W9-CLO-451

Extensions and Absolutes of Hausdorff Spaces

Jack R. Porter R. Grant Woods

Extensions and Absolutes of Hausdorff Spaces

With 27 Illustrations

Springer-Verlag
New York Berlin Heidelberg
London Paris Tokyo

Jack R. Porter
Department of Mathematics
The University of Kansas
Lawrence, KS 66045-2142
USA

R. Grant Woods
Department of Mathematics
University of Manitoba
Winnipeg, Manitoba R3T 2N2
Canada

Mathematics Subjects Classification (1980): 54-02, 54C20

Library of Congress Cataloging-in-Publication Data
Porter, Jack R.
Extensions and absolutes of Hausdorff spaces.
Bibliography: p.
Includes index.
1. Hausdorff compactifications. 2. Linear
topological spaces. I. Woods, R. Grant. II. Title.
QA611.23.P67 1987 515.7'3 87-32734

Camera-ready copy provided by the authors.
Printed and bound by Quinn-Woodbine, Woodbine, New Jersey.
Printed in the United States of America.

9 8 7 6 5 4 3 2 1

ISBN 0-387-96212-3 Springer-Verlag NewYork Berlin Heidelberg
ISBN 3-540-96212-3 Springer-Verlag Berlin Heidelberg NewYork

To Peggy and Sheila

PREFACE

An **extension** of a topological space X is a space that contains X as a dense subspace. The construction of extensions of various sorts — compactifications, realcompactifications, H-closed extensions — has long been a major area of study in general topology. A ubiquitous method of constructing an extension of a space is to let the "new points" of the extension be ultrafilters on certain lattices associated with the space. Examples of such lattices are the lattice of open sets, the lattice of zero-sets, and the lattice of clopen sets.

A less well-known construction in general topology is the "absolute" of a space. Associated with each Hausdorff space X is an extremally disconnected zero-dimensional Hausdorff space EX, called the **Iliadis absolute** of X, and a perfect, irreducible, Θ-continuous surjection from EX onto X. A detailed discussion of the importance of the absolute in the study of topology and its applications appears at the beginning of Chapter 6. What concerns us here is that in most constructions of the absolute, the points of EX are certain ultrafilters on lattices associated with X. Thus extensions and absolutes, although conceptually very different, are constructed using similar tools.

The purpose of this book is to undertake a systematic study of the extensions and absolutes of Hausdorff spaces. A secondary theme is to show that by investigating the structure of certain lattices of subsets of a space, we obtain powerful tools that allow us to build many sorts of extensions and to construct absolutes in a variety of ways. Hence another purpose of this book is to demonstrate the power of lattice-theoretic concepts when applied to topology.

Chapter 1 is devoted to a discussion of some topics from elementary topology. In Chapter 2 we develop the basic concepts of lattices, filters, and convergence. We then discuss linearly ordered topological spaces, and finish the chapter with a discussion of ordinal and cardinal numbers. In Chapter 3 we discuss Boolean algebras and the Stone duality theorem. We finish by introducing Martin's axiom and some of its topological consequences.

Chapters 1 to 3 are preparatory material. In Chapters 4 to 7 we undertake a detailed study of the central matter of the book. Chapter 4 begins with a discussion of the general theory of extensions, and then embarks upon a detailed discussion of different methods of constructing compactifications. The chapter closes with a discussion of H-closed spaces. In Chapter 5 we investigate those topological properties P for which each "suitable" space has a "largest" extension with P. At the end there is a detailed discussion of realcompact spaces and extensions. Chapter 6 is devoted to the construction of the Iliadis and Banaschewski absolutes of a space. Chapter 7 studies the various types of H-closed extensions that a space may have, and discusses their interrelationships.

Chapter 8 is a collection of four essentially unrelated topics. After a brief introduction in 8.1, we investigate when the absolute of an extension is "the same" as the extension of the absolute. In 8.2 we are concerned with the commuting of absolutes with various H-closed extensions; in 8.3 we study when the Iliadis absolute commutes with extension properties containing realcompactness. In 8.4 we generalize the notion of absolutes and develop a theory of "covers" of spaces that is in many ways analogous to the theory of extensions discussed in Chapters 4 and 5. In 8.5 we relate real-valued continuous functions on a space to those on its absolute.

Chapter 9 provides a brief introduction to abstract category theory, and then concentrates on interpreting the construction of extensions and absolutes discussed in previous chapters in a category-theoretic light.

Each chapter is followed by a lengthy collection of problems. Some are routine verifications; others are harder. A number of hints are provided to aid the reader in solving the more difficult problems. The presence of these problems will, we hope, make this book useful as a text for a graduate course in topology. We have often referred in the body of the text to results appearing in problems from previous chapters, and we encourage the reader to attempt as many problems as possible.

We have not tried to give an exhaustive description of "who proved what," but some historical comments along these lines, together with guides to further reading, appear in the "Notes."

The heart of this book is Chapters 4 to 7, together with 8.4 (Chapter 9 is in part a retrospective, couched in the language of category theory, of what happened earlier). Readers with a good knowledge of topology who wish to reach the central ideas as soon as possible are advised to start at Chapter 4, and refer back to Chapters 1, 2, and 3 when necessary for notation, terminology, and basic results.

We do assume that the reader is familiar with the basic ideas of general topology. Some topics of particular relevance to our later work are reviewed in Chapter 1. The texts by Willard [Wi], Dugundji [Du], and Engelking [En] are excellent references for topological ideas not discussed here.

Although general topology in recent years has become increasingly dependent on axiomatic set theory, we do not assume that the reader has a detailed knowledge of this subject. It will suffice for the reader to know that the axioms of set theory can be formulated precisely in the language of mathematical logic. The formulation that we will implicitly use is the Zermelo-Frankel axioms, together with the Axiom of Choice (henceforth abbreviated ZFC). We do not state these axioms explicitly in this book. The reader who wishes to pursue these ideas is referred to the text by Kunen [Ku], which gives a rigorous treatment of those aspects of set theory most useful in topology.

We do assume that the reader is familiar with the distinction between a set and a proper class, with transfinite induction, and with the basic facts about ordinal and cardinal numbers. However, in 2.7 we give a rapid review (without proofs) of those facts about ordinal and cardinal numbers that are needed in this book.

This is not a book about set-theoretic topology, and we will not often be concerned with topological questions whose answers depend upon which "relatively consistent" model of ZFC is being assumed. Most of the theorems we discuss are "real" theorems in the sense that they are derivable from the axioms of ZFC, and do not require any additional set-theoretic assumptions for their proofs. There is one major exception to this, however; at the end of Chapter 3 we include a discussion of the continuum hypothesis and Martin's axiom. The reason for doing this is that our previous work on Boolean algebras and filters has equipped us to formulate the various versions of Martin's axiom, and prove their equivalence, with relatively little additional work. It also provides the reader with an opportunity to see one of the major tools of modern set-theoretic topology at work. Exercises using Martin's axiom and/or the continuum hypothesis are scattered throughout the problem sections of Chapter 3 and subsequent chapters.

Finally, we wish to acknowledge the extensive and valuable assistance we have received from our students, colleagues, and typists. Melvin Henriksen, Mohan Tikoo, and Johannes Vermeer read portions of an early version of the manuscript and provided us with many useful suggestions. A number of the problems reflect Vermeer's valuable contributions to the theory of H-closed spaces and absolutes. Beverly Diamond read Chapters 4 and 5, and the problems for Chapter 5, and made several valuable contributions to the final form of this portion of the book. Above all, we would like to express our sincere thanks and gratitude to Alan Dow, who read essentially the entire book to the end of Chapter 7, did virtually all of the problems, and protected us from numerous errors (in both style and substance). We appreciate the work done by our typists — Sharon Gumm, Carol Johnson, Susan Levine, Beverly Preiss, Everly Scherko, and Edith Despins — who had to endure our sloppy handwriting and numerous changes of mind. It has been a pleasure to work with such helpful and generous colleagues.

Needless to say, whatever errors remain are the sole responsibility of the authors.

CONTENTS

Preface v

Chapter 1 Topological background 1
1.1 Notation and terminology from elementary set theory 1
1.2 Notation and terminology for elementary topological concepts .. 4
1.3 C(X) as a lattice-ordered ring 9
1.4 Tychonoff spaces, zero-sets, and cozero-sets 11
1.5 Clopen sets and zero-dimensional spaces 16
1.6 Continuous functions 18
1.7 Product spaces and evaluation maps 20
1.8 Perfect functions 28
1.9 C- and C*-embedding 35
1.10 Normal spaces 43
1.11 Pseudocompact spaces 50
Problems 55

Chapter 2 Lattices, filters, and topological spaces 74
2.1 Posets and lattices 74
2.2 Regular open sets, regular closed sets, and semiregular spaces .. 81
2.3 Filters on a lattice 91
2.4 More lattice properties 100
2.5 Completions of lattices and ordered topological spaces 105
2.6 Ordinals, cardinals, and spaces of ordinals 121
Problems 133

Chapter 3 Boolean algebras 155
3.1 Definition and basic properties 155
3.2 Stone's representation and duality theorems 169
3.3 Atomless, countable Boolean algebras 180
3.4 Completions of Boolean algebras 186
3.5 The continuum hypothesis and Martin's Axiom 196
Problems 214

Chapter 4 Extensions of spaces 238
4.1 Basic concepts 238
4.2 Compactifications 246
4.3 One-point compactifications 252
4.4 Wallman compactifications 255
4.5 Gelfand compactifications 261
4.6 The Stone-Čech compactification 277
4.7 Zero-dimensional compactifications 290
4.8 H-closed spaces 297
Problems 316

Chapter 5 Maximum *P*-extensions 362
5.1 Introductory remarks 362
5.2 *P*-regular and *P*-compact spaces 365
5.3 Characterizations of extension properties 367
5.4 E-compact spaces 373
5.5 Examples of E-compactness 376
5.6 Tychonoff extension properties 380
5.7 Zero-dimensional extension properties 382
5.8 Hausdorff extension properties 383
5.9 More on Tychonoff and zero-dimensional extension properties 385
5.10 Two examples of maximum *P*-extensions 391
5.11 Realcompact spaces and extensions 396
Problems 407

Chapter 6 Extremally disconnected spaces and absolutes 440
6.1 Introduction 440
6.2 Characterization of extremally disconnected spaces 444
6.3 Examples of extremally disconnected spaces 450
6.4 Extremally disconnected spaces and zero-dimensionality 451
6.5 Irreducible functions 452
6.6 The construction of the Iliadis absolute 457
6.7 The uniqueness of the absolute 463
6.8 The construction of EX as a space of open ultrafilters 466
6.9 Elementary properties of EX 473
6.10 Examples of absolutes 479
6.11 The Banaschewski absolute 484
Problems 495

Chapter 7 H-closed extensions 531
7.1 Strict and simple extensions 531
7.2 The Fomin extension 539

7.3 One-point H-closed extensions 543
7.4 Partitions of $\sigma X \setminus X$ 547
7.5 Minimal Hausdorff spaces 551
7.6 p-maps 557
7.7 An equivalence relation on H(X) 566
Problems 575

Chapter 8 Further properties and generalization of absolutes 612
8.1 Introduction 612
8.2 Absolutes and H-closed extensions 612
8.3 Absolutes and extension properties 620
8.4 Covers of topological spaces 639
8.5 Completions of C(X) vs. C(EX) 661
Problems 672

Chapter 9 Categorical interpretations of absolutes and extensions 691
9.1 Introduction 691
9.2 Categories, functors, natural transformations, and subcategories 692
9.3 Topological categories 700
9.4 Morphisms 701
9.5 Products and coproducts 711
9.6 Reflective and epireflective subcategories 716
9.7 Coreflections 723
9.8 Projective covers 730
Problems 745

Notes 765

Bibliography 780

List of Symbols 800

Index 815

CHAPTER 1

TOPOLOGICAL BACKGROUND

In this chapter we do two things. First, we introduce notation and terminology that we will use throughout the book. Second, we discuss a number of topological ideas that fall into a "grey zone"; they do not form part of the central subject matter of this book (extensions and absolutes), and they sometimes do not appear in a typical graduate level course in point-set topology. A familiarity with these ideas is necessary to what follows, so a detailed discussion of them is given here. The topologically sophisticated reader may wish to skip this material and to refer to it when the need arises.

Throughout this book, all hypothesized topological spaces are assumed to be Hausdorff unless it is explicitly stated otherwise. Thus the word "space" means "Hausdorff topological space." We will occasionally repeat this assumption for emphasis.

1.1 Notation and terminology from elementary set theory

In this section we introduce some of the set-theoretic notation and terminology that we will be using.

(a) If X and Y are two sets, then F(X,Y) will denote the set of functions whose domain is X and whose range is Y. If $Z \subseteq Y$, then evidently F(X,Z) is a subset of F(X,Y).

(b) If $A \subseteq X$ and $f \in F(X,Y)$, then $f|A$, called the **restriction of f to A**, is the function in $F(A,Y)$ defined as follows:
$f|A = \{(x,f(x)) : x \in A\}$.

(c) The cardinality of a set X is denoted by $|X|$. The **power set** of X (i.e., the set of all subsets of X) is denoted by $\mathbb{P}(X)$. The cardinality of $\mathbb{P}(X)$ is denoted by either $2^{|X|}$ or $\exp(|X|)$, depending on typographical convenience. Thus if α is a cardinal number, 2^{α} (or $\exp \alpha$) denotes the cardinality of ″(X), where X is a set with $|X| = \alpha$. (A more comprehensive discussion of cardinal numbers appears in 2.5(ℓ).) The smallest cardinal larger than α is denoted by α^{+}, We will use the well-known "aleph" notation ($\aleph_0$, $\aleph_1$, etc.) for cardinals.

(d) If A is a subset of a set X, then $\chi_{A,X}$ is that function in $F(X,\{0,1\})$ defined as follows: $\chi_{A,X}(x) = 1$ if $x \in A$ and $\chi_{A,X}(x) = 0$ if $x \in X \setminus A$. We call $\chi_{A,X}$ the **characteristic function** of A. If the identity of X is clear from the context, we may write "χ_A" instead of "$\chi_{A,X}$".

(e) Let $f \in F(X,Y)$. If $f[X] = Y$ we call f a **surjection**, or say that f is an "**onto**" function, or say that f takes X **onto** Y. A one-to-one surjection is called a **bijection**.

(f) Let $\{X_i : i \in I\}$ be a set of sets, indexed by the set I. The **product** of $\{X_i : i \in I\}$, denoted $\Pi\{X_i : i \in I\}$, is defined to be the set $\{f \in F(I, \cup\{X_i : i \in I\}) : \text{for each } i \in I,\ f(i) \in X_i\}$. If $f \in \Pi\{X_i : i \in I\}$ and $f(i) = x_i$ for each $i \in I$, we will often denote f by $\langle x_i \rangle_{i \in I}$, thereby generalizing in the standard way the usual notation used to describe countable sequences. If $j \in I$ the jth **projection function** on $\Pi\{X_i : i \in I\}$, denoted Π_j, is the member of $F(\Pi\{X_i : i \in I\}, X_j)$ defined as follows: $\Pi_j(f) = f(j)$ for each $f \in \Pi\{X_i : i \in I\}$. More generally, if S is a subset of I, we define a function $\Pi_S \in F(\Pi\{X_i : i \in I\}, \Pi\{X_i : i \in S\})$ as follows: $\Pi_S(f) = f|S$ for each $f \in \Pi\{X_i : i \in I\}$.

(g) If $f \in F(X,Y)$ and $A \subseteq Y$ then $f^{\leftarrow}[A]$ is defined to be the set $\{x \in X : f(x) \in A\}$ (we reserve the symbol "f^{-1}" for a different purpose; see 1.3 below).

(h) If $\mathcal{C}$ is a set of sets, then $\cup\mathcal{C}$ will denote the union of all the sets in $\mathcal{C}$; thus $\cup\mathcal{C} = \cup\{C : C \in \mathcal{C}\}$. Similarly $\cap\mathcal{C}$ denotes $\cap\{C : C \in \mathcal{C}\}$, and so on.

(i) We assume the reader is familiar with the notation of a partially ordered set. We will use the term "poset" as an abbreviation for "partially ordered set."

(j) If A is a set, then "id_A" is used to denote the identity function in $F(A,A)$; thus $id_A(x) = x$ for each $x \in A$.

(k) The notation "$A \subseteq B$" means "A is a subset of B", while "$A \subset B$" means "A is a proper subset of B".

1.2 Notation and terminology for elementary topological concepts

We now review some terminology from elementary topology.

We will use special symbols for certain frequently occurring topological spaces. These are summarized in the following table.

(a) Table of Symbols

Space	Symbol
closed unit interval	[0,1]
real numbers	$\mathbb{R}$
rational numbers	$\mathbb{Q}$
discrete space of cardinality α	$D(\alpha)$
two-point discrete space	$\mathbf{2}$

(The elements of $\mathbf{2}$ are the integers 0 and 1).

(b) **Definition**. The symbol $\mathbb{N}$ denotes the set of positive integers. It is regarded as a discrete topological space, and we use $\mathbb{N}$ in preference to $D(\aleph_0)$. We also use $\mathbb{N}$ as an indexing set when enumerating countable sets.

The following notation is used when discussing families of continuous functions.

(c) **Definition**.

(1) Let X and Y be spaces. Then $C(X,Y)$ denotes the set of

all continuous functions from X into Y. If $Z \subseteq Y$ we regard $C(X,Z)$ as a subset of $C(X,Y)$.

(2) $C(X,\mathbb{R})$ is denoted by $C(X)$.

(3) If $r \in \mathbb{R}$ and X is some space under discussion, then $\underline{r}$ denotes the "constant function" in $C(X)$ that sends each point of X to r.

We will need some notation for the topology of a space.

(d) <u>**Definition**</u>. Let X be a space.

(1) The collection of open subsets of X is denoted by $\tau(X)$.

(2) An **open base for** X (or, a **base for** $\tau(X)$) is a subcollection $\mathcal{C}$ of $\tau(X)$ such that each nonempty member of $\tau(X)$ can be written as a union of sets belonging to $\mathcal{C}$.

(3) A **closed base for** X is a subcollection $\mathcal{A}$ of $\{X\backslash V : V \in \tau(X)\}$ such that each closed subset of X can be written as an intersection of sets belonging to $\mathcal{A}$.

(4) Occasionally we will wish to specify the topology of a space, and on such occasions we will talk about "the space (X,τ)". In such instances τ denotes $\tau(X)$.

(5) If $A \subseteq X$ then the closure of A in X is denoted by $c\ell_X A$, the interior of A in X is denoted by $\mathrm{int}_X A$, and the boundary of A in X is denoted $\mathrm{bd}_X A$. If there is no possibility of confusion, these concepts will be denoted by $c\ell A$, int A, and bd A, respectively.

(6) The set of neighborhoods of a point p is a space X will be denoted $\mathcal{N}(p)$ (or $\mathcal{N}(p,X)$ if the identity of X is ambiguous).

The following sorts of functions between spaces will be of

interest.

(e) __Definition.__ Let X and Y be spaces, and let $f \in F(X,Y)$. Then:

(1) f is an **open** (respectively **closed**) function from X to Y if f[V] is an open (respectively closed) subset of Y whenever V is an open (respectively closed) subset of X. (Note that such functions need not be either continuous or surjective.);

(2) f is an **embedding** of X in Y if f is an homeomorphism from X onto f[X].

(f) __Definition.__ A topological property P is **hereditary** (respectively **open-hereditary**, respectively **closed-hereditary**) if every subspace (respectively every open subspace, respectively every closed subspace) of a space with P has P.

Examples of hereditary topological properties are the T_i separation properties ($i = 1, 2, 3$), and the property of being metrizable. Compactness is closed-hereditary but not open-hereditary, while separability is open-hereditary but not closed-hereditary.

(g) __Definition.__ A **G_δ-set** of a space X is a subset of X that can be written as the intersection of countable many open subsets of X. An **F_σ-set** of X is the complement of a G_δ-set.

(h) __Definition.__

(1) Let $\{X_i : i \in I\}$ be a set of spaces. The **sum**, or **free**

union, of $\{X_i : i \in I\}$, denoted $\oplus\{X_i : i \in I\}$, is the topological space X whose underlying set is $\cup\{X_i \times \{i\} : i \in I\}$, and whose topology is defined as follows: $V \in \tau(X)$ if $\{x \in X_i : (x,i) \in V\} \in \tau(X_i)$ for each $i \in I$.

(2) If $\{X_i : i \in I\}$ and $\{Y_i : i \in I\}$ are two sets of spaces and $f_i \in F(X_i,Y_i)$ for each $i \in I$, then $\oplus\{f_i : i \in I\}$ is defined to be the function $f \in F(\oplus\{X_i : i \in I\}, \oplus\{Y_i : i \in I\})$ defined as follows: If $(x,j) \in \oplus\{X_i : i \in I\}$ then $f((x,j)) = (f_j(x),j)$ (for each $j \in I$).

We close this section with definitions of certain sorts of covers of topological spaces, and a brief discussion of spaces defined in terms of such covers.

(i) <u>Definition</u>.

(1) A **cover** of a space X is a collection of subsets of X whose union is X.

(2) An **open cover** of X is a cover whose members are open subsets of X.

(j) <u>Definition</u>. Let $\mathcal{C}$ and $\mathcal{D}$ be covers of a space X. Then:

(1) $\mathcal{D}$ is said to **refine** $\mathcal{C}$, or $\mathcal{C}$ is **refined by** $\mathcal{D}$, or $\mathcal{D}$ is a **refinement of** $\mathcal{C}$ (denoted $\mathcal{D} < \mathcal{C}$) for each $D \in \mathcal{D}$ there is some $C \in \mathcal{C}$ such that $D \subseteq C$;

(2) $\mathcal{D}$ is a **subcover** of $\mathcal{C}$ if $\mathcal{D} \subseteq \mathcal{C}$;

(3) $\mathcal{C}$ is a **partition** of X if $A, B \in \mathcal{C}$ and $A \cap B \neq \emptyset$ implies $A = B$. Thus $\mathcal{C}$ is a family of pairwise disjoint sets whose union is X.

If the refinement (subcover, partition) consists of open subsets of X, we call it an **open refinement** (subcover, partition).

(k) Definition. Let C be a collection of subsets of a space X. Then:

(1) C is **locally finite** if each point of X has a neighborhood that intersects only finitely many members of C;

(2) C is **point finite** if each point of X belongs to only finitely many members of C.

Obviously every locally finite collection is point finite, but it is easy to find examples of point finite collections that are not locally finite.

(l) Definition. A space X is **paracompact** (respectively **metacompact**) if every open cover has a locally finite (respectively point finite) open refinement.

We shall not be much concerned with metacompactness or paracompactness in this book, and we will not spend time developing the properties of these spaces, which can be found in [Wi], [Du], or [En]. One place in which these spaces do arise is in 5.11(l), where we prove a theorem that implies that every normal metacompact (and hence every regular paracompact) space is realcompact.

1.3 C(X) as a lattice-ordered ring

We can impose an algebraic structure on $C(X)$ as follows:

(a) **Definition**. If $f,\ g \in F(X,\mathbb{R})$ and $\alpha \in \mathbb{R}$, we define αf, $f + g$, and $f \cdot g$ in $F(X,\mathbb{R})$ as follows:

$$(f+g)(x) = f(x) + g(x) \qquad \text{for each } x \in X$$

$$(\alpha f)(x) = \alpha(f(x)) \qquad \text{for each } x \in X$$

$$(f \cdot g)(x) = f(x)g(x) \qquad \text{for each } x \in X$$

(We will usually write "fg" instead of "f·g".)

It is straightforward to verify that if f and g belong to $C(X)$, and $\alpha \in \mathbb{R}$, then αf, $f + g$, and $fg \in C(X)$ (see 1B). From the field-theoretic properties of $\mathbb{R}$ one can readily deduce that $(C(X),+,\cdot)$ is a ring. If $f \in C(X)$ and $f(x) \neq 0$ for each $x \in X$, then the multiplicative inverse f^{-1} of f exists in $C(X)$, and $f^{-1}(x) = 1/f(x)$ for each $x \in X$ (see 1B(5)).

In a manner similar to the above, the order structure of $\mathbb{R}$ can be used to impose an order structure on $C(X)$.

(b) **Definition**. Let $f,\ g \in C(X)$. We say that $f \leqslant g$ if $f(x) \leqslant g(x)$ for each $x \in X$. Then $(C(X),\leqslant)$ is a poset, as is easily checked. We can define functions $f \vee g$ and $f \wedge g$ as follows:

$(f \vee g)(x) = \max \{f(x), g(x)\}$ for each $x \in X$

$(f \wedge g)(x) = \min \{f(x), g(x)\}$ for each $x \in X$

It is straightforward to prove that $f \vee g$ and $f \wedge g$ belong to $C(X)$ (See 1B). In Chapter 2 we will see that the poset $(C(X),\leqslant)$ is a lattice, where $\sup\{f,g\} = f \vee g$ and $\inf\{f,g\} = f \wedge g$ (as defined above); see 2.1(f)(5) for details. One special case of this construction that will be important to us is the following.

If $f \in C(X)$ and $r, s \in \mathbb{R}$, with $r \leqslant s$, let $g = \underline{r} \vee (\underline{s} \wedge f)$. Then

$$g(x) = \begin{cases} r & \text{if } f(x) \leqslant r \\ f(x) & \text{if } r < f(x) < s \\ s & \text{if } f(x) \geqslant s, \end{cases}$$

and $g \in C(X)$.

Suppose $f \in C(X)$. Evidently $(f \vee (-1)f)(x) = |f(x)|$ for each $x \in X$. If we define $|f| \in F(X,\mathbb{R})$ by the equation $|f|(x) = |f(x)|$ for each $x \in X$, we see that $|f| = f \vee (-f)$ and so $|f| \in C(X)$. Evidently $\underline{0} \leqslant |f|$ for each $f \in C(X)$.

The order-theoretic and ring-theoretic properties of $C(X)$ interact "nicely"; to be precise, $C(X)$, endowed with the above algebraic and order-theoretic structures, is a "lattice-ordered ring". These ideas are discussed in detail in 4.5 when we construct "Gelfand compactifications" of Tychonoff spaces using certain subalgebras of $C(X)$.

1.4 Tychonoff spaces, zero-sets, and cozero-sets

(a) **Definition**. A space X is called a **Tychonoff space** if, whenever A is a closed subspace of X and $p \in X \setminus A$, there exists $f \in C(X)$ such that $f[A] \subseteq \{0\}$ and $f(p) = 1$.

Tychonoff spaces are also called **completely regular** (Hausdorff) spaces by many authors. Note that the function f in the above definition may be chosen so that $\underline{0} \leqslant f \leqslant \underline{1}$, for if f did not have this property, one could replace it by the function $\underline{0} \vee (f \wedge \underline{1})$.

(b) **Definition**. A **zero-set** of a space X is a subset of X of the form $f^{\leftarrow}(0)$, where $f \in C(X)$. The set $f^{\leftarrow}(0)$ is also denoted by $Z(f)$. The set of all zero-sets of X is denoted by $\mathcal{Z}(X)$.

(c) **Definition**. A **cozero-set** of a space X is a set of the form $X \setminus Z(f)$, where $f \in C(X)$. The set $X \setminus Z(f)$ is denoted by $coz(f)$; the set of all cozero-sets of X is denoted by $coz(X)$.

Evidently $coz(f) = \{x \in X : f(x) \neq 0\}$.

(d) **Definition**. If $f \in C(X)$ the **support** of f, denoted $spt(f)$, is defined to be $c\ell_X(coz(f))$.

(e) **Proposition**. The following are equivalent for a space X:

(1) X is Tychonoff

(2) $\mathcal{Z}(X)$ is a closed base for X

(3) coz(X) is an open base for X

Proof.

(1) ⇒ (2) Let A be a closed subset of X and let $p \in X \setminus A$. By hypothesis there exists $f_p \in C(X)$ such that $f_p(p) = 1$ and $f_p[A] \subseteq \{0\}$. Thus $A \subseteq Z(f_p) \subseteq X \setminus \{p\}$. It follows that $A = \cap\{Z(f_p) : p \in X \setminus A\}$ and so (2) is true.

(2) ⇒ (1) Let A be a closed subset of X and let $p \in X \setminus A$. As A is the intersection of the zero-sets containing it, there exists $f \in C(X)$ such that $A \subseteq Z(f)$ and $p \notin Z(f)$. Let $g = \underline{(f(p))}^{-1} \cdot f$. Then $g \in C(X)$, $g[A] \subseteq \{0\}$, and $g(p) = 1$, and so X is Tychonoff. Obviously (2) and (3) are equivalent.

■

The following technical result will be used frequently.

(f) Proposition. Let $Z \in \mathcal{Z}(X)$ and let $r > 0$. Then there exists $f \in C(X)$ such that $Z = Z(f)$ and $\underline{0} \leq f \leq \underline{r}$.

Proof. Let $Z = Z(g)$ for $g \in C(X)$, and define f to be $|g| \wedge \underline{r}$.

■

We now develop some basic facts about zero-sets.

(g) Proposition. If M is a metric space then every closed

subset of M is a zero-set of M.

Proof. Let d be the metric on M and let A be a nonempty closed subset of M. Define $f_A \in F(M,\mathbb{R})$ as follows: $f_A(x) = \inf\{d(x,a) : a \in A\}$. It is straightforward to show that $f_A \in C(M)$ and that $Z(f_A) = A$.

■

(h) **Proposition**. Let X and Y be spaces, let $f \in C(X,Y)$, and let $Z \in \mathcal{Z}(Y)$. Then $f^{\leftarrow}[Z] \in \mathcal{Z}(X)$.

Proof. There exists $g \in C(Y)$ such that $Z = Z(g)$. Then $g \circ f \in C(X)$ and $f^{\leftarrow}[Z] = Z(g \circ f)$.

■

(i) **Theorem**. Let X be a space. Then

(1) $\mathcal{Z}(X)$ is closed under finite unions and countable intersection.

(2) coz(X) is closed under finite intersections and countable unions.

Proof. If $f, g \in C(X)$ then $Z(f) \cup Z(g) = Z(fg)$. This fact, together with a finite induction, shows that Z(X) is closed under finite unions.

Let $\{Z_i : i \in \mathbb{N}\}$ be a countable subset of $\mathcal{Z}(X)$. By 1.4(f) there exists, for each $i \in \mathbb{N}$, a function $f_i \in C(X)$ such that $\underline{0} \leqslant f_i \leqslant \underline{2}^{-i}$ and $Z_i = Z(f_i)$. Using the Weierstrass M-test, we see that $\sum_{i=1}^{\infty} f_i(x)$ is a convergent infinite series for each $x \in X$, and so we can define $f \in F(X,\mathbb{R})$ by putting $f(x) = \sum_{i=1}^{\infty} f_i(x)$.

We now show that f is continuous. Let $x_0 \in X$; to show that f is continuous at x_0 we must show that, given $\delta > 0$, there exists an open subset $V(\delta)$ of X such that $x_0 \in V(\delta)$ and

$$f[V(\delta)] \subseteq (f(x_0) - \delta,\ f(x_0) + \delta).$$

Choose $i(0) \in \mathbb{N}$ such that $\sum_{i=i(0)}^{\infty} 2^{-i} < \frac{\delta}{4}$. As $\sum_{i=1}^{i(0)-1} f_i \in C(X)$ (see 1.3), there is an open subset $V(\delta)$ of X such that $x_0 \in V(\delta)$ and such that

$$\left|\sum_{i=1}^{i(0)-1} f_i(x_0) - \sum_{i=1}^{i(0)-1} f_i(x)\right| < \frac{\delta}{4}$$

for each $x \in V(\delta)$. Thus if $n \geqslant i(0)$ and $x \in V(\delta)$ it follows that

$$\left|\sum_{i=1}^{n} f_i(x_0) - \sum_{i=1}^{n} f_i(x)\right| \leqslant$$

$$\leqslant \left|\sum_{i=1}^{i(0)-1} f_i(x_0) - \sum_{i=1}^{i(0)-1} f_i(x)\right|$$

$$+ \sum_{i=i(0)}^{n} f_i(x_0) + \sum_{i=i(0)}^{n} f_i(x)$$

$$< \frac{\delta}{4} + \frac{\delta}{4} + \frac{\delta}{4} = \frac{3}{4}\delta$$

Taking the limit as $n \to \infty$, we see that

$$|f(x_0) - f(x)| \leq \tfrac{3}{4}\delta < \delta$$

whenever $x \in V(\delta)$; thus $f \in C(X)$. Evidently $Z(f) = \bigcap\{Z(f_i) : i \in \mathbb{N}\}$ and so $Z(X)$ is closed under countable intersections. Part (2) is an immediate consequence of part (1).

■

(j) **Theorem**. The following are equivalent for a subspace S of a space X.

(1) $S \in Z(X)$

(2) There exists a closed subset A of $\mathbb{R}$ and $f \in C(X)$ such that $S = f^{\leftarrow}[A]$

(3) There exists a metric space M, a closed subset B of M, and a function $f \in C(X,M)$ such that $S = f^{\leftarrow}[B]$.

Proof

(1) ⇒ (2) Evidently $S = Z(f)$ for some $f \in C(X)$; choose A to be $\{0\}$.

(2) ⇒ (3) Obvious (since $\mathbb{R}$ is a metric space).

(3) ⇒ (1) By 1.4(g) $B \in Z(M)$, so by 1.4(h) $f^{\leftarrow}[B] \in Z(X)$.

■

1.5 Clopen sets and zero-dimensional spaces

(a) **Definition.**

(1) A **clopen subset** of a space X is a subset of X that is both open and closed in X. The set of clopen subsets of X is denoted by $B(X)$.

(2) A space X is **zero-dimensional** if $B(X)$ is an open base for X.

We record some obvious but important facts about clopen sets. Their proofs are all trivial.

(b) **Proposition.** Let X be a space. Then:

(1) $\{\emptyset,X\} \subseteq B(X)$ and $\{\emptyset,X\} = B(X)$ iff X is connected.

(2) $B(X) = \mathbb{P}(X)$ iff X is discrete.

(3) $B(X)$ is closed under formation of finite unions, finite intersections, and complements.

(4) X is zero-dimensional iff $B(X)$ is a base for the closed subsets of X.

Our next result describes the relationship between $B(X)$ and C(X).

(c) **Proposition.** Let X be a space. Then:

(1) If $A \subseteq X$ then $\chi_A \in C(X)$ iff $A \in B(X)$

(2) $B(X) \subseteq Z(X)$

(3) If X is zero-dimensional then X is Tychonoff

(4) Zero-dimensionality is hereditary.

Proof

(1) As noted in 1.1(a) and 1.1(d), $\chi_A \in F(X,\mathbb{R})$. If $\chi_A \in C(X)$ then A is closed in X as $A = \chi_A^{\leftarrow}(1)$, and $X \setminus A$ is closed in X as $X \setminus A = \chi_A^{\leftarrow}(0)$. Thus $A \in B(X)$. Conversely, suppose $A \in B(X)$. If $S \subseteq \mathbb{R}$ then $\chi_A^{\leftarrow}[S]$ is one of $\emptyset$, X, A, or $X \setminus A$, each of which is open in X, so $\chi_A \in C(X)$.

(2) If $A \in B(X)$ then $\chi_A \in C(X)$; since $A = \chi_A^{\leftarrow}(1)$, by 1.4(j), $A \in Z(X)$.

(3) This follows from (2) above and 1.4(e).

(4) This is obvious.

■

(d) Examples of zero-dimensional spaces

Obviously any discrete space X is zero-dimensional; $B(X) = \mathbb{P}(X)$ in this case. As a slightly less trivial example, let $X \cup \{\omega\}$ denote the one-point compactification of the infinite discrete space X. Then $B(X \cup \{\omega\}) = \{A \subseteq X : A \text{ is finite}\} \cup \{B \subseteq X \cup \{\omega\} : \omega \in B \text{ and } X \setminus B \text{ is finite}\}$. Let K be closed in $X \cup \{\omega\}$ and suppose $p \notin K$. If $p \in X$ then $p \in \{p\} \subseteq X \setminus K$, and $\{p\} \in B(X \cup \{\omega\})$. If $p = \omega$ then K must be a finite subset of X and therefore a member of $B(X \cup \{\omega\})$; thus $(X \cup \{\omega\}) \setminus K \in B(X \cup \{\omega\})$ also. We see that $(X \cup \{\omega\}) \setminus K$ is a union of members of $B(X \cup \{\omega\})$, and so $X \cup \{\omega\}$ is zero-dimensional.

Other zero-dimensional spaces are $\mathbb{Q}$ and $\mathbb{R} \setminus \mathbb{Q}$, as the reader can easily verify. In Chapters 2 and 3, we will discuss zero-dimensional spaces in much greater detail, and give many more examples of such spaces.

1.6 Continuous functions

In this section we collect a miscellany of technical, useful properties of continuous functions.

Let $\mathcal{C}$ be a cover of a space X. Suppose $f_C \in C(C,Y)$ for each $C \in \mathcal{C}$ and suppose that $f_C|C \cap D = f_D|C \cap D$ for each C, $D \in \mathcal{C}$. Evidently there is a unique function $f \in F(X,Y)$ such that $f|C = f_C$ for each $C \in \mathcal{C}$. An important problem in topology is to determine conditions on $\mathcal{C}$ that ensure that f is continuous; that is, to determine when it is possible to go from "local" to "global" continuity. Theorems such as 1.6(a) and 1.6(c) below, which assert the continuity of f in situations of the above sort, are called "pasting" theorems.

(a) **Theorem**. Let $\{V_i : i \in I\}$ be an open cover of the space X. Let Y be a space, let $f \in F(X,Y)$, and suppose that $f|V_i \in C(V_i,Y)$ for each $i \in I$. Then $f \in C(X,Y)$.

Proof. Let W be open in Y. Then

$$\begin{aligned} f^{\leftarrow}[W] &= \cup\{f^{\leftarrow}[W] \cap V_i : i \in I\} \\ &= \cup\{(f|V_i)^{\leftarrow}[W] : i \in I\}. \end{aligned}$$

By hypothesis $(f \mid V_i)^{\leftarrow}[W]$ is open in V_i, and hence open in X, for each $i \in I$. The conclusion follows.

■

(b) **Lemma**. Let X and Y be spaces and let $f \in F(X,Y)$. Let $\mathcal{F}$ be a finite collection of closed subsets of X such that $\bigcup \mathcal{F} = X$. Suppose that $f \mid F \in C(F,Y)$ for each $F \in \mathcal{F}$. Then $f \in C(X,Y)$.

Proof. Let B be a closed subset of Y. By hypothesis $(f \mid F)^{\leftarrow}[B]$ is a closed subset of F for each $F \in \mathcal{F}$, and hence is closed in X. But

$$\begin{aligned} f^{\leftarrow}[B] &= \bigcup\{f^{\leftarrow}[B] \cap F : F \in \mathcal{F}\} \\ &= \bigcup\{(f \mid F)^{\leftarrow}[B] : F \in \mathcal{F}\}, \end{aligned}$$

so as $\mathcal{F}$ is a finite collection of closed subsets of X, $f^{\leftarrow}[B]$ is closed in X. Thus $f \in C(X,Y)$.

■

(c) **Theorem**. Let X be a union of a locally finite collection $\mathcal{C}$ of closed subsets of X. Let Y be a space, let $f \in F(X,Y)$, and suppose that $f \mid S \in C(S,Y)$ for each $S \in \mathcal{C}$. Then $f \in C(X,Y)$.

Proof. Let $p \in X$. There is a finite subcollection $\mathcal{F}$ of $\mathcal{C}$, and an open subset V of X, such that $p \in V$ and $V \cap C = \emptyset$ for each $C \in \mathcal{C} \setminus \mathcal{F}$. Let $F = \bigcup \mathcal{F}$; then F is a closed subset of X and

$V \subseteq F$. By hypothesis and the previous lemma, $f|F \in C(F,Y)$. Suppose W is open in Y and that $f(p) \in W$. There exists an open set U of X such that $p \in U \cap F \subseteq (f|F)^{\leftarrow}[W]$. Thus $p \in U \cap V \subseteq f^{\leftarrow}[W]$ and so f is continuous at p. Thus $f \in C(X,Y)$.

■

(d) **Theorem**. Let X be a dense subspace of a space Y, let Z be a space, and let $f,g \in C(Y,Z)$. If $f|X = g|X$ then $f = g$.

Proof. Let $A = \{y \in Y : f(y) = g(y)\}$. By hypothesis A contains X and thus is dense in Y. Hence it suffices to prove that A is closed in Y.

Let $y_0 \in Y \setminus A$. Then $f(y_0) \neq g(y_0)$ so there are disjoint open sets U and V of Z such that $f(y_0) \in U$ and $g(y_0) \in V$. Then $f^{\leftarrow}[U] \cap g^{\leftarrow}[V]$ is an open set of Y containing y_0 and disjoint from A. It follows that A is closed in Y.

■

1.7 Product spaces and evaluation maps

In this section we review the definition and basic properties of product spaces, and then discuss the evaluation map from a space into a product of spaces.

(a) **Definition**. Let X be a set, let $\{Y_i : i \in I\}$ be a set of topological spaces, and let $F_i \subseteq F(X,Y_i)$ for each $i \in I$. The **weak topology** induced on X by the family $\cup\{F_i : i \in I\}$ is defined to be the topology τ on X for which $\{f_i^{\leftarrow}[V_i] : f_i \in F_i, V_i$ is open in Y_i, and $i \in I\}$ is a subbase. That is, τ is the smallest topology on X for which $F_i \subseteq C((X,\tau),Y_i)$ for each $i \in I$.

(b) Lemma. Let X have the weak topology induced by $\cup\{F_i : i \in I\}$ as described above. Let Z be a space and let $g \in F(Z,X)$. Then $g \in C(Z,X)$ iff $f_i \circ g \in C(Z,Y_i)$ for each $f_i \in F_i$ and each $i \in I$.

Proof. If $g \in C(Z,X)$, then obviously $f_i \circ g \in C(Z,Y_i)$ for each $i \in I$. To prove the converse, let $i \in I$, let V_i be open in Y_i, and let $f_i \in F_i$. Then $g^{\leftarrow}[f_i^{\leftarrow}[V_i]] = (f_i \circ g)^{\leftarrow}[V_i]$, which is open in X as $f_i \circ g \in C(Z,Y_i)$. Thus X has an open subbase whose inverse images under g are open in Z. It follows that $g \in C(Z,X)$.

■

(c) Definition. Let $\{X_i : i \in I\}$ be a set of spaces. The **product topology** on the product set $\Pi\{X_i : i \in I\}$ is defined to be the weak topology induced on the product set by the set of projection functions $\{\Pi_i : i \in I\}$ (see 1.1(f)). We call $\Pi\{X_i : i \in I\}$, equipped with the product topology, the **product space**, or just the **product**, of $\{X_i : i \in I\}$. Explicitly, $\{\cap\{\Pi_i^{\leftarrow}[V_i] : i \in F\} : F$ is a finite subset of I and V_i is open in X_i for each $i \in F\}$ is an open base for the product space. If X is a space and $X_i = X$ for each $i \in I$, then

$\Pi\{X_i : i \in I\}$ is denoted by X^I.

The following result is well-known; part (2) is an immediate consequence of 1.7(b) and characterizes product spaces; see 1E.

(d) **Proposition.** Let X be the product of $\{X_i : i \in I\}$.

(1) Each projection function Π_i is an open map from X onto X_i.

(2) Let Y be a space and let $f \in F(Y,X)$. Then $f \in C(Y,X)$ iff $\Pi_i \circ f \in C(Y,X_i)$ for each $i \in I$.

Corresponding to products of spaces, we can define products of functions.

(e) **Definition.** Let $\{X_i : i \in I\}$ and $\{Y_i : i \in I\}$ be two sets of spaces and let $f_i \in F(X_i,Y_i)$ for each $i \in I$. Let $\Pi\{X_i : i \in I\}$ be denoted by X and $\Pi\{Y_i : i \in I\}$ be denoted by Y. The **product function** $\Pi\{f_i : i \in I\}$ is the member of F(X,Y) defined as follows:

$$\Pi\{f_i : i \in I\} (\langle x_i \rangle_{i \in I}) = \langle f_i(x_i) \rangle_{i \in I}$$

(f) **Proposition.** Let $\{X_i : i \in I\}$, $\{Y_i : i \in I\}$, $\{f_i : i \in I\}$, X and Y be as in the preceding definition. If $f_i \in C(X_i,Y_i)$ for each $i \in I$, then $\Pi\{f_i : i \in I\} \in C(X,Y)$.

Proof. Let f denote $\Pi\{f_i : i \in I\}$. Let Π_j^X denote the jth

projection function from X onto X_j, and define Π_j^Y similarly. A straightforward computation shows that

$$f^{\leftarrow}[\Pi_j^{Y\leftarrow}[V_j]] = (\Pi_j^X)^{\leftarrow}[f_j^{\leftarrow}[V_j]]$$

for each $j \in I$ and each subset V_j of Y_j. Thus inverse images under f of subbasic open sets of Y are open sets of X, and so f is continuous. ■

The following technical result will be useful in 1.8(g) when we discuss products of perfect functions.

(g) **Theorem**. Let $\{X_i : i \in I\}$ be a set of spaces, let K_i be a compact subspace of X_i for each $i \in I$, and let W be an open subset of $\Pi\{X_i : i \in I\}$ that contains $\Pi\{K_i : i \in I\}$. Then there exists a finite subset F of I, and open subsets U_i of X_i for each $i \in F$, such that $\Pi\{K_i : i \in I\} \subseteq \cap\{\Pi_i^{\leftarrow}[U_i] : i \in F\} \subseteq W$.

Proof. First suppose $|I| = 2$; let $I = \{1,2\}$. Fix $z \in K_2$. For each $y \in K_1$ there exist open sets $V_1(y)$ of X_1 and $V_2(y)$ of X_2 such that $(y,z) \in V_1(y) \times V_2(y) \subseteq W$. Then $K_1 \subseteq \cup\{V_1(y) : y \in K_1\}$ so by the compactness of K_1 there is a finite subset S of K_1 such that $K_1 \subseteq \cup\{V_1(y) : y \in S\}$. Let $T_1(z) = \cup\{V_1(y) : y \in S\}$ and $T_2(z) = \cap\{V_2(y) : y \in S\}$. Then $K_1 \times \{z\} \subseteq T_1(z) \times T_2(z) \subseteq W$ for each $z \in K_2$. As K_2 is compact and contained in $\cup\{T_2(z) : z \in K_2\}$, there is a finite subset Q of K_2 such that $K_2 \subseteq \bigcup\{T_2(z) : z \in Q\}$. Let $U_1 = \bigcap\{T_1(z) : z \in Q\}$ and $U_2 = \bigcup\{T_2(z) : z \in Q\}$. Then $K_1 \times K_2 \subseteq U_1 \times U_2 \subseteq W$, and the theorem is proved for the case in which $|I| = 2$.

Now assume I is finite. We prove the theorem in this case by induction on $|I|$. Let $|I| = k$ and assume the theorem holds for any product of k-1 spaces, where $k \geqslant 3$. Let $j \in I$. By identifying the space $\Pi\{X_i : i \in I\}$ (respectively $\Pi\{K_i : i \in I\}$) with the product $(\Pi\{X_i : i \in I\setminus\{j\}\}) \times X_j$ (respectively $(\Pi\{K_i : i \in I\setminus\{j\}\}) \times K_j$) and applying the special case above, we can find open sets V of $\Pi\{X_i : i \in I\setminus\{j\}\}$ and U_j of X_j such that

$$\Pi\{K_i : i \in I\} \subseteq V \times U_j \subseteq W \tag{1}$$

By our induction hypothesis we can find open subsets U_i of X_i (for each $i \in I\setminus\{j\}$) such that

$$\Pi\{K_i : i \in I\setminus\{j\}\} \subseteq \Pi\{U_i : i \in I\setminus\{j\}\} \subseteq V \tag{2}$$

Combining (1) and (2) we obtain

$$\Pi\{K_i : i \in I\} \subseteq \Pi\{U_i : i \in I\} \subseteq W,$$

and we have proved the theorem for the case in which I is finite.

Now we prove the theorem for an arbitrary index set I. By the compactness of $\Pi\{K_i : i \in I\}$ there exists a finite collection

$\{F_j : 1 \leqslant j \leqslant n\}$ of finite subsets of I, and a collection of sets $\{E(s) : s \in \cup\{F_j : 1 \leqslant j \leqslant n\}\}$ such that E(s) is an open subset of X_s for each $s \in \cup\{F_j : 1 \leqslant j \leqslant n\}$ and

$$(3) \qquad \Pi\{K_i : i \in I\} \subseteq \cup\{\cap\{\Pi_s^{\leftarrow}[E(s)] : s \in F_j\} : 1 \leqslant j \leqslant n\} \subseteq W$$

Let $F = \cup\{F_j : 1 \leqslant j \leqslant n\}$. Let $G = \cup\{\{\langle x_s\rangle_{s\in F} : x_s \in E(s)$ for each $s \in F_j\} : 1 \leqslant j \leqslant n\}$; G is an open subset of $\Pi\{X_i : i \in F\}$. If $t \in I\setminus F$, define X_t to be $\Pi\{X_i : i \in I\setminus F\}$, and define K_t to be $\Pi\{K_i : i \in I\setminus F\}$. Consider the product space $\Pi\{X_i : i \in F \cup \{t\}\}$. This product space can be identified with $\Pi\{X_i : i \in I\}$, its subspace $\Pi\{K_i : i \in F \cup \{t\}\}$ can be identified with $\Pi\{K_i : i \in I\}$, and with these identifications (3) becomes

$$\Pi\{K_i : i \in F \cup \{t\}\} \subseteq G \times X_t \subseteq W$$

Using the fact that our theorem is true for finite index sets, we see that there exist open sets $\{U_j : j \in F\}$ such that $\Pi\{K_i : i \in F \cup \{t\}\} \subseteq \cap\{\Pi_j^{\leftarrow}[U_j] : j \in F\} \subseteq G \times X_t \subseteq W$. Then $\Pi\{K_i : i \in I\} \subseteq \cap\{\Pi_j^{\leftarrow}[U_j] : j \in F\} \subseteq W$ as claimed.

■

Just as we defined hereditary topological properties in 1.2, we now define productive topological properties.

(h) **Definition**. Let α be a cardinal. A topological property *P* is called **α-productive** if, whenever $\{X_i : i \in I\}$ is a set of spaces each of which has P, and $|I| < \alpha$, then $\Pi\{X_i : i \in I\}$ has P. If P is α-productive for every α, then P is said to be **productive.** If P is $\aleph_0$-productive then P is said to be **finitely productive.**

Compactness is a productive property; so is the property of being Tychonoff (see 1F(3)). Metrizability is $\aleph_1$-productive, and separability is $\left(2^{\aleph_0}\right)^+$-productive (see, for example, 30), but neither is productive. Local compactness is finitely productive, but a product of infinitely many nonempty locally compact spaces is locally compact iff all but finitely many of the spaces are compact. The property of being normal, and the Lindelof property, are not even finitely productive.

One way to construct extensions of a space, which we will use several times in Chapters 4 and 5, is to embed a space in a product space. The embedding function will usually be an "evaluation map"; we introduce and discuss these now.

(i) <u>Definition</u>.

(1) Let X be a space, let $\{X(i) : i \in I\}$ be a set of spaces, and let $F_i \subseteq C(X,X(i))$ for each $i \in I$. Let $F = \cup\{F_i : i \in I\}$. For each $i \in I$ and each $f \in F_i$, let X_f be homeomorphic to $X(i)$. Let Y denote the product space $\Pi\{X_f : f \in F\}$. The **evaluation map** e_F corresponding to F is the member of $F(X,Y)$ defined as follows: $\Pi_f \circ e_F = f$ for each $f \in F$. (In other words $e_F(x) = \langle f(x)\rangle_{f \in F}$ for each $x \in X$.)

(2) The family F defined above **separates points and closed sets of X** if for each closed subset A of X and each $p \in X \setminus A$, there exists $i \in I$ and $f \in F_i$ such that $f(p) \notin c\ell_{X(i)} f[A]$.

(j) **Embedding Theorem**. Let X, $\{X(i) : i \in I\}$, F_i, F, Y and e_F be as in the preceding definition. Then:

(1) $e_F \in C(X,Y)$, and

(2) If F separates points and closed sets of X, then e_F is an embedding.

Proof

(1) Since $\Pi_f \circ e_F = f \in C(X,X(i))$ for each $i \in I$ and each $f \in F_i$, then e_F is continuous by 1.7(d).

(2) We must show that e_F is a one-to-one open function from X onto $e_F[X]$. If x_0 and x_1 are distinct points of X then there exists $i \in I$ and $f \in F_i$ such that $f(x_0) \notin c\ell_{X(i)}\{f(x_1)\}$. Thus $f(x_0) \neq f(x_1)$ so $\Pi_f \circ e_F(x_0) \neq \Pi_f \circ e_F(x_1)$. Hence $e_F(x_0) \neq e_F(x_1)$ so e_F is one-to-one.

Now suppose that V is open in X and that $x_0 \in V$. There exists $i \in I$ and $f \in F_i$ such that $f(x_0) \notin c\ell_{X(i)} f[X \setminus V]$. We now claim that

$$(1) \qquad e_F(x_0) \in e_F[X] \cap \Pi_f^{\leftarrow}[X(i) \setminus c\ell_{X(i)} f[X \setminus V]] \subseteq e_F[V]$$

which will show that $e_F[V]$ is open in $e_F[X]$. Obviously $\Pi_f(e_F(x_0)) = f(x_0) \in X(i) \setminus c\ell_{X(i)} f[X \setminus V]$ and so $e_F(x_0) \in$

$\Pi_f^{\leftarrow}[X(i)\backslash c\ell_{X(i)}f[X\backslash V]]$. Next, if $x_1 \in X$ and $e_F(x_1) \in \Pi_f^{\leftarrow}[X(i)\backslash c\ell_{X(i)}f[X\backslash V]]$, then $f(x_1) = \Pi_f \circ e_F(x_1) \in X(i)\backslash c\ell_{X(i)}f[X\backslash V]$. It follows that $x_1 \in V$ and so (1) holds. Thus e_F is an embedding.

■

1.8 Perfect functions

Perfect functions are ubiquitous in the theory of extensions, and also form an integral part of the definition of the absolute of a space. Consequently we shall be working with perfect functions throughout this book. Here we collect the most important properties of these functions.

(a) **Definition**. Let X and Y be spaces and let $f \in F(X,Y)$.

(1) f is a **compact** function if $f^{\leftarrow}(y)$ is a compact subset of X for each $y \in Y$.

(2) f is a **perfect** function if f is a closed, compact function.

Note that we do not assume that perfect functions are either continuous or onto. However, if $f : X \to Y$ is perfect then $f[X]$ will be a closed subset of Y. The following theorem provides a class of examples of perfect functions.

(b) **Theorem**. Continuous functions with compact domain are perfect.

__Proof__. Let X be compact, let Y be any space, and let $f \in C(X,Y)$. If A is a closed subset of X it is compact, so $f[A]$ is a compact subset of Y and hence closed in Y. If $y \in Y$, then $f^{\leftarrow}(y)$ is closed in X and hence is compact.

■

Perfect continuous functions preserve (in both directions) most topological properties that are described in terms of the existence of certain kinds of open subcovers or refinements of a given open cover of a certain type. (A typical theorem of this sort appears in 1J.) The reason that perfect continuous functions preserve such properties is that they transfer certain open covers of their domain to open covers of their range in a standard fashion, as described in the next theorem.

(c) __Theorem__. Let X and Y be spaces, and let $f \in F(X,Y)$. The following are equivalent.

(1) f is perfect

(2) For each $y \in Y$, $f^{\leftarrow}(y)$ is a closed subset of X, and if $\mathcal{C}$ is an open cover of X that is closed under finite unions, then $\{Y \setminus f[X \setminus C] : C \in \mathcal{C}\}$ is an open cover of Y.

__Proof__

(1) ⇒ (2) Let f be perfect. Since $f^{\leftarrow}(y)$ is compact for each $y \in Y$, $f^{\leftarrow}(y)$ is closed in X. Let $\mathcal{C}$ be an open cover of X that is closed under finite unions, and let $f^{*}[\mathcal{C}] = \{Y \setminus f[X \setminus C] : C \in \mathcal{C}\}$.

If $C \in \mathcal{C}$, then since f is a closed function the set $Y \setminus f[X \setminus C]$ is an open set of Y. Thus it suffices to show that $f^*[\mathcal{C}]$ covers Y.

Let $y \in Y$. If $y \notin f[X]$ then $y \in Y \setminus f[X \setminus C]$ for each $C \in \mathcal{C}$. If $y \in f[X]$, then since f is perfect $f^{\leftarrow}(y)$ is a compact subset of X. Thus there is a finite subset $\mathcal{F}$ of $\mathcal{C}$ such that $f^{\leftarrow}(y) \subseteq \bigcup\{C : C \in \mathcal{F}\} = C(y)$. By hypothesis $C(y) \in \mathcal{C}$, and evidently $y \in Y \setminus f[X \setminus C(y)]$. Thus $f^*[\mathcal{C}]$ covers Y.

(2) $\Rightarrow$ (1) Let V be open in X. Then $\{X,V\}$ is an open cover of X that is closed under finite unions, so by hypothesis $\{Y, Y \setminus f[X \setminus V]\}$ is an open cover of Y. Thus $Y \setminus f[X \setminus V]$ is open in Y, and so f is a closed function.

Now let $y \in Y$. If $y \in Y \setminus f[X]$ then $f^{\leftarrow}(y) = \emptyset$ and hence is compact. Otherwise let $\mathcal{C}$ be a collection of open sets that covers $f^{\leftarrow}(y)$, and let $\mathcal{C}_1$ denote the set of all unions of finite subcollections of $\mathcal{C} \cup \{X \setminus f^{\leftarrow}(y)\}$. As $f^{\leftarrow}(y)$ is a closed subset of X, $\mathcal{C}_1$ is an open cover of X that is closed under finite unions. Thus by hypothesis there exists $C \in \mathcal{C}_1$ such that $y \in Y \setminus f[X \setminus C]$. Therefore $f^{\leftarrow}(y) \subseteq C$. There must exist a finite subcollection $\mathcal{F}$ of $\mathcal{C}$ such that $C \subseteq (X \setminus f^{\leftarrow}(y)) \cup [\bigcup \mathcal{F}]$. Thus $f^{\leftarrow}(y) \subseteq \bigcup \mathcal{F}$. It follows that $f^{\leftarrow}(y)$ is compact, and so f is perfect.

■

(d) **Lemma**. Let X and Y be spaces, and let $f : X \to Y$ be perfect. If K is a compact subset of Y then $f^{\leftarrow}[K]$ is a compact subset of X.

Proof. Let $\mathcal{U}$ be a collection of open sets of X that covers $f^{\leftarrow}[K]$, and let $\mathcal{C}$ be the collection of all unions of finite subcollections of $\mathcal{U}$. If $k \in K$ then $f^{\leftarrow}(k) \subseteq \bigcup\mathcal{U}$, so as $f^{\leftarrow}(k)$ is compact there exists $C(k) \in \mathcal{C}$ such that $f^{\leftarrow}(k) \subseteq C(k)$. Thus $k \in Y\setminus f[X\setminus C(k)]$. Hence $K \subseteq \bigcup\{Y\setminus f[X\setminus C] : C \in \mathcal{C}\}$ and so as K is compact, there exist $C_1, \ldots, C_n \in \mathcal{C}$ such that $K \subseteq \bigcup\{Y\setminus f[X\setminus C_i] : i = 1 \text{ to } n\}$. It follows that $f^{\leftarrow}[K] \subseteq \bigcup\{C_i : i = 1 \text{ to } n\}$, and as each C_i is a union of a finite subfamily of $\mathcal{U}$, $f^{\leftarrow}[K]$ is covered by a finite subfamily of $\mathcal{U}$. Thus $f^{\leftarrow}[K]$ is compact.

■

(e) **Theorem**. A composition of perfect functions is perfect.

Proof. Let X, Y, Z be spaces. Let $f : X \to Y$ and $g : Y \to Z$ be perfect functions. Then $g \circ f$ is a closed function as g and f are. If $z \in Z$, then $g^{\leftarrow}(z)$ is a compact subset of Y, so by the preceding lemma $(g \circ f)^{\leftarrow}(z) = f^{\leftarrow}(g^{\leftarrow}(z))$ is a compact subset of X. Thus $g \circ f$ is perfect.

■

Restrictions of perfect functions are not in general perfect, but in two important special cases they are.

(f) **Proposition**. Let $f : X \to Y$ be a perfect function.

(1) If A is a closed subset of X then $f|A$ is perfect (whether regarded as a function to Y or as a function to $f[A]$).

(2) If $B \subseteq Y$ then $f|f^{\leftarrow}[B]$ is a perfect function from $f^{\leftarrow}[B]$ to B.

<u>Proof</u>

(1) If C is closed in A then it is closed in X and so $(f \mid A)[C] = f[C]$ is closed in Y (and in $f[A]$). If $y \in Y$ then $(f \mid A)^{\leftarrow}(y) = A \cap f^{\leftarrow}(y)$, which is a closed subset of the compact space $f^{\leftarrow}(y)$, and hence is compact.

(2) If C is closed in X then $(f \mid f^{\leftarrow}[B])[C \cap f^{\leftarrow}[B]] = f[C] \cap B$, which is closed in B as f is closed. Thus $f \mid f^{\leftarrow}[B] : f^{\leftarrow}[B] \to B$ is a closed function. If $b \in B$ then $(f \mid f^{\leftarrow}[B])^{\leftarrow}(b) = f^{\leftarrow}(b)$, which is compact. Thus $f \mid f^{\leftarrow}[B]$ is a compact function, and therefore is perfect.

■

We now show that a product of perfect functions is a perfect function.

(g) **<u>Theorem</u>**. Let $\{X_i : i \in I\}$ and $\{Y_i : i \in I\}$ be two sets of spaces and let f_i be a perfect function from X_i to Y_i for each $i \in I$. Then $\Pi\{f_i : i \in I\}$ is a perfect function from $\Pi\{X_i : i \in I\}$ to $\Pi\{Y_i : i \in I\}$.

<u>Proof</u>. Denote $\Pi\{X_i : i \in I\}$, $\Pi\{Y_i : i \in I\}$, and $\Pi\{f_i : i \in I\}$ respectively by X, Y, and f. If $y = \langle y_i \rangle_{i \in I} \in Y$, then $f^{\leftarrow}(y) = \Pi\{f_i^{\leftarrow}(y_i) : i \in I\}$, which is compact as each $f_i^{\leftarrow}(y_i)$ is.

Now suppose A is a closed subset of X and that $y = \langle y_i \rangle_{i \in I} \in Y \setminus f[A]$. Then $\Pi\{f_i^{\leftarrow}(y_i) : i \in I\} = f^{\leftarrow}(y) \subseteq X \setminus A$. By 1.7(g) there exists a finite subset F of I, and open sets U_i of X_i, for each $i \in F$, such that $f^{\leftarrow}(y) \subseteq \cap\{\Pi_i^{\leftarrow}[U_i] : i \in F\} \subseteq X \setminus A$.

Thus $f_i^{\leftarrow}(y_i) \subseteq U_i$ for each $i \in F$; equivalently $y_i \in Y_i \setminus f_i[X_i \setminus U_i] = V_i$ for each $i \in F$. As f_i is a closed function, each V_i is open in Y_i and $f_i^{\leftarrow}[V_i] \subseteq U_i$. Thus, $y \in \bigcap\{\Pi_i^{\leftarrow}[V_i] : i \in F\}$. Furthermore, if $z = \langle z_i \rangle_{i \in I} \in \bigcap\{\Pi_i^{\leftarrow}[V_i] : i \in F\}$ then $z_i \in Y_i \setminus f_i[X \setminus U_i]$ for each $i \in F$. Thus $f_i^{\leftarrow}(z_i) \subseteq U_i$ for each $i \in F$, and so $f^{\leftarrow}(z) \subseteq \bigcap\{\Pi_i^{\leftarrow}[U_i] : i \in F\} \subseteq X \setminus A$. It follows that $z \in Y \setminus f[A]$. Hence $\bigcap\{\Pi_i^{\leftarrow}[V_i] : i \in F\}$ is a Y-neighborhood of y disjoint from f[A]. It follows that f[A] is closed in Y and so f is perfect.

■

Perfect continuous surjections preserve some separation properties. In particular:

(h) **Theorem**. A perfect continuous image of a regular space is regular.

Proof. Let $f : X \to Y$ be a perfect continuous surjection and let X be regular. Let A be a closed subset of Y and let $y \in Y \setminus A$. Then $f^{\leftarrow}(y)$ is a compact subset of X that is disjoint from the closed subset $f^{\leftarrow}[A]$. As X is regular, for each $x \in f^{\leftarrow}(y)$ there exist disjoint open sets U(x) and V(x) of X such that $x \in U(x)$ and $f^{\leftarrow}[A] \subseteq V(x)$. As $f^{\leftarrow}(y)$ is compact, there exists a finite subset F of $f^{\leftarrow}(y)$ such that $f^{\leftarrow}(y) \subseteq \bigcup\{U(x) : x \in F\} = U$. Let $V = \bigcap\{V(x) : x \in F\}$. Then V is open, $f^{\leftarrow}[A] \subseteq V$, and $U \cap V = \emptyset$. Because f is a closed surjection and $U \cap V = \emptyset$, it follows that $Y \setminus f[X \setminus U]$ and $Y \setminus f[X \setminus V]$ are disjoint neighborhoods in Y of y and A respectively. Hence Y is regular.

■

See 1J for the "other half" of this result.

Our final result will be fundamental when we study the interaction of perfect functions and extensions of topological spaces.

(i) **Theorem.** Let X and Y be spaces. If $f \in C(X,Y)$, S is dense in X, and $f|S$ is a perfect function from S onto $f[S]$, then $f[X\setminus S] \subseteq Y\setminus f[S]$.

Proof. Suppose not; then there exists $x \in X\setminus S$ such that $f(x) \in f[S]$. Let $T = S \cup \{x\}$. Since $f|S$ is a compact function, $f^{\leftarrow}(f(x)) \cap S$ is a compact set K and thus closed in T. As X is Hausdorff there exists an open subset U of T such that $K \subseteq U$ and $x \notin c\ell_T U$. Now $c\ell_T U \cup c\ell_T(S\setminus U) = T$, since S is dense in X and thus in T; hence

$$f(x) \in f[c\ell_T(S\setminus U)] \subseteq c\ell_{f[S]} f[S\setminus U] = f[S\setminus U] ;$$

the last equality follows since $f|S$ is a closed function. Thus $K\setminus U \neq \emptyset$, which is a contradiction. The theorem follows.

■

Another useful (and surprising) property of perfect functions appears in 8.4(d).

1.9 C- and C^*-embedding

In this section we investigate when certain types of real-valued continuous functions can be continuously extended from a subspace of a space X to all of X. Most of the results in this section may be found in Chapter 1 of [GJ].

(a) __Definition__. If X is a space then $C^*(X)$ is defined to be $\{f \in C(X) :$ there exists $M \in \mathbb{R}$ such that $|f(x)| \leq M$ for each $x \in X\}$.

Members of $C^*(X)$ are called **bounded** continuous functions. Evidently $f \in C^*(X)$ iff $f \in C(X)$ and there exists $m \in \mathbb{R}$ such that $|f| \leq \underline{m}$. It is routine to verify that $C^*(X)$ is a subring of $C(X)$; as we will see in Chapter 2, $C^*(X)$ is also a sublattice of $C(X)$.

(b) __Definition__. Let A be a subspace of a space X. A is said to be **C^*-embedded** (respectively **C-embedded**) in X if for each $g \in C^*(A)$ (respectively, $g \in C(A)$) there exists $f \in C^*(X)$ (respectively, $C(X)$) such that $f|A = g$.

(c) __Proposition__.

(1) Each C-embedded subspace of a space X is C^*-embedded in X.

(2) If A is C^*-embedded (respectively, C-embedded) in B and B is C^*-embedded (respectively, C-embedded) in X, then A is C^*-embedded (respectively, C-embedded) in X.

Proof.

(1) Let A be a C-embedded subspace of X and let $f \in C^*(A)$. Then $f \in C(A)$ so there exists $g \in C(X)$ such that $g|A = f$. There exists $m \in \mathbb{R}$ such that $|f| \leqslant \underline{m}$. Let $h = (-\underline{m} \vee g) \wedge \underline{m}$. Then $h \in C^*(X)$ and $h|A = f$, so A is C^*-embedded in X.

(2) This is obvious.

■

A fundamental problem of topology is to characterize the C^*-embedded and C-embedded subspaces of a given topological space, or, alternatively, to characterize those topological spaces all of whose subspaces of a certain kind are C^*-embedded (or C-embedded) in X. The notion of two subsets of a space being completely separated from one another is closely related to this problem.

(d) Definition. The subsets A and B of the space X are **completely separated in** X if there exists $f \in C^*(X)$ such that $\underline{0} \leqslant f \leqslant \underline{1}$, $f[A] \subseteq \{0\}$, and $f[B] \subseteq \{1\}$. We say that f **completely separates** A **and** B **in** X.

(e) Proposition. Let X be a space.

(1) If $A \in \mathcal{B}(X)$ then A and $X \backslash A$ are completely separated in X.

(2) If A and B are completely separated in X, and if $C \subseteq B$, then A and C are completely separated in X.

(3) X is Tychonoff iff each closed subset of X is completely separated from each singleton set disjoint from it.

Proof.

(1) $\chi_A \in C^*(X)$ and completely separates A and $X \setminus A$ in X.

(2) Obvious.

(3) See 1.4(a).

■

(f) **Theorem.** Two subsets of X are completely separated in X iff they are contained in disjoint zero-sets of X.

Proof. Let A, B $\subseteq$ X. Suppose $A \subseteq Z_1$, $B \subseteq Z_2$, $Z_1 \cap Z_2 = \emptyset$, and $Z_1, Z_2 \in Z(X)$. By 1.4(f) there exist $f_1, f_2 \in C(X)$ such that $\underline{0} \leqslant f_i \leqslant \underline{1}$ and $Z_i = Z(f_i)$ $(i = 1,2)$. Define $g \in F(X,\mathbb{R})$ by:

$$g(x) = \frac{[f_1(x)]^2}{[f_1(x)]^2 + [f_2(x)]^2}$$

Since $Z(f_1) \cap Z(f_2) = \emptyset$, g is well-defined and continuous. Evidently $\underline{0} \leqslant g \leqslant \underline{1}$ and $g[Z(f_1)] \subseteq \{0\}$ while $g[Z(f_2)] \subseteq \{1\}$. Thus g completely separates A and B in X.

Conversely, suppose A and B are completely separated in X. There exists $f \in C^*(X)$ such that $f[A] \subseteq \{0\}$ and $f[B] \subseteq \{1\}$. Then $Z(f)$ and $f^{\leftarrow}(1)$ are the required disjoint zero-sets.

■

(g) **Corollary.** The following are equivalent for subsets A and B of a space X.

(1) A and B are completely separated in X.

(2) There exists $f \in C(X)$, and distinct numbers a and b in $\mathbb{R}$, such that $f[A] \subseteq \{a\}$ and $f[B] \subseteq \{b\}$.

(3) There exists $f \in C(X)$ such that $cl_{\mathbb{R}} f[A] \cap cl_{\mathbb{R}} f[B] = \emptyset$.

Proof

(1) $\Rightarrow$ (2) Take $a = 0$ and $b = 1$ and then apply the definition of complete separation.

(2) $\Rightarrow$ (3) This is obvious.

(3) $\Rightarrow$ (1) By 1.4(j) $f^{\leftarrow}[cl_{\mathbb{R}} f[A]]$ and $f^{\leftarrow}[cl_{\mathbb{R}} f[B]]$ are disjoint zero-sets of X containing A and B respectively. Hence by the preceding theorem A and B are completely separated in X.

■

The most powerful criterion for deciding whether a subspace of X is C^*-embedded in X is given by Urysohn's extension theorem.

(h) **Theorem** (Urysohn's extension theorem). A subspace S of a space X is C^*-embedded in X iff any two completely separated sets in S are completely separated in X.

Proof. Suppose that S is C^*-embedded in X and that A and B are completely separated sets in S. Then there exists $f \in C^*(S)$ such that $\underline{0} \leq f \leq \underline{1}$ and $f[A] \subseteq \{0\}$, $f[B] \subseteq \{1\}$. By hypothesis there exists $g \in C^*(X)$ such that $g|S = f$. Then $\underline{0} \vee (g \wedge \underline{1}) \in C^*(X)$ and completely separates A and B in X.

Conversely, suppose that any two completely separated sets in S are completely separated in X, and suppose that $f \in C^*(S)$. Find

$M \in \mathbb{R}$ such that $|f| \leq \underline{M}$. For each positive integer n define r_n to be $(M/2)(2/3)^n$. Then $|f| \leq 3\underline{r}_1$. Let $f_1 = f$, let $A_1 = \{s \in S : f_1(s) \leq -r_1\}$, and let $B_1 = \{s \in S : f_1(s) \geq r_1\}$. By 1.9(f) and 1.4(j), A_1 and B_1 are completely separated in S, so by hypothesis there exists $h_1 \in C^*(X)$ such that $\underline{0} \leq h_1 \leq \underline{1}$, $h_1[A_1] \subseteq \{0\}$, and $h_1[B_1] \subseteq \{1\}$. Let $g_1 = 2\underline{r}_1 h_1 - \underline{r}_1$; then $g_1 \in C^*(X)$ and $g_1[A_1] \subseteq \{-r_1\}$, $g_1[B_1] \subseteq \{r_1\}$, and $|g_1| \leq \underline{r}_1$.

Let $n > 1$. Proceeding inductively, suppose that for each $k < n$ we have defined $f_k \in C^*(S)$, subsets A_k and B_k of S, and $g_k \in C^*(X)$, such that:

(1) $|f_k| \leq 3\underline{r}_k$

(2) $A_k = \{s \in S : f_k(s) \leq -r_k\}$

(3) $B_k = \{s \in S : f_k(s) \geq r_k\}$

(4) $|g_k| \leq \underline{r}_k$, $g_k[A_k] \subseteq \{-r_k\}$, $g_k[B_k] \subseteq \{r_k\}$.

Let $f_n = f_{n-1} - g_{n-1}|S$. Then $f_n \in C^*(S)$ and by (1) and (4), $|f_n| \leq 3\underline{r}_n$. Define A_n to be $\{s \in S : f_n(s) \leq -r_n\}$ and B_n to be $\{s \in S : f_n(s) \geq r_n\}$. Arguing as we did when we constructed g_1, we can find $g_n \in C^*(X)$ such that $|g_n| \leq \underline{r}_n$, $g_n[A_n] \subseteq \{-r_n\}$ and $g_n[B_n] \subseteq \{r_n\}$. This completes our inductive construction of f_n, g_n, A_n and B_n.

By the Weierstrass M-test and the convergence of $\sum_{n=1}^{\infty} r_n$, it follows that if $x \in X$ then $\sum_{n=1}^{\infty} g_n(x)$ converges to a real number which we denote by $g(x)$, thereby defining $g \in F(X,\mathbb{R})$. Proceeding exactly as in the proof of 1.4(i), we can show that $g \in C(X)$. As $\sum_{n=1}^{\infty} r_n = M$, we see that $|g| \leq \underline{M}$;

thus $g \in C^*(X)$. Since

$$g_1 + \dots + g_n | S = (f_1 - f_2) + (f_2 - f_3) + \dots + (f_n - f_{n+1})$$
$$= f_1 - f_{n+1}$$

and as $\lim_{n\to\infty} f_n(s) = 0$ (by (1) and the fact that $\lim_{n\to\infty} r_n = 0$), it follows that $g(s) = f_1(s) = f(s)$ for each $s \in S$. Thus $g|S = f$ and S is C^*-embedded in X.

■

(i) **Corollary**. Let K be a compact subspace of the Tychonoff space X. Then

(1) K is completely separated in X from each closed set of X disjoint from it, and

(2) K is C^*-embedded in X.

Proof

(1) Let A be closed in X and disjoint from K. For each $x \in K$ there exists $f_x \in C^*(X)$ such that $\underline{0} \leq f_x \leq \underline{1}$, $f_x(x) = 0$, and $f_x[A] \subseteq \{1\}$. By compactness there exist finitely many points $x(1), \dots, x(n) \in K$ such that $K \subseteq \bigcup\{f_{x(i)}^{\leftarrow}[[0,\frac{1}{2}]] : i = 1 \text{ to } n\} = B$. By 1.4(h,i), B and $\bigcap\{f_{x(i)}^{\leftarrow}(1) : i = 1 \text{ to } n\}$ are disjoint zero-sets of X containing K and A respectively. Thus by 1.9(f) K and A are completely separated in X.

(2) Let A and B be completely separated in K. By 1.9(f) there exist disjoint zero-sets Z_1 and Z_2 in $Z(K)$ such that $A \subseteq Z_1$

and $B \subseteq Z_2$. As Z_2 is compact (being closed in K), by (1) above Z_2 is completely separated in X from Z_1, since Z_1 is closed in X and disjoint from Z_2. Thus K is C^*-embedded in X by Urysohn's extension theorem.

■

By 1.9(c)(1) if S is C-embedded in a space X, then S is C^*-embedded in X; the next result gives a necessary and sufficient condition for the converse to be true.

(j) **Theorem**. Let S be a C^*-embedded subspace of X. Then S is C-embedded in X iff S is completely separated in X from every zero-set of X disjoint from S.

Proof. Suppose S is C-embedded in X and let $Z \in Z(X)$ such that $Z \cap S = \emptyset$. Let $f \in C(X)$ such that $Z = Z(f)$. Define $g \in F(S,\mathbb{R})$ by $g(s) = \frac{1}{f(s)}$ for each $s \in S$. Then $g \in C(S)$, so by hypothesis there exists $k \in C(X)$ such that $k|S = g$. Let $q = fk$. Then $q \in C(X)$, $q[Z] \subseteq \{0\}$, and $q[S] \subseteq \{1\}$. Thus S and Z are completely separated in X.

Conversely, suppose that S is completely separated in X from each zero-set of X disjoint from it. Let h be a homeomorphism from $\mathbb{R}$ onto $(-1,1)$ and let $f \in C(S)$. Then $h \circ f \in C^*(S)$, so as S is C^*-embedded in X there exists $g \in C^*(X)$ such that $g|S = h \circ f$. Let $Z = \{x \in X : |g(x)| \geq 1\}$. If $s \in S$ then $g(s) = h(f(s)) \in (-1,1)$ so $Z \cap S = \emptyset$. By 1.4(h) $Z \in Z(X)$, so Z and S are completely separated in X. Thus there exists $k \in C(X)$ such that $\underline{0} \leqslant k \leqslant \underline{1}$, $k[Z] \subseteq \{0\}$, and $k[S] \subseteq \{1\}$. Thus $h^{\leftarrow} \circ (gk) \in C(X)$ and if $s \in S$,

$$\begin{aligned} h^{\leftarrow} \circ (gk)(s) &= h^{\leftarrow}(g(s)k(s)) \\ &= h^{\leftarrow}(h(f(s))) \\ &= f(s) \end{aligned}$$

Thus $h^{\leftarrow} \circ (gk) | S = f$, and S is C-embedded in X.

■

(k) <u>Corollary</u>. Every compact subspace of a Tychonoff space X is C-embedded in X.

<u>Proof</u>. See 1.9(i) and 1.9(j).

■

Associated with the concept of C^*-embedding is the related (but weaker) notion of z-embedding.

(l) <u>Definition</u>. A subspace S of a space X is **z-embedded** in X if, for each $Z \in Z(S)$, there exists $T \in Z(X)$ such that $Z = S \cap T$.

(m) <u>Proposition</u>. If S is a subspace of X that is C^*-embedded in X, then S is z-embedded in X.

<u>Proof</u>. Let $Z \in Z(S)$. By 1.4(f) there exists $f \in C^*(S)$ such that $Z = Z(f)$. By hypothesis there exists $g \in C^*(X)$ such that

$g|S = f$. Then $Z = Z(g) \cap S$ and $Z(g) \in Z(X)$.

■

However, a subspace of a space can be z-embedded without being C^*-embedded. Every subspace of a metric space X is z-embedded in X, but only closed subspaces are C^*-embedded in X.

1.10 Normal spaces

(a) Definition. A **normal space** is a (Hausdorff) space in which each pair of disjoint closed subsets is contained in a pair of disjoint open subsets.

Compact spaces and metric spaces are familiar classes of normal spaces. Our goal in this section is to characterize normal spaces as being those spaces in which each closed subset is C^*-embedded (equivalently, C-embedded). We do this in Theorem 1.10(g) below. In approaching this problem we introduce the notion of a "separating chain" in a space X; this will provide us with a very general method of constructing members of C(X).

(b) Definition. A **separating chain** in a space X is a family $\mathcal{V}$ of subsets of X with the following properties:

(1) $\mathcal{V}$ is a countable family of open subsets of X that is totally ordered by inclusion.

(2) If U, V $\in \mathcal{V}$ and U $\subsetneq$ V, then there exists W $\in \mathcal{V}$ such

that $c\ell_X U \subseteq W \subseteq c\ell_X W \subseteq V$.

(3) $\cap V = \emptyset$ and $\cup V = X$.

(c) **Definition**. A **Q-indexing** of a separating chain V in X is a function α from the set $\mathbb{Q}$ of rationals onto V such that $c\ell_X \alpha(r) \subseteq \alpha(s)$ whenever $r < s$.

(d) **Lemma**. Any separating chain can be $\mathbb{Q}$-indexed.

Proof. Enumerate $\mathbb{Q}$ as $\{r_i : i \in \mathbb{N}\}$ and the separating chain V as $\{V_i : i \in \mathbb{N}\}$. If $V = \{\emptyset, X\}$ define α as follows: $\alpha(r) = \emptyset$ if $r \leqslant 0$ and $\alpha(r) = X$ if $r > 0$. Obviously α is a $\mathbb{Q}$-indexing of V.

If $V \neq \{\emptyset, X\}$ without loss of generality we may assume that $\emptyset \neq V_1 \neq X$. We will define α inductively. First, let $\alpha(r_1) = V_1$.

Let $n \in \mathbb{N}$ and suppose inductively that we have chosen $\{r_{i(j)} : 1 \leqslant j \leqslant n\} \subseteq \mathbb{Q}$ and $\{V_{j(i)} : 1 \leqslant i \leqslant n\} \subseteq V$, and have defined a function $\alpha_n : \{r_i : 1 \leqslant i \leqslant n\} \cup \{r_{i(j)} : 1 \leqslant j \leqslant n\} \to \{V_j : 1 \leqslant j \leqslant n\} \cup \{V_{j(i)} : 1 \leqslant i \leqslant n\}$ with the following properties:

(a) $\alpha_n(r_i) = V_{j(i)}, \quad 1 \leqslant i \leqslant n$

(b) $\alpha_n(r_{i(j)}) = V_j, \quad 1 \leqslant j \leqslant n$

(c) If $r,s \in \{r_i : 1 \leqslant i \leqslant n\} \cup \{r_{i(j)} : 1 \leqslant j \leqslant n\}$ and $r < s$, then $c\ell_X \alpha_n(r) \subseteq \alpha_n(s)$.

We will now define $r_{i(n+1)}$, $V_{j(n+1)}$, and α_{n+1}.

If $r_{n+1} \in \{r_{i(j)} : 1 \leqslant j \leqslant n\}$ - say $r_{n+1} = r_{i(k)}$ - then let $V_{j(n+1)}$ be V_k and define $\alpha_{n+1}(r_{n+1})$ to be V_k. If $r_{n+1} \notin \{r_{i(j)} : 1 \leqslant j \leqslant n\}$, let

$$S = \{r \in \{r_i : 1 \leqslant i \leqslant n\} \cup \{r_{i(j)} : 1 \leqslant j \leqslant n\} : r < r_{n+1}\}$$

and

$$T = \{r \in \{r_i : 1 \leqslant i \leqslant n\} \cup \{r_{i(j)} : 1 \leqslant j \leqslant n\} : r > r_{n+1}\}.$$

As V is a chain, $\{\alpha_n(r) : r \in S\}$ has a largest member V, and $\{\alpha_n(r) : r \in T\}$ has a smallest member W. Evidently $V \subseteq W$. If $V = W$, by (c) above $V \in B(X)$, and we define $\alpha_n(r_{n+1})$ to be V. We let $V_{j(n+1)} = V$ in this case. If $V \neq W$, there exists $V_{j(n+1)} \in V$ such that $c\ell_X V \subseteq V_{j(n+1)} \subseteq c\ell_X V_{j(n+1)} \subseteq W$, and we define $\alpha_{n+1}(r_{n+1})$ to be $V_{j(n+1)}$.

Now we will define $r_{i(n+1)}$ and $\alpha_{n+1}(r_{i(n+1)})$. If $V_{n+1} \in \{V_j : 1 \leqslant j \leqslant n\} \cup \{V_{j(i)} : 1 \leqslant i \leqslant n\}$, choose $r_{i(n+1)}$ to be one of the elements of $\alpha_n^{\leftarrow}(V_{n+1})$ and define $\alpha_{n+1}(r_{i(n+1)})$ to be V_{n+1}. If $V_{n+1} \notin \{V_j : 1 \leqslant j \leqslant n\} \cup \{V_{j(i)} : 1 \leqslant i \leqslant n+1\} = W_n$, let $A = \{V \in W_n : V \subset V_{n+1}\}$ and $B = \{V \in W_n : V_{n+1} \subset V\}$. Let $Q(A) = \{r \in \{r_i : 1 \leqslant i \leqslant n\} \cup \{r_{i(j)} : 1 \leqslant j \leqslant n+1\} : \alpha_n(r)$ or $\alpha_{n+1}(r) \in A\}$ and let $Q(B) = \{r \in \{r_i : 1 \leqslant i \leqslant n\} \cup \{r_{i(j)} : 1 \leqslant j \leqslant n\} : \alpha_n(r)$ or $\alpha_{n+1}(r) \in B\}$. Then $Q(A)$ has a largest element $r(A)$, $Q(B)$ has a smallest element $r(B)$, and $r(A) < r(B)$. Choose $r_{i(n+1)}$ to be any rational in $(r(A), r(B))$ and define $\alpha_{n+1}(r_{i(n+1)})$ to be V_{n+1}. We now define $\alpha_{n+1} : \{r_i : 1 \leqslant i \leqslant n+1\} \cup \{r_{i(j)} : 1 \leqslant j \leqslant n+1\} \to \{V_j : 1 \leqslant j \leqslant n+1\} \cup \{V_{j(i)} : 1 \leqslant i \leqslant n+1\}$ by saying that α_{n+1} extends α_n and is defined at $r_{i(n+1)}$ and r_{n+1} as described above. It is evident that our inductive assumptions are again satisfied, so our inductive step is complete. Evidently $\cup\{\alpha_n : n \in \mathbb{N}\}$ is the desired

$\mathbb{Q}$-indexing of $\mathcal{V}$.

■

(e) **Lemma**. Let $\mathcal{V}$ be a separating chain in X, and let α be a $\mathbb{Q}$-indexing of $\mathcal{V}$. If $x \in X$, let $f(x) = \inf\{r \in \mathbb{Q} : x \in \alpha(r)\}$. Then f is a well-defined member of C(X).

Proof. Let $x \in X$. As $\cup\mathcal{V} = X$, it follows that $\{r \in \mathbb{Q} : x \in \alpha(r)\} \neq \emptyset$. As $\cap\mathcal{V} = \emptyset$, there exists $r_0 \in \mathbb{Q}$ such that $x \notin \alpha(r_0)$. If $r < r_0$ then $c\ell_X\alpha(r) \subseteq \alpha(r_0)$ and $x \notin \alpha(r)$. Then $\{r \in \mathbb{Q} : x \in \alpha(r)\}$ is bounded below by r_0. Thus $\inf\{r \in \mathbb{Q} : x \in \alpha(r)\}$ is a well-defined real number, and so f is a well-defined function.

The set of open intervals of $\mathbb{R}$ with rational endpoints forms a base for the open sets of $\mathbb{R}$, so to show that f is continuous it suffices to show that if $r_0, s_0 \in \mathbb{Q}$ and $r_0 < s_0$, then $f^{\leftarrow}[(r_0,s_0)]$ is an open set of X. Our first step is to show that $f^{\leftarrow}[(r_0,\infty)]$ is open. Suppose that $x \in f^{\leftarrow}[(r_0,\infty)]$. There exist $r_1, r_2 \in \mathbb{Q}$ such that $r_0 < r_2 < r_1 < f(x)$. Thus $r_1 < \inf\{r \in \mathbb{Q} : x \in \alpha(r)\}$ so $x \notin \alpha(r_1)$. Thus $x \notin c\ell_X\alpha(r_2)$ and so $x \in X\backslash c\ell_X\alpha(r_2) \subseteq f^{\leftarrow}[(r_0,\infty)]$. Also, $f^{\leftarrow}[(-\infty,s_0)] = \{x \in X : \inf\{r \in \mathbb{Q} : x \in \alpha(r)\} < s_0\} = \cup\{\alpha(s) : s < s_0\}$. It follows that $f \in C(X)$.

■

(f) **Proposition**. The following are equivalent for two subsets A and B of a space X.

(1) A and B are completely separated in X.

(2) There exists a separating chain $\mathcal{V}$ in X, and sets U and V of $\mathcal{V}$, such that $A \subseteq U \subseteq c\ell_X U \subseteq V \subseteq X\backslash B$.

Proof

(1) $\Rightarrow$ (2) If A and B are completely separated in X, there exists $f \in C^*(X)$ such that $f[A] \subseteq \{0\}$, $f[B] \subseteq \{1\}$, and $\underline{0} \leqslant f \leqslant \underline{1}$. If $r \in \mathbb{Q}$ define V(r) as follows:

$$V(r) = \begin{cases} f^{\leftarrow}[[0,r)] & \text{if } 0 < r < 1 \\ \varnothing & \text{if } r \leqslant 0 \\ X & \text{if } r \geqslant 1 \end{cases}$$

Then $\{V(r) : r \in \mathbb{Q}\}$ is a separating chain in X and $A \subseteq f^{\leftarrow}[[0,\frac{1}{2})] \subseteq c\ell_X f^{\leftarrow}[[0,\frac{1}{2})] \subseteq f^{\leftarrow}[[0,\frac{3}{4})] \subseteq X\backslash B$.

(2) $\Rightarrow$ (1) Let $\mathcal{V}$ be a separating chain in X containing sets U and V such that $A \subseteq U \subseteq c\ell_X U \subseteq V \subseteq X\backslash B$. Let α be a $\mathbb{Q}$-indexing of $\mathcal{V}$ and suppose that $U = \alpha(r)$, $V = \alpha(s)$, for rationals r and s. If $U = V$ then $U \in \mathcal{B}(X)$ so by 1.9(e) A and B are completely separated in X. If $U \neq V$ then $r < s$. Consider the function f constructed in 1.10(e). Evidently $c\ell_{\mathbb{R}} f[A] \subseteq (-\infty,r]$ while $c\ell_{\mathbb{R}} f[B] \subseteq [s,+\infty)$ so by 1.9(g) A and B are completely separated in X.

■

We now turn to a consideration of complete separation in normal spaces.

(g) **Theorem**. The following are equivalent for a space X:

(1) X is normal.

(2) Disjoint closed subsets of X are completely separated in X.

(3) Each closed subset of X is C^*-embedded in X.

(4) Each closed subset of X is C-embedded in X.

<u>Proof</u>

(1) $\Rightarrow$ (2) Let A and B be disjoint closed subsets of X. We will inductively define a separating chain $\mathcal{V}$ in X that contains sets U and V such that $A \subseteq U \subseteq c\ell_X U \subseteq V \subseteq X \backslash B$. Then we will apply the preceding proposition.

If $r < 0$ let $V(r) = \emptyset$; if $r > 1$ let $V(r) = X$. Now enumerate $\mathbb{Q} \cap [0,1]$ as $\{r_n : n \in N\}$, with $r_0 = 0$ and $r_1 = 1$. As X is normal, there exist open sets U_0 and U_1 of X such that $A \subseteq U_0 \subseteq c\ell_X U_0 \subseteq U_1 \subseteq B$. We define V(0) to be U_0, and V(1) to be U_1.

Let $n(0) \geqslant 1$ and inductively assume that for each $n \leqslant n(0)$ we have defined $V(r_n)$ so that $V(r_n)$ is open in X and $c\ell_X V(r_m) \subseteq V(r_n)$ whenever $m,n \leqslant n(0)$ and $r_m < r_n$. Let $\max\{r_k : k \leqslant n(0)$ and $r_k < r_{n(0)+1}\} = r$ and let $\min\{r_k : k \leqslant n(0)$ and $r_k > r_{n(0)+1}\} = s$ (the sets in question are nonempty since $r_0 = 0$ and $r_1 = 1$). Then $r < s$ so $c\ell_X V(r) \subseteq V(s)$. Hence by normality there exists an open subset U of X such that $c\ell_X V(r) \subseteq U \subseteq c\ell_X U \subseteq V(s)$. Let $V(r_{n(0)+1}) = U$. This completes the inductive step.

One can easily check that the family $\{V(r) : r \in \mathbb{Q}\}$ is a separating chain of the desired form. Hence by the preceding proposition A and B are completely separated in X.

(2) $\Rightarrow$ (3) Let A be closed in X and let B and C be subsets of A that are completely separated in A. By 1.9(f) B and C are contained respectively in disjoint zero-sets Z_1 and Z_2 of A. As A is closed in X, Z_1 and Z_2 are closed in X. By hypothesis Z_1 and Z_2 are completely separated in X. Thus B and C are completely separated in X. It follows from 1.9(h) that A is C^*-embedded in X.

(3) $\Rightarrow$ (4) Let A be closed in X. If $Z \in \mathcal{Z}(X)$ and $Z \cap A =$

⇒ (1) Compact spaces are normal, by 1.10(h) normal spaces
noff, and subspaces of Tychonoff spaces are Tychonoff.

■

1.11 Pseudocompact spaces

this final section of Chapter 1 we introduce the class of
pact spaces and develop a number of characterizations of
ces. When we investigate realcompact spaces in Chapter 5
ave occasion to use these results. Although the concept of
pactness makes sense for all (Hausdorff) spaces, it is most
en working with Tychonoff spaces. However, we shall begin
ng the notion of "feeble compactness" and developing its
. Then we introduce pseudocompactness and show that for
spaces (although not in general), pseudocompactness and
npactness are equivalent.

Definition. A space X is **feebly compact** if every locally
ection of open subsets of X is finite.

give a number of characterizations of feebly compact

Theorem. The following conditions on a space X are
t:

X is feebly compact.

Every locally finite family of pairwise disjoint open subsets

$\emptyset$, then since $\chi_{A,A\cup Z} \in C(A\cup Z)$, Z and A
$A\cup Z$. As $A\cup Z$ is closed it is C^*-embedded
A are completely separated in X. But A
C^*-embedded in X. so by 1.9(j) A is C-embe

(4) ⇒ (1) Let A and B be disjoin
above, $\chi_{A,A\cup B} \in C(A\cup B)$. By hypothesis
that $f|A\cup B = \chi_{A,A\cup B}$. Then $f^{\leftarrow}[($
disjoint open sets of X containing A and
that X is normal.

(h) **Corollary**. Normal spaces are T

(i) **Corollary**. The following are eq

(1) X is Tychonoff

(2) X is (homeomorphic to) a subsp

(3) X is (homeomorphic to) a dense
space.

Proof

(1) ⇒ (2) Let X be Tychonoff.
$C(X,[0,1])$ separates points and closed sets
is an embedding of X in the compact spac

(2) ⇒ (3) If X is a subspace of
$c\ell_K X$ is a compact space containing X as

are T

pseud
these
we wi
pseud
useful
by de
proper
Tychon
feeble

finite

spaces.

equival

of X is finite.

(3) If $\{V_n : n \in \mathbb{N}\}$ is a decreasing family of nonempty open subsets of X, then $\bigcap\{c\ell_X V_n : n \in \mathbb{N}\} \neq \emptyset$.

(4) Every countable open cover of X has a finite subfamily whose union is dense in X.

Proof

(2) ⇒ (1) Suppose (1) fails, and let $\{C_n : n \in \mathbb{N}\}$ be a countably infinite, locally finite collection of open subsets of X. Let $k_1 = 1$. Choose $x_1 \in C_1$ and find an open subset V_1, and $k_2 \in \mathbb{N}$, such that $x_1 \in V_1$ and $V_1 \cap C_k = \emptyset$ whenever $k \geq k_2$. Inductively assume that $\{k(i) : i = 1 \text{ to } m\}$ is an increasing sequence of integers, and $\{V_i : i = 1 \text{ to } m - 1\}$ a sequence of open sets, such that $C_{k(i)} \cap V_i \neq \emptyset$ and $i < j$ implies $V_i \cap C_{k(j)} = \emptyset$. Choose $x_m \in C_{k(m)}$ and find an open set V_m, and an integer $k(m+1) > k(m)$, such that $x_m \in C_{k(m)} \cap V_m$ and $V_m \cap C_k = \emptyset$ whenever $k \geq k(m+1)$. Then $\{V_m \cap C_{k(m)} : m \in \mathbb{N}\}$ is an infinite pairwise disjoint locally finite family of open sets, and (2) fails.

(1) ⇒ (3) Suppose (3) fails, and $\{V_n : n \in \mathbb{N}\}$ is a decreasing sequence of nonempty open subsets of X such that $\bigcap\{c\ell_X V_n : n \in \mathbb{N}\} = \emptyset$. If $x \in X$ there exists $n(x) \in \mathbb{N}$ such that $n \geq n(x)$ implies $x \notin c\ell_X V_n$. Then $X \setminus c\ell_X V_n$ meets only finitely many of $\{V_n : n \in \mathbb{N}\}$, and contains x. Thus $\{V_n : n \in \mathbb{N}\}$ is an infinite, locally finite family of open sets and (1) fails.

(3) ⇒ (4) Let $\{C_j : j \in \mathbb{N}\}$ be a countable open cover of X and let $V_n = X \setminus c\ell_X[\bigcup\{C_j : 1 \leq j \leq n\}]$. Then $\{V_n : n \in \mathbb{N}\}$ is a decreasing sequence of open sets and $\bigcap\{c\ell_X V_n : n \in \mathbb{N}\} \subseteq$

$X \setminus \bigcup\{C_j : j \in \mathbb{N}\} = \emptyset$. Thus by hypothesis $V_k = \emptyset$ for some $k \in \mathbb{N}$. Then $\bigcup\{C_j : 1 \leqslant j \leqslant k\}$ is dense in X.

(4) $\Rightarrow$ (2) Suppose (2) fails. Then there exists an infinite, locally finite family $\{V_n : n \in \mathbb{N}\}$ of pairwise disjoint open sets of X. For each $p \in X$, fix an open neighborhood $U(p)$ of p such that $\{n \in \mathbb{N} : U(p) \cap V_n \neq \emptyset\}$ is a finite subset $F(p)$ of $\mathbb{N}$. Let $\mathcal{F}$ denote the set of finite subsets of $\mathbb{N}$, and for each $F \in \mathcal{F}$, let $W(F) = \bigcup\{U(p) : p \in X \text{ and } F(p) = F\}$. Then $\{W(F) : F \in \mathcal{F}\}$ is a countable open cover $\mathcal{C}$ of X. Let $\{F_1, \ldots, F_n\}$ be a finite subset of $\mathcal{F}$. Choose $k \in \mathbb{N} \setminus \bigcup\{F_i : 1 \leqslant i \leqslant n\}$, and suppose that $q \in \bigcup\{W(F_i) : 1 \leqslant i \leqslant n\} \cap V_k$. Find $j \in \{1, \ldots, n\}$ such that $q \in W(F_j) \cap V_k$. Then there exists $p \in X$ such that $q \in U(p)$ and $F(p) = F_j$. As $k \notin F_j$, it follows that $U(p) \cap V_k = \emptyset$, which is a contradiction. Thus $\bigcup\{W(F_i) : 1 \leqslant i \leqslant n\} \cap V_k = \emptyset$, and so no finite subset of $\mathcal{C}$ has a union dense in X. Thus (4) fails.

■

It is an immediate consequence of the preceding theorem that every compact space is feebly compact. In 1Q(8) we exhibit a feebly compact, noncompact space. Some of the spaces of ordinals discussed at the end of Chapter 2 are also feebly compact but not compact.

(c) __Definition__. A space S is **pseudocompact** if every continuous real-valued function on X is bounded, i.e., if $C(X) = C^*(X)$.

(d) __**Theorem**__

(1) Every feebly compact space is pseudocompact.

(2) A Tychonoff space is feebly compact iff it is

pseudocompact.

<u>Proof</u>

(1) Suppose X is not pseudocompact, and find $f \in C(X) \setminus C^*(X)$. By replacing f by $|f|$ if necessary, we may assume that $f \geq \underline{0}$. As $f \notin C^*(X)$, there exists a sequence $\{r_n : n \in \mathbb{N}\}$ of positive real numbers such that $r_{n+1} \geq r_n + 1$ for each $n \in \mathbb{N}$, and $\{r_n : n \in \mathbb{N}\} \subseteq f[X]$. Then $\{f^{\leftarrow}[(r_n - \frac{1}{3}, r_n + \frac{1}{3})] : n \in \mathbb{N}\}$ is an infinite, locally finite collection of nonempty open subsets of X. Hence X is not feebly compact.

(2) By the above it suffices to show that if the Tychonoff space X is not feebly compact, then it is not pseudocompact. Let $\{V_n : n \in \mathbb{N}\}$ be an infinite, pairwise disjoint locally finite collection of nonempty open subsets of X (by 1.11(b) such a collection exists if X is not feebly compact). As X is Tychonoff, there exists a collection $\{C_n : n \in \mathbb{N}\}$ of nonempty cozero-sets of X such that $c\ell_X C_n \subseteq V_n$. Choose $p_n \in C_n$; because X is Tychonoff for each $n \in \mathbb{N}$ there exists $f_n \in C(c\ell_X C_n)$ such that $f_n(p_n) = n$ and $f_n[c\ell_X C_n \setminus C_n] \subseteq \{0\}$. Define $f \in F(X,\mathbb{R})$ as follows:

$$f(x) = \begin{cases} f_n(x) & \text{if } x \in C_n \\ 0 & \text{if } x \in X \setminus \cup\{C_n : n \in \mathbb{N}\} \end{cases}$$

Now $\{c\ell_X C_n : n \in \mathbb{N}\} \cup \{X \setminus \cup\{C_n : n \in \mathbb{N}\}\}$ is a locally finite closed cover $\mathcal{C}$ of X, and $f|A \in C(A)$ for each $A \in \mathcal{C}$. Thus by 1.6(c), $f \in C(X)$. As $f(p_n) = n$ for each $n \in \mathbb{N}$, it follows that $f \notin C^*(X)$. Thus X is not pseudocompact.

■

(e) <u>**Corollary**</u>. The following are equivalent for a Tychonoff space X.

(1) X is pseudocompact

(2) X is feebly compact

(3) Each locally finite family of pairwise disjoint open subsets of X is finite.

(4) If $\{V_n : n \in \mathbb{N}\}$ is a decreasing sequence of nonempty open subsets of X then $\cap\{c\ell_X V_n : n \in \mathbb{N}\} \neq \emptyset$.

(5) Each countable open cover of X has a finite subfamily whose union is a dense subspace of X.

<u>Proof</u>. This is an immediate consequence of 1.11(b) and 1.11(d).

■

There are (Hausdorff) spaces that are pseudocompact but not feebly compact; see 1U for an example.

Chapter 1 - Problems

1A. Elementary properties of closures

(1) If X is dense in T and V is open in T, show that $c\ell_T V = c\ell_T(V \cap X)$.

(2) Let $A \subseteq B \subseteq X$. Prove that $c\ell_B A = B \cap c\ell_X A$. If B is closed in X, prove that $c\ell_B A = c\ell_X A$.

1B. Properties of C(X)

Let X be any space. Let f, g, h $\in F(X,\mathbb{R})$.

(1) Define $a : \mathbb{R} \times \mathbb{R} \to \mathbb{R}$ and $m : \mathbb{R} \times \mathbb{R} \to \mathbb{R}$ as follows:

$$a(r,s) = r + s$$

$$m(r,s) = rs$$

Prove that a and m are continuous.

(2) If f, g $\in$ C(X) and $r > 0$ prove that f + g, fg, and $|f|^r$ are continuous.

(3) Prove these formulae:

$$f \vee g = (2)^{-1}(f + g + |f - g|)$$
$$f \wedge g = 2^{-1}(f + g - |f - g|)$$
$$fg = 4^{-1}((f+g)^2 - (f-g)^2)$$
$$f + (g \wedge h) = (f+g) \wedge (f+h)$$

$$f + (g \vee h) = (f+g) \vee (f+h)$$

$$f \wedge (g \vee h) = (f \wedge g) \vee (f \wedge h)$$

$$f \vee (g \wedge h) = (f \vee g) \wedge (f \vee h)$$

(4) Prove that if $f, g \in C(X)$ then $f \vee g$, $f \wedge g \in C(X)$.

(5) If $f \in C(X)$ and $Z(f) = \emptyset$, prove that f^{-1} is continuous.

(6) (The Weierstrass M-test) Prove that if $\{f_n : n \in \mathbb{N}\} \subseteq F(X,\mathbb{R})$, and if $|f_n| \leq \underline{M}_n$ for each $n \in \mathbb{N}$, and if $\sum_{n=1}^{\infty} M_n$ converges, then $\sum_{n=1}^{\infty} f_n(x)$ converges uniformly.

(7) Find a necessary and sufficient condition on a Tychonoff space X such that the ring C(X) is a field.

(8) If $Z \in \mathcal{Z}(X)$ and C is a proper closed subset of $\mathbb{R}$, show that there exists $f \in C(X)$ such that $Z = f^{\leftarrow}[C]$.

1C. Another proof of 1.6(d)

We present another proof that if X and Y are spaces, $f, g \in C(X,Y)$, and $\{x \in X : f(x) = g(x)\}$ is dense in X, then $f = g$.

(1) Let $\Delta_Y = \{(y,y) : y \in Y\}$ (Δ_Y is called the diagonal of $Y \times Y$). Prove that Δ_Y is closed in $Y \times Y$. (Hint: Use the fact that Y is Hausdorff.)

(2) Define $h \in F(X,Y \times Y)$ by $h(x) = (f(x),g(x))$. Prove that $h \in C(X,Y \times Y)$.

(3) Show that $h^{\leftarrow}(\Delta_Y) = \{x \in X : f(x) = g(x)\}$.

(4) Show that $f = g$. (Hint: $h^{\leftarrow}(\Delta_Y)$ is closed.)

1D. <u>Theorem 1.8(i) for T_1 spaces</u>

Show that if the spaces X and Y of 1.8(i) are assumed only to be T_1 spaces, the conclusion of 1.8(i) need not follow, even if f is a homeomorphism. (Hint: Let X = Y = the positive integers with the cofinite topology. Let $f(2n) = f(2n-1) = n$ for each $n \in X$.)

1E. <u>A characterization of product spaces</u>

Let X be the product of the spaces $\{X_i : i \in I\}$. Prove that the product topology on X is characterized by this property: for each space Y and each function $f \in F(Y,X)$, $f \in C(Y,X)$ iff $\Pi_i \circ f \in C(Y,X_i)$ for each $i \in I$. (Hint: By 1.7(d)(2), the product topology τ on X has this property. Now, let σ be a topology on the set X such that (X,σ) has the property. Show that $\tau = \sigma$ by selecting appropriate spaces Y and functions f.)

1F. <u>Elementary properties of product spaces</u>

Let $\{X_i : i \in I\}$ be a family of nonempty spaces and let $X = \Pi\{X_i : i \in I\}$. Let $A_i \subseteq X_i$ for each $i \in I$.

(1) Show that $c\ell_X(\Pi\{A_i : i \in I\}) = \Pi\{c\ell_{X_i} A_i : i \in I\}$.

(2) If F is a finite subset of I, show that $\text{int}_X(\Pi\{A_i : i \in F\} \times \Pi\{X_i : i \in I\setminus F\}) = \Pi\{\text{int}_{X_i} A_i : i \in F\} \times \Pi\{X_i : i \in I\setminus F\}$. Is the restriction that F be finite necessary?

(3) If each X_i is Tychonoff (respectively zero-dimensional), show that X is Tychonoff (respectively zero-dimensional).

1G. Density character and $C^*(X)$.

The **density character** of a space X, denoted d(X), is defined to be the larger of $\aleph_0$ and $\min\{|S| : S \text{ is a dense subset of } X\}$ (see 2N for more information about d(X)). Prove that $|C^*(X)| \leq 2^{d(X)}$. (Note: A detailed discussion of the cardinality $|Y|$ of a set Y appears in 2.6.)

1H. The Sorgenfrey line.

Let $\mathcal{B} = \{[a,b) : a, b \in \mathbb{R} \text{ and } a < b\}$.

(1) Prove that $\mathcal{B}$ is an open base for a topology τ on the real line. (The space $(\mathbb{R},\tau)$ is called the **Sorgenfrey line**, and is henceforth denoted by S.)

(2) Prove that S is zero-dimensional (and hence Tychonoff).

(3) Prove that $|C^*(S \times S)| = 2^{\aleph_0}$. (Hint: Use 1G.)

(4) Prove that $S \times S$ has a closed discrete subspace of cardinality $2^{\aleph_0}$. Combine this with (3) above, and 1.10(g), to prove that $S \times S$ is not normal.

(5) Prove that S is Lindelöf, and conclude that neither normality nor the Lindelöf property is productive. (Hint: Let $S = \cup\{[a_i,b_i) : i \in I\}$ and let $M = S \setminus \cup\{(a_i,b_i) : i \in I\}$. For each $x \in M$ find $j(x) \in I$ and $r(x) > x$ such that $[x,r(x)) \subset [a_{j(x)},b_{j(x)})$. If x and y are distinct points of M, show that $[x,r(x)) \cap [y,r(y)) = \emptyset$, and note that as $\mathbb{R}$ is separable, M must be countable. Now use the fact that $\mathbb{R}$ is hereditarily Lindelöf.)

(6) Show that a compact subspace A of S contains no increasing sequence. (Hint: Assume A contains an

increasing sequence $\{a_n : n \in \mathbb{N}\}$. If $\{a_n : n \in \mathbb{N}\}$ has no supremum, let $\mathcal{C} = \{(-\infty,a_1)\} \cup \{[a_n,a_{n+1}) : n \in \mathbb{N}\}$ and if $b = \sup\{a_n : n \in \mathbb{N}\}$, let $\mathcal{C} = \{(-\infty,a_1)\} \cup \{[a_n,a_{n+1}) : n \in \mathbb{N}\} \cup \{[b,\infty)\}$. Show $\mathcal{C}$ is an open cover of A with no finite subcover.)

(7) Show that a compact subset A of S is countable. (Hint: For each $p \in A$, use (6) to find some $n_p \in \mathbb{N}$ such that $(p - n_p^{-1},p) \cap A = \emptyset$. First show that $\{(p - n_p^{-1},p) : p \in A\}$ is a family of pairwise disjoint intervals. Use this fact to deduce that for $n \in \mathbb{N}$, $A_n = \{p \in A : n = n_p\}$ is countable. Now show that $A = \bigcup\{A_n : n \in \mathbb{N}\}$.)

1I. The Nemitskii plane

Define a topology $\mathcal{T}$ on $\{(x,y) \in \mathbb{R}^2 : y \geq 0\}$ as follows. If $y > 0$, a neighborhood base at (x,y) is identical to the neighborhood base at (x,y) in the usual Euclidean topology of $\mathbb{R}^2$. If $x \in \mathbb{R}$ and $r > 0$, let $V_r(x) = \{(u,v) \in \mathbb{R}^2 : (u-x)^2 + (v-r)^2 < r^2\} \cup \{(x,0)\}$, and let $\{V_r(x) : r > 0\}$ be a neighborhood base at $(x,0)$.

(1) Prove that $\mathcal{T}$ as described is a Tychonoff topology on $\{(x,y) \in \mathbb{R}^2 : y \geq 0\}$. (Hint: If $r > 0$ and $x_0 \in \mathbb{R}$, define $f_{x_0,r} : \{(x,y) \in \mathbb{R}^2 : y \geq 0\} \to \mathbb{R}$ to be 1 at $(x_0,0)$, 0 outside $V_r(x_0)$, and "linear on $V_r(x_0)$".) Let Γ denote the resulting topological space, and let $\Gamma_0 = \{(x,0) : x \in \mathbb{R}\}$.

(2) Calculate the value of $|C^*(\Gamma)|$ (use 1G).

(3) Calculate the value of $|C^*(\Gamma_0)|$, and infer that Γ is not normal.

(4) Show that each closed subset of Γ is a G_δ-set of Γ (see 1K).

1J. <u>Preservation of covering properties by perfect maps.</u>

(1) Let $f : X \to Y$ be a perfect continuous surjection. Use 1.8(c) to give quick proofs that X has P iff Y has P, where P is local compactness or the Lindelöf property.

(2) Let $f : X \to Y$ be a perfect continuous surjection. Prove that X is regular iff Y is regular (see 1.8(h)).

1K. <u>Closed G_δ-sets</u>

Recall that a G_δ-set of a space is any subset that is the intersection of countably many open subsets of the space.

(1) Prove that a closed subset A of a normal space X is a G_δ-set of X iff $A \in Z(X)$.

(2) Let X be a Tychonoff space in which each singleton set is a G_δ-set. Prove that each C-embedded subset of X is closed in X.

(3) Give an example of a Tychonoff space with a closed G_δ-set that is not a zero-set of the space.

(4) Characterize the C-embedded subspaces of a metric space.

1L. $\mathbf{2}$-embedding and $\mathbb{N}$-embedding

Let X be a zero-dimensional space and let S be a subspace of X. Then S is said to be $\mathbf{2}$-embedded in X if for each $f \in C(S,\mathbf{2})$ there exists $F \in C(X,\mathbf{2})$ such that $F|S = f$. If we replace $\mathbf{2}$ by $\mathbb{N}$ in the above we obtain the definition of S being $\mathbb{N}$-embedded in X. We say that S is C^*-$\mathbb{N}$-embedded in X if, for each $f \in C(S,\mathbb{N})$ for which $f[S]$ is finite, there exists $F \in C(X,\mathbb{N})$ such that $F[X]$ is finite and $F|S = f$.

(1) Show that $C(X,\mathbf{2})$ separates points and closed subsets of X.

(2) Prove that X can be embedded as a subspace of a compact zero-dimensional space. (Hint: Use 1.7(j).)

(3) Prove that S is $\mathbf{2}$-embedded in X iff S is C^*-$\mathbb{N}$-embedded in X.

(4) Prove the following "discrete version" of Urysohn's extension theorem: S is $\mathbf{2}$-embedded in X iff any pair of subsets of S that are $\mathbf{2}$-separated in S are $\mathbf{2}$-separated in X (A and B are $\mathbf{2}$-separated in S if there exists $f \in C(S,\mathbf{2})$ such that $f[A] \cap f[B] = \varnothing$).

(5) Prove the following "discrete version" of 1.9(j). Suppose S is $\mathbf{2}$-embedded in X. Then S is $\mathbb{N}$-embedded in X iff whenever F is the intersection of countably many clopen subsets of X and disjoint from S, there exists a clopen subset T of X such that $F \subseteq T$ and $S \subseteq X \setminus T$. (See 4.7 for further results on $\mathbf{2}$-embedding.)

1M. Projection maps vs. perfect maps

Let $\{X_i : i \in I\}$ be a set of spaces with product X. Show that the jth projection map $\Pi_j : X \to X_j$ is perfect iff $\Pi\{X_i : i \in I\setminus\{j\}\}$ is compact.

1N. The space ψ

(1) Show that there exists an infinite family M of infinite subsets of $\mathbb{N}$ such that the intersection of any two is finite, and such that M is maximal with respect to this property. (Hint: Use Zorn.)

(2) Let $\psi = \mathbb{N} \cup M$, and let $B = \{\{n\} : n \in \mathbb{N}\} \cup \{\{M\} \cup S : M \in M$ and S is a cofinite subset of M$\}$. Show that B is an open base for a locally compact Hausdorff topology on ψ.

(3) Show that $|M| > \aleph_0$.

(4) Show that $M \in Z(\psi)$ but M is not C^*-embedded in ψ. (Hint: Let N be a countably infinite subset of M and U an open set of ψ such that $U \cap M = N$. Show that $c\ell_\psi U \cap (M\setminus N) \neq \emptyset$. Now apply Urysohn's extension theorem 1.9(h).)

(5) Show that each subset of ψ is a G_δ-set, but ψ is not normal.

1O. C^*-embedding in normal spaces.

(1) Let X be a normal space and S a subspace of X. Prove

that the following are equivalent:

(a) S is C^*-embedded in X.

(b) If $Z_1, Z_2 \in Z(S)$ and $Z_1 \cap Z_2 = \emptyset$ then $c\ell_X Z_1 \cap c\ell_X Z_2 = \emptyset$.

(2) If X is Tychonoff but not normal, show that for some subspace S of X the equivalence of (1) is guaranteed to fail.

Note. If S is a dense subspace of the Tychonoff space X, then (a) and (b) are equivalent; see 4.6(h).

1P. Countably compact spaces.

A space is **countably compact** if each countable open cover of it has a finite subcover.

(1) Prove that X is countably compact iff X has no countably infinite closed discrete subspace.

(2) Prove that continuous images, and closed subspaces, of countably compact spaces are countably compact.

(3) Prove that every countably compact space is feebly compact, but that the converse is false. (See 1N.)

(4) Prove that the product of a countably compact space and a compact space is countably compact.

Note. Examples of countably compact noncompact spaces appear in 2.6(q).

1Q. Properties of feebly compact and pseudocompact spaces.

(1) Prove that if the space Y is the continuous image of a feebly compact space, then Y is feebly compact.

(2) Prove that if V is an open subset of the feebly compact space X, then $c\ell_X V$ is feebly compact.

(3) Prove that if X is a feebly compact dense subspace of the space T, then T is feebly compact.

(4) Prove each of the above statements with each instance of "space" replaced by "Tychonoff space," and each instance of "feebly compact" replaced by "pseudocompact".

(5) Show that every pseudocompact metric space is compact.

(6) Let Y be a dense subspace of the non-pseudocompact Tychonoff space T. Show that there exists a countably infinite discrete subspace D of Y such that D is closed and C-embedded in T. (Hint: Let $f \in C(T)\setminus C^*(T)$ and inductively construct $D = (d_n)_{n\in\mathbb{N}} \subseteq Y$ such that $|f(d_{n+1})| \geq |f(d_n)| + 1$ for each $n \in \mathbb{N}$. Prove that $f[D]$ is C-embedded in $\mathbb{R}$. If $g \in C(D)$, use this to extend $g\circ(f^{\leftarrow}|f[D])$ to $k \in C(\mathbb{R})$, and consider $k\circ f$.)

(7) Show that a Tychonoff space is pseudocompact iff it contains no closed C-embedded countably infinite discrete subspace.

(8) Show that ψ (see 1N) is pseudocompact but not countably compact.

(9) Show that pseudocompactness is not closed-hereditary.

1R. Weak normality properties

A Tychonoff space X is called **δ-normally separated** (respectively, **weakly δ-normally separated**) if whenever $Z \in Z(X)$ and A is closed in X (respectively $A = c\ell_X V$ where V is open in X) and $Z \cap A = \emptyset$, then Z and A are completely separated in X.

(1) Prove that every pseudocompact space is weakly δ-normally separated.

(2) Prove that a Tychonoff space is countably compact iff it is pseudocompact and δ-normally separated.

1S. Functions defined using separating chains

In the proof of 1.10(e) show that it is not necessarily true that $f^{\leftarrow}[(r_0,\infty)] = X \setminus c\ell_X \alpha(r_0)$; that is, find a space X, a separating family V on X, a $\mathbb{Q}$-indexing α for V, and a point $p \in X$ such that $p \in X \setminus c\ell_X \alpha(0)$ and $0 = \inf\{s : p \in \alpha(s)\}$. (Hint: Use subspaces of $\mathbb{R}$ to form X.)

1T. k-spaces and k-maps

For a space X, let kX denote X with the topology whose closed sets are $C = \{C \subseteq X : C \cap K$ is closed in K for each compact subset K of X$\}$. A space Y is a **k-space** if $Y = kY$.

(1) Let $U \subseteq X$. Show that the following are equivalent:

(a) U is open in kX,

(b) $kX \setminus U \in \mathcal{C}$ and

(c) $U \cap K$ is open in K for each compact subset K of X.

(2) Show that the topology of kX is finer than the topology of X and infer that kX is a Hausdorff space.

(3) Let $A \subseteq X$. Show that A is a compact subset of X iff A is a compact subset of kX.

(4) For any space X, prove that $k(kX) = kX$; in particular, prove that kX is a k-space.

(5) Prove that locally compact spaces and first countable spaces are k-spaces.

(6) A function $f \in C(X,Y)$ is called a **k-map** if $f^{\leftarrow}[A]$ is compact for each compact $A \subseteq Y$. Prove that a perfect continuous function is a k-map.

(7) Prove that $id_X : kX \to X$ is a k-map.

(8) Let X and Y be spaces and let $f \in C(X,Y)$. Define $kf : kX \to kY$ by $kf(x) = f(x)$. Prove that kf is continuous.

(9) Let X and Y be spaces. If $f : X \to Y$ is a k-map, prove that $kf : kX \to kY$ is perfect and continuous.

1U. <u>A pseudocompact space that is not feebly compact</u>

Let $\{X(n) : n \in \mathbb{N}\}$ be a countably infinite partition of the closed unit interval [0,1] into dense subsets. Let τ denote the usual topology on [0,1], and let σ denote the topology of [0,1] for which the family $\tau \cup \{X(2n-1) : n \in \mathbb{N}\} \cup \{X(2n-1) \cup X(2n) \cup X(2n+1) : n \in \mathbb{N}\}$ is a sub-base. Let Y =

$([0,1],\sigma)$.

(1) Show that Y is Hausdorff but not regular.

(2) Let $f \in C(Y)$, let $a, b \in \mathbb{R}$ and let $c, d \in Y$. Show that if $(c,d) \cap f^{\leftarrow}[(a,b)] \neq \emptyset$, then $(c,d) \cap f^{\leftarrow}[(a,b)] \cap X(k) \neq \emptyset$ for each $k \in \mathbb{N}$. (Hint: Suppose $(c,d) \cap f^{\leftarrow}[(a,b)] \cap X(k_0) \neq \emptyset$. First prove that $(c,d) \cap f^{\leftarrow}[(a,b)] \cap X(k_0-1) \neq \emptyset \neq (c,d) \cap f^{\leftarrow}[(a,b)] \cap X(k_0+1)$. Consider separately the case when k_0 is even and when k_0 is odd. The latter is more involved, and utilizes the regularity of $\mathbb{R}$.)

(3) Show that if $m \in \mathbb{N}$, the subspace topology that $X(m)$ inherits from Y is the same as the subspace topology that $X(m)$ inherits from the usual topology τ on $[0,1]$.

(4) Prove that Y is pseudocompact. (Hint: If $f \in C(Y)\setminus C^*(Y)$ and $f \geqslant \underline{0}$, choose $\{a_k : k \in \mathbb{N}\}$ and $\{b_k : k \in \mathbb{N}\}$ such that $a_k < b_k < a_{k+1} - 1$ and $f^{\leftarrow}[(a_k,b_k)] \neq \emptyset$ for each $k \in \mathbb{N}$. Choose $x_k \in f^{\leftarrow}[(a_k,b_k)]$ and $p \in [0,1]$ such that p is a limit point of $\{x_k : k \in \mathbb{N}\}$ in $([0,1],\tau)$. Let $p \in X(m)$ and use (2) to find a sequence in $X(m)$ that converges to p. Consider $f(p)$ and use (3) to derive a contradiction.)

(5) Prove that Y is not feebly compact.

1V. <u>Hilbert space</u>

For $x \in \mathbb{R}^{\mathbb{N}}$, the infinite series $\Sigma\{x(n) : n \in \mathbb{N}\}$ will be abbreviated as $\Sigma x(n)$. Let $X = \{x \in \mathbb{R}^{\mathbb{N}} : \Sigma x(n)^2 < \infty\}$.

For $x, y \in X$, define $d(x,y) = \left[\Sigma(x(n) - y(n))^2\right]^{1/2}$.

(1) Show that X is a metric space. (Hint: You may assume Holder's inequality, i.e., for each $x, y \in X$, $\left[\Sigma x(n)y(n)\right]^2 \leqslant \left[\Sigma x(n)^2\right]\left[\Sigma y(n)^2\right]$.)

(2) Show that X is separable. (Hint: Show that $D = \{x \in X \cap \mathbb{Q}^{\mathbb{N}} : \text{for some } n \in \mathbb{N},\ x(m) = 0 \text{ for } m \geqslant n\}$ is dense in X.) Let X' be X with the subspace topology introduced from the space $\mathbb{R}^{\mathbb{N}}$ with the product topology.

(3) Prove that $\mathcal{T}(X') \subseteq \mathcal{T}(X)$.

Let $Y = X \cap \mathbb{Q}^{\mathbb{N}}$ and $Y' = X' \cap \mathbb{Q}^{\mathbb{N}}$.

(4) Show that Y is totally disconnected. (Hint: Use the fact that Y is totally disconnected and that $\mathcal{T}(Y') \subseteq \mathcal{T}(Y)$.)

For $n \in \mathbb{N}$, let $\mathbb{Q}^n = \{x \in \mathbb{Q}^{\mathbb{N}} : x(m) = 0 \text{ for } m \geqslant n + 1\}$.

(5) Show that $\mathcal{T}(Y \cap \mathbb{Q}^n) = \mathcal{T}(Y' \cap \mathbb{Q}^n)$ for $n \in \mathbb{N}$.

(6) Prove that Y is not zero-dimensional. (Hint: Define $\hat{0} \in Y$ by $\hat{0}(n) = 0$ for $n \in \mathbb{N}$. Let U be a bounded, open Y-neighborhood of $\hat{0}$. Since $U \cap \mathbb{Q}^1$ is bounded and open in $\mathbb{Q}^1$, there are points $r^1 = (r_1,0,0,\ldots) \in U \cap \mathbb{Q}^1$ and $s^1 = (s_1,0,0,\ldots) \in \mathbb{Q}^1 \setminus U$ such that $d(r^1,s_1) < 1$. Hence, $d(r^1,Y \setminus U) < 1$. Since $U \cap \mathbb{Q}^2$ is bounded

and open in $\mathbb{Q}^2$, show that there is a point $r^2 = (r_1,r_2,0,0,\ldots) \in U \cap \mathbb{Q}^2$ such that $d(r^2,Y\setminus U) < 1/2$. Continue by induction to obtain a point $(r_1,r_2,\ldots,r_3\ldots) \in c\ell_Y U \cap c\ell_Y(Y\setminus U)$.)

1W. P-points and P-spaces

A point p of a space X is called a **P-point** of X if every G_δ-set of X containing p is a neighborhood of p. The set of P-points of X is denoted by P(X). If P(X) = X then X is called a **P-space**.

(1) Show that each regular P-space is zero-dimensional. (Hint: If $p \in V$, where V is open in the P-space X, find an open set V_1 of X such that $p \in V_1 \subseteq c\ell_X V_1 \subseteq V$. Proceed inductively and intersect.)

(2) Show that a Tychonoff space is a P-space iff $Z(X) = B(X)$.

(3) Let m be an uncountable cardinal, let X be a set of cardinality m, and let $x_0 \in X$. Topologize X as follows: every point except x_0 is isolated, and $\{C \subseteq X : x_0 \in C \text{ and } |X\setminus C| \leqslant \aleph_0\}$ is a neighborhood base at x_0. Show that X is a non-discrete P-space.

(4) Show that feebly compact regular P-spaces are finite. (Hint: See 1.11(b). Inductively produce an infinite, pairwise disjoint collection of cozero-sets.)

(5) Show that the product of finitely many P-spaces is a P-space, and that subspaces of P-spaces are P-spaces.

(6) Let (X,τ) be a space. Show that the set of G_δ-sets of

(X,τ) is a base for a topology τ_δ on X, and that (X,τ_δ) is a P-space; also, show that if (X,τ) is regular, then so is (X,τ_δ).

(7) Denote the space (X,τ_δ) described in (6) by X_δ. Show that if Y is a P-space and if $f \in C(Y,X)$ then there exists $g \in C(Y,X_\delta)$ such that $j \circ g = f$, where $j : X_\delta \to X$ is the (continuous) function defined to be the identity on the underlying set.

(8) If m is an uncountable cardinal, show that $(\{0,1\}^m)_\delta$ is a zero-dimensional P-space without isolated points.

1X. Σ-products

Let $\{X_a : a \in A\}$ be a nonempty family of nonempty spaces. For $c \in \Pi\{X_a : a \in A\}$, let $\Sigma(c)$ denote $\{x \in \Pi\{X_a : a \in A\}: |\{a \in A : x_a \neq c_a\}| \leq \aleph_0\}$. A **$\Sigma$-product of** $\{X_a : a \in A\}$ is a subspace $\Sigma(c)$ for some $c \in \Pi\{X_a : a \in A\}$. Fix $b \in \Pi\{X_a : a \in A\}$. If $\emptyset \neq T \subseteq A$, let $\Pi_T : \Pi\{X_a : a \in A\} \to \Pi\{X_a : a \in T\}$ be the usual projection function, and for $x \in \Pi\{X_a : a \in A\}$ let $S_T(x) = \{y \in \Sigma(b) : y_a = x_a \text{ for } a \in T\}$. Let Y be a space and $f : \Sigma(b) \to Y$ a continuous function. Throughout this problem, suppose each X_a is separable and each singleton subset of Y is a G_δ-set of Y.

(1) If $x \in \Sigma(b)$, prove there is a countable subset $T \subseteq A$ such that $f[S_T(x)] = \{f(x)\}$. (Hint: Use that $\{f(x)\}$ is a G_δ-set of Y.)

For each $x \in \Sigma(b)$, let $T(x)$ denote a countable subset of A (there may be many) such that $f[S_{T(x)}(x)] =$

$\{f(x)\}$ and $\{a \in A : x_a \neq b_a\} \subseteq T(x)$. Let $B(1) = \{b\}$ and $R(1) = T(b)$. Since $R(1)$ is countable, the product space $S_{A \setminus R(1)}(b)$ is separable and contains a countable dense subspace $B(2)$ such that $B(1) \subseteq B(2)$. Since $R(2) = \bigcup\{T(x) : x \in B(2)\}$ is countable, $S_{A \setminus R(2)}(b)$ is separable and contains a countable dense subspace $B(3)$ such that $B(3) \supseteq B(2)$. Continue by induction and let $B = \bigcup\{B(n) : n \in \mathbb{N}\}$ and $R = \bigcup\{R(n) : n \in \mathbb{N}\}$.

(2) Prove that B is dense in $S_{A \setminus R}(b)$ and $\bigcup\{S_R(x) : x \in B\}$ is dense in $\Sigma(b)$.

(3) Prove that $f[B]$ is dense in $f[\Sigma(b)]$. (Hint: Use (2).)

For each $x \in \Sigma(b)$, let $\bar{x}$ denote the unique element in $S_{A \setminus R}(b)$ such that $x_a = \bar{x}_a$ for $a \in R$.

(4) For each $x \in \Sigma(b)$, prove $f(x) = f(\bar{x})$. (Hint: If G and H are basic open subsets in the product space $\Pi\{X_a : a \in A\}$ such that $x \in G$ and $\bar{x} \in H$, show there is some $z \in B$ such that $S_R(z) \cap G \neq \emptyset$ and $z \in H$. Show that $f[G \cap \Sigma(b)] \cap f[H \cap \Sigma(b)] \neq \emptyset$. Use that Y is Hausdorff to conclude that $f(x) = f(\bar{x})$.)

Define $g : \Pi\{X_a : a \in R\} \to Y$ by $g(x) = f(z)$ where z is any element of $\Sigma(b)$ such that $x = \Pi_R(z)$. By (4), g is well-defined.

(5) Show that g is continuous and $(g \circ \Pi_R) | \Sigma(b) = f$.

(6) Use (5) to show that f has a continuous extension $F : \Pi\{X_a : a \in A\} \to Y$. In particular, conclude that $\Sigma(b)$ is C-embedded in $\Pi\{X_a : a \in A\}$.

1Y. <u>The "Jones Machine" for constructing regular non-Tychonoff spaces</u>

(1) Suppose that H and K are disjoint closed subsets of a space E which cannot be put inside disjoint open subsets of E. Suppose that V is open in E and $H \subseteq V$. Show that H and $K \cap c\ell_E V$ cannot be put inside disjoint open subsets of E.

(2) Let X be a non-normal space. Suppose that H and K are disjoint closed subsets of X such that whenever U is open in X and $H \subseteq U$, then $K \cap c\ell_X U \neq \emptyset$. Let $Z = (X \times \mathbb{N}) \cup \{\infty\}$, topologized as follows: $X \times \mathbb{N}$ has the usual product topology and is open in Z, and $\{Z \setminus (X \times \{1,\ldots,n\}) : n \in \mathbb{N}\}$ is an open neighborhood base at ∞. Prove that Z is Hausdorff, and is regular (respectively, Tychonoff) if X is.

(3) Define a partition E on Z as follows:

$$\begin{aligned}
E_1 &= \{\{(x,n), (x,n+1)\} : x \in H \text{ and } n \text{ is even}\}\\
E_2 &= \{\{(x,n), (x,n+1)\} : x \in K \text{ and } n \text{ is odd}\}\\
E_3 &= \{\{x\} : x \in Z \setminus \cup\{P : P \in E_1 \cup E_2\}\}\\
E &= E_1 \cup E_2 \cup E_3
\end{aligned}$$

If $z \in Z$, define $q(z)$ to be the unique member of E containing z. Thus $q : Z \to E$ is surjective; give E the quotient topology induced by q, and denote E, thus topologized, by J(X). If $A \subseteq X$, denote $q[A \times \{n\}]$ by

A_n. (Intuitively, $J(X)$ is obtained from Z by "gluing $X \times \{n\}$ to $X \times \{n+1\}$ along H" if n is even, and "gluing $X \times \{n\}$ to $X \times \{n+1\}$ along K" if n is odd.) Prove that $J(X)$ is Hausdorff, that each X_n is closed in $J(X)$ and homeomorphic to X, and that $q : Z \to J(X)$ is a perfect continuous surjection.

(4) Prove that $J(X)$ is regular iff X is regular (see 1J).

(5) Prove that $J(X)$ is not Tychonoff. (Hint: Suppose that $f \in C(J(X))$, $\underline{0} \leqslant f \leqslant \underline{1}$, and that $f[X_1] = \{1\}$ while $f(\infty) = 0$. Let $\{t_i : i \in \mathbb{N}\}$ be a monotone increasing sequence in $(0,1/2)$ converging to $1/2$. For each $i \in \mathbb{N}$ let $V_i = f^{\leftarrow}[[0,t_i)]$ and $A_i = f^{\leftarrow}[[0,t_i)]$. Find an even integer n such that $H_n \subseteq V_1$. Use (1) to show that H_n and $K_n \cap A_1$ $(=K_{n-1} \cap A_1)$ cannot be put inside disjoint open sets of $J(X)$. Verify that $K_{n-1} \cap A_1 \subseteq V_2$, and deduce (using (1)) that $H_{n-1} \cap A_2$ and K_{n-1} cannot be put inside disjoint open sets of $J(X)$. Continue through at most n steps and show that $f[X_1] \cap [0,1/2] \neq \varnothing$, which is a contradiction.)

(6) Let $DJ(X)$ be the quotient space obtained from $J(X) \times \{0,1\}$ by identifying $(x,0)$ with $(x,1)$ for each $x \in X_1$. Show that $f(\infty,0) = f(\infty,1)$ for each $f \in C(DJ(X))$.

Note that we have shown how to construct from any given regular non-normal space X a regular, non-Tychonoff space $J(X)$ that will evidently share many of the topological properties of X.

CHAPTER 2

LATTICES, FILTERS, AND TOPOLOGICAL SPACES

There is a close relation between partially ordered sets and topological spaces. The theme of this chapter is a development of this relationship. In a first course in topology, orders are often used to develop important tools such as filters and nets and to define a topology on some spaces, such as the space of real numbers. In this chapter, we start by defining orders and discussing several of the ways in which they are associated with spaces. In the middle of the chapter, we introduce spaces whose topologies are defined by an order, and prove that such spaces are hereditarily normal. The chapter closes with a summary of the properties of well-ordered sets and ordinal numbers, and an investigation of the order topology on an ordinal number.

2.1 Posets and Lattices

In this section, we define some of the fundamental concepts and terms used in the theory of ordered sets and give a few examples to illustrate these concepts. The reader already familiar with partially ordered sets and lattices may wish to bypass this section.

(a) <u>Definition</u>

(1) A **partial order** on a set A is a binary relation "$\leq$" on A,

i.e., a subset of $A \times A$, that satisfies the following axioms:

(01) **Reflexive.** If $a \in A$, then $a \leqslant a$.

(02) **Antisymmetric.** If $a, b \in A$, $a \leqslant b$, and $b \leqslant a$, then $a = b$.

(03) **Transitive.** If $a, b, c \in A$, $a \leqslant b$ and $b \leqslant c$, then $a \leqslant c$.

The pair $(A,\leqslant)$ is called a **partially ordered set** (usually abbreviated to **poset**).

(2) If $(A,\leqslant)$ is a poset and $B \subseteq A$, then $\leqslant_B$ is used to denote $\leqslant \cap (B \times B)$.

(b) <u>**Examples**</u>

(1) Let X be a set and $\mathfrak{a} \subseteq \mathbb{P}(X)$. For $B, C \in \mathfrak{a}$, define $B \leqslant C$ if $B \subseteq C$. It is easily verified that $\leqslant$ satisfies 01, 02, and 03; thus, $(\mathfrak{a},\leqslant)$ is a partially ordered set. We say $\mathfrak{a}$ is **partially ordered by inclusion** and write $(\mathfrak{a},\subseteq)$ if we want to emphasize the order.

(2) As before, let $\mathbb{R}$ denote the set of all the real numbers. Define $a \leqslant b$ to mean that the real number a is less than or equal to the real number b in the usual sense. It is clear that $(\mathbb{R},\leqslant)$ is a partially ordered set.

(c) <u>**Notation**</u>. Let $(A,\leqslant)$ be a poset, $a, b \in A$, and $B \subseteq A$.

(1) If $a \leqslant b$ and $a \neq b$, we write $a < b$.

(2) The notation (a,b) (respectively $(a,b]$, $[a,b)$, $[a,b]$) is used to denote $\{c \in A : a < c < b\}$ (respectively $\{c \in A : a < c \leqslant b\}$, $\{c \in$

$A : a \leqslant c < b\}$, $\{c \in A: a \leqslant c \leqslant b\})$.

(3) The element $a \in A$ is called an **upper** (respectively, **lower**) **bound** of B if $a \geqslant b$ (respectively, $a \leqslant b$) for all $b \in B$. We denote this by writing $a \geqslant B$ (respectively, $a \leqslant B$).

A subset of a poset may have no upper (or lower) bounds or may have many upper or lower bounds; e.g., in $(\mathbb{R},\leqslant)$ in 2.1(b)(2), $(\pi,7.6]$ has an infinite number of upper and lower bounds but $\mathbb{R}$ itself has no upper nor lower bounds.

(4) The notation (a,∞) (respectively, $[a,\infty)$, $(-\infty,a)$, $(-\infty,a]$) is used to denote $\{c \in A : a < c\}$ (respectively, $\{c \in A : a \leqslant c\}$, $\{c \in A : c < a\}$, $\{c \in A : c \leqslant a\}$).

(5) An element $a \in A$ is a **least upper bound** (respectively, **greatest lower bound**) of B if $a \geqslant B$ (respectively, $a \leqslant B$) and if $b \geqslant a$ (respectively, $b \leqslant a$) whenever $b \geqslant B$ (respectively, $b \leqslant B$). It is straightforward (see 2B) to show that there is at most one least upper bound and one greatest lower bound of a nonempty subset $B \subseteq A$. If a is the least upper bound (respectively, greatest lower bound) of B, we write $a = \bigvee B = \sup B = \text{lub}(B)$ (respectively, $a = \bigwedge B = \inf B = \text{glb}(B)$). The reader should note that for arbitrary $B \subseteq A$, it may be true that neither $\bigvee B$ nor $\bigwedge B$ exists; on the other hand, it may be possible for $\bigvee B$ (respectively, $\bigwedge B$) to exist but $\bigvee B \notin B$ (respectively, $\bigwedge B \notin B$).

(6) If $\bigvee A$ exists (respectively, $\bigwedge A$ exists), then $\bigvee A$ (respectively, $\bigwedge A$) is an upper (respectively, lower) bound for A; in such cases, $\bigvee A$ (respectively, $\bigwedge A$) is called the **maximum** (respectively, **minimum**) element of A. Sometimes the maximum (respectively, minimum) element of A when it exists is denoted by 1 (respectively, 0). Thus, the poset $(A,\leqslant)$ could be written as $(A,\leqslant,0,1)$ to emphasize the

existence of a maximum and minimum element. In particular, the example $(\mathbb{P}(X),\subseteq)$ of a poset in 2.1(b)(1) might be written as $(\mathbb{P}(X),\subseteq,\emptyset,X)$ as $\emptyset = \bigwedge\mathbb{P}(X)$ and $X = \bigvee\mathbb{P}(X)$. A poset is called **bounded** if it possesses both maximum and minimum elements.

(7) The poset $(A,\leqslant)$ is a **lattice** if for every pair of elements $a, b \in A$, $\bigvee\{a,b\}$ and $\bigwedge\{a,b\}$ exist. We write $a \vee b$ for $\bigvee\{a,b\}$ and $a \wedge b$ for $\bigwedge\{a,b\}$; $a \vee b$ is called the **join** of $\{a,b\}$ and $a \wedge b$ is called the **meet** of $\{a,b\}$. It is left to the reader (see 2C) to verify that if $a, b, c \in A$, and if $(A,\leqslant)$ is a lattice, then the following properties hold:

(L0) $a\leqslant b$ iff $a \vee b = b$ iff $a \wedge b = a$

(L1) $a = a \vee a = a \wedge a$

(L2) $a \wedge b = b \wedge a$, $a \vee b = b \vee a$

(L3) $(a \wedge b) \wedge c = a \wedge (b \wedge c)$, $(a \vee b) \vee c = a \vee (b \vee c)$

(L4) $a \wedge (a \vee b) = a$, $a \vee (a \wedge b) = a$

Also, the converse is presented in 2D; that is, if A is a set with two binary operations $\wedge$ and $\vee$ and satisfying (L1) - (L4), then by using (L0) as a definition of $\leqslant$, $(A,\leqslant)$ is a poset whose join and meet are $\vee$ and $\wedge$, respectively. Sometimes, a lattice will be denoted as $(A,\wedge,\vee)$; in the latter case $(A,\wedge,\vee)$ will also be considered a poset with the order relation defined by (L0).

(d) <u>**Definition**</u>. A poset $(A,\leqslant)$ is an **upper semilattice** (respectively, **lower semilattice**) if for $a, b \in A$, $a \vee b$ (respectively, $a \wedge b$) exists. An upper (respectively, lower) semilattice is **complete** if for all $\emptyset \neq B \subseteq A$, $\bigvee B$ (respectively, $\bigwedge B$) exists. A lattice is **complete** if it is complete both as an upper and a lower semilattice. (Sometimes we will use the convention that $\bigvee\emptyset$ is the smallest member of A and $\bigwedge\emptyset$ is the largest member of A.)

(e) **Theorem**. A complete upper (respectively, lower) semilattice with a minimum (respectively, maximum) element is a complete lattice.

Proof. Suppose $(A,\leqslant)$ is a complete upper semilattice and $0 = \bigwedge A$. Let $\emptyset \neq B \subseteq A$ and $C = \{c \in A : c \leqslant B\}$. Since $0 \in C$, then $C \neq \emptyset$ and $d = \bigvee C$ exists. We will show $\bigwedge B$ exists and, in fact, $d = \bigwedge B$. Now, for $b \in B$, since $C \leqslant b$, then $d = \bigvee C \leqslant b$. Thus, $d \leqslant B$. Suppose $e \leqslant B$. Then $e \in C$ implying $e \leqslant \bigvee C = d$. This shows that $d = \bigwedge B$ and completes the proof that A is a complete lattice. A similar proof works when $(A,\leqslant)$ is a complete lower semilattice with a maximum element. Thus $(A,\leqslant)$ is a complete lattice.

■

(f) **Examples**

(1) Let X be a set. By 2.1(b)(1), $(\mathbb{P}(X),\subseteq)$ is a poset. If $\emptyset \neq \mathcal{B} \subseteq \mathbb{P}(X)$, then it is clear that $\bigvee\mathcal{B} = \bigcup\mathcal{B}$ and $\bigwedge\mathcal{B} = \bigcap\mathcal{B}$. So, $(\mathbb{P}(X),\subseteq)$ is a complete lattice.

(2) Let X be a nonempty set and $\mathcal{P} = \{P : P \text{ is a partition of } X\}$. For $P_1, P_2 \in \mathcal{P}$, define $P_1 \leqslant P_2$ if P_1 refines P_2, i.e., for $A \in P_1$, there is some $B \in P_2$ such that $A \subseteq B$. It is easy to show that $(\mathcal{P},\leqslant)$ is a poset with $\{X\}$ as the maximum element and $\{\{x\} : x \in X\}$

as the minimum element. Let $\emptyset \neq \mathcal{B} \subseteq \mathcal{P}$. Let $Q = \{\bigcap\{f(P) : P \in \mathcal{B}\} : f \in \Pi\{P : P \in \mathcal{B}\}\} \setminus \{\emptyset\}$. Now, we will show that $Q = \bigwedge\mathcal{B}$. Let $x \in X$. For each $P \in \mathcal{B}$, there is a set $A_P \in P$ such that $x \in A_P$. Define $f \in \Pi\{P : P \in \mathcal{B}\}$ by $f(P) = A_P$. Then $x \in \bigcap\{A_P : P \in \mathcal{B}\} \in Q$ and so Q covers X. Suppose $f, g \in \Pi\{P : P \in \mathcal{B}\}$ and $f \neq g$. For some $R \in \mathcal{B}$, $f(R) \neq g(R)$. Thus, $f(R) \cap g(R) = \emptyset$ implying $(\bigcap\{f(P) : P \in \mathcal{B}\}) \cap (\bigcap\{g(P) : P \in \mathcal{B}\}) = \emptyset$. This shows that $Q \in \mathcal{P}$. If $R \in \mathcal{B}$ and $\bigcap\{f(P) : P \in \mathcal{B}\} \neq \emptyset$, then since $f(R) \in R$, $\bigcap\{f(P) : P \in \mathcal{B}\} \subseteq f(R)$. Thus, $Q \leqslant R$ and so $Q \leqslant \mathcal{B}$. Let $T \in \mathcal{P}$ and $T \leqslant \mathcal{B}$. Then for each $P \in \mathcal{B}$, $T \leqslant P$. Let $A \in T$. For each $P \in \mathcal{B}$, there is some $A_P \in P$ such that $A \subseteq A_P$. Define $f \in \Pi\{P : P \in \mathcal{B}\}$ by $f(P) = A_P$ for $P \in \mathcal{B}$. Then $A \subseteq \bigcap\{f(P) : P \in \mathcal{B}\} \in Q$. Thus, $T \leqslant Q$. This shows that $Q = \bigwedge\mathcal{B}$; thus $\mathcal{P}$ is a complete lower semilattice. Since $\mathcal{P}$ has a maximum element, then by 2.1(e), $\mathcal{P}$ is a complete lattice.

(3) Let X be a space and $\tau(X)$ the set of open sets of X. Then $(\tau(X),\subseteq)$ is a complete lattice for if $\emptyset \neq \mathcal{U} \subseteq \tau$, then $\bigvee\mathcal{U} = \bigcup\mathcal{U}$ and $\bigwedge\mathcal{U} = \mathrm{int}(\bigcap\mathcal{U})$. Similarly, the set of closed sets of X with inclusion as the order relation forms a complete lattice.

(4) Let X be a space and $Z(X)$ the set of zero sets defined in 1.4. Then $(Z(X),\subseteq)$ is a lattice when partially ordered by inclusion for if $Z(f), Z(g) \in Z(X)$ where $f, g \in C(X)$, then $Z(f) \wedge Z(g) = Z(f) \cap Z(g) = Z(f^2+g^2)$ and $Z(f) \vee Z(g) = Z(f) \cup Z(g) = Z(f \cdot g)$ (see 1.4(i)). Now, $X \in Z(X)$ is the maximum element of $Z(X)$ and $\emptyset$ is the minimum element of $Z(X)$. If $\emptyset \neq \mathcal{B} \subseteq Z(X)$, then by 1.4(i), $\bigwedge\mathcal{B}(=\bigcap\mathcal{B})$ exists if $\mathcal{B}$ is countable. However, in general, $Z(X)$ is not a complete lattice as is shown in 2E.

(5) Let X be a space and $C(X)$ the set of continuous functions

from X to $\mathbb{R}$ introduced in 1.3. As in 1.3, if f, g $\in$ C(X), then f $\leqslant$ g is defined to mean that $f(x) \leqslant g(x)$ for all $x \in X$. Now, $(C(X),\leqslant)$ is a poset (01 - 03 are easily verified). Given f, g $\in$ C(X), let h $\in$ $F(X,\mathbb{R})$ be defined by

$$h(x) = \max\{f(x), g(x)\} \text{ for } x \in X \quad \text{(compare to 1.3)}$$

If h $\in$ C(X), then it is straightforward to show that $h = f \vee g$. But $h = (\underline{2})^{-1}(f + g + |f-g|)$ and is continuous by 1B(3). Likewise, $f \wedge g$ exists, is continuous, and is defined by

$$(f \wedge g)(x) = \min\{f(x), g(x)\} \text{ for } x \in X.$$

Thus, $(C(X),\leqslant)$ is a lattice. In a similar fashion it follows that $(C^*(X),\leqslant)$ is a lattice. However, in 2F, we exhibit a space X for which neither C(X) nor $C^*(X)$ is a complete lattice. For more on completeness properties of C(X), see §8.5.

(6) Let X be a space and $\mathcal{B}(X)$ be the set of clopen sets of X defined in 1.5. As noted in 1.5(b)(3), $(\mathcal{B}(X),\subseteq)$, partially ordered by inclusion, is a lattice; for if A, B $\in \mathcal{B}(X)$, then $A \vee B = A \cup B$ and $A \wedge B = A \cap B$. Now, we will show that $\mathcal{B}(\mathbb{Q})$ is not complete. Let $U = (0,\infty)$ and $\mathcal{U} = \{C \in \mathcal{B}(\mathbb{Q}) : C \subseteq U\}$. Then $U = \cup\mathcal{U}$. Assume $\vee\mathcal{U}$ exists. Then $\vee\mathcal{U} \supseteq \cup\mathcal{U}$ and since $\vee\mathcal{U}$ is clopen, then $\vee\mathcal{U} \supseteq [0,\infty)$. Also, for each $r < 0$, there is a clopen set C_r such that $r \in C_r \subseteq (-\infty,0)$. Hence, $\mathbb{Q}\setminus C_r$ is a clopen set and $\mathbb{Q}\setminus C_r \supseteq C$ for all $C \in \mathcal{U}$. This implies that $\mathbb{Q}\setminus C_r \supseteq \vee\mathcal{U}$. Since this is true for all $r \in (-\infty,0)$, then $\vee\mathcal{U} \subseteq [0,\infty)$. So, $\vee\mathcal{U} = [0,\infty)$ which is impossible as $\vee\mathcal{U}$ must be open.

2.2 Regular open sets, regular closed sets, and semiregular spaces

In this section two lattices of subsets of a space X, namely $\mathcal{RO}(X)$ and $\mathcal{R}(X)$, are defined and studied. Each lattice plays an important role in understanding extensions and absolutes. Also, these lattices are used to define a coarser topology on a given space. Before we define these lattices, we need some preliminary results.

(a) Proposition. Let X be a space, $U \subseteq X$ be open, $C \subseteq X$ be closed, and $A, B \subseteq X$. Then:

(1) $\mathrm{int}_X(c\ell_X(\mathrm{int}_X(c\ell_X A)) = \mathrm{int}_X(c\ell_X A)$ and $c\ell_X(\mathrm{int}_X(c\ell_X(\mathrm{int}_X A))) = c\ell_X(\mathrm{int}_X A)$,

(2) if $A \subseteq B$, then $\mathrm{int}_X(c\ell_X A) \subseteq \mathrm{int}_X(c\ell_X B)$ and $c\ell(\mathrm{int}_X A) \subseteq c\ell_X(\mathrm{int}_X B)$,

(3) $c\ell_X(U \cap A) = c\ell_X(U \cap c\ell_X A)$,

(4) $\mathrm{int}_X(c\ell_X(U \cap A)) = \mathrm{int}_X(c\ell U) \cap \mathrm{int}_X(c\ell_X A)$, and

(5) $c\ell_X(\mathrm{int}_X(C \cup A)) = c\ell_X(\mathrm{int}_X C) \cup c\ell_X(\mathrm{int}_X A))$.

Proof. The proof is left as an exercise (see 2G).

■

(b) Definition. An open set U (respectively, a closed set A) in a space X is called **regular open** (respectively, **regular closed**) if $U = \mathrm{int}_X(c\ell_X U)$ (respectively, $A = c\ell_X(\mathrm{int}_X A)$). We denote by $\mathcal{RO}(X)$ (respectively, $\mathcal{R}(X)$) the set of all regular open (respectively, regular closed) subsets of X. Thus, by 2.2(a), $RO(X) = \{int_X(c\ell_X A) : A \subseteq X\}$ and $R(X) = \{c\ell_X(int_X A) : A \subseteq X\}$.

It is easy to verify that when partially ordered by inclusion, $(RO(X),\subseteq)$ and $(R(X),\subseteq)$ are posets. Much more is true as indicated in the next result.

(c) **Proposition**. For a space X, $(RO(X),\subseteq)$ and $(R(X),\subseteq)$ are complete lattices.

Proof. We will only prove that $(RO(X),\subseteq)$ is a complete lattice, as the proof that $(R(X),\subseteq)$ is a complete lattice is quite similar. If $U, V \in RO(X)$, then since $U \cap V \in RO(X)$ by 2.2(a), it is immediate that $U \wedge V = U \cap V$. Also, if $\emptyset \neq \mathcal{B} \subseteq RO(X)$, then $int_X(c\ell_X(\cup\mathcal{B}) \supseteq B$ for all $B \in \mathcal{B}$; if $C \in RO(X)$ and $C \supseteq B$ for all $B \in \mathcal{B}$, then by 2.2(a)(1) and (2), $C = int_X(c\ell_X C) \supseteq int_X(c\ell_X \cup\mathcal{B}))$. Thus, $int_X(c\ell_X)\cup\mathcal{B})) = \vee\mathcal{B}$. Since $\emptyset \in RO(X)$ and $\emptyset$ is the minimum element of $RO(X)$, then by 2.1(e), $(RO(X),\subseteq)$ is a complete lattice.

■

A common error is to assume that $U \cup V \in RO(X)$ whenever $U, V \in RO(X)$; however, note that in the space $\mathbb{R}$, $(0,1), (1,2) \in RO(\mathbb{R})$ but $(0,1) \cup (1,2) \notin RO(\mathbb{R})$.

By 2.2(c), we have associated with an arbitrary space X two complete lattices, namely, $(RO(X),\subseteq)$ and $(R(X),\subseteq)$. If $U, V \in RO(X)$ and $\emptyset \neq \mathcal{B} \subseteq RO(X)$, then $U \vee V = int_X(c\ell_X(U \cup V))$, $U \wedge V = U \cap V$, $\vee\mathcal{B} = int_X(c\ell_X(\cup\mathcal{B}))$ and $\wedge\mathcal{B} = int_X(c\ell_X(\cap\mathcal{B}))$. Note that by using 2.2(a)(4), it follows that the intersection of two (and by induction, any finite number) of regular open sets is regular open; so, if $B_1,\ldots,B_n \in RO(X)$, then $\bigwedge\{B_i : 1 \leq i \leq n\} = \bigcap\{B_i : 1 \leq i \leq n\}$. However, the intersection of an arbitrary family of regular open sets may not be regular open or even open (for example, consider the subsets $\{(-1/n,1/n) : n \in \mathbb{N}\}$ of $\mathbb{R}$). Similarly, if $A, B \in R(X)$ and $\emptyset \neq \mathcal{B} \subseteq R(X)$, then $A \vee B = A \cup B$, $A \wedge B = c\ell_X(int_X(A \cap B))$, $\bigwedge\mathcal{B} = c\ell_X(int_X(\bigcup\mathcal{B}))$, and $\bigvee\mathcal{B} = c\ell_X(int_X(\bigcap\mathcal{B}))$. **The reader should be aware** that these equalities for suprema and infima in $RO(X)$ and $R(X)$ are frequently used in this book without reference.

(d) **Proposition**. Let X be a space. Then $RO(X)$ is an open base for a Hausdorff topology on the underlying set of X that is contained in the original topology on X.

Proof. By 2.2(a)(4), $RO(X)$ is closed under finite intersections. Obviously $\{\emptyset,X\} \subseteq RO(X)$; so, $RO(X)$ is an open base for a topology $\tau(s)$ on X. If $p, q \in X$ and $p \neq q$ there are disjoint open sets S and T of X such that $p \in S$ and $q \in T$; thus $int_X(c\ell_X S)$ and $int_X(c\ell_X T)$ are disjoint members of $\tau(s)$ containing p and q respectively, so $(X,\tau(s))$ is Hausdorff. Since $RO(X)$ is a subcollection of the open sets of the original topology of X, obviously $\tau(s)$ is contained in this original topology.

■

(e) **Definition**.

(1) A space X is **semiregular** if $RO(X)$ forms a base for the open sets of X.

(2) If X is a space, then X(s) will denote the underlying set of

X equipped with topology $\tau(s)$ described above. We call X(s) the **semiregularization** of X.

The motivation for the above terminology is evident from the following proposition.

(f) Proposition. Let X be a space, let U be open in X, and let A be closed in X. Then:

(1) X(s) is a Hausdorff space,

(2) $c\ell_X U = c\ell_{X(s)} U$ and $\text{int}_X A = \text{int}_{X(s)} A$,

(3) $\text{int}_X(c\ell_X U) = \text{int}_{X(s)}(c\ell_{X(s)} U)$ and $c\ell_X(\text{int}_X A) = c\ell_{X(s)}(\text{int}_{X(s)} A)$,

(4) $RO(X(s)) = RO(X)$, $R(X(s)) = R(X)$, and $B(X) = B(X(s))$,

(5) X is semiregular iff $X(s) = X$, and

(6) $(X(s))(s) = X(s)$.

Proof.

(1) See 2.2(d) above.

(2) Since X(s) has a coarser topology than that of X, then $c\ell_X U \subseteq c\ell_{X(s)} U$. If $x \notin c\ell_X U$, there is an open set V in X such that $x \in V$ and $V \cap U = \varnothing$. By 2.2(a), $\text{int}_X(c\ell_X V) \cap \text{int}_X(c\ell_X U) = \varnothing$. Since $\text{int}_X(c\ell_X V)$ is an open neighborhood of x in X(s) and $U \subseteq \text{int}_X(c\ell_X U)$, then $x \notin c\ell_{X(s)} U$. This shows that $c\ell_X U = c\ell_{X(s)} U$. Since $X \backslash A$ is open in X, then $c\ell_X(X\backslash A) = c\ell_{X(s)}(X\backslash A)$; hence

$X\backslash int_X A = X\backslash int_{X(s)}A$. This shows that $int_X A = int_{X(s)}A$.

(3) The proof is straightforward using (2).

(4) By 2.2(a), $RO(X) = \{int_X(c\ell_X U) : U$ is open in $X\}$. Since the topology of X(s) is coarser than the topology of X, it follows that $RO(X(s)) = \{int_{X(s)}(c\ell_{X(s)}U) : U$ is open in $X\}$. By (3), $RO(X) = RO(X(s))$. Since $R(X) = \{c\ell_X(int_X A) : A \subseteq X\}$ by 2.2(a) and the fact that $c\ell_X(int_X A) = c\ell_X(int_X(X\backslash(X\backslash A))) = X\backslash int_X(c\ell_X(X\backslash A))$, it follows that $R(X) = \{X\backslash int_X(c\ell_X B) : B \subseteq X\} = \{X\backslash V : V \in RO(X)\}$. Thus, it follows that $R(X) = R(X(s))$. It immediately follows that $B(X) = B(X(s))$.

(5) This is obvious.

(6) This follows immediately from the first half of (4).

■

(g) **Proposition**.

(1) A regular space is semiregular.

(2) If $f \in C(X,Y)$ where X is a space and Y is regular and if $f_s \in F(X(s),Y)$ is defined by $f_s(x) = f(x)$ for all $x \in X$, then f_s is continuous.

Proof. To prove (1), it suffices to show that $RO(X)$ forms a base for the open sets of the regular space X. If $x \in X$ and V is an open neighborhood of x, there is an open set U containing x such that $c\ell_X U \subseteq V$. Thus, $x \in int_X(c\ell_X U) \subseteq V$ and $int_X(c\ell_X U) \in RO(X)$ by 2.2(a). To prove (2), let $x \in X$ and V be an open neighborhood in Y of f(x). There is an open neighborhood U of f(x) in Y such that $c\ell_Y U \subseteq V$. Since f is continuous, there is an open neighborhood W of x such that $f[W] \subseteq U$. By continuity of f, $f[c\ell_X W] \subseteq c\ell_Y f[W] \subseteq c\ell_Y U \subseteq V$. So, $f[\mathrm{int}_X(c\ell_X W)] \subseteq V$. Since $RO(X) = \{\mathrm{int}_X(c\ell_X U) : U \text{ open in } X\}$ is an open basis for $X(s)$, then f_s is continuous.

■

To show that 2.2(g) cannot be strengthened much, some examples of semiregular spaces are needed.

(h) **Examples**.

(1) Let $X = \mathbb{R} \cup \{p_-, p_+\}$ where $\mathbb{R} \cap \{p_-, p_+\} = \emptyset$ and $p_- \neq p_+$. A subset $U \subseteq X$ is defined to be open in X if $U \cap \mathbb{R}$ implies in the usual topology on $\mathbb{R}$, and $p_+ \in U$ (respectively, $p_- \in U$) and there is some $n \in \mathbb{N}$ such that $\cup\{(-m-1,-m) \cup (m,m+1) : m \geqslant n$ and m even (respectively, m odd)$\} \subseteq U$. It is straightforward to verify that this is a valid definition of a topology on X. For $n \in \mathbb{N}$, let $V_n = \{p_+\} \cup \cup\{(-m-1,-m) \cup (m,m+1) : m \geqslant n \text{ and } m \text{ even}\}$. Then V_n is open in X and $\mathrm{int}_X(c\ell_X V_n) = \mathrm{int}_X(\{p_+\} \cup \cup\{[-m-1,-m] \cup [m,m+1] : m \geqslant n \text{ and } m \text{ even}\}) = \{p_+\} \cup \cup\{(-m-1,-m) \cup (m,m+1) : m \geqslant n \text{ and } m \text{ even}\} = V_n$. Similarly define W_n to be $\{p_-\} \cup \cup\{(-m-1,-m) \cup (m,m+1) : m \geqslant n \text{ and } m \text{ odd}\}$; then W_n is open in X and $\mathrm{int}_X(c\ell_X W_n) = W_n$. Also, $V_1 \cap W_1 = \emptyset$. If $x \in \mathbb{R}$ and $t > 0$, then $(x-t,x+t)$ is an open neighborhood of x in X and $\mathrm{int}_X(c\ell_X(x-t,x+t)) = \mathrm{int}_X[x-t,x+t] = (x-t,x+t)$. Thus, it easily follows that X is semiregular and Hausdorff. For any $n, k \in \mathbb{N}$, $c\ell_X V_n \cap c\ell_X W_k \neq \emptyset$ which implies that X is not regular. This illustrates that the converse of 2.2(g)(1) is false.

(2) Consider the following subspace Y of the space X of 2.2(h)(1): $Y = \{p_+\} \cup \cup\{[m,m+1] : m \geqslant 0, m \text{ even}\}$. Let $f \in F(Y,X)$

be defined by $f(y) = y$ for $y \in Y$. Then f is the inclusion function and is continuous. Note that $int_Y(c\ell_Y(V_n \cap Y)) = int_Y(\{p_+\} \cup \bigcup\{[m,m+1] : m \geqslant n, m \text{ even}\}) = \{p_+\} \cup \bigcup\{[m,m+1] : m \geqslant n, m \text{ even}\}$. This shows that Y(s) has a strictly coarser topology than that of Y. In fact Y(s) is a compact space whereas Y is not even regular as p_+ and the closed set $\mathbb{N}$ cannot be separated by disjoint open sets. Also, the function $f_s \in F(Y(s),X)$ (defined in 2.2(g)(2)) is not continuous which illustrates that regularity in 2.2(g)(2) cannot be replaced by semiregularity. Moreover, this illustrates that if $f \in C(Y,X)$, then f_s may not belong to $C(Y(s),X(s))$.

(3) Let $\mathbb{R}'$ be $\mathbb{R}$ with the topology for which $\{\mathbb{Q}\} \cup \{(a,b) : a, b \in \mathbb{R}, a < b\}$ is a subbase; i.e., we obtain the topology of $\mathbb{R}'$ by enlarging the usual topology on $\mathbb{R}$ so that $\mathbb{Q}$ is open. If $a, b \in \mathbb{R}$ and $a < b$, then $int_{\mathbb{R}'}(c\ell_{\mathbb{R}'}((a,b) \cap \mathbb{Q})) = int_{\mathbb{R}'}([a,b]) = (a,b)$. Since $(a,b) \not\subseteq \mathbb{Q}$, then $\mathbb{Q}$ contains no members of $RO(\mathbb{R}')\backslash\{\emptyset\}$, and so $\mathbb{R}'$ is not semiregular. It is easy to verify that $\mathbb{R}'(s) = \mathbb{R}$. In $\mathbb{R}'$, both $\mathbb{Q}$ and $\mathbb{R}'\backslash\mathbb{Q}$ have the usual topology inherited from $\mathbb{R}$; hence, both $\mathbb{Q}$ and $\mathbb{R}'\backslash\mathbb{Q}$ are regular (and, hence, semiregular) subspaces of $\mathbb{R}'$. This shows that the union of two semiregular spaces may not be semiregular as $\mathbb{R}' = \mathbb{Q} \cup (\mathbb{R}'\backslash\mathbb{Q})$.

(4) Let $\mathbb{R}''$ be $\mathbb{R}$ with the topology generated by $\{\mathbb{R}\backslash\mathbb{Q}\} \cup \{(a,b) : a, b \in \mathbb{R}, a < b\}$. As above, $(\mathbb{R}'')(s) = \mathbb{R}$. Also, $\mathbb{R}'$ and $\mathbb{R}''$ are not homeomorphic spaces as $\mathbb{R}'$ has a countable dense open set, but $\mathbb{R}''$ does not. However, the semiregularizations of $\mathbb{R}'$ and $\mathbb{R}''$ are the same space, namely $\mathbb{R}$.

As noted in 2.2(h)(2), the subspace Y of the semiregular space X is not semiregular; so, in general, semiregularity is not a hereditary property. However, we do have the following hereditary properties.

(i) <u>**Proposition**</u>.

(1) Semiregularity is open-hereditary (defined in 1.2).

(2) If D is a dense subspace of a space X, then $D(s)$ is a dense subspace of $X(s)$; in particular, semiregularity is hereditary on dense sets.

<u>Proof</u>.

(1) Let X be a semiregular space and U be an open subset of X. Let $p \in U$ and W be a regular open subset of X such that $p \in W \subseteq U$. Now, $\mathrm{int}_U(c\ell_U W) = \mathrm{int}_U(U \cap c\ell_X W) = \mathrm{int}_X(U \cap c\ell_X W)$ as U is open in X. But $\mathrm{int}_X(U \cap c\ell_X W) = \mathrm{int}_X(c\ell_X W) \cap \mathrm{int}_X U = W \cap U = W$. Thus, U is a semiregular space.

(2) First note that if T is an open set in X, then $c\ell_D(T \cap D) = c\ell_X(T \cap D) \cap D = (c\ell_X T) \cap D$ (the last equality follows from 2.2(a)(3)). Let W be an open set in X. To show that $RO(D)$ is an open base for the subspace topology that D inherits from $X(s)$, it suffices to show that $\mathrm{int}_D(c\ell_D(W \cap D)) = (\mathrm{int}_X(c\ell_X W)) \cap D$. Since $(\mathrm{int}_X(c\ell_X W)) \cap D \subseteq (c\ell_X W) \cap D = c\ell_D(W \cap D)$ (as noted above) and $(\mathrm{int}_X(c\ell_X W)) \cap D$ is open in D, then $(\mathrm{int}_X(c\ell_X W)) \cap D \subseteq \mathrm{int}_D(c\ell_D(W \cap D))$. Conversely suppose U is an open set in X such that $U \cap D \subseteq c\ell_D(W \cap D)$ ($= (c\ell_X W) \cap D$ as noted above). Then $c\ell_X U = c\ell_X(U \cap D) \subseteq c\ell_X((c\ell_X W) \cap D) \subseteq c\ell_X W$. So, $U \subseteq \mathrm{int}_X(c\ell_X U) \subseteq \mathrm{int}_X(c\ell_X W)$, which implies that $U \cap D \subseteq (\mathrm{int}_X(c\ell_X W)) \cap D$. Hence, $\mathrm{int}_D(c\ell_D(W \cap D)) \subseteq (\mathrm{int}_X(c\ell_X W)) \cap D$. This shows that $\mathrm{int}_D(c\ell_D(W \cap D)) = (\mathrm{int}_X(c\ell_X W)) \cap D$ and completes the proof that $D(s)$ is a subspace of $X(s)$. Since $X(s)$ has a coarser topology

than that of X, then D(s) is also dense in X(s).

■

We now investigate products of semiregular spaces.

(j) **Proposition**. The product of nonempty spaces is semiregular iff each coordinate space is semiregular.

Proof. Let $\{X_i : i \in I\}$ be a family of nonempty spaces and let $X = \Pi\{X_i : i \in I\}$. For each $i \in I$, the projection function $\Pi_i : X \to X_i$ is continuous, open and onto. Let $x \in U$ where U is open in X_i. There is a point $y \in X$ such that $\Pi_i(y) = x$. Thus $y \in \Pi_i^{\leftarrow}[U] = U \times \Pi\{X_j : j \in J\}$ where $J = I \setminus \{i\}$. Now, suppose X is semiregular. There is a regular open neighborhood V of y such that $y \in V \subseteq \Pi_i^{\leftarrow}[U]$. There is a finite set $F \subseteq I$ and open sets W_j in X_j for $j \in F$ such that the open set $W = \Pi\{W_j : j \in F\} \times \Pi\{X_j : j \in K\}$ for $K = I \setminus F$ has the property that $y \in W \subseteq V \subseteq \Pi_i^{\leftarrow}[U]$. Now, $int_X(c\ell_X W) \subseteq int_X(c\ell_X V) = V \subseteq \Pi_i^{\leftarrow}[U]$. But $int_X(c\ell_X W) = int_X(\Pi\{c\ell_X W_j : j \in F\} \times \Pi\{X_j : j \in K\}) = \Pi\{int_{X_j}(c\ell_X W_j) : j \in F\} \times \Pi\{X_j : j \in K\}$ (see 1F). Thus, $x \in W_i \subseteq int_{X_i}(c\ell_{X_i} W_i) \subseteq U$. Since $int_{X_i}(c\ell_{X_i} W_i) \in RO(X_i)$, it follows that X_i is semiregular. Conversely, suppose X_j is semiregular for each $j \in I$. Let $y \in X$ and T be a basic open neighborhood of y in X; then $T = \Pi\{T_i : i \in F\} \times \Pi\{X_i : i \in K\}$ where $K = I \setminus F$, F is a finite subset of I, and T_i is an open set in X_i for each $i \in F$. Now, $\Pi_i(y) \in T_i$ and there exists $V_i \in RO(X_i)$ such that $\Pi_i(y) \in V_i \subseteq T_i$ for each $i \in F$. Let $V = \Pi\{V_i : i \in F\} \times \Pi\{X_i : i \in K\}$. Then $y \in V \subseteq U$ and

$\text{int}_X(c\ell_X V) = \text{int}_X(\Pi\{c\ell_{X_i} V_i : i \in F\} \times \Pi\{X_i : i \in K\}) = \Pi\{(\text{int}_{X_i}(c\ell_{X_i} V_i) : i \in F\} \times \Pi\{X_i : i \in K\} = \Pi\{V_i : i \in F\} \times \Pi\{X_i : i \in K\} = V$. So, $V \in RO(X)$; this completes the proof that X is semiregular.

■

We conclude this section with an example that shows semiregularity is not preserved by perfect, continuous, open surjections.

(k) **Example**. Let $X = \mathbb{R}^2 \cup \{p_+, p_-\}$. A subset U of X is defined to be open if $U \cap \mathbb{R}^2$ is open in $\mathbb{R}^2$ (with the usual plane topology) and if $p_+ \in U$ (respectively, $p_- \in U$) implies that there exists $n \in \mathbb{N}$ such that $\{(x,y) : x > n, y > 0\} \subseteq U$ (respectively, $\{(x,y) : x > n, y < 0\} \subseteq U$). For $n \in \mathbb{N}$, let $V_n = \{p_+\} \cup \{(x,y) : x > n, y > 0\}$ and $W_n = \{p_-\} \cup \{(x,y) : x > n, y < 0\}$. Now, $\text{int}_X(c\ell_X V_n) = \text{int}_X(\{p_+\} \cup \{(x,y) : x \geq n, y \geq 0\}) = \{p_+\} \cup \{(x,y) : x > n, y > 0\} = V_n$. Likewise, $\text{int}_X(c\ell_X W_n) = W_n$. Since $\{V_n : n \in \mathbb{N}\}$ (respectively, $\{W_n : n \in \mathbb{N}\}$) is a neighborhood base of p_+ (respectively, p_-), it easily follows that X is semiregular. Now consider two subspaces: $Y = \{p_+\} \cup \{(x,y) : x \in \mathbb{R}, y \geq 0\}$ and $Z = \{p_-\} \cup \{(x,y) : x \in \mathbb{R}, y \leq 0\}$. Define $f : X \to Y$ by $f(p_+) = f(p_-) = p_+$, and $f((x,y)) = (x, |y|)$. It is straightforward to check that f is continuous and onto and $f|Y$ and $f|Z$ are homeomorphisms. If C is closed in X, then $C \cap Y$ and $C \cap Z$ are closed; hence, $f[C \cap Y]$ and $f[C \cap Z]$ are closed in Y since $f|Y$ and $f|Z$ are homeomorphisms. Thus, $f[C] = f[C \cap Y] \cup f[C \cap Z]$ is closed in Y. This shows f is a closed function. If $y \in Y$, then $f^{\leftarrow}(y)$ is either a singleton or a doubleton. Hence, f is compact. To show f is open, it suffices to show f[B] is open for elements B of an open base. Now, for each $n \in \mathbb{N}$, $f[V_n] = f[W_n] = V_n$ is open. Suppose $(x,y) \in \mathbb{R}^2$ and $y \neq 0$. Let $t > 0$ such that $(x-t,x+t) \times (y-t,y+t) \cap \mathbb{R} \times \{0\} = \emptyset$. Then $f[(x-t,x+t) \times (y-t,y+t)] = (x-t,x+t) \times (|y|-t,|y|+t)$ is open. If $t > 0$, then $f[(x-t,x+t) \times (-t,t)] = (x-t,x+t) \times [0,t)$ is open in Y. This shows that f is open. So, we have shown that f is perfect, open, continuous and onto and that X is a semiregular space. It remains to show that Y is not semiregular. Now V_1 is an open neighborhood in Y of p_+. Let W be an open neighborhood of p_+ in Y such that $W \subseteq V_1$. Then there is some $n \in \mathbb{N}$ such that $V_n \subseteq W$. But $\text{int}_Y(c\ell_Y V_n) = \text{int}_Y(\{p_+\} \cup \{(x,y) : x \geqslant n, y \geqslant 0\}) = \{p_+\} \cup \{(x,y) : x > n, y \geqslant 0\}$. So, $(n+1,0) \in \text{int}_Y(c\ell_Y V_n) \subseteq \text{int}_Y(c\ell_Y W)$ and $(n+1,0) \notin V_1$. Thus, V_1 contains no regular open neighborhood of p_+. This completes the proof that Y is not semiregular.

2.3 Filters on a lattice

There are many lattices (see 2.1(f) and 2.2(b)) associated with a space X and filters on these lattices (precisely defined in 2.3(a) below) are useful in studying extensions and absolutes of X. It is the aim of this section to define and introduce some of the basic properties of filters on a lattice. Throughout this section, L will denote a subset of $\mathbb{P}(X)$ such that

(a) $\emptyset, X \in L$, and

(b) $(L, \subseteq)$ is a lattice with respect to set inclusion "$\subseteq$".

This does **not** necessarily mean that if A, B $\in L$, then $A \wedge B = A \cap B$ and $A \vee B = A \cup B$. For as we saw in 2.2(c), if $L = RO(X)$, then $A \vee B = \text{int}_X(c\ell_X(A \cup B))$, which, in general, is not the same as $A \cup B$.

(a) <u>Definition</u>

(1) A subset $F \subseteq L$ is a L-**filter base** if F satisfies:

(FB1) $\emptyset \neq F$, and

(FB2) if $F_1, F_2 \in F$, there is some $F_3 \in F$ such that $\emptyset \neq F_3 \subseteq F_1 \wedge F_2$.

Note that $\emptyset \notin F$ which implies that $F \neq L$.

(2) If an L-filter base F satisfies

(FB3) if $F_1 \in F$ and $F_1 \subseteq F_2 \in L$, then $F_2 \in F$,

then F is called a L-**filter**.

(b) <u>Proposition</u>. If F is a L-filter and $F_1, F_2 \in F$, then $F_1 \wedge F_2 \in F$.

<u>Proof</u>. Combine (FB2) and (FB3).

■

(c) <u>Definition</u>

(1) If F is a L-filter base, define $\hat{F}$ to be $\{A \in L : A \supseteq F$ for some $F \in F\}$. It is easily seen that $\hat{F}$ is a L-filter; $\hat{F}$ is said to be **generated** by F.

(2) A L-**ultrafilter** U is a maximal element in the set of all

L-filters when partially ordered by inclusion; in particular, if F is a L-filter and $F \supseteq U$, then $F = U$.

(3) Let F be a L-filter base. The set $\bigcap\{c\ell_X F : F \in F\}$ is called the **adherence** of F and denoted by $a(F)$. Clearly, $a(F) = a(\hat{F})$. If $a(F) \neq \emptyset$, F is said to be **fixed**; otherwise, F is called **free**. The L-filter base F is said to **converge** to some $p \in X$ if $N(p) \cap L \subseteq \hat{F}$ (recall $N(p)$ is the set of neighborhoods of p in X). Let $c(F)$ denote the set of convergent points of F. If there is a possibility of confusion, $a(F)$ will be denoted by $a_X(F)$ and $c(F)$ by $c_X(F)$.

(4) A $\mathbb{P}(X)$-filter (respectively, $\mathbb{P}(X)$-filter base, $\mathbb{P}(X)$-ultrafilter) is called a **filter** (respectively, **filter base**, **ultrafilter**) on X. A $\tau(X)$-filter (respectively, $\tau(X)$-filter base, $\tau(X)$-ultrafilter) is called an **open filter** (respectively, **open filter base**, **open ultrafilter**) on X. Similarly, we define **closed** (respectively, **regular open**, **regular closed**, **zero-set**, **clopen**) **filters**, **filter bases**, and **ultrafilters**. The term zero-set filter (respectively, zero-set filter base, zero-set ultrafilter) is abbreviated to **z-filter** (respectively, **z-filter base**, **z-ultrafilter**).

(5) If $A \in L$ and F is a L-filter base, we say A **meets** F if $A \wedge B \neq \emptyset$ for all $B \in F$; otherwise, we say A **misses** F. If F and G are L-filter bases, we say F **meets** G if $F \wedge G \neq \emptyset$ for all $F \in F$ and $G \in G$; otherwise, we say F **misses** G. We now list some general properties common to all L-filters.

(d) <u>**Proposition**</u>. Let X be a space, $A \subseteq X$, and F a L-filter base. Then:

(1) $\hat{F} = \bigcap\{G : G \text{ is a } L\text{-filter and } F \subseteq G\}$,

(2) F is contained in some L-ultrafilter,

(3) F is a L-ultrafilter iff for each $A \in L$, if A meets F,

then $A \in \mathcal{F}$,

(4) if $\mathcal{U}$ and $\mathcal{V}$ are distinct $\mathcal{L}$-ultrafilters, then there are sets $U \in \mathcal{U}$ and $V \in \mathcal{V}$ such that $U \wedge V = \emptyset$, and

(5) if $\{\mathcal{F}_i : i \in I\}$ is a nonempty family of $\mathcal{L}$-filters, then $\cap\{\mathcal{F}_i \; i \in I\}$ is a $\mathcal{L}$-filter.

Proof. Left as an exercise (2H) for the reader.

■

(e) **Proposition**. Let $\mathcal{F}$ be a $\mathcal{L}$-filter base where $\mathcal{L}$ is one of $\mathbb{P}(X)$, $\tau(X)$, $\mathcal{RO}(X)$, or $\mathcal{R}(X)$. Then $\mathcal{F}$ is a $\mathcal{L}$-ultrafilter iff

(1) $\mathcal{L} = \mathbb{P}(X)$ and for each $A \subseteq X$, $A \in \mathcal{F}$ or $X \setminus A \in \mathcal{F}$, or

(2) $\mathcal{L} = \tau(X)$ or $\mathcal{RO}(X)$ and for each $A \in \mathcal{L}$, $A \in \mathcal{F}$ or $X \setminus c\ell_X A \in \mathcal{F}$, or

(3) $\mathcal{L} = \mathcal{R}(X)$ and for each $A \in \mathcal{L}$, $A \in \mathcal{F}$ or $X \setminus \mathrm{int}_X A \in \mathcal{F}$.

Proof. The proofs for (1), (2) and (3) are quite similar. We will prove (2) and leave the proofs of (1) and (3) as an exercise (see 2H) for the reader. Suppose $\mathcal{F}$ is an $\mathcal{L}$-ultrafilter and $A \in \mathcal{L}$. If A meets $\mathcal{F}$, then $A \in \mathcal{F}$ by 2.3(d)(3). If $A \cap F = \emptyset$ for some $F \in \mathcal{F}$, then $F \subseteq X \setminus c\ell_X A$ and $X \setminus c\ell_X A \in \mathcal{F}$. Conversely, suppose for each $A \in \mathcal{L}$, $A \in \mathcal{F}$ or $X \setminus c\ell_X A \in \mathcal{F}$. Now, $\mathcal{F}$ is contained in some $\mathcal{L}$-ultrafilter $\mathcal{U}$ by 2.3(d)(2). Let $A \in \mathcal{U}$. If $A \notin \mathcal{F}$, then $X \setminus c\ell_X A \in \mathcal{F} \subseteq \mathcal{U}$. This is impossible as $A \in \mathcal{U}$ and $A \cap (X \setminus c\ell_X A) = \emptyset$. So, $A \in \mathcal{F}$. This shows $\mathcal{U} \subseteq \mathcal{F}$. Hence, $\mathcal{U} = \mathcal{F}$.

■

Proposition 2.3(e)(3) is not true when L is the lattice of closed sets. An example is developed in 2H to illustrate this point.

(f) **Proposition**. Let $L = \mathbb{P}(X)$, $\tau(X)$, $RO(X)$, or $R(X)$ and let F be a L-filter base and U be a L-ultrafilter. Then:

(1) $c(F) \subseteq a(F)$,

(2) $c(U) = a(U)$,

(3) $c(F)$ contains at most one point, and

(4) if X is compact and $a(F) = \{p\}$ where $p \in X$, then $c(F) = \{p\}$.

Proof

(1) Let $p \in c(F)$. Then $N(p) \cap L \subseteq \hat{F}$. Let $F \in F$ and $U \in N(p)$ be open. If F is a filter base or open filter base, then $U \in \hat{F}$. So, $U \cap F = U \wedge F \neq \emptyset$ and $p \in a(F)$. If F is a regular open filter base, then $int_X(c\ell_X U) \in N(p) \cap L \subseteq \hat{F}$ and $\emptyset \neq (int_X(c\ell_X U)) \cap F = int_X(c\ell_X(U \cap F))$ by 2.2(a). So, $U \cap F \neq \emptyset$ which implies that $p \in a(F)$. Suppose F is a regular closed filter base; then $c\ell_X U \in N(p) \cap L \subseteq \hat{F}$. If $U \cap F = \emptyset$, then $int_X((c\ell_X U) \cap F) = \emptyset$ implying $(c\ell_X U) \wedge F = \emptyset$. So, $c\ell_X U \notin \hat{F}$, a contradiction. Hence, $U \cap F \neq \emptyset$ and $p \in a(F)$.

(2) By (1), $c(U) \subseteq a(U)$. Suppose $p \in a(U)$ and let $F = N(p) \cap L$. Then F is a L-filter base. Since F meets U, then by 2.3(d)(3), $F \subseteq U$. Thus, $p \in c(U)$. This completes the proof that $a(U) = c(U)$.

(3) Suppose $p \in c(F)$. Then $N(p) \cap L \subseteq \hat{F}$. Let $q \in X$ and $q \neq p$. There are disjoint open sets $U \in N(p)$ and $V \in N(q)$.

Thus, when $L = \mathbb{P}(X)$, $\tau(X)$, or $RO(X)$, then $\mathrm{int}_X(c\ell_X U) \in N(p) \cap L$ and $\mathrm{int}_X(c\ell_X V) \in N(q) \cap L$. By 2.2(a), $\mathrm{int}_X(c\ell_X U) \cap \mathrm{int}_X(c\ell_X V) = \emptyset$. Hence, $N(q) \cap L \not\subseteq \hat{F}$, i.e., $q \notin c(F)$. If F is a regular closed filter base, then $c\ell_X U \in N(p) \cap L$ and $c\ell_X V \in N(q) \cap L$. But $c\ell_X U \wedge c\ell_X V = \emptyset$. So, $N(p) \cap L \subseteq F$ and $q \notin c(F)$. Thus, $c(F)$ contains at most one point.

(4) By (3), it suffices to show that F converges to p. Let U be an open neighborhood of p. Since $X \setminus U$ is compact and $a(F) = \{p\}$, then it is easy to find $F_1, \ldots, F_n \in F$ for some positive integer n such that $F_1 \cap \ldots \cap F_n \subseteq U$. If $L = \mathbb{P}(X)$, $\tau(X)$, or $RO(X)$, then $F_1 \cap \ldots \cap F_n = F_1 \wedge \ldots \wedge F_n$; if L is $R(X)$, then $F_1 \cap \ldots \cap F_n \supseteq F_1 \wedge \ldots \wedge F_n$. So, if $F \in F$ such that $F \subseteq F_1 \wedge \ldots \wedge F_n$, then $F \subseteq U$. Hence, F converges to p.

■

If L is the lattice of closed sets, then the above proposition fails, as the following example shows.

(g) **Example**. Let X be the space defined in 2.2(h)(1). Let $F = \{\mathbb{N} \setminus \{1, 2, \ldots, n\} : n \in \mathbb{N}\}$. Then F is a closed filter base and is contained in some closed ultrafilter U by 2.3(d)(2). Clearly, $a(U) = \cap U \subseteq \cap F = \emptyset$. If U is an open neighborhood of p_+, then $c\ell_X U \in N(p_+) \cap L$ and for some $n \in \mathbb{N}$, $\mathbb{N} \setminus \{1, 2, \ldots, n\} \subseteq c\ell_X U$. Thus, $c\ell_X U \in U$. This shows that $p_+ \in c(U)$. Likewise, $p_- \in c(U)$. Thus, $|c(U)| = 2$ and $|a(U)| = 0$. Note that F is not a regular closed filter base as $\mathrm{int}_X(\mathbb{N} \setminus \{1, 2, \ldots, n\}) = \emptyset$.

(h) **Proposition**. If U is an L-ultrafilter where $L = \mathbb{P}(X)$,

$\tau(X)$, $RO(X)$, $\{X\backslash U : U \in \tau(X)\}$, or $R(X)$, then $|a(\mathcal{U})| \leqslant 1$.

Proof. If $\mathcal{L}$ is not the lattice of closed sets, then the result follows from 2.3(f)(2), (3). If $\mathcal{L}$ is the lattice of closed sets and $p, q \in a(\mathcal{U})$, then $\{p\} \in \mathcal{U}$ and $\{q\} \in \mathcal{U}$ by 2.3(d)(3). Thus, $\{p\} \cap \{q\} \neq \emptyset$ so $p = q$.

■

(i) **Proposition**. For the space X let $\mathcal{L}$ be $\mathbb{P}(X)$, $\tau(X)$, or $\{X\backslash U : U \in \tau(X)\}$. If Y is a space, $f \in C(X,Y)$, f maps onto Y, and $\mathcal{F}$ is an $\mathcal{L}$-filter base on Y, then $f^{\leftarrow}[\mathcal{F}] = \{f^{\leftarrow}[F] : F \in \mathcal{F}\}$ is a $\mathcal{L}$-filter base on X and $f^{\leftarrow}[a(\mathcal{F})] \supseteq a(f^{\leftarrow}[\mathcal{F}])$.

Proof. It is easily seen that $f^{\leftarrow}[\mathcal{F}]$ is an $\mathcal{L}$-filter base on X. Let $p \in a(f^{\leftarrow}[\mathcal{F}])$ and $F \in \mathcal{F}$. Let U be an open neighborhood of $f(p)$. There is an open neighborhood V of p such that $f[V] \subseteq U$. By our choice of p, $V \cap f^{\leftarrow}[F] \neq \emptyset$ which implies that $U \cap F \supseteq f[V] \cap F \neq \emptyset$. This shows that $f(p) \in a(\mathcal{F})$ and so $p \in f^{\leftarrow}[a(\mathcal{F})]$.

■

Now, we restrict our attention to the case when $\mathcal{L} = \mathbb{P}(X)$.

(j) **Proposition**. Let $\mathcal{F}$ be a filter on X. Then $\mathcal{F} = \cap\{\mathcal{U} : \mathcal{U}$ is an ultrafilter on X and $\mathcal{U} \supseteq \mathcal{F}\}$.

Proof. Let $\mathcal{G} = \cap\{\mathcal{U} : \mathcal{U}$ is an ultrafilter on X and $\mathcal{U} \supseteq \mathcal{F}\}$; by 2.3(d), $\mathcal{G}$ is a filter on X. Clearly, $\mathcal{G} \supseteq \mathcal{F}$. Let $A \in \mathcal{G}$. Suppose $X\backslash A$ meets $\mathcal{F}$. Then $\mathcal{H} = \{(X\backslash A) \cap F : F \in \mathcal{F}\}$ is a filter base on X and is contained in some ultrafilter V on X. But, $F \subseteq \hat{H} \subseteq V$. Since $X\backslash A \in H$, then $X\backslash A \in V$. But $A \in G \subseteq V$, a contradiction as $X\backslash A \in V$. This shows that $X\backslash A$ misses F; that is, there is some $F \in F$ such that $(X\backslash A) \cap F = \emptyset$. Hence, $F \subseteq A$ which implies that $A \in F$.

■

In 2.3(j) we saw that if $L = \mathbb{P}(X)$, then an L-filter is the intersection of all the L-ultrafilters containing it. For other lattices L, this may not be true; in fact, it fails even when L is the lattice of open subsets of X. The best that can happen for an arbitrary open filter is recorded in the next result.

(k) **Proposition.** Let F be an open filter on a space X. Then:

(1) if $G = \bigcap\{U : U$ is an open ultrafilter on X and $U \supseteq F\}$, then $G = \{T \subseteq X : T$ is open and $\text{int}_X(c\ell_X T) \in F\}$ and

(2) F is contained in a unique open ultrafilter iff there exists an open ultrafilter U such that $\{\text{int}_X(c\ell_X U) : U \in U\} \subseteq F$.

Proof

(1) By 2.3(d), G is an open filter on X. Let $G' = \{T \subseteq X : T$ is open and $\text{int}_X(c\ell_X T) \in F\}$. Let $T \in G$. If $X\backslash c\ell_X T$ meets F, then $\{F \cap (X\backslash c\ell_X T) : F \in F\}$ is an open filter base and is contained in some open ultrafilter V. So, $F \subseteq V$ and $X\backslash c\ell_X T \in V$. But $T \in G \subseteq V$, which is impossible as $T \cap (X\backslash c\ell_X T) = \emptyset$. Thus, $X\backslash c\ell_X T$ misses F; that is, there is some $F \in F$ such that $F \cap (X\backslash c\ell_X T) = \emptyset$. Hence, $F \subseteq c\ell_X T$ implying $F \subseteq$

$\text{int}_X(c\ell_X T)$. This shows that $\text{int}_X(c\ell_X T) \in \mathcal{F}$ and $T \in \mathcal{G}'$. So, $\mathcal{G} \subseteq \mathcal{G}'$. Conversely, suppose $T \in \mathcal{G}'$. Then $\text{int}_X(c\ell_X T) \in \mathcal{F}$. Let $\mathcal{U}$ be an open ultrafilter containing $\mathcal{F}$. Then $\text{int}_X(c\ell_X T) \in \mathcal{U}$. By 2.3(e)(2), $T \in \mathcal{U}$ or $X\backslash c\ell_X T \in \mathcal{U}$. But $(X\backslash c\ell_X T) \cap (\text{int}_X(c\ell_X T)) = \emptyset$, and so $T \in \mathcal{U}$. This shows that $T \in \mathcal{G}$ and $\mathcal{G}' \subseteq \mathcal{G}$. Thus, $\mathcal{G}' = \mathcal{G}$.

(2) Suppose $\mathcal{F}$ is contained in an unique open ultrafilter $\mathcal{U}$; then by (1), $\mathcal{U} = \{T \subseteq X : T \text{ is open and } \text{int}_X(c\ell_X T) \in \mathcal{F}\}$. Thus, $\{\text{int}_X(c\ell_X U) : U \in \mathcal{U}\} \subseteq \mathcal{F}$. Conversely, suppose $\{\text{int}_X(c\ell_X U) : U \in \mathcal{U}\} \subseteq \mathcal{F}$ for some open ultrafilter $\mathcal{U}$. Let $\mathcal{V}$ be an open ultrafilter such that $\mathcal{F} \subseteq \mathcal{V}$. Let $U \in \mathcal{U}$. Then $\text{int}_X(c\ell_X U) \in \mathcal{F} \subseteq \mathcal{V}$. Since $(X\backslash c\ell_X U) \cap \text{int}_X(c\ell_X U) = \emptyset$, then by 2.3(e)(2), $U \in \mathcal{V}$. So, $\mathcal{U} \subseteq \mathcal{V}$. Since $\mathcal{U}$ is maximal, then $\mathcal{U} = \mathcal{V}$. This shows that $\mathcal{F}$ is contained in an unique open ultrafilter.

■

To show that the situations in 2.3(j) and 2.3(k) differ, we must find a space Y and an open filter $\mathcal{F}$ on Y such that $\mathcal{F} \neq \mathcal{G}$ (where $\mathcal{G} = \bigcap \{\mathcal{U} : \mathcal{U} \text{ is an open ultrafilter and } \mathcal{U} \supseteq \mathcal{F}\}$).

(l) **Example.** Let Y be the space in Example 2.2(h)(2). Now $\{\{p_+\} \cup \bigcup\{[m,m+1] : m \geqslant n, \ m \text{ even}\} : n \in \mathbb{N}\}$ is an open filter base and generates an open filter $\hat{\mathcal{F}}$. But, $\{p_+\} \cup \bigcup\{(m,m+1) : m \geqslant 0\} \in \mathcal{G}\backslash\hat{\mathcal{F}}$. So, $\hat{\mathcal{F}} \neq \mathcal{G}$.

In studying the interplay of the filters on L and the topology on X when $L = Z(X)$ (respectively, $B(X)$), one usually assumes that X is a Tychonoff (respectively, zero-dimensional) space.

(m) **Proposition.** Suppose X is Tychonoff (respectively, zero-dimensional), $L = Z(X)$ (respectively, $B(X)$), F is a L-filter base, and U a L-ultrafilter. Then:

(1) $c(F) \subseteq a(F)$,

(2) $c(U) = a(U)$,

(3) $|c(F)| \leq 1$, and

(4) when $L = B(X)$, then F is a clopen ultrafilter iff for each $A \in L$, $A \in F$ or $X \setminus A \in F$.

Proof. The proof of these results is similar to the proof of the corresponding results in 2.3(e,f) and is left as an exercise (see 2H) for the reader.

■

2.4 More lattice properties

In this section, we study four specific lattice properties - sublattices, homomorphisms, set representations, and distributivity. This provides the groundwork for investigating the very special type of lattice introduced in the next chapter. We begin by considering sublattices of a lattice.

(a) **Definition.** Let $(L,\vee,\wedge)$ and $(L^*,\vee^*,\wedge^*)$ be lattices. Then $(L^*,\vee^*,\wedge^*)$ is a **sublattice** of $(L,\vee,\wedge)$ if $L^* \subseteq L$ and for $a, b \in L^*$, $a \vee^* b = a \vee b$ and $a \wedge^* b = a \wedge b$. If the induced orders on L and L^* are denoted by $\leq$ and $\leq^*$, respectively, then an easy consequence of (LO) in 2.1(c)(7) is that for a, b $\in L^*$, a $\leqslant^*$ b iff a $\leqslant$ b.

For a space X, consider the lattice $(\mathbb{P}(X),\cup,\cap)$. Both $(\tau(X),\vee,\wedge)$ and $(B(X),\vee,\wedge)$ are sublattices of $(\mathbb{P}(X),\cup,\cap)$ as $\vee = \cup$ and $\wedge = \cap$. On the other hand, $(RO(X),\vee,\wedge)$ is not necessarily a sublattice of $(P(X),\cup,\cap)$ for if U, V $\in RO(X)$, it may happen that $U \cup V \neq U \vee V$ $(= \text{int}_X c\ell_X(U \cup V))$ even though $U \cap V = U \wedge V$ (see 2.2(c)).

A lattice $(L,\vee,\wedge)$ is called a **ring of subsets of a set** X if $(L,\vee,\wedge)$ is a sublattice of $(\mathbb{P}(X),\cup,\cap)$. In this case, if A, B $\in L$, then $A \vee B = A \cup B$ and $A \wedge B = A \cap B$; thus, $(L,\vee,\wedge)$ could be written as $(L,\cup,\cap)$ or $(L,\subseteq,\cup,\cap)$. In particular, $(\tau(X),\cup,\cap)$ is a ring of subsets of X, but, in general, as noted above, $(RO(X),\vee,\wedge)$ is not necessarily a ring of subsets of X.

Of course, we are interested in knowing when an abstract lattice $(L,\vee,\wedge)$ is the same as a ring of subsets of some set X. The first step is to define the word "same". We begin by defining various types of homomorphisms.

(b) <u>Definition</u>. Let $(L,\leqslant,\vee,\wedge)$ and $(L^*,\leqslant^*,\vee^*,\wedge^*)$ be lattices and $f \in F(L,L^*)$. The function f is:

(1) an **order homomorphism** if $a \leqslant b$ implies $f(a) \leqslant^* f(b)$ for each a, b $\in L$, and **order isomorphism** if f is a bijection and both f and $f^{\leftarrow}$ are order homomorphisms (only posets are needed in these definitions),

(2) a **join** (respectively, **meet**) **homomorphism** if $f(a \vee b) = f(a) \vee^* f(b)$ (respectively, $f(a \wedge b) = f(a) \wedge^* f(b)$) for each a, b $\in L$,

(3) a **lattice homomorphism** if f is both a join and a meet

homomorphism, and

(4) a **lattice isomorphism** if f is a bijection and both f and $f^{\leftarrow}$ are lattice homomorphisms.

Here are some useful facts about homomorphisms.

(c) **Proposition**. Let $(L,\leqslant,\vee,\wedge)$ and $(L',\leqslant',\vee',\wedge')$ be lattices and $f \in F(L,L')$. Then

(1) if f is a join or meet homomorphism, then f is an order homomorphism,

(2) if f is a bijection and both f and $f^{\leftarrow}$ are order homomorphisms, then f is a lattice isomorphism, and

(3) if f is a bijection and a join (respectively, meet) homomorphism, then $f^{\leftarrow}$ is also a join (respectively, meet) homomorphism. Hence, a bijective join or meet homomorphism is a lattice isomorphism.

Proof

(1) Suppose f is a join homomorphism and a, b $\in L$ such that $a \leqslant b$. Then $b = a \vee b$ which implies $f(b) = f(a) \vee' f(b)$. Thus, $f(a) \leqslant' f(b)$. A similar proof works when f is a meet homomorphism.

(2) By symmetry, it suffices to show that f is a lattice homomorphism. Let a, b $\in L$. Since $a \vee b \geqslant a$ and $a \vee b \geqslant b$, then $f(a \vee b) \ '\!\geqslant f(a)$ and $f(a \vee b) \ '\!\geqslant f(b)$, implying $f(a \vee b) \ '\!\geqslant f(a) \vee' f(b)$. Let $c' = f(a) \vee' f(b)$. Since f is onto, there is some c $\in L$ such that $f(c) = c'$. Since $f(c) \ '\!\geqslant f(a)$ and $f(c) \ '\!\geqslant f(b)$, then $c = f^{\leftarrow}(c') \geqslant a$ and $c \geqslant b$. So, $c \geqslant a \vee b$, which implies that

$c' = f(c) \;'\geqslant f(a \vee b)$. Thus, $f(a \vee b) = c' = f(a) \vee' f(b)$. Similarly, $f(a \wedge b) = f(a) \wedge' f(b)$.

(3) Suppose f is a bijection and a join homomorphism. Let a, b $\in L$. Since f is onto, it suffices to show $f^{\leftarrow}(f(a) \vee' f(b)) = f^{\leftarrow}(f(a)) \vee f^{\leftarrow}(f(b))$. But $f^{\leftarrow}(f(a) \vee' f(b)) = f^{\leftarrow}(f(a \vee b)) = a \vee b$ (since f is one-to-one) and $a \vee b = f^{\leftarrow}(f(a)) \vee f^{\leftarrow}(f(b))$. A similar proof shows that if f is a bijective meet homomorphism, then $f^{\leftarrow}$ is a meet homomorphism. Thus, if f is a bijective meet or join homomorphism then $f^{\leftarrow}$ is a meet of join homomorphism, respectively. By (1), f and $f^{\leftarrow}$ are order homomorphisms, and by (2) f is a lattice isomorphism.

■

(d) **Definition**. A lattice $(L,\leqslant,\vee,\wedge)$ is "the same as" a ring $(L',\subseteq,\cup,\cap)$ of subsets of a set X if there is a lattice isomorphism $f \in F(L,L')$; we say $(L',\subseteq,\cup,\cap)$ is a **representation** of $(L,\leqslant,\vee,\wedge)$ and $(L,\leqslant,\vee,\wedge)$ has a **set representation**.

We are interested in identifying those lattices which have set representations because in such cases we often are able to topologize the set in question in a natural way by using the given lattice.

(e) **Definition**. A lattice L satisfying

(L5) for a, b, c $\in L$, $a \wedge (b \vee c) = (a \wedge b) \vee (a \wedge c)$
and $a \vee (b \wedge c) = (a \vee b) \wedge (a \vee c)$

is called **distributive**.

(f) **Proposition**. Let $(L,\vee,\wedge)$ be a distributive lattice and a, $a_1, \ldots, a_n \in L$ for some $n \in \mathbb{N}$. Then

(1) $a \wedge (\vee\{a_k : 1 \leqslant k \leqslant n\}) = \vee\{a \wedge a_k : 1 \leqslant k \leqslant n\}$,

(2) $a \vee (\wedge\{a_k : 1 \leqslant k \leqslant n\}) = \wedge\{a \vee a_k : 1 \leqslant k \leqslant n\}$, and

(3) if $\emptyset \neq S \subseteq L$, then the smallest sublattice of L containing S — i.e., the sublattice of L generated by S — is the subset $\{\vee\{\wedge\{s_{ij} : 1 \leqslant i \leqslant m_j\} : 1 \leqslant j \leqslant n\} : s_{ij} \in S,\ n, m_1, \ldots, m_n \in \mathbb{N}\}$.

Proof. Use L5 and induction.

■

Since "$\cup$" and "$\cap$" satisfy (L5), then a lattice of the form $(L,\subseteq,\cup,\cap)$ is distributive; hence, a lattice with a set representation is also distributive. Surprisingly, the converse is true. This result is known as the Birkhoff-Stone Theorem and the proof is divided into a number of intermediate steps in the exercises. (See 2I).

(g) **Theorem (Birkhoff-Stone)**. A lattice is distributive iff it has a set representation.

2.5 Completions of lattices and ordered topological spaces.

In this section we construct two "completions" of a lattice. Then we show how to convert a linearly ordered set into a topological space, and discuss the properties of such a space. We combine these ideas and prove that every linearly ordered topological space is hereditarily normal.

(a) <u>**Definition**</u>

(1) A **linearly ordered set** is a poset $(L,\leqslant)$ satisfying:

(L6) (Trichotomy) if x, y $\in L$ then $x \leqslant y$ or $y \leqslant x$. (The term "linearly ordered set" will sometimes be abbreviated to "loset".)

(2) A poset $(P,\leqslant)$ is **lattice-complete** if it is a lattice and is complete (as a lattice) (see 2.1(d).)

(3) A poset $(P,\leqslant)$ is **conditionally complete** if for each nonempty subset S of P that is bounded above (respectively, below), $\vee S$ (respectively, $\wedge S$) exists in $(P,\leqslant)$.

Evidently any linearly ordered set is a lattice. We will employ the notation of 2.1(c); thus if $(L,\leqslant)$ is linearly ordered and a, b $\in L$ then (a,b) denotes $\{x \in L : a < x \text{ and } x < b\}$, etc. Also we denote $\{x \in L : x > a\}$ and $\{x \in L : x < a\}$ by $(a,+\infty)$ and $(-\infty,a)$ respectively. If there is any ambiguity concerning the identity of the set L or the ordering $\leqslant$, we write "$(a,b)_{(L,\leqslant)}$" rather than "(a,b)", etc. We also denote $(-\infty,a) \cup \{a\}$ by $(-\infty,a]$ and so on.

We wish to show that every linearly ordered set can be embedded "densely" in a lattice-complete linearly ordered set. We begin by proving a more general result, namely that every poset $(P,\leqslant)$ can be "densely embedded" in a complete lattice $(D(P),\subseteq)$ and that this embedding (as defined below) is essentially unique. If $(P,\leqslant)$ is linearly ordered, we will see that $(D(P),\subseteq)$ is also linearly ordered.

(b) **Theorem**. Let $(P,\leqslant)$ be a poset. Then:

(1) There exists a complete lattice $(D(P),\subseteq)$ and a one-to-one function $j : P \to D(P)$ such that:

(a) $x < y$ iff $j(x) \subset j(y)$

(b) if $z \in D(P)$ then $z = \wedge\{j(x) : x \in P \text{ and } j(x) \supseteq z\}$
$= \vee\{j(x) : x \in P \text{ and } j(x) \subseteq z\}$.

(Here we adopt the convention that $\wedge\varnothing = \vee D(P)$ and $\vee\varnothing = \wedge D(P)$; recall $D(P)$ has a largest and a smallest element.)

(2) If $(L,\leqslant)$ is another complete lattice and if $k : P \to L$ is a one-to-one function satisfying (a) and (b) above, then there is a lattice isomorphism $i : D(P) \to L$ such that $i \circ j = k$.

(3) If $(P,\leqslant)$ is linearly ordered then so is $(D(P),\subseteq)$.

(4) If $(P,\leqslant)$ is linearly ordered, and if $a, b \in D(P)$ and $(a,b)_{(D(P),\subseteq)} \neq \varnothing$, then there exists $x \in P$ such that $j(x) \in (a,b)_{(D(P),\subseteq)}$.

(5) If $x \in P$ and $j(x) \in (a,b)_{(D(P),\subseteq)}$, then there exist s_0, $y_0 \in P$ such that $j(x) \in (j(s_0),j(y_0))_{(D(P),\subseteq)} \subseteq (a,b)_{(D(P),\subseteq)}$.

Proof

(1) If $S \subseteq P$, let $S^\wedge$ denote the set of all lower bounds of S; thus $S^\wedge = \{x \in P : \text{for each } s \in S,\ x \leqslant s\}$. Let $D(P) = \{S^\wedge : S \subseteq P\}$, and partially order D(P) by inclusion. Then $(D(P),\subseteq)$ is a poset; we will show it is a complete lattice. Let $\{S_i : i \in I\} \subseteq \mathbb{P}(P)$. It is straightforward to show that $(\cup\{S_i : i \in I\})^\wedge = \cap\{S_i^\wedge : i \in I\} = \wedge\{S_i^\wedge : i \in I\}$. (Here infima and suprema refer to $(D(P),\subseteq)$). Let $R = \{y \in P : y \text{ is an upper bound of } \cup\{S_i^\wedge : i \in I\}\}$. We claim that $\vee\{S_i^\wedge : i \in I\} = R^\wedge$. Obviously $\cup\{S_i^\wedge : i \in I\} \subseteq R^\wedge$. Suppose that $\cup\{S_i^\wedge : i \in I\} \subseteq T^\wedge$. If $t \in T$, it follows that $s \leqslant t$ for each $s \in \cup\{S_i^\wedge : i \in I\}$. Thus $T \subseteq R$, and it follows that $R^\wedge \subseteq T^\wedge$. Hence $(D(P),\subseteq)$ is a complete upper semilattice. Note that $\emptyset^\wedge = \{x \in P : \text{if } y \in \emptyset \text{ then } x \leqslant y\} = P$, which is the largest member of D(P). If P has a largest member 1 note also that $\{1\}^\wedge = P$. If P has a smallest member 0, then $P^\wedge = \{0\}$; if P has no smallest member, then $P^\wedge = \emptyset$. Either way $P^\wedge$ is the smallest member of D(P). It now follows from 2.1(e) that $(D(P),\subseteq)$ is a complete lattice.

Define $j: P \to D(P)$ by $j(x) = \{x\}^\wedge$ for each $x \in P$. If $x < y$ then obviously $\{x\}^\wedge \subseteq \{y\}^\wedge$ (note $y \in \{y\}^\wedge \setminus \{x\}^\wedge$) so $j(x) \subset j(y)$; conversely if $j(x) \subset j(y)$ then $x \in j(y)$ so $x < y$. Thus j is one-to-one and (a) holds.

To verify (b), let $S^\wedge \in D(P)$. We want to show that $S^\wedge = \wedge\{j(x) : x \in P \text{ and } j(x) \supseteq S^\wedge\}$. To do that, we must show that if $T \subseteq P$ and $T^\wedge \subseteq \{x\}^\wedge$ whenever $S^\wedge \subseteq \{x\}^\wedge$, then $T^\wedge \subseteq S^\wedge$. So, let us suppose that $S^\wedge \subseteq \{x\}^\wedge$ implies $T^\wedge \subseteq \{x\}^\wedge$. This means that if x is an upper bound for $S^{\wedge}$, then it is an upper bound for $T^{\wedge}$. But every member of S is an upper bound for $S^{\wedge}$, so every member of S is an upper bound of $T^{\wedge}$. This means that each member of $T^{\wedge}$ is a lower bound for S, i.e., $T^{\wedge} \subseteq S^{\wedge}$.

Now we show that $S^{\wedge} = \bigvee\{j(x) : x \in P \text{ and } j(x) \subseteq S^{\wedge}\}$. To do this we must show that if $T \subseteq P$ and $\{x\}^{\wedge} \subseteq T^{\wedge}$ whenever $\{x\}^{\wedge} \subseteq S^{\wedge}$, then $S^{\wedge} \subseteq T^{\wedge}$. So, let us suppose that $\{x\}^{\wedge} \subseteq S^{\wedge}$ implies $\{x\}^{\wedge} \subseteq T^{\wedge}$. It is easily checked that $\{x\}^{\wedge} \subseteq S^{\wedge}$ iff $x \in S^{\wedge}$. Thus we are supposing that $x \in S^{\wedge}$ implies $x \in T^{\wedge}$, which means $S^{\wedge} \subseteq T^{\wedge}$. Thus we have verified (b), and so (1) holds.

To prove (2), let $(L,\leqslant)$ and k be as hypothesized. Define $i : D(P) \to L$ as follows:

$$i(S^{\wedge}) = \bigwedge\{k(x) : x \in P \text{ and } j(x) \supseteq S^{\wedge}\}$$

As L is complete, $i(S^{\wedge})$ is well-defined. It is relatively straightforward to verify that i is a lattice isomorphism for which $i \circ j = k$; we leave the details to the reader.

(3) Suppose $(P,\leqslant)$ is linearly ordered, and let $S, T \in \mathbb{P}(P)$. Suppose that $y \in T^{\wedge} \setminus S^{\wedge}$. Then $y \leqslant t$ for each $t \in T$, but there exists $s \in S$ such that $y \nleqslant s$. As $(P,\leqslant)$ is linearly ordered, it follows that $s < y$. If $z \in S^{\wedge}$, then $z \leqslant s < y \leqslant t$ for each $t \in T$. Hence $z \in T^{\wedge}$, and it follows that $S^{\wedge} \subseteq T^{\wedge}$. Thus $(D(P),\subseteq)$ is linearly ordered.

To prove (4) and (5), assume $(S^\wedge,T^\wedge)_{(D(P),\subseteq)} \neq \emptyset$. There exists $U \subseteq P$ such that $S^\wedge \subset U^\wedge \subset T^\wedge$. Let $z_0 \in U^\wedge \setminus S^\wedge$ and $y_0 \in T^\wedge \setminus U^\wedge$. Then there exists $s_0 \in S$ such that $s_0 < z_0$, and of course $z_0 \leqslant u$ for each $u \in U$. Similarly there exists $u_0 \in U$ such that $u_0 < y_0$. Then

$$S^\wedge \subseteq (-\infty,s_0] \subset (-\infty,z_0] \subseteq U^\wedge \subset (-\infty,y_0] \subseteq T^\wedge$$

Thus $j(z_0) \in (S^\wedge,T^\wedge)$ and (4) holds. If $U^\wedge = j(x)$ for some $x \in P$, obviously $j(x) \in (j(s_0),j(y_0))_{(D(P),\subseteq)} \subseteq (S^\wedge,T^\wedge)_{(D(P),\subseteq)}$. This shows that (5) holds.

■

The complete lattice $(D(P),\subseteq)$ (or one possessing the characterizing properties described in (1) above) is called the **Dedekind-MacNeille completion** of $(P,\leqslant)$. We now describe the conditional completion $(K(P),\subseteq)$ of $(P,\leqslant)$ as follows. If P has no largest element (respectively, smallest element), delete the largest element (respectively, the smallest element) from D(P). The resulting set is K(P), and its order is the inclusion order inherited from D(P). The conditional completion $(K(P),\subseteq)$ can be characterized as follows:

(c) **Theorem**. Let $(P,\leqslant)$ be a poset. Then

(1) The conditional completion $(K(P),\subseteq)$ is a conditionally complete poset, i.e., if S is a nonempty subset of K(P) that is bounded above (respectively, below) in K(P), then $\bigvee S$ (respectively, $\bigwedge S$) exists in K(P).

(2) If $(P,\leqslant)$ is a lattice then so is $(K(P),\subseteq)$.

(3) There exists a one-to-one function $j : P \to K(P)$ such that $x < y$ iff $j(x) \subset j(y)$.

(4) If $z \in K(P)$ then $\{j(x) : x \in P \text{ and } j(x) \supseteq z\} \neq \emptyset \neq \{j(x) : x \in P \text{ and } j(x) \subseteq z\}$, and $z = \bigwedge\{j(x) : x \in P \text{ and } j(x) \supseteq z\} = \bigvee\{j(x) : x \in P \text{ and } j(x) \subseteq z\}$.

(5) The pair $(K(P),j)$ is unique in the sense that if M is another conditionally complete poset and $k : P \to M$ is an order isomorphism onto $k[P]$ satisfying (4), then there is an order isomorphism $m : K(P) \to M$ such that $k = m \circ j$.

Proof

(1) If P has a largest (respectively, smallest) member than so does $K(P)$, and the completeness of $D(P)$ ensures that $\bigvee S$ (respectively, $\bigwedge S$) exists in $K(P)$. If P has no largest member, then the largest member of $D(P)$, namely $P = \emptyset^{\wedge}$, does not belong to $K(P)$. Thus if $S \subseteq K(P)$ and $T^{\wedge}$ is an upper bound of S in $K(P)$, then $T^{\wedge} \neq \emptyset^{\wedge}$. Thus the supremum of S in $D(P)$ belongs to $K(P)$, and hence is the supremum of S in $K(P)$. A similar argument shows that if P has no smallest member and S is bounded below in $K(P)$, it has an infimum in $K(P)$.

(2) If P has a largest (respectively, smallest) member and $S^{\wedge}$, $T^{\wedge} \in K(P)$, then $S^{\wedge} \vee T^{\wedge}$ (respectively, $S^{\wedge} \wedge T^{\wedge}$) exists in $D(P)$ and hence in $K(P)$. If P has no largest member, then $S^{\wedge} \neq P \neq T^{\wedge}$ so $S \neq \emptyset \neq T$. Choose $s \in S$ and $t \in T$; then $S^{\wedge} \cup T^{\wedge} \subseteq \{s \vee t\}^{\wedge} \neq P$ (as $s \vee t$ is not the largest member of P). Thus $\{S^{\wedge},T^{\wedge}\}$ is bounded above in $K(P)$ and hence has a supremum in $K(P)$ by (1). If P has no smallest member then $S^{\wedge} \neq \emptyset \neq T^{\wedge}$ so there exists $y \in S^{\wedge}$ and $z \in T^{\wedge}$. Then $\{y \wedge z\}^{\wedge} \subseteq S^{\wedge} \cap T^{\wedge}$

and $\{y \wedge z\}^{\wedge}$ is **not** the smallest member of D(P). Thus $\{S^{\wedge},T^{\wedge}\}$ is bounded below in K(P) and hence has an infimum in K(P) by (1).

(3) This follows readily from (1)(i) of 2.5(b), once one notes that if $x \in P$ then $\{x\}^{\wedge} \in K(P)$ whether or not P has a largest or smallest member.

(4) Let $z \in K(P)$; then $z = S^{\wedge}$ for some $S \subset P$. If P has a largest member 1 (respectively, smallest member 0) then $\{1\}^{\wedge} \in \{j(x) : x \in P \text{ and } j(x) \supseteq z\}$ (respectively, $\{0\}^{\wedge} \in \{j(x) : x \in P \text{ and } j(x) \subseteq z\}$). If P has no largest member then $P \notin K(P)$ so $S^{\wedge} \neq \emptyset^{\wedge} = P$, so $S \neq \emptyset$. If $s \in S$, then $j(s) = \{s\}^{\wedge} \supseteq S^{\wedge} = z$. Hence $\{j(x) : x \in P \text{ and } j(x) \supseteq z\}$ (which, as noted in (3) above, is a subset of K(P)), is nonempty. If P has no smallest member then $S^{\wedge} \neq \emptyset$ so there exists $y \in S^{\wedge}$. Thus $j(y) = \{y\}^{\wedge} \subseteq S^{\wedge}$, hence $\{j(x) : x \in P \text{ and } j(x) \subseteq z\}$ is a nonempty subset of K(P). The remainder of (4) follows from 2.5(b)(1)(ii).

(5) By adjoining a largest and smallest element to M (if M does not already possess them), we produce a complete lattice L satisfying the description in 2.5(b)(2). We can apply 2.5(b)(2) and readily check that if we let $m = i \mid K(P)$, then m has the required properties.

■

We now convert a linearly ordered set into a topological space.

(d) **Lemma**. Let $(L,\leqslant)$ be a linearly ordered set (containing more than one element). Let $\mathcal{B} = \{(a,b) : a, b \in L \text{ and } a < b\} \cup \{(-\infty,a) : a \in L\} \cup \{(b,+\infty) : b \in L\}$. Then $\mathcal{B}$ is a basis for a Hausdorff topology on L.

<u>Proof</u>. It is routine to show that B is closed under finite intersections, and that the union of all members of B is L. Let a, b $\in$ L such that a < b. If there is c $\in$ L such that a < c < b, then $(-\infty,c)$ and (c,∞) are disjoint open neighborhoods of a and b respectively. If (a,b) = $\emptyset$, then $(-\infty,b)$ and (a,∞) are disjoint open neighborhoods of a and b, respectively. Thus, L is Hausdorff.

■

(e) <u>Definition</u>

(1) Let $(L,\leqslant)$ be a linearly ordered set. The topology induced on the set L by the collection B defined above is called the **order topology** on L (induced by $\leqslant$); we will denote it by $\tau(\leqslant)$.

(2) A topological space (X,τ) is called an **ordered topological space** if there is a linear order $\leqslant$ on X such that $\tau = \tau(\leqslant)$.

(3) An **interval** in a linearly ordered set $(L,\leqslant)$ (or in the ordered topological space $(L,\tau(\leqslant))$) is any subset S of L with the following property: if x, y $\in$ S then $(x,y) \subseteq S$. An **open interval** is an interval belonging to $\tau(\leqslant)$.

Note that a bounded open interval in a linearly ordered space $(L,\tau(\leqslant))$ need not be of the form (a,b), where a, b $\in$ L. For example, consider the linearly ordered space of rational numbers, and the open interval $\{x \in \mathbb{Q} : \sqrt{2} < x < \sqrt{3}\}$.

(f) <u>Proposition</u>. Let $(L,\leqslant)$ and $(M,\leqslant)$ be two linearly ordered sets. Then an order isomorphism from $(L,\leqslant)$ onto $(M,\leqslant)$ is a homeomorphism from $(L,\tau(\leqslant))$ onto $(M,\tau(\leqslant))$.

<u>Proof</u>. The proof of the proposition is straightforward and left to the reader as an exercise (see 2J).

■

(g) **<u>Lemma</u>**. Let $(L,\leqslant)$ be a linearly ordered set. Then

(1) if $a, b \in L$ and $a \leqslant b$, then (a,b) is an open interval,

(2) if $\{I(\alpha) : \alpha \in A\}$ is a set of intervals of L and $\cap\{I(\alpha) : \alpha \in A\} \neq \emptyset$, then $\cup\{I(\alpha) : \alpha \in A\}$ is an interval and

(3) if $V \in \tau(\leqslant)$, then V can be written as a union of pairwise disjoint open intervals of L.

(4) The linearly ordered set $(L,\leqslant)$ is lattice-complete iff the linearly ordered space $(L,\tau(\leqslant))$ is compact.

<u>Proof</u>

(1) The proof is easy and left to the reader (see 2J).

(2) Let $x \in \cap\{I(\alpha) : \alpha \in A\}$ and let $\{y,z\} \subseteq \cup\{I(\alpha) : \alpha \in A\}$ where $y < z$. Find $\alpha_1, \alpha_2 \in A$ such that $y \in I(\alpha_1)$ and $z \in I(\alpha_2)$. If $y \leqslant x \leqslant z$ then $(y,x) \subset I(\alpha_1)$ and $(x,z) \subset I(\alpha_2)$ so $(y,z) \subseteq (y,x) \cup \{x\} \cup (x,z) \subseteq I(\alpha_1) \cup I(\alpha_2) \subseteq \cup\{I(\alpha) : \alpha \in A\}$. If $x < y$ then $(y,z) \subseteq (x,z) \subseteq I(\alpha_2)$. A similar argument works when $z < x$. It follows that $\cup\{I(\alpha) : \alpha \in A\}$ is an interval.

(3) Let $x, y \in V$, and let $I(x) = \cup\{I : I$ is an open interval and $x \in I \subseteq V\}$. By (1), $x \in I(x)$ and by (2), $I(x)$ is an open interval and $I(x) \subseteq V$. If $I(x) \cap I(y) \neq \emptyset$, then by (2), $I(x) \cup I(y)$ is an open interval. Since $x \in I(x) \cup I(y) \subseteq V$, then by definition of $I(x)$, $I(x) \cup I(y) \subseteq I(x)$. Thus, $I(y) \subseteq I(x)$. By symmetry, $I(x) \subseteq I(y)$. So, if $I(x) \cap I(y) \neq \emptyset$, then $I(x) = I(y)$. This shows that $\{I(x) : x \in V\}$ is a

pairwise disjoint collection of open intervals whose union is V.

(4) Suppose $(L,\leq)$ is lattice-complete. Let S be a collection of closed subsets of L with the finite intersection property. We must show that $\cap S \neq \emptyset$.

Let $A = \{a \in L : [a,+\infty) \cap [\cap F] \neq \emptyset$ for each finite subcollection F of $S\}$. Note that $\wedge L$ (which exists as L is lattice-complete) belongs to A, so $A \neq \emptyset$. Let $b = \vee A$; we will show that $b \in \cap S$. Suppose not; then there exists $S_0 \in S$ and x_0, $y_0 \in L$ such that $b \in (x_0,y_0) \subseteq L \setminus S_0$. (If $b = \vee L$ (respectively, $\wedge L$), then $y_0 = +\infty$ (respectively, $x_0 = -\infty$)). By definition of b there exists $a_0 \in A$ such that $a_0 > x_0$. Thus $[a_0,+\infty)$ meets the intersection of each finite subfamily of S. As $y_0 \notin A$, there is a finite subfamily F of S such that $[y_0,+\infty) \cap [\cap F] = \emptyset$. But $[a_0,+\infty) \cap S_0 \cap [\cap F] \subseteq [y_0,+\infty) \cap [\cap F] = \emptyset$, which is a contradiction. It follows that $\cap S \neq \emptyset$ and so $(L,\tau(\leq))$ is compact.

Conversely, if $(L,\tau(\leq))$ is not lattice-complete, let A be a nonempty subset with no supremum (say). Then $\{(-\infty,a) : a \in A\} \cup \{(b,+\infty) : b > a$ for each $a \in A\}$ is an open cover of $(L,\tau(\leq))$ with no finite subcover, so $(L,\tau(\leq))$ is not compact.

■

(h) **Theorem**. Every linearly ordered space is (homeomorphic to) a dense subspace of a compact linearly ordered space.

Proof. Let $(L,\leq)$ be a linearly ordered set and let $(D(L),\subseteq)$ be the lattice-complete linearly ordered set constructed in 2.5(b). Denote $\{j[V] : V \in \tau(\leq)\}$ by $j[\tau(\leq)]$, where $j : L \to D(L)$ is as defined in

2.5(b). By 2.5(b)(1) and 2.5(f), $(L,\tau(\leqslant))$ and $(j[L],j[\tau(\leqslant)]$ are homeomorphic linearly ordered spaces. Let σ denote the subspace topology induced on $j[L]$ by the topology $\tau(\subseteq)$ on the linearly ordered space $(D(L),\tau(\subseteq))$. We will show that $j[\tau(<)] = \sigma$ and that $(j[L],\sigma)$ is dense in $(D(L),\tau(\subseteq))$. Since by 2.5(g) $(D(L),\tau(\subseteq))$ is compact, this will prove that $(L,\tau(\leqslant))$ is homeomorphic to a dense subspace of a compact linearly ordered space.

Basic open sets of $j[\tau(\leqslant)]$ are of the form $\{j(z) : z \in L$ and $j(x) \subset j(z) \subset j(y)\}$, where $x, y \in L$. Such a set is of the form $(j(x),j(y))_{(D(L),\subseteq)} \cap j[L]$, which obviously belongs to σ. Thus $j[\tau(\leqslant)] \subseteq \sigma$. Conversely, let $(a,b)_{(D(L),\subseteq)}$ be a basic open set of $\tau(\subseteq)$. Then $(a,b)_{(D(L),\subseteq)} \cap j[L]$ would be a basic open set of σ. If $x \in L$ and $j(x) \in (a,b)_{(D(L),\subseteq)} \cap j[L]$, by 2.5(b)(5) there exist $s_0, y_0 \in L$ such that $j(x) \in (j(s_0),j(y_0))_{(D(L),\subseteq)} \subseteq (a,b)_{(D(L),\subseteq)}$. Now $j[(s_0,y_0)_{(L,\leqslant)}] = (j(s_0),j(y_0))_{(D(L),\subseteq)} \cap j[L] \in j[\tau(\leqslant)]$. Thus

$$j(x) \in j[(s_0,y_0)_{(L,\leqslant)}] \subseteq (a,b)_{(D(L),\subseteq)} \cap j[L].$$

It follows that each basic open set of σ of the above type can be written as a union of members of $j[\tau(\leqslant)]$. Thus $\sigma \subseteq j[\tau(\leqslant)]$, and so $\sigma = j[\tau(\leqslant)]$. By 2.5(b)(4), $(j[L],\sigma)$ is dense in $(D(L),\tau(\subseteq))$ and so the proof is complete.

■

(i) <u>Corollary</u>. Every linearly ordered space is Tychonoff.

<u>Proof</u>. This follows from 2.5(h) and 1.10(i).

■

Our next goal is to show that every linearly ordered space is hereditarily normal. It follows from 2.5(h) that it will suffice to show that every compact linearly ordered space is hereditarily normal. As we can infer from the general definition of hereditary topological properties given in 1.2(f), a space is said to be hereditarily normal if each of its subspaces is normal. Obviously metric spaces and countable Tychonoff spaces are hereditarily normal (countable spaces are Lindelöf, and Lindelöf Tychonoff spaces are normal).

(j) **Lemma**. The following are equivalent for a space X:

(1) X is hereditarily normal,

(2) each open subspace of X is normal, and

(3) if A and B are subsets of X such that $A \cap c\ell_X B = B \cap c\ell_X A = \emptyset$, then there are disjoint open subsets U and V of X such that $A \subseteq U$ and $B \subseteq V$.

Proof

(1) $\Rightarrow$ (2) Obvious.

(2) $\Rightarrow$ (3) Let $W = X \backslash (c\ell_X A \cap c\ell_X B)$. Then W is an open subset of X, and by hypothesis $A \cup B \subseteq W$. Obviously $c\ell_W A \cap c\ell_W B = W \cap c\ell_X A \cap c\ell_X B = \emptyset$. As W is normal by hypothesis, there are disjoint open sets U and V of W such that $c\ell_W A \subseteq U$ and $c\ell_W B \subseteq V$. Then U and V are disjoint open sets of X and $A \subseteq U$, $B \subseteq V$.

(3) $\Rightarrow$ (1) Let $S \subseteq X$ and let A and B be disjoint closed subsets of S. Thus $A = S \cap c\ell_X A$ and $B = S \cap c\ell_X B$. Thus $A \cap$

$c\ell_X B = (S \cap c\ell_X A) \cap c\ell_X B = A \cap B = \emptyset$, and similarly $B \cap c\ell_X A = \emptyset$. By hypothesis there exist disjoint open subsets U and V of X with $A \subseteq U$, $B \subseteq V$. Thus A and B are contained respectively in the disjoint open subsets $U \cap S$ and $V \cap S$ of S. It follows that S is normal.

∎

(k) **Lemma**. In a compact linearly ordered space L each open interval can be written in one of the forms $\emptyset$, L, (p,q), $(-\infty,q)$, or $(p,+\infty)$, where p, $q \in L$.

Proof. Let $\emptyset \neq I$ be an open interval in the compact linearly ordered space L. Let $H = \{x \in L : \text{for each } y \in I,\ x > y\}$. Let $K = \{x \in L : \text{for each } y \in I,\ x < y\}$. Suppose $H \neq \emptyset \neq K$. As L is compact, $\vee K$ and $\wedge H$ exist (i.e., are elements of L) by 2.5(g)(4). We claim that $I = (\vee K, \wedge H)$.

Suppose $x \in I$. Find a, $b \in L$ such that $x \in (a,b) \subseteq I$ (as I is open). If $r > a$ and $r \in K$, then $r < x$ (as $x \in I$), so $r \in (a,x) \subset (a,b) \subseteq I$. Thus $r < r$ (as $r \in K$), which is a contradiction. Thus a is an upper bound for K, and so $\vee K \leqslant a < x$. A similar argument shows that $x < b \leqslant \wedge H$, and so $I \subseteq (\vee K, \wedge H)$. Conversely, suppose $z \in (\wedge K, \vee H)$. We will show that $z \in I$. Since $z < \wedge H$, there exists some $y_0 \in I$, such that $z < y_0$. Likewise, there is some $x_0 \in I$ such that $x_0 < z$. So, $z \in (x_0,y_0)$. But $(x_0,y_0) \subseteq I$. So, $z \in I$. Hence $(\vee K, \wedge H) \subseteq I$ and so $(\vee K, \wedge H) = I$.

If $K = \emptyset \neq H$ then $I = (-\infty, \wedge H)$; if $K \neq \emptyset = H$ then $I = (\vee K, +\infty)$. If $K = H = \emptyset$, then $I = L$. The proofs of these assertions are similar to the above.

■

For the remainder of this section, when we talk of an open interval (a,b), we will allow the possibility that $a = -\infty$ or $b = +\infty$.

(l) **Lemma**. Let $\{(r_j,t_j) : j \in J\}$ be a pairwise disjoint collection of nonempty open intervals of the ordered space L. Then:

(1) Let $b \in c\ell_L[\cup\{(r_j,t_j) : j \in J\}] \setminus \cup\{c\ell_L(r_j,t_j) : j \in J\}$. If $b \in (c,d)$ then there exists $j_0 \in J$ such that $(r_{j_0},t_{j_0}) \subseteq (c,d)$.

(2) Let $x \in L$. Then $|\{j \in J : x \in c\ell_L(r_j,t_j)\}| \leqslant 2$.

Proof. The proof is rather lengthy but straightforward. In 2J we invite the reader to construct this proof.

■

(m) **Theorem**. A compact linearly ordered space is hereditarily normal.

Proof. Let A and B be subsets of the compact linearly ordered space L such that $A \cap c\ell_L B = B \cap c\ell_L A = \emptyset$. By 2.5(j) it suffices to show that there are disjoint open sets V and W of L such that $A \subseteq V$, $B \subseteq W$, and $V \cap W = \emptyset$. By 2.5(g)(3) and 2.5(k) we can write

$$L\backslash(c\ell_L A \cap c\ell_L B) = \bigcup\{(p_s,q_s) : s \in J\}$$

where $\{(p_s,q_s) : s \in J\}$ is a pairwise disjoint collection of nonempty open intervals of L. (Note this is where we invoke the compactness of L.) By hypothesis $A \cup B \subseteq \bigcup\{(p_s,q_s) : s \in J\}$. Let $J_1 = \{s \in J : (p_s,q_s) \cap B = \emptyset\}$, $J_2 = \{s \in J : (p_s,q_s) \cap A = \emptyset\}$, and $J_3 = \{s \in J : (p_s,q_s) \cap A \neq \emptyset \neq (p_s,q_s) \cap B\}$.

Let $s \in J_3$. We will construct disjoint open subsets V(s) and W(s) of L such that $(p_s,q_s) \cap A \subseteq V(s) \subseteq (p_s,q_s)$ and $(p_s,q_s) \cap B \subseteq W(s) \subseteq (p_s,q_s)$. Using 2.5(g)(3) and 2.5(k), we write $(p_s,q_s)\backslash cl_L B$ as $\bigcup\{(r_j,t_j) : j \in I\}$ where $\{(r_j,t_j) : j \in I\}$ is a pairwise disjoint collection of nonempty open intervals of L. Let $K = \{j \in I : (r_j,t_j) \cap A \neq \emptyset\}$. Since $A \cap c\ell_X B = \emptyset$ it follows that $A \cap (p_s,q_s) \subseteq \bigcup\{(r_j,t_j) : j \in K\}$. For each $b \in (p_s,q_s) \cap B$ let $K(b) = \{j \in K : b \in c\ell_L(r_j,t_j)\}$. By 2.5(l)(2), $|K(b)| \leqslant 2$, and by 2.5(l)(1), together with the fact that $b \notin c\ell_L A$, it follows that $b \notin c\ell_L[\bigcup\{(r_j,t_j) : j \in K\backslash K(b)\}]$. Let $E = \{r_j : j \in K\} \cup \{t_j : j \in K\}$. If $b \in B \cap E$ choose G(b) to be an open subset of L such that $b \in G(b) \subseteq c\ell_L G(b) \cap X\backslash A$. If $b \in B\backslash E$ let $G(b) = L$. For each $b \in B \cap (p_s,q_s)$, let

$$W(s,b) = [(p_s,q_s)\backslash c\ell_L[\bigcup\{(r_j,t_j) : j \in K\backslash K(b)\}]] \cap G(b)$$

Then $b \in W(s,b) \subseteq (p_s,q_s)$ and W(s,b) is open in X. Put $W(s) = \bigcup\{W(s,b) : b \in B \cap (p_s,q_s)\}$. Then $B \cap (p_s,q_s) \subseteq W(s) \subseteq (p_s,q_s)$.

If $a \in A \cap (p_s,q_s)$, there is a unique $j(a) \in K$ such that $a \in (r_{j(a)},t_{j(a)})$. Define V(s,a) by

$$V(s,a) = (r_{j(a)},t_{j(a)})\backslash \bigcup\{c\ell_L G(b) : b \in [r_{j(a)},t_{j(a)}] \cap B\}$$

Put $V(s) = \bigcup\{V(s,a) : a \in A \cap (p_s,q_s)\}$. Then $V(s)$ is open and $A \cap (p_s,q_s) \subseteq V(s) \subseteq (p_s,q_s)$.

We now claim that $V(s) \cap W(s) = \emptyset$. For if $u \in V(s) \cap W(s)$, there exist $a \in A \cap (p_s,q_s)$ and $b \in B \cap (p_s,q_s)$ such that $u \in V(s,a) \cap W(s,b)$. Thus $u \in (r_{j(a)},t_{j(a)}) \setminus c\ell_L[\bigcup\{(r_j,t_j) : j \in K \setminus K(b)\}]$ which implies $j(a) \in K(b)$ and $b \in \{r_{j(a)},t_{j(a)}\}$. Since $u \in V(s,a)$, then $u \notin \bigcup\{c\ell_X G(b) : b \in \{r_{j(a)},t_{j(a)}\}\}$ yet $u \in W(s,b) \subseteq G(b)$, which is a contradiction. Thus our claim holds.

Now let $V = \bigcup\{(p_s,q_s) : s \in J_1\} \cup [\bigcup\{(V(s) : s \in J_3\}]$ and let $W = \bigcup\{(p_s,q_s) : s \in J_2\} \cup [\bigcup\{W(s) : s \in J_3\}]$. It is straightforward to show that $A \subseteq V$, $B \subseteq W$ and $V \cap W = \emptyset$. Hence by 2.5(j), L is hereditarily normal.

■

(n) **Theorem**. Every linearly ordered space is hereditarily normal.

Proof. By 2.5(h) a linearly ordered space L has a compactification Y that is a linearly ordered space. By the previous theorem Y and all its subspaces are normal, so in particular L and all its subspaces are normal.

■

2.6 Ordinals, Cardinals and Spaces of Ordinals

In this section we give a rapid summary of the theory of ordinal and cardinal numbers. We follow this with a brief study of the topological properties of ordinals as linearly ordered spaces. Proofs of purely order-theoretic results are not included; the reader is referred to Dugundji [Du], Monk [Mo], or Hrbacek and Jech [HrJ] for proofs. Proofs of the topological assertions are included.

(a) **Definition**. A partially ordered set $(A,\leqslant)$ is a **well-ordered set** (abbreviated as "**woset**") if every nonempty subset has a least element.

The set $A = \{\emptyset, \{\emptyset\}, \{\emptyset,\{\emptyset\}\}\}$ which is partially ordered by set inclusion "$\subseteq$" and the subset $\mathbb{N} \cup \{0\}$ of $\mathbb{R}$ which is partially ordered by the usual order on $\mathbb{R}$ are wosets.

Recall from 2.1(a) that if $(A,\leqslant)$ is a poset and $B \subseteq A$, then $\leqslant_B$ is used to denote $(B \times B) \cap \leqslant$. The proof of the following is trivial.

(b) **Proposition**. Let $(A,\leqslant)$ be a woset. Then:

(1) $(A,\leqslant)$ is a linearly ordered set and

(2) if $B \subseteq A$, then $(B,\leqslant_B)$ is a woset.

We now define a class of "canonical" wosets with the property that every woset is order isomorphic to a unique canonical woset. These canonical wosets are called the ordinal numbers, more precisely, the von Neumann ordinals. In what follows, we give the definition and a brief outline of the salient features of ordinal numbers.

(c) <u>**Definition**</u>

(1) An **ordinal number** is a set α satisfying these two properties:

(a) If $x, y \in \alpha$, then either $x \in y$, $y \in x$, or $x = y$, and

(b) if $x \in y$ and $y \in \alpha$, then $x \in \alpha$.

(2) If α is an ordinal number, define $\leqslant$ on α by $x \leqslant y$ iff $x = y$ or $x \in y$.

Evidently, ordinal numbers are sets whose elements are themselves sets. Intuitively, an ordinal number is a chain of sets such that each set and its members belong to the next set up in the chain. As examples, we list five ordinal numbers:

(a) $\emptyset$,

(b) $\{\emptyset\}$,

(c) $\{\emptyset,\{\emptyset\}\}$,

(d) $\{\emptyset,\{\emptyset\},\{\emptyset,\{\emptyset\}\}\}$, and

(e) $\{\emptyset,\{\emptyset\},\{\emptyset,\{\emptyset\}\}, \{\emptyset,\{\emptyset\},\{\emptyset,\{\emptyset\}\}\}\}$

We now list some of the basic properties of ordinal numbers. Henceforth, we will use the term "ordinal" rather than "ordinal number".

(d) <u>**Theorem**</u>. Let α and β be ordinals. Then:

(1) $(\alpha,\leqslant)$ is a woset,

(2) $\alpha \cup \{\alpha\}$ is an ordinal,

(3) if $\alpha \subseteq \beta \subseteq \alpha \cup \{\alpha\}$, then $\beta = \alpha$ or $\beta = \alpha \cup \{\alpha\}$,

(4) $\alpha \subset \beta$ iff $\alpha \in \beta$,

(5) the least element of $(\alpha,\leqslant)$ is $\varnothing$,

(6) $\alpha = \{\gamma : \gamma$ is an ordinal and $\gamma \subset \alpha\}$,

(7) either $\alpha \in \beta$, $\alpha = \beta$, or $\beta \in \alpha$,

(8) if $\{\alpha_i : i \in I\}$ is a nonempty set of ordinals, then $\bigcup\{\alpha_i : i \in I\}$ and $\bigcap\{\alpha_i : i \in I\}$ are ordinals and $\bigcap\{\alpha_i : i \in I\} \in \{\alpha_i : i \in I\}$,

(9) if $\varnothing \neq A \subset \alpha$, then $\bigcup A \in \alpha$ or $\bigcup A = \alpha$,

(10) if $\{\alpha_i : i \in \mathbb{N}\}$ is a set of ordinals such that $\alpha_{i+1} \leqslant \alpha_i$ for $i \in \mathbb{N}$, then there exists some $n \in \mathbb{N}$ such that $\alpha_n = \alpha_{n+i}$ for each $i \in \mathbb{N}$,

(11) if $(W,\leqslant)$ is a woset, there is a unique ordinal α such that W and α are order isomorphic,

(12) if $\leqslant$ is the order on $\mathbb{N} \cup \{0\}$ inherited from the usual order on $\mathbb{R}$, then $(\mathbb{N} \cup \{0\},\leqslant)$ is a woset.

(e) <u>**Definition**</u>

(1) If α is an ordinal, then $\alpha \cup \{\alpha\}$ is called the **successor ordinal** (see 2.6(d)(2)) **of** α and is denoted by $\alpha + 1$.

(2) An ordinal β is called a **successor ordinal** if there is an ordinal α such that $\beta = \alpha + 1$.

(3) By 2.6(d)(11) and (12), $(\mathbb{N} \cup \{0\},\leqslant)$ is order isomorphic to unique ordinal with is denoted by ω or ω_0. Under this isomorphism, 0 corresponds to $\varnothing$, 1 to $\{\varnothing\} = \varnothing \cup \{\varnothing\}$, 2 to $\{\varnothing,\{\varnothing\}\} = \{\varnothing\} \cup \{\{\varnothing\}\}$, etc. So, sometimes, we denote $\varnothing$ by 0, $\{\varnothing\}$ by 1, $1 \cup \{\varnothing\}$ by 2, etc. Thus, $\omega = \{0,1,2,3,\ldots\}$.

(4) A **limit ordinal** is a nonzero ordinal that is not a successor

ordinal.

If α is a limit ordinal and $\gamma \in \alpha$, then $\gamma + 1 \leqslant \alpha$. But since α is not a successor ordinal, then $\alpha \neq \gamma + 1$; hence, $\gamma + 1 \in \alpha$. It follows that $\bigvee\alpha = \bigvee\{\gamma : \gamma \in \alpha\} = \bigvee\{\gamma+1 : \gamma \in \alpha\} = \alpha$. Thus, every nonzero ordinal α is either a successor ordinal (i.e., $\alpha = \beta + 1$ for some ordinal β) or the union of ordinals preceding α (i.e., $\alpha = \bigvee\alpha$). In particular, ω is a nonzero limit ordinal and $\omega = \bigvee\omega$.

So, at this point, we have two methods of obtaining ordinals - the successor operation and the union operation. In fact, all ordinals can be obtained from $\emptyset$ by iteration of these two methods. This fact is formalized in our next result.

(f) **Theorem**. (Transfinite Induction). Let A be a set and for each ordinal α, let $P(\alpha)$ be a set. Then:

(1) if

(a) $P(0) \subseteq A$,

(b) for all ordinals β, $P(\beta) \subseteq A$ implies $P(\beta+1) \subseteq A$,

and

(c) for all limit ordinals β, $P(\gamma) \subseteq A$ for all $\gamma \in \beta$ implies $P(\beta) \subseteq A$;

then $P(\alpha) \subseteq A$ for all ordinals α and

(2) if for each ordinal β, $P(\gamma) \subseteq A$ for all $\gamma \in \beta$ implies $P(\beta) \subseteq A$, then $P(\alpha) \subseteq A$ for all ordinals α.

(g) **Theorem**. (Transfinite Construction). Let α be an ordinal, A a set, $B = \bigcup\{F(\beta,A) : \beta \in \alpha\}$, and $g \in F(B,A)$. Then there is a unique function $f \in F(\alpha,A)$ such that $f(\beta) = g(f \mid \beta)$ for all $\beta \in \alpha$.

This last theorem allows us to construct functions with domain α recursively; that is, for $\beta \in \alpha$, $f(\beta)$ can be determined by $\{f(\gamma) : \gamma \in \beta\}$.

For example, consider the process of defining the sets $\{A_n : n \in \omega+1\}$ as follows: $A_0 = \mathbb{Q}$, $A_{n+1} = \{e^x : x \in A_n\}$ for $0 \leq n < \omega$, and $A_\omega = \bigcup\{A_\alpha : \alpha \in \omega\}$. What we are doing, when we index our family of sets, is to define a function f from the index set $\omega+1$ into $\mathbb{P}(\mathbb{R})$. This is, in fact, a recursive construction that proceeds in the following manner. Let $\alpha = \omega+1$ and $A = \mathbb{P}(\mathbb{R})$. Let $h \in B = \bigcup\{F(\beta,A) : \beta < \omega+1\}$, and suppose the domain of h is β, where $\beta \in \omega+1$. Define $g(h) = \mathbb{Q}$ if $\beta = 0$ (and hence $h = \emptyset$), $g(h) = \{e^x : x \in h(\bigcup\beta)\}$ if $\beta \in \omega$, and $g(h) = \bigcup\{h(\gamma) : \gamma \in \beta\}$ if $\beta = \omega$. It is easy to verify that the function f, whose existence and uniqueness is given by 2.6(g), has the property that $f(\gamma) = A_\gamma$ for all $\gamma \in \omega+1$; thus f is the required "indexing function".

We now turn our attention towards constructing the cardinal numbers. First, a few preliminary results and definitions are needed.

(h) **Proposition**. Let A, B be sets. Then:

(1) A has a well-ordering defined on it (this is equivalent to the axiom of choice),

(2) if $\leq$ is a well-ordering defined on A, then there is a unique ordinal α and a unique function $f \in F(A,\alpha)$ which is an order

isomorphism,

(3) if α is an ordinal, $f \in F(A,\alpha)$ is a bijection, and $\leqslant$ is defined on A by stipulating that $a \leqslant b$ iff $f(a) \leqslant f(b)$, then $(A,\leqslant)$ is a woset and f is an order isomorphism,

(4) if A is a set, then $\{\alpha : \alpha$ is an ordinal and there is a bijection $f \in F(A,\alpha)\}$ (denoted as ord(A)) is a nonempty set,

(5) $\mathrm{ord}(A) = \mathrm{ord}(B)$ iff there is a bijection $f \in F(A,B)$,

(6) if $\alpha \in \mathrm{ord}(A)$, then $\mathrm{ord}(A) = \mathrm{ord}(\alpha)$, and

(7) if $\mathrm{ord}(A) \neq \mathrm{ord}(B)$, then $\mathrm{ord}(A) \cap \mathrm{ord}(B) = \emptyset$.

(i) <u>**Definition**</u>

(1) An ordinal α is called an **initial ordinal** if $\alpha = \bigcap\mathrm{ord}(\alpha)$ (note that by 2.6(d)(8) and 2.6(h)(4), if A is a set, then $\bigcap\mathrm{ord}(A) \in \mathrm{ord}(A)$).

(2) An ordinal α is called **finite** if $\mathrm{ord}(\alpha) = \{\alpha\}$. An ordinal which is not finite is called **infinite**.

(j) <u>**Proposition**</u>. Let α and β be ordinals. Then:

(1) $\omega = \{\gamma : \gamma$ is a finite ordinal$\}$,

(2) α is infinite iff $\alpha \geqslant \omega$,

(3) if $\alpha \geqslant \omega$, then $\bigcup\mathrm{ord}(\alpha)$ is an initial ordinal,

(4) if A is a set, then $\bigcap\mathrm{ord}(A)$ is an initial ordinal, and

(5) if α and β are initial ordinals and $\alpha < \beta$, then $\bigcup\mathrm{ord}(\alpha) \leqslant \beta$.

Let β be infinite, initial ordinal. Then $A = \{\alpha : \alpha$ is an initial infinite ordinal and $\alpha < \beta\}$ is a woset by 2.6(b) (as $A \subseteq \beta+1$) and by 2.6(d)(11), there is a unique ordinal γ such that A and γ

are order isomorphic. We denote β by ω_γ.

(k) **Proposition**. Let α, β be infinite, initial ordinals. The following are true:

(1) $\omega_0 = \omega$ is the first, infinite initial ordinal.

(2) $\omega_{\alpha+1} = \bigvee \text{ord}(\omega_\alpha)$

(3) if γ is a limit ordinal, then $\omega_\gamma = \bigvee\{\omega_\xi : \xi < \gamma$ and ξ is an ordinal$\}$, and

(4) $\omega_\alpha \leqslant \omega_\beta$ iff $\alpha \leqslant \beta$.

(l) **Cardinal numbers**. We have seen that among the ordinals, there are certain ordinals, called initial ordinals which are split into two camps - the finite initial ordinals and the infinite initial ordinals. All of the finite ordinals, i.e., members of ω, are initial ordinals. The infinite initial ordinals are written in the form ω_β where β is an ordinal. If A is a set, then $\bigwedge\text{ord}(A)$ is an initial ordinal by 2.6(j)(4). Thus, for two sets A and B, it follows by 2.6(h) that $\bigwedge\text{ord}(A) = \bigwedge\text{ord}(B)$ iff there is a bijection between the two sets A and B.

Sometimes when we work with an arbitrary set A, we are only interested in those sets B for which there is a bijection between A and B instead of, for example, being interested in the different well-orderings on A. Thus, if A is infinite and $\omega_\alpha = \bigwedge\text{ord}(A)$, then we are interested in those sets B such that $\omega_\alpha = \bigwedge\text{ord}(B)$; in this case, we are in effect ignoring the order structure on ω_α. In this situation we write $\aleph_\alpha$ instead of ω_α and write $|A| = \aleph_\alpha$. In particular, $|\omega| = \aleph_0$. If A is finite, we define $|A| = \bigwedge\text{ord}(A)$; so, $|A| \in \omega$. The ordinal $|A|$ is called the **cardinal** or **cardinality** of A and is used to focus our attention on $\bigwedge\text{ord}(A)$ as a non-ordered set. In particular, if $n \in \omega$, then $|n| = n$ and if $\alpha \geq \omega$ and $\omega_\beta = \bigcap \text{ord}(\alpha)$, then $|\alpha| = \aleph_\beta$. The elements of ω plus the $\aleph_\beta$'s where β is an ordinal are called **cardinal numbers** (usually shorten to **cardinals**). Thus, the cardinals coincide with the initial ordinals but indicate to the reader than the order structure of the ordinals is not being used. If m is a finite cardinal, then m^+ is used to denote m+1 and if $m = \omega_\alpha$ for some ordinal α, then m^+ is used to denote $\omega_{\alpha+1}$.

Cardinal addition, multiplication, and exponentiation are defined as follows:

(m) **Definition**. Let $\{m_i : i \in I\}$ be a set of cardinals.

(1) $\Sigma\{m_i : i \in I\}$ is defined to be $|\bigcup\{A_i : i \in I\}|$, where $\{A_i : i \in I\}$ is a pairwise disjoint collection of sets such that $|A_i| = m_i$ for each $i \in I$.

(2) $\Pi\{m_i : i \in I\}$ is defined to be $|\Pi\{B_i : i \in I\}|$ where $\{B_i : i \in I\}$ is a collection of sets such that $|B_i| = m_i$ for each $i \in I$.

(n) **Definition**. Let m and k be two cardinal numbers. Then m^k is defined to be $|F(K,M)|$, where K and M are sets for which $|K| = k$ and $|M| = m$.

Although the definitions above are given in terms of specific collections of sets $\{A_i : i \in I\}$, $\{B_i : i \in I\}$, K, and M, it is not difficult to show that these definitions are unambiguous.

We now list some of the properties for cardinal addition,

multiplication, and exponentiation. Note that $\Sigma\{m_i : i \in \{1,2\}\}$ is denoted by $m_1 + m_2$, etc.

(o) **Theorem**. Let m, n, p be cardinals, $\{m_i : i \in I\}$ a nonempty family of cardinals, and A a nonempty set. Then:

(1) $m + n = n + m$ and $mn = nm$,

(2) $m + (n+p) = (m+n) + p$ and $m(np) = (mn)p$,

(3) if m or n is infinite, then $m + n = mn = m \vee n$,

(4) if $2 \leqslant m \leqslant n$ and $\aleph_0 \leqslant n$, then $m^n = 2^n$,

(5) if $m \geqslant \aleph_0$, then $m^m \geqslant m^+$,

(6) $\wedge\{m_i : i \in I\} \in \{m_i : i \in I\}$ and $\vee\{m_i : i \in I\}$ is a cardinal,

(7) if $|I| \geqslant \aleph_0$, then $\Sigma\{m_i : i \in I\} = |I| \cdot (\vee\{m_i : i \in I\})$,

(8) (König). if $\{n_i : i \in I\}$ is a family of cardinals such that $m_i < n_i$ for each $i \in I$, then $\Sigma\{m_i : i \in I\} < \Pi\{n_i : i \in I\}$, and

(9) (Hausdorff). if m and n are infinite, then $(m^+)^n = (m^n)m^+$.

(p) **Definition**

(1) Let $\aleph_\alpha$ be a cardinal. The **cofinality** of $\aleph_\alpha$, denoted by $cf(\aleph_\alpha)$, is defined to be $\wedge\{m : m$ is a cardinal and ω_α can be written as $\vee\{S_i : i \in I\}$ where $|I| = m$ and $|S_i| < \aleph_\alpha$ for each $i \in I\}$. (It is immediate that $cf(\aleph_\alpha) \leqslant \aleph_\alpha$).

(2) $\aleph_\alpha$ is called **regular** if $cf(\aleph_\alpha) = \aleph_\alpha$ and is called **singular** if $cf(\aleph_\alpha) < \aleph_\alpha$.

(3) The cofinality of ω_α, denoted by $cf(\omega_\alpha)$, is defined to be the cofinality of $\aleph_\alpha$. Note that $cf(\omega_\alpha) = \wedge\{|S| : S \subseteq$

ω_α and for each $\gamma \in \omega_\alpha$, there is some $s \in S$ such that $\gamma \leqslant s\}$.

(4) A subset S of α is **cofinal** in α if $\alpha = \bigvee S$. A subset of α that is not cofinal is called **bounded**.

For example, $cf(\omega_1) = \aleph_1$; so, $\aleph_1$ is regular. However, $cf(\aleph_\omega) = \aleph_0 < \aleph_\omega$ since $\omega_\omega = \bigvee\{\omega_i : i \in \omega\}$; thus, $\aleph_\omega$ is singular.

In 2.5, we saw that any linearly ordered set could be viewed as a hereditarily normal topological space when given the order topology. Now, we study the topological properties of wosets.

(q) **Theorem**. Let α be an ordinal and give α and ω_α the order topology defined in 2.5(e). Then:

(1) α is compact iff α is a successor or $\alpha = 0$,

(2) if α is a limit ordinal, then α is a dense subset of the compact space $\alpha + 1$,

(3) if $cf(\omega_\alpha) = \aleph_0$, then ω_α is σ-compact (i.e., the union of countably many compact subspaces) and if $cf(\omega_\alpha) > \aleph_0$, then ω_α is countably compact,

(4) if $cf(\omega_\alpha) > \aleph_0$ and if A and B are disjoint closed subsets of ω_α, then $\bigvee A \in \omega_\alpha$ or $\bigvee B \in \omega_\alpha$, and therefore either A or B is compact,

(5) if $f \in C(\omega_\alpha)$ where $cf(\omega_\alpha) > \aleph_0$, then there exists $\beta \in \omega_\alpha$ and $r \in \mathbb{R}$ such that $f(\gamma) = r$ whenever $\beta \leqslant \gamma < \alpha$, and

(6) if $cf(\omega_\alpha) > \aleph_0$, then ω_α is a dense C-embedded subspace of $\omega_\alpha + 1$.

Proof

(1) If α is not a successor ordinal and $\alpha \neq 0$, then $\{\{\gamma : \gamma < \beta\} : \beta < \alpha\}$ is an open cover of α with no finite subcover; so, α is not compact. Clearly α is compact when $\alpha = 0$; so, suppose $\alpha = \beta + 1$ for some ordinal β. By 2.5(g), it suffices to show that $\beta + 1$ is lattice-complete. If $\emptyset \neq S \subseteq \beta + 1$, then $\cap S \in S \subseteq \alpha$ by 2.6(d)(8), and $\cap S$ is the infimum of S. By 2.1(e), $\beta + 1$ is lattice-complete and α is compact.

(2) Note that $(\alpha+1)\setminus\alpha = \{\alpha\}$. Basic neighborhoods of α in $\alpha + 1$ are of the form $(\beta,\alpha]$, where $\beta \in \alpha$. As α is not a successor, then $\beta + 1 \in (\beta,\alpha] \cap \alpha$. Thus, α is dense in $\alpha + 1$.

(3) If $cf(\omega_\alpha) = \aleph_0$, then by 2.6(p) there is a countable subset $\{\beta_i : i \in \omega\}$ of ω_α such that $\omega_\alpha = \cup\{\beta_i : i \in \omega\}$. Since $\beta_i \in \omega_\alpha$, then $\beta_i + 1 \subseteq \omega_\alpha$; hence, $\omega_\alpha = \cup\{\beta_i + 1 : i \in \omega\}$. By (1), ω_α is the countable union of compact subspaces. If $cf(\omega_\alpha) > \aleph_0$, let $\{\beta_i : i \in \omega\} \subseteq \omega_\alpha$. Then by 2.6(p) and 2.6(d)(9), $\cup\{\beta_i : i \in \omega\} \in \omega_\alpha$ as $cf(\omega_\alpha) > \aleph_0$. So, $\{\beta_i : i \in \omega\} \subseteq (\cup\{\beta_i : i \in \omega\}) + 1 \subseteq \omega_\alpha$. But $(\cup\{\beta_i : i \in \omega\}) + 1$ is compact by (1); thus, $\{\beta_i : i \in \omega\}$ has an accumulation point in $(\cup\{\beta_i : i \in \omega\}) + 1$ and, hence, in ω_α. This shows ω_α is countably compact. (In fact, it shows that ω_α is $\aleph_0$-bounded; see 5.6(c).)

(4) By 2.6(d)(9), $\cup A \in \omega_\alpha$ or $\cup A = \omega_\alpha$ and $\cup B \in \omega_\alpha$ or $\cup B = \omega_\alpha$. Assume $\cup A = \cup B = \omega_\alpha$. Inductively, choose $\{\alpha_i : i \in \omega\} \subseteq A$ and $\{\beta_i : i \in \omega\} \subseteq B$ such that $\alpha_n < \beta_n < \alpha_{n+1}$ for each $n \in \omega$. As $cf(\omega_\alpha) > \aleph_0$, $\cup\{\alpha_i : i \in \omega\} = \cup\{\beta_i : i \in \omega\} \in \omega_\alpha$. Let $t = \cup\{\alpha_i : i \in \omega\}$. Since A and B are closed, then $t \in A \cap B$, which contradicts the hypothesis. So, $\cup A \in \omega_\alpha$ or $\cup B \in \omega_\alpha$.

(5) By (3), ω_α is countably compact. Hence, for each $\gamma < \omega_\alpha$, the closed subspace $\omega_\alpha \setminus [0,\gamma)$ of ω_α is countably compact; so, $f[\omega_\alpha \setminus [0,\gamma)]$ is a countably compact subset of $\mathbb{R}$ and, thus, compact. Evidently, $\{f[\omega_\alpha \setminus [0,\gamma)] : \gamma < \omega_\alpha\}$ has the finite intersection property, and so $\bigcap \{f[\omega_\alpha \setminus [0,\gamma)] : \gamma < \omega_\alpha\} = K \neq \emptyset$. Let $r \in K$. For each $n \in \mathbb{N}$, $f^{\leftarrow}[\mathbb{R} \setminus (r - 1/n, r + 1/n)] = T_n$ and $f^{\leftarrow}(r)$ are disjoint closed subsets of ω_α. As $\bigvee f^{\leftarrow}(r) = \omega_\alpha$, then by (4), $\bigvee T_n = \sigma_n \in \omega_\alpha$. Since $cf(\omega_\alpha) > \aleph_0$, then $\bigvee \{\sigma_n : n \in \mathbb{N}\} = \beta \in \omega_\alpha$. Let $\gamma \geqslant \beta$. It follows that $f(\gamma) = r$.

(6) Let $f \in C^*(\omega_\alpha)$. By (2), ω_α is dense in $\omega_\alpha + 1$. By (5), there exists some $r \in \mathbb{R}$ and $\beta \in \omega_\alpha$ such that for $\gamma \geqslant \beta$, $f(\gamma) = r$. Define $F : \omega_\alpha + 1 \to \mathbb{R}$ by $F|\omega_\alpha = f$ and $F(\omega_\alpha) = r$. By 1.6(a), $F \in C^*(\omega_\alpha + 1)$.

■

Chapter 2 - Problems

2A. Duality for Posets. Let $(A,\leqslant)$ be a poset. For a, b $\in$ A, define a $\leqslant'$ b if b $\leqslant$ a.

(1) Show that $(A,\leqslant')$ is a poset.

(2) If $\emptyset \neq B \subseteq A$ and $a \in A$ is an upper bound (respectively, least upper bound) of B in $(A,\leqslant)$, prove that a is a lower bound (respectively, greatest lower bound) of b in $(A,\leqslant')$.

Note: Using this result, it is straightforward to show that results about upper bounds (respectively, least upper bounds) also hold for lower bounds (respectively, greatest least bounds).

2B. Bounds of Posets.

(1) Let $(A,\leqslant)$ be a poset and $\emptyset \neq B \subseteq A$. If $a_1, a_2 \in A$ are least upper bounds (respectively, greatest lower bounds) of B, then show $a_1 = a_2$.

(2) Give an example of a poset $(A,\leqslant)$ and a subset B of A such that B has no upper bound.

(3) Give an example of a poset $(A,\leqslant)$ and a subset B of A such that B has an upper bound but no least upper bound.

2C. Lattices.

(1) Let $(A,\leqslant)$ be an upper semilattice. Show that $(A,\leqslant)$ satisfies the following properties for a, b, c $\in$ A:

$(\vee$ - L0) $a \leqslant b$ iff $a \vee b = b$

$(\vee$ - L1) $a = a \vee a$

$(\vee$ - L2) $a \vee b = b \vee a$

$(\vee$ - L3) $(a \vee b) \vee c = a \vee (b \vee c)$

Note. By using the duality property given in 2A(2), it follows that a lower semilattice satisfies the meet statements corresponding to $(\vee$ - L0) -- $(\vee$ - L3). Combining these results for upper and lower semilattices, it follows that a lattice $(A,\leqslant)$ satisfies (L0) -- (L3) in 2.1(c)(7).

(2) Show that a lattice $(A,\leqslant)$ satisfies:

(L4) for a, b $\in$ A, $a \wedge (a \vee b) = a = a \vee (a \wedge b)$.

2D. Lattices to Posets. Let A be a set and let $\vee$ and $\wedge$ be two binary operations on A (i.e., there are functions $f, g : A \times A \to A$ such that f(a,b) is denoted by $a \vee b$ and g(a,b) is denoted by $a \wedge b$). Furthermore, suppose $(A, \vee, \wedge)$ satisfies (L1) --

(L4). Let a, b $\in$ A.

(1) Show $a \vee b = b$ iff $a \wedge b = a$

(2) Define a relation $\leqslant$ on A using (L0) of 2.1(c)(7), i.e., define $a \leqslant b$ to mean $a \vee b = b$. Show that $(A,\leqslant)$ satisfy (01) -- (03) of 2.1(a) and that relative to $\leqslant$, $\sup\{a,b\} = a \vee b$ and $\inf\{a,b\} = a \wedge b$.

2E. Z(X) is not lattice-complete. Let $\mathbb{R}_d$ be the set of real numbers with the discrete topology. Suppose A, B $\subseteq \mathbb{R}$ such that $|A| = |B| = |\mathbb{R}|$, $A \cap B = \emptyset$, and $\mathbb{R} = A \cup B$. Let Y be the one-point compactification of $\mathbb{R}_d$ where the point at infinity is denoted as p, i.e., $\{p\} = Y \setminus \mathbb{R}_d$.

(1) If $g \in C(Y)$ and $p \in Z(g)$, show there is a countable subset $T \subseteq \mathbb{R}$ such that $Z(g) = Y \setminus T$.

(2) For each finite set $D \subseteq A$ (respectively, $E \subseteq B$), let $g_D = \chi_{Y \setminus D, Y}$ and $h_E = \chi_{E,Y}$. Show that $g_D, h_E \in C^*(Y)$, $Z(g_D) = D$, and $Z(h_E) = Y \setminus E$.

(3) Suppose $Z = \vee\{Z(g_D) : D \subseteq A,\ D \text{ finite}\}$ exists in Z[Y]. Using that $Z(g_D) \subseteq Z \subseteq Z(h_E)$ for all finite subsets $D \subseteq A$ and $E \subseteq B$, show that $Z = A \cup \{p\}$ and that this is impossible by (1). Thus Z(Y) is not a complete lattice.

(4) If C is a countable subset of $\mathbb{R}_d$, show that $\{\{c\} : c \in C\}$ is a countable family of Z(Y) with no supremum in Z(Y). Thus Z(Y) is not even countably complete.

2F. <u>$C(X)$ and $C^*(X)$ need not be conditionally complete</u>. Let Y be the same space as in 2E above and for a finite set $D \subseteq A$, let $g_D = \chi_{Y \setminus D}$. Now, $\{g_D : D \text{ finite}, D \subseteq A\}$ is bounded below by χ_Y and above by $\chi_\emptyset$. Suppose $g = \bigwedge\{g_D : D \text{ finite}, D \subseteq A\}$ exists in $C(Y)$ or $C^*(Y)$. Show that $Z(g) = \bigvee\{Z(g_D) : D \text{ finite}, D \subseteq A\}$ (in $Z[Y]$). Of course this is impossible by 2E. This shows that $C(Y)$ and $C^*(Y)$ are not conditionally complete and, hence, not complete. What analogous conclusion can be drawn using 2E(4)?

2G. <u>Semiregular spaces and semiregularization of products</u>.

(1) Prove 2.2(a).

(2) Find a subspace $A \subseteq \mathbb{R}$ such that A, $\text{int}_{\mathbb{R}}A$, $c\ell_{\mathbb{R}}A$, $\text{int}_{\mathbb{R}}(c\ell_{\mathbb{R}}A)$, $c\ell_{\mathbb{R}}(\text{int}_{\mathbb{R}}A)$, $\text{int}_{\mathbb{R}}(c\ell_{\mathbb{R}}(\text{int}_{\mathbb{R}}A))$, and $c\ell_{\mathbb{R}}(\text{int}_{\mathbb{R}}(c\ell_{\mathbb{R}}A))$ are seven different sets.

(3) Let X be a space and Y be the one-point compactification of $\mathbb{N}$ where $Y \setminus \mathbb{N} = \{p\}$. Enlarge the product topology on $X \times Y$ be making $\{(x,n)\}$ open for each $x \in X$ and $n \in \mathbb{N}$. Let Z be $X \times Y$ with this topology. Prove:

(a) X is homeomorphic to $X \times \{p\}$,

(b) $X \times \{p\}$ is a closed, nowhere dense subspace of Z, and

(c) Z is semiregular (and Hausdorff).

Note: This shows that every Hausdorff space can be embedded as a closed, nowhere dense subspace of a semiregular space.

(4) Let $\{X_i : i \in I\}$ be a family of nonempty spaces and $Y = \Pi\{X_i : i \in I\}$. Prove $Y(s) = \Pi\{X_i(s) : i \in I\}$.

2H. Filters and ultrafilters.

(1) Prove 2.3(d). (Hint: Zorn's Lemma is needed.)

(2) Prove the characterizations of ultrafilters given in 2.3(e)(1) and 2.3(e)(3).

(3) Consider the subspaces $Y = \{1/n : n \in \mathbb{N}\} \cup \{0\}$ and $X = \{1/n : n \in \mathbb{N}\}$ of $\mathbb{R}$; X is discrete and Y is compact. Let $\mathcal{G} = \{\{1/n : n \geq m\} : m \in \mathbb{N}\}$. By 2.3(d)(2), $\mathcal{G}$ is contained in some ultrafilter $\mathcal{U}$ on X. Let $\mathcal{F} = \{A \subseteq Y : A \text{ is closed and } A \cap X \in \mathcal{U}\}$. Prove:

(a) $\mathcal{F}$ is a closed filter on Y.

(b) if A is closed in Y, then $A \in \mathcal{F}$ or $Y \setminus \mathrm{int}_Y A \in \mathcal{F}$, and

(c) $\mathcal{F}$ is not a closed ultrafilter on Y. (Hint: Show $\{0\} \notin \mathcal{F}$.)

Note. This shows that 2.3(e)(3) is false in general when $\mathcal{L}$ is the lattice of closed subsets.

(4) If F is a closed filter on a regular space X, prove that $c(F) \subseteq a(F)$.

(5) If X is a space such that for every $\mathbb{P}(X)$-filter base $\mathbb{F}$ on X, $c(\mathbb{F}) = \{p\}$ whenever $a(\mathbb{F}) = \{p\}$ and $p \in X$, prove X is compact. Thus the converse of 2.3(f)(4) is true when L is $\mathbb{P}(X)$.

(6) If X is the space described in 2.2(h)(1), L is $\tau(X)$, $RO(X)$, or $R(X)$, and F is an L-filter such that $a(F) = \{p\}$ for some $p \in X$, show $c(F) = \{p\}$. As noted in 2.2(h)(1), X is not regular. Hence, X is not compact, and the converse of 2.3(f)(4) is false when L is $\tau(X)$, $RO(X)$, or $R(X)$.

(7) Prove 2.3(m).

2I. Proof of the Birkhoff-Stone Theorem. By the comments preceding 2.4(g), we know that a lattice with a set representation is distributive. So, for the remainder of this exercise, we will suppose that $(L,\leqslant,\wedge,\vee)$ is distributive.

(1) If L has no minimum element, let $L' = L \cup \{0\}$ (assuming that $0 \notin L$) and define $a \leqslant' b$ if $a, b \in L$ and $a \leqslant b$ or $a = 0$. Prove $(L',\leqslant')$ is a distributive lattice with 0 as the minimum element and $(L,\leqslant)$ is a sublattice.

(2) If L has no maximum element, prove L is a sublattice

of a distributive lattice with a maximum element 1.

(3) If $(L,\leqslant,\vee,\wedge)$ is a sublattice of a lattice $(L',\leqslant',\vee',\wedge')$ and $(L',\leqslant',\vee',\wedge')$ has a set representation, show that $(L,\leqslant,\vee,\wedge)$ has a set representation.

Henceforth, we will assume by (1), (2), and (3) that $(L,\leqslant,\vee,\wedge)$ is a distributive lattice with minimum element 0 and maximum element 1. A subset $F \subseteq L$ is an L-**filter** if

(a) $0 \notin F$ and $F \neq \emptyset$,

(b) if a, b $\in F$, then a $\wedge$ b $\in F$, and

(c) if a $\in F$ and b $\in L$ such that a $\leqslant$ b, then b $\in F$.

Note that this definition extends the corresponding definition in 2.3(a)(2).

(4) Let $0 \neq a \in L$ and $F_a = \{b \in L : b \geqslant a\}$. Show F_a is an L-filter.

An L-filter F is called **prime** if, when a, b $\in L$ and a $\vee$ b $\in F$, then a $\in F$ or b $\in F$.

(5) Let F be an L-filter and a $\in L$ such that a $\notin F$. Let $S = \{G : G$ is an L-filter, $F \subseteq G$, and a $\notin G\}$ be partially ordered by set inclusion. Apply Zorn's lemma to $(S,\subseteq)$ to obtain a maximal element $M \in S$. Note that M is an L-filter, a $\notin M$, and if G is an

$\mathcal{L}$-filter, $G \supsetneq M$, and $a \notin G$ then $G = M$.

(6) Let M be a maximal element in (5). Show M is a prime $\mathcal{L}$-filter. (Hint: Let $b, c \in L$ such that $b \vee c \in M$ and $M \cap \{b,c\} = \emptyset$. Let $G_b = \{d \in L : d \geqslant (b \wedge m)$ for some $m \in M\}$ and show G_b is an $\mathcal{L}$-filter such that $M \subsetneq G_b$. So, there is an element $m_1 \in M$, such that $a \geqslant b \wedge m_1$. Likewise there is some $m_2 \in M$ such that $a \geqslant c \wedge m_2$. So, $(b \wedge m_1) \vee (c \wedge m_2) \leqslant a$. To obtain a contradiction, use the distributive law to show that $(b \wedge m_1) \vee (c \wedge m_2) \in M$.)

(7) Let $X = \{F : F$ is a prime $\mathcal{L}$-filter$\}$. Define $f \in F(L,\mathbb{P}(X))$ by $f(a) = \{F \in X : a \in F\}$. For $a, b \in L$, show $f(a \wedge b) = f(a) \cap f(b)$, $f(a \vee b) = f(a) \cup f(b)$, $f(0) = \emptyset$, and $f(1) = X$.

(8) If $a, b \in L$ and $a \neq b$, note that $a \nleqslant b$ or $b \nleqslant a$. Now, use (4), (5) and (6) to show that $f(a) \neq f(b)$.

(9) Consider the sublattice $(f[L],\cup,\cap)$ of $(\mathbb{P}(X),\cup,\cap)$. Use (7) and (8) in conjunction with 2.4(c)(3) to deduce that $f \in F(L,f[L])$ is a lattice isomorphism.

Remark. By (9), we have completed the other half of the proof of the Birkhoff-Stone Theorem that a lattice L has a set representation iff L is distributive.

2J. Ordered Spaces

(1) Prove 2.5(f), and show that the converse is false. (Hint: Consider $\mathbb{N}$ with the discrete topology.)

(2) Prove 2.5(g)(1).

(3) Prove 2.5(l).

2K. Subsets of Posets. Let $(A,\leq)$ be a poset and $B \subseteq A$.

(1) Prove $(B,\leq_B)$ is a poset.

(2) Find an example of a lattice $(A,\leq)$ and a subset $B \subseteq A$ such that $(B,\leq_B)$ is **not** a lattice.

(3) Find an example of a lattice $(A,\leq)$ and a subset $B \subseteq A$ such that $(B,\leq_B)$ is a lattice and there are elements $b_1, b_2 \in B$ such that $\sup_A\{b_1,b_2\} \neq \sup_B\{b_1,b_2\}$.

(4) If $(A,\leq)$ is a loset, prove $(B,\leq_B)$ is a loset.

(5) Prove 2.6(b)(2).

2L. Subspaces of ordered spaces. Let $(L,\leq)$ be a loset and $(L,\tau(\leq))$ the induced ordered space (see 2.5(e)). Let $M \subseteq L$. Let $\tau(\leq)_M$ denote the subspace topology on M induced by $\tau(\leq)$.

(1) Prove $\tau(\leq_M) \subseteq \tau(\leq)_M$.

(2) Find a woset $(L,\leq)$ and subset $M \subseteq L$ such that $\tau(\leq_M) \neq \tau(\leq)_M$. (Hint: First solve the problem for a

loset, e.g., when $L = \mathbb{R}$. Then extend the idea to a woset.)

(3) A set M is **<-dense** in L if for each a, b ∈ L such that a < b, there is a c ∈ M with a < c < b. Prove $\tau(\leqslant_M) = \tau(\leqslant)_M$ whenever M is <-dense in L.

(4) Show that there exists a loset $(L,\leqslant)$ and a subset M of L that is dense in the space $(L,\tau(\leqslant))$ but for which $\tau(\leqslant)_M \neq \tau(\leqslant_M)$. (Hint: Let $L = (-\infty,0] \cup [1,\infty)$ with the order induced by $\mathbb{R}$). Compare this to (3).

2M. Cantor's characterization of $\mathbb{Q}$.

(1) Prove that a countable loset $(X,\leqslant)$ is orderisomorphic to a subset of $(\mathbb{Q},\leqslant)$. (Hint: Let $X = \{x_n : n \in \omega\}$ and $X_k = \{x_n : n \leqslant k\}$ for $k \in \omega$. Show that a one-to-one order-preserving function $f \in F(X_k,\mathbb{Q})$ has a one-to-one order-preserving extension to a function $g \in F(X_{k+1},\mathbb{Q})$.)

(2) Let $(L,\leqslant)$ be a loset and a, b ∈ L such that a < b. Then the open interval (a,b) is called a **gap** in L if (a,b) = ∅. Prove that a countable loset $(X,\leqslant)$ without gaps and without greatest and least elements is order isomorphic to $(\mathbb{Q},\leqslant)$. (Hint: Let $X = \{x_n : n \in \omega\}$, index $\mathbb{Q}$ as $\{r_n : n \in \omega\}$, let $X_k = \{x_n : n \leqslant k\}$, and $Q_k = \{r_n : n \leqslant k\}$ for $k \in \omega$. As in 2M(1), show that a one-to-one order-preserving function $f \in F(A,B)$, where $X_k \subseteq A \subseteq X$ and $Q_k \subseteq B \subseteq \mathbb{Q}$, has a

one-to-one, order-preserving extension $g \in F(A',B')$ where A' and B' are sets satisfying $A \cup X_{k+1} \subseteq A' \subseteq X$ and $B \cup \mathbb{Q}_{k+1} \subseteq B' \subseteq \mathbb{Q}$.)

Note: 2M(2) shows that $(\mathbb{Q},\leqslant)$ is characterized by being a countable loset without gaps and without greatest and least elements. Also, 2M(1) shows that if $(L,\leqslant)$ is a countable loset, then one can assume that $L \subseteq \mathbb{Q}$ with the order $\leqslant$ inherited from the usual ordering on $\mathbb{Q}$.

2N. Cardinal Functions. Let X be an infinite space.

(1) If μ is a cardinal, let

$[X]^{<\mu} = \{A \subseteq X : |A| < \mu\}$,

$[X]^{\leqslant\mu} = \{A \subseteq X : |A| \leqslant \mu\}$, and

$[X]^{\mu} = [X]^{\leqslant\mu} \setminus [X]^{<\mu}$.

(a) Prove $|[X]^{<\aleph_0}| = |X|$.

(b) Prove $|[X]^{\mu}| \leqslant |[[X]^{\leqslant\mu}]^{\leqslant\mu}| \leqslant |X|^{\mu}$.

(2) Density Character. From 1G, recall that $d(X) = \min\{|D| : D$ is a dense subset of $X\}$ and $d(X)$ is called the **density character** of X.

(a) Prove $|X| \leqslant 2^{\mu}$ where $\mu = 2^{d(X)}$. (Hint: Remember that all spaces are Hausdorff.)

(b) If Y is a space, $f \in C(X,Y)$, $f[X]$ is dense in Y, and D is dense in X, prove $f[D]$ is dense in Y.

(c) Prove that $|RO(X)| \leqslant 2^{d(X)}$.

(3) Weight. The **weight** of the space X, denoted w(X), is defined to be $\min\{|\mathcal{B}| : \mathcal{B}$ is an open base for X$\}$.

(a) Prove that $|X| \leqslant 2^{w(X)}$.

(b) If $\mathcal{B}$ is a base for X, prove there is a base $\mathcal{B}' \subseteq \mathcal{B}$ such that $w(X) = |\mathcal{B}'|$.

(c) If X is a compact, 0-dimensional space, prove that $w(X) = |B(X)|$.

(4) Character. For $p \in X$, define $\chi(p,X) = \min\{|\mathcal{B}| : \mathcal{B}$ is a neighborhood base at p$\}$. The **character** of a space X, denoted $\chi(X)$, is defined to be $\sup\{\chi(p,X) : p \in X\}$.

(a) Prove that $|X| \leqslant d(X)^{\chi(X)}$. (Hint: Let D be a dense subset of X such that $|D| = d(X)$ and let $\mu = \chi(X)$. For each $x \in X$, there is a neighborhood base $\mathcal{B}(x)$ at x such that $|\mathcal{B}(x)| \leqslant \mu$. Choose $D_x \subseteq D$ with $|D_x| \leqslant \mu$ and $B \cap D_x \neq \emptyset$ for each $B \in \mathcal{B}(x)$. Define $f : X \to [[D]^{\leqslant\mu}]^{\leqslant\mu}$ by $f(x) = \{B \cap D_X : B \in \mathcal{B}(X)\}$. Show f is one-to-one and use 2N(1).)

(b) If $A \subseteq X$, prove that $|c\ell_X A| \leqslant |A|^{\chi(X)}$.

(5) π-weight. A family $\mathcal{B}$ of nonempty open sets of X is a

π-base if for each nonempty open set U in X, there is some B $\in$ $\mathcal{B}$ such that B $\subseteq$ U. The **π-weight** of a space X is defined as $\pi w(X) = \min\{|\mathcal{B}| : \mathcal{B}$ is a π-base for X$\}$.

(a) Give an example of an infinite space Y such that $\aleph_0 = \pi w(Y) < w(Y)$. (Hint: See 1N.)

(b) Prove that $d(X) \leqslant \pi w(X) \leqslant w(X)$.

(c) Prove that $\pi w(X) \leqslant d(X) \cdot \chi(X)$.

(6) Cellularity. The **cellularity** of a space X is defined as $c(X) = \sup\{|\mathcal{F}| : \mathcal{F}$ is a family of pairwise disjoint, nonempty open sets of X$\}$.

(a) Prove $c(X) \leqslant d(X)$.

(b) If X is metrizable, prove that $w(X) = c(X)$. (Hint: First prove that $d(X) = w(X)$ and then use $1/n$-covers to prove that $c(X) = d(X)$.)

A space with $c(X) \leqslant \aleph_0$ is said to satisfy the **countable chain condition** (abbreviated as *ccc*).

(7) Lindelöf Degree. For a cardinal μ, X is said to be **μ-Lindelöf** if every open cover of X has a subcover of cardinaltiy $\leqslant \mu$. Thus "Lindelöf" is the same as "$\aleph_0$-Lindelöf". The **Lindelöf degree** of a space X is defined as $L(X) = \min\{\mu : X$ is μ-Lindelöf$\}$.

(a) If $A \subseteq X$ is closed, prove $L(A) \leqslant L(X)$.

(b) If X is a metric space, prove $L(X) = w(X)$.

20. Arhangel'skii's Theorem. Let X be an infinite space, and let $H : \mathbb{P}(X) \to X$ be a choice function, i.e., $H(A) \in A$ for $A \in \mathbb{P}(X)\setminus\{\emptyset\}$. For each $x \in X$, let $\{B(x,\alpha) : \alpha < \chi(X)\}$ be an open neighborhood base at x. For $A \subseteq X$ and $f \in F(A,\chi(X))$, let $G(A,f) = \bigcup\{B(x,f(x)) : x \in A\}$.

(1) For $A \subseteq X$, prove

$$|\{G(C,f) : C \subseteq A \subseteq G(C,f),\ f \in F(C,\chi(X)),$$

and

$$|C| \leqslant L(A)\}| \leqslant \chi(X)^{L(A)} \cdot |A|^{L(A)}.$$

Let $\mu = \chi(X) \cdot L(X)$. Recursively define a family $\{A_\alpha : \alpha < \mu^+\}$ of subsets of X as follows:

(a) $A_0 = \{H(\emptyset)\}$ and

(b) if $\alpha < \mu^+$ and A_γ is defined for $\gamma < \alpha$, let $S_\alpha = c\ell_X(\bigcup\{A_\gamma : \gamma < \alpha\})$. Then let $A_\alpha = c\ell_X(B_\alpha \cup [\bigcup\{A_\gamma : \gamma < \alpha\}])$ where $B_\alpha = \{H(X\setminus G(C,f)) : C \subseteq S_\alpha \subseteq G(C,f),\ f \in F(C,\chi(X)),\ \text{and}\ |C| \leqslant L(S_\alpha)\}$.

Let $B = \bigcup \{A_\alpha : \alpha < \mu^+\}$.

(2) If $\alpha < \beta < \mu^+$, prove $A_\alpha \subseteq A_\beta$.

(3) Prove $|A_\alpha| \leqslant 2^\mu$. (Hint: Prove by transfinite induction and use 2N(4)(b) and 2O(1), noting that S_α is closed.)

(4) Prove B is closed and $|B| \leqslant 2^\mu$. (Hint: To prove B is closed use that $\chi(X) < \mu^+$.)

(5) (Arhangel'skii's Theorem). Prove $|X| \leqslant 2^\mu$ where $\mu = L(X) \cdot \chi(X)$. (Hint: Assume there is some $x \in X \setminus B$. Since $L(B) \leqslant L(X) \leqslant \mu$ (by 2N(7)(a)), show there is a subset $C \subseteq B$ and a function $f \in F(C, \chi(X))$ such that $B \subseteq G(C,f)$, $x \notin G(C,f)$, and $|C| \leqslant L(X) \leqslant \mu$. Since $C \subseteq B$ and $|C| \leqslant \mu$, show there is some $\alpha < \mu^+$ such that $C \subseteq A_\alpha$. Now, $G(C,f) \neq X$ and, hence $H(X \setminus G(C,f)) \in A_{\alpha+1}$. Show this is impossible as $H(X \setminus G(C,f)) \in X \setminus G(C,f)$ and $A_{\alpha+1} \subseteq B \subseteq G(C,f)$.)

2P. Ulam-measurable cardinals. Let D be an infinite set with the discrete topology. An **Ulam-measure** μ on D is a function $\mu : \mathbb{P}(D) \to 2$ satisfying these properties:

(a) $\mu(\{d\}) = 0$ for all $d \in D$,

(b) $\mu(D) = 1$, and

(c) if $\{A_n : n \in \omega\} \subseteq \mathbb{P}(D)$ and $A_n \cap A_m = \varnothing$ for $n \neq m$, then $\mu(\bigcup\{A_n : n \in \omega\}) =$

$$\Sigma\{\mu(A_n) : n \in \omega\}.$$

(1) If μ is an Ulam-measure on D and $\mathcal{U} = \{A \subseteq D : \mu(A) = 1\}$, prove $\mathcal{U}$ is a free ultrafilter on D with the **countable intersection property** (i.e., if $\{U_n : n \in \omega\} \subseteq \mathcal{U}$, then $\bigcap\{U_n : n \in \omega\} \neq \emptyset$.

(2) If $\mathcal{U}$ is a free ultrafilter on D with the countable intersection property, prove there is an Ulam-measure μ on D.

An infinite cardinal ν is said to be **Ulam-measurable** if there is an Ulam-measure on ν.

(3) Prove $\aleph_0$ is not Ulam-measurable.

(4) If ν and λ are infinite cardinals, $\nu < \lambda$, and λ is not Ulam-measurable, prove ν is not Ulam-measurable.

(5) If ν is an infinite cardinal and ν is not Ulam-measurable, prove $\lambda = 2^{\nu}$ is not Ulam-measurable. (Hint: Assume there is an Ulam-measure μ on 2^{ν}. For $x \in \nu$, let $P_x = \{A \subseteq \nu : x \in A\}$ and $M = \{x \in \nu : \mu(P_x) = 1\}$. For $x \in \nu$, define $F_x = P_x$ if $x \notin M$ and $F_x = \mathbb{P}(\nu)\setminus P_x$ if $x \in M$. Recursively define G_x to be $F_x \setminus \bigcup\{F_y : y < x\}$ for each $x \in \nu$. If $A \in \nu$, define $p(A)$ to be $\mu(\bigcup\{G_x : x \in A\})$. Observing that $\mu(G_x) = 0$ for $x \in \nu$ and that $\bigcup\{G_x : x \in \nu\} = \mathbb{P}(\nu)\setminus\{M\}$, show that p is a Ulam-measure on ν).

2Q. z-ultrafilters vs. closed ultrafilters. Let X be a Tychonoff space, $\mathcal{A}$ a z-ultrafilter on X, and $F(\mathcal{A}) = \{F$ closed in $X : Z \cap F \neq \emptyset$ for all $Z \in \mathcal{A}\}$. Obviously, $\mathcal{A} \subseteq F(\mathcal{A})$.

(1) Show there is a closed ultrafilter on X that contains $\mathcal{A}$.

(2) Suppose that $F(\mathcal{A})$ has the finite intersection property. Show there is precisely one closed ultrafilter on X that contains $\mathcal{A}$, namely $F(\mathcal{A})$.

(3) If X is normal, show that $F(\mathcal{A})$ is the unique closed ultrafilter on X containing $\mathcal{A}$.

(4) If X is δ-normally separated (see 1R) but not normal, show there exists a z-ultrafilter on X which is contained in two distinct closed ultrafilters. (Hint: Let A and B be disjoint closed sets that are not completely separated. Let $G = \{Z_1 \cap Z_2 : A \subseteq Z_1, B \subseteq Z_2$, and $Z_1, Z_2 \in Z(X)\}$. Show that

(a) G is a free z-filter on X,

(b) if $Z \in G$, then $Z \cap A \neq \emptyset$ and $Z \cap B \neq \emptyset$, and

(c) if $Z \in Z(X)$ and $Z \cap A = \emptyset$ or $Z \cap B = \emptyset$, then for some $Z' \in G$, $Z \cap Z' = \emptyset$. Examine any z-ultrafilter containing G.)

(Note that if X is not δ-normally separated, there may be no z-ultrafilter contained in distinct closed ultrafilters; see 2R(4).)

(5) Show the following are equivalent for a δ-normally separated space X:

(a) X is normal,

(b) $F(\mathcal{Q})$ has the finite intersection property for each z-ultrafilter $\mathcal{Q}$ on X,

(c) for each z-ultrafilter $\mathcal{Q}$, $F(\mathcal{Q})$ is a closed filter on X, and

(d) for each z-ultrafilter $\mathcal{Q}$ on X, $F(\mathcal{Q})$ is the unique closed ultrafilter on X that contains $\mathcal{Q}$.

(6) Show that even if X is normal, it is not necessarily true that $F(\mathcal{Q}) = \{F \text{ closed in } X : Z \subseteq F \text{ for some } Z \in \mathcal{Q}\}$. (Hint: Consider the space ω_1.)

2R. The Tychonoff Plank. Let $\omega_1 + 1$ have the order topology (see 2.6(q)). Then $\omega_1 + 1$ and $\omega + 1$ are compact by 2.6(q)(1), and hence $(\omega_1+1) \times (\omega+1)$ is compact. The subspace $T = (\omega_1+1) \times (\omega+1)\setminus\{(\omega_1,\omega)\}$ is called the **Tychonoff plank.** Let $A = \{\omega_1\} \times \omega$ and $B = \omega_1 \times \{\omega\}$.

(1) If U is an open set and $\{\omega_1\} \times [n,\omega) \subseteq U$ for some $n \in \omega$, prove there is some $\alpha \in \omega_1$ such that $[\alpha,\omega_1) \times [n,\omega) \subseteq U$ and $[\alpha,\omega_1) \times \{\omega\} \subseteq cl_T U$.

(Hint: For each $m \in [n,\omega)$, show there is some $\alpha_m \in \omega_1$ such that $(\alpha_m,\omega_1] \times \{m\} \subseteq U$. Since $cf(\omega_1) > \aleph_0$, find some $\alpha \in \omega_1$ such that $(\alpha,\omega_1] \times [n,\omega) \subseteq U$. Show that $[\alpha,\omega_1) \times \{\omega\} \subseteq c\ell_T U$.)

(2) If U is an open set and $[\alpha,\omega_1) \times \{\omega\} \subseteq U$ for some $\alpha \in \omega_1$, prove that $\{\omega_1\} \times [n,\omega) \subseteq c\ell_T U$ for some $n \in \omega$.

(3) Show that T is 0-dimensional but not normal. (Hint: A and B cannot be put in disjoint open sets of T.)

(4) Show that T is pseudocompact and C-embedded in $(\omega_1+1) \times (\omega+1)$. (Hint: Let $f \in C(T)$. Use 2.6(q)(5) plus the technique developed in the hint for (1) above to find some $\alpha \in \omega_1$ such that f is constant on each $(\alpha,\omega_1] \times \{n\}$ for $n \in \omega$ and on $(\alpha,\omega_1) \times \{\omega\}$.)

(5) Let $A = \{Z(f \mid T) : f \in C((\omega_1+1) \times (\omega+1))$ and $f((\omega_1+1,\omega+1)) = 0\}$. Show that A is a z-ultrafilter on T that is not contained in two distinct closed ultrafilters on T (cf., 2Q(4).)

2S. The Tychonoff Spirals

(1) Let T, A, and B be as described in 2R. Feed T into the "Jones machine" described in 1Y, using A for H, B for K, and T for X. Verify that the "output spaces" J(T) and DJ(T) are regular but not Tychonoff.

(2) Let S_n be the subspace $\bigvee\{T_i : 1 \leqslant i \leqslant n\}$ of J(T)

(see 1Y(3) for notation). Let $\{p_+,p_-\} \cap S_n = \emptyset$ and let $Y_n = \{p_+,p_-\} \cup S_n$. Suppose $n \geqslant 2$. Topologize Y_n as follows: S_n has the subspace topology inherited from $J(T)$, and if $U \subseteq Y_n$ then U is defined to be open in Y_n if $U \cap S_n$ is open in S_n and whenever $p_+ \in U$ (respectively, $p_- \in U$), there is some $\alpha \in \omega_1$ and $m \in \omega$ such that $[\alpha,\omega_1) \times [m,\omega] \times \{1\} \subseteq U$ (respectively, $[\alpha,\omega_1) \times [m,\omega] \times \{n\} \subseteq U$). Show that Y_2 is a semiregular space and that if U, V are neighborhoods of p_+, p_-, respectively, in Y_2, then $c\ell_{Y_2}U \cap c\ell_{Y_2}V \neq \emptyset$. In particular, this shows that Y_2 is not regular. For all $n \geqslant 2$, show that Y_n is not regular.

2T. Another Plank. Let $\omega_1 + 1$ have the order topology (see 2.5(e)). By 2.6(q)(1), $\omega_1 + 1$ is compact; so, $(\omega_1+1) \times (\omega_1+1)$ is compact. Let $P = (\omega_1+1) \times (\omega_1+1)\setminus\{(\omega_1,\omega_1)\}$, $A = \{\omega_1\} \times \omega_1$, $B = \{(\alpha,\alpha) : \alpha \in \omega_1\}$, and $C = \omega_1 \times \{\omega_1\}$.

(1) Prove P is a zero-dimensional space.

(2) If U is an open set in P and $A \subseteq U$ or $C \subseteq U$, prove $B \cap c\ell_P U \neq \emptyset$. (Hint: Suppose $A \subseteq U \subseteq \{\{\alpha\} \times (\alpha,\omega) : \alpha < \omega_1\}$. Show that for each $\alpha \in \omega_1$, there is some $f(\alpha) \in \omega_1$ such that $f(\alpha) > \alpha$ and $(f(\alpha),\alpha) \in U$. Let $\alpha_1 = f(\alpha)$ and $\alpha_{n+1} = f(\alpha_n)$ for $n \geqslant 1$. Then $\{\alpha_n : n \in \mathbb{N}\}$ is an increasing sequence in ω_1; let

$\beta = \sup\{\alpha_n : n \in \mathbb{N}\}$. Show that $(\beta,\beta) \in B \cap c\ell_P U$.)

(3) If U is an open set and $B \subseteq U$, prove $A \cap c\ell_P U \neq \emptyset$ and $C \cap c\ell_P U \neq \emptyset$.

(4) Show that neither P nor $\omega_1 \times (\omega_1+1)$ is normal.

(5) Prove that ω_1 and $\omega_1 + 1$ are normal spaces. Thus we have another example showing that the product of normal spaces in not necessarily normal; the first example is in 1H.

(6) Prove that P is pseudocompact and C-embedded in $(\omega_1+1) \times (\omega_1+1)$.

2U. Directed sets and Inverse Limits. A poset $(P,\leqslant)$ is **directed** if for a, b $\in$ P, there is some c $\in$ P such that $a \leqslant c$ and $b \leqslant c$. Let $\{X_a : a \in P\}$ be a family of nonempty spaces indexed over a directed set $(P,\leqslant)$ and $\{f_{ba} : a, b \in P, a \leqslant b\}$ be a family of functions where $f_{ba} \in C(X_b,X_a)$. The functions f_{ba} where a, b $\in$ P and $a \leqslant b$ are sometimes called **bonding maps**. Now, $(\{X_a\},f_{ba},P)$ is called an **inverse system** if

(a) for a $\in$ P, $f_{aa} = id_{X_a}$ and

(b) for a, b, c $\in$ P, $a \leqslant b$ and $b \leqslant c$, $f_{ba} \circ f_{cb} = f_{ca}$.

Let $Y = \Pi\{X_a : a \in P\}$ and Π_a be the projection function from Y to X_a. The **inverse limit** of $(\{X_a\},f_{ba},P)$ is the space

$X_\infty = \{y \in Y : f_{ba}(\Pi_b(y)) = \Pi_a(y)$ where $a, b \in P$ and $a \leq b\}$. Let $f_a = \Pi_a | X_\infty$ for $a \in P$.

(1) For $a \leq b$, prove $f_a = f_{ba} \circ f_b$ and $\{f_a^{\leftarrow}[U] : a \in P$, U open in $X_a\}$ is a base for X_∞.

(2) Prove X_∞ is a space (i.e., is Hausdorff) and is closed in Y.

(3) If X_a is compact for each $a \in P$, prove X_∞ is compact and nonempty.

(4) If each f_{ba} is open, prove f_a is open.

Note: The space X_∞ described above is frequently denoted by $\varprojlim(\{X_a\}, f_{ba}, P)$ or simply $\varprojlim\{X_a\}$.

CHAPTER 3

BOOLEAN ALGEBRAS

In this chapter, a special type of lattice, called a Boolean algebra, is investigated. The set of clopen subsets of an arbitrary space is shown to be a Boolean algebra with respect to unions and intersections, and any Boolean algebra is shown to be isomorphic to the set of clopen sets of a unique compact, zero-dimensional space. This correspondence establishes a duality between the class of Boolean algebras and the class of compact, zero-dimensional spaces. This duality result is used to show that certain (atomless) countable Boolean algebras are isomorphic to each other and to the set of clopen sets of the Cantor space. In 3.4 we focus our attention on complete Boolean algebras (Boolean algebras in which arbitrary suprema and infima exist) and show that every Boolean algebra is isomorphic to a subalgebra of a complete Boolean algebra. We close the chapter by discussing Martin's Axiom, and some topological and combinatorial applications of it, in 3.5.

3.1 Definition and basic properties

In 2.4, we investigated when a lattice $(L,\vee,\wedge)$ is lattice isomorphic to a sublattice of the lattice $(\mathbb{P}(X),\cup,\cap)$ for some set X, i.e., when $(L,\vee,\wedge)$ has a set representation. In this section, we go one step further and investigate when $(L,\vee,\wedge)$ is lattice isomorphic to the lattice of clopen sets of some topological space.

(a) <u>Definition</u>. The lattice $(L,\vee,\wedge)$ has a **topological representation** if $(L,\vee,\wedge)$ is lattice isomorphic to $(B(X),\cup,\cap)$ for some space X.

Suppose $(L,\vee,\wedge)$ has a topological representation $(B(X),\cup,\cap)$ for some space. Since a topological representation is a special type of set representation, then $(L,\vee,\wedge)$ is a distributive lattice. Also, since $(B(X),\cup,\cap)$ has a maximum element X and a minimum element $\emptyset$, then $(L,\vee,\wedge)$ is bounded. Moreover, if $A \in B(X)$, then $X \setminus A \in B(X)$; in terms of lattice notation, for each $A \in B(X)$ there exists an element $B \in B(X)$ such that $A \cup B = X$ and $A \cap B = \emptyset$. This gives rise to the following definition.

(b) <u>Definition</u>. Let $(L,\vee,\wedge)$ be a bounded lattice with 0 and 1 denoting the minimum and maximum elements, respectively, and let $a \in L$. Then $b \in L$ is called a **complement** of a if $a \vee b = 1$ and $a \wedge b = 0$. A **complemented lattice** is a bounded lattice in which every element has a complement.

(c) <u>Proposition</u>. If $(L,\vee,\wedge)$ is a distributive, bounded lattice, then each element in L has at most one complement.

<u>Proof</u>. Let $a \in L$ and suppose b and c are complements of a. Then $c = c \wedge 1 = c \wedge (a \vee b) = (c \wedge a) \vee (c \wedge b) = 0 \vee (c \wedge b) = c \wedge b \leqslant b$. By symmetry, $b \leqslant c$. So, $b = c$.

■

(d) **Definition**

(1) If $(L,\vee,\wedge)$ is a complemented lattice in which each element $a \in L$ has a unique complement, then the unique complement of a is denoted by a'.

(2) A complemented, distributive lattice is called a **Boolean algebra**.

When the complement operation is unique and it is important to emphasize it, we will denote the lattice by $(L,\vee,\wedge,')$ (or $(L,\leq,\vee,\wedge,')$, $(L,\vee,\wedge,',0,1)$,...).

By the comments in the paragraph preceding 3.1(b), a lattice with a topological representation is a Boolean algebra. The converse of this statement is also true and will be proven in the next section. In the remainder of this section we will investigate some examples and properties of Boolean algebras.

(e) **Examples**

(1) For a set X, $(\mathbb{P}(X),\cup,\cap)$ is a distributive, bounded lattice. If $A \in \mathbb{P}(X)$, then $X \setminus A = A'$ is the complement of $\mathbb{P}(X)$. Note that complements are unique since $(\mathbb{P}(X),\cup,\cap)$ is distributive. So, $(\mathbb{P}(X),\cup,\cap,')$ is a Boolean algebra.

(2) For a space X, $(B(X),\cup,\cap)$, the set of clopen sets of X, is also a Boolean algebra with $A' = X \setminus A$ for $A \in B(X)$.

(3) We saw in 2.2(c) that if X is a space then $(RO(X),\vee,\cap)$ is a bounded lattice. If $U \in RO(X)$, then $U \wedge (X \setminus c\ell_X U) =$

$U \cap (X \setminus c\ell_X U) = \emptyset$ and $U \vee (X \setminus c\ell_X U) = \text{int}_X c\ell_X(U \cup (X \setminus c\ell_X U)) = \text{int}_X X = X$. Since $\text{int}_X c\ell_X(X \setminus c\ell_X U) = X \setminus c\ell_X \text{int}_X c\ell_X U = X \setminus c\ell_X U$ (by 2.2(a)), it follows that $X \setminus c\ell_X U \in RO(X)$; thus, $RO(X)$ is a complemented lattice. If U, V, W $\in RO(X)$, then $U \wedge (V \vee W) = U \cap (\text{int}_X c\ell_X(V \cup W)) = \text{int}_X c\ell_X(U \cap (V \cup W))$ (by 2.2(a)) $= \text{int}_X c\ell_X((U \cap V) \cup (U \cap W)) = (U \wedge V) \vee (U \wedge W)$; similarly, $U \vee (V \wedge W) = (U \vee V) \wedge (U \vee W)$. So, $RO(X)$ is a distributive lattice. Thus, $RO(X)$ is a Boolean algebra and for $U \in RO(X)$, $U' = X \setminus c\ell_X U$. It is important to note in this example that $U' \neq X \setminus U$, in general.

(4) For a space X, $(R(X),\vee,\wedge)$ is a bounded lattice. By a proof similar to the proof of (3) (which the reader is invited to supply; see 3B), $R(X)$ is a Boolean algebra and if $A \in R(X)$, then $A' = X \setminus \text{int}_X A = c\ell_X(X \setminus A)$.

(5) Let X be a set and $CF(X) = \{F \subseteq X : F$ is finite or $X \setminus F$ is finite$\}$. Then $(CF(X),\cup,\cap)$ is a sublattice of $(\mathbb{P}(X),\cup,\cap)$. Hence, $CF(X)$ is a distributive lattice. Since $\emptyset$, $X \in CF(X)$, it follows that $CF(X)$ is bounded. Also, if $A \in CF(X)$, then $X \setminus A$ is the complement A'. Thus, $CF(X)$ is a Boolean algebra.

(6) For a space X, $(\tau(X),\cup,\cap)$ is the complete lattice of open sets in X. $\tau(X)$ is bounded and distributive. In general, $\tau(X)$ is not complemented for if $U \in \tau(X)$ and U' existed (it would be unique by 3.1(c)), then $U \cap U' = \emptyset$ and $U \cup U' = X$. Hence, $U' = X \setminus U$. This would mean that $X \setminus U$ is open, and would imply that U is closed. But in a connected space X, the only clopen sets are $\emptyset$ and X. So, in general, $(\tau(X),\cup,\cap)$ is not a Boolean algebra.

(f) **Notation**. If $(B,\vee,\wedge,')$ is a Boolean algebra and a, b $\in$

B, then $a \backslash b$ is used to denote $a \wedge b'$ (when using the notation for complementation, it will be clear from the context whether it is Boolean algebra complementation or set-theoretic complementation) and a'' is used to denote $(a')'$.

(g) **Proposition**. Let $(B, \vee, \wedge, ')$ be a Boolean algebra and a, $b \in B$. Then:

(1) (De Morgan Laws) $(a \wedge b)' = a' \vee b'$ and $(a \vee b)' = a' \wedge b'$,

(2) $a'' = a$, $0' = 1$, and $1' = 0$,

(3) $a \wedge b = (a' \vee b')'$ and $a \vee b = (a' \wedge b')'$,

(4) $a \leqslant b$ iff $b' \leqslant a'$, $a \wedge b = 0$ iff $b \leqslant a'$, and $a \vee b = 1$ iff $a' \leqslant b$,

(5) $a \backslash b = 0$ iff $a \leqslant b$ and $a \backslash b = 1$ iff $a = 1$ and $b = 0$, and

(6) if $\{a_k : 1 \leqslant k \leqslant n\} \subseteq B$, then $a \backslash \vee\{a_k : 1 \leqslant k \leqslant n\} = \wedge\{a \backslash a_k : 1 \leqslant k \leqslant n\}$ and $a \backslash \wedge\{a_k : 1 \leqslant k \leqslant n\} = \vee\{a \backslash a_k : 1 \leqslant k \leqslant n\}$.

Proof

(1) Since $(a \wedge b) \wedge (a' \vee b') = (a \wedge b \wedge a') \vee (a \wedge b \wedge b') = (0 \wedge b) \vee (a \wedge 0) = 0$, $(a \wedge b) \vee (a' \vee b') = (a \vee a' \vee b') \wedge (b \vee a' \vee b') = (1 \vee b') \wedge (1 \vee a') = 1$, and complementation is unique, then $(a \wedge b)' = a' \vee b'$. A similar proof shows $(a \vee b)' = a' \wedge b'$.

(2) Since $a' \wedge a = 0$ and $a' \vee a = 1$, then $a'' = a$ by uniqueness of complementation of a'. Likewise, since $0 \wedge 1 = 0$ and $0 \vee 1 = 1$, then $0' = 1$ and $1' = 0$.

(3) By (1) and (2), $(a' \vee b')' = a'' \wedge b'' = a \wedge b$

and $(a' \wedge b')' = a'' \vee b'' = a \vee b$.

(4) Note that $a \leqslant b$ iff $a \wedge b = a$. By (1) and (2), $a \wedge b = a$ iff $a' = (a \wedge b)' = a' \vee b'$. But $a' = a' \vee b'$ iff $a' \geqslant b'$. If $a \wedge b = 0$, then $a' = a' \vee 0 = a' \vee (a \wedge b) = (a' \vee a) \wedge (a' \vee b) = 1 \wedge (a' \vee b) = a' \vee b$. Thus, $b \leqslant a'$. If $b \leqslant a'$, then $a \wedge b \leqslant a \wedge a' \leqslant 0$. Hence $a \wedge b = 0$. Now, $a \vee b = 1$ iff $0 = 1' = (a \vee b)' = a' \wedge b'$. But $0 = a' \wedge b'$ iff $a' \leqslant b''$ iff $a' \leqslant b$.

(5) By definition, $a\backslash b = 0$ iff $a \wedge b' = 0$. By (2) and (4), $a \wedge b' = 0$ iff $a \leqslant b$. If $a = 1$ and $b = 0$, then clearly $a\backslash b = 1$. If $a\backslash b = 1$, then $1 = a \wedge b' \leqslant a$; hence $a = 1$. Also, $1 = a \wedge b' \leqslant b'$ and so $b' = 1$ which implies that $b = 1' = 0$.

(6) By (1) and induction, it follows that $(\wedge\{a_k : 1 \leqslant k \leqslant n\})' = \vee \{a_k' : 1 \leqslant k \leqslant n\}$ and $(\vee\{a_k : 1 \leqslant k \leqslant n\})' = \wedge\{a_k' : 1 \leqslant k \leqslant n\}$. Now (6) follows from 3.1(f).

■

Note that in a Boolean algebra $(B,\vee,\wedge,')$, it follows by 3.1(g)(3), that "$\vee$" is determined by "$\wedge$" and "$'$" and "$\wedge$" is determined by "$\vee$" and "$'$".

(h) <u>**Definition**</u>. Let $(B,\vee,\wedge,')$ be a Boolean algebra and $A \subseteq B$. A is a **Boolean subalgebra** if $\{0,1\} \subseteq A$ and $a, b \in A$ implies $a \wedge b$, $a \vee b$, $a' \in A$.

Note that, in particular, a Boolean subalgebra $(A,\vee,\wedge,')$ is a sublattice of $(B,\vee,\wedge,')$.

(i) <u>**Proposition**</u>. Let $A \subseteq B$ where $(B,\vee,\wedge,')$ is a Boolean algebra. Then:

(1) A is a Boolean subalgebra if $\{0,1\} \subseteq A$ and $a, b \in A$ imply $a' \in A$ and $a \vee b \in A$ (or $a \wedge b \in A$), and

(2) If $\mathcal{A}$ is a nonempty family of Boolean subalgebras of B, then $\cap\mathcal{A}$ is a Boolean subalgebra of B.

Proof. Now (1) is an immediate consequence of 3.1(g)(3). The proof of (2) is left to the reader (see 3C).

■

(j) **Definition**. If $E \subseteq B$ where $(B,\vee,\wedge,')$ is a Boolean algebra, then $\langle E\rangle$ denotes the intersection of all the Boolean subalgebras of B containing E. By 3.1(i)(2), $\langle E\rangle$ is a Boolean subalgebra of $(B,\vee,\wedge,')$ and is the smallest Boolean subalgebra of $(B,\vee,\wedge,')$ containing E. Sometimes we say $\langle E\rangle$ is the Boolean subalgebra of $(B,\vee,\wedge,')$ **generated** by E. If $E = \emptyset$, then $\langle E\rangle = \{0,1\}$ which is the smallest Boolean subalgebra of B.

(k) **Examples**.

(1) If X is a space, then $B(X)$ is a Boolean subalgebra of $R(X)$, $RO(X)$, and $\mathbb{P}(X)$. If $C \in B(X)$, then $\{\emptyset,C,X\setminus C,X\}$ is a Boolean subalgebra of $B(X)$; in fact, $\langle\{C\}\rangle = \{\emptyset,C,X\setminus C,X\}$. On the other hand, if $C, D \in B(X)$ such that $(C\setminus D) \cup (D\setminus C) \neq \emptyset$, and either $\emptyset \neq C \cap D$ or $C \cup D \neq X$, then $A = \{\emptyset,C,D,X\}$ is not a Boolean subalgebra of $B(X)$ or $\mathbb{P}(X)$. However, $(A,\subseteq)$ as a poset is a complemented, distributive lattice and, hence, a Boolean algebra with operations other than those of $\cap$, $\cup$ and $'$ of $B(X)$.

(2) In general, for a space X, $R(X)$ and $RO(X)$ are not

Boolean subalgebras of $\mathbb{P}(X)$. Consider $C = (a,b) \subseteq \mathbb{R}$ where $a, b \in \mathbb{R}$ and $a < b$. In $RO(\mathbb{R})$, $C' = (-\infty,a) \cup (b,\infty)$; however, in $\mathbb{P}(\mathbb{R})$, $C' = (-\infty,a] \cup [b,\infty)$.

(3) Consider the subsets $\mathcal{A} = \{(a,b) : a, b \in \mathbb{R}, a < b\}$ and $\mathcal{B} = \mathcal{A} \cup \{(-\infty,a) : a \in \mathbb{R}\} \cup \{(a,\infty) : a \in \mathbb{R}\}$ of $RO(\mathbb{R})$. Clearly, $<\mathcal{A}> \subseteq <\mathcal{B}> \subseteq RO(\mathbb{R})$. Now $C \in <\mathcal{A}>\backslash\{\emptyset,\mathbb{R}\}$ iff for some positive integer $n \geqslant 2$ and subset $\{a_2,a_3,\ldots,a_n\} \subseteq \mathbb{R}$ where $a_1 < a_2 < \ldots < a_n$, $C = (-\infty,a_1) \cup (a_1,a_2) \cup \ldots \cup (a_n,\infty)$ or $C = (a_1,a_2) \cup \ldots \cup (a_{n-1},a_n)$. Thus, $<\mathcal{A}> \neq <\mathcal{B}>$. Likewise, $<\mathcal{B}> \neq RO(\mathbb{R})$ as $\cup \{(2n,2n+1) : n \in \mathbb{N}\} \in RO(\mathbb{R})\backslash<\mathcal{B}>$. The Boolean subalgebra generated by $\{(\sqrt{2},\pi)\}$ in $RO(\mathbb{R})$ is

$$<\{(\sqrt{2},\pi)\}> = \{\emptyset,(\sqrt{2},\pi),(-\infty,\sqrt{2}) \cup (\pi,\infty),\mathbb{R}\}.$$

(I) **Proposition**. Let $(B,\vee,\wedge,')$ be a Boolean algebra. Then:

(1) If $A \subseteq B$ is a Boolean subalgebra and $b \in B$, then $<A \cup \{b\}> = \{(a_0 \wedge b) \vee (a_1 \wedge b') : a_0, a_1 \in A\}$ and

(2) if $\emptyset \neq E \subseteq B$ and $E' = \{a' : a \in E\}$, then $<E> = \{\vee\{(\wedge\{b : b \in F_i\}) : 1 \leqslant i \leqslant n\} : n \in \mathbb{N}$ and F_i is a finite set of $E \cup E'$ for $1 \leqslant i \leqslant n\}$.

Proof

(1) Let $C = \{(a_0 \wedge b) \vee (a_1 \wedge b') : a_0, a_1 \in A\}$. Clearly, $C \subseteq <A \cup \{b\}>$. Since $a = (a \wedge b) \vee (a \wedge b')$ and $b = (1 \wedge b) \vee (0 \wedge b')$, then $A \cup \{b\} \subseteq C$. To show $<A \cup \{b\}> \subseteq C$, it suffices by 3.1(i)(2) to show that C is a Boolean subalgebra of $(B,\vee,\wedge,')$. Since $\{0,1\} \subseteq A$, it follows that $\{0,1\} \subseteq C$. So by 3.1(i)(1), it suffices to show C is closed under "$\vee$" and "$'$". Let $a_0, a_1, a_2, a_3 \in A$.

Now, $[(a_0 \wedge b) \vee (a_1 \wedge b')] \vee [(a_2 \wedge b) \vee (a_3 \wedge b')] = [(a_0 \vee a_2) \wedge b] \vee [(a_1 \vee a_3) \wedge b'] \in C$ as $a_0 \vee a_2 \in A$ and $a_1 \vee a_3 \in A$. Also, $[(a_0 \wedge b) \vee (a_1 \wedge b')]' = (a_0' \vee b') \wedge (a_1' \vee b) = (a_0' \wedge a_1') \vee (a_0' \wedge b) \vee (a_1' \wedge b') \vee (b \wedge b') \in C$ as $A \subseteq C$ and C is closed under "$\vee$". This completes the proof that C is a Boolean subalgebra.

(2) The proof of (2) is similar to the proof of (1), but uses the generalized De Morgan Laws and distribution laws of 3F. The details are left to the reader.

■

(m) **Definition**. A function $f \in F(A,B)$ between Boolean algebras A and B is a **Boolean homomorphism** if f is a lattice homomorphism and for $a \in A$, $f(a') = (f(a))'$. By 3.1(g)(3), f is a Boolean homomorphism iff for a, $b \in A$, $f(a') = f((a))'$ and $f(a \vee b) = f(a) \vee f(b)$ (or $f(a \wedge b) = f(a) \wedge f(b)$). If f is a Boolean homomorphism, then $f(1) = f(0') = f(0)'$ which implies that $f(0) = f(0 \wedge 1) = f(0) \wedge f(0)' = 0$; similarly, $f(1) = 1$. The function f is a **Boolean isomorphism** if f is a bijection and both f and $f^{\leftarrow}$ are Boolean homomorphisms.

(n) **Proposition**. Let C and B be Boolean algebras.

(1) If $f : C \to B$ is an order isomorphism, then f is a Boolean isomorphism.

(2) If $f : C \to B$ is a bijective Boolean homomorphism, then f is a Boolean isomorphism.

(3) Let A be a Boolean subalgebra of C and $b \in C$. Suppose $f \in F(A \cup \{b\},B)$ and $f|A \in F(A,B)$ is a Boolean homomorphism. Let

$a \in A$. If $a \wedge b = 0$ implies $f(a) \wedge f(b) = 0$ and $a \wedge b' = 0$ implies $f(a) \wedge (f(b))' = 0$, then f has a unique extension to a Boolean homomorphism $g \in F(\langle A \cup \{b\}\rangle, B)$.

Proof

(1) By 2.4(c)(2), f is a lattice isomorphism. So, it remains to show that $f(c') = (f(c))'$ for $c \in C$. Since f is an order isomorphism, it follows that $f(0) = 0$ and $f(1) = 1$. For $c \in C$, $0 = f(0) = f(c) \wedge f(c')$ and $1 = f(1) = f(c) \vee f(c')$. Hence, $f(c')$ is the complement of $f(c)$; so, $f(c') = (f(c))'$.

(2) By 2.4(c)(1) and 2.4(c)(3), f is an order isomorphism. By (1) above, f is a Boolean isomorphism.

(3) Let $c \in \langle A \cup \{b\}\rangle$. By 3.1(l), there are elements $a_0, a_1 \in A$ such that $c = (a_0 \wedge b) \vee (a_1 \wedge b')$. The obvious extension of f to c is to define g(c) to be $(f(a_0) \wedge f(b)) \vee (f(a_1) \vee f(b)')$, but the problem is whether such a "definition" is well defined. That is, if c also equals $(a_2 \wedge b) \vee (a_3 \wedge b')$, then does $(f(a_0) \wedge f(b)) \vee (f(a_1) \wedge f(b)') = (f(a_2) \wedge f(b)) \vee (f(a_3) \wedge f(b)')$? To show this, first note the following:

$$\begin{aligned}
0 &= c \wedge c' \\
&= [(a_0 \wedge b) \vee (a_1 \wedge b')] \wedge [(a_2 \wedge b) \vee (a_3 \wedge b')]' \\
&= [(a_0 \wedge b) \vee (a_1 \wedge b')] \wedge [(a_2' \vee b') \wedge (a_3' \vee b)] \\
&= [(a_0 \wedge b) \wedge (a_2' \vee b') \wedge (a_3' \vee b)] \vee \\
&\quad [(a_1 \wedge b') \wedge (a_2' \vee b') \wedge (a_3' \wedge b)].
\end{aligned}$$

This implies that

$$0 = (a_0 \wedge b) \wedge (a_2' \vee b') \wedge (a_3' \vee b) \text{ and}$$
$$0 = (a_1 \wedge b') \wedge (a_2' \vee b') \wedge (a_3' \vee b).$$

But $(a_0 \wedge b) \wedge (a_2' \vee b') \wedge (a_3' \vee b) = (a_0 \wedge a_2' \wedge b)$ and so $a_0 \wedge a_2' \wedge b = 0$. Similarly, $a_1 \wedge a_3' \wedge b' = 0$. By the hypothesis and since $f|A$ is a Boolean homomorphism, $0 = f(a_0 \wedge a_2') \wedge f(b) = f(a_0) \wedge f(a_2)' \wedge f(b)$; similarly, $0 = f(a_1) \wedge f(a_3)' \wedge f(b)'$. By reversing the above steps, we obtain the following:

$$0 = [(f(a_0) \wedge f(b)) \vee (f(a_1) \wedge f(b)')] \wedge [(f(a_2) \wedge f(b)) \vee (f(a_3) \wedge f(b)')]'$$

By 3.1(g)(4), $(f(a_0) \wedge f(b)) \vee (f(a_1) \wedge f(b)') \leqslant (f(a_2) \wedge f(b)) \vee (f(a_3) \wedge f(b)')$; by symmetry, $(f(a_2) \wedge f(b)) \vee (f(a_3) \wedge f(b)') \leqslant (f(a_0) \wedge f(b)) \vee (f(a_1) \wedge f(b)')$. This shows that g is well-defined and it easily follows that g is a Boolean homomorphism. To show the uniqueness of g, suppose g and h are Booleans homomorphisms that extend f. If $c \in \langle A \cup \{b\}\rangle$, then $c = (a_0 \wedge b) \vee (a_1 \wedge b')$ for some a_0, a_1, $\in A$. So, $g(c) = (g(a_0) \wedge g(b)) \vee (g(a_1) \wedge g(b)') = (f(a_0) \wedge f(b)) \vee (f(a_1) \wedge f(b)') = (h(a_0) \wedge h(b)) \vee (h(a_1) \wedge h(b)') = h(c)$. Thus, $g = h$.

■

We will now give an application of 3.1(n). First we need to define the concept of a complete Boolean algebra.

(o) <u>**Definition**</u>. A Boolean algebra B is **complete** if B is complete as a lattice.

(p) **Proposition**. Let A be a subalgebra of a Boolean algebra C and let $h \in F(A,B)$ be a Boolean homomorphism where B is a complete Boolean algebra. Then h has a Boolean homomorphism extension $H \in F(C,B)$.

Proof. First, we will show that if $c \in C$, then h has a Boolean homomorphism extension $g \in F(\langle A \cup \{c\}\rangle,B)$. By 3.1(n) it suffices to find some $b \in B$ such that for $a \in A$, if $a \wedge c = 0$, then $h(a) \wedge b = 0$ and if $a \wedge c' = 0$, then $h(a) \wedge b' = 0$. Let $S = \{a \in A : a \leqslant c\}$ and $b = \bigvee\{h(a) : a \in S\}$. Suppose $a_1 \wedge c = 0$ for some $a_1 \in A$. Then $c \leqslant a_1'$. If $a \in S$, then $a \leqslant c$. So, $a \leqslant a_1'$ which implies that $h(a) \leqslant h(a_1') = h(a_1)'$ for all $a \in S$. So, $b \leqslant h(a_1)'$ which implies that $b \wedge h(a_1) = 0$ by 3.1(g)(4). Suppose $a \wedge c' = 0$. Then $a \leqslant c$. So, $a \in S$ and $h(a) \leqslant b$. Thus, $h(a) \wedge b' = 0$. By 3.1(n), there is a Boolean homomorphism $g \in F(\langle A \cup \{c\}\rangle,B)$ such that $g|A = f$ and $g(c) = b$.

Now let $C = \{c_\alpha : \alpha < \kappa\}$ be an indexing of the elements of C with respect to some cardinal κ and let $A_\alpha = \langle A \cup \{c_\beta : \beta \leqslant \alpha\}\rangle$ for $\alpha < \kappa$. By the above fact, since $A_0 = \langle A \cup \{c_0\}\rangle$, it follows that h has a Boolean homomorphism extension $h_0 \in F(A_0,B)$. Suppose $\beta < \kappa$ and for each ordinal $\alpha < \beta$, suppose a Boolean homomorphism $h_\alpha \in F(A_\alpha,B)$ is defined so that $h_\alpha|A = h$ and for $\gamma < \alpha$, $h_\alpha|A_\gamma = h_\gamma$. If $\beta = \alpha + 1$, then since $A_\beta = \langle A_\alpha \cup \{c_\beta\}\rangle$, by the above remarks there is a Boolean homomorphism $h_\beta \in F(A_\beta,B)$ which extends h_α. If β is a limit ordinal, then $D = \bigcup\{A_\alpha : \alpha < \beta\}$ is a Boolean subalgebra (see 3E). Define $g \in F(D,B)$ by $g(d) =$

$h_\alpha(d)$ if $d \in A_\alpha$; g is a Boolean homomorphism (see 3E) that extends h_α for $\alpha < \beta$. Now, $A_\beta = \langle D \cup \{c_\beta\}\rangle$ and by the above fact, there is a Boolean homomorphism $h_\beta \in F(A,B)$ which extends h_α for $\alpha < \beta$ and $h_\beta | A = h$. Thus, by transfinite induction, there is a Boolean homomorphism $h_\alpha \in F(A_\alpha,B)$ for $\alpha < \kappa$ such that $h_\alpha = h_\beta | A_\alpha$ for $\alpha < \beta < \kappa$ and $h_\alpha | A = h$. Now $C = \bigcup\{A_\alpha : \alpha < \kappa\}$. Define $H \in F(C,B)$ by $H(c) = h_\alpha(c)$ where $c \in A_\alpha$. Then H is a Boolean algebra homomorphism (see 3E) that extends h.

■

As Boolean algebras are lattices, we can define filters (and related concepts) on Boolean algebras as we did for lattices in Chapter 2. Explicitly, we make the following definitions.

(q) __Definition__. Let $(B,\vee,\wedge,')$ be a Boolean algebra.

(1) A family $F \subseteq B$ is a **B-filter base** (or a **filter base on** B) if

(a) $F \neq \emptyset$ and

(b) if a, b $\in F$, there is some $c \in F$ such that $0 \neq c \leqslant a \wedge b$.

(2) A B-filter base F is a **B-filter** if, in addition, F satisfies

(c) if $a \in F$ and $a \leqslant b \in B$, then $b \in F$.

(3) A B-filter F is a **B-ultrafilter** if, whenever G is a B-filter and $G \supseteq F$, then $G = F$; F is a **prime B-filter** if for a, b $\in$

B, $a \vee b \in F$ implies $a \in F$ or $b \in F$. (Prime filters on lattices and posets are defined in 2I.)

Note that $1 \in F$ for every B-filter F.

(r) **Proposition**. Let F be a B-filter where $(B, \vee, \wedge, ')$ is a Boolean algebra. Then:

(1) if $a, b \in F$, then $a \wedge b \in F$.

(2) F is contained in some B-ultrafilter and

(3) the following are equivalent:

(a) F is a B-ultrafilter,

(b) for $a \in B$, if $a \wedge b \neq 0$ for all $b \in F$, then $a \in F$,

(c) F is a prime B-filter, and

(d) for $a \in B$, $a \in F$ or $a' \in F$.

Proof. The proof of (1) is similar to the proof of 2.3(b) and the proof of (2) is similar to the proof of 2.3(d)(2). We now prove (3).

(a) implies (b). Suppose $a \in B$ and $a \wedge b \neq 0$ for all $b \in F$. Let $U = \{c \in B : c \geq a \wedge b$ for some $b \in F\}$. Since $b \in F$ implies $b \geq a \wedge b$, it follows that $F \subseteq U$, and since $1 \in F$ and $a \geq a \wedge 1$, it follows that $a \in U$. So, $U \neq \emptyset$. Also, $c \geq a \wedge b > 0$ implies $c \neq 0$ and $0 \notin U$. If $c_1 \geq a \wedge b_1$ and $c_2 \geq a \wedge b_2$ where $b_1, b_2 \in F$, then $c_1 \wedge c_2 \geq (a \wedge b_1) \wedge (a \wedge b_2) = a \wedge (b_1 \wedge b_2)$. Thus, $c_1 \wedge c_2 \in U$ as $b_1 \wedge b_2 \in F$. If $d \geq c \geq a \wedge b$, then $d \geq a \wedge b$ which implies that $d \in U$. This shows that U is a B-filter. Since F is maximal among the B-filters and $F \subseteq U$, it follows that $F = U$. Thus, $a \in F$.

(b) implies (c). Suppose $a \vee b \in F$. Assume both $a \notin F$ and $b \notin F$. There are elements $c_1, c_2 \in F$ such that $a \wedge c_1 = 0$ and $b \wedge c_2 = 0$. Now, $(a \vee b) \wedge (c_1 \wedge c_2) \in F$. But $(a \vee b) \wedge (c_1 \wedge c_2) = (a \wedge c_1 \wedge c_2) \vee (b \wedge c_1 \wedge c_2) = 0 \vee 0 = 0$ implying $0 \in F$, a contradiction. Thus, $a \in F$ or $b \in F$.

(c) implies (d). Since $a \vee a' = 1 \in F$, it follows that $a \in F$ or $a' \in F$.

(d) implies (a). Let U be a B-ultrafilter containing F and let $a \in U$. Now $a \in F$ or $a' \in F$. If $a' \in F$, then $a' \in U$ and $a \wedge a' = 0 \in U$, which is a contradiction. Thus, $a \in F$. Hence, $U \subseteq F$. So, F is a B-ultrafilter.

■

There is a close relation between filters and Boolean homomorphisms. This relationship is developed in 3G.

3.2 Stone's Representation and Duality Theorems

In this section we establish a one-to-one correspondence between the class (technically, the category) of Boolean algebras and Boolean algebra homomorphisms, and the class of compact zero-dimensional spaces and continuous functions between them. This correspondence is called the **Stone duality theorem**. One aspect of it — namely the fact that every Boolean algebra is isomorphic to $B(X)$ for some suitably chosen compact zero-dimensional space X — is called the **Stone representation theorem.** These results are named after Marshall Stone, whose 1937 paper first introduced them (see the Notes). The Stone duality theorem can be presented in a categorical setting; the categorical version of this theorem (and others) is presented in Chapter 9.

We will assume for the remainder of this book that the elements 0 and 1 of each hypothesized Boolean algebra B are distinct; otherwise, we would have a trivial Boolean algebra $B = \{0\}$ which contains little information.

(a) Definition. Let B be a Boolean algebra. Put $S(B) = \{\mathcal{U} : \mathcal{U}$ is a B-ultrafilter$\}$. For $a \in B$, let $\lambda(a) = \{\mathcal{U} \in S(B) : a \in \mathcal{U}\}$. If more than one Boolean algebra B is involved, we write λ_B for λ. Clearly, $\lambda \in F(B,\mathbb{P}(S(B)))$.

(b) Proposition. Let B be a Boolean algebra and $a, b \in B$. Then:

(1) $\lambda(0) = \emptyset$ and $\lambda(1) = S(B)$,

(2) $\lambda(a \vee b) = \lambda(a) \cup \lambda(b)$,

(3) $\lambda(a \wedge b) = \lambda(a) \cap \lambda(b)$, and

(4) $\lambda(a') = S(B) \setminus \lambda(a)$.

Proof.

(1) If $\mathcal{U}$ is a B-ultrafilter, then $0 \notin \mathcal{U}$ and $1 \in \mathcal{U}$; hence $\lambda(0) = \emptyset$ and $\lambda(1) = S(B)$.

(2) Since a B-ultrafilter $\mathcal{U}$ is prime (see 3.1(r)) it follows that $a \vee b \in \mathcal{U}$ iff $a \in \mathcal{U}$ or $b \in \mathcal{U}$. Thus, (2) follows.

(3) Since a B-ultrafilter $\mathcal{U}$ is a filter, it follows that $a \wedge b \in$

U iff $a \in U$ and $b \in U$. Thus, (3) follows.

(4) Since $\lambda(a) \cap \lambda(a') = \lambda(a \wedge a') = \lambda(0) = \emptyset$ and $\lambda(a) \cup \lambda(a') = \lambda(a \vee a') = \lambda(1) = S(B)$, it follows that $\lambda(a') = S(B) \setminus \lambda(a)$.

■

By 3.2(b), $\{\lambda(a) : a \in B\}$ is a base for the open sets of a topology on S(B). This leads to the following.

(c) Definition. Let B be a Boolean algebra. The set S(B), equipped with the topology for which $\{\lambda(a) : a \in B\}$ is an open base, is called the **Stone space** of B.

Note that since $\lambda(a) = S(B) \setminus \lambda(a')$ for each $a \in B$, $\lambda(a)$ is clopen in S(B). This means that $\{\lambda(a) : a \in B\} \subseteq \mathcal{B}(S(B))$ and $\lambda \in F(B, \mathcal{B}(S(B)))$.

(d) Theorem. (Stone's representation theorem). Let B be a Boolean algebra. Then

(1) S(B) is a compact, zero-dimensional space.

(2) $\{\lambda(a) : a \in B\} = \mathcal{B}(S(B))$ and

(3) λ is a Boolean isomorphism from B onto $\mathcal{B}(S(B))$.

Proof

(1) First, we show S(B) is Hausdorff. Let $U, V \in S(B)$ be such that $U \neq V$. Since $V \not\subseteq U$, there is an element $b \in V \setminus U$. Since $b \notin U$, it follows by 3.1(r) that $b' \in U$. By 3.2(b), $\lambda(b') \cap \lambda(b) = \emptyset$. As $U \in \lambda(b')$ and $V \in \lambda(b)$, it follows that S(B) is Hausdorff. As noted above, since $\lambda(a)$ is clopen for all $a \in$

B, S(B) is zero-dimensional. To show that S(B) is compact, we show that if $\mathcal{F}$ is a closed filter on S(B), then $\bigcap\mathcal{F} \neq \emptyset$. Let $\mathcal{G} = \{a \in B : \lambda(a) \supseteq F$ for some $F \in \mathcal{F}\}$. As $\{\lambda(a) : a \in B\}$ is obviously a closed base for S(B), it follows that $\bigcap\mathcal{F} = \bigcap\{\lambda(a) : a \in \mathcal{G}\}$. If $a_1, a_2 \in \mathcal{G}$, there exist $F_i \in \mathcal{F}$ such that $\lambda(a_i) \supseteq F_i$ $(i = 1,2)$. Thus $\lambda(a_1 \wedge a_2) = \lambda(a_1) \cap \lambda(a_2) \supseteq F_1 \cap F_2 \neq \emptyset$. As $\mathcal{F}$ is a closed filter, $F_1 \cap F_2 \in \mathcal{F}$; so $a_1 \wedge a_2 \in \mathcal{G}$. It quickly follows that $\mathcal{G}$ is a filter on B, and hence by 3.1(r) is contained in some $\mathcal{U} \in$ S(B). Thus $\mathcal{U} \in \lambda(a)$ for each $a \in \mathcal{G}$, so $\mathcal{U} \in \bigcap\{\lambda(a) : a \in \mathcal{G}\} = \bigcap\mathcal{F}$. Thus S(B) is compact.

(2) As noted in the remark preceding this theorem, we know that $\{\lambda(a) : a \in B\} \subseteq \mathcal{B}(S(B))$. Conversely, suppose $C \in \mathcal{B}(S(B))$. Since C is open and $\{\lambda(a) : a \in B\}$ is a base, there exists $D \subseteq B$ such that $C = \bigcup\{\lambda(a) : a \in D\}$. Since C is closed and thus compact, there is a finite set $F \subseteq D$ such that $C = \bigcup\{\lambda(a) : a \in F\}$. But by 3.2(b) and a finite induction, $\bigcup\{\lambda(a) : a \in F\} = \lambda(\vee\{a : a \in F\})$. So, $C \in \{\lambda(b) : b \in B\}$ and $\mathcal{B}(S(B)) \subseteq \{\lambda(b) : b \in B\}$.

(3) By 3.2(b), λ is a Boolean homomorphism and by (2), λ is onto. Let $a, b \in B$ such that $a \neq b$. Then $a \nleq b$ or $b \nleq a$. Suppose $a \nleq b$; then by 3.1(g)(5), $a\backslash b \neq 0$. Now $\mathcal{F} = \{c \in B : c \geqslant a\backslash b\}$ is a B-filter and is contained in some B-ultrafilter $\mathcal{U}$. So, $a\backslash b \in \mathcal{U}$ which implies that $\mathcal{U} \in \lambda(a\backslash b)$. Since $a \geqslant a\backslash b = a \wedge b'$ and $b' \geqslant a\backslash b$, it follows that $a, b' \in \mathcal{U}$. So, $\mathcal{U} \in \lambda(a)$ and $\mathcal{U} \notin \lambda(b)$. Hence, $\lambda(a) \neq \lambda(b)$. This shows λ is a bijection. By 3.1(n)(2), λ is a Boolean isomorphism.

■

Thus, by Stone's representation theorem, a Boolean algebra B has a topological representation as the Boolean algebra of clopen sets of a compact zero-dimensional space, namely S(B), the **Stone space** of B.

(e) **Definition**. Let $f : A \to B$ be a Boolean homomorphism between two Boolean algebras A and B. Define $\lambda(f) \in F(S(B),\mathbb{P}(A))$ by $\lambda(f)(\mathcal{U}) = \{a \in A : f(a) \in \mathcal{U}\}$.

(f) **Proposition**. If $f : A \to B$ is a Boolean homomorphism where A and B are Boolean algebras, then:

(1) $\lambda(f) \in F(S(B),S(A))$,

(2) for $a \in A$, $\lambda(f)^{\leftarrow}[\lambda_A(a)] = \lambda_B(f(a))$ and $\lambda(f)[\lambda_B(f(a))] = \lambda_A(a) \cap \lambda(f)[S(B)]$,

(3) $\lambda(f)$ is continuous and closed,

(4) f is one-to-one iff $\lambda(f)$ is onto S(A),

(5) f is onto iff $\lambda(f)$ is one-to-one, and

(6) f is a Boolean isomorphism iff $\lambda(f)$ is a homeomorphism.

Proof

(1) For $\mathcal{U} \in S(B)$, we need to show $\lambda(f)(\mathcal{U}) = \{a \in A : f(a) \in \mathcal{U}\}$ is an A-ultrafilter. Since $f(1) = 1 \in \mathcal{U}$, it follows that $1 \in \lambda(f)(\mathcal{U})$ and $\lambda(f)(\mathcal{U}) \neq \varnothing$. If $f(a) \in \mathcal{U}$, then $f(a) \neq 0$, which implies that $a \neq 0$. So, $0 \notin \lambda(f)(\mathcal{U})$. If $a, b \in \lambda(f)(\mathcal{U})$, then $f(a), f(b) \in \mathcal{U}$ and, hence, $f(a \wedge b) = f(a) \wedge f(b) \in \mathcal{U}$. This shows $a \wedge b \in \lambda(f)(\mathcal{U})$. Also, if $a \in \lambda(f)(\mathcal{U})$ and $b \geq a$, then $f(b) \geq f(a)$ and $f(a) \in \mathcal{U}$. So, $f(b) \in \mathcal{U}$, which implies $b \in \lambda(f)(\mathcal{U})$. Thus $\lambda(f)(\mathcal{U})$ is an A-filter. If $a \vee b \in \lambda(f)(\mathcal{U})$, then $f(a \vee b) = f(a) \vee f(b) \in \mathcal{U}$. Since $\mathcal{U}$ is prime, either $f(a) \in \mathcal{U}$ or $f(b) \in \mathcal{U}$ which implies that a

$\in \lambda(f)(U)$ or $b \in \lambda(f)(U)$. Thus, $\lambda(f)(U)$ is a prime A-filter and by 3.1(r), $\lambda(f)(U) \in S(A)$.

(2) If $U \in \lambda_B(f(a))$, then $f(a) \in U$ which implies that $a \in \lambda(f)(U)$. So, $\lambda(f)(U) \in \lambda_A(a)$ and $U \in (\lambda(f))^{\leftarrow}[\lambda_A(a)]$. This shows $\lambda_B(f(a)) \subseteq (\lambda(f))^{\leftarrow}[\lambda_A(a)]$. Conversely, suppose $U \in (\lambda(f))^{\leftarrow}[\lambda_A(a)]$. Then $\lambda(f)(U) \in \lambda_A(a)$ which implies that $a \in \lambda(f)(U)$ and $f(a) \in U$. So, $U \in \lambda_B(f(a))$. This completes the proof that $\lambda_B(f(a)) = (\lambda(f))^{\leftarrow}[\lambda_A(a)]$ and implies that $\lambda(f)[\lambda_B(f(a))] \subseteq \lambda_A(a) \cap \lambda(f)[S(B)]$. If $V \in \lambda_A(a) \cap \lambda(f)[S(B)]$, then $V = \lambda(f)(U)$ for some $U \in S(B)$. Since $a \in V = \lambda(f)(U)$, then $f(a) \in U$. Thus, $U \in \lambda_B(f(a))$ and $V = \lambda(f)(U) \in \lambda(f)[\lambda_B(f(a))]$. So, $\lambda_A(a) \cap \lambda(f)[S(B)] \subseteq \lambda(f)[\lambda_B(f(a))]$.

(3) To show $\lambda(f)$ is continuous, it suffices to show $\lambda(f)^{\leftarrow}[C]$ is open for each $C \in B(S(A))$. By 3.2(d)(2), it suffices to show $\lambda(f)^{\leftarrow}[\lambda_A(a)]$ is open for each $a \in A$. But by (2), $\lambda(f)^{\leftarrow}[\lambda_A(a)] = \lambda_B(f(a)) \in B(S(B))$. So, $\lambda(f)$ is continuous. Since $S(B)$ is compact and $S(A)$ is Hausdorff, $\lambda(f)$ is also a closed function.

(4) Suppose f is one-to-one. Let $U \in S(A)$ and define V to be $\{b \in B : b \geqslant f(a)$ for some $a \in U\}$. If $a \in U$, then $a \neq 0$; so, $f(a) \neq 0$ as f is one-to-one. So, $0 \notin V$. Also, if $b_1 \geqslant f(a_1)$ and $b_2 \geqslant f(a_2)$ where $a_1, a_2 \in U$, then $b_1 \wedge b_2 \geqslant f(a_1) \wedge f(a_2) = f(a_1 \wedge a_2)$ and $a_1 \wedge a_2 \in U$. Thus, $b_1 \wedge b_2 \in V$. If $b \in V$ and $d \geqslant b \geqslant f(a)$ for some $a \in U$, then $d \geqslant f(a)$ which implies that $d \in V$. This shows that V is a B-filter. Now, V is contained in some B-ultrafilter W. Since $\{f(a) : a \in U\} \subseteq W$, it follows that $\lambda(f)(W) \supseteq U$. Since U is maximal, then $\lambda(f)(W) = U$. Thus, $\lambda(f)$ is onto. Conversely, suppose $\lambda(f)$ is onto and $a, b \in A$ such that $a \neq b$. Since λ_A is one-to-one, $\lambda_A(a) \neq \lambda_A(b)$. Thus, $\lambda_A(a) \setminus \lambda_A(b) \neq$

$\emptyset$ or $\lambda_A(b)\setminus\lambda_A(a) \neq \emptyset$. If $U \in \lambda_A(a)\setminus\lambda_A(b)$, then $a \in U$ and $b \notin U$. Hence, $a \in U$ and $b' \in U$. Since $\lambda(f)$ is onto, there is some $V \in S(B)$ such that $\lambda(f)(V) = U$. Thus, $f(a) \in V$ and $f(b') \in V$. Since V is a B-filter, $0 \neq f(a) \wedge f(b') = f(a) \wedge f(b)'$. This shows $f(a) \neq f(b)$. A similar proof shows that if $\lambda_A(b)\setminus\lambda_A(a) \neq \emptyset$, then $f(a) \neq f(b)$. Hence, f is one-to-one.

(5) Suppose f is onto. Let $U, V \in S(B)$ such that $U \neq V$. Let $b \in U\setminus V$. For some $a \in A$, $f(a) = b$. Thus, $a \in \lambda(f)(U)$ and $a \notin \lambda(f)(V)$ implying $\lambda(f)(U) \neq \lambda(f)(V)$. Conversely, suppose $\lambda(f)$ is one-to-one and $b \in B$. Thus, by (3), $\lambda(f) \in F(S(B),S(A))$ is an embedding, i.e., $\lambda(f) \in F(S(B),\lambda(f)[S(B)])$ is a homeomorphism. So, $\lambda(f)[\lambda_B(b)] = U \cap \lambda(f)[S(B)]$ for some open set U in S(A). For some $C \subseteq A$, $U = \bigvee\{\lambda_A(a) : a \in C\}$. Since $U \supseteq \lambda(f)[\lambda_B(b)]$ and $\lambda(f)[\lambda_B(b)]$ is closed in S(B) and hence compact, there is a finite set $F \subseteq C$ such that $\lambda(f)[\lambda_B(b)] \subseteq \bigvee\{\lambda_A(a) : a \in F\} = \lambda_A(\bigvee F) \subseteq U$. Let $a_1 = \bigvee F$. Then $\lambda(f)[\lambda_B(b)] = \lambda_A(a_1) \cap \lambda(f)[S(B)]$. By (2), $\lambda(f)[\lambda_B(b)] = \lambda(f)[\lambda_B(f(a_1))]$. Since $\lambda(f)$ is one-to-one, $\lambda_B(b) = \lambda_B(f(a_1))$. Hence $f(a_1) = b$ which implies that f is onto.

(6) If f is a Boolean isomorphism, then by (3), (4) and (5), $\lambda(f)$ is a homeomorphism. If $\lambda(f)$ is a homeomorphism, then by (4) and (5), f is a bijective, Boolean homomorphism. By 3.1(n)(2), f is a Boolean isomorphism.

■

We will now show that the dual to 3.2(f) is true. In other words, if we are given compact zero-dimensional spaces S and T, and if $f \in C(S,T)$, we show that there is a "natural" Boolean homomorphism $\nu f : \mathcal{B}(T) \to \mathcal{B}(S)$ that is defined in terms of f and whose properties reflect those of f.

(g) **Definition**. Let X be a compact zero-dimensional space and let $x \in X$. Then $\nu \in F(X,\mathbb{P}(B(X)))$ is defined as follows: $\nu(x) = \{C \in B(X) : x \in C\}$. If more than one space X is under discussion, we write ν_X for ν.

(h) **Proposition**. For a compact, zero-dimensional space X, $\nu \in F(X,S(B(X)))$ and ν is a homeomorphism.

Proof. First we show that if $x \in X$, then $\nu(x)$ is a $B(X)$-ultrafilter. It is easily verified that $\nu(x)$ is a $B(X)$-filter. Suppose $C_1 \cup C_2 \in \nu(x)$ where $C_1, C_2 \in B(X)$. Then $x \in C_1$ or $x \in C_2$. Hence, $C_1 \in \nu(x)$ or $C_2 \in \nu(x)$. Thus, by 3.1(r)(3), $\nu(x)$ is a $B(X)$-ultrafilter. Next, we show ν is continuous. Since $\{\lambda(C) : C \in B(X)\}$ is a base for $S(B(X))$, it suffices to show $\nu^{\leftarrow}[\lambda(C)]$ is open for $C \in B(X)$. Now, $\nu^{\leftarrow}[\lambda(C)] = \{x \in X : \nu(x) \in \lambda(C)\} = \{x \in X : C \in \nu(x)\} = \{x \in X : x \in C\} = C$. So, ν is continuous. If $x, y \in X$ and $x \neq y$ then as X is zero-dimensional, there is a clopen set $C \in B(X)$ such that $x \in C$ and $y \in C' = X \backslash C$. So, $\nu(x) \in \lambda(C)$ and $\nu(y) \in \lambda(C')$. Since $\lambda(C) \cap \lambda(C') = \lambda(C \cap C') = \lambda(\emptyset) = \emptyset$, it follows that $\nu(x) \neq \nu(y)$. Thus, ν is one-to-one. To show ν is onto, let $\mathcal{U} \in S(B(X))$. Then, $\mathcal{U}$ is a $B(X)$-ultrafilter. Since X is compact, there is some point $x \in \bigcap\mathcal{U}$. This implies $\mathcal{U} \subseteq \nu(x)$. Since $\mathcal{U}$ is maximal, then $\mathcal{U} = \nu(x)$. This completes the proof that ν is a continuous bijection. Since X is compact and $S(B(X))$ is Hausdorff, it follows that ν is closed. So, ν is a homeomorphism.

■

(i) **Definition**. Let X and Y be compact, zero-dimensional spaces and let $f \in C(X,Y)$. Define $\nu(f) \in F(B(Y),B(X))$ by $\nu(f)(C) = f^{\leftarrow}[C]$.

(j) **Proposition**. Let $f \in C(X,Y)$ where X and Y are compact, zero-dimensional spaces. Then:

(1) $\nu(f) \in F(B(Y),B(X))$ is a Boolean homomorphism,

(2) f is one-to-one iff $\nu(f)$ is onto,

(3) f is onto iff $\nu(f)$ is one-to-one, and

(4) f is a homeomorphism iff $\nu(f)$ is a Boolean isomorphism.

Proof. The proofs of (1), (3), and (4) are left to the reader (see 3H).

(2) Suppose f is one-to-one. To show $\nu(f)$ is onto, let $C \in B(X)$. Since f is one-to-one, continuous, and closed, it follows that f is an embedding of X into Y. So, $f[C]$ is open in $f[X]$ and $f[C] = U \cap f[X]$ for some open set U in Y. Now, $U = \bigcup A$ for some $A \subseteq B(Y)$. Since $f[C]$ is compact, there is a finite set $F \subseteq A$ such that $f[C] \subseteq \bigcup F$. So, $f[C] = (\bigcup F) \cap f[X]$. Now, $\bigcup F \in B(Y)$ and $f^{\leftarrow}[\bigcup F] = C$ as f is one-to-one. So, $\nu(f)(\bigcup F) = C$. Conversely, suppose $\nu(f)$ is onto and let x and y be distinct points of X. There is a clopen set $C \in B(X)$ such that $x \in C$ and $y \in X \setminus C$. Since $\nu(f)$ is onto, there is a clopen set $D \in B(Y)$ such that $\nu(f)(D) = C$. So, $f^{\leftarrow}[D] = C$ and $f(x) \in D$. Also, $f(y) \in D' = Y \setminus D$. Thus, $f(x) \neq f(y)$.

■

If A is a Boolean algebra, then by 3.2(d)(3), λ_A is a Boolean isomorphism from A onto $B(S(A))$ and if X is a compact, zero-dimensional space, then by 3.2(h), ν_X is a homeomorphism between X and $S(B(X))$. This duality between Boolean algebras and compact, zero-dimensional spaces is shown, in the next result, to preserve continuous functions and Boolean homomorphisms. The results 3.2(f) and 3.2(j) together with the next result, are collectively called the Stone Duality Theorem.

(k) **Proposition.**

(1) If $f \in C(X,Y)$ where X and Y are compact, zero-dimensional spaces, then $\lambda(\nu(f)) \circ \nu_X = \nu_Y \circ f$ (see the accompanying diagram).

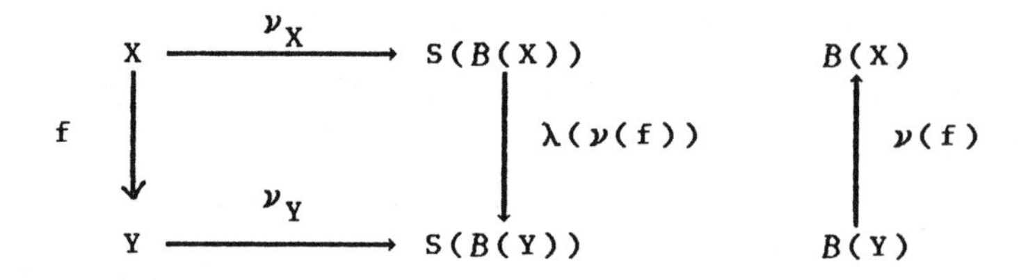

(2) If $f \in F(A,B)$ is a Boolean homomorphism between Boolean algebras A and B, then $\nu(\lambda(f)) \circ \lambda_A = \lambda_B \circ f$ (see the accompanying diagram).

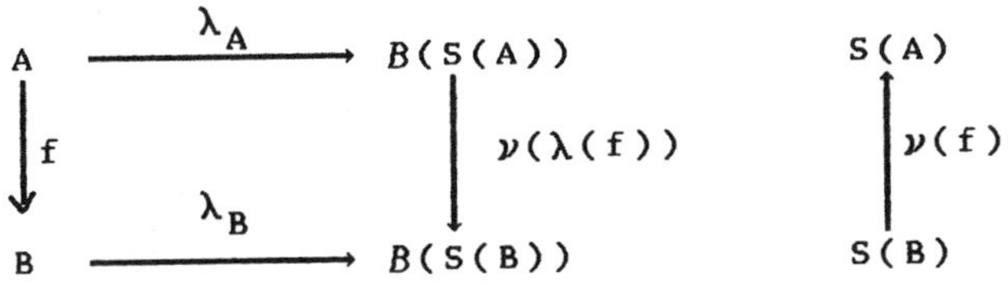

$$\begin{array}{ccc} A & \xrightarrow{\lambda_A} & \mathcal{B}(S(A)) \\ \downarrow f & & \downarrow \nu(\lambda(f)) \\ B & \xrightarrow{\lambda_B} & \mathcal{B}(S(B)) \end{array} \qquad \begin{array}{c} S(A) \\ \uparrow \nu(f) \\ S(B) \end{array}$$

<u>Proof</u>

(1) Let $x \in X$. Then $(\lambda(\nu(f)) \circ \nu_X)(x) = \lambda(\nu(f))(\mathcal{U})$ where $\mathcal{U} = \{C \in \mathcal{B}(X) : x \in C\}$. But $\lambda(\nu(f))(\mathcal{U}) = \{D \in \mathcal{B}(Y) : \nu(f)(D) \in \mathcal{U}\} = \{D \in \mathcal{B}(Y) : f^{\leftarrow}[D] \in \mathcal{U}\} = \{D \in \mathcal{B}(Y) : x \in f^{\leftarrow}[D]\} = \{D \in \mathcal{B}(Y) : f(x) \in D\} = \nu_Y(f(x)) = (\nu_Y \circ f)(x)$.

(2) Let $a \in A$. $(\nu(\lambda(f)) \circ \lambda_A)(a) = \nu(\lambda(f))(\lambda_A(a)) = (\lambda(f))^{\leftarrow}[\lambda_A(a)] = \{\mathcal{U} : \lambda(f)(\mathcal{U}) \in \lambda_A(a)\} = \{\mathcal{U} : a \in \lambda(f)(\mathcal{U})\} = \{\mathcal{U} : f(a) \in \mathcal{U}\} = \lambda_B(f(a)) = (\lambda_B \circ f)(a)$.

■

We now use Stone's Duality Theorem to convert a Boolean algebra concept into a topological concept.

(l) <u>**Definition**</u>. An element a in a Boolean algebra B is an **atom** if $0 < a$ and if $0 < b \leq a$ implies $b = a$. B is called **atomic** if for $0 < b \in B$, there is an atom $a \in B$ such that $a \leq b$. B is called **atomless** if B has no atoms.

(m) <u>**Proposition**</u>. Let B be a Boolean algebra. Then:

(1) $\{\lambda(a) : a \text{ is an atom in } B\} = \{\{p\} : p \text{ is isolated in } S(B)\}$,

(2) B is atomic iff S(B) has a dense set of isolated points, and

(3) B is atomless iff S(B) has no isolated points.

Proof. The proofs of (2) and (3) easily follow from (1).

(1) Suppose a is an atom. Then since $0 < a$, $\lambda(a) \neq \emptyset$. Let $p \in \lambda(a)$ and $q \in S(B)\setminus\{p\}$. There is a clopen set $C \in \mathcal{B}(S(B))$ such that $p \in C$ and $q \in S(B)\setminus C$. There is some $b > 0$ such that $\lambda(b) = C \cap \lambda(a)$. Since $\lambda(b) \subseteq \lambda(a)$, it follows that $b \leqslant a$. But a is an atom which implies that $b = a$. In particular, $q \notin \lambda(a) \cap C = \lambda(a)$. So, $\lambda(a) = \{p\}$. Conversely, suppose p is isolated in $S(B)$. Then $\{p\} \in \mathcal{B}(S(B)) = \{\lambda(b) : b \in B\}$ and $\lambda(a) = \{p\}$ for some $a \in B$. Then $0 < a$ and if $0 < b \leqslant a$, then $\emptyset \neq \lambda(b) \subseteq \lambda(a)$. This implies $\lambda(b) = \{p\} = \lambda(a)$. Thus, $b = a$ and a is an atom.

■

3.3 Atomless, countable Boolean algebras

In this section, an application of Stone's Duality Theorem is given by showing that all atomless, countable Boolean algebras are Boolean isomorphic. To accomplish this, we first prove that two second countable, compact, zero-dimensional spaces without isolated points are homeomorphic.

We will assume that the reader is acquainted with the topological result that a compact space is second countable (i.e., has a countable open base) iff it is metrizable.

(a) **Definition**. Let $\mathcal{C}$ and $\mathcal{D}$ be covers of a space. For each $C \in \mathcal{C}$, we denote $\{D \in \mathcal{D} : D \subseteq C\}$ by $\mathcal{D}(C)$.

(b) **Lemma**. An open cover $\mathcal{C}$ of a compact, zero-dimensional space X has a refinement $\mathcal{D}$ which is a finite partition of nonempty clopen sets.

Proof. Since X is zero-dimensional, $\mathcal{C}$ can be refined by a cover $\mathcal{E}$ of nonempty clopen sets. Since X is compact, $\mathcal{E}$ has a finite subcover, say $\mathcal{F}$. Now, $\mathcal{F}$ refines $\mathcal{C}$ and $\mathcal{F}$ is a finite cover of nonempty clopen sets but may not be a partition. If $\mathcal{F} = \{F_1, F_2, \ldots, F_n\}$ for some $n \in \mathbb{N}$, let $D_1 = F_1$ and $D_k = F_k \setminus \cup\{F_j : 1 \leqslant j \leqslant k-1\}$ for $2 \leqslant k \leqslant n$. Let $\mathcal{D} = \{D_\ell : 1 \leqslant \ell \leqslant n\} \setminus \{\emptyset\}$; then $\cup\mathcal{D} = \cup\mathcal{F} = X$ and $\mathcal{D}$ is a refinement of $\mathcal{F}$ and, hence, of $\mathcal{C}$. The elements of $\mathcal{D}$ are nonempty clopen sets and are pairwise disjoint. Thus, $\mathcal{D}$ is a finite partition of nonempty clopen sets.

■

It is useful to note that if $\mathcal{D}$ is a partition of X, then for each $x \in X$, there is a unique $D \in \mathcal{D}$ such that $x \in D$.

(c) **Lemma**. If C is a nonempty clopen set in a zero-dimensional space X without isolated points, then for each $n \in \mathbb{N}$, C can be partitioned into n nonempty clopen sets $D_1, \ldots, D_n$.

Proof. Proceed by induction. If $n = 1$, let $D_1 = C$ and we are done. Suppose C can be partitioned into nonempty clopen sets $D_1, \ldots, D_k$. Since D_k is nonempty and clopen and since X contains no isolated points, D_k is infinite. Choose distinct points x and y of D_k. Since X is zero-dimensional, there is a clopen set E in X such that $x \in E$ and $y \in X \setminus E$. So, $D_k = (D_k \cap E) \cup (D_k \setminus E)$. Thus, D_k can be partitioned into two nonempty clopen sets $D_k \cap E$ and $D_k \setminus E$. So, C can be partitioned into k + 1 nonempty clopen sets $D_1, \ldots, D_{k-1}$, $D_k \cap E$, $D_k \setminus E$. This completes the proof.

■

Finally, we are ready to prove the main lemma.

(d) **Lemma**. If X is a zero-dimensional, compact, metric space without isolated points, then there is a sequence $\{\mathcal{C}_n : n \in \mathbb{N}\}$ of partitions of nonempty clopen sets of X and a subsequence $n_1 < n_2 < n_3 < \ldots$ satisfying:

(1) $\mathcal{C}_n > \mathcal{C}_{n+1}$ for $n \in \mathbb{N}$,

(2) $|\mathcal{C}_n| = 2^n$,

(3) if $C \in \mathcal{C}_n$, then $|\mathcal{C}_{n+1}(C)| = 2$ (recall $\mathcal{C}_{n+1}(C)$ is defined in 3.3(a)), and

(4) for $C \in \mathcal{C}_{n_i}$, diam $C \leqslant 1/2^i$ (where diam C, the diameter of C, is defined to be $\sup\{d(x,y) : x, y \in C\}$ and d denotes the metric on X).

<u>Proof</u> **(Step 1).** Let $\mathcal{C}$ be an open cover of X by sets C such that diam(C) $\leqslant 1/2$ (such a $\mathcal{C}$ exists since X is compact). By 3.3(b), $\mathcal{C}$ has a refinement $\mathcal{C}'$ which is a finite partition of nonempty clopen sets. Let n_1 be the least integer $k \in \mathbb{N}$ such that $|\mathcal{C}'| \leqslant 2^k$. By using 3.3(c) if necessary, we can find a finite partition $\mathcal{C}_{n_1}$ of nonempty clopen sets such that $|\mathcal{C}_{n_1}| = 2^{n_1}$ and $\mathcal{C}_{n_1}$ refines $\mathcal{C}'$. If $n_1 > 1$, then as there are an even number of elements of

$\mathcal{C}_{n_1}$, there is an onto function $f \in F(\mathcal{C}_{n_1},K)$ where $K = \{n \in \mathbb{N} : n \leqslant 2^{n_1-1}\}$ and $|f^{\leftarrow}(n)| = 2$ for all $n \in K$. Define $\mathcal{C}_{n_1-1} = \{\{A \cup B : A, B \in f^{\leftarrow}(n)\} : n \in K\}$. Thus, $\mathcal{C}_{n_1}$ is a refinement of $\mathcal{C}_{n_1-1}$, $|\mathcal{C}_{n_1-1}| = 2^{n_1-1}$, and $\mathcal{C}_{n_1-1}$ is a partition of X consisting of nonempty clopen sets. If $n_1 - 1 > 1$, then continue by induction to obtain the initial sequence $\{\mathcal{C}_1,\mathcal{C}_2,\ldots,\mathcal{C}_{n_1}\}$ of partitions of nonempty clopen sets satisfying (1) through (4) for the initial subsequence n_1. (Thus we constructed $\mathcal{C}$, then $\mathcal{C}_{n_1}$, then $\mathcal{C}_{n_1-1}$, $\mathcal{C}_{n_1-2}$, ..., $\mathcal{C}_1$, in that order).

Step 2. Let $\mathcal{C}$ be an open refinement of $\mathcal{C}_{n_1}$ such that diam $C \leqslant 1/4$ for $C \in \mathcal{C}$. By 3.3(a), $\mathcal{C}$ has a refinement $\mathcal{C}'$ which is a finite partition of nonempty clopen sets. Let ℓ be the least integer $k \in \mathbb{N}$ such that $k > n_1$ and $|\mathcal{C}'| \leqslant 2^k$. Also, $\mathcal{C}'$ refines $\mathcal{C}_{n_1}$. For each $C \in \mathcal{C}_{n_1}$, $|\mathcal{C}'(C)| \leqslant 2^{\ell}$. By using 3.3(c), we can find a partition $\mathcal{C}''$ of nonempty clopen sets such that $\mathcal{C}''$ refines $\mathcal{C}'$ and for $C \in \mathcal{C}_{n_1}$, $|\mathcal{C}_{n_1}(C)| = 2^{\ell}$. Hence, $|\mathcal{C}''| = 2^{n_2}$ where $n_2 = n_1 + \ell$. Since $n_2 > n_1$, there is an onto function $g \in F(\mathcal{C}_{n_2},L)$ where $L = \{n \in \mathbb{N} : n \leqslant 2^{n_2-1}\}$ and for $n \in L$, $|g^{\leftarrow}(n)| = 2$ and $g^{\leftarrow}(n) \subseteq \mathcal{C}_{n_2}(C)$ for some $C \in \mathcal{C}_{n_2}$. Define $\mathcal{C}_{n_2-1} = \{\bigcup g(n) : n \in L\}$. Now, $\mathcal{C}_{n_1} > \mathcal{C}_{n_2-1} > \mathcal{C}_{n_2}$, $|\mathcal{C}_{n_2-1}| = 2^{n_2-1}$, $|\mathcal{C}_{n_2}(C)| = 2$ for $C \in \mathcal{C}_{n_2-1}$, and $\mathcal{C}_{n_2-1}$ is a partition of nonempty clopen sets. Continue by induction to obtain $\mathcal{C}_{n_1+1} > \mathcal{C}_{n_1+2} > \ldots > \mathcal{C}_{n_2}$. Now, the initial sequence $\{\mathcal{C}_1,\ldots,\mathcal{C}_{n_2}\}$ of partitions satisfies (1) through (4) for the initial subsequence $n_1 < n_2$.

Step 3. The final step is to continue by induction by repeating Step 2. In doing so, we obtain a subsequence $n_1 < n_2 < \ldots$ and partition $\mathcal{C}_1 > \mathcal{C}_2 > \ldots$ satisfying (1) through (4).

■

(e) **Theorem**. Two second countable, compact, zero-dimensional spaces without isolated points are homeomorphic.

Proof. Let X be a second countable, compact, zero-dimensional space without isolated points. Note that X is metrizable; we fix a metric on X that induces the topology of X. We will show X is homeomorphic to $2^{\mathbb{N}}$ (denoted by Y in this proof). Let $\{\mathcal{C}_n : n \in \mathbb{N}\}$ be the sequence of partitions of nonempty clopen sets constructed in 3.3(d) and $n_1 < n_2 < \ldots$ the associated subsequence. Let $\mathcal{C}_0 = \{X\}$. For $n \geq 1$ and $C \in \mathcal{C}_{n-1}$, we have that $|\mathcal{C}_n(C)| = 2$; we index $\mathcal{C}_n(C)$ by $\{D_0^{\,C}, D_1^{\,C}\}$. Let $p \in X$ and for $n \geq 0$, let C_n be the unique element of $\mathcal{C}_n$ such that $p \in C_n$. Since $C_n = D_0^{\,C_n} \cup D_1^{\,C_n}$, it follows that $C_{n+1} = D_0^{\,C_n}$ or $C_{n+1} = D_1^{\,C_n}$. Define $f_p \in Y$ by $f_p(n) = i$ if $p \in D_i^{\,C_{n-1}}$ for $n \geq 1$. Now, define $e \in F(X,Y)$ by $e(p) = f_p$. We prove, in a series of steps, that e is a homeomorphism.

Claim: **e is onto.** Suppose $f \in Y$. First, we inductively define sets $M_f(n) \in \mathcal{C}_n$ for $n \geq 1$ where $M_f(1) \supseteq M_f(2) \ldots$. Let $M_f(1) = D_{f(1)}{}^X$ and $M_f(n+1) = D_{f(n+1)}{}^{M_f(n)}$. Since $\{M_f(n) : n \in \mathbb{N}\}$ is a decreasing sequence of closed sets in a compact space X, there is a point $p \in \bigcap\{M_f(n) : n \in \mathbb{N}\}$. By our definition of f_p, it is clear that $e(p) = f_p = f$.

Claim: **e is one-to-one**. If $p, q \in X$, $p \neq q$, let $\mathrm{diam}(p,q) = t > 0$. There is some $k \in \mathbb{N}$ such that $1/2^k < t$. There exists $C \in \mathcal{C}_{n_k}$ such that $p \in C$. Since $\mathrm{diam}(C) \leq 1/2^k$, it follows that $q \notin C$. Let ℓ be the least integer in $\mathbb{N} \cup \{0\}$ such that p and q are in distinct sets of $\mathcal{C}_i$; so, $\ell \geq 1$, for some $C \in \mathcal{C}_{\ell-1}$, $\{p,q\} \subseteq C$. Since $C = D_0{}^C \cup D_1{}^C$, $p \in D_i{}^C$ and $q \in D_{1-i}{}^C$ where $i \in \{0,1\}$. Thus, $f_p(\ell) = i \neq 1 - i = f_q(\ell)$. So, $e(p) \neq e(q)$.

Claim: **e is continuous.** For $n \in \mathbb{N}$, let $\Pi_n \in F(Y,\mathbf{2})$ be the nth-projection function. To show e is continuous, it suffices to show $\Pi_n \circ e$ is continuous for $n \in \mathbb{N}$. Now, $(\Pi_n \circ e)^{\leftarrow}(i) = \bigcup\{D_i{}^C : C \in \mathcal{C}_{n-1}\}$, and thus is open for $i = 0$ or $i = 1$. So, e is continuous. Since X is compact and Y is Hausdorff, e is closed. Thus, e is a homeomorphism between X and $\mathbf{2}^{\mathbb{N}}$.

■

Of course, the space $Y = \mathbf{2}^{\mathbb{N}}$ used in the proof of 3.3(e) will be recognized by the reader as the Cantor space. Since the Cantor space is a second countable, compact, zero-dimensional space without isolated points, the path of the proof of 3.3(e) is most reasonable. The next result is an immediate corollary to the proof of 3.3(e).

(f) **Corollary**. A second countable, compact, zero-dimensional space without isolated points is homeomorphic to the Cantor space.

Now we convert this result into the language of Boolean algebras by using Stone's Duality Theorem.

(g) **Theorem**. A countable, atomless Boolean algebra is Boolean isomorphic to $\mathcal{B}(\mathbf{2}^{\mathbb{N}})$.

Proof. Let B be a countable, atomless Boolean algebra. By 3.2(d) and (m), S(B) is a compact, zero-dimensional space without isolated points. Since B is Boolean isomorphic to $\mathcal{B}$(S(B)) by 3.2(d), S(B) is second countable. By 3.3(f), S(B) is homeomorphic to $\mathbf{2}^{\mathbb{N}}$. By 3.2(j), $\mathcal{B}$(S(B)) is Boolean isomorphic to $\mathcal{B}(\mathbf{2}^{\mathbb{N}})$. Hence, B is Boolean isomorphic to $\mathcal{B}(\mathbf{2}^{\mathbb{N}})$.

■

3.4 Completions of Boolean algebras

Recall (see 3.1(o)) that a Boolean algebra B is complete if B is complete as a lattice. In this section, we show that every Boolean algebra is a subalgebra of many complete Boolean algebras and that among these complete Boolean algebras, there is a "smallest one."

(a) **Definition**. A subset A of a Boolean algebra B is a

complete subalgebra of B if A is a subalgebra of B and for each nonempty subset S of A, if $\bigvee S$ (respectively, $\bigwedge S$) exists in B, then $\bigvee S \in A$ (respectively, $\bigwedge S \in A$). Note that B is a complete subalgebra of B and that a "complete subalgebra" need not be a complete Boolean algebra.

(b) **Proposition**. Let $\mathcal{A}$ be a nonempty family of complete subalgebras of a Boolean algebra B. Then $\bigcap\mathcal{A}$ is a complete subalgebra of B.

Proof. By 3.1(i), $\bigcap\mathcal{A}$ is a subalgebra of B. Suppose $\emptyset \neq S \subseteq \bigcap\mathcal{A}$ and $\bigvee S$ (respectively, $\bigwedge S$) exists in B. Then for each $A \in \mathcal{A}$, $\bigvee S \in A$ (respectively, $\bigwedge S \in A$). Thus $\bigvee S \in \bigcap\mathcal{A}$ (respectively, $\bigwedge S \in \bigcap\mathcal{A}$).

■

Let S be a nonempty subset of a Boolean algebra B and let $\mathcal{A} = \{A \subseteq B : S \subseteq A$ and A is a complete subalgebra of B$\}$. Note that $\mathcal{A} \neq \emptyset$ as $B \in \mathcal{A}$. By 3.4(b), $\bigcap\mathcal{A}$ is a complete subalgebra and, hence, is the smallest complete subalgebra of B containing S. We denote $\bigcap\mathcal{A}$ by $\langle\langle S\rangle\rangle$.

We will now introduce the concept of a "Boolean embedding" which is similar to the concept of a "topological embedding". First, a result is needed.

(c) **Proposition**. Let $f \in F(A,B)$ be a Boolean homomorphism between Boolean algebras A and B. Then:

(1) f[A] is a subalgebra of B, and

(2) if f is one-to-one, then f is a Boolean isomorphism from A onto f[A].

Proof.

(1) Let a, b $\in$ A. Since f(a) $\vee$ f(b) = f(a $\vee$ b), f(a) $\wedge$ f(b) = f(a $\wedge$ b), f(a)$'$ = f(a$'$), f(0) = 0 and f(1) = 1, it follows that f[A] is a subalgebra of B.

(2) Since f $\in$ F(A,f[A]) is a bijection and f is a Boolean homomorphism, then by 3.1(n)(2), f is a Boolean isomorphism from A onto f[A].

■

A one-to-one Boolean homomorphism f $\in$ F(A,B) between Boolean algebras A and B is called a **Boolean monomorphism**. By 3.4(c), a Boolean monomorphism g $\in$ F(A,B) is a Boolean isomorphism from A onto f[A]. Thus, a Boolean monomorphism is similar to the concept of "topological embedding".

(d) Definition

(1) Let A, B be Boolean algebras. A pair (B,i) is a **completion** of a Boolean algebra A if i $\in$ F(A,B) is a Boolean monomorphism, B is a complete Boolean algebra, and <<i[A]>> = B. If (B,i) is a completion of A, frequently we will identify A with i[A] and assume that A is a subalgebra of B and i is the inclusion function.

(2) A Boolean homomorphism f $\in$ F(A,B) between Boolean algebras is **complete** if, whenever $\emptyset \neq S \subseteq A$ and $\bigvee S$ (respectively, $\bigwedge S$) exists in A, then $f(\bigvee S) = \bigvee\{f(a) : a \in S\}$ (respectively, $f(\bigwedge S) = \bigwedge\{f(a) : a \in S\}$).

(3) A subalgebra A of B is said to be **regular** if $i \in F(A,B)$ is complete where i is the inclusion function.

(4) A subset S of B is said to be **dense** in B if for $0 < b \in B$, there is some $a \in S$ such that $0 < a \leqslant b$.

(e) **Proposition**. Let X be a zero-dimensional space and let $i \in F(\mathcal{B}(X), \mathcal{RO}(X))$ be the inclusion function. Then:

(1) $(\mathcal{RO}(X), i)$ is a completion of $\mathcal{B}(X)$,

(2) $\mathcal{B}(X)$ is dense and regular in $\mathcal{RO}(X)$, and

(3) for each completion (A,j) of $\mathcal{B}(X)$, there is a Boolean monomorphism $h \in F(\mathcal{RO}(X), A)$ such that $h \circ i = j$.

Proof

(1) First note that i is one-to-one and if $B, C \in \mathcal{B}(X)$, then $int_X c\ell_X(B \cup C) = B \cup C$, and $X \setminus c\ell_X B = X \setminus B$. So, i is a Boolean monomorphism. By 2.2(c) and 3.1(e)(3), $\mathcal{RO}(X)$ is a complete Boolean algebra. Let $\emptyset \neq U \in \mathcal{RO}(X)$. Since X is zero-dimensional, $U = \bigcup\{i(V) \in \mathcal{B}(X) : V \subseteq U\}$, so, $U = \bigvee\{i(V) \in \mathcal{B}(X) : V \subseteq U\}$. Thus, $\langle\langle \mathcal{B}(X) \rangle\rangle = \mathcal{RO}(X)$.

(2) Suppose $\emptyset \neq U \in \mathcal{RO}(X)$. Since X is zero-dimensional, there is some $V \in \mathcal{B}(X)$ such that $\emptyset \neq V \subseteq U$. So, $\mathcal{B}(X)$ is dense in $\mathcal{RO}(X)$. Suppose $\emptyset \neq S \subseteq \mathcal{B}(X)$ and $\bigvee S$ exists in $\mathcal{B}(X)$. Now, $\bigvee S$ is clopen and $\bigvee S \supseteq \bigcup S$. So, $\bigvee S \supseteq int_X c\ell_X(\bigcup S)$. Also, since X is zero-dimensional, $\bigvee S \subseteq \bigcap\{U \in \mathcal{B}(X) : \bigcup S \subseteq U\} = c\ell_X(\bigcup S)$. Since $\bigvee S$ is clopen, $\bigvee S = c\ell_X(\bigcup S) = int_X c\ell_X(\bigcup S)$. This shows that $i(\bigvee S) = \bigvee i[S]$. A similar proof works for infima. Hence $\mathcal{B}(X)$ is regular in $\mathcal{RO}(X)$.

(3) Let (A,j) be a completion of $\mathcal{B}(X)$. By 3.1(p), j has a

Boolean homomorphism extension $h \in F(RO(X),A)$; so, $h \circ i = j$. To show h is one-to-one, suppose $h(U) = h(V)$ where $U, V \in RO(X)$ and $U \neq V$. Then $U \setminus c\ell_X V \neq \emptyset$ or $V \setminus c\ell_X U \neq \emptyset$. Suppose $U \setminus c\ell_X V \neq \emptyset$. Then $h(U \setminus c\ell_X V) = h(U \wedge V') = h(U) \wedge h(V)' = 0$. There is some $W \in B(X)$ such that $\emptyset \neq W \subseteq U \setminus c\ell_X V$. Now, $h(W) = h(i(W)) = j(W) \neq 0$ as j is one-to-one. So, $0 < h(W) \leqslant h(U \setminus c\ell_X V) = 0$ which is a contradiction. Thus, h is a Boolean monomorphism.

■

If A is a Boolean algebra, $\emptyset \neq S \subseteq A$, and $|S| \geqslant \aleph_0$, then it follows easily from 3.1(ℓ)(2) that $|\langle S \rangle| = |S|$; however, it is not necessarily true that $|\langle\langle S \rangle\rangle| = |S|$. If X is a zero-dimensional space, then for each $U \in RO(X)$, $U = \vee S$ where $S = \{V \in B(X) : V \subseteq U\}$. This shows that $|\langle\langle B(X) \rangle\rangle| = |RO(X)| \leqslant |\mathbb{P}(B(X))|$. We now give a couple of examples to show it is possible for $|RO(X)| = |\mathbb{P}(B(X))|$ and for $|\langle\langle S \rangle\rangle| > |S|$.

(f) **Examples**

(1) Let X be an infinite compact zero-dimensional metric space. Thus X is second countable and has a countable open base $\{V_n : n \in \mathbb{N}\}$. If $B \in B(X)$, then there exists, by the compactness of B, a finite subset $F(B)$ of $\mathbb{N}$ such that $B = \cup\{V_n : n \in F(B)\}$. Thus, $B \to F(B)$ is a one-to-one function from $B(X)$ into the finite subsets of $\mathbb{N}$. Hence, $|B(X)| = \aleph_0$. As X is infinite, we can find an infinite family $\mathcal{Q}$ of nonempty, pairwise disjoint open subsets of X. For $F \subseteq \mathcal{Q}$, let $V(F) = \text{int}_X c\ell_X(\cup F)$. Then $V(F) \in RO(X)$ and it is easily seen that $F \to V(F)$ is a one-to-one function from $\mathbb{P}(\mathcal{Q})$ to $RO(X)$.

Thus, $2^{\aleph_0} \leqslant 2^{|\alpha|} \leqslant |RO(X)|$. But $|RO(X)| \leqslant |\mathbb{P}(B(X))| = 2^{\aleph_0}$. Thus, $|RO(X)| = 2^{\aleph_0}$.

(2) Given an infinite cardinal τ, we will find a complete Boolean algebra B and a countable subset $S \subseteq B$ such that $B = \langle\langle S\rangle\rangle$ and $|B| \geqslant \tau$. Let $D(\tau)$ be τ with the discrete topology and let $X = D(\tau)^{\mathbb{N}}$. Viewed as a set, $X = F(\mathbb{N},D(\tau))$; viewed as a product space, X is zero-dimensional. Let $B = RO(X)$. Then, B is a complete Boolean algebra. For m, $n \in \mathbb{N}$, let $T(m,n) = \{f \in X : f(m) < f(n)\}$ and $S = \{T(m,n) : m, n \in \mathbb{N}\}$ (here we are regarding τ as an ordinal and using the standard ordering on ordinals.) We will show

(a) $|B| \geqslant \tau$ and

(b) $B = \langle\langle S\rangle\rangle$.

First, we show (a). For $\alpha \in \tau$, $\Pi_3^{\leftarrow}[\{\alpha\}] \in RO(X)$ and if $\beta \in \tau\setminus\{\alpha\}$, then $\Pi_3^{\leftarrow}[\{\alpha\}] \cap \Pi_3^{\leftarrow}[\{\beta\}] = \emptyset$. Thus, $\tau = |\{\Pi_3^{\leftarrow}[\{\alpha\}] : \alpha < \tau\}| \leqslant |B|$. To show (b), first we show that $S \subseteq B(X)$. Now, for m, $n \in \mathbb{N}$, $T(m,n) = \bigvee\{\Pi_m^{\leftarrow}[[0,\alpha)] \cap \Pi_n^{\leftarrow}[\{\alpha\}] : \alpha < \tau\}$ and $T(m,n)$ is open. If $f \notin T(m,n)$, then $f(m) \geqslant f(n)$. But $f \in U$ where $U = \Pi_m^{\leftarrow}[\{f(m)\}] \cap \Pi_n^{\leftarrow}[\{f(n)\}]$ and $U \cap T(m,n) = \emptyset$. So, $T(m,n) \in B(X) \subseteq B$. Next note that $\mathcal{B} = \langle\{\Pi_n^{\leftarrow}[\{\alpha\}] : n \in \mathbb{N}, \alpha < \tau\}\rangle$ is a subbase for X. So, to show $\langle\langle S\rangle\rangle = RO(X)$, it suffices to show $\Pi_n^{\leftarrow}[\{\alpha\}] \in \langle\langle S\rangle\rangle$ for $n \in \mathbb{N}$ and $\alpha < \tau$. Now, $\Pi_n^{\leftarrow}[\{\alpha\}] = \Pi_n^{\leftarrow}[[0,\alpha]]\setminus\Pi_n^{\leftarrow}[[0,\alpha)]$; so, we will show the clopen sets $\Pi_n^{\leftarrow}[[0,\alpha]]$ and $\Pi_n^{\leftarrow}[[0,\alpha)] \in \langle\langle S\rangle\rangle$ (recall that $D(\tau)$ has the discrete topology). Let $\alpha < \tau$. By induction, assume that for each $m \in \mathbb{N}$ and each $\beta < \alpha$, $\Pi_m^{\leftarrow}[[0,\beta)]$ and $\Pi_m^{\leftarrow}[[0,\beta]] \in \langle\langle S\rangle\rangle$. Note that $\Pi_m^{\leftarrow}[[0,\alpha)] = \bigvee\{\Pi_m^{\leftarrow}[[0,\beta]] : \beta < \alpha\}$ and $\Pi_m^{\leftarrow}[[0,\alpha)]$ is clopen. So,

$\Pi_m^{\leftarrow}[[0,\alpha)] = \mathrm{int}_X c\ell_X(\cup\{\Pi_m^{\leftarrow}[[0,\beta]] : \beta < \alpha\}) \in \langle\langle S\rangle\rangle$. Finally, to show $\Pi_m^{\leftarrow}[[0,\alpha]] \in \langle\langle S\rangle\rangle$ for all $m \in \mathbb{N}$, fix $n \in \mathbb{N}$ and for $m \in \mathbb{N}$, let $C_m = (X\setminus T(m,n)) \cup \Pi_m^{\leftarrow}[[0,\alpha)]$. Then C_m is clopen and $C_m \in \langle\langle S\rangle\rangle$. It is easy to verify that

$$\Pi_n^{\leftarrow}[[0,\alpha]] \subseteq \cap \{C_m : m \in \mathbb{N}\},$$

which implies that $\Pi_n^{\leftarrow}[[0,\alpha]] \subseteq \mathrm{int}_X c\ell_X(\cap\{C_m : m \in \mathbb{N}\}) = \mathrm{int}_X(\cap\{C_m : m \in \mathbb{N}\})$ (since $\cap\{C_m : m \in \mathbb{N}\}$ is closed). Let $g \in \mathrm{int}_X(\cap\{C_m : m \in \mathbb{N}\})$. Then for some $k \geq n$, $g \in V \subseteq \cap\{C_m : m \in \mathbb{N}\}$ where $V = \cap\{\Pi_m^{\leftarrow}[\{g(m)\}] : m \leq k\}$. Assume $g(n) > \alpha$. Define $h \in X$ by $h(m) = g(m)$ for $m \leq k$ and $h(m) = \alpha$ for $m > k$. Then $h \in V$. Since $V \subseteq \cap\{C_m : m \in \mathbb{N}\}$, it follows that $V \subseteq C_{k+1}$. But $h(k+1) = \alpha < g(n) = h(n)$ which implies that $h \notin C_{k+1}$, a contradiction. Thus, $g(n) \leq \alpha$. This shows that $g \in \Pi_n^{\leftarrow}[[0,\alpha]]$ and that $\mathrm{int}_X(\cap\{C_m : m \in \mathbb{N}\}) = \Pi_n^{\leftarrow}([0,\alpha])$. Hence, $\Pi_n^{\leftarrow}[[0,\alpha]] \in \langle\langle S\rangle\rangle$. This completes the induction and the proof that $\Pi_n^{\leftarrow}[\{\alpha\}] \in \langle\langle S\rangle\rangle$ for $n \in \mathbb{N}$ and $\alpha < \tau$.

■

Since a Boolean algebra B is determined by its order structure and an order structure on B is simply a subset of B × B, it follows that the number of non-isomorphic countable Boolean algebras is $\leq 2^{\aleph_0}$. By the result in 3.4(f)(2), it follows that there is a countable Boolean algebra whose collection of non-isomorphic completions is not a set. So, there is not, necessarily, a largest completion of a fixed Boolean algebra. However, there is a completion which is "minimal" in the sense of 3.4(e). Here is a characterization of such completions.

(g) **Proposition**. Let (B,i) be a completion of a Boolean subalgebra A where $i \in F(A,B)$ is the inclusion function. Then A is dense in B iff for each completion (C,j) of A, there is a Boolean monomorphism $h \in F(B,C)$ such that $h \circ i = j$.

Proof. Suppose A is dense in B. The proof for this direction is identical to the method of the proof of 3.4(e)(3). To show the converse, let (B,i) be a completion of A such that for each completion (C,j) of A, there is a Boolean monomorphism $h \in F(B,C)$ such that $h \circ i = j$. By 3.4(e) and Stone's Representation Theorem, there is a completion (C,j) of A such that $j[A]$ is dense in C. So, there is a Boolean monomorphism $h \in F(B,C)$ such that $h \circ i = j$. Let $b \in B$ such that $0 < b$. Then $0 = h(0) < h(b)$. There is an element $a \in A$ such that $0 < j(a) \leqslant h(b)$. But $j(a) = h(i(a))$ and $0 < h(i(a)) \leqslant h(b)$. Since h is an order isomorphism, it follows that $0 < i(a) \leqslant b$. Thus, $A = i[A]$ is dense in B.

■

(h) **Definition**. A completion B of a Boolean subalgebra A is called a **minimal completion** if A is dense in B.

Before establishing the unqiueness of minimal completions of a Boolean algebra, we want to compare the concepts of a Dedekind-MacNeille completion (introduced in 2.5) and a minimal completion of a Boolean algebra. First, a result about denseness is needed.

(i) **Proposition**. Let A be a subalgebra of a Boolean algebra B. Then A is dense in B iff for each $b \in B$, $b = \bigwedge\{s \in A : s \geq b\} = \bigvee\{s \in A : s \leq b\}$.

Proof. For b in B, let $S^b = \{s \in A : s \geq b\}$ and $S_b = \{s \in A : s \leq b\}$. If $b \in A$, then, clearly, $b = \bigwedge S^b = \bigvee S_b$. Suppose A is dense in B and $b \in B \setminus A$. Then $0 < b < 1$. Note that b is an upper bound of S_b, and assume that b is not the least of the upper bounds of S_b. Thus, there is some $c \in B$ such that $b > c$ and $c \geq s$ for each $s \in S_b$. It follows from 3.1(g)(5) that $b \wedge c' = b \setminus c > 0$. By denseness of A, there is some $a \in A$ such that $b \wedge c' \geq a > 0$. So, $c' \geq a$ and $b \geq a$ which implies $a \in S_b$. As $a \in S_b$, $a \leq c$; hence $a \leq c \wedge c' = 0$, a contradiction. This completes the proof that $b = \bigvee S_b$. It is left as an exercise (see 3F) to show that $\bigwedge S^b = (\bigvee S_{b'})'$. So, $\bigwedge S^b = (b')' = b$. Conversely, suppose for each b in B, $b = \bigvee S_b$. Let $b \in B$ and $b > 0$. Since $b = \bigvee S_b$, there is some $a \in S_b$ such that $0 < a \leq b$. Hence, A is dense in B.

■

(j) **Corollary**. Let A be a Boolean algebra.

(1) A minimal completion of A is also the Dedekind-MacNeille completion of A.

(2) Minimal completions of A are unique in this sense: if (B_1,i_1) and (B_2,i_2) are minimal completions of A, there is a Boolean isomorphism $h \in F(B_1,B_2)$ such that $h \circ i_1 = i_2$.

(3) Every Boolean algebra has a minimal completion.

(4) The Dedekind-MacNeille completion of a Boolean algebra is also a Boolean algebra.

Proof. Now, (1) follows from 2.5(b) and 3.4(i), (2) from 2.5(b)(2) and 3.1(n)(1), (3) from 3.4(e) and Stone's Representation Theorem that every Boolean algebra is isomorphic to $B(X)$ for some compact zero-dimensional space X, and (4) from (1), (3), and 2.5(b)(2).

■

By Stone's Representation Theorem, a Boolean algebra B is Boolean isomorphic to $B(X)$ for some compact zero-dimensional space X. But $RO(X)$ is the unique minimal completion of $B(X)$ and, hence, of B.

We are at a crossroads where results about compact zero-dimensional spaces can be obtained either by using Boolean algebras and then applying Stone duality, or by using topological methods. For the results we wish to obtain, the topological method gives generalizations (usually in the setting of regular spaces) of the results obtained by Boolean algebras and Stone duality (where the setting would be compact zero-dimensional spaces). In particular, generalizations of the "Stone duals" of the last few theorems in this section appear in Chapter 6. In the problems for Chapter 6, the reader will be invited to translate some of the results in Chapter 6 into the language of Boolean algebras.

The theory of Boolean algebras is an important tool in studying extensions and absolutes. For example, in 4.7, Boolean algebras are used to construct all the compact zero-dimensional extensions of a zero-dimensional space and in 6.6, Boolean algebras are used to construct the absolute of a space.

3.5 The Continuum Hypothesis and Martin's Axiom

As mentioned in the Preface, the reader should be aware that most mathematicians use a set theory which is based on the Zermelo-Fraenkel Axioms plus the Axiom of Choice, abbreviated ZFC. In what follows, it is not necessary that the reader be familiar with the Zermelo-Fraenkel Axioms, but the reader must know and be able to use the Axiom of Choice and some of its equivalences.

There are statements, say S, expressible in the language of set theory, for which neither S nor the negation of S (denoted as $\neg S$) can be proven using the axioms of ZFC. Such a statement S is said to be **independent of ZFC**. The term "axiom" is frequently used to describe such a statement, since it can consistently be used together with ZFC as a starting point for set theory (assuming, of course, that ZFC is itself consistent). Since S cannot be proved from ZFC, this new theory will yield more theorems and examples than could be deduced from ZFC alone. In this section, two important axioms are introduced - the Continuum Hypothesis and Martin's Axiom. Since one important formulation of Martin's Axiom is expressed in the language of Boolean algebras, we have waited until now to introduce this topic.

(a) <u>The Continuum Hypothesis</u>. The axiom best known among mathematicians is the **Continuum Hypothesis** (denoted as CH) which is defined as follows:

$$\text{CH} : 2^{\aleph_0} = \aleph_1$$

When CH or ¬CH (recall that ¬CH is the symbol for the negation of CH) is used in proving a result, we will indicate this by including [CH] or [¬CH] as part of the statement.

Another statement often used as an axiom in conjunction with ZFC is the **Generalized Continuum Hypothesis** (denoted as GCH), defined as follows:

$$\text{GCH : For every ordinal } \alpha,\ 2^{\aleph_\alpha} = \aleph_{\alpha+1}$$

It is clear that CH is the special case of GCH in which $\alpha = 0$.

Before Martin's Axiom can be formulated, we need to extend the definition of three concepts to the setting of posets.

(b) <u>Definition</u>. Let $(P,\leq)$ be a poset and $x, y \in P$.

(1) The elements x and y are said to be **compatible** if there is some $z \in P$ such that $z \leq x$, $z \leq y$, and if $\wedge P$ exists, then $z \neq \wedge P$. If the elements x and y are not compatible, we write $x \perp y$.

(2) A subset $E \subseteq P$ is called an **antichain** if for distinct elements $x, y \in E$, $x \perp y$ and $\wedge P \notin E$ if $\wedge P$ exists.

(3) P is said to satisfy the **countable chain condition** (abbreviated as *ccc*) if every antichain is countable. (See [Ku., pg. 53] for a comment on this strangely-chosen terminology.)

(4) A Boolean algebra $(B,\leq)$ is said to **satisfy the** ***ccc*** if $(B,\leq)$, as a poset, satisfies the *ccc*.

(5) A space X is said to satisfy the *ccc* if the poset $(\tau(X),\subseteq)$ satisfies the *ccc*. Note that this definition coincides with the definition given in 2N(6). An antichain in a space X is an antichain in $(\tau(X),\subseteq)$, i.e., a family of pairwise disjoint, nonempty open

sets.

(c) **Proposition**

(1) For each set X, $(\mathbb{P}(X),\subseteq)$ has the *ccc* iff X is countable,

(2) for each space X, a subset $E \subseteq C^*(X)$ (respectively, $E \subseteq C(X)$) is an antichain in $(C^*(X),\leqslant)$ (respectively, $(C(X),\leqslant)$) (the partial order on C(X) and $C^*(X)$ is defined in 2.1(f)(5)) iff E is a singleton; in particular, both $C^*(X)$ and C(X) have the *ccc*,

(3) if $(L,\leqslant)$ is a loset, then $E \subseteq L$ is an antichain iff E is a singleton and $\wedge L \notin E$ if $\wedge L$ exists; in particular, a loset has the *ccc*,

(4) the space ω_1 with the order topology does not satisfy the *ccc*, and

(5) each antichain E of the poset $(P,\leqslant)$ is contained in a maximal antichain.

Proof

(1) For nonempty sets $A, B \subseteq X$, note that $A \perp B$ iff $A \cap B = \varnothing$. Hence, $\{\{x\} : x \in X\}$ is an antichain. So, (1) follows.

(2) If $f, g \in C^*(X)$ (or C(X)), note that $f \wedge g \in C^*(X)$ (or C(X)) by 2.1(f)(5). Hence, for $E \subseteq C^*(X)$ (or C(X)), E is an antichain iff E is a singleton.

(3) If $a, b \subseteq L$, then $a \wedge b = \min\{a,b\}$. So, $E \subseteq L$ is an antichain iff E is a singleton and $\wedge L \notin E$ if $\wedge L$ exists.

(4) The set $\{\{\alpha\} : a \in \omega_1$ and α a successor ordinal$\}$ is an uncountable antichain in $\tau(\omega_1)$.

(5) This Zorn's Lemma argument is left as an exercise for the

reader (see 3P).

∎

(d) <u>Definition</u>. Let $(P.\leqslant)$ be a poset. A subset $F \subseteq P$ is a **P-filter** or a **filter on P** if

(1) $\varnothing \neq F$ and $\wedge P \notin F$ if $\wedge P$ exists,

(2) if $x, y \in F$, there is some $r \in F$ such that $r \leqslant x$ and $r \leqslant y$, and

(3) if $x \in F$ and $x \leqslant y \in P$, then $y \in F$.

Note that this definition is equivalent to the definition of a filter given in 2.3(a) if $(P,\leqslant)$ is a lattice of sets.

A maximal filter on P is called a **P-ultrafilter** or an **ultrafilter on P**.

(e) <u>Proposition</u>. Let $(P,\leqslant)$ be a poset. A P-filter is contained in a P-ultrafilter.

<u>Proof</u>. The proof is similar to the proof of 2.3(d)(2) which was left as an exercise in 2H.

∎

(f) <u>Definition</u>. Let $(P,\leqslant)$ be a poset.

(1) A subset $D \subseteq P$ is **dense** if $\wedge P \notin D$ when $\wedge P$ exists and for each $x \in P$, there is some $d \in D$ such that $d \leqslant x$. Note that this definition coincides with the definition given in 3.4(d)(4) when $(P,\leqslant)$ is a Boolean algebra.

(2) If $x \in P$, $\hat{x}$ will denote the set $\{y \in P : y \leqslant x\}$.

Before introducing Martin's Axiom, we present a useful fact

about dense subsets of posets which will be needed later.

(g) **Proposition**. If $(P,\leq)$ is a poset and $p, q \in P$, then $D(p,q) = \{r \in P \setminus \{\Lambda P\} : r \perp p, r \perp q, \text{ or } r \in \hat{p} \cap \hat{q}\}$ is dense.

Proof. Let $t \in P$. Suppose that if $s \in P$ and $s \leq t$, then $s \not\perp p$ and $s \not\perp q$. Thus, $t \not\perp p$, so there is some $s \in \hat{t} \cap \hat{p}$. Since $s \leq t$, it follows that $s \not\perp q$. So, there is some $w \in \hat{s} \cap \hat{q}$. Now, $w \in \hat{p} \cap \hat{q}$ and $w \leq t$. This shows that $D(p,q)$ is dense in P.

■

(h) **Martin's Axiom**. Let κ be an infinite cardinal. The **κ-Martin's Axiom** is this statement:

$MA(\kappa)$: If X is a compact space with *ccc* and $\{U_\alpha : \alpha < \kappa\}$ is a family of open, dense sets of X, then $\bigcap\{U_\alpha : \alpha < \kappa\} \neq \emptyset$.

The alert reader will notice that $MA(\aleph_0)$ is a special case of the Baire Category Theorem, in which the "*ccc*" hypothesis is not needed. Also, it is clear that if κ and m are infinite cardinals, $\kappa \leq m$, and $MA(m)$ is true, then $MA(\kappa)$ is also true. If X is the unit interval with the usual topology, $X = \{p_\alpha : \alpha < 2^{\aleph_0}\}$, and $U_\alpha = X \setminus \{p_\alpha\}$ for $\alpha < 2^{\aleph_0}$, then $\bigcap\{U_\alpha : \alpha < 2^{\aleph_0}\} = \emptyset$. So, $MA(\kappa)$ is false for $\kappa \geq 2^{\aleph_0}$. Thus the assumption that $MA(\kappa)$ is true implies that $\aleph_0 \leq \kappa < 2^{\aleph_0}$; henceforth, the use of the symbol "$MA(\kappa)$" will mean that κ is a cardinal and $\aleph_0 \leqslant \kappa < 2^{\aleph_0}$. **Martin's Axiom** is this statement:

MA: For each infinite cardinal $\kappa < 2^{\aleph_0}$, MA(κ) is true.

If CH is true, then $\aleph_0 \leqslant \kappa < 2^{\aleph_0}$ implies that $\kappa = \aleph_0$ and, hence, that MA is true. On the other hand, if MA($\aleph_1$) is true, then $\aleph_0 \leqslant \aleph_1 < 2^{\aleph_0}$ and $\neg$CH is true.

Our next result gives characterizations of Martin's Axiom in terms of posets and complete Boolean algebras.

(i) **Theorem**. Let κ be an infinite cardinal. The following are equivalent:

(1) MA(κ).

(2) If $(B,\leqslant)$ is a Boolean algebra with the *ccc* property and if $\{D_\alpha : \alpha < \kappa\}$ is a family of dense subsets of B, then there is a filter F on B such that $F \cap D_\alpha \neq \emptyset$ for each $\alpha < \kappa$.

(3) If $(P,\leqslant)$ is a poset with *ccc*, $|P| \leqslant \kappa$, and $\{D_\alpha : \alpha < \kappa\}$ is a family of dense subsets of P, then there is a filter F on P such that $F \cap D_\alpha \neq \emptyset$ for each $\alpha < \kappa$.

(4) If $(P,\leqslant)$ is a poset with *ccc* and $\{D_\alpha : \alpha < \kappa\}$ is a family of dense sets, then there is a filter F on P such that $F \cap D_\alpha \neq \emptyset$ for each $\alpha < \kappa$.

Proof

(1) $\Rightarrow$ (2) Let B be a Boolean algebra with the *ccc* property, and let $\{D_\alpha : \alpha < \kappa\}$ be a family of dense sets. Let X be the Stone space of B, i.e., X = S(B) (see 3.2(c)), and let $\lambda : B \rightarrow$

$B(X)$ be the Boolean isomorphism described in 3.2. First, we show that the compact, zero-dimensional space X satisfies the *ccc* property. Let C be an antichain of nonempty open sets in X. For each $U \in C$, there is a nonempty clopen set $V_U \subseteq U$. Let $C' = \{V_U : U \in C\}$. Thus, $|C| = |C'|$. Since λ is an isomorphism, $\{\lambda^{\leftarrow}(V) : V \in C'\}$ is an antichain in B and hence is countable. Thus C' and C are countable families. This shows that X satisfies *ccc*. For $\alpha < \kappa$, let $E_\alpha = \bigcup\{\lambda(a) : a \in D_\alpha\}$. It is immediate that E_α is open and dense in X. By MA(κ), there is some $p \in \bigcap\{E_\alpha : \alpha < \kappa\}$. Now, $F = \{a \in B : p \in \lambda(a)\}$ is a filter on B. For $\alpha < \kappa$, since $p \in E_\alpha$, there is some $a \in D_\alpha$ such that $p \in \lambda(a)$. Thus, $a \in F \cap D_\alpha$.

(2) $\Rightarrow$ (3) Suppose $(P,\leqslant)$ is a poset satisfying *ccc*, $\{D_\alpha : \alpha < \kappa\}$ is a family of dense sets, and $|P| \leqslant \kappa$. Suppose $\bigwedge P$ exists; since D_α is dense in $P\setminus\{\bigwedge P\}$ for $\alpha < \kappa$ and F is a filter on P iff F is a filter on $P\setminus\{\bigwedge P\}$, we can assume $\bigwedge P$ does not exist. Note that if $x, y \in P$ and $z \in \hat{x} \cap \hat{y}$, then $\hat{z} \subseteq \hat{x} \cap \hat{y}$. So, $\{\hat{x} : x \in P\}$ is an open base for a topology (not necessarily Hausdorff) on P. Let $B = RO(P)$. Then B is a Boolean algebra (by the same argument as in 3.1(e)(3)). Let C be an antichain in B. For each $U \in B$, there is some $p_U \in P$ such that $\mathrm{int}_P c\ell_P \hat{p}_U \subseteq U$. If $U, V \in C$ and $U \neq V$, then $U \cap V = \varnothing$; so, the function $U \to p_U$ is one-to-one and $\hat{p}_U \cap \hat{p}_V \subseteq U \cap V = \varnothing$. Therefore, $p_U \perp p_V$ and $\{p_U : U \in C\}$ is an antichain in P. Since P satisfies *ccc*, $\{p_U : U \in C\}$, and therefore C, is countable and B satisfies *ccc*. Let $D = \{D_\alpha : \alpha < \kappa\} \cup \{D(p,q) : p, q \in P\}$. By 3.5(g), D is a family of dense subsets of P. Evidently $|D| = \kappa$ (since $|P| \leqslant \kappa$). Fix $D \in D$ and

let $E = \{\text{int}_P c\ell_P \hat{p} : p \in D\}$. To show E is dense in B, let $U \in B \setminus \{\emptyset\}$. There is some $q \in P$ such that $\text{int}_P c\ell_P \hat{q} \subseteq U$. For some $p \in D$, $p \leqslant q$. So, $\text{int}_P c\ell_P \hat{p} \subseteq U$ and, hence, E is dense in B. Specifically, let $E = E_\alpha$ when $D = D_\alpha$ for $\alpha < \kappa$ and $E = E(p,q)$ when $D = D(p,q)$ for $p, q \in P$. By (2), there is a filter F on B such that $F \cap E_\alpha \neq \emptyset$ for $\alpha < \kappa$ and $F \cap E(p,q) \neq \emptyset$ for $p, q \in P$. Let $G = \{p \in P : \text{int}_P c\ell_P \hat{p} \in F\}$. Clearly, $G \cap D_\alpha \neq \emptyset$ for $\alpha < \kappa$ and $G \cap D(p,q) \neq \emptyset$ for $p, q \in P$. It remains to show that G is a filter on P. Obviously, $G \neq \emptyset$; by our initial assumption, $\bigwedge P$ does not exist. If $p \in G$, $p \leqslant q$, and $q \in P$, then $\hat{p} \subseteq \hat{q}$ and $\text{int}_P c\ell_P \hat{p} \subseteq \text{int}_P c\ell_P \hat{q}$. Since F is a filter and $\text{int}_P c\ell_P \hat{p} \in F$, it follows that $\text{int}_P c\ell_P \hat{q} \in F$ and $q \in G$. Suppose $p, q \in G$. Then $\text{int}_P c\ell_P \hat{p} \cap \text{int}_P c\ell_P \hat{q} = \text{int}_P c\ell_P (\hat{p} \cap \hat{q}) \in F$ by 2.2(a). Let $r \in D(p,q) \cap G$. If $r \perp p$, then $\hat{r} \cap \hat{p} = \emptyset$ as $\bigwedge P$ does not exist. By 2.2(a), $\text{int}_P c\ell_P \hat{r} \cap \text{int}_P c\ell_P \hat{p} = \emptyset$ which is impossible as F is a filter. Likewise $r \perp q$. So, $r \in \hat{p} \cap \hat{q}$ and $r \in G$. This completes the proof that G is a filter.

(3) $\Rightarrow$ (4) Suppose $(P,\leqslant)$ is a poset satisfying *ccc* and $\{D_\alpha : \alpha < \kappa\}$ is a family of dense sets. As in the argument of (2) $\Rightarrow$ (3), we can assume that $\bigwedge P$ does not exist. For each $\alpha < \kappa$, use the Axiom of Choice to define a function $f_\alpha : P \to D_\alpha$ such that $f_\alpha(p) \leqslant p$ for $p \in P$, and use the Axiom of Choice to define a function $f : P \times P \to P$ such that $f(p,q) \in \hat{p} \cap \hat{q}$ if $\hat{p} \cap \hat{q} \neq \emptyset$. Let $f(p,q)$ be chosen arbitrarily in P if $\hat{p} \cap \hat{q} = \emptyset$. Let $q \in P$ and $S_0 = \{q\}$. If S_n is defined for $n \in \omega$, define $S_{n+1} = S_n \cup f[S_n \times S_n] \cup (\bigcup\{f_\alpha[S_n] : \alpha < \kappa\})$. Let $S = \bigcup\{S_n : n \in \omega\}$. Then $|S| \leqslant \kappa$, $f[S \times S] \subseteq S$, and $f_\alpha[S] \subseteq S$ for $\alpha < \kappa$. Now, S is a poset and if C is an antichain in S, then C is an antichain in P as $f[S \times S] \subseteq$

S. So, S satisfies *ccc*. Since $f_\alpha[S] \subseteq S$ for $\alpha < \kappa$, it follows that $S \cap D_\alpha$ is dense in S. By (3) for $(S,\leqslant)$, there is a filter F on S such that $F \cap (S \cap D_\alpha) \neq \emptyset$ for each $\alpha < \kappa$. Hence, $G = \{p \in P : p \geqslant r$ for some $r \in F\}$ is a filter on P and $G \cap D_\alpha \neq \emptyset$ for $\alpha < \kappa$.

(4) $\Rightarrow$ (1) Let X be a compact space satisfying *ccc*. Obviously $MA(\kappa)$, applied to X, is true if X is finite; so, suppose X is infinite. Suppose for $\alpha < \kappa$, D_α is an open, dense set. Let $P = \tau(X)\setminus\{\emptyset\}$. Then $(P,\subseteq)$ is a poset and $\bigwedge P$ does not exist. Also, for U, $V \in P$, $U \perp V$ iff $U \cap V = \emptyset$. So, P satisfies *ccc*. For $\alpha < \kappa$, let $E_\alpha = \{U \in P : c\ell_X U \subseteq D_\alpha\}$. Then E_α is dense in P. By (4) for P, there is a filter F on P such that $F \cap E_\alpha \neq \emptyset$ for $\alpha < \kappa$. Now, F is an open filter on X; since X is compact, there is some $x \in \bigcap\{c\ell_X U : U \in F\}$. For $\alpha < \kappa$, let $U \in F \cap E_\alpha$. Then $p \in c\ell_X U \subseteq D_\alpha$. Hence, $x \in \bigcap\{D_\alpha : \alpha < \kappa\}$.

■

We are now ready to give a topological application of Martin's Axiom.

(j) **Theorem**. $[MA(\kappa)]$. Suppose X is compact and satisfies *ccc*, and suppose $\aleph_0 < \kappa$ and κ is a regular cardinal. If $\mathcal{U}$ is a family of open sets in X and $|\mathcal{U}| = \kappa$, there is a subfamily $\mathcal{G} \subseteq \mathcal{U}$ such that $|\mathcal{G}| = \kappa$ and $\bigcap\mathcal{G} \neq \emptyset$.

Proof. Let $\mathcal{U} = \{U_\alpha : \alpha < \kappa\}$ and, for $\alpha < \kappa$, let $V_\alpha = \bigcup\{U_\gamma : \alpha < \gamma < \kappa\}$. For $\alpha < \beta < \kappa$, $V_\alpha \supseteq V_\beta$ and $c\ell_X V_\alpha \supseteq$

$c\ell_X V_\beta$. Assume for each $\alpha < \kappa$, there is some $\beta > \alpha$ such that $c\ell_X V_\alpha \supsetneq c\ell_X V_\beta$; hence, $V_\alpha \backslash c\ell_X V_\beta \neq \emptyset$. For $\alpha < \kappa$, let $\beta(\alpha)$ be the least ordinal such that $V_\alpha \backslash c\ell_X V_{\beta(\alpha)} \neq \emptyset$. Define $\{\mu(\alpha) : \alpha < \kappa\} \subseteq \kappa$ inductively as follows: $\mu(0) = 0$ and $\mu(\alpha) = \sup\{\beta(\mu(\delta)) : \delta < \alpha\} + 1$ for each $\alpha < \kappa$. Since κ is regular, $\mu(\alpha) < \kappa$ for each $\alpha < \kappa$. It is straightforward to check that $\{V_{\mu(\alpha)} \backslash c\ell_X V_{\beta(\mu(\alpha))} : \alpha < \kappa\}$ is an uncountable antichain of open sets, contradicting the assumption that X satisfies the *ccc*. So, there is a least $\alpha < \kappa$ such that $c\ell_X V_\beta = c\ell_X V_\alpha$ for all $\beta > \alpha$. Let $Y = c\ell_X V_\alpha$. If $\mathcal{W}$ is an antichain of open sets in Y, then $\{W \cap V_\alpha : W \in \mathcal{W}\}$ is an antichain of open sets in X; hence, $\mathcal{W}$ is countable. So, the compact space Y satisfies *ccc*. By MA(κ) applied to Y, there is some point $p \in \cap\{V_\beta : \beta > \alpha\}$. Then $\{U \in \mathcal{U} : p \in U\}$ is the required $\mathcal{G}$.

■

(k) **Definition**

(1) A space is **perfect** if each of its closed sets is a G_δ-set. (Note: Some authors call a space perfect if it has no isolated points. We will not use that terminology in this book.)

(2) A space is **perfectly normal** if it is perfect and normal.

(3) A space X is **hereditarily separable** if every subspace of X is separable.

Part (3) of the next result is another topological consequence of Martin's Axiom.

(1) **Theorem**

(1) A countably compact space X in which every point is a G_δ is first countable.

(2) A perfect countably compact space X satisfies the *ccc*.

(3) $[MA(\aleph_1)]$. A perfect compact space X is hereditarily separable.

Proof.

(1) Let $p \in X$. There is a family $\{G_n : n \in \omega\}$ of open sets such that $\bigcap\{G_n : n \in \omega\} = \{p\}$. Since X is regular, we can inductively find a family $\{U_n : n \in \omega\}$ of open sets such that $c\ell_X U_{n+1} \subseteq U_n$ for $n \in \omega$ and $\{p\} = \bigcap\{U_n : n \in \omega\}$. Thus, $\bigcap\{c\ell_X U_n : n \in \omega\} = \{p\}$. Let U be an open neighborhood of p. Then $\{X \setminus c\ell_X U_n : n \in \omega\}$ is a countable open cover of the countably compact subspace $X \setminus U$. Since $\{X \setminus c\ell_X U_n : n \in \omega\}$ is an increasing sequence, there is some $n \in \omega$ for which $X \setminus U \subseteq X \setminus c\ell_X U_n$. Hence, $U_n \subseteq U$. This shows that X is first countable at p.

(2) Assume that $\{U_\alpha : \alpha < \omega_1\}$ is an antichain of open sets. Let $C = c\ell_X(\bigcup\{U_\alpha : \alpha < \omega_1\}) \setminus \bigcup\{U_\alpha : \alpha < \omega_1\}$ and choose $x_\alpha \in U_\alpha$ for each $\alpha < \omega_1$.

Since X is perfect, there is a family $\{V_n : n \in \omega\}$ of open sets such that $C = \bigcap\{V_n : n \in \omega\}$. If $\{x_\alpha : \alpha < \omega_1\} \setminus V_n$ is a finite set for each $n \in \omega$, then $\{x_\alpha : \alpha < \omega_1\} \setminus C$ ($= \{x_\alpha : \alpha < \omega_1\}$) is countable which is a contradiction. Hence, for some $n \in$

ω, $\{x_\alpha : \alpha < \omega_1\}\setminus V_n$ is infinite. But $\{x_\alpha : \alpha < \omega_1\}\setminus V_n$ is a closed infinite discrete subspace contradicting the assumption that X is countably compact. This shows that X satisfies the *ccc*.

(3) Assume X contains a nonseparable subspace. Thus, there is a subset $\{p_\alpha : \alpha < \omega_1\}$ such that

$$c\ell_X\{p_\beta : \beta < \alpha\} \neq c\ell_X\{p_\beta : \beta < \omega_1\}$$

for all $\alpha < \omega_1$. Let $Y = c\ell_X\{p_\beta : \beta < \omega_1\}$ and $W_\alpha = Y\setminus c\ell_X\{p_\beta : \beta < \alpha\}$. Since the property of being perfect is easily verified to be closed-hereditary, it follows that Y is perfect and compact. By (2) above Y satisfies the *ccc*. By 3.5(j), there is an uncountable subfamily $\mathcal{G}$ of $\{W_\alpha : \alpha < \omega_1\}$ such that $\cap\mathcal{G} \neq \emptyset$. Since $\{W_\alpha : \alpha < \omega_1\}$ is a decreasing family, it follows that $\cap\mathcal{G} = \cap\{W_\alpha : \alpha < \omega_1\}$. So, there exists $p \in X$ such that $p \in \cap\{W_\alpha : \alpha < \omega_1\}$. As Y is first countable at p, there is a countable open neighborhood base $\{U_n : n \in \omega\}$ for p. Let $D_n = \{\alpha < \omega_1 : U_n \subseteq W_\alpha\}$ for $n < \omega$. Then $\omega_1 = \cup\{D_n : n \in \omega\}$ as $\{U_n : n \in \omega\}$ is neighborhood base of p. Thus, for some $m \in \omega$, $|D_m| = \aleph_1$ and D_m is cofinal in ω_1. Hence, $U_m \subseteq \cap\{W_\alpha : \alpha < \omega_1\}$ which is impossible as $\{p_\alpha : \alpha < \omega_1\}$ is dense in Y. This completes the proof that Y is separable.

■

(m) **Theorem**. Let κ be a cardinal such that $\aleph_0 \leqslant \kappa < c$. Then MA($\kappa$) is equivalent to this statement:

If X is a space satisfying *ccc* such that $\{p \in X : p$ has a compact neighborhood$\}$ is dense in X and $\mathcal{U}$ is a family of open, dense subsets of X such that $|\mathcal{U}| = \kappa$, then $\cap\mathcal{U}$ is dense in X.

Proof. The proof of one direction is trivial. To prove the converse, suppose X satisfies *ccc* and $D = \{p \in X : p$ has a compact neighborhood$\}$ is dense in X. Let $\mathcal{U}$ be a family of open, dense subsets of X such that $|\mathcal{U}| = \kappa$. Let W be a nonempty open set in X. There is a point $p \in W \cap D$ and an open set V such that $p \in V$ and $Y = c\ell_X(V \cap W)$ is compact. Since X satisfies *ccc*, so does Y. Also, for $U \in \mathcal{U}$, $U \cap V \cap W$ is open and dense in Y. By MA(κ) for Y, $\emptyset \neq \cap\{U \cap V \cap W : U \in \mathcal{U}\} \subseteq W \cap (\cap\mathcal{U})$. This shows that $\cap\mathcal{U}$ is dense in X.

■

We conclude this section with some applications of the poset version of Martin's Axiom. Our next result is due to Solovay.

(n) **Theorem**. [MA(κ)]. Let $\mathcal{A}$ and $\mathcal{B}$ be subsets of $\mathbb{P}(\omega)$ such that $|\mathcal{A}| \leq \kappa$, $|\mathcal{B}| \leq \kappa$, and $|A \setminus \cup K| = \aleph_0$ for $A \in \mathcal{A}$ and finite $K \subseteq \mathcal{B}$. Then there is a subset $S \subseteq \omega$ such that $|A \cap S| = \aleph_0$ for $A \in \mathcal{A}$ and $|B \cap S| < \aleph_0$ for $B \in \mathcal{B}$.

Proof. Let $P = \{(c,C) : c \subset \omega, |c| < \aleph_0, C \subset \mathcal{B}, |C| < \aleph_0\}$ and define $(c,C) \leq (d,D)$ if $d \subseteq c$, $D \subseteq C$, and $B \cap c \subseteq d$ for each $B \in D$. Clearly, $(P,\leq)$ is a poset. If $\mathcal{C}$ is an uncountable subset of P, there is some finite subset $c \subset \omega$ such that $\{(c,C) : (c,C) \in \mathcal{C}\}$ is uncountable. If (c,C_1) and $(c,C_2) \in \mathcal{C}$, then $(c,C_1 \cup C_2) \leq (c,C_i)$ for $i = 1, 2$. This shows that P satisfies the *ccc*.

For $A \in \mathcal{A}$ and $n \in \omega$, let $D_{nA} = \{(c,C) \in P : |c \cap A| \geq n\}$. To show D_{nA} is dense in P, let $(d,D) \in P$. Since $|A \setminus \cup D| = \aleph_0$, let $c \subset A \setminus \cup D$ such that $|c| = n$. Then $(c \cup d, D) \in D_{nA}$. If $B \in D$, then $c \cap B = \emptyset$. So, $B \cap (c \cup d) = B \cap d \subseteq d$. Thus, $(c \cup d, D) \leq (d,D)$. This completes the proof that D_{nA} is dense in P. For $B \in \mathcal{B}$, let $D_B = \{(c,C) \in P : B \in C\}$. If $(d,D) \in P$, then $(d, D \cup \{B\}) \in D_B$ and $(d, D \cup \{B\}) \leq (d,D)$; hence, D_B is dense in P.

By MA(κ), there is a filter F on P such that $F \cap D_{nA} \neq \emptyset$ and $F \cap D_B \neq \emptyset$ for $A \in \mathcal{A}$, $B \in \mathcal{B}$, and $n \in \omega$. Let $S = \cup\{c : (c,C) \in F\}$. If $A \in \mathcal{A}$ and $n \in \omega$, then there is some $(c,C) \in D_{nA} \cap F$. So, $|c \cap A| \geq n$. Hence, $|S \cap A| \geq n$ for all $n \in \omega$, i.e., $|S \cap A| = \aleph_0$. If $B \in \mathcal{B}$, there is some $(d,D) \in D_B \cap F$. If we show that $B \cap S \subseteq d$, it will follow that $|B \cap S| < \aleph_0$. Let $p \in B \cap S$. There is some $(c,C) \in F$ such that $p \in C$. Since F is a filter, there is some $(e,E) \in F$ such that $(e,E) \leq (d,D)$ and $(e,E) \leq (c,C)$. Since $B \in D$, it follows that $B \cap e \subseteq d$. But $c \subseteq e$. So, $p \in B \cap c \subseteq B \cap e$ and $B \cap c \subseteq d$.

■

Using 3.5(n) as our motivation, we define the following combinatorial statement for $\aleph_0 \leq \kappa \leq 2^{\aleph_0}$:

(o) **Definition**. SL(κ) is the statement:

Let $\mathcal{A}$ and $\mathcal{B}$ be subsets of $\mathbb{P}(\omega)$ such that $|\mathcal{A}| < \kappa$, $|\mathcal{B}| < \kappa$, and $|A \setminus \cup K| = \aleph_0$ for $A \in \mathcal{A}$ and finite $K \subseteq \mathcal{B}$. There is an infinite subset $S \subseteq \omega$ such that $|A \cap S| = \aleph_0$ for $A \in \mathcal{A}$ and $|B \cap S|$

$< \aleph_0$ for $B \in \mathcal{B}$.

Clearly, if $\aleph_0 \leqslant \kappa \leqslant \mu \leqslant 2^{\aleph_0}$, then SL($\mu$) implies SL($\kappa$). Also, by 3.5(n), MA($\kappa$) implies SL($\kappa^+$). In particular, SL($\aleph_0$) and SL($\aleph_1$) are true in ZFC. The statement SL($2^{\aleph_0}$), like MA, is independent of ZFC.

(p) **Proposition**. [SL(κ)]. Let $\mathcal{A}$ and $\mathcal{B}$ be families of infinite subsets of ω such that $|\mathcal{A}| < \kappa$, $|\mathcal{B}| < \kappa$, and $|A \cap B| < \aleph_0$ for $A \in \mathcal{A}$ and $B \in \mathcal{B}$. Then there is an infinite subset $S \subseteq \omega$ such that $|A \cap S| = \aleph_0$ for $A \in \mathcal{A}$ and $|B \cap S| < \aleph_0$ for $B \in \mathcal{B}$.

Proof. This is an easy consequence of SL(κ).

■

(q) **Definition**. If $\aleph_0 \leqslant \kappa \leqslant 2^{\aleph_0}$, we define P($\kappa$) to be the following statement:

P(κ): Let $\mathcal{F} \subseteq \mathbb{P}(\omega)$ such that $|\mathcal{F}| < \kappa$ and $|\bigcap \mathcal{K}| = \aleph_0$ for each nonempty finite set $\mathcal{K} \subseteq \mathcal{F}$. There is an infinite subset $S \subseteq \omega$ such that $S \setminus F$ is finite for all $F \in \mathcal{F}$.

Applications of P(κ) appear in 3U.

(r) **Proposition**. For $\aleph_0 \leqslant \kappa \leqslant 2^{\aleph_0}$, SL($\kappa$) iff P($\kappa$).

Proof. Suppose SL(κ) is true and $F \subseteq P(\omega)$ such that $|F| < \kappa$ and $|\cap K| = \aleph_0$ for each nonempty finite set $K \subseteq F$. Now, let $A = \{\omega\}$ and $B = \{\omega \setminus F : F \in F\}$. For $A \in A$ and finite $K \subseteq B$, $|A \setminus \cup K| = \aleph_0$. By SL($\kappa$), there is an infinite set $S \subseteq \omega$ such that $|S \cap (\omega \setminus F)| = |S \setminus F| < \aleph_0$ for all $F \in F$. Conversely, suppose P(κ) is true and $\mathcal{A}$ and B are subsets of $\mathbb{P}(\omega)$, $|\mathcal{A}| < \kappa$, $|B| < \kappa$, and $|A \setminus \cup K| = \aleph_0$ for $A \in \mathcal{A}$ and finite $K \subseteq B$. Let $S = \{S \subset \omega : S \text{ is finite}\}$. For $A \in \mathcal{A}$, let $A' = \{S \in S : S \cap A \neq \emptyset\}$ and $\mathcal{A}' = \{A' : A \in \mathcal{A}\}$. For $B \in B$, let $B' = \{S \in S : S \cap B \neq \emptyset\}$ and $B' = \{S \setminus B' : B \in B\}$. For $n \in \omega$, let $E_n = \{S \in S : S \cap \{0,1,\ldots,n\} \neq \emptyset\}$ and $E = \{S \setminus E_n : n \in \omega\}$. Let $F = \mathcal{A}' \cup B' \cup E$. Clearly, $|F| < \kappa$ and $F \subseteq \mathbb{P}(S)$. For m, n, k $\in \omega$, let $\{A_0,\ldots,A_n\} \subseteq \mathcal{A}$ and $\{B_0,\ldots,B_k\} \subseteq B$ and note that $E_0 \subseteq E_1 \subseteq \ldots \subseteq E_m$. For each $j \leqslant n$, let $T_j = A_j \setminus \cup\{B_i : i \leqslant k\}$. By hypothesis, T_j is an infinite set. So, it follows that $\{S \in S : S \cap T_j \neq \emptyset$ for all $j \leqslant n$, $S \subseteq \cup\{T_j : j \leqslant n\}$, and $S \cap \{0, 1, \ldots, m\} = \emptyset\}$ is infinite. This implies that $|\cap\{A_j' : j \leqslant n\} \cap (\cap\{S \setminus B_i' : i \leqslant k\}) \cap (S \setminus E_m)| = \aleph_0$. Since $|S| = \aleph_0$, it follows that P(κ) is true when ω is replaced by S (and $F \subseteq \mathbb{P}(S)$). Thus, by P($\kappa$), there is an infinite subset $C \subseteq S$ such that for $A \in \mathcal{A}$, $B \in B$, and $n \in \omega$, $C \setminus A'$, $C \setminus (S \setminus B')$, and $C \setminus (S \setminus E_n)$ are finite sets. Therefore, $C \cap B'$ and $C \cap E_n$ are finite sets. Let $C = \cup C$. Since C is infinite, C is infinite. Assume $C \cap A$ is finite for some $A \in \mathcal{A}$. Let $C \setminus A' = \{S_0, \ldots, S_m\}$; for some $n \in \omega$, $(C \cap A) \cup (\cup\{S_i : i \leqslant m\}) \subseteq \{0, 1, \ldots, n\}$. If $S \in$

$\mathcal{C}\setminus E_n$, then $S \neq S_i$ for $i \leq m$ and $S \cap (C \cap A) = \emptyset$. Also, $S \in \mathcal{C} \subseteq A' \cup (\cup\{S_i : i \leq n\})$ implies $S \in A'$ and $\emptyset \neq S \cap A \subseteq S \subseteq C$. Thus $\emptyset \neq S \cap A \subseteq S \cap (C \cap A)$, which is a contradiction. This shows $|C \cap A| = \aleph_0$ for $A \in A$. For $B \in \mathcal{B}$, let $\mathcal{C} \cap B' = \{S_0, \ldots, S_n\}$. Thus, $C \cap B \subseteq \cup\{S_i : i \leq n\}$ and $C \cap B$ is a finite set. This completes the proof that $SL(\kappa)$ holds.

■

One consequence of Solovay's result is this next fact due to Silver.

(s) **Theorem**. $[SL(\kappa^+)]$. If $X \subseteq \mathbb{R}$ and $|X| = \kappa$, then every subset of X is a G_δ subset of X.

Proof. Let $Z \subseteq X$. For each $n \in \omega$, let $\mathcal{B}_n = \{(m + (k/2^n)\ ,\ m + (k+2)/2^n) : m \text{ an integer}, k = 0, 1, \ldots, 2^n\}$, let $\mathcal{B} = \cup\{\mathcal{B}_n : n \in \omega\}$. Then $\mathcal{B}$ is a countable open base for $\mathbb{R}$ such that if $x, y \in \mathbb{R}$, $x \neq y$, then $|\{U \in \mathcal{B} : x, y \in U\}| < \aleph_0$. Let $\mathcal{B} = \{U_i : i \in \omega\}$. For $p \in X$, let $\omega(p) = \{i \in \omega : p \in U_i\}$. Let $A = \{\omega(p) : p \in X\setminus Z\}$ and $\mathcal{C} = \{\omega(p) : p \in Z\}$. By $SL(\kappa^+)$, there is some $S \subseteq \omega$ such that $|\omega(p) \cap S| < \aleph_0$ for $p \in X\setminus Z$ and $|\omega(p) \cap S| = \aleph_0$ for $p \in Z$. For $n \in \omega$, define W_n to be $\cup\{U_i : i \in S, i \geq n\}$. Since $|\omega(p) \cap S| = \aleph_0$ for $p \in Z$, there is some $i \in \omega(p) \cap S$ such that $i \geq n$. So, $p \in W_n$. This shows that $Z \subseteq W_n$ for all $n \in \omega$. If $p \in X\setminus Z$, then $|\omega(p) \cap S| < \aleph_0$. So, there exists some n such that $p \notin W_n$. Hence, $Z = X \cap [\cap\{W_n : n \in \omega\}]$ and so Z is a G_δ subset of X.

■

An easy consequence of Silver's result is the next corollary.

(t) **Corollary**. The following hold:

(1) $[SL(\kappa^+)]$ $2^\kappa = 2^{\aleph_0}$.

(2) $[SL(2^{\aleph_0})]$ $2^{\aleph_0}$ is a regular cardinal.

Proof

(1) Since $\aleph_0 \leqslant \kappa$, evidently $2^{\aleph_0} \leqslant 2^\kappa$. Let X be a subset of $\mathbb{R}$ such that $|X| = \kappa$. Let $B = \{U_i : i \in \omega\}$ be the countable open base of $\mathbb{R}$ used in the proof of 3.5(s), and for $p \in X$, let $\omega(p) = \{i \in \omega : p \in U_i\}$. For each $Z \subseteq X$, use $SL(\kappa^+)$ (see proof of 3.5(s)), to find a subset S_Z of ω such that $|S_Z \cap \omega(p)| = \aleph_0$ for $p \in Z$ and $|S_Z \cap \omega(p)| < \aleph_0$ for $p \in X\setminus Z$.

Evidently, if $T \subseteq X$ and $T \neq Z$, then $S_Z \neq S_T$. Thus the function from $\mathbb{P}(X)$ to $\mathbb{P}(\omega)$ taking Z to S_Z is one-to-one, and it follows that $2^\kappa \leqslant 2^{\aleph_0}$.

(2) Assume that $\kappa = cf(2^{\aleph_0}) < 2^{\aleph_0}$. Then there is a subset $S \subseteq 2^{\aleph_0}$ such that $2^{\aleph_0} = \cup\{[0,\alpha) : \alpha \in S\}$ and $|S| = \kappa$. Hence, $2^{\aleph_0} = |\cup S| \leqslant \Sigma\{|\alpha| : \alpha \in S\} < \Pi\{2^{\aleph_0} : \alpha \in S\} = (2^{\aleph_0})^\kappa$. The strict inequality follows from 2.6(o)(8). But $(2^{\aleph_0})^\kappa = 2^{\aleph_0 \kappa} = 2^\kappa = 2^{\aleph_0}$ (the latter equality follows from (1) above). This shows that $2^{\aleph_0} < (2^{\aleph_0})^\kappa = 2^{\aleph_0}$ which is a contradiction.

■

Chapter 3 - Problems

3A. $\tau(X)$ as a Boolean algebra. For a space X, prove that $(\tau(X),\cup,\cap)$ is a Boolean algebra iff X is discrete.

3B. $R(X)$ as a Boolean algebra. Let X be a space.

(1) Provide the details of the proof of 3.1(e)(4), i.e., prove that if X is a space, then $(R(X),\vee,\wedge)$ is a Boolean algebra.

(2) Prove that the function $f : RO(X) \rightarrow R(X)$ defined by $f(U) = c\ell_X U$, for $U \in RO(X)$, is a Boolean isomorphism. If $A \in R(X)$, show that $f^{\leftarrow}(A) = \text{int}_X A$.

(3) Let X be a dense subspace of a space T and let $A \in R(X)$. Prove that $c\ell_T A \in R(T)$.

(4) With X and T as above, define $g : R(X) \rightarrow R(T)$ by letting $g(A) = c\ell_T A$. Use 3.1(n) to prove that g is a Boolean isomorphism, and verify that $g^{\leftarrow}(B) = B \cap X$ if $B \in R(T)$.

3C. Intersection of Boolean subalgebras. If $\mathcal{a}$ is a nonempty family of Boolean subalgebras of B, prove that $\cap\mathcal{a}$ is a Boolean subalgebra of B.

3D. Lattice homomorphism vs. Boolean homomorphism. Let $f \in F(A,B)$ be a lattice homomorphism between two Boolean algebras

A and B such that $f(0) = 0$ and $f(1) = 1$. Prove that f is a Boolean homomorphism.

3E. Unions of Boolean subalgebras

(1) Give an example of a Boolean algebra B and two Boolean subalgebras A_1 and A_2 such that $A_1 \cup A_2$ is not a Boolean subalgebra of B.

(2) Let B be a Boolean algebra, and let $\mathfrak{a}$ be a nonempty chain of Boolean subalgebras of B. Prove $\cup\mathfrak{a}$ is a Boolean subalgebra of B.

(3) Let B and C be Boolean algebras, and let $\mathfrak{a}$ be a nonempty chain of Boolean subalgebras of B. For $A \in \mathfrak{a}$, let $h_A \in F(A,C)$ be a Boolean homomorphism such that if $A_1, A_2 \in \mathfrak{a}$ and $A_1 \subseteq A_2$, then $h_{A_1} = h_{A_2} \mid A_1$. Define h by $h(a) = h_A(a)$ for $A \in \mathfrak{a}$ and $a \in A$. Show that h is a function, i.e., $h \in F(\cup\mathfrak{a},C)$, and a Boolean homomorphism.

3F. Generalized Distributive and De Morgan's Laws. Let B be a Boolean algebra, let $F = \{b_k : k \in K\}$ and $E = \{e_{ij} : (i,j) \in I \times J\}$ be subsets of B where I and J are nonempty, finite sets and K is an arbitrary indexing set. Prove:

(1) $\wedge_I(\vee_J e_{i,j}) = \vee_{f \in F(I,J)}(\wedge_I e_{i,f(i)})$,

(2) $\vee_I(\wedge_J e_{i,j}) = \wedge_{f \in F(I,J)}(\vee_I e_{i,f(i)})$, and

(3) if $\vee\{b_k : k \in K\}$ exists, then $\wedge\{b_k{}' : k \in K\} =$

$(\vee\{b_k : k \in K\})'$.

3G. Boolean homomorphisms and B-filters. Let B be a Boolean algebra and 2 denote the Boolean algebra $\{0,1\}$ with $0 < 1$.

(1) If $f : B \to 2$ is a surjective Boolean homomorphism, show that $f^{\leftarrow}(1)$ is a B-ultrafilter.

(2) If $\mathcal{U}$ is a B-ultrafilter, define $f_B \in F(B,2)$ by $f_B(a) = 1$ if $a \in \mathcal{U}$ and $f_B(a) = 0$ if $a \notin \mathcal{U}$. Show that f_B is a surjective Boolean homomorphism.

(3) Let $X(B) = \{f \in F(B,2) : f$ is a surjective Boolean homomorphism$\}$. Since $X(B) \subseteq 2^B$, where 2 has the discrete topology, $X(B)$ can be considered as a space. Prove that $X(B)$ is a compact, zero-dimensional space by proving it is closed in 2^B.

(4) For $a \in B$, let $\pi_a \in F(2^B,2)$ be the projection function defined by $\pi_a(f) = f(a)$. Define $k \in F(B,\mathcal{B}(X(B)))$ by $k(a) = \pi_a^{\leftarrow}(1) \cap X(B)$. Prove that k is a Boolean isomorphism.

(5) Let Y be a compact, zero-dimensional space. For $f \in X(\mathcal{B}(Y))$, show that $\cap f^{\leftarrow}(1)$ is a singleton. So, $\cap f^{\leftarrow}(1) = \{y_f\}$ for some $y_f \in Y$.

(6) Define $G \in F(X(\mathcal{B}(Y)),Y)$ by $G(f) = y_f$. Prove that G is a homeomorphism.

3H. Compact, zero-dimensional spaces to Boolean algebras. Prove (1), (3), and (4) of 3.2(j).

3I. Generated Boolean Algebras. Let A and B be Boolean algebras, let $E \subseteq B$ such that $B = \langle E \rangle$, and let $f, g \in F(B,A)$ be Boolean homomorphisms.

(1) If $f|E = g|E$, prove that $f = g$.

(2) If $A = \langle f[E] \rangle$, prove that f is onto.

(3) If E is infinite, prove that $|B| = |E|$.

(4) If E is finite, prove that $|B| \leqslant 2^{(2^{|E|})}$.

(5) Let $h \in F(E,A)$ have the property that $h(b') = h(b)'$ whenever both $b, b' \in E$. Let $E' = \{b' : b \in E\}$. Extend the domain of h to $E \cup E'$ by defining $h(b')$ to be $h(b)'$ for $b \in E$. Prove h can be extended to a Boolean homomorphism $H \in F(B,A)$ iff for each nonempty finite set $F \subseteq E \cup E'$ such that $\bigwedge F = 0$, $\bigwedge\{h(b) : b \in F\} = 0$. (Hint: Repeat the proof of 3.1(n) using 3.1(l)(2), the De Morgan Laws of 3.1(g)(6), and the generalized distribution laws of 3F.)

(6) Let $H \subseteq A$ be such that $\langle H \rangle = A$ and let $h \in F(E,H)$ be an onto function. Suppose for finite subsets $F_1, F_2 \subseteq E$, $\bigwedge F_1 \leqslant \bigvee F_2$ iff $\bigwedge h[F_1] \leqslant \bigvee h[F_2]$. Prove B and A are isomorphic Boolean algebras.

3J. Dense Subalgebras and Minimal Completions

(1) Let A be a dense subalgebra of a Boolean algebra B, $a \in A$, and $b \in B$. Prove:

(a) if $a < b$ (respectively, $b < a$), there is some $a_1 \in$

A such that $a < a_1 \leq b$ (respectively, $b \leq a_1 < a$),

(b) if $\emptyset \neq S \subseteq A$ and $a = \bigvee S$ (respectively, $a = \bigwedge S$) exists in A, then $\bigvee S$ (respectively, $\bigwedge S$) exists in B and equals a, and

(c) if $0 < b < 1$ and $S = \{a \in A : 0 < a \leq b\}$ (respectively, $S = \{a \in A : b \leq a < 1\}$), then $b = \bigvee S$ (respectively, $b = \bigwedge S$).

(2) Let (B,i) be a completion of a Boolean algebra A where $i \in F(A,B)$ is the inclusion function. Prove that if A is dense in B, then i is complete.

(3) Let (B,i) be a minimal completion of a Boolean algebra A. Prove that A is dense and regular in B.

3K. <u>Free Boolean Algebras and Cantor κ-spaces.</u> A Boolean algebra B is **free** if there is a subset $E \subseteq B$ such that

(a) $B = \langle E \rangle$ and

(b) for each Boolean algebra C, every function $f \in F(E,C)$ has a Boolean homomorphism extension $\bar{f} \in F(B,C)$, i.e., $\bar{f}$ is a Boolean homomorphism and $\bar{f} \upharpoonright E = f$.

The set E is called a **set of free generators** for B.

(1) If E_i is a set of free generators for Boolean algebras B_i, $i = 1, 2$, then prove B_1 and B_2 are Boolean

isomorphic iff $|E_1| = |E_2|$. (Hint: If $f \in F(E_1,E_2)$ is a bijection and $g = f^{\leftarrow}$, extend to $\bar{f}$ and $\bar{g}$ and use 3I(1) to show $\bar{f} \circ \bar{g} = 1_{B_2}$ and $\bar{g} \circ \bar{f} = 1_{B_1}$. Now show $\bar{f}$ is a Boolean isomorphism. Conversely, suppose B_1 and B_2 are isomorphic. Use 3I(3) to handle the infinite case. If $|E_1| < |E_2| < \aleph_0$, then there is a function $f \in F(E_2,E_1)$ which is onto but not one-to-one. Then f has a Boolean homomorphism extension $\bar{f} \in F(B_2,B_1)$ which is onto but not one-to-one. Since B_1 and B_2 are finite, $|B_2| > |B_1|$.)

For each cardinal κ, let C_κ denote the compact zero-dimensional space $\mathbf{2}^\kappa$. The space C_κ is called the **Cantor κ-space**; $C_{\aleph_0}$ is the well-known Cantor space discussed in 3.3. For each $\alpha < \kappa$, let $\Pi_\alpha \in F(C_\kappa,\mathbf{2})$ be the projection function onto the αth coordinate space.

(2) Prove $B(C_\kappa)$ is a free Boolean algebra with $E = \{\Pi_\alpha^{\leftarrow}(\{1\}) : \alpha < \kappa\}$ as a set of free generators. (Hint: Use 3I(5).)

Note: It follows from (1) and (2) that a Boolean algebra B is free iff $B \cong B(C_\kappa)$ for some cardinal $\kappa > 0$.

(3) Suppose B is a Boolean algebra, $E \subseteq B$, and $B = \langle E \rangle$. Prove that E is a set of free generators of B if for each nonempty finite set $A \subseteq E \cup E'$, $\wedge A = 0$ implies $e, e' \in A$ for some $e \in E$.

(4) Let B be a Boolean algebra satisfying *ccc*. Let α

be an ordinal and $\{b_\beta : \beta < \alpha\} \subseteq B$ such that if $\beta < \gamma < \alpha$, then $b_\beta < b_\gamma$. Prove $|\alpha| \leqslant \aleph_0$.

(5) Let B be a Boolean algebra. Prove B satisfies *ccc* iff $S(B)$ is a *ccc* space.

Note that in 3R, it is proven that every free Boolean algebra satisfies *ccc*.

3L. Factor Algebras. An **ideal** of a Boolean algebra B is a subset J of B such that $\{a' : a \in J\}$ is a filter on B. If $a, b \in B$, we denote $(a \wedge b') \vee (a' \wedge b)$ by $a \Delta b$.

(1) Let J be an ideal of a Boolean algebra B. Define a relation $\equiv$ on B as follows: for $a, b \in B$, $a \equiv b$ if $a \Delta b \in J$. Prove that $\equiv$ is an equivalence relation on B.

(2) If J is an ideal of the Boolean algebra B, let B/J denote the set of equivalence classes on B induced by $\equiv$, and let $[a]$ denote the equivalence class containing the element a of B. Show that if $\wedge$, $\vee$, and $'$ are defined on B/J by:

$$[a] \vee [b] = [a \vee b],$$
$$[a] \wedge [b] = [a \wedge b], \text{ and}$$
$$[a]' = [a'].$$

then $\vee$, $\wedge$, and $'$ are well-defined operations on B/J; furthermore, show that $(B/J, \vee, \wedge, ')$ is a Boolean algebra such that $[a] \leqslant [b]$ iff $b \wedge a' \in J$.

The Boolean algebra B/J is called the **factor algebra** or **quotient algebra** of B with respect to J.

(3) Show that $\phi : B \to B/J$ defined by $\phi(b) = [b]$ is a Boolean algebra homomorphism from B onto B/J, and that $J = \phi^{\leftarrow}(0)$.

(4) Prove that $S(B/J)$ is homeomorphic to the subspace $S(B) \setminus \bigcup\{\lambda(b) : b \in J\}$ of $S(B)$. (Hint: Dualize (3).)

(5) Let $\phi : B_1 \to B_2$ be a Boolean algebra homomorphism from B_1 into B_2. Prove that $\phi[B_1]$ is a subalgebra of B_2 and

(a) $\{b \in B_1 : \phi(b) = 0\}$ is an ideal $J(\phi)$ of B_1,

(b) the factor algebra $B_1/J(\phi)$ is isomorphic to $\phi[B_1]$, and

(c) $S(\phi[B_1])$ is homeomorphic to $S(B_1)\setminus\bigcup\{\lambda(b) : b \in B_1 \text{ and } \phi(b) = 0\}$.

(6) Prove that every ideal of a Boolean algebra is contained in a maximal ideal.

(7) Let B be a Boolean algebra and $J \subseteq B$

(a) Prove that J is a maximal ideal in B iff $B\setminus J$ is an ultrafilter on B.

(b) Let J be an ideal of B. Prove that B/J is isomorphic to the Boolean algebra 2 described in 3G iff J is a maximal ideal.

3M. The Boolean Algebra $\mathbb{P}(\kappa)/\mathrm{FN}$. Let κ be an infinite cardinal, $\mathrm{FN} = \{A \subset \kappa : A \text{ finite}\}$, and $A, B \in \mathbb{P}(\kappa)$.

(1) Prove FN is an ideal in the Boolean algebra $(\mathbb{P}(\kappa), \cup, \cap)$.

By 3L, $\mathbb{P}(\kappa)/\mathrm{FN}$ is a Boolean algebra satisfying $[A] \vee [B] = [A \cup B]$, $[A] \wedge [B] = [A \cap B]$, $[A]' = [\kappa \setminus A]$, and $[A] \leqslant [B]$ iff $A \setminus B$ is finite.

(2) Prove that $|[A]| = \kappa$, $\bigvee(\mathbb{P}(\kappa)/\mathrm{FN}) = \mathbb{P}(\kappa)$, and $|\mathbb{P}(\kappa)/\mathrm{FN}| = 2^{\kappa}$.

(3) Prove that $\mathbb{P}(\kappa)/\mathrm{FN}$ does not satisfy *ccc*. (Hint: It suffices to prove that $\mathbb{P}(\aleph_0)/\mathrm{FN}$ does not satisfy *ccc*. Let $f : \aleph_0 \to \mathbb{Q}$ be a bijection, and for each $x \in \mathbb{R}$, choose a nonconstant sequence $S_x \subseteq \mathbb{Q}$ that converges to x. Show $\{[f^{\leftarrow}[S_x]] : x \in \mathbb{R}\}$ is an uncountable antichain in $\mathbb{P}(\aleph_0)/\mathrm{FN}$.

(4) Show that $\mathbb{P}(\kappa)/\mathrm{FN}$ is not a complete Boolean algebra. (Hint: Let $\{E_n : n \in \mathbb{N}\}$ be a pairwise disjoint family of infinite sets. Assume $[A] = \bigvee\{[E_n] : n \in \mathbb{N}\}$. Let $p_n \in E_n \cap A$ for each $n \in \mathbb{N}$, and let $B = A \setminus \{p_n : n \in \mathbb{N}\}$. Show that $\bigvee\{[E_n] : n \in \mathbb{N}\} \leqslant [B] < [A]$.)

3N. The Boolean Algebra L/Z. In this problem it is assumed that the reader knows some basic facts about the set L of Lebesgue measurable subsets of [0,1] (the measure is denoted by μ); in particular, the reader should know enough measure theory to verify that $(L,\cup,\cap,',\subseteq)$ is a Boolean algebra. Let $Z = \{C \in L : \mu(C) = 0\}$ and let A, B $\in L$.

(1) Prove that Z is an ideal in L. By 3L, L/Z is a Boolean algebra where $[A] \vee [B] = [A \cup B]$, $[A] \wedge [B] = [A \cap B]$, $[A'] = [[0,1]\setminus A]$, and $[A] \leqslant [B]$ iff $\mu(A\setminus B) = 0$.

(2) Prove that $|L/Z| = 2^{\aleph_0}$. (Hint: Let $U_n = (1/n+1,1/n)$ for $n \in \mathbb{N}$. If D, E $\subseteq \mathbb{N}$ and D $\neq$ E, show that $[\cup\{U_n : n \in D\}] \neq [\cup\{U_n : n \in E\}]$.)

(3) Show that L/Z satisfies the *ccc* property. (Hint: If $\{[A_\alpha] : \alpha < \omega_1\}$ is an antichain, then for $n \in \mathbb{N}$, let $S_n = \{\alpha < \omega_1 : \mu(A_n) > 1/n\}$. Show that $\omega_1 = \cup\{S_n : n \in \mathbb{N}\}$ and deduce that S_n is uncountable for some $n \in \mathbb{N}$. Let $\{[B_k] : 1 \leqslant k \leqslant n + 1\} \subseteq S_n$. Show that $\mu(\cup\{B_k : 1 \leqslant k \leqslant n + 1\}) = \Sigma\{\mu(B_k) : 1 \leqslant k \leqslant n + 1\} \geqslant 1 + 1/n$ and this is impossible.)

(4) Show that L/Z is complete. (Hint: Let $A = \{[A_i] : i \in I\} \subseteq L/Z$. Show that there is an antichain B which is maximal with respect to this property: if $[B] \in B$, there is some $i \in I$ such that $[B] \leqslant [A_i]$. Use (3) to deduce that B is countable. Show that $\vee B$ exists and $\vee B = \vee \mathfrak{a}$.)

30. <u>Density of Product Spaces</u>. Let κ be an infinite cardinal and $m = 2^\kappa$. Let $C_\kappa = 2^\kappa$ be the Cantor κ-space introduced in 3K.

(1) Prove that $w(C_\kappa) = \kappa$ (see 2N(3) for definition of weight).

Let $D = D(\kappa)$, i.e., κ with the discrete topology. By (1), there is a base $\mathcal{B}$ for C_κ such that $|\mathcal{B}| = \kappa$. Let $\mathfrak{a} = \{(B_1,\ldots,B_n;\ \alpha_1,\ldots,\alpha_n) : n \in \mathbb{N},\ B_i \in \mathcal{B},\ \alpha_i < \kappa$ for $1 \leqslant i \leqslant n$ and $B_1, B_2, \ldots, B_n$ are pairwise disjoint and nonempty.

(2) Prove that $|\mathfrak{a}| = \kappa$.

Define $f \in F(\mathfrak{a}, D^m)$ as follows: for $\beta < m$,

$$f(B_1,\ldots,B_n;\ \alpha_1,\ldots,\alpha_n)(\beta) = \begin{cases} \alpha_i \text{ if } \beta \in B_i \text{ for some } i \leqslant n. \\ 0 \text{ otherwise} \end{cases}$$

(3) Prove that $f[\mathfrak{a}]$ is dense in D^m and conclude that $d(D^m) = \kappa$.

(4) (Hewitt-Pondiczery). Let $\{X_a : a \in A\}$ be a family of spaces such that $d(X_a) \leqslant \kappa$ for each $a \in A$ and $|A| \leqslant m$. Prove that $d(\Pi\{X_a : a \in A) \leqslant \kappa$. (Hint: First, show it suffices to assume that $|A| = m$. For each $a \in A$, find a function $g_a \in F(D, X_a)$ such

that $g_a[D]$ is dense in X_a. Let $g = \Pi\{g_a : a \in A\}$. By 1.7(f), g is continuous. Use 2N(2)(b) to deduce that it suffices to show that $d(D^m) = \kappa$.)

(5) Prove that $d(C_\kappa) \geqslant \min\{\beta : \beta$ is a cardinal, $2^\beta \geqslant \kappa\}$. (Hint: Use (1) and 2N(2)(c).)

3P. Properties of Spaces Satisfying *ccc*. First recall from 2N that if a space X is metrizable, then X has the *ccc* iff X is Lindelöf iff X is second countable.

(1) Show that the property of satisfying *ccc* is hereditary on open subspaces. Also, give an example to show that *ccc* is not hereditary on closed spaces.

(2) Prove that the continuous image of a space satisfying *ccc* is a space satisfying *ccc*.

A space is called **weakly Lindelöf** if every open cover has a countable subfamily whose union is dense.

(3) Prove that a space satisfies *ccc* iff every open subspace is weakly Lindelöf.

(4) Prove that a weakly Lindelöf paracompact space is Lindelöf.

(5) Prove 3.5(c)(5); i.e., that an antichain in a poset is contained in a maximal antichain.

Note that additional properties of the product of spaces satisfying *ccc* are discussed in 3R, 3S, and 3T.

3Q. Denseness in Topologies and Posets

(1) Let X be a space and $D \subseteq \tau(X)\setminus\{\emptyset\}$. Show that D is dense in the poset $(\tau(X),\subseteq)$ iff D is a π-base for X (see 2N(5) for the definition of a π-base).

(2) Let X be a set and $D \subseteq \mathbb{P}(X)\setminus\{\emptyset\}$. Prove that D is dense in $(\mathbb{P}(X),\subseteq)$ iff $\{\{x\} : x \in X\} \subseteq D$.

(3) Let X be a nonempty space. Find a countable dense subset D of $C^*(X)$ (relative to the partial order on $C^*(X)$ defined in 2.1(f)(5)).

(4) Let X be a normal but not a countably compact space, and let D be a dense subset of C(X) (partially ordered as described in 2.1(f)(5)). Show that $|D| > \aleph_0$.

Let κ and m be nonzero cardinals and $FN(\kappa,m) = \{f \subseteq \kappa \times m : f$ is a function and $|\text{dom } f| < \aleph_0\}$ (dom f is used to denote the domain of f). Define $f \leqslant g$ if $f \supseteq g$, i.e., if f extends g. It is obvious that $FN(\kappa,m)$ is a poset.

(5) If $m \leqslant \aleph_0$, prove that $FN(\kappa,m)$ satisfies *ccc*.

(6) For $\alpha \in \kappa$, prove that $D_\alpha = \{f \in FN(\kappa,m) : \alpha \in \text{dom } f\}$ is dense in $FN(\kappa,m)$.

(7) If $\kappa \geqslant \aleph_0$ and $\beta \in m$, prove that $R_\beta = \{f \in FN(\kappa,m) : \beta \in \text{ran } f\}$ is dense in $FN(\kappa,m)$ (ran f is used to denote the range of f, i.e., $\{f(\alpha) : \alpha < \kappa\}$).

(8) If F is a filter on $FN(\kappa,m)$, prove that $\cup F \in F(S,m)$ for some $S \subseteq \kappa$; in particular, if F is an ultrafilter,

prove that $S = \kappa$.

(9) If $\kappa \geqslant \aleph_0$ and $\mathcal{F}$ is a filter on $FN(\kappa, m)$ such that $D_\alpha \cap \mathcal{F} \neq \emptyset$ for $\alpha \in \kappa$ and $R_\beta \cap \mathcal{F} \neq \emptyset$ for $\beta \in m$, prove that $\bigcup\mathcal{F} \in F(\kappa, m)$ and $\bigcup\mathcal{F}$ is onto.

3R. __Δ-systems and Products of *ccc* Spaces.__ Let κ be a cardinal. A family $\mathcal{A} \subseteq \mathbb{P}(\kappa)$ is called a **Δ-system** if there is a subset $r \subseteq \kappa$ such that $\alpha \cap \beta = r$ for all $\alpha, \beta \in \mathcal{A}$, $\alpha \; G \; \beta$. The subset r is called the **root** of the Δ-system. Clearly a family of pairwise disjoint sets is a Δ-system.

(1) If $\emptyset \neq \mathcal{A} \subseteq \mathbb{P}(\kappa)$, prove there is a family $\mathcal{B} \subseteq \mathcal{A}$ which is maximal with respect to this property: if $A, B \in \mathcal{B}$, $A \neq B$, then $A \cap B = \emptyset$. Such a set $\mathcal{B}$ is called a **maximal family of pairwise disjoint subsets for $\mathcal{A}$**.

(2) Let $n \in \omega\backslash\{0\}$ and let $\mathcal{A} \subseteq \mathbb{P}(\kappa)$ be an uncountable collection of sets of cardinality n. Prove there is an uncountable Δ-system $\mathcal{B} \subseteq \mathcal{A}$. (Hint: Prove by induction. When $n = 1$, note that $\mathcal{A}$ is a Δ-system of pairwise disjoint sets. Assume the result is true for all $k < n$. Let $\mathcal{C}$ be a maximal family of pairwise disjoint subsets for $\mathcal{A}$. If $\mathcal{C}$ is uncountable, we are done. Suppose $\mathcal{C}$ is countable. Show (i) for some $C \in \mathcal{C}$, $\{A \in \mathcal{A} : A \cap C \neq \emptyset\}$ is uncountable and (ii) for some $x \in C$, $\{A \in \mathcal{A} : x \in A\}$ is uncountable. Now, apply the inductive hypothesis to $\{A\backslash\{x\} : x \in A \in \mathcal{A}\}$.)

(3) (Δ-system Lemma for finite sets.) If $\mathcal{A} \subseteq \mathbb{P}(\kappa)$ is

an uncountable collection of finite sets, then there is an uncountable Δ-system $\mathcal{B} \subseteq \mathcal{Q}$. (Hint: Show that for some $n \in \mathbb{N}$, $\{A \in \mathcal{Q} : |A| = n\}$ is uncountable; now, apply (2).)

(4) Let $\{X_\alpha : \alpha < \kappa\}$ be a family of spaces. Prove that $\Pi\{X_\alpha : \alpha < \kappa\}$ is a *ccc* space iff for each finite set $F \subseteq \kappa$, $\Pi\{X_\alpha : \alpha \in F\}$ is a *ccc* space. (Hint: Suppose $\Pi\{X_\alpha : \alpha \in F\}$ is a *ccc* space for each finite set $F \subseteq \kappa$. Let $\mathcal{U}$ be an uncountable family of basic nonempty pairwise disjoint open sets in $\Pi\{X_\alpha : \alpha < \kappa\}$. So, for $U \in \mathcal{U}$, there is a finite set $F_U \subseteq \kappa$ and nonempty open sets $U_i \subseteq X_i$ for $i \in F_U$ such that $U = \Pi\{U_i : i \in F_U\} \times \Pi\{X_\alpha : \alpha \in \kappa \setminus F_U\}$. Note that $U \cap V = \emptyset$ iff there is some $i \in F_U \cap F_V$ such that $U_i \cap V_i = \emptyset$; in particular, if F is a finite subset of κ, then $\{U \in \mathcal{U} : F_U \subseteq F\}$ is countable by hypothesis. Thus, if $\{F_U : U \in \mathcal{U}\}$ is countable then $\mathcal{U}$ is countable. So, $\{F_U : U \in \mathcal{U}\}$ is uncountable. By the Δ-system Lemma, there is an uncountable $\mathcal{V} \subseteq \mathcal{U}$ and a finite set $F \subseteq \kappa$ such that $F_U \cap F_V = F$ for $U, V \in \mathcal{V}$, $U \neq V$. This implies that $\{(\Pi V_i : i \in F_V \cap F\} : V \in \mathcal{V}\}$ is an uncountable family of nonempty, pairwise disjoint open sets in $\Pi\{X_i : i \in F\}$, a contradiction.)

(5) Show that the product of separable spaces is a *ccc* space.

(6) If $\kappa > 2^{\aleph_0}$, show that the Cantor κ-space C_κ (defined in 3K) is compact and satisfies *ccc* but is

not separable. (Hint: Use 3O.)

(7) Prove that a free Boolean algebra satisfies *ccc*. (Hint: Use (5) and the results of 3K.)

3S. MA and the Product of *ccc* Spaces

(1) Let X be a space satisfying *ccc* and $\mathcal{U} \subseteq \tau(X)\setminus\{\emptyset\}$ such that $|\mathcal{U}| = \kappa$, where κ is an uncountable cardinal.

(a) Let $\mathcal{U} = \{U_\alpha : \alpha < \kappa\}$ and $V_\alpha = \cup\{U_\beta : \beta > \alpha\}$ for $\alpha < \kappa$. Show there is some $\alpha < \kappa$ such that $c\ell_X V_\beta = c\ell_X V_\alpha$ for all $\beta > \alpha$. (Hint: See the proof of 3.5(j).)

(b) Let α be the least ordinal satisfying (a). Let P $= (\tau(X)\setminus\{\emptyset\}) \cap \mathbb{P}(V_\alpha)$. Show that $(P,\subseteq)$ is a poset satisfying *ccc*.

(c) For $\alpha \leqslant \beta < \kappa$, show that $D_\beta = \{U \in P : U \subseteq U_\gamma \text{ for some } \gamma > \beta\}$ is dense in P.

(d) [MA(κ)]. If κ is a regular cardinal, show there is some $\mathcal{W} \subseteq \mathcal{U}$ such that $|\mathcal{W}| = \kappa$ and $\mathcal{W}$ has the finite intersection property. (Hint: Let F be a filter in P such that $D_\beta \cap F \neq \emptyset$ for $\alpha \leqslant \beta < \kappa$. Show that F is an open filter base on X. Let G be the open filter generated by F and $\mathcal{W} = G \cap \mathcal{U}$. Clearly, $\mathcal{W}$ has the finite intersection property. Show $\{\gamma \in \kappa : U_\gamma \in \mathcal{W}\}$ is cofinal in κ; hence, $|\mathcal{W}| = \kappa$.)

(2) [MA($\aleph_1$)]. Prove that the product of spaces satisfying *ccc* is a space satisfying *ccc*. (Hint: Note, by 3R(4), that it suffices to prove that if X and Y are spaces satisfying *ccc*, then $X \times Y$ satisfies *ccc*. Assume $X \times Y$ does not satisfy *ccc*. Show there are families $\{U_\alpha : \alpha < \omega_1\}$ and $\{V_\alpha : \alpha < \omega_1\}$ of nonempty open sets in X and Y, respectively, such that $\{U_\alpha \times V_\alpha : \alpha < \omega_1\}$ is an antichain in $X \times Y$. Apply (1) above twice to obtain an uncountable $A \subseteq \omega_1$ such that $\{U_\alpha : \alpha \in A\}$ and $\{V_\alpha : \alpha \in A\}$ each has the finite intersection property. For $\alpha, \beta \in A$, $\alpha \neq \beta$, $(U_\alpha \times V_\alpha) \cap (U_\beta \times V_\beta) = (U_\alpha \cap U_\beta) \times (V_\alpha \cap V_\beta) \neq \emptyset$, a contradiction.)

3T. <u>CH and the Product of *ccc* Spaces</u>. If S is a set let $[S]^2 = \{A \in \mathbb{P}(S) : |A| = 2\}$. Let $K \subseteq [\omega_1]^2$ and $X_K = \{A \subseteq \omega_1 : A$ is maximal with respect to the property that $[A]^2 \subseteq K\}$. For a finite set $F \subseteq \omega_1$, let $\hat{F} = \{A \in X_K : F \subseteq A\}$. For $\alpha < \omega_1$, let $K_\alpha = K \cap [\alpha]^2$ and $K(\alpha) = \{\beta < \alpha : \{\alpha, \beta\} \subseteq K\}$.

(1) If $\emptyset \neq A \subseteq \omega_1$ and $[A]^2 \subseteq K$, prove that $A \subseteq B$ for some $B \in X_K$.

(2) If $\{\alpha, \beta\} \notin K$, prove that $\widehat{\{\alpha\}} \cap \widehat{\{\beta\}} = \emptyset$.

(3) If $F_1, F_2 \subset \omega_1$ are finite, prove that $\hat{F}_1 \cap \hat{F}_2 = \widehat{F_1 \cup F_2}$.

(4) If $F \subset \omega_1$ is finite, prove that $\hat{F} = \emptyset$ iff $[F]^2 \setminus K \neq \emptyset$.

(5) Prove that X_K with the topology generated by the open base $\{\hat{F} : F \subset \omega_1$ is finite$\}$ is a zero-dimensional Hausdorff space and $w(X_K) \leq \aleph_1$.

(6) Suppose $\widehat{\{\alpha\}} \neq \emptyset$ for each $\alpha \in \omega_1 \setminus \omega$. Prove that X_K satisfies *ccc* iff, whenever $\mathcal{C}$ is a collection of pairwise disjoint finite subsets of ω_1 such that

(a) $[F]^2 \subseteq K$ for every $F \in \mathcal{C}$ and

(b) $[F_1 \cup F_2]^2 \setminus K \neq \emptyset$ whenever $F_1, F_2 \in \mathcal{C}$ and $F_1 \neq F_2$, it follows that $|\mathcal{C}| \leq \aleph_0$.

(Hint: Use the Δ-system Lemma in 3R to prove one direction.)

Let $\mathfrak{a} = \{\mathcal{F} \subseteq \mathbb{P}(\omega_1) : \mathcal{F}$ is an antichain, $|\mathcal{F}| = \aleph_0$, and $F \in \mathcal{F}$ implies $|F| < \aleph_0\}$.

(7) Prove that $|\mathfrak{a}| = 2^{\aleph_0}$. (Hint: Show that $|\mathfrak{a}| = \aleph_1^{\aleph_0}$ and note that $2^{\aleph_0} \leq \aleph_1^{\aleph_0} \leq (2^{\aleph_0})^{\aleph_0} = 2^{\aleph_0}$.)

For notational purposes, let $\mathfrak{a} = \{\mathcal{F}_\mu : \mu < 2^{\aleph_0}\}$ and $\mathcal{F}_\mu = \{F_\mu^{\ n} : n < \omega\}$. When we assume CH, then $\mathfrak{a}$ will be indexed over ω_1, i.e., $\mathfrak{a} = \{\mathcal{F}_\mu : \mu < \omega_1\}$.

(8) [CH]. If K satisfies, for $\alpha < \omega_1$,

(a) $K(\alpha) \neq \emptyset$ if $\alpha \geqslant \omega$ and

(b) if $\beta < \alpha$, $\bigcup\{F_\beta^n : n \in \omega\} \subseteq \alpha$, $F \subseteq \alpha \setminus \bigcup\{F_\beta^n : n < \omega\}$ is finite, and $W_{\beta,F} = \{n : [F_\beta^n \cup F]^2 \subseteq K_\alpha\}$ is infinite, then $\{n \in W_{\beta,F} : F_\beta^n \subseteq K(\alpha)\}$ is infinite,

then prove $\{\hat{\alpha}\} \neq \emptyset$ for $\alpha \in \omega_1 \setminus \omega$ and X_K satisfies *ccc*. (Hint: Let C be an uncountable collection of pairwise disjoint finite subsets of ω_1 satisfying 6(a) and 6(b). For some $\beta < \omega_1$, $F_\beta \subseteq C$. Show that there exists $\alpha < \omega_1$ and $G \in C$ such that $\alpha > \beta$, $\bigcup F_\beta \subseteq \alpha$, and $G \cap \alpha = \emptyset$. Let $G = \{\alpha_1, \ldots, \alpha_n\}$ where $n \in \omega$ and $\alpha_1 < \alpha_2 < \ldots < \alpha_n$. Apply 8(b) n times where $F = \emptyset$, $\{\alpha_1\}$, $\{\alpha_1, \alpha_2\}$, ..., $\{\alpha_1, \ldots, \alpha_{n-1}\}$ and $\alpha = \alpha_1, \alpha_2, \ldots, \alpha_n$, respectively, to obtain that $\{n : [F_\beta^n \cup G]^2 \subseteq K\}$ is infinite, yielding a contradiction to 6(b).)

(9) If $K, M \subseteq [\omega_1]^2$ are disjoint such that X_K, X_M satisfy the *ccc* and $\{\hat{\alpha}\} \neq \emptyset$ in both X_K and X_M for each $\alpha \in \omega_1 \setminus \omega$, then prove that $X_K \times X_M$ does not satisfy the *ccc*. (Hint: Look at $\{\{\hat{\alpha}\} \times \{\hat{\alpha}\} : \alpha \in \omega_1 \setminus \omega\}$.)

(10) Let $\{S_n : n < \omega\} \subseteq \mathfrak{a}$ and $\bigcup(\bigcup S_n) \subseteq \alpha$. So, $|S_n| = \aleph_0$ and $S_n = \{S_n^m : m \in \omega\}$.

(a) Let $D \subseteq \bigcup S_n$ be a finite subset of pairwise disjoint sets and let $n < \omega$. Show there is an m

$< \omega$ such that S_n^m is disjoint from each element of D.

(b) Show there are disjoint sets A, B such that $A \cup B \subseteq \alpha$ and for $n < \omega$, $\{m : S_n^m \subseteq A\}$ and $\{m : S_n^m \subseteq B\}$ are infinite. (Hint: Use 10(a) to construct for $n < \omega$, a finite subset $D_n \subseteq \bigcup\{S_m : m < \omega\}$ consisting of pairwise disjoint sets such that $|D_n \cap S_m| \geq n$ for $m \leq n$ and $D_n \subseteq D_m$ for $n \leq m$. Construct A and B as subsets of $\bigcup\{D_n : n < \omega\}$.)

(11) [CH]. Find disjoint subsets $K^0, K^1 \subseteq [\omega_1]^2$ which satisfy 8(a) and (b). (Hint: For $n \in \omega$, let $K^i(n) = \emptyset$ for $i = 0, 1$. Let $\alpha \geq \omega$ and suppose $(K^i)_\alpha$ is defined for $i = 0, 1$ so that $K^i(\beta)$ satisfy 8(a) and (b). Look at all the triples (β,F,i) where $\beta < \alpha$, $i \in \{0,1\}$, $\bigcup\{F_\beta^n : n < \omega\} \subseteq \alpha$, $F \subseteq \alpha \setminus \bigcup\{F_\beta^n : n < \omega\}$ is finite, and $\{n : [F_\beta^n \cup F]^2 \subseteq (K^i)_\alpha\}$ is infinite. Since α is countable and F is a finite subset of a countable set, then the number of triples is countable. For each triple (β,F,i), let $S_{\beta,F}^i = \{F_\beta^n : [F_\beta^n \cup F]^2 \leq (K^i)_\alpha\}$; let S be the collection of all $S_{\beta,F}^i$'s. Let $K^0(\alpha)$ and $K^1(\alpha)$ be the disjoint sets provided by (10). Show that $(K^0)_\alpha$, $(K^1)_\alpha$ satisfy 8(a) and (b).)

Note that in this problem, by assuming CH is true, we have shown that the product of spaces with *ccc* is not necessarily a space satisfying *ccc*. Compare this result with the conclusion of 3S where it is shown, by assuming $MA(\aleph_1)$, that the product of spaces with *ccc* is a space with *ccc*. Thus, the statement "the product of spaces with *ccc* is a space with *ccc*" is independent of the usual axioms of set theory.

3U. <u>The Combinatorial Principle</u> $BF(\kappa)$. In 3.5 we proved that one consequence of $MA(\kappa)$ is the combinatorial statement $P(\kappa)$, showed that $P(\kappa)$ is equivalent to another combinatorial statement $SL(\kappa)$, and derived a topological property of $\mathbb{R}$ using $P(\kappa)$. In this problem, we derive a combinatorial consequence of $P(\kappa)$, which we will denote by $BF(\kappa)$. In 3V, we exhibit a typical topological application of $BF(\kappa)$.

(1) Let $F \subseteq F(\omega,\omega)$. Show there exists a subset $G \subseteq F(\omega,\omega)$ such that: (i) $|F| = |G|$, (ii) if $f \in F$, there is some $g \in G$ such that $f(n) \leqslant g(n)$ for all $n \in \omega$, and (iii) if $g \in G$ and $n \leqslant m$, then $g(n) \leqslant g(m)$.

Let $F \subseteq F(\omega,\omega)$. For each $f \in F$, define $A(f)$ to be $\{(m,n) \in \omega \times \omega : n > f(m)\}$. For each $r \in \omega$, define $C(r)$ to be $\{(m,n) \in \omega \times \omega : m > r\}$. Let $S = \{A(f) : f \in F\} \cup \{C(r) : r \in \omega\}$.

(2) Prove that S has the finite intersection property.

(3) $[P(\kappa)]$. Let $F \subseteq F(\omega,\omega)$ with $|F| < \kappa$. Show there exists a subset $B \subseteq \omega \times \omega$ with these two properties: (i) for each $m \in \omega$, $\{n \in \omega : (m,n) \in$

B} is finite, and (ii) for each $f \in F$, there is some $m(f) \in \omega$ such that if $(m,n) \in B$ and $m \geqslant m(f)$, then $n > f(n)$. (Hint: Apply $P(\kappa)$ to $\omega \times \omega$ and use (2).)

For $\aleph_0 \leqslant \kappa \leqslant 2^{\aleph_0}$, $BF(\kappa)$ is this statement: $BF(\kappa)$: If $F \subset F(\omega,\omega)$ and $|F| < \kappa$, then there is some $g \in F(\omega,\omega)$ such that for each $f \in F$, there is some $m(f) \in \omega$ such that $g(n) > f(n)$ whenever $n \geqslant m(f)$.

(4) Show that $P(\kappa)$ implies $BF(\kappa)$. (Hint: Build g by using the B constructed in (3).)

Note. Although $MA(\kappa)$ implies $P(\kappa^+)$ and $P(\kappa)$ implies $BF(\kappa)$, neither implication is reversible; see the Notes.

3V. Uncountable Discrete Subspaces of Lindelf Spaces Whose Points are G_δ's. Let L be a Lindelöf space whose singletons are G_δ-sets, and suppose that D is an uncountable discrete subspace of L. We ask whether or not it is possible for $L \setminus D$ to be countable. In this problem, we show the answer is independent of the usual axioms of set theory. First, assuming CH, we produce a particular L and the required subspace D such that $L \setminus D$ is countable. Then, assuming $MA(\aleph_1)$ (actually, $BF(\aleph_2)$, as discussed in 3U), we show that no such example can exist.

A **Lusin set** is an uncountable subspace of $\mathbb{R}$ which intersects each closed nowhere dense subset of $\mathbb{R}$ in a countable set.

(1) [CH]. Prove that $\mathbb{R}$ contains a dense Lusin set which is a subset of $\mathbb{R}\setminus\mathbb{Q}$. (Hint: Use CH to enumerate the closed nowhere dense subsets of $\mathbb{R}$ as $\{K(\alpha) : \alpha < \omega_1\}$. Inductively, choose $\{p(\alpha) : \alpha < \omega_1\} \subseteq \mathbb{R}\setminus\mathbb{Q}$ so that $p(\alpha) \neq p(\beta)$ if $\alpha \neq \beta$ and $p(\alpha) \notin \bigcup\{K(\beta) : \beta < \alpha\}$; this requires the use of the Baire category theorem and the second countability of $\mathbb{R}$.)

Let S denote the Lusin set chosen in (1). Let $L = S \cup \mathbb{Q}$; $U \subseteq L$ is defined to be open if $q \in U \cap \mathbb{Q}$ implies $L \cap (q - \frac{1}{n}, q + \frac{1}{n}) \subseteq U$ for some $n \in \mathbb{N}$. In particular, the points of S are discrete in L.

(2) Prove that L is a first countable, regular Lindelöf space with an uncountable dense discrete subspace D such that $L\setminus D$ is countable. (Hint: To prove the Lindelöf property, note that an open subset of $\mathbb{R}$ that contains $\mathbb{Q}$ is of the form $\mathbb{R}\setminus K(\alpha)$.)

(3) [$BF(\kappa^+)$]. Suppose $F \subseteq F(\omega,\omega)$ with $|F| = \kappa$ where κ is a regular, uncountable cardinal Show there exists $G \subseteq F$ and $h \in F(\omega,\omega)$ such that $|G| = \kappa$ and $f \leqslant h$ for all $f \in G$. (Hint: If g witnesses $BF(\kappa^+)$ (see 3U), then for each $f \in F$ there is a least $n(f) \in \omega$ such that $n \geqslant n(f)$ implies $f(n) \leqslant g(n)$. For $n \in \omega$, let $S(n) = \{f \in F : n(f) = n\}$. Find some $m \in \omega$ such that $|S(m)| = \kappa$, and find $G \subseteq S(m)$ such that $|G| = \kappa$ and $s(n) = t(n)$ for each $s, t \in G$ and $n \leqslant$

m. Now, define h.)

(4) [BF($\aleph_2$)]. Prove there does not exist a Lindelöf space X whose points are G_δ's and which has an uncountable discrete subspace D such that X\D is countable. (Hint: Assume such a space X exists. First, show we can assume that $|D| = \aleph_1$. Let X\D = $\{x(n) : n \in \omega\}$ and $\{G_n(m) : n \in \omega\}$ be a decreasing family of open sets of X such that $\{x_n\} = \bigcap\{G_n(m) : m \in \omega\}$. For $x \in D$ and $n \in \omega$, choose $f_x \in F(\omega,\omega)$ such that $x \notin G_n(f_x(n))$. Apply (3) to $\{f_x : x \in D\}$ to obtain a $\mathcal{G}$ and h, and use h to define an open cover of X with no countable subcover.)

CHAPTER 4

EXTENSIONS OF SPACES

The reader may recall that a space Y is called an **extension** of a space X if X is a dense subspace of Y. One of the reasons for studying extensions is the possibility of shifting a problem concerning a space X to a problem concerning an extension Y of X where Y is a "nicer" space than X and the "shifted" problem can be solved. Thus, an important goal in extension theory is to generate "nice" extensions of a fixed space X. After we have defined and developed some of the basic notions of extension theory, we will proceed to generate all the compact extensions of a Tychonoff space and all the compact, zero-dimensional extensions of a zero-dimensional space. In the final section of the chapter, we study certain "nice" extensions of an arbitrary (Hausdorff) space, namely the H-closed extensions.

4.1 Basic Concepts

In this section we define when two extensions of a fixed space are equivalent and show that "up to equivalence" the collection of extensions of a fixed space is a set.

(a) **Definition**.

(1) Let Y_1 and Y_2 be extensions of spaces X_1 and X_2, respectively. Let $f \in F(X_1,X_2)$. A function $F \in F(Y_1,Y_2)$ is said to

extend f if $F|X_1 = f$. The function F is called an **extension** of f.

(2) If Y is an extension of X, the space $Y \setminus X$ is called the **remainder** of X (in Y).

(b) <u>Proposition</u>. Let Y_1 and Y_2 be extensions of spaces X_1 and X_2, respectively, and let $f \in C(X_1,X_2)$. Then f has at most one extension $F \in C(Y_1,Y_2)$.

<u>Proof</u>. This is a corollary of 1.6(d).

■

(c) <u>Proposition</u>. If Y is an extension of X, then $|Y| \leq 2^{2^{|X|}}$.

<u>Proof</u>. For each $p \in Y$ let $0^p = \{U \cap X : U$ is open in Y and $p \in U\}$. Now 0^p is an open filter on X. If $p, q \in Y$ and $p \neq q$, then since Y is Hausdorff, and because X is dense in Y, there are disjoint nonempty sets $U \in 0^p$ and $V \in 0^q$. Thus, $p \to 0^p$ is one-to-one. Since $0^p \in \mathbb{P}(\mathbb{P}(X))$, it follows that $|Y| \leq |\mathbb{P}(\mathbb{P}(X))| = 2^{2^{|X|}}$.

■

(d) <u>Definition</u>

(1) Two extensions Y_1 and Y_2 of X are said to be **equivalent** if there is a homeomorphism $h : Y_1 \to Y_2$ such that $h(x) = x$ for each $x \in X$. The fact that Y_1 and Y_2 are equivalent extensions of X is indicated by writing $Y_1 \equiv_X Y_2$.

(2) Let X be a space and $E(X)$ a collection of extensions of

X such that distinct members of $E(X)$ are non-equivalent and any extension of X is equivalent to some extension in $E(X)$. If Y is an extension of X and $Y \equiv_X Z$ where $Z \in E(X)$, we identify Y and Z and write $Y \in E(X)$ to denote this identification. Thus, we regard $E(X)$ as "the collection of extensions of X."

In this book equivalent extensions of a given space are identified and considered to be the same extension.

The next result is an immediate consequence of 4.1(c).

(e) Corollary. For a space X, $E(X)$ is a set.

We now define an order relation on $E(X)$.

(f) Definition. For a space X and Y, $Z \in E(X)$, define Y to be **projectively larger than** Z, written as $Y \geqslant Z$ (or $Z \leqslant Y$) if there is a continuous function $f:Y \to Z$ such that $f|X = id_X$.

This relation on $E(X)$ is an order relation as we prove in the next result.

(g) Proposition. For a space X, $(E(X),\leqslant)$ is a complete upper semilattice.

Proof. First, we show $(E(X),\leqslant)$ is a poset. The proofs that $\leqslant$ is a reflexive, transitive relation are immediate. To show that $\leqslant$ is antisymmetric, suppose $Y \leqslant Z$ and $Z \leqslant Y$ where Y, $Z \in E(X)$. There are continuous functions $f:Z \to Y$ and $g:Y \to Z$ such that f(x)

$= g(x) = x$ for $x \in X$. Now, $f \circ g: Y \to Y$ is continuous and $f \circ g | X = id_X = id_Y | X$. By 4.1(b), $f \circ g = id_Y$. Similarly, $g \circ f = id_Z$. Thus, $g = f^{\leftarrow}$, and f is a homeomorphism. So, $Y \equiv_X Z$. Finally, we show $E(X)$ has arbitrary suprema. Let $\emptyset \neq S \subseteq E(X)$ and define $e: X \to \Pi S (= \Pi\{Y : Y \in S\})$ by $(\Pi_Y \circ e)(x) = x$ for all $Y \in S$. By the Embedding Theorem 1.7(j), e is an embedding of X into ΠS. Let $Z = c\ell_{\Pi S} e[X]$ and identify x with $e(x)$ for $x \in X$. Thus, Z is an extension of X and $Z \in E(X)$. For $Y \in S$, let $f_Y = \Pi_Y | Z$ where $\Pi_Y : \Pi S \to Y$ is the projection function onto the coordinate space Y. Then $f_Y : Z \to Y$ is a continuous function and $f_Y(x) = \Pi_Y(e(x)) = x$ for each $x \in X$. So, $Z \geq Y$ for all $Y \in S$. Suppose $W \in E(X)$ and $W \geq Y$ for all $Y \in S$. To show $Z = \bigvee S$ in $E(X)$, we need to show $W \geq Z$. Since for $Y \in S$, $W \geq Y$, there is a continuous function $g_Y : W \to Y$ such that $g_Y(x) = x$ for $x \in X$. Now, define $h: W \to \Pi S$ by $\Pi_Y(h(w)) = g_Y(w)$ for $w \in W$ and $Y \in S$. Since $\Pi_Y \circ h = g_Y$ is continuous for each $Y \in S$, then h is continuous. Also, $\Pi_Y \circ h(x) = g_Y(x) = x = \Pi_Y \circ e(x)$ for all $Y \in S$. So, $h | X = e | X$ and $h[X] = e[X]$. Since $h[W] = h[c\ell_W X] \subseteq c\ell_{\Pi S} h[X] = c\ell_{\Pi S} e[X] = Z$, then $W \geq Z$. This completes the proof that $Z = \bigvee S$ in $E(X)$.

■

(h) **Definition**. Let $\emptyset \neq \mathcal{Q} \subseteq E(X)$ for a space X. An extension $Y \in E(X)$ is a **projective maximum** in $\mathcal{Q}$ if $Y \in \mathcal{Q}$ and $Y \geq Z$ for all $Z \in \mathcal{Q}$. (Note that $Y \geq \bigvee \mathcal{Q}$, and since $Y \in \mathcal{Q}$, it follows that $\bigvee \mathcal{Q} \geq Y$. So, $Y \equiv_X \bigvee \mathcal{Q}$. Hence if $\mathcal{Q}$ has a projective maximum, it is unique and is the extension $\bigvee \mathcal{Q}$.)

The next results identify the projective maximum of $E(X)$.

(i) **Proposition**. For each space X, the projective maximum of $E(X)$ is X.

Proof. Since $X \in E(X)$ and $X \geqslant Y$ for all $Y \in E(X)$ (as witnessed by the function that embeds X into Y), then X is the projective maximum of $E(X)$.

■

(j) **Definition**

(1) Let P be a class of spaces closed under homeomorphism, i.e., if $X \in P$ and X is homeomorphic to Y, then $Y \in P$; such classes are called **replete**. We will usually refer to such classes as "**topological properties**". We will use interchangeably the phrases "$X \in P$" and "X has the property P". Recall our convention that all hypothesized spaces are assumed to be Hausdorff; thus any hypothesized topological property is assumed to include the property of being Hausdorff, unless this is explicitly denied. For example, if we say "let P be connectedness", we mean "let P be the class of connected Hausdorff spaces".

Another convention that we will adopt is that **if *P* is a hypothesized topological property, then there exists a space with *P* that contains more than one point.**

(2) For a space X, let $P(X) = \{Y \in E(X) : Y \in P\}$; the set $P(X)$ is called the set of ***P*-extensions** of X.

Recall from Chapter 1 that a class P of spaces is closed-hereditary if $A \in P$ whenever $X \in P$ and A is a closed subset of X; P is productive if for each set A of spaces in P, $\Pi A (= \Pi\{X : X \in A\}) \in P$.

(k) **Proposition**. Let X be a space and P a topological property that is both closed-hereditary and productive. If $P(X) \neq \emptyset$, then $P(X)$ is a complete upper semilattice of $E(X)$; in particular, $P(X)$ has a projective maximum.

Proof. Let $\emptyset \neq S \subseteq P(X)$. Then $Z = \bigvee S$ exists in $E(X)$. Since $Z \geq Y$ for all $Y \in S$, it suffices to show that $Z \in P(X)$. But, by the proof of 4.1(g), $Z \equiv_X c\ell_{\Pi S} e[X]$ where $e: X \to \Pi S$ is as in 4.1(g). Since P is productive, $\Pi S \in P$. Since S is closed-hereditary, it follows that $c\ell_{\Pi S}[X] \in P(X)$. Hence, $Z \in P(X)$.

■

We now begin to investigate the problem of when a given continuous function defined on a space X has a continuous extension to a given member of $E(X)$. This investigation, in one form or another, will occupy us for a majority of the remainder of this book.

(l) **Proposition**. Let $Y \in E(X)$ for a space X, let Z be a regular space, and let $f \in C(X,Z)$. The following are equivalent:

(1) There exists a function $F \in C(Y,Z)$ such that $F|X = f$, and

(2) for each $y \in Y$, the filter $\mathcal{F}_y = \{A \subseteq Z : A \supseteq f[U]$ for some $U \in 0^y\}$ converges (where $0^y = \{W \cap X : W$ is open in Y and $y \in W\}$).

Proof

(1) $\Rightarrow$ (2) Suppose F exists and $y \in Y$. We will show $\mathcal{F}_y$ converges to $F(y)$. Let W be an open neighborhood of $F(y)$ in Z. There is an open neighborhood U of y such that $F[U] \subseteq W$. Thus, $f[U \cap X] = F[U \cap X] \subseteq W$ and $f[U \cap X] \in \mathcal{F}_y$. Thus, $\mathcal{F}_y$ converges to $F(y)$.

(2) $\Rightarrow$ (1) Suppose for each $y \in Y$, $\mathcal{F}_y$ converges to some point. By 2.3(f), $\mathcal{F}_y$ converges to a unique point which we denote by $F(y)$. So, $F:Y \to Z$ is a function. If $x \in X$ and W is an open neighborhood of $f(x)$, there is an open set U of X with $x \in U$ and $f[U] \subseteq W$. If V is an open set in Y such that $V \cap X = U$, then $f[V \cap X] \in \mathcal{F}_x$. Thus, $\mathcal{F}_x$ converges to $f(x)$ (and no other point by 2.3(f)) and $F|X = f$. To show F is continuous, let $y \in Y$ and let W be an open neighborhood of $F(y)$. As Z is regular, there is an open subset V of Z such that $F(y) \in V \subseteq c\ell_Z V \subseteq W$. Since $\mathcal{F}_y$ converges to $F(y)$, there is an open set U of Y such that $y \in U$ and $f[U \cap X] \subseteq V$. Let $p \in U$ and T be an open set of Z such that $F(p) \in T$. There is a Y-open neighborhood R of p such that $R \subseteq U$ and $f[R \cap X] \subseteq T$. Since $R \cap X \neq \emptyset$, it follows that $\emptyset \neq f[R \cap X] \subseteq V$; thus $T \cap V \neq \emptyset$. This shows that $F(p) \in c\ell_Z V \subseteq W$ and $F[U] \subseteq W$. So, F is continuous.

■

The regularity of Z cannot be omitted from the hypotheses in the above; see 4L(8).

A consequence of this result is the following result of Taimanov.

(m) **Theorem**. Let X be a space, $Y \in E(X)$, Z a compact space, and $f \in C(X,Z)$. Then there is a continuous function $F:Y \to Z$ such that $F|X = f$ iff for every pair of disjoint closed sets B and C in Z, $c\ell_Y f^{\leftarrow}[B] \cap c\ell_Y f^{\leftarrow}[C] = \emptyset$.

Proof. Suppose there is a continuous function $F:Y \to Z$ such that $F|X = f$. Since $F^{\leftarrow}[B]$ is closed and $f^{\leftarrow}[B] \subseteq F^{\leftarrow}[B]$, it follows that $c\ell_Y f^{\leftarrow}[B] \subseteq F^{\leftarrow}[B]$. Since $B \cap C = \emptyset$, we conclude that $\emptyset = F^{\leftarrow}[B] \cap F^{\leftarrow}[C] \supseteq c\ell_Y f^{\leftarrow}[B] \cap c\ell_Y f^{\leftarrow}[C]$. Conversely, suppose for every pair of disjoint closed sets B and C in Z, $c\ell_Y f^{\leftarrow}[B] \cap c\ell_Y f^{\leftarrow}[C] = \emptyset$. Let $y \in Y$. By 4.1(l), it suffices to show that $F = \{A \subseteq Z : A \supseteq f[U] \text{ for some } U \in 0^y\}$ converges. Since Z is compact, then $\emptyset \neq \cap\{c\ell_Z f[U] : U \in 0^y\} = a(F)$. Assume $p, q \in a(F)$ such that $p \neq q$. There are open sets W and V such that $p \in W$, $q \in V$, and $c\ell_Z W \cap c\ell_Z V = \emptyset$. Thus, $c\ell_Y f^{\leftarrow}[c\ell_Z W] \cap c\ell_Y f^{\leftarrow}[c\ell_Z V] = \emptyset$. Since $W \cap f[U] \neq \emptyset$ for all $U \in 0^y$, it follows that $U \cap f^{\leftarrow}[W] \neq \emptyset$ for all $U \in 0^y$. Thus, $y \in c\ell_Y f^{\leftarrow}[W]$. Likewise, $y \in c\ell_Y f^{\leftarrow}[V]$. So, $y \in c\ell_Y f^{\leftarrow}[W] \cap c\ell_Y f^{\leftarrow}[V] \subseteq c\ell_Y f^{\leftarrow}[c\ell_Z W] \cap c\ell_Y f^{\leftarrow}[c\ell_Z V] = \emptyset$, which is a contradiction.

Thus, $\{p\} = a(F_y)$ for some $p \in Z$. Let W be an open neighborhood of p. Since $Z \setminus W$ is compact and $Z \setminus W \subseteq \cup\{Z \setminus c\ell_Z F : F \in F_y\}$, it follows that there is a finite family $S \subseteq F_y$ such that $Z \setminus W \subseteq \cup\{Z \setminus c\ell_Z F : F \in S\} \subseteq Z \setminus \cap S$. But $\cap S \in F_y$ and $\cap S \subseteq$

W. This shows F_y converges to p. By 4.1(l), the conclusion follows. ■

Another consequence of 4.1(l) is the next result.

(n) **Proposition**. Let $Y \in E(X)$ for a space X and let Z be a regular space. Let $g \in F(Y,Z)$ such that for each $y \in Y$, $g|(X \cup \{y\})$ is continuous. Then g is continuous.

Proof. Let $0^y = \{W \cap X : W \text{ is open in } Y \text{ and } y \in W\}$. Note that $0^y = \{W \cap X : W \text{ open in } X \cup \{y\} \text{ and } y \in W\}$. Since $g|(X \cup \{y\})$ is continuous, then by 4.1(l), $\{A \subseteq Z : A \supseteq (g|X)[U] \text{ for some } U \in 0^y\}$ converges to $g(y)$. Thus, by the other direction of 4.1(l), g is continuous. ■

4.2 Compactifications

In this section, we begin our discussion of compact extensions. A compact extension of a space X is called a **compactification** of X. The theory of compactifications is one of the more active areas of study and research in general topology. Here, we will show that if the family of compactifications of a given space is nonempty then it has a projective maximum, and we will construct that projective maximum.

Let K denote the class of all compact spaces. Thus K is a replete class that is both productive and closed-hereditary. The next result is an immediate consequence of 4.1(k).

(a) **Proposition**. For a space X, $K(X)$ (which is $\{Y \in E(X) : Y \in K\}$; see 4.1(j)(2)) is a complete upper semilattice and has a projective maximum if $K(X) \neq \emptyset$.

(b) **Definition**. If $K(X) \neq \emptyset$ for a space X, then the projective maximum of X is denoted by βX (so, $\beta X = \bigvee K(X)$) and is called the **Stone-Čech compactification** of X.

We saw in 1.10(h) that if X is a space, then $K(X) \neq \emptyset$ iff X is Tychonoff. We now show that if X and Y are Tychonoff and $f \in C(X,Y)$, then there exists some $F \in C(\beta X, \beta Y)$ such that $F|X = f$.

(c) **Proposition**. If X and Y are Tychonoff spaces and if $f \in C(X,Y)$, there is a unique continuous extension $F \in C(\beta X, \beta Y)$ such that $F|X = f$.

Proof. Let $C = C(X,[0,1])$ where $[0,1]$ is the unit interval of $\mathbb{R}$ with the topology induced by $\mathbb{R}$. Define $e:X \to [0,1]^C$ by setting $e(x)(f) = f(x)$ for each $f \in C$. Since X is Tychonoff, then by the Embedding Theorem 1.7(j), e is an embedding. By the Tychonoff Product Theorem for compact spaces, $[0,1]^C$ is compact. Identify X and $e[X]$ by identifying x with $e(x)$ for $x \in X$, and let $cX = c\ell e[X]$ (closure relative to $[0,1]^C$). Evidently, cX is a compactification of X. Construct cY similarly using $C(Y,[0,1]) = D$. We first show that there exists $G \in C(cX,cY)$ such that $G|X = f$; then we show that $cX \equiv_X \beta X$. Note that as Y and $e[Y]$ are now identified, we can identify the function f with the function that takes e(x) to e(f(x)) for each x $\in$ X.

Define $f^*:D \to C$ by $f^*(g) = g \circ f$. Define $f^{**}:[0,1]^C \to [0,1]^D$ by $f^{**}(h) = h \circ f^*$. To show f^{**} is continuous, we will show $\Pi_g \circ f^{**}$ is continuous for each $g \in D$ where $\Pi_g:[0,1]^D \to [0,1]$ is the gth projection function. Let $h \in [0,1]^C$. Then $(\Pi_g \circ f^{**})(h) = \Pi_g(f^{**}(h)) = f^{**}(h)(g) = (h \circ f^*)(g) = h(f^*(g)) = h(g \circ f) = \Pi_{g \circ f}(h)$. Thus, $\Pi_g \circ f^{**} = \Pi_{g \circ f}$ and so f^{**} is continuous. Next we will show that $f^{**}[e[X]] \subseteq e[Y]$. Let $x \in X$ and $g \in D$. Then $f^{**}(e(x))(g) = (e(x) \circ f^*)(g) = e(x)(f^*(g)) = e(x)(g \circ f) = (g \circ f)(x) = g(f(x)) = e(f(x))(g)$, which implies that $f^{**}(e(x)) = e(f(x)) \in e[Y]$. Thus, $f^{**}[cX] = f^{**}[c\ell e[X]] \subseteq c\ell f^{**}[e[X]] \subseteq c\ell e[Y] = cY$. Let $G = f^{**} \mid cX$. Then $G:cX \to cY$ is continuous and $G \mid X = f$. Now, let $Y = \beta X$. Then Y is compact and, hence, $Y = cY$. Let $i:X \to \beta X$ be the inclusion function. Then by the above result, there is a continuous function $G:cX \to c(\beta X) = \beta X$ such that $G \mid X = i$. Thus, $G(x) = i(x) = x$ for $x \in X$. So, $cX \geq \beta X$. Since $\beta X = \bigvee K(X)$, then $\beta X \geq cX$. Thus, $\beta X \equiv_X cX$. Therefore, by the above result, if $f:X \to Y$ is continuous, there is a continuous function $F:\beta X \to \beta Y$ such that $F \mid X = f$. By 4.1(b), f is unique.

■

(d) <u>Definition</u>. If X and Y are Tychonoff spaces and $f \in C(X,Y)$, then the unique continuous function $F:\beta X \to \beta Y$ such that $F \mid X = f$ is denoted by βf, and is called the **Stone extension** of f.

(e) <u>Corollary</u>. Let X be a Tychonoff space. Then:

(1) If K is a compact space and $f \in C(X,K)$, then f has a

unique continuous extension $F:\beta X \to K$, and

(2) X is C^*-embedded in βX.

Proof. By 4.2(c) there exists $\beta f \in C(\beta X, \beta K)$ such that $\beta f \mid X = f$. But $\beta K = K$ since K is compact, so βf is the desired F. Evidently, (2) follows from (1), for if $f \in C^*(X)$, let $K = c\ell_{\mathbb{R}} f[X]$ and apply (1).

■

We next present an extremely useful characterization of perfect continuous surjections (between Tychonoff spaces) in terms of the properties of extensions of such surjections to compactifications.

(f) **Theorem**. Let X and Y be Tychonoff spaces and let $f:X \to Y$ be a continuous surjection. The following are equivalent:

(1) f is perfect,

(2) If $\alpha X \in K(X)$, $\delta Y \in K(Y)$, and if there exists $F \in C(\alpha X, \delta Y)$ such that $F \mid X = f$, then $F^{\leftarrow}[Y] = X$ (i.e., $F[\alpha X \setminus X] = \delta Y \setminus Y$), and

(3) $(\beta f)^{\leftarrow}[Y] = X$ (i.e., $\beta f[\beta X \setminus X] = \beta Y \setminus Y$).

Proof

(1) ⇒ (2) Obviously $X \subseteq F^{\leftarrow}[Y]$, and it is an immediate consequence of 1.8(i) that $F^{\leftarrow}[Y] \subseteq X$.

(2) ⇒ (3) Obvious

(3) ⇒ (1) Let $y \in Y$. Then $(\beta f)^{\leftarrow}(y) \subseteq X$, so $(\beta f)^{\leftarrow}(y) = f^{\leftarrow}(y)$. Thus $f^{\leftarrow}(y)$ is compact. If A is a closed subset of X, there exists a compact subset H of βX such that $A = H \cap X$. Thus $f[A] =$

$f[H \cap X]$. But $(\beta f)[H] = f[H \cap X] \cup (\beta f)[H \setminus X]$. By hypothesis $(\beta f)[H \setminus X] \subseteq \beta Y \setminus Y$, and so $f[H \cap X] = Y \cap (\beta f)[H]$. Thus $f[A]$ is closed in Y. It follows that f is perfect.

■

We can generalize the previous theorem slightly to work with functions that are not surjective.

(g) **Corollary**. Let X and Y be Tychonoff spaces. Let $f \in C(X,Y)$ and assume that $f[X]$ is a closed subset of Y. The following are equivalent:

(1) f is perfect.

(2) If $\alpha X \in K(X)$, $\delta Y \in K(Y)$, and if there exists $F \in C(\alpha X, \delta Y)$ such that $F|X = f$, then $F^{\leftarrow}[Y] = X$ and $F[\alpha X \setminus X] \subseteq \delta Y \setminus Y$.

(3) $(\beta f)^{\leftarrow}[Y] = X$ and $\beta f[\beta X \setminus X] \subseteq \beta Y \setminus Y$.

Proof

(1) ⇒ (2) By 1.8(f)(2), f is a perfect continuous surjection from X onto $f[X]$. Evidently $c\ell_{\delta Y} f[X]$ is a compactification of $f[X]$ and $F \in C(\alpha X, c\ell_{\delta Y} f[X])$. Hence by (f) above $F^{\leftarrow}[c\ell_{\delta Y} f[X] \setminus f[X]] \subseteq \alpha X \setminus X$. As $f[X]$ is closed in Y, it follows that $Y \cap c\ell_{\delta Y} f[X] = f[X]$ and so $F^{\leftarrow}[Y] = X$. From (f), $F[\alpha X \setminus X] = c\ell_{\delta Y} f[X] \setminus f[X] \subseteq \delta Y \setminus Y$.

(2) ⇒ (3) Obvious.

(3) ⇒ (1) Apply the argument used in (f) above to prove (3) ⇒ (1), replacing Y by $f[X]$ and βY by $c\ell_{\beta Y} f[X]$. (Note that we need the hypothesis that $f[X]$ is closed in Y in order to prove that f:X

$\rightarrow Y$ is a closed function.)

■

We finish this section by giving an important special case of Taimanov's theorem (4.1(m)).

(h) **Theorem**. Let αX, $\gamma X \in K(X)$. The following are equivalent:

(1) $\alpha X \geq \gamma X$.

(2) If $Z_1, Z_2 \in Z(X)$ and $cl_{\gamma X}Z_1 \cap cl_{\gamma X}Z_2 = \emptyset$, then $cl_{\alpha X}Z_1 \cap cl_{\alpha X}Z_2 = \emptyset$.

(3) If A and B are disjoint closed subsets of γX, then $cl_{\alpha X}(A \cap X) \cap cl_{\alpha X}(B \cap X) = \emptyset$.

Proof

(1) $\Rightarrow$ (2) By hypothesis the embedding function $i:X \rightarrow \gamma X$ can be continuously extended to αX. By 4.1(m), $cl_{\alpha X}i^{\leftarrow}[cl_{\gamma X}Z_1] \cap cl_{\alpha X}i^{\leftarrow}[cl_{\gamma X}Z_2] = \emptyset$. Evidently $i^{\leftarrow}[cl_{\gamma X}Z_j] = Z_j$ $(j = 1,2)$.

(2) $\Rightarrow$ (3) As γX is normal, there exist disjoint zero-sets H and K of γX such that $A \subseteq H$ and $B \subseteq K$. Then $A \cap X \subseteq H \cap X$, $B \cap \subseteq K \cap X$, and $cl_{\gamma X}(H \cap X) \cap cl_{\gamma X}(K \cap X) = \emptyset$. Hence by (2), $cl_{\alpha X}(H \cap X) \cap cl_{\alpha X}(K \cap X) = \emptyset$. Our conclusion follows.

(3) $\Rightarrow$ (1) Let $i:X \rightarrow \gamma X$ be the embedding function. Our hypothesis says that if A and B are disjoint closed sets of cX, then $cl_{\alpha X}i^{\leftarrow}[A] \cap cl_{\alpha X}i^{\leftarrow}[B] = \emptyset$. It follows from 4.1(m) that $\alpha X \geq \gamma X$.

■

4.3 One-point Compactifications

In this section, we show that if X is a locally compact space, then $\mathcal{K}(X)$ is nonempty, has a smallest element, and is a complete lattice. Recall that a space X is **locally compact** if every point has a compact neighborhood. We need another basic fact about locally compact spaces.

(a) Proposition. Let A be a subspace of a locally compact space X. Then A is locally compact iff A is open in $c\ell_X A$.

Proof. The proof is left to the reader (see 4A). ■

(A generalization of this is given in 7.3(b).)

(b) Definition. For a space X, an element $Y \in \mathcal{K}(X)$ is called a **one-point compactification** of X if $Y \setminus X$ is a singleton.

(c) Proposition

(1) If a space X has a one-point compactification, then X is locally compact and not compact.

(2) If a space X is locally compact and not compact, then X has a one-point compactification Y such that $Z \geq Y$ for all $Z \in \mathcal{K}(X)$.

Proof

(1) Let Y be a one-point compactification of X. Then $Y \setminus X$ is closed and, hence, X is open in Y. By 4.3(a) X is locally compact. As it is a proper dense subspace of Y, X is not compact.

(2) Let $Y = X \cup \{\infty\}$ where $\infty \notin X$. It is easy to verify that the collection $\{W \subseteq Y : W \cap X$ is open in X and if $\infty \in W$, then $X \setminus W$ is compact$\}$ is a compact Hausdorff topology on Y such that X is a dense subspace of Y. So, Y is a one-point compactification. Let $Z \in K(X)$ and define $f: Z \to Y$ by $f(x) = x$ for $x \in X$ and $f(z) = \infty$ for $z \in Z \setminus X$. Since X is locally compact, it follows from 4.3(a) that X is open in Z. Since X is also open in Y and $f|X = id_X$, then f is continuous at points of X. Let $z \in Z \setminus X$ and $f(z) \in W$ for some W open in Y. Since $f(z) = \infty \in W$, it follows that $X \setminus W$ is compact. Thus, $T = Z \setminus (X \setminus W)$ is open in Z and $z \in Z \setminus X \subseteq T$. Also, $f[T] \subseteq W$. So, f is continuous. This shows that $Z \geq Y$.

■

Since a compact space is Tychonoff and the property of being Tychonoff is hereditary, the next result is immediate.

(d) **Corollary**. A locally compact space is Tychonoff.

We next obtain some interesting characterizations of local compactness.

(e) **Proposition**. For a noncompact, Tychonoff space X, the following are equivalent:

(1) $K(X)$ is a complete lattice,

(2) X is locally compact,

(3) X has a one-point compactification,

(4) X is open in every compactification, and

(5) X is open in some compactification.

Proof. By 4.3(c), (2) and (3) are equivalent. Clearly the equivalence of (2), (4), and (5) follows from 4.3(a).

(2) $\Rightarrow$ (1) Let Y be the one-point compactification of X which is constructed in 4.3(c). Then $Z \geq Y$ for all $Z \in K(X)$. Since $Y \in K(X)$, it follows that $Y = \wedge K(X)$. By 2.1(e) and 4.2(a), $K(X)$ is a complete lattice.

(1) $\Rightarrow$ (3) Let $Z = \wedge K(X)$. Since X is not compact, it follows that $X \neq Z$ and $Z \setminus X \neq \emptyset$. In particular, $|Z \setminus X| \geq 1$. Now, we will show that $|Z \setminus X| \leq 1$. Let $p, q \in Z \setminus X$ and let Y be the set $(Z \setminus \{p,q\}) \cup \{r\}$ where $r \notin Z$. Let $f:Z \to Y$ be defined as follows: $f(z) = z$ if $z \neq p, q$ and $f(p) = f(q) = r$. Give Y the quotient topology induced by f. Since $\{p,q\}$ is compact, then it is easily shown that Y is a Hausdorff space and $f|Z \setminus \{p,q\}$ is a homeomorphism. Also, since $\{p,q\} \cap X = \emptyset$, then X is a subspace of Y. Since X is dense in Z and f is continuous, then $f[X] = X$ is dense in Y. As Z is compact, then $Y = f[Z]$ is compact. Hence Y is a compactification of X and $Z \geq Y$. But $Z = \wedge K(X)$ implies $Y \geq Z$. This means there is a continuous function $g:Y \to Z$ such that $g|X = id_X$. Since $g \circ f:Z \to Z$ is continuous and $g \circ f|X = id_X$, then by 4.1(b), $g \circ f = id_Z$. So, $p = (g \circ f)(p) = (g \circ f)(q) = q$. Hence, $|Z \setminus X| \leq 1$ and Z is a one-point compactification of X.

■

Finally we show that the one-point compactification of a locally compact space is unique (up to equivalence).

(f) **Proposition**. A noncompact, locally compact space X has a unique one-point compactification.

Proof. Let Z be a one-point compactification of X and let $Y = \wedge\mathcal{K}(X)$ (Y exists by 4.3(e)). So, there is a continuous function $f: Z \to Y$ such that $f|X = id_X$. By 1.8(i), $f[Z\backslash X] = Y\backslash X$. Hence, $Y\backslash X$ is a singleton. This implies that f is a bijection. Since f is closed by 1.8(b), then f is a homeomorphism and $Z \equiv_X Y$.

■

4.4 Wallman Compactifications

From 4.2, we know that if X is a Tychonoff space, then X has at least one compactification, namely βX (which is the largest one). If X is also locally compact, than X has a one-point compactification (which is the smallest one). In this section, we show that certain closed bases of X give rise to (or generate) compactifications of X.

Recall from 2.4(a) that a collection $\mathcal{L}$ of subsets of X is a **ring of sets** of X if for A, B $\in \mathcal{L}$, $A \cup B \in \mathcal{L}$ and $A \cap B \in \mathcal{L}$. Note that a ring of sets of X is a lattice on X. Hence we can talk about filters on a ring of sets.

(a) **Definition**

(1) A **Wallman base** L on a space X is a ring of sets of X satisfying: (i) $\emptyset$, X $\in L$, (ii) L is a closed base for X, (iii) if A $\in L$ and x $\in$ X\A, there is a B $\in L$ such that x $\in$ B and A $\cap$ B $= \emptyset$, and (iv) if A, B $\in L$ such that A $\subseteq$ X\B, then there are C, D $\in L$ such that A $\subseteq$ X\C $\subseteq$ D $\subseteq$ X\B.

(2) If L is a Wallman base on X, let $w_L X = \{F : F$ is a L-ultrafilter on X$\}$. For A $\in L$, let $S(A) = \{F \in w_L X : A \in F\}$.

(b) **Example**. If X is a Tychonoff space and $L = Z(X)$, then by 1.4 and 2.1(f)(4), L is a ring of sets satisfying (i), (ii) and (iii) of (a)(1). If Z_1 and Z_2 are disjoint elements of $Z(X)$, then by 1.9(f) there exists f $\in C^*(X)$ such that $f[Z_1] \subseteq \{0\}$, $f[Z_2] \subseteq \{1\}$, and $f[X] \subseteq [0,1]$. Now, $C = f^{\leftarrow}([0,\frac{1}{2}])$ and $D = f^{\leftarrow}([\frac{1}{2},1])$ are zero-sets such that $Z_2 \subseteq X\backslash C \subseteq D \subseteq X\backslash Z_1$.

We will now show that each Wallman base on X generates a compactification of X.

(c) **Lemma**. Let L be a Wallman base on X and A, B $\in L$. Then:

(1) $S(\emptyset) = \emptyset$ and $S(X) = w_L X$,

(2) $S(A \cap B) = S(A) \cap S(B)$,

(3) $S(A \cup B) = S(A) \cup S(B)$, and

(4) $S(A) \subseteq (w_L X)\backslash S(B)$ iff $A \subseteq X\backslash B$.

Proof. This straightforward proof is left to the reader (see 4B).

■

It follows that $\{S(A) : A \in L\}$ is a closed base for the closed sets of a topology on $w_L X$. We topologize $w_L X$ using this closed base.

(d) **Proposition**. If L is a Wallman base on X, then $w_L X$ is a compact space.

Proof. First, we will show that $w_L X$ is Hausdorff. Let U, $V \in w_L X$ such that $U \neq V$. Now $U \setminus V \neq \emptyset$. Let $A \in U \setminus V$. By 2.3(d), $A \cap B = \emptyset$ for some $B \in V$. So, there are C, D $\in L$ such that $A \subseteq X \setminus C \subseteq D \subseteq X \setminus B$. Now, $U \in S(A) \subseteq w_L X \setminus S(C)$. Since $B \subseteq X \setminus D$, then $V \in S(B) \subseteq w_L X \setminus S(D)$. Since $(w_L X \setminus S(C)) \cap (w_L X \setminus S(D)) = w_L X \setminus (S(C) \cup S(D)) = w_L X \setminus S(C \cup D) = w_L X \setminus S(X) = w_L X \setminus w_L X = \emptyset$, it follows that $w_L X$ is Hausdorff. To show $w_L X$ is compact, suppose that F is a filter of closed sets on $w_L X$. For each $F \in F$, $F = \cap\{S(A) : F \subseteq S(A), A \in L\}$; hence $G = \{S(A) : F \subseteq S(A)$ for some $F \in F$ and $A \in L\}$ is closed under finite intersections and $\cap G = \cap F$. Let $B = \{A \in L : S(A) \in G\}$. If A, B $\in B$, then $S(A), S(B) \in G$ and $\emptyset \neq S(A) \cap S(B) = S(A \cap B)$ which implies that $\emptyset \neq A \cap B \in B$. If $C \in L$ and $S(C) \supseteq S(D)$ where $S(D) \in G$, then $S(C) \in G$. So, $C \in B$. Hence, B is a L-filter and is contained in some L-ultrafilter W by 2.3(d). Since $B \subseteq W$, it follows that $W \in S(A)$ for all $A \in B$; thus $W \in \cap G = \cap F$ and $\cap F \neq \emptyset$. This shows $w_L X$ is compact.

■

Our next goal is to embed X as a dense subspace of $w_L X$. First, a lemma is needed.

(e) **Lemma**. Let L be a Wallman base on a space X. For $x \in X$, let $\mathcal{U}_x = \{A \in L : x \in A\}$. Then $\mathcal{U}_x \in w_L X$.

Proof. Clearly, $\mathcal{U}_x$ is a L-filter. Let $\mathcal{U}$ be a L-ultrafilter such that $\mathcal{U} \supseteq \mathcal{U}_x$. Assume $A \in \mathcal{U} \backslash \mathcal{U}_x$. Then $x \notin A$. By 4.4(a)(1)(iii), there is some $B \in L$ such that $x \in B$ and $A \cap B = \varnothing$. But $B \in \mathcal{U}_x \subseteq \mathcal{U}$, so $A, B \in \mathcal{U}$ and $A \cap B \neq \varnothing$, which is a contradiction. So, $\mathcal{U} = \mathcal{U}_x$.

■

We define a function $e: X \to w_L X$ by $e(x) = \mathcal{U}_x$.

(f) **Theorem. (Frink-Wallman).** Let L be a Wallman base on a space X. Then:

(1) $e: X \to w_L X$ is a dense embedding and

(2) for $A, B \in L$, $c\ell e[A \cap B] = c\ell e[A] \cap c\ell e[B]$ where closure is with respect to $w_L X$.

Proof

(1) First, we show e is one-to-one. Suppose $x, y \in X$ and $x \neq y$. Since L is a closed base and $\{y\}$ is closed, there is some $A \in L$ such that $y \in A$ and $x \notin A$. Thus $A \in \mathcal{U}_y \backslash \mathcal{U}_x$ and so $e(x) \neq$

e(y). As e is one-to-one, to show e is an embedding it suffices to show for $A \in L$, $e[A] = S(A) \cap e[X]$. Clearly, $e[A] \subseteq S(A) \cap e[X]$. If $U_x \in S(A)$, then $A \in U_x$. Hence $x \in A$, so $S(A) \cap e[X] \subseteq e[A]$. This shows that e is an embedding. Finally, to show that e[X] is dense in $w_L X$, it suffices to show for $A \in L$, $S(A) = c\ell e[A]$ (closure in $w_L X$). Since $e[A] \subseteq S(A)$, it follows that $c\ell e[A] \subseteq S(A)$ as $S(A)$ is closed in $w_L X$. Let $U \in S(A)$ and $U \in w_L X \setminus S(B)$ where $B \in L$. Now, $A \in U$ and $B \notin U$. So, there is some $C \in U$ such that $B \cap C = \emptyset$, by 2.3(d). Thus, $C \subseteq X \setminus B$ and by 4.4(c), $S(C) \subseteq w_L X \setminus S(B)$. Let $x \in A \cap C$. Then $C \in U_x$ and $U_x \in S(C)$. So, $U_x \in e[A] \cap (w_L X \setminus S(B))$. This shows that $S(A) \subseteq c\ell e[A]$. Hence, $w_L X = S(X) = c\ell e[X]$.

(2) If $C \in L$, by (1) $S(C) = c\ell e[C]$ (closure in $w_L X$); hence $c\ell e[A \cap B] = S(A \cap B) = S(A) \cap S(B) = c\ell e[A] \cap c\ell e[B]$.

■

Thus, for each Wallman base L on a space X, $w_L X \in K(X)$ and $K(X) \neq \emptyset$. We call $w_L X$ the **Wallman compactification** of X with respect to L. Hence we have the next result.

(g) **Corollary**. If a space X has a Wallman base, then X is Tychonoff.

On the other hand, if X is Tychonoff, then $L = Z(X)$ is a Wallman base by 4.4(b); so, $w_L X$ is a compactification when $L = Z(X)$ and X is Tychonoff.

(h) **Theorem**. Let $L = Z(X)$ for a Tychonoff space X. Then

$\beta X \equiv_X w_L X$.

<u>Proof</u>. Let $Y = w_L X$. Since $\beta X \geq Y$, it suffices to show $Y \geq \beta X$. Let Z_1 and Z_2 be disjoint zero-sets of X. Since $Z_1, Z_2 \in L$, it follows from 4.4(f) that $c\ell_Y Z_1 \cap c\ell_Y Z_2 = c\ell_Y(Z_1 \cap Z_2) = \emptyset$. By 4.2(h) it follows that $Y \geq \beta X$.

■

(i) **<u>Theorem</u>**. Let $Y \in K(X)$ for a Tychonoff space X. The following are equivalent:

(1) $\beta X \equiv_X Y$,

(2) if $Z_1, Z_2 \in Z(X)$ and $Z_1 \cap Z_2 = \emptyset$, then $c\ell_Y Z_1 \cap c\ell_Y Z_2 = \emptyset$, and

(3) if $Z_1, Z_2 \in Z(X)$, then $c\ell_Y(Z_1 \cap Z_2) = c\ell_Y Z_1 \cap c\ell_Y Z_2$.

<u>Proof</u>. Clearly, (3) implies (2). The proof that (2) implies (1) is the same as the proof of 4.4(h). Finally, to show (1) implies (3), let $\beta X \equiv_X Y$. Then $Y \equiv_X w_{Z(X)} X$ by 4.4(h). Thus by 4.4(f), if $Z_1, Z_2 \in Z(X)$, then $c\ell_Y(Z_1 \cap Z_2) = c\ell_Y Z_1 \cap c\ell_Y Z_2$.

■

According to 4.4(h) we can regard to points of $\beta X \backslash X$ as being free z-ultrafilters on X with the fixed z-ultrafilters corresponding to points of X. We will make frequent use of this construction of βX later.

A compactification Y of a Tychonoff space X is said to be of **Wallman type** if there is a Wallman base L on X such that $Y \equiv_X$

$w_L X$. A famous problem in topology was to determine if every compactification is of Wallman type. The problem was solved when a compactification of an uncountable discrete space which is not Wallman was found (see the Notes).

4.5 Gelfand compactifications

In 4.2 and 4.4, the Stone-Čech compactification βX of a Tychonoff space X was constructed by two methods. The first method involved embedding X in a product of closed intervals, and the second method involved constructing the Wallman compactification $w_{Z(X)}X$. In this section, βX is constructed by using maximal ideals of $C^*(X)$. In fact, we show that all compactifications of a Tychonoff space can be constructed by using the set of maximal ideals of certain subrings of $C^*(X)$.

For a space X and $f \in C^*(X)$, define $\|f\|$ to be $\sup\{|f(x)| : x \in X\}$.

If $r \in \mathbb{R}$, recall that $\underline{r}$ is used to denote the constant function from X to $\mathbb{R}$ defined by $\underline{r}(x) = r$ for each $x \in X$. Let $\underline{\mathbb{R}} = \{\underline{r} : r \in \mathbb{R}\}$. If $f, g \in C^*(X)$, let $d(f,g) = \|f-g\|$.

(a) **<u>Proposition</u>**. Let X be a space, $f, g \in C^*(X)$, and $r \in \mathbb{R}$. Then:

(1) $\|f\| \geq 0$,

(2) $\|f\| = 0$ iff $f = \underline{0}$,

(3) $\|f+g\| \leq \|f\| + \|g\|$,

(4) $\|\underline{r}f\| = |r| \, \|f\|$,

(5) $\|fg\| \leqslant \|f\| \cdot \|g\|$, and

(6) d is a complete metric on $C^*(X)$.

Proof. This is left as an exercise for the reader (see 4E(1)).

■

(b) **Definition**

(1) Conditions (1) — (4) in the above proposition can be expressed by saying that $\| \ \|$ is a **norm** on $C^*(X)$. Conditions (1) — (5) are equivalent to the assertion that $C^*(X)$ is a **normed ring** with respect to $\| \ \|$.

(2) If $\mathfrak{a} \subseteq C^*(X)$, then d is a metric on $\mathfrak{a}$; $\mathfrak{a}$ is said to be **complete** if $\mathfrak{a}$ is a complete metric space relative to d.

For a commutative ring R with identity 1, recall that a subset $S \subseteq R$ is a subring iff $0, 1 \in S$ and for $a, b \in S$, both $a - b$ and $ab \in S$. Recall that an ideal I in R is maximal iff for $r \in R \setminus I$, there are elements $s \in R$ and $t \in I$ such that $r \cdot s + t = 1$.

Our first goal is to establish the Stone-Weierstrass property for compact spaces. First, a lemma is needed.

(c) **Lemma**. For a space X, let $\mathfrak{a} \subseteq C^*(X)$ be a complete subring of $C^*(X)$ such that $\underline{\mathbb{R}} \subseteq \mathfrak{a}$. Then $\mathfrak{a}$ is a sublattice of $C^*(X)$.

Proof. Since, for $f, g \in \mathfrak{a}$, $f \vee g = \underline{2}^{-1}(f + g + |f - g|)$ and $f \wedge g = f + g - (f \vee g)$ (see 1B(3)), it suffices to show that $|f| \in \mathfrak{a}$ when $f \in \mathfrak{a}$. Since f is bounded, it suffices to show $|f|$

$\in \mathcal{A}$ when $\|f\| < 1$. From calculus, we know that $(1-t)^{1/2}$ has a binomial expansion $\Sigma_{n=0}^{\infty} c_n t^n$ which converges uniformly to $(1-t)^{1/2}$ in the interval $[0,1]$ (see 4C). Thus, for $\epsilon > 0$, there is a polynomial $p(t)$ such that $|(1-t)^{1/2} - p(t)| < \epsilon$ for $0 \leq t \leq 1$. Thus, for $t = 1 - y^2$, $|(y^2)^{1/2} - p(1-y^2)| < \epsilon$ for $-1 \leq y \leq 1$. But $|y| = (y^2)^{1/2}$. So, since $-1 < f(x) < 1$ for all $x \in X$, then $||f(x)| - p(1-(f(x))^2)| < \epsilon$. But $p \circ (1-f^2) \in \mathcal{A}$ since p is a polynomial and $\mathcal{A}$ is a subring with $\mathbb{R} \subseteq \mathcal{A}$. Since $\mathcal{A}$ is complete and $\||f| - p \circ (1-f^2)\| \leq \epsilon$, then $|f| \in \mathcal{A}$.

■

(d) <u>**Definition**</u>. A subset $\mathcal{A} \subseteq C^*(X)$, is said to **separate points** of the space X if for $p, q \in X$ such that $p \neq q$, there is some $f \in \mathcal{A}$ such that $f(p) \neq f(q)$.

(e) <u>**Theorem**</u>. **(Stone-Weierstrass)** Let X be a compact space and let $\mathcal{A} \subseteq C^*(X)$ be a complete subring which contains $\mathbb{R}$. If $\mathcal{A}$ separates points, then $\mathcal{A} = C^*(X)$.

<u>Proof</u>. Let $f \in C^*(X)$ and $p, q \in X$. Since $\mathcal{A}$ is a subring of $C^*(X)$ that separates points and $\mathbb{R} \subseteq \mathcal{A}$, it follows that there is a function $g_{pq} \in \mathcal{A}$ such that $g_{pq}(p) = f(p)$ and $g_{pq}(q) = f(q)$. Let $\epsilon > 0$, $U_{pq} = \{x \in X : g_{pq}(x) - f(x) < \epsilon\}$, and $V_{pq} = \{x \in X : g_{pq}(x) - f(x) > -\epsilon\}$. Let q be fixed. As X is compact, there is a finite set $F_q \subseteq X$ such that $X = \bigvee\{U_{pq} : p \in F_q\}$. Let $V_q = \bigcap\{V_{pq} : p \in F_q\}$ and $g_q = \bigwedge\{g_{pq} : p \in F_q\}$. By 4.5(c), $g_q \in \mathcal{A}$. If $x \in X$, there exists $r \in F_q$ such that $x \in U_{rq}$ and $g_q(x) \leq g_{rq}(x) < f(x) + \epsilon$.

Hence, $g_q < f + \underline{\epsilon}$ for all $q \in X$. If $y \in V_q$, there is some $t \in F_q$ such that $g_q(y) = g_{tq}(y) > f(y) - \epsilon$. Since $q \in V_q$, there is a finite set $F \subseteq X$ such that $X = \bigcup\{V_q : q \in F\}$. Let $g = \bigvee\{g_q : q \in F\}$. Now, $g \in \mathfrak{a}$ by 4.5(c). For $x \in X$, there is some $s \in F$ such that $x \in V_s$ and $g(x) \geqslant g_s(x) > f(x) - \epsilon$. Hence, $g > f - \underline{\epsilon}$. Since $g_q < f + \underline{\epsilon}$ for all $q \in X$, it follows that $g < f + \underline{\epsilon}$. This shows $\|f - g\| < \epsilon$. Since $\mathfrak{a}$ is complete, it follows that $f \in \mathfrak{a}$. Thus $\mathfrak{a} = C^*(X)$.

■

(f) <u>Definition</u>

(1) A ring R equipped with a partial order $\leqslant$ is called a **partially ordered ring** if for a, b, c $\in$ R, $a \geqslant c$ implies $a + b \geqslant c + b$ and $a \geqslant 0$, $b \geqslant 0$ imply $ab \geqslant 0$.

(2) A partially ordered ring in which the partial order is a linear order is called an **ordered ring**.

(3) An ordered field F (i.e., as a ring, it is ordered) is **Archimedian** if the set of integers is cofinal in F, that is, if for each $x \in F$, there is some positive integer $n \in F$ ($=1 + 1 + \ldots + 1$ (n times) for some $n \in \mathbb{N}$) such that $-n < x < n$.

(4) Let $\mathfrak{a}$ be a subring of $C^*(X)$ for a space X such that $\underline{\mathbb{R}} \subseteq \mathfrak{a}$. We denote the set of maximal ideals of $\mathfrak{a}$ by $m_{\mathfrak{a}}X$. If $f \in \mathfrak{a}$, define $S(f)$ to be $\{M \in m_{\mathfrak{a}}X : f \in M\}$.

(g) <u>Lemma</u>. An ordered field F is Archimedian iff F is isomorphic to a subfield of $\mathbb{R}$ (with the isomorphism preserving both the order-theoretic and algebraic structure of F).

Proof. This well-known algebraic result is left as an exercise for the reader (see 4D).

■

Now we are able to derive properties of certain complete subrings of $C^*(X)$ which will make the main theorem in this section more accessible.

(h) **Proposition**. Let X be a space and $\mathfrak{a} \subseteq C^*(X)$ be a complete subring such that $\mathfrak{a} \supseteq \underline{\mathbb{R}}$. Let f, g, h $\in \mathfrak{a}$ and let M $\in$ $m_{\mathfrak{a}}X$. Then:

(1) if $n \in \mathbb{N}$ and $n > \|f\|$, then $(\underline{n}-f)^{-1} \in \mathfrak{a}$,

(2) if $1 > \|\underline{1}-f\|$, then $f^{-1} \in \mathfrak{a}$,

(3) if $f \geqslant \underline{\epsilon} > \underline{0}$ for some $\epsilon \in \mathbb{R}$, then $f^{-1} \in \mathfrak{a}$,

(4) if $|f| \geqslant |g|$ and $f \in M$, then $g \in M$,

(5) Define a relation $\leqslant$ on the quotient field $\mathfrak{a}/M$ as follows: $h + M \leqslant f + M$ if there is some $g \in \mathfrak{a}$ such that $g \geqslant \underline{0}$ and $f + M = h + g + M$. Then $\leqslant$ is a well-defined partial order on $\mathfrak{a}/M$.

(6) if f, g $\in$ M, then $|g| \in M$ and $f \vee g \in M$,

(7) $(f+M) \vee (g+M)$ exists and equals $(f \vee g) + M$,

(8) $f + M \geqslant \underline{0} + M$ iff $f + M = |f| + M$, and

(9) there is a unique order isomorphism between $\mathfrak{a}/M$ and $\mathbb{R}$.

Proof

(1) If $|t| \leqslant \alpha < 1$, the sequence of partial sums $p_n(t) = \Sigma_{k=0}^{n} t^k$ converges uniformly to $(1-t)^{-1}$. If $n > \|f\|$, then for $x \in X$, $n^{-1}|f(x)| \leqslant n^{-1}\|f\| < 1$. But $p_n \circ (\underline{n}^{-1} \cdot f) \in \mathfrak{a}$. Since $\mathfrak{a}$

is complete, it follows that $(\underline{1}-\underline{n}^{-1}\cdot f)^{-1} \in \mathfrak{a}$. So, $(\underline{n}-f)^{-1} = \underline{n}^{-1}(\underline{1}-\underline{n}^{-1}f)^{-1} \in \mathfrak{a}$.

(2) By (1), $(\underline{1}-(\underline{1}-f))^{-1} = f^{-1} \in \mathfrak{a}$.

(3) There is some $n \in \mathbb{N}$ such that $\underline{n} > f \geqslant \underline{\epsilon} > \underline{0}$. So, $\underline{1} > f\cdot(\underline{n})^{-1} \geqslant \underline{\epsilon}\ \underline{n}^{-1} > \underline{0}$ which implies that

$$\underline{0} \leqslant \underline{1} - \underline{f}\ \underline{n}^{-1} \leqslant \underline{1} - \underline{\epsilon}\ \underline{n}^{-1} < \underline{1}.$$

By (2), $(\underline{n}^{-1}\cdot f)^{-1} = \underline{n}\cdot f^{-1} \in \mathfrak{a}$. So, $f^{-1} \in \mathfrak{a}$.

(4) Suppose $|f| \geqslant |g|$ and $g \notin M$. Since $\mathfrak{a}/M$ is a field, there is some $h \in \mathfrak{a}$ such that $\underline{1} + M = (g+M)(h+M) = gh + M$, which implies that $\underline{1} + M = g^2h^2 + M$. So, for some $k \in M$, $g^2h^2 + k = \underline{1}$. Since $g^2 = |g|^2 \leqslant |f|^2 = f^2$, it follows that $f^2h^2 + k \geqslant g^2h^2 + k = \underline{1}$. As $f^2h^2 + k \in \mathfrak{a}$, we see by (3) that $j = (f^2h^2+k)^{-1} \in \mathfrak{a}$. Thus, $(f^2h^2+k)j = \underline{1}$ and $f(fh^2j) + M = \underline{1} + M$. This shows $f \notin M$.

(5) First we must show that $\leqslant$ is well-defined on $\mathfrak{a}/M$. Suppose $f_1 + M = f_2 + M$, $h_1 + M = h_2 + M$, and $g \in \mathfrak{a}$ such that $g \geqslant \underline{0}$ and $f_1 + M = g + h_1 + M$. So, $f_2 + M = f_1 + M = g + (h_1+M) = g + (h_2+M)$. Hence, $\leqslant$ is well-defined on $\mathfrak{a}/M$. Next, we show that $\leqslant$ is a partial ordering on the quotient field $\mathfrak{a}/M$. Since $\underline{0} \in M$ and $f + M = f + \underline{0} + M$, then $f + M \geqslant f + M$ and $\leqslant$ is reflexive. Suppose $f + M \leqslant g + M$ and $g + M \leqslant h + M$. There are elements $k, l \in \mathfrak{a}$ such that $k \geqslant \underline{0}$, $l \geqslant \underline{0}$, $f + k + M = g + M$, and $g + l + M = h + M$. So, $f + (k+l) + M = h + M$; since $k + l \geqslant \underline{0}$, then $h + M \geqslant f + M$. This shows $\leqslant$ is transitive. Suppose $f + M \leqslant g + M$ and $g + M \leqslant f + M$. There are functions $k, l \in \mathfrak{a}$ such that $k \geqslant \underline{0}$, $l \geqslant \underline{0}$, $f + k + M = g + M$ and $g + l + M = f + M$. Thus, $k + l \in M$. Since $\underline{0} \leqslant k \leqslant k + l$, then by (4), $k \in$

M. Thus, $f + M = f + k + M = g + M$ and $\leqslant$ is antisymmetric. This shows that $\leqslant$ is a partial ordering on $\mathfrak{a}/M$. To show $\mathfrak{a}/M$ is a partially ordered field, suppose $f + M \geqslant g + M$; we want to show $f + h + M \geqslant g + h + M$. There is some $k \in \mathfrak{a}$ such that $k \geqslant \underline{0}$ and $f + M = g + k + M$. So, $f + h + M = g + h + k + M$. Hence, $f + h + M \geqslant g + h + M$. Finally, we need to show that if $f + M \geqslant \underline{0} + M$ and $g + M \geqslant \underline{0} + M$, then $fg + M \geqslant \underline{0} + M$. There are elements $k, \ell \in M$ such that $k \geqslant \underline{0}$, $\ell \geqslant \underline{0}$, $f + M = k + M$, and $g + M = \ell + M$. Now, $fg + M = (f+M)(g+M) = (k+M)(\ell+M) = k\ell + M$. Since $k\ell \geqslant \underline{0}$, then $fg + M \geqslant \underline{0} + M$. Thus, $\mathfrak{a}/M$ is a partially ordered field.

(6) Since $||g|| \leqslant |g|$, it follows from (4) that $|g| \in M$ if $g \in M$. So, if $f, g \in M$, then $|f| + |g| \in M$. Since $-(|f| + |g|) \leqslant f \vee g \leqslant |f| + |g|$, it follows that $\underline{0} \leqslant |f \vee g| \leqslant |f| + |g|$. By (4), $f \vee g \in M$.

(7) Clearly $f \vee g \geqslant f$ and $f \vee g \geqslant g$; thus $(f \vee g) + M \geqslant f + M$ and $(f \vee g) + M \geqslant g + M$. Suppose, for some $h \in \mathfrak{a}$, $h + M \geqslant f + M$ and $h + M \geqslant g + M$. There are elements $k, \ell \in \mathfrak{a}$ such that $k \geqslant \underline{0}$, $\ell \geqslant \underline{0}$, and $h + M = f + k + M = g + \ell + M$. So, there are elements $m_1, m_2 \in M$ such that $h - f + m_1 = k \geqslant \underline{0}$ and $h - g + m_2 = \ell \geqslant \underline{0}$. Thus, $h + m_1 \geqslant f$ and $h + m_2 \geqslant g$. Since $h + (m_1 \vee m_2) \geqslant h + m_1$ and $h + (m_1 \vee m_2) \geqslant h + m_2$, it follows that $h + (m_1 \vee m_2) \geqslant (h+m_1) \vee (h+m_2) \geqslant f \vee g$. By (6), $m_1 \vee m_2 \in M$. So, $h + M \geqslant f \vee g + M$. This shows $f \vee g + M = (f+M) \vee (g+M)$.

(8) Suppose $f + M \geqslant \underline{0} + M$. Then $\underline{0} + M \geqslant -f + M$ and $f + M = (f+M) \vee (-f+M) = f \vee (-f) + M = |f| + M$. Conversely, suppose $f + M = |f| + M$. Then $f + M \geqslant \underline{0} + M$ as $|f| \geqslant \underline{0}$.

(9) Since $(f - |f|)(f + |f|) = \underline{0} \in M$, then $f - |f| \in M$ or $f + |f| \in M$. Hence, $f + M = |f| + M$ or $-f + M = |f| + M$. By (8), either $f + M \geq \underline{0} + M$ or $-f + M = |-f| + M$ and $-f + M \geq \underline{0} + M$, i.e., $f + M \leq \underline{0} + M$. So $\mathcal{Q}/M$ is an ordered field. Since f is bounded, there is some $n \in \mathbb{N}$ such that $f \leq |f| \leq \underline{n}$. So, $-\underline{n} + M \leq f + M \leq \underline{n} + M$ and $\mathcal{Q}/M$ is Archimedian. By 4.5(g), $\mathcal{Q}/M$ is isomorphic (as a field and as a poset) to a subfield F of $\mathbb{R}$; let $\phi : \mathcal{Q}/M \to F$ be an order isomorphism. Then $\phi(\underline{0}+M) = 0$ and $\phi(\underline{n}+M) = n$ for $n \in \mathbb{Z}$. Also, for $r \in \mathbb{Q}$, $\phi(\underline{r}+M) = r$. Let $s \in \mathbb{R}$ and $T = \{r \in \mathbb{Q} : r \leq s\}$. So, $\phi(\underline{s}+M) \geq \phi(\underline{r}+M) = r$ for all $r \in T$. Hence, $\phi(\underline{s}+M) \geq s$. Similarly, $\phi(\underline{s}+M) \leq s$. So, $\phi(\underline{s}+M) = s$. Thus, $F = \mathbb{R}$, $\mathcal{Q}/M$ is order isomorphic to $\mathbb{R}$, and $\mathcal{Q}/M = \{\underline{s} + M : s \in \mathbb{R}\}$. If $\psi : \mathcal{Q}/M \to \mathbb{R}$ is an order isomorphism, then by repeating the above argument with "ϕ" replaced by "ψ", we get that $\psi(\underline{s}+M) = s$ for all $s \in \mathbb{R}$. Since $\phi(\underline{s}+M) = \psi(\underline{s}+M)$ for all $s \in \mathbb{R}$ and $\mathcal{Q}/M = \{\underline{s} + M : s \in \mathbb{R}\}$, then $\phi = \psi$.

■

Our next goal is to define a compact topology on $m_{\mathcal{Q}}X$ so that X can be densely embedded in $m_{\mathcal{Q}}X$.

(i) **Lemma**. Let $\mathcal{Q}$ be a subring of $C^*(X)$ for a space X such that $\mathbb{R} \subseteq \mathcal{Q}$. Let $f, g \in \mathcal{Q}$. Then:

(1) $S(\underline{1}) = \emptyset$ and $S(\underline{0}) = m_{\mathcal{Q}}X$,

(2) $S(f) \cup S(g) = S(fg)$,

(3) $S(f) \cap S(g) \subseteq S(f^2 + g^2)$, and

(4) $\{S(f) : f \in \mathcal{Q}\}$ is closed base for a topology on $m_{\mathcal{Q}}X$ (called the **Stone topology**).

(See 4.5(f)(4) for notation.)

<u>Proof</u>. Parts (1) – (3) are an exercise in algebra and are left to the reader (see 4E). Part (4) is an immediate consequence of (1) – (3).

■

There is a similarity between the Stone space of a Boolean algebra and the Stone topology defined on the set $m_{\mathfrak{a}}X$ of maximal ideals of a subring $\mathfrak{a}$ of $C^*(X)$. An equivalent way of constructing the Stone space S(B) of a Boolean algebra B is to define a topology on the set of maximal ideals of B where B is considered as a Boolean ring. This approach using maximal ideals is the usual construction of S(B) in an algebraic setting. The topology on S(B) uses $\{S(a) : a \in B\}$ as a closed base where $S(a) = \{M : M$ is maximal ideal on B and $a \in M\}$. From this viewpoint, the Stone space of a Boolean algebra and the Stone topology defined on a set of maximal ideals of a subring of $C^*(X)$ are quite similar.

(j) **<u>Theorem</u>**. Let X be a space and $\mathfrak{a} \subseteq C^*(X)$ be a complete subring such that $\mathfrak{a} \supseteq \mathbb{R}$. Then $m_{\mathfrak{a}}X$ is a compact Hausdorff space and for $f, g \in \mathfrak{a}$, $\{M \in m_{\mathfrak{a}}X : f + M > g + M\}$ is open in $m_{\mathfrak{a}}X$.

<u>Proof</u>. Let $\mathcal{F}$ be a filter of closed sets of $m_{\mathfrak{a}}X$. Let $\mathcal{G} = \{S(f) : f \in \mathfrak{a}$ and $S(f) \supseteq F$ for some $F \in \mathcal{F}\}$. Since $\{S(f) : f \in \mathfrak{a}\}$ is a closed base, it follows that $\cap\mathcal{F} = \cap\mathcal{G}$. We will show that $\cap\mathcal{G} \neq \emptyset$. Let $B = \{f \in \mathfrak{a} : S(f) \in \mathcal{G}\}$. Since $S(f) \neq \emptyset$ for all $S(f)$

$\in \mathcal{G}$, it follows that $f \neq \underline{1}$ for all $f \in B$. Let C be the ideal generated by B and M a maximal ideal containing C. Then $M \in m_{\mathfrak{a}}X$ and for $g \in B \subseteq M$, $M \in S(g)$. So, $M \in \cap\mathcal{G} = \cap\mathcal{F}$. Thus, $\cap\mathcal{F} \neq \emptyset$. This shows that $m_{\mathfrak{a}}X$ is compact. Before showing $m_{\mathfrak{a}}X$ is Hausdorff, we will show that for $f, g \in \mathfrak{a}$, $\{M \in m_{\mathfrak{a}}X : f + M > g + M\}$ is open. First, note that $\{M \in m_{\mathfrak{a}}X : f + M \geqslant g + M\} = \{M \in m_{\mathfrak{a}}X : f - g + M \geqslant \underline{0} + M\} = \{M \in m_{\mathfrak{a}}X : f - g + M = |f - g| + M\}$ by 4.5(h)(8). But $\{M \in m_{\mathfrak{a}}X : f - g + M = |f - g| + M\} = \{M \in m_{\mathfrak{a}}X : f - g - |f - g| \in M\} = S(f - g - |f - g|)$. Thus, $\{M \in m_{\mathfrak{a}}X : f + M > g + M\} = m_{\mathfrak{a}}X \backslash \{M \in m_{\mathfrak{a}}X : f + M \leqslant g + M\} = m_{\mathfrak{a}}X \backslash S(g - f - |g - f|)$ is open. To show $m_{\mathfrak{a}}X$ is Hausdorff, suppose $M_1, M_2 \in m_{\mathfrak{a}}X$ and $M_1 \neq M_2$. There is some $f \in M_1 \backslash M_2$. Since $f \notin M_2$, there is some $g \in \mathfrak{a}$ such that $fg + M_2 = \underline{1} + M_2$. Now, $M_2 \in \{M \in m_{\mathfrak{a}}X : fg + M > \underline{1/2} + M\}$ and $M_1 \in \{M \in m_{\mathfrak{a}}X : fg + M < \underline{1/2} + M\}$. This shows that $m_{\mathfrak{a}}X$ is Hausdorff.

■

(k) **Definition**. Let X be a space. A subring $\mathfrak{a}$ of $C^*(X)$ is called a **regular subring** if $\mathfrak{a}$ is a complete (with respect to the sup norm metric) subring of $C^*(X)$ such that: (i) $\mathfrak{a} \supseteq \underline{\mathbb{R}}$, and (ii) $Z(\mathfrak{a}) = \{Z(f) : f \in \mathfrak{a}\}$ is a closed base for X.

If $x \in X$, and if we have specified the regular subring $\mathfrak{a}$, we will denote $\{f \in \mathfrak{a} : f(x) = 0\}$ by M_x. Note that if $C^*(X)$ has a regular subring, it follows that X is Tychonoff.

(l) **Lemma**. Let $\mathfrak{a}$ be a regular subring of $C^*(X)$, and let $x \in X$. Then $M_x \in m_{\mathfrak{a}}X$.

Proof. The function $\hat{x} : \mathcal{A} \to \mathbb{R}$ defined by $\hat{x}(f) = f(x)$ is easily checked to be a ring homomorphism; note that $\hat{x}[\mathcal{A}] = \mathbb{R}$ since $\mathbb{R} \subseteq \mathcal{A}$. Recall from algebra that $(\hat{x})^{\leftarrow}(0)$ (called the kernel of $\hat{x}$) is a maximal ideal. Clearly, $(\hat{x})^{\leftarrow}(0) = M_x$.

■

If $\mathcal{A}$ is a regular subring of $C^*(X)$, define $\lambda : X \to m_{\mathcal{A}}X$ by $\lambda(x) = M_x$. We are now ready to present one of the main results in this section.

(m) **Theorem.** **(Gelfand)** Let $\mathcal{A}$ be a regular subring of $C^*(X)$ for a space X. Then:

(1) $\lambda : X \to m_{\mathcal{A}}X$ is a dense embedding,

(2) each $f \in \mathcal{A}$, viewed as a function defined on $\lambda[X]$, can be extended to a unique $f^e \in C^*(m_{\mathcal{A}}X)$; that is, if $f \in \mathcal{A}$, there exists a unique $f^e \in C^*(m_{\mathcal{A}}X)$ such that $f^e | \lambda[X] = f \circ \lambda^{\leftarrow}$,

(3) if $\mathcal{A}^e = \{f^e : f \in \mathcal{A}\}$, then $\mathcal{A}^e = C^*(m_{\mathcal{A}}X)$ and $\mathcal{A} = \{f \in C^*(X) : f = F|X$ for some $F \in C^*(m_{\mathcal{A}}X)\}$, and $f \to f^e$ is a ring isomorphism from $\mathcal{A}$ onto $C^*(m_{\mathcal{A}}X)$, and

(4) for $f \in \mathcal{A}$, $S(f) = Z(f^e)$, the zero-set of f^e in $m_{\mathcal{A}}X$.

Proof

(1) First, we show λ is one-to-one. Suppose $x, y \in X$ and $x \neq y$. Since $Z(\mathcal{A})$ is a closed base for X, there is some $f \in \mathcal{A}$ such that $f(x) = 0$ and $f(y) = 1$. Thus, $f \in M_x \setminus M_y$. So, $\lambda(x) \neq \lambda(y)$. Let $f \in \mathcal{A}$. To show λ is continuous and $\lambda : X \to \lambda[X]$ is closed, it suffices to show (since λ is one-to-one) $\lambda[Z(f)] = \lambda[X] \cap$

$S(f)$. Let $x \in Z(f)$. Then $f(x) = 0$ and $f \in M_x$. So, $\lambda(x) = M_x \in S(f)$, and this shows $\lambda[Z(f)] \subseteq \lambda[X] \cap S(f)$. Now, let $\lambda(x) \in \lambda[X] \cap S(f)$. Thus $M_x \in S(f)$; hence $f(x) = 0$ and $x \in Z(f)$. Hence, $\lambda[X] \cap S(f) = \lambda[Z(f)]$. Finally, we need to show that $\lambda[X]$ is dense in $m_{\mathfrak{a}}X$. Suppose $\lambda[X] \subseteq S(f)$. Then $X = \lambda^{\leftarrow}[\lambda[X]] \subseteq \lambda^{\leftarrow}[S(f)] = Z(f)$. Hence, $f = \underline{0}$ and $S(f) = m_{\mathfrak{a}}X$. Since $\{S(f) : f \in \mathfrak{a}\}$ is a closed base for $m_{\mathfrak{a}}X$, then $m_{\mathfrak{a}}X = \cap\{S(f) : f \in \mathfrak{a}$ and $S(f) \supseteq \lambda[X]\} = c\ell\lambda[X]$ (closure in $m_{\mathfrak{a}}X$). Thus, $\lambda[X]$ is dense in $m_{\mathfrak{a}}X$.

(2) For each $M \in m_{\mathfrak{a}}X$, there is a unique order isomorphism j_M from $\mathfrak{a}/M$ onto $\mathbb{R}$ by 4.5(h)(9). Define $f^e(M)$ to be $j_M(f+M)$; thus $f + M = \underline{f^e(M)} + M$. We need to show f^e is continuous and $f^e(M_x) = f(x)$ for $x \in X$. Let $r \in \mathbb{R}$. Then $(f^e)^{\leftarrow}[(r,\infty)] = \{M \in m_{\mathfrak{a}}X : f^e(M) > r\} = \{M \in m_{\mathfrak{a}}X : f + M > \underline{r} + M\}$ (as j_M is an order isomorphism), which is open by 4.5(j). Similarly, $(f^e)^{\leftarrow}[(-\infty,r)]$ is open. Hence, f^e is continuous. For $x \in X$, recall that $\hat{x} : \mathfrak{a} \to \mathbb{R}$ defined by $\hat{x}(f) = f(x)$ is a ring homomorphism onto $\mathbb{R}$ and $\hat{x}^{\leftarrow}(0) = M_x$. This ring homomorphism induces an order isomorphism between $\mathfrak{a}/M_x$ and $\mathbb{R}$. Since such order isomorphisms are unique by 4.5(h)(9), then $f^e(M_x) = \hat{x}(f) = f(x)$. By 4.1(b), f has at most one continuous extension from $m_{\mathfrak{a}}X$ to $\mathbb{R}$.

(3) Clearly, $\underline{\mathbb{R}} \subseteq \mathfrak{a}^e \subseteq C^*(m_{\mathfrak{a}}X)$. Let $f, g \in \mathfrak{a}$. Since $(f^e + g^e)|X = f + g = (f + g)^e|X$, then by 4.1(b), $f^e + g^e = (f + g)^e$. Similarly, $f^e g^e = (fg)^e$. So, $\mathfrak{a}^e$ is a subring of $C^*(m_{\mathfrak{a}}X)$. Let $h \in C^*(m_{\mathfrak{a}}X)$ be the limit of $\{f_n{}^e : n \in \mathbb{N}\}$ where $f_n \in \mathfrak{a}$ (this is relative to the metric defined in 4.5(b)). Then $h|X$ is a limit of $f_n{}^e|X$. So, $h|X \in \mathfrak{a}$. Thus, $h = (h|X)^e \in \mathfrak{a}^e$ and $\mathfrak{a}^e$ is complete. Suppose $M_1, M_2 \in m_{\mathfrak{a}}X$ such that $M_1 \neq M_2$. Let $f \in$

$M_1 \backslash M_2$. Then $f^e(M_1) = 0$ as $f + M_1 = \underline{0} + M_1$ and $f^e(M_2) \neq 0$ as $f + M_2 \neq \underline{0} + M_2$. So, $\mathfrak{a}^e$ separates points of $m_{\mathfrak{a}}X$. By the Stone-Weierstrass Theorem 4.5(e), $\mathfrak{a}^e = C^*(m_{\mathfrak{a}}X)$. Clearly, $\mathfrak{a} \subseteq \{f \in C^*(X) : f = F|X$ for some $F \in C^*(m_{\mathfrak{a}}X)\}$. Suppose $F \in C^*(m_{\mathfrak{a}}X)$; then $F = g^e$ for some $g \in \mathfrak{a}$. Since $F|X = g^e|X = g \in \mathfrak{a}$, then $\mathfrak{a} \supseteq \{f \in C^*(X) : f = F|X$ for some $F \in C^*(m_{\mathfrak{a}}X)\}$.

(4) Let $M \in m_{\mathfrak{a}}X$. Then $M \in S(f)$ iff $f \in M$. But $f \in M$ iff $f + M = \underline{0} + M$ iff $f^e(M) = 0$. Finally, $f^e(M) = 0$ iff $M \in Z(f^e)$.

■

(n) **Definition**. A compactification Y of a Tychonoff space X is called a **Gelfand compactification** if $Y \equiv_X m_{\mathfrak{a}}X$ for some regular subring $\mathfrak{a}$ of $C^*(X)$.

Unlike the situation for Wallman compactifications, it is true that every compactification is a Gelfand compactification; this is shown in the next theorem. Problem 4G addresses the question of when the closed base $Z(\mathfrak{a})$ associated with the regular subring $\mathfrak{a}$ is a Wallman base, and the relationship between $m_{\mathfrak{a}}X$ and $w_{Z(\mathfrak{a})}X$.

(o) **Theorem**. Let X be a Tychonoff space and $T \in K(X)$. Let $\mathfrak{a} = \{f|X : f \in C^*(T)\}$. Then:

(1) $\mathfrak{a}$ is a regular subring of $C^*(X)$.

(2) $m_{\mathfrak{a}}X = \{M^p : p \in T\}$ where $M^p = \{f|X : f \in C^*(T)$ and $f(p) = 0\}$, and for $x \in X$, $M_x = M^x$, and

(3) $T \equiv_X m_{\mathfrak{a}}X$.

Proof. The proof of (1) is left to the reader (see 4E(2)).

(2) Let $p \in T$. Clearly, M^p is an ideal in $\mathfrak{a}$. Let $g \in C^*(T)$. Then $g|X \in \mathfrak{a}$. Suppose $g|X \notin M^p$. Then $g(p) \neq 0$. So, $p \notin Z_T(g)$. There is some $f \in C^*(T)$ such that $f \geq \underline{0}$, $f(p) = 0$, and $Z_T(f) \cap Z_T(g) = \emptyset$. Thus, $f|X \in M^p$. Now, $f + g^2 \in C^*(T)$ and $Z_T(f + g^2) = \emptyset$. So, $(f + g^2)^{-1} \in C(T) = C^*(T)$. Let $h = (f + g^2)^{-1}$. Then $\underline{1} = h(f + g^2) = hf + g(gh)$ implying $\underline{1} = (hf)|X + (g|X)((gh)|X)$. But $(hf)|X \in M^p$ and $(gh)|X \in \mathfrak{a}$. This shows that M^p is maximal. Conversely, suppose $M \in m_{\mathfrak{a}}X$. Let $M^e = \{f \in C^*(T) : f|X \in M\}$. Clearly, M^e is an ideal in $C^*(T)$. Let $g \notin M^e$ where $g \in C^*(T)$. Then $g|X \in \mathfrak{a}$ and $g|X \notin M$. So, there are f, $h \in C^*(T)$ such that $f|X \in M$, $h|X \in \mathfrak{a}$ and $(h|X)(g|X) + f|X = \underline{1}$. Hence, $(hg+f)|X = \underline{1}$. By 4.1(b), $hg + f = \underline{1}$. So, M^e is a maximal ideal in $C^*(T)$. Consider $F = \{Z_T(f) : f \in M^e\}$. If $f, g \in M^e$, then $Z_T(f) \cap Z_T(g) = Z_T(f^2 + g^2) \in F$ as $f^2 + g^2 \in M^e$. Also, $Z_T(f) \neq \emptyset$ for all $f \in M^e$. So, F is a family of closed sets of T with the finite intersection property. Since T is compact, it follows that there is some $p \in \cap F$. So, $M^e \subseteq M_p (= \{f \in C^*(T) : f(p) = 0\})$, and since M^e is maximal, then $M^e = M_p$. Thus, $M = \{f|X : f \in M^e\} = \{f|X : f \in M_p\} = M^p$. For $x \in X$, it easily follows that $M_x = M^x$.

(3) Define $\psi : T \to m_{\mathfrak{a}}X$ by $\psi(p) = M^p$. By (2), ψ is onto. If $p, q \in T$ and $p \neq q$, then there is some $f \in C^*(T)$ such that $f(p) = 0$ and $f(q) = 1$. Thus, $f|X \in M^p \setminus M^q$. So, $\psi(p) \neq \psi(q)$ and ψ is one-to-one. By (2), for $x \in X$, $\psi(x) = M^x = M_x$ $(= \lambda(x)$; see 4.5(m)(1)). So, $\psi|X = id_X$ if we identify X and the dense subspace $\lambda[X]$ of $m_{\mathfrak{a}}X$. For $f \in C^*(T)$, $\psi[Z_T(f)] = \psi[\{p \in T : f(p) = 0\}] =$

$\{M^p : f(p) = 0\}$. Now, $f(p) = 0$ iff $f|X \in M^p$ iff $M^p \in S(f|X)$. So, $\{M^p : f(p) = 0\} = S(f|X)$. Since $\psi[Z_T(f)] = S(f|X)$ and ψ is a bijection, it follows that ψ is both continuous and closed. This shows ψ is a homeomorphism and completes the proof that $T \equiv_X m_{\mathfrak{a}}X$.

■

(p) **Corollary**. Let X be a Tychonoff space. Then:

(1) $\beta X \equiv_X m_{\mathfrak{a}}X$ where $\mathfrak{a} = C^*(X)$,

(2) βX is the only compactification in which X is C^*-embedded, and

(3) if $X \subseteq T \subseteq \beta X$, then $\beta T \in K(X)$ and $\beta T \equiv_X \beta X$.

Proof

(1) By 4.5(o)(3), $\beta X \equiv_X m_{\mathfrak{a}}X$ where $\mathfrak{a} = \{f \in C^*(X) : f = F|X$ for some $F \in C^*(\beta X)\}$. But by 4.2(e), X is C^*-embedded in βX; hence, $\mathfrak{a} = C^*(X)$.

(2) Suppose X is C^*-embedded in a compactification Y of X. By 4.5(o), $Y \equiv_X m_{\mathfrak{a}}X$ where $\mathfrak{a} = \{f : f = F|X$ and $F \in C^*(Y)\}$. But X is C^*-embedded in Y; so, $\mathfrak{a} = C^*(X)$. By (1), $\beta X \equiv_X m_{\mathfrak{a}}X$. Thus, $Y \equiv_X \beta X$.

(3) By (2), we only need to show X is C^*-embedded in βT. Since X is C^*-embedded in βX, then X is C^*-embedded in T. But T is C^*-embedded in βT. Hence, X is C^*-embedded in βT.

■

For a Tychonoff space X, let $GB(X)$ denote the collection of regular subrings of $C^*(X)$. Clearly, $GB(X)$ is a poset with respect to inclusion. Our next result shows there is an order isomorphism between $GB(X)$ and $K(X)$.

(q) **Theorem**. Let X be a Tychonoff space. Define ψ : $GB(X) \to K(X)$ by $\psi(B) = m_B X$. Then ψ is an order isomorphism.

Proof. By 4.5(o)(3), ψ is onto. Suppose $B, C \in GB(X)$ and $\psi(B) \equiv_X \psi(C)$. Let $\mathfrak{a}_B = \{f|X : f \in C^*(m_B X)\}$ and $\mathfrak{a}_C = \{f|X : f \in C^*(m_C X)\}$. Clearly $\mathfrak{a}_B = \mathfrak{a}_C$ as $m_B X \equiv_X m_C(X)$. By 4.5(m) $B = \mathfrak{a}_B = \mathfrak{a}_C = C$. This shows that ψ is one-to-one. Let Y $= \psi(B)$ and Z $= \psi(C)$. Then $B = \mathfrak{a}_B$ and $C = \mathfrak{a}_C$ (defined above). Suppose $Y \geq Z$. Then there is a continuous function $f : Y \to Z$ such that $f|X = id_X$. If $g \in C = \mathfrak{a}_C$ and g^Z is the continuous extension of g to $C^*(Z)$, then $g^Z \circ f \in C^*(Y)$ and $g^Z \circ f|X = g$. So, $g \in \mathfrak{a}_B = B$. Hence, $C \subseteq B$. Conversely, suppose $C = \mathfrak{a}_C \subseteq \mathfrak{a}_B = B$. Let A and B be disjoint closed sets in Z. There is a continuous function $h \in C^*(Z)$ such that $h[A] \subseteq \{0\}$, $h[B] \subseteq \{1\}$, and $h[Z] \subseteq [0,1]$. Since $h|X \in \mathfrak{a}_C \subseteq \mathfrak{a}_B$, then there is some $g \in C^*(Y)$ such that $g|X = h|X$. So, $g[A \cap X] = h[A \cap X] \subseteq \{0\}$ and $g[B \cap X] = h[B \cap X] \subseteq \{1\}$. This shows $g[c\ell_Y(A \cap X)] \subseteq \{0\}$ and $g[c\ell_Y(B \cap X)] \subseteq \{1\}$ which implies that $c\ell_Y(A \cap X) \cap c\ell_Y(B \cap X) = \emptyset$. Thus $Y \geq Z$ by 4.2(h).

■

Since X is C^*-embedded in βX, each member of $C^*(X)$ extends to a unique member of $C^*(\beta X)$ (uniqueness follows from 1.6(d)). We introduce some notation to cope with this.

(r) **Definition**. If $f \in C^*(X)$, let f^β denote the unique $g \in C^*(\beta X)$ for which $g|X = f$.

Note that if $f \in C(X)$, then the function βf described in 4.2(d) is a well-defined member of $C(\beta X, \beta \mathbb{R})$, while f^β exists only if $f \in C^*(X)$. If f^β **does** exist, however, then $f^\beta(x) = \beta f(x)$ for each $x \in \beta X$. We use these distinct notations to emphasize that $f^\beta \in C^*(\beta X)$ if $f \in C^*(X)$.

4.6 The Stone-Čech Compactification

In this section, we focus our attention on the Stone-Čech compactification of a Tychonoff space. The methods of Wallman and Gelfand are used to develop more characterizations of the Stone-Čech compactification. First, we look at the internal structure of the Stone-Čech compactification as revealed by the Wallman and Gelfand models.

We now have several methods for constructing βX; each yields an equivalent compactification of X, but the points in these various equivalent compactifications vary in nature. In one case they are maximal ideals, in another case z-ultrafilters, and in yet another case they are points in a certain product space. In order to avoid specifying the identity of the underlying set of the Stone-Čech compactification, in the ensuing discussion we shall think of βX as being an abstract space which happens to be a compactification of X that is equivalent to each of $w_{Z(X)}X$, $m_{\mathcal{Q}}X$ and $c\ell_Y e[X]$, where $Y = [0,1]^{C^*(X)}$ and e is the evaluation map.

Because $\beta X \equiv_X m_{C^*(X)}X$, we can index the maximal ideals in $C^*(X)$ as follows: $m_{C^*(X)}X = \{(M^*)^p : p \in \beta X\}$.

(a) **Proposition**. Let X be a Tychonoff space. Then:

(1) for $p \in \beta X$, $M^{*p} = \{f \in C^*(X) : f^\beta(p) = 0\}$,

(2) for $x \in X$, $M^{*x} = M_x = \{f \in C^*(X) : f(x) = 0\}$, and

(3) $\{Z(f^\beta) : f \in C^*(X)\}$ is a closed base for βX.

Proof. Evidently (1) and (2) follow from 4.5(o)(2), and (3) follows from 4.5(m)(4).

■

Because $\beta X \equiv_X w_{Z(X)}X$, we can index the set of all z-ultrafilters on X as follows: $w_{Z(X)}X = \{A^p : p \in \beta X\}$.

(b) **Proposition**. Let X be a Tychonoff space. Then:

(1) for $p \in \beta X$, $A^p = \{Z \in Z(X) : p \in c\ell_{\beta X}Z\}$,

(2) for $x \in X$, $A^x = \{Z \in Z(X) : x \in Z\}$,

(3) $\{c\ell_{\beta X}Z : Z \in Z(X)\}$ is a closed base for βX,

(4) for $Z \in Z(X)$, $c\ell_{\beta X}Z = Z \cup \{A \in \beta X \backslash X : Z \in A\}$,

and

(5) for $p \in \beta X$, A^p is the unique z-ultrafilter on X which converges (as a filter base on βX) to p in βX.

Proof. Let $Z \in Z(X)$. In the proof of 4.4(f)(1), it is shown that $S(Z) = c\ell_Y Z$ where $Y = w_L X$ and $L = Z(X)$. Thus, $Y \equiv_X \beta X$.

So, $S(Z) = c\ell_{\beta X}Z$. Thus, (3) follows. Also, $p \in c\ell_{\beta X}Z$ iff $A^p \in S(Z)$ and $A^p \in S(Z)$ iff $Z \in A^p$. So, (1) and (4) follow. Now, (2) follows from (1). To show (5), first note that for $q \in \beta X$, $\{q\} = \cap\{c\ell_{\beta X}Z : q \in c\ell_{\beta X}Z\}$ by (3). But $\cap\{c\ell_{\beta X}Z : q \in c\ell_{\beta X}Z\} = \cap\{c\ell_{\beta X}Z : Z \in A^q\} = a_{\beta X}(A^q)$ (viewing A^q as a filter base on βX). By 2.3(f)(4), A^q converges to q. Thus, A^p is the unique z-ultrafilter on X converging on p.

■

For a function $f \in C^*(X)$ for a Tychonoff space X, it seems reasonable, in view of 4.6(a)(3) and 4.6(b)(3), to compare $Z(f^\beta)$ and $c\ell_{\beta X}Z(f)$. Since $Z(f^\beta)$ is closed and $Z(f^\beta) \supseteq Z(f)$, then $Z(f^\beta) \supseteq c\ell_{\beta X}Z(f)$. In general, these sets are not equal as demonstrated in the next example.

(c) **Example**. Let $f \in C^*(\mathbb{N})$ be defined by $f(n) = 1/n$. Then $Z(f) = \emptyset$ and $c\ell_{\beta\mathbb{N}}Z(f) = \emptyset$. If $p \in \beta\mathbb{N}\setminus\mathbb{N}$, then $f^\beta(p) \geq 0$. Now A^p is a z-ultrafilter on the discrete space $\mathbb{N}$; hence A^p is an ultrafilter on $\mathbb{N}$. Since $p \notin \mathbb{N}$, it follows that A^p is a free ultrafilter on $\mathbb{N}$, and each $F \in A^p$ is an infinite subset of $\mathbb{N}$. So, $0 \in c\ell_{\mathbb{R}}f[F]$. But $f^\beta[A^p] = f[A^p] = \{f[F] : F \in A^p\}$, which converges to $f^\beta(p)$. Thus, $f^\beta(p) \in \cap\{c\ell_{\mathbb{R}}f[F] : F \in A^p\} = \{0\}$. So, $f^\beta(p) = 0$. Hence, $Z(f^\beta) = \beta\mathbb{N}\setminus\mathbb{N} \neq \emptyset = c\ell_{\beta\mathbb{N}}Z(f)$.

On the other hand, there is this relationship between the zero-sets of βX and the βX-closure of zero-sets of X.

(d) **Proposition**. Let X be a Tychonoff space X. Each zero-set of βX is the countable intersection of the βX-closures of zero-sets of X.

Proof. Let $F \in C^*(\beta X)$. Then $Z(F) = \cap\{F^{\leftarrow}[[(-n^{-1},n^{-1})]] : n \in \mathbb{N}\}$. Let $f = F|X$. Then $F^{\leftarrow}[(-n^{-1},n^{-1})]$ is an open set in βX. Also, $F^{\leftarrow}[(-n^{-1},n^{-1})] \cap X = f^{\leftarrow}[(-n^{-1},n^{-1})]$. Now $c\ell_{\beta X}f^{\leftarrow}[(-n^{-1},n^{-1})] = c\ell_{\beta X}(F^{\leftarrow}[(-n^{-1},n^{-1})] \cap X) = c\ell_{\beta X}(F^{\leftarrow}[(-n^{-1},n^{-1})] \subseteq F^{\leftarrow}[[-n^{-1},n^{-1}]]$. So, $F^{\leftarrow}[[-(n+1)^{-1},(n+1)^{-1}]] \subseteq c\ell_{\beta X}f^{\leftarrow}[[-n^{-1},n^{-1}]] \subseteq F^{\leftarrow}[[-n^{-1},n^{-1}]]$ and $Z(F) = \cap\{c\ell_{\beta X}f^{\leftarrow}[[-n^{-1},n^{-1}]] : n \in \mathbb{N}\}$. The conclusion follows as $f^{\leftarrow}[[-n^{-1},n^{-1}]] \in Z(X)$.

■

It is natural to conjecture about the relationship (if any) between A^p and M^{*p} where $p \in \beta X$. When we compare 4.6(a) and 4.6(b) it becomes evident that there is a one-to-one correspondence between the maximal ideals of $C^*(X)$ and the z-ultrafilters on X. In the following two results we make that relationship explicit. Our formulation of the relationship has the advantage that it does not refer to the space βX or its properties, but stays entirely "within X".

(e) **Theorem**. Let X be a Tychonoff space.

(1) If M is a maximal ideal in $C^*(X)$, let $\mathcal{Q}(M) = \{Z \in Z(X) :$ if $f \in M$ then $0 \in c\ell_{\mathbb{R}}f[Z]\}$. Then $\mathcal{Q}(M)$ is a z-ultrafilter on X.

(2) Let $\mathcal{Q}$ be a z-ultrafilter on X. Let $M(\mathcal{Q}) = \{f \in C^*(X) :$

if $S \in \mathcal{Q}$ then $0 \in c\ell_{\mathbb{R}} f[S]\}$. Then $M(\mathcal{Q})$ is a maximal ideal of $C^*(X)$.

Proof

(1) It is routine to show that $\emptyset \notin \mathcal{Q}(M)$ and that if $Z_1 \in \mathcal{Q}(M)$ and $Z_1 \subseteq Z_2 \in Z(X)$, then $Z_2 \in \mathcal{Q}(M)$.

Now suppose that Z_1 and Z_2 are in $\mathcal{Q}(M)$. Thus for each $f \in M$,

$$(*) \qquad 0 \in c\ell_{\mathbb{R}} f[Z_1] \cap c\ell_{\mathbb{R}} f[Z_2]$$

We will assume that $Z_1 \cap Z_2 \notin \mathcal{Q}(M)$ and derive a contradiction. If $Z_1 \cap Z_2 \notin \mathcal{Q}(M)$, there exists $f_0 \in M$ such that $0 \notin c\ell_{\mathbb{R}} f_0[Z_1 \cap Z_2]$. Hence there exists $\epsilon > 0$ such that $f_0^{\leftarrow}[[-\epsilon,\epsilon]] \cap Z_1 \cap Z_2 = \emptyset$. We can assume that $\epsilon < 1$. Thus by 1.9(f) there exists $g \in C^*(X)$ such that $g[f_0^{\leftarrow}[[-\epsilon,\epsilon]] \cap Z_1] = \{0\}$ and $g[Z_2] = \{1\}$. Obviously $g \notin M$ by (*). As M is a maximal ideal, there exists $a \in C^*(X)$ and $w \in M$ such that $w + ag = \underline{1}$. Then $w^2 + f_0^2 \in M$. If $x \in f_0^{\leftarrow}[[-\epsilon,\epsilon]] \cap Z_1$, then $g(x) = 0$ and $(w^2+f_0^2)(x) \geqslant [w(x)]^2 = [w(x) + a(x)g(x)]^2 = 1$ (as $g(x) = 0$). If $x \in Z_1 \setminus f_0^{\leftarrow}[[-\epsilon,\epsilon]]$ then $(w^2+f_0^2)(x) \geqslant [f_0(x)]^2 \geqslant \epsilon^2$. Thus $(w^2+f_0^2)[Z_1] \subseteq [\epsilon^2,+\infty)$, which contradicts (*). Hence $Z_1 \cap Z_2 \in \mathcal{Q}(M)$ and $\mathcal{Q}(M)$ is a z-filter.

Now suppose that $Z_0 \in Z(X) \setminus \mathcal{Q}(M)$. As above, there exists $f_0 \in M$ and $\epsilon > 0$ so that $f_0^{\leftarrow}[[-\epsilon,\epsilon]] \cap Z_0 = \emptyset$. We will prove that $f_0^{\leftarrow}[[-\epsilon,\epsilon]] \in \mathcal{Q}(M)$; by 2.3(d)(3) this will show that $\mathcal{Q}(M)$ is a z-ultrafilter.

If $f_0^{\leftarrow}[[-\epsilon,\epsilon]] \notin \mathcal{Q}(M)$ there exists $g \in M$ and $\delta > 0$ such that $g^{\leftarrow}[[-\delta,\delta]] \cap f_0^{\leftarrow}[[-\epsilon,\epsilon]] = \emptyset$. Let $\sigma = \min\{\delta^2,\epsilon^2\}$. A routine calculation shows that $f_0^2 + g^2 \geq \underline{\alpha}$. Thus $f_0^2 + g^2$ is invertible in $C^*(X)$, and also belongs to the (proper) ideal M, which is a contradiction.

(2) If f is invertible in $C^*(X)$ then there exists $\delta > 0$ such that $|f| \geq \underline{\delta}$. Thus $0 \notin c\ell_{\mathbb{R}}f[X]$ so $f \notin M$. Hence if M is an ideal, it is proper.

Let $f \in M(\mathcal{a})$ and $g \in C^*(X)$. Thus there exists $K > 0$ such that $|g| \leq \underline{K}$. Let $S \in \mathcal{a}$ and let $\epsilon > 0$ be given; as $f \in M(\mathcal{a})$ there exists $x_0 \in S$ such that $|f(x_0)| < \epsilon/K$. It follows that $|fg(x_0)| < \epsilon$. As ϵ was arbitrary, $0 \in c\ell_{\mathbb{R}}(fg[S])$, and as S was arbitrary, $fg \in M(\mathcal{a})$.

Now suppose that $f, g \in M(\mathcal{a})$ and $f - g \notin M(\mathcal{a})$. We will derive a contradiction. As $f - g \notin M(\mathcal{a})$ there exists $S_0 \in \mathcal{a}$ such that $0 \notin c\ell_{\mathbb{R}}[(f - g)[S_0]]$. Hence there exists $\epsilon > 0$ such that $|f(x) - g(x)| > \epsilon$ whenever $x \in S_0$. If $f^{\leftarrow}[[-\frac{\epsilon}{4}, \frac{\epsilon}{4}]] \cap S_0 \notin \mathcal{a}$, by the maximality of $\mathcal{a}$ there exists $S_1 \in \mathcal{a}$ such that $f^{\leftarrow}[[-\frac{\epsilon}{4}, \frac{\epsilon}{4}]] \cap S_0 \cap S_1 = \emptyset$. Thus $0 \notin c\ell_{\mathbb{R}}f[S_0 \cap S_1]$, contradicting the assumption that $f \in M(\mathcal{a})$ and the fact that $S_0 \cap S_1 \in \mathcal{a}$. Thus $f^{\leftarrow}[[\frac{\epsilon}{4}, \frac{\epsilon}{4}]] \cap S_0 \in \mathcal{a}$. A similar argument shows that $g^{\leftarrow}[[-\frac{\epsilon}{4}, \frac{\epsilon}{4}]] \cap S_0 \in \mathcal{a}$. Thus there exists some $y \in f^{\leftarrow}[[-\frac{\epsilon}{4}, \frac{\epsilon}{4}]] \cap g^{\leftarrow}[[-\frac{\epsilon}{4}, \frac{\epsilon}{4}]] \cap S_0$. But then $|f(y) - g(y)| \leq |f(y)| + |g(y)| \leq \frac{\epsilon}{2}$, contradicting our earlier result. Thus $f - g \in M(\mathcal{a})$ and so $M(\mathcal{a})$ is a proper ideal.

It remains to show that $M(\mathcal{a})$ is maximal. Suppose $h \notin M(\mathcal{a})$. Arguing as above, we see there exists $\epsilon > 0$ and $S_0 \in \mathcal{a}$ such that $h^{\leftarrow}[[-\epsilon, \epsilon]] \cap S_0 = \emptyset$. Hence by 1.9(f) there exists $r \in C^*(X)$ such that $r[S_0] = \{1\}$, $r[h^{\leftarrow}[[-\epsilon, \epsilon]]] = \{0\}$; and $\underline{0} \leq r \leq \underline{1}$. Define $g \in F(X,\mathbb{R})$ as follows:

$$g(x) = \begin{cases} \frac{r(x)}{h(x)} & \text{if } x \notin \text{int}_X Z(r) \\ 0 & \text{if } x \in Z(r). \end{cases}$$

This is a well-defined function (since $h^{\leftarrow}[(-\epsilon,\epsilon)] \subseteq \text{int}_X Z(r)$) and is continuous by 1.6(c). It is easily checked that $|g(x)| \leqslant 1/\epsilon$ for every $x \in X$, and so $g \in C^*(X)$. Finally, it is evident that $r = hg$ (for if $x \in \text{int}_X Z(r)$ then $r(x) = 0 = g(x)$). Define f to be $\underline{1} - r$; then $f \in C^*(X)$. We claim that $f \in M(\mathcal{a})$; for if $S \in \mathcal{a}$ then $S \cap S_0 \neq \emptyset$, and if $y \in S \cap S_0$ then $f(y) = 1 - r(y) = 0$. Thus $0 \in c\ell_{\mathbb{R}} f[S]$ and $f \in M(\mathcal{a})$. Evidently $\underline{1} = f + hg$, and from this we infer that $M(\mathcal{a})$ is a maximal ideal.

■

(f) Corollary. The mapping $M \to \mathcal{a}(M)$ is a bijection from the set of all maximal ideals of $C^*(X)$ onto the set of z-ultrafilters of X. (See 4.6(e) for notation.)

Proof. It is routine to check that $M(\mathcal{a}(M)) \supseteq M$ and $\mathcal{a}(M(\mathcal{a})) \supseteq \mathcal{a}$ for each maximal ideal M and each z-ultrafilter $\mathcal{a}$. The maximality of maximal ideals and z-ultrafilters then yields the result.

■

In 4C we invite the reader to work out the correspondence between the maximal ideals of C(X) and the z-ultrafilters on X, and to construct βX as a "maximal ideal space" of C(X).

We are now ready to state and prove our main characterization theorem for the Stone-Čech compactification of a Tychonoff space.

(g) **Theorem**. The Stone-Čech compactification βX of a Tychonoff space X is unique (up to equivalence) among the compactifications of X with respect to each of the following properties:

(1) βX is the projective maximum of $K(X)$,

(2) X is C^*-embedded in βX,

(3) every continuous function from X to a compact space has a continuous extension to βX,

(4) if $Z_1, Z_2 \in Z(X)$, then $c\ell_{\beta X}Z_1 \cap c\ell_{\beta X}Z_2 = c\ell_{\beta X}(Z_1 \cap Z_2)$,

(5) disjoint zero-sets in X have disjoint closures in βX,

(6) completely separated sets in X have disjoint closures in βX, and

(7) each point of βX is the limit of a unique z-ultrafilter on X (viewed as a filter base on βX).

Proof. Clearly βX satisfies (1) by definition, (2) and (3) by 4.2(c), (4) and (5) by 4.4(i), (6) by 4.4(i) and 1.9(f), and (7) by 4.6(b). Next we establish the uniqueness of βX with respect to the various properties.

(1) If $Y \in K(X)$, then $\beta X \geq Y$. If Y is maximal, then $\beta X \equiv_X Y$.

(2,3) The uniqueness of βX with respect to (2) follows from 4.5(p)(2). If $Y \in K(X)$ and Y satisfies (3), then Y satisfies (2). So, by 4.5(p)(2), $Y \equiv_X \beta X$.

(4,5,6) The uniqueness of βX with respect to (4) and (5) follows from 4.4(i). If $Y \in K(X)$ and Y satisfies (6), then by 1.9(f), Y satisfies (5). So, $Y \equiv_X \beta X$ by 4.4(i).

(7) Suppose $Y \in K(X)$ and Y satisfies (7). Let $f : \beta X \to Y$ be the continuous function such that $f|X = id_X$. If $p \in \beta X$, then A^p converges to p. Since f is continuous, then $f(A^p) = \{f[Z] : Z \in A^p\} = A^p$ converges to $f(p)$. If $f(q) = f(p)$, then A^q and A^p converge to $f(p)$. But Y satisfies (7). So, $A^q = A^p$ and $p = q$. Hence, f is one-to-one. That is, f is a continuous bijection. Since βX is compact and Y is Hausdorff, then f is closed. So, f is a homeomorphism. This shows $Y \equiv_X \beta X$.

■

(h) **Corollary**. The following are equivalent for the dense subspace S of the Tychonoff space X.

(1) S is C^*-embedded in X.

(2) If Z_1 and Z_2 are disjoint zero-sets of S, then $c\ell_X Z_1 \cap c\ell_X Z_2 = \emptyset$.

Proof

(1) ⇒ (2) By the transitivity of C^*-embedding, S is C^*-embedded and dense in βX; so $\beta X \equiv_S \beta S$ by 4.6(g)(2). Thus by 4.6(g)(5) $c\ell_{\beta X} Z_1 \cap c\ell_{\beta X} Z_2 = \emptyset$, so $c\ell_X Z_1 \cap c\ell_X Z_2 = \emptyset$.

(2) ⇒ (1) Suppose that S is C^*-embedded in $S \cup \{p\}$ for each $p \in X \setminus S$. Then by 4.1(n) for each $f \in C^*(S)$ there exists $F \in C(X)$ such that $F|S = f$, and as $F|S \in C^*(S)$ it follows that $F \in C^*(X)$. Thus S would be C^*-embedded in X.

So, suppose $p \in X \setminus S$ and let $S(p)$ denote the collection of closed neighborhoods of p in $S \cup \{p\}$. If $f \in C^*(S)$, then $c\ell_{\mathbb{R}} f[A \cap X]$ is a compact nonempty subset of $\mathbb{R}$, and $\{c\ell_{\mathbb{R}} f[A \cap S] : A \in \mathcal{S}(p)\}$ has the finite intersection property. Thus $\cap\{c\ell_{\mathbb{R}} f[A \cap X] : A \in \mathcal{S}(p)\} \neq \emptyset$. Note that if $s \in \cap\{c\ell_{\mathbb{R}} f[A \cap S] : A \in \mathcal{S}(p)\}$ and $\delta > 0$, then $p \in c\ell_{S \cup \{p\}} f^{\leftarrow}[[s - \delta, s + \delta]]$; to see this, note that if $A \in \mathcal{S}(p)$ then $(s - \delta, s + \delta) \cap f[A \cap S] \neq \emptyset$ and so $A \cap f^{\leftarrow}[(s - \delta, s + \delta)] \neq \emptyset$.

Choose $r \in \cap\{c\ell_{\mathbb{R}} f[A \cap S] : A \in \mathcal{S}(p)\}$ and define $F : S \cup \{p\} \to \mathbb{R}$ as follows: $F|S = f$ and $F(p) = r$. Obviously F is continuous at each point of S; we must show that F is continuous at p. Let $\delta > 0$ be given. We claim there exists $A_0 \in \mathcal{S}(p)$ such that $f[A_0 \cap S] \subseteq (r - \delta, r + \delta)$. For if this were not the case, then $c\ell_{\mathbb{R}} f[A \cap S] \backslash (r - \frac{3\delta}{4}, r + \frac{3\delta}{4})$ is a nonempty compact subset of $\mathbb{R}$ for each $A \in \mathcal{S}(p)$. As $\{c\ell_{\mathbb{R}} f[A \cap S] \backslash (r - \frac{3\delta}{4}, r + \frac{3\delta}{4}) : A \in \mathcal{S}(p)\}$ has the finite intersection property, there exists $s \in \cap\{c\ell_{\mathbb{R}} f[A \cap S] \backslash (r - \frac{3\delta}{4}, r + \frac{3\delta}{4}) : A \in \mathcal{S}(p)\}$. As noted in the previous paragraph, it follows that $p \in c\ell_{S \cup \{p\}} f^{\leftarrow}[[s - \frac{\delta}{4}, s + \frac{\delta}{4}]] \cap c\ell_{S \cup \{p\}} f^{\leftarrow}[[r - \frac{\delta}{4}, r + \frac{\delta}{4}]]$. As $f^{\leftarrow}[[s - \frac{\delta}{4}, s + \frac{\delta}{4}]]$ and $f^{\leftarrow}[[r - \frac{\delta}{4}, r + \frac{\delta}{4}]]$ are disjoint zero-sets of S, this is in contradiction to our hypothesis (2). Thus there exists $A_0 \in \mathcal{S}(p)$ such that $f[A_0 \cap S] \subseteq (r - \delta, r + \delta)$. Thus $F[A_0] \subseteq (r - \delta, r + \delta)$ and F is continuous at p. As f was arbitrarily chosen from $C^*(S)$, it follows that S is C^*-embedded in $S \cup \{p\}$ and (1) follows.

■

One consequence of the Gelfand method of constructing βX for a Tychonoff space X is the fact that the topology of βX is determined by the algebraic properties of $C^*(X)$.

(i) **Theorem**

(1) The function $\Phi : C^*(X) \to C^*(\beta X)$ defined by $\Phi(f) = f^\beta$ is a ring isomorphism.

(2) Let X and Y be Tychonoff spaces. Then $C^*(X)$ and $C^*(Y)$ are ring isomorphic iff βX and βY are homeomorphic.

Proof

(1) This is a straightforward proof (see 4H).

(2) Suppose $C^*(X)$ and $C^*(Y)$ are ring isomorphic. Let $\mathcal{A} = C^*(X)$ and $\mathcal{B} = C^*(Y)$. Since $m_{\mathcal{A}}X$ and $m_{\mathcal{B}}X$ are constructed using only the algebraic properties of $\mathcal{A}$ and $\mathcal{B}$, respectively, then $m_{\mathcal{A}}X$ and $m_{\mathcal{B}}X$ are homeomorphic. Hence, βX and βY are homeomorphic. Conversely, suppose βX and βY are homeomorphic. Then $C^*(\beta X)$ and $C^*(\beta Y)$ are ring isomorphic. By (1), $C^*(X)$ and $C^*(Y)$ are ring isomorphic.

■

Now we give a necessary and sufficient condition for the closure in βX of a subspace of a Tychonoff space X to be the Stone-Čech compactification of the subspace.

(j) **Proposition.** Let $S \subseteq X$ where X is Tychonoff. Then S is C^*-embedded in X iff $\beta X \equiv_S c\ell_{\beta X}X$.

Proof. Suppose S is C^*-embedded in X. It suffices to show S is C^*-embedded in $c\ell_{\beta X}S$ as $c\ell_{\beta X}S$ is a compactification of S. Let

$f \in C^*(S)$; then there is some $F \in C^*(X)$ such that $F|S = f$. Now, $F^\beta | cl_{\beta X}S \in C^*(cl_{\beta X}S)$ and $F^\beta | cl_{\beta X}S$ extends f. So, S is C^*-embedded in $cl_{\beta X}S$. Conversely, suppose $\beta S \equiv_S cl_{\beta X}S$. Then S is C^*-embedded in $cl_{\beta X}S$. By 1.9(k), $cl_{\beta X}S$ is C^*-embedded in βX. Hence, S is C^*-embedded in βX; this implies S is C^*-embedded in X.

■

Occasionally in later chapters it will be useful to have an explicit description of the Stone extension βf of a function $f \in C(X,Y)$ in terms of z-ultrafilters on X and Y. We provide such a description in the following theorem. In this theorem, we regard the points of $\beta X \setminus X$ as free z-ultrafilters on the space X. The points of $\beta Y \setminus Y$ (respectively, Y) are identified with free (respectively, fixed) z-ultrafilters on Y.

(k) **Theorem**. Let X and Y be Tychonoff spaces, let $f \in C(X,Y)$, let $\mathcal{a} \in \beta X \setminus X$, and let $\mathcal{F} = \{Z \in Z(Y) : f^{\leftarrow}[Z] \in \mathcal{a}\}$. Then:

(1) $\mathcal{F}$ is a z-filter on Y,

(2) $\mathcal{F}$ is contained in a unique z-ultrafilter $f\mathcal{a}$ on Y,

(3) $|\bigcap\{cl_{\beta Y}F : F \in \mathcal{F}\}| = 1$, and

(4) $\beta f(\mathcal{a}) = f\mathcal{a}$ and $\{f\mathcal{a}\} = \bigcap\{cl_{\beta Y}Z : Z \in Z(Y)$ and $f^{\leftarrow}[Z] \in \mathcal{a}\}$.

Proof

(1) This follows in a straightforward fashion from the fact that $\mathcal{a}$ is a z-filter on X.

(2) By 2.3(d) F is contained in at least one z-ultrafilter on Y. Suppose that G and H are distinct z-ultrafilters on Y, each containing F. By 2.3(d) there exist G ∈ G and H ∈ H such that G ∩ H = ∅. By 1.9(f) there exists $g \in C^*(X)$ such that $G \subseteq g^{\leftarrow}(0)$, $H \subseteq g^{\leftarrow}(1)$, and $\underline{0} \leqslant g \leqslant \underline{1}$. Let $L = g^{\leftarrow}[[0,1/2]]$ and $M = g^{\leftarrow}[[1/2,1]]$. Then L, M ∈ Z(Y) by 1.4(j) and $f^{\leftarrow}[L] \cup f^{\leftarrow}[M] = X$. By 1.4(h), $f^{\leftarrow}[L]$ and $f^{\leftarrow}[M]$ are in Z(X). Since $\mathcal{Q}$ is a z-ultrafilter on X, either $f^{\leftarrow}[L] \in \mathcal{Q}$ or $f^{\leftarrow}[M] \in \mathcal{Q}$ (see 2I(6)). Suppose $f^{\leftarrow}[L] \in \mathcal{Q}$; then L ∈ F, and so L ∈ H (as $F \subseteq H$). Thus L ∩ H ≠ ∅ (as H is a z-filter), which contradicts the definitions of L and g. It follows that F is contained in a unique z-ultrafilter on Y, which we denote by $f\mathcal{Q}$.

(3) Because F has the finite intersection property it follows from the compactness of βY that $\bigcap\{c\ell_{\beta Y}F : F \in F\} \neq \emptyset$. If p_1 and p_2 were distinct points of $\bigcap\{c\ell_{\beta Y}F : F \in F\}$, it would follow from 1.4(j) that there exist $Z_i \in Z(\beta Y)$ (i = 1,2) such that $p_i \in int_{\beta Y}Z_i$ and $Z_1 \cap Z_2 = \emptyset$. Then $p_i \in c\ell_{\beta Y}(Z_i \cap Y)$ and $Z_i \cap Y \in Z(Y)$. Evidently $\{Z_i \cap F : F \in F\}$ is a filter base on Z(Y) and hence is contained in a z-ultrafilter on Y (i = 1,2) (see 2.3(d)). As $Z_1 \cap Z_2 = \emptyset$, this contradicts (2). Hence $|\bigcap\{c\ell_{\beta Y}F : F \in F\}| = 1$.

(4) If F ∈ F then F ∈ $f\mathcal{Q}$ so $f\mathcal{Q} \in c\ell_{\beta Y}F$ (see 4.6(b)(4)). Thus by (3), $\{f(\mathcal{Q})\} = \bigcap\{c\ell_{\beta Y}Z : Z \in Z(Y)$ and $f^{\leftarrow}[Z] \in \mathcal{Q}\}$. If Z ∈ F then $f^{\leftarrow}[Z] \in \mathcal{Q}$ so $\mathcal{Q} \in c\ell_{\beta X}f^{\leftarrow}[Z]$. Thus $\beta f(\mathcal{Q}) \in (\beta f)[c\ell_{\beta X}f^{\leftarrow}[Z]] \subseteq c\ell_{\beta Y}Z$. Hence $\beta f(\mathcal{Q}) \in \bigcap\{c\ell_{\beta Y}Z : Z \in F\}$, and so $\beta f(\mathcal{Q}) = f\mathcal{Q}$.

■

If Y is compact the preceding result simplifies slightly.

(l) **Corollary**. Let X be Tychonoff and Y be compact. If $f \in C(X,Y)$ and $\mathcal{a} \in \beta X \setminus X$, then $\beta f(\mathcal{a})$ is the unique point in the set $\cap\{Z \in Z(Y) : f^{\leftarrow}[Z] \in \mathcal{a}\}$.

Proof. Use (k) above and note that if $Z \in Z(Y)$, then $c\ell_{\beta Y} Z = Z$.

■

Note that the z-filter F discussed in (k) above need not be a z-ultrafilter. To see this, let X be $\mathbb{N}$, let Y be the one-point compactification $\mathbb{N} \cup \{\infty\}$ of $\mathbb{N}$, and let f be the obvious embedding of X in Y. If $\mathcal{a}$ is a free z-ultrafilter on $\mathbb{N}$, then $F = \{A \cup \{\infty\} : A \in \mathcal{a}\}$ which is not a z-ultrafilter on Y because $\{\infty\} \in Z(Y) \setminus F$.

4.7 Zero-dimensional Compactifications

The theory of zero-dimensional compactifications of a zero-dimensional space is parallel to the theory of compactifications of a Tychonoff space. Let K_0 denote the class of all zero-dimensional compact spaces. We can apply the techniques and results about Boolean algebras that we obtained in Chapter 3 to this class K_0.

Note that K_0 is a replete class that is productive and closed-hereditary. By 4.1(k), we have this result.

(a) **Proposition**. For a space X, $K_0(X)$ (= $\{Y \in E(X) : Y \in$

K_0)) is a complete upper semilattice and has a projective maximum if $K_0(X) \neq \emptyset$.

Let X be zero-dimensional and let $\mathcal{B} \subseteq \mathcal{B}(X)$ be a Boolean subalgebra such that $\mathcal{B}$ is an open basis for X. The Stone space $S(\mathcal{B})$, denoted as $c_{\mathcal{B}}X$, is compact and zero-dimensional. For $x \in X$, let $\mathcal{U}_x = \{B \in \mathcal{B} : x \in B\}$. Recall that the underlying set of $S(\mathcal{B})$ is the collection of ultrafilters on $\mathcal{B}$, and that the topology of $S(\mathcal{B})$ is generated by $\{\lambda(B) : B \in \mathcal{B}\}$ where $\lambda(B) = \{\mathcal{U} \in S(\mathcal{B}) : B \in \mathcal{U}\}$.

(b) **Proposition**. Let X be a zero-dimensional space and let $\mathcal{B} \subseteq \mathcal{B}(X)$ be a Boolean algebra such that $\mathcal{B}$ is an open base for X. Let Z be a space. Then:

(1) for $x \in X$, $\mathcal{U}_x \in c_{\mathcal{B}}X$,

(2) the function $\psi : X \to c_{\mathcal{B}}X$ defined by $\psi(x) = \mathcal{U}_x$ is a dense embedding,

(3) if $B_1, B_2 \in \mathcal{B}$ are disjoint, then $c\ell_Y[B_1] \cap c\ell_Y[B_2] = \emptyset$, where $Y = c_{\mathcal{B}}X$, and

(4) $K_0(Z) \neq \emptyset$ iff Z is zero-dimensional.

Proof. It is easily checked that $\mathcal{B}$ is a Wallman base for X, and if $B \in \mathcal{B}$ then $S(B) = \lambda(B)$ (see 4.4(a)(2) for the definition of $S(B)$). It quickly follows that the Stone space $S(\mathcal{B})$ is just $w_{\mathcal{B}}X$. Thus (1), (2), and (3) follow immediately from 4.4(e), (f). We now prove (4).

If $K_0(Z) \neq \emptyset$, then Z is a subspace of a compact, zero-dimensional space. Since zero-dimensionality is hereditary, Z is zero-dimensional. Conversely, is Z is zero-dimensional, then $\mathcal{B} = B(Z)$ is a Boolean subalgebra of $B(Z)$ and $\mathcal{B}$ is an open base for Z. By (2), $c_{\mathcal{B}}Z \in K_0(Z)$. So $K_0(Z) \neq \emptyset$.

■

As usual, we identify $\psi[X]$ with X by associating $\mathcal{U}_x$ with x for each $x \in X$. Thus, we think of X as being a dense subspace of $c_{\mathcal{B}}X$. Now $K_0(X) \neq \emptyset$ if X is a zero-dimensional space. Thus by 4.7(a), $K_0(X)$ has a projective maximum which we denote as $\beta_0 X$; in other words, $\beta_0 X = \vee K_0(X)$. At this point, the reader may have noticed that $B(X)$ is the largest element in $\{\mathcal{B} : \mathcal{B}$ is a Boolean subalgebra of $B(X)$ and is an open base for X$\}$ and may be conjecturing that $\beta_0 X \equiv_X c_{B(X)}X$.

(c) **Proposition**. For a zero-dimensional space X, $\beta_0 X \equiv_X c_{B(X)}X$ and $c\ell_{\beta_0 X}B \in B(\beta_0 X)$ for each $B \in B(X)$.

Proof. Let $Y = c_{B(X)}X$. By definition of $\beta_0 X$, $\beta_0 X \geq Y$. To show the converse is also true, let $i : X \to \beta_0 X$ be the inclusion function and let A and B be disjoint closed sets in $\beta_0 X$. Since $\beta_0 X$ is compact and zero-dimensional, there is a clopen set U in $\beta_0 X$ such that $A \subseteq U$ and $U \cap B = \emptyset$. Let $V = \beta_0 X \setminus U$. Thus, $U \cap X$, $V \cap X \in B(X)$. Since $(U \cap X) \cap (V \cap X) = \emptyset$, then by 4.7(b)(3), $c\ell_Y(U \cap X) \cap c\ell_Y(V \cap X) = \emptyset$. It follows that $c\ell_Y(A \cap X) \cap c\ell_Y(B \cap X) = \emptyset$. Thus $Y \geq \beta_0 X$ by 4.2(h). The second part of the statement is a consequence of applying 4.7(b)(3) to B and $X \setminus B$.

■

It follows from the preceding proposition and the definition of $c_B X$ that $\beta_0 X$ can be identified with the Stone space of $B(X)$. The points of $\beta_0 X$ can thus be thought of as clopen ultrafilters on X, with the fixed clopen ultrafilters corresponding to the points of X. This is analogous to our representation of the points of βX as z-ultrafilters on X (see the comments following 4.4(i)).

Our next result gives a characterization of $\beta_0 X$ is terms of a product of copies of $\mathbf{2}$.

(d) **Proposition**. Let X and Y be zero-dimensional spaces and $\mathcal{a} = C(X,\mathbf{2})$. Define $e_X : X \to \Pi_{\mathcal{a}}\mathbf{2}$ $(=\mathbf{2}^{\mathcal{a}})$ by $e_X(x)(g) = g(x)$ for $x \in X$ and $g \in \mathbf{2}^{\mathcal{a}}$. Then:

(1) e_X is an embedding and $\beta_0 X \equiv_X c\ell e_X[X]$ (closure relative to $\mathbf{2}^{\mathcal{a}}$), and

(2) if $f \in C(X,Y)$, then f has a unique continuous extension $\beta_0 f \in C(\beta_0 X, \beta_0 Y)$.

<u>Proof</u>. Since X is zero-dimensional, it follows that $\mathcal{a}$ separates points and closed sets; so, by the Embedding Theorem 1.7(j), e_X is an embedding. Let $cX = c\ell e_X[X]$ (closure relative to $\mathbf{2}^{\mathcal{a}}$). So, cX is a zero-dimensional compactification of X. In order to show that $cX \equiv_X \beta_0 X$, one shows that if $\mathcal{C} = C(Y,\mathbf{2})$ and $cY = c\ell e_Y[Y]$ (closure relative to $\mathbf{2}^{\mathcal{C}}$) and $f \in C(X,Y)$, there is a unique continuous function $\beta_0 f \in C(cX,cY)$, such that $\beta_0 f|X = f$. The remainder of the proof is entirely analogous to the proof used in 4.2(c).

■

(e) **Definition**. Let Y be an extension of X. X is **$\mathbf{2}$-embedded** in Y is each continuous function $f : X \to \mathbf{2}$ has a continuous extension $F : Y \to \mathbf{2}$ (see 1L).

Now, $\mathbf{2}$-embeddedness for $\mathcal{K}_0(X)$ is the analogue of C^*-embeddedness for $\mathcal{K}(X)$. Also, "$\mathbf{2}$-embeddedness characterizes $\beta_0 X$" in the following sense:

(f) **Corollary**. Let X be a zero-dimensional space. $\beta_0 X$ is the unique zero-dimensional compactification of X in which X is $\mathbf{2}$-embedded (up to equivalence)..

Proof. First, we show that X is $\mathbf{2}$-embedded in $\beta_0 X$. By 4.7(d), if $f \in C(X,\mathbf{2})$, there is a continuous extension $F \in C(\beta_0 X, \beta_0 \mathbf{2})$. But $\beta_0 \mathbf{2} = \mathbf{2}$ as $\mathbf{2}$ is compact. So, X is $\mathbf{2}$-embedded in $\beta_0 X$. Now, let $Y \in \mathcal{K}_0(X)$ such that X is $\mathbf{2}$-embedded in Y. By definition of $\beta_0 X$, $\beta_0 X \geq Y$. To show $Y \geq \beta_0 X$, let B and C be disjoint closed sets of $\beta_0 X$. There exists $A \in \mathcal{B}(\beta_0 X)$ such that $B \subseteq A$ and $C \subseteq \beta_0 X \backslash A$. Then $\chi_{A \cap X} \in C(X,\mathbf{2})$. So there exists $F \in C(Y,\mathbf{2})$ such that $F|X = \chi_{A\cap X}$. Obviously, $c\ell_Y(B \cap X) \subseteq F^{\leftarrow}(1)$ and $c\ell_Y(A \cap X) \subseteq F^{\leftarrow}(0)$. It follows from 4.2(h) that $Y \geq \beta_0 X$.

■

As in the case of Gelfand bases for compactifications, the Boolean subalgebras of $\mathcal{B}(X)$ that are open bases generate all the zero-dimensional compactifications of X. Also, other facts about compactifications have analogues for zero-dimensional compactifications. Some of the facts are presented in the exercises (see 4I and 4J).

If X is zero-dimensional, then βX exists and $\beta X \geq \beta_0 X$. We now determine when $\beta X \equiv_X \beta_0 X$.

(g) **Proposition**. Let X be zero-dimensional. Then $\beta X \in K_0(X)$ (i.e., $\beta X \equiv_X \beta_0 X$) iff for every pair of disjoint zero-sets A and B in X, there is a clopen set $C \subseteq X$ such that $A \subseteq C$ and $B \cap C = \emptyset$.

Proof. Suppose $\beta X \in K_0(X)$. Then $c\ell_{\beta X}A \cap c\ell_{\beta X}B = \emptyset$. Since βX is compact and zero-dimensional, there is a clopen set W in βX such that $c\ell_{\beta X}A \subseteq W$ and $W \cap c\ell_{\beta X}B = \emptyset$. Let $C = W \cap X$. Then C is clopen in X, $A \subseteq C$, and $C \cap B = \emptyset$. Conversely, suppose for each pair of disjoint zero sets A and B, there is a clopen set C such that $A \subseteq C$ and $B \cap C = \emptyset$. Now, C, $X \backslash C \in B(X)$ and by 4.7(b,c), $c\ell_Y C \cap c\ell_Y(X \backslash C) = \emptyset$ where $Y = \beta_0 X$. Thus, $c\ell_Y A \cap c\ell_Y B = \emptyset$. By a characterization of βX in 4.6(g), $\beta_0 X \equiv_X \beta X$. So, $\beta X \in K_0(X)$.

■

(h) **Definition**. A Tychonoff space X is **strongly zero-dimensional** if βX is zero-dimensional.

Thus, by 4.7(g), X is strongly zero-dimensional iff X is zero-dimensional and disjoint sets in X are contained in disjoint clopen sets. An example of a zero-dimensional space X such that $\beta X \not\equiv_X \beta_0 X$ is given in 4V. We finish this section by giving another criterion for a zero-dimensional space to be strongly zero-dimensional. First we need a lemma.

(i) **Lemma**. Let A and B be disjoint closed subsets of the zero-dimensional Lindelöf space X. Then there exists $C \in B(X)$ such that $A \subseteq C$ and $B \cap C = \emptyset$. In particular X is strongly zero-dimensional.

Proof. For each $x \in X$ there exists $U(x) \in B(X)$ such that either $U(x) \cap A = \emptyset$ or $U(x) \cap B = \emptyset$. As X is Lindelöf there is a countable subset $\{x_n : n \in \mathbb{N}\}$ of X such that $X = \cup\{U(x_n) : n \in \mathbb{N}\}$. Let $V(k) = U(x_k) \setminus \cup\{U(x_i) : i < k\}$ for each $k \in \mathbb{N}$. Then $V(k) \in B(X)$, $\cup\{V(k) : k \in \mathbb{N}\} = X$, and if $j \neq k$ then $V(j) \cap V(k) = \emptyset$. Let $C = \cup\{V(k) : V(k) \cap B = \emptyset\}$. It is straightforward to verify that C has the desired properties. That X is strongly zero-dimensional now follows from (g) above.

■

(j) **Theorem**. The following are equivalent for a Tychonoff space X:

(1) X is strongly zero-dimensional

(2) each zero-set of X is the intersection of countably many clopen subsets of X.

Proof

(1) ⇒ (2) As X is z-embedded in βX, if $Z \in Z(X)$ there exists $S \in Z(\beta X)$ such that $S \cap X = Z$. As βX is zero-dimensional, $\beta X \setminus S$ is the union of clopen subsets of βX; as $\beta X \setminus S$ is an F_σ-set of βX and hence Lindelöf, there exists $\{A_n : n \in \mathbb{N}\} \subseteq$

$B(\beta X)$ such that $\beta X \setminus S = \cup\{A_n : n \in \mathbb{N}\}$. Then $Z = \cap\{X \setminus A_n : n \in \mathbb{N}\}$ and $X \setminus A_n \in B(X)$.

(2) ⇒ (1) Let Z_1 and Z_2 be disjoint zero-sets of X. Let $Z_1 = \cap\{A_n : n \in \mathbb{N}\}$ and $Z_2 = \cap\{B_n : n \in \mathbb{N}\}$, where each A_n and B_n is in $B(X)$. Evidently X is zero-dimensional by hypothesis, so $\beta_0 X$ exists; let $L = \beta_0 X \setminus \cap\{c\ell_{\beta_0 X} A_n \cap c\ell_{\beta_0 X} B_n : n \in \mathbb{N}\}$. We see that $X \setminus L = Z_1 \cap Z_2 = \emptyset$, so $X \subseteq L \subseteq \beta_0 X$. Evidently L is an F_σ-set of $\beta_0 X$ and hence is Lindelöf. As $L \subseteq \beta_0 X$ it follows that L is zero-dimensional; hence by (i) above L is strongly zero-dimensional. Now $L \cap [\cap\{c\ell_{\beta_0 X} A_n : n \in \mathbb{N}\}]$ and $L \cap [\cap\{c\ell_{\beta_0 X} B_n : n \in \mathbb{N}\}]$ are disjoint zero-sets of L, as each is the intersection of countably many clopen subsets of L. Hence by 4.7(g) there exists $C \in B(L)$ such that $L \cap [\cap\{c\ell_{\beta_0 X} A_n : n \in \mathbb{N}\}] \subseteq C$ and $L \cap [\cap\{c\ell_{\beta_0 X} B_n : n \in \mathbb{N}\}] \cap C = \emptyset$. Thus $C \cap X \in B(X)$, $Z_1 \subseteq C \cap X$, and $Z_2 \cap (C \cap X) = \emptyset$. Hence by 4.7(g) X is strongly zero-dimensional.

■

4.8 H-closed Spaces

In the previous sections, we have presented some of the basic information about compactifications and zero-dimensional compactifications. But a space must be Tychonoff in order to have a compactification. We will now introduce the concept of an H-closed space; such spaces are similar to compact spaces, but every Hausdorff space can be densely embedded in an H-closed space. A complete chapter (Chapter 7) is devoted to studying H-closed extensions. In fact, in many ways, compact spaces are to Tychonoff spaces as H-closed spaces are to Hausdorff spaces.

(a) **Definition**. A space X is **H-closed** if X is closed in every space containing X as a subspace. (In fact, "H-closed" is an abbreviation for "Hausdorff-closed" - closed in Hausdorff spaces.)

Every H-closed space is feebly compact (see 1.11(a) and the following result). We can characterize H-closed spaces as follows.

(b) **Proposition**. For a space X, the following are equivalent:

(1) X is H-closed,

(2) for every open cover of X, there is a finite subfamily whose union is dense in X,

(3) every open filter on X has nonvoid adherence, and

(4) every open ultrafilter on X converges.

Proof

(1) $\Rightarrow$ (3) Let F be an open filter in X such that $a(F) = \emptyset$. Let $Y = X \cup \{F\}$ and define a set $U \subseteq Y$ to be open if $U \cap X$ is open in X and $F \in U$ implies $U \cap X \in F$. It is easily checked that this is a valid definition of a topology on Y. Since X is Hausdorff and open in Y, it follows that every pair of distinct points in X can be separated by disjoint open sets in Y. Suppose $p \in X$. Since $a_X(F) = \emptyset$, there is an open set $U \in F$ and an open neighborhood V of p such that $U \cap V = \emptyset$. Now, V and $U \cup \{F\}$ are disjoint open sets in Y and contain p and F, respectively. This shows Y is Hausdorff. Since X is not closed in Y, it follows that X

is not H-closed.

(3) ⇒ (4) Let $\mathcal{U}$ be an open ultrafilter on X. Then $a_X(\mathcal{U}) \neq \emptyset$. By 2.3(f), $a_X(\mathcal{U}) = c_X(\mathcal{U})$. So, $\mathcal{U}$ converges.

(4) ⇒ (3) Suppose $\mathcal{F}$ is an open filter on X. By 2.3(d), $\mathcal{F}$ is contained in some open ultrafilter $\mathcal{U}$. Since $\mathcal{F} \subseteq \mathcal{U}$, then $a_X(\mathcal{U}) \subseteq a_X(\mathcal{F})$. But $\emptyset \neq c_X(\mathcal{U})$ by (4) and by 2.3(f), $c_X(\mathcal{U}) = a_X(\mathcal{U})$. Hence, $a_X(\mathcal{F}) \neq \emptyset$.

(3) ⇒ (1) Suppose X is not H-closed. Then there is a space Y such that X is a subspace and $c\ell_Y X \neq X$. Let $p \in c\ell_Y X \setminus X$ and $\mathcal{F} = \{U \cap X : U \text{ open in } Y \text{ and } p \in U\}$. Then $\mathcal{F}$ is an open filter on X. Since Y is Hausdorff, $a_X(\mathcal{F}) = \emptyset$.

(2) ⇒ (3) Suppose $\mathcal{F}$ is a free open filter on X. Then $\{X\setminus c\ell_X U : U \in \mathcal{F}\}$ is an open cover of X. A finite subfamily is of the form $\{X\setminus c\ell_X U : U \in A\}$ where A is a finite subset of $\mathcal{F}$. Now $c\ell_X(\bigcup\{X\setminus c\ell_X U : U \in A\}) = \bigcup\{X\setminus \text{int}_X c\ell_X U : U \in A\} = X\setminus\bigcap\{\text{int}_X c\ell_X U : U \in A\} \subseteq X\setminus\bigcap\{U : U \in A\} \neq X$ as $\bigcap\{U : U \in A\} \neq \emptyset$ since $\mathcal{F}$ is a filter. So, no finite subfamily of $\{X\setminus c\ell_X U : U \in \mathcal{F}\}$ has a dense union in X.

(3) ⇒ (2) Let $\mathcal{C}$ be an open cover of X such that for each finite set $A \subseteq \mathcal{C}$, $X \neq c\ell_X(\bigcup A)$. Let $\mathcal{F} = \{U : U \text{ open and } U \supseteq X\setminus c\ell_X(\bigcup A) \text{ for some finite } A \subseteq \mathcal{C}\}$. Now, $\mathcal{F}$ is an open filter on X, and $a_X(\mathcal{F}) = \bigcap\{c\ell_X U : U \in \mathcal{F}\} \subseteq \bigcap\{c\ell_X(X\setminus c\ell_X(\bigcup A)) : A \subseteq \mathcal{C} \text{ is finite}\} \subseteq \bigcap\{c\ell_X(X\setminus c\ell_X V) : V \in \mathcal{C}\} \subseteq X\setminus\bigcup\mathcal{C} = X\setminus X = \emptyset$. So, $\mathcal{F}$ is a free open filter on S.

■

(c) **Corollary**. A space X is H-closed and regular iff X is compact.

__Proof__. Clearly, a compact space is H-closed and regular. Conversely, suppose X is H-closed and regular. Let $\mathcal{C}$ be an open cover of X. For each $x \in X$, there is an open set $U(x) \in \mathcal{C}$ such that $x \in U(x)$. Since X is regular, there is an open set V(x) containing x such that $c\ell_X V(x) \subseteq U(x)$. By (b) above $\{V(x) : x \in X\}$ has a finite subfamily $\{V(x) : x \in A\}$ such that $X = c\ell_X(\bigcup\{V(x) : x \in A\}) = \bigcup\{c\ell_X V(x) : x \in A\}$. But $c\ell_X V(x) \subseteq U(x)$. So, $X = \bigcup\{U(x) : x \in A\}$. Hence, $\mathcal{C}$ has a finite subcover. This shows X is compact.

■

Our next result shows there are noncompact H-closed spaces.

(d) __**Example**__. The subset $Y = \{(1/n,1/m) : n \in \mathbb{N}, |m| \in \mathbb{N}\} \cup \{(1/n,0) : n \in \mathbb{N}\}$ of $\mathbb{R}^2$ is given the subspace topology inherited from the usual topology on the plane $\mathbb{R}^2$. Let $X = Y \cup \{p^+,p^-\}$. A subset $U \subseteq X$ is defined to be open if $U \cap Y$ is open in Y and if $p^+ \in U$ (respectively, $p^- \in U$) implies that there is some $r \in \mathbb{N}$ such that $\{(1/n,1/m) : n \geqslant r, m \in \mathbb{N}\} \subseteq U$ (respectively, $\{(1/n,1/m) : n \geqslant r, -m \in \mathbb{N}\} \subseteq U$). It is easily seen that this is a valid definition of a Hausdorff topology on X. Let $\mathcal{C}$ be an open cover of X. There are open sets U^+, U^- in $\mathcal{C}$ containing p^+, p^-, respectively. There is some $r \in \mathbb{N}$ such that $X \setminus c\ell_X(U^+ \cup U^-) \subseteq D$ where $D = \{(1/n,1/m) : n \leqslant r, |m| \in \mathbb{N}\} \cup \{(1/n,0) : n \leqslant r\}$. But D is compact. So, there is a finite family $\mathcal{Q} \subseteq \mathcal{C}$ such that $D \subseteq \bigcup\mathcal{Q}$. Thus, $X = c\ell_X(U^+ \cup U^-) \cup (\bigcup\mathcal{Q}) = c\ell_X(U^+ \cup U^- \cup (\bigcup\mathcal{Q}))$. This shows X is H-closed by 4.8(b). Since $\{(1/n,0) : n \in \mathbb{N}\}$ is a closed, discrete infinite subspace of X, X is not compact.

We now show that regular closed subsets of H-closed spaces are H-closed.

(e) **Proposition**. Let X be H-closed and $U \subseteq X$ be open. Then $c\ell_X U$ is H-closed.

Proof. Let $A = c\ell_X U$ and $\mathcal{F}$ be an open filter on A. Then $\{F \cap U : F \in \mathcal{F}\}$ is an open filter base on X. Let $\mathcal{G} = \{W \subseteq X : W$ open in X and $W \supseteq F \cap U$ for some $F \in \mathcal{F}\}$. Then $\mathcal{G}$ is an open filter on X and $\emptyset \neq a_X(\mathcal{G}) \subseteq \cap\{c\ell_X(F \cap U) : F \in \mathcal{F}\} = \cap\{c\ell_A(U \cap F) : F \in \mathcal{F}\} \subseteq \cap\{c\ell_A F : F \in \mathcal{F}\} = a_A(\mathcal{F})$. By 4.8(b), A is H-closed.

■

In contrast to the above proposition, we now show that being H-closed is **not** a closed-hereditary property.

(f) **Example**. In 4.8(d), the closed subspace $A = \{(1/n,0) : n \in \mathbb{N}\}$ is infinite and discrete (and, hence, regular). Thus, A is not compact, and by 4.8(c), A is not H-closed.

Before looking at other properties of H-closed spaces, we need to define a concept which is a slight generalization of continuity and is very useful in studying spaces which are not regular.

(g) **Definition**. Let X and Y be spaces, $f \in F(X,Y)$, and $x_0 \in X$.

(1) f is **Θ-continuous at** x_0 if for each open neighborhood V of $f(x_0)$, there is an open neighborhood U of x_0 such that $f[c\ell_X U] \subseteq c\ell_Y V$.

(2) f is **Θ-continuous** if f is Θ-continuous at each point of X.

(3) The set of Θ-continuous functions from X to Y is denoted by $\Theta C(X,Y)$.

(4) f is a **Θ-homeomorphism** if f is a bijection and both f and $f^{\leftarrow}$ are Θ-continuous.

We warn the reader that Θ-continuity differs from continuity in the following important way. If $f : X \to Y$ is Θ-continuous, then it is **not** necessarily true that $f : X \to f[X]$ is Θ-continuous. An example will be given in 4.8(i) to illustrate this point.

In the following proposition, we collect a number of basic facts about Θ-continuity which will be used in the remainder of this chapter and, especially, in Chapters 6, 7, and 9.

(h) <u>Proposition</u>. Let X, Y and Z be spaces and $f \in F(X,Y)$ and $g \in F(Y,Z)$. Then:

(1) if f is Θ-continuous at $x_0 \in X$ and g is Θ-continuous at $f(x_0)$, then $g \circ f$ is Θ-continuous at x_0,

(2) $C(X,Y) \subseteq \Theta C(X,Y)$,

(3) if Y is regular, then $C(X,Y) = \Theta C(X,Y)$,

(4) if $A \subseteq X$, and $f \in \Theta C(X,Y)$, then $f|A \in \Theta C(A,Y)$,

(5) if D is dense in Y, $f \in \Theta C(X,Y)$, and $f[X] \subseteq D$, then $f \in \Theta C(X,D)$,

(6) if f is a Θ-continuous surjection and X is H-closed, then Y is H-closed,

(7) the identity function $id: X(s) \to X$ is a θ-homeomorphism

and,

(8) X is H-closed iff X(s) is H-closed.

Proof

(1) Suppose $g \circ f(x_0) \in W$ for some open set W in Z. Since g is θ-continuous at $f(x_0)$, there is an open subset V of Y such that $f(x_0) \in V$ and $g[c\ell_Y V] \subseteq c\ell_Z W$. Since f is θ-continuous at x_0, it follows that there is an open subset U of X such that $x_0 \in U$ and $f[c\ell_X U] \subseteq c\ell_Y V$. Thus, $g \circ f[c\ell_X U] \subseteq g[c\ell_Y V] \subseteq c\ell_Z W$ and $g \circ f$ is θ-continuous at x_0.

(2) Obvious.

(3) By (2), we only need to show $\theta C(X,Y) \subseteq C(X,Y)$. Let $f \in \theta C(X,Y)$ and $x_0 \in X$. Let V be an open neighborhood of $f(x_0)$. Since Y is regular, there is an open neighborhood W of $f(x_0)$ such that $c\ell_Y W \subseteq V$. By θ-continuity of f, there is an open set U of X such that $x_0 \in U$ and $f[c\ell_X U] \subseteq c\ell_Y W$. Hence, $f[U] \subseteq V$. This shows that f is continuous at x_0. So, $f \in C(X,Y)$.

(4) Let $x_0 \in A$ and V be an open neighborhood of $f(x_0)$ in Y. Since $f : X \to Y$ is θ-continuous, there is an open set U such that $x_0 \in U$ and $f[c\ell_X U] \subseteq c\ell_Y V$. Now, $x_0 \in U \cap A$ and $c\ell_A(U \cap A) \subseteq c\ell_X U$. So, $(f \mid A)[c\ell_A(U \cap A)] \subseteq c\ell_Y V$.

(5) Let $x_0 \in X$ and let V be an open neighborhood of $f(x_0)$ in D. There is an open set W in Y such that $W \cap D = V$. Since $f : X \to Y$ is θ-continuous and $f(x_0) \in W$, there is an open set U such that $x_0 \in U$ and $f[c\ell_X U] \subseteq c\ell_Y W$. So, $f[c\ell_X U] \subseteq c\ell_Y W \cap f[X] \subseteq [c\ell_Y(W \cap D)] \cap D = c\ell_D(V)$. Thus, $f \in \theta C(X,D)$.

(6) Let $\mathcal{C}$ be an open cover of Y. For each $x \in X$, there is

an open set $V(x) \in \mathcal{C}$ such that $f(x) \in V(x)$. Since f is Θ-continuous, there is an open set $U(x)$ in X such that $x \in U(x)$ and $f[c\ell_X U(x)] \subseteq c\ell_Y V(x)$. Since X is H-closed, there is a finite set $F \subseteq X$ such that $X = c\ell_X \bigcup\{U(x) : x \in F\} = \bigcup\{c\ell_X U(x) : x \in F\}$. Now, $Y = f[X] = \bigcup\{f[c\ell_X U(x)] : x \in F\} \subseteq \bigcup\{c\ell_Y(V(x)) : x \in F\}$. By 4.8(b), Y is H-closed.

(7) Since id is a bijection and $id^{\leftarrow} : X \rightarrow X(s)$ is Θ-continuous by 2.2(d), we only need to show $id : X(s) \rightarrow X$ is Θ-continuous. Let $x_0 \in X(s)$ and $id(x_0) = x_0 \in V$ for some open set V in X. Let $U = int_{X(s)} c\ell_{X(s)} V$; U is an open neighborhood of x_0 in X(s). By 2.2(f), $c\ell_{X(s)} U = c\ell_{X(s)} int_{X(s)} c\ell_{X(s)} V = c\ell_X int_X c\ell_X V = c\ell_X V$. Hence, $id[c\ell_{X(s)} U] = c\ell_{X(s)} U \subseteq c\ell_X V$. Thus, id is Θ-continuous.

(8) (8) follows from (6) and (7).

■

Now, as promised, we give an example of spaces Z and X and a Θ-continuous function $f : Z \rightarrow X$ such that $f : Z \rightarrow f[Z]$ is not Θ-continuous. This function also provides an example of a Θ-continuous function that is not continuous. Also, the importance of "onto" in 4.8(h)(6) is demonstrated as Z is compact and f[Z] is not H-closed!

(i) **Example**. Let Z be the one-point compactification of $\mathbb{N}$ which exists by 4.3(f). Let $Z \backslash \mathbb{N} = \{p\}$. Let X be the space described in 4.8(d). Define $f \in F(Z,X)$ by $f(n) = (1/n,0)$ and $f(p) = p^+$. First we show $f : Z \rightarrow X$ is Θ-continuous. Now, f is Θ-continuous (and continuous) at $n \in Z$ since $\{n\}$ is clopen in Z. To show f is

Θ-continuous at p, let V be an open neighborhood of $f(p) = p^+$ in X. There is an open set $T \subseteq V$ where $T = \{(1/n,1/m) : n \geqslant r$ and $m \in \mathbb{N}\} \cup \{p^+\}$ for some $r \in \mathbb{N}$. Let $W = \{p\} \cup \{n \in \mathbb{N} : n \geqslant r\}$. Then $f[c\ell_Z W] = f[W] = \{p^+\} \cup \{1/n,0) : n \geqslant r\} \subseteq c\ell_X T \subseteq c\ell_X V$. So, $f : Z \to X$ is Θ-continuous. Also, $f[Z] = \{p^+\} \cup \{(1/n,0) : n \in \mathbb{N}\}$ is a discrete, infinite subspace; so f[Z] is not compact. Since f[Z] is regular, we infer from 4.8(c) that f[Z] is not H-closed. Since Z is compact it follows from 4.8(h)(6) that $f : Z \to f[Z]$ is not Θ-continuous. This also shows $f : Z \to X$ is not continuous for if $f : Z \to X$ were continuous, then $f : Z \to f[Z]$ would be continuous and, hence by 4.8(h)(2), $f : X \to f[Z]$ would be Θ-continuous.

(j) Definition. A space X is **Urysohn** if for $p, q \in X$ with $p \neq q$, there are open sets U and V such that $p \in U$, $q \in V$, and $c\ell_X U \cap c\ell_X V = \emptyset$.

Clearly, a regular space is Urysohn and an Urysohn space is Hausdorff. We leave it to the reader to verify these two facts and to find an example of a space which is not Urysohn and an Urysohn space which is not regular (see 4K).

(k) **Corollary**. A space X is compact iff X is H-closed, semiregular, and Urysohn.

Proof. Clearly, a compact space is H-closed, semiregular, and Urysohn. Conversely, suppose X is H-closed, semiregular, and Urysohn. By 4.8(c), it suffices to show X is regular. Let $A \subseteq X$ be closed and $x \in X \setminus A$. Since X is semiregular, there is a regular open set U such that $x \in U \subseteq X \backslash A$. Let $V = X \backslash c\ell_X U$. Then $A \subseteq c\ell_X V$ and $x \notin c\ell_X V$. By 4.8(e), $c\ell_X V$ is H-closed. For each $p \in c\ell_X V$, there are open neighborhoods $W(p)$ of p and $T(p)$ of x such that $c\ell_X W(p) \cap c\ell_X T(p) = \emptyset$. There is a finite set $F \subseteq c\ell_X V$ such that $c\ell_X V \subseteq \bigcup\{c\ell_X W(p) : p \in F\}$. Let $T = \bigcap\{T(p) : p \in F\}$. Then $x \in T$ and $c\ell_X T \cap (\bigcup\{c\ell_X W(p) : p \in F\}) = \emptyset$. So, $(c\ell_X T) \cap A = \emptyset$. This shows X is regular.

■

Now we investigate products of H-closed spaces.

(l) **Proposition**. Let $\{X_a : a \in A\}$ be a family of nonvoid spaces. Then $\Pi\{X_a : a \in A\}$ is H-closed iff X_a is H-closed for each $a \in A$.

Proof. Let $Y = \Pi\{X_a : a \in A\}$ and for each $a \in A$, let $\Pi_a : Y \to X_a$ be the ath projection. Suppose Y is H-closed. Since Π_a is a continuous surjection, then by 4.8(h)(6), X_a is H-closed. Conversely, suppose X_a is H-closed for each $a \in A$. Let $\mathcal{U}$ be an open ultrafilter on Y. Since Π_a is open and continuous, then $\Pi_a[\mathcal{U}] = \{\Pi_a[U] : U \in \mathcal{U}\}$ is an open filter on X_a. To show $\Pi_a[\mathcal{U}]$ is an open ultrafilter on X_a, let W be an open set in X_a such that $W \cap \Pi_a[U] \neq \emptyset$ for all $U \in \mathcal{U}$. Hence, $\Pi_a^{\leftarrow}[W] \cap U \neq \emptyset$ for all $U \in \mathcal{U}$. By 2.3(d), $\Pi_a^{\leftarrow}[W] \in \mathcal{U}$. Hence, $W = \Pi_a[\Pi_a^{\leftarrow}[W]] \in \Pi_a[\mathcal{U}]$. So, by 2.3(d), $\Pi_a[\mathcal{U}]$ is an open ultrafilter on X_a. By 4.8(b), $\Pi_a[\mathcal{U}]$ converges to some point $x_a \in X_a$. Let $x = \langle x_a \rangle_{a \in A}$. Since $\Pi_a[\mathcal{U}]$ converges to $\Pi_a(x)$ for all $a \in A$, it easily follows that $\mathcal{U}$ converges to x. Thus, by 4.8(b), Y is H-closed.

■

In 1924, Alexandroff and Urysohn [AU] asked if every space can be embedded in an H-closed space. In 1930, Tychonoff [Ty] answered this in the affirmative. In contrast to the analogous situation for compact spaces, however, the closure of the embedded subspace may not be H-closed. The question of whether a space could be **densely** embedded in an H-closed space remained open for ten more years until Katetov [Ka_1] and Stone [Sto] answered it.

(m) **Definition**. Let H be the class of all H-closed spaces, and let X be a space. By 4.1(j), $H(X) = \{Y \in E(X) : Y \text{ is H-closed}\}$ is a set of H-closed extensions of X such that no two are equivalent extensions of X and each H-closed extension of X is equivalent to some $Y \in H(X)$.

We want to investigate if $H(X)$ has a projective maximum when $H(X) \neq \emptyset$. Unfortunately, we cannot use the proof that $H(X)$ has a projective maximum by applying 4.1(k) for, even though H is productive, H is not closed-hereditary by 4.8(f). So, we will use a direct approach; for each space X we will build an H-closed extension κX of it, and then show that κX is the projective maximum of $H(X)$. This is the method used by Katetov.

Let X be a space and let $\kappa X = X \cup \{\mathcal{U} : \mathcal{U} \text{ is a free open ultrafilter on } X\}$. It is easily verified that $\{U : U \text{ open in } X\} \cup \{U \cup \{\mathcal{U}\} : U \in \mathcal{U}, \mathcal{U} \in \kappa X \setminus X\}$ is an open base for some topology on κX. We topologize κX in this way.

(n) **Theorem**. Let X be a space. Then:

(1) κX is an H-closed extension of X, and X is open in κX,

(2) if $Y \in \mathcal{H}(X)$, there is a unique continuous function $f : \kappa X \to Y$ such that $f|X = id_X$, i.e., $\kappa X \geq Y$, and

(3) if $Z \in \mathcal{H}(X)$ and $Z \geq Y$ for all $Y \in \mathcal{H}(X)$, then $\kappa X \equiv_X Z$; in particular, $\kappa X = \bigvee \mathcal{H}(X)$.

Proof

(1) Clearly, X is an open subspace of κX. Since X is open in κX and X is Hausdorff, it follows that distinct points of X can be separated by disjoint open sets in κX. If $x \in X$ and $\mathcal{U} \in \kappa X \setminus X$, then there are open sets U and V in X such that $U \in \mathcal{U}$, $x \in V$, and $U \cap V = \emptyset$. Now, V and $U \cup \{\mathcal{U}\}$ are disjoint open neighborhoods of x and $\mathcal{U}$, respectively, in κX. Likewise, if $\mathcal{U}, \mathcal{V} \in \kappa X \setminus X$ such that $\mathcal{U} \neq \mathcal{V}$, then by 2.3(d), there are disjoint open sets $U \in \mathcal{U}$ and $V \in \mathcal{V}$. Then, $U \cup \{\mathcal{U}\}$ and $V \cup \{\mathcal{V}\}$ are disjoint, open neighborhoods of $\mathcal{U}$ and $\mathcal{V}$, respectively, in κX. This shows κX is Hausdorff. Since a basic neighborhood of $\mathcal{U} \in \kappa X \setminus X$ is of the form $U \cup \{\mathcal{U}\}$ where $U \in \mathcal{U}$, then $(U \cup \{\mathcal{U}\}) \cap X = U \neq \emptyset$ (as $U \in \mathcal{U}$). So, X is dense in κX. To show κX is H-closed, assume $\mathcal{W}$ is a free open ultrafilter on κX. Then $X \in \mathcal{W}$ as X is dense and open in κX, and $\mathcal{U} = \{W \cap X : W \in \mathcal{W}\}$ is a free open ultrafilter on X. Now, $\mathcal{U} \subseteq \mathcal{W}$ and for $U \in \mathcal{U}$, $U \subseteq U \cup \{\mathcal{U}\}$. This implies that $U \cup \{\mathcal{U}\} \in \mathcal{W}$. So, $\mathcal{W}$ converges to $\mathcal{U}$, contradicting the assumption that $\mathcal{W}$ is free. This shows that κX is H-closed.

(2) Suppose Y is an H-closed extension of X. Let $\mathcal{U} \in \kappa X \setminus X$ and $\mathcal{F} = \{W \subseteq Y : W \text{ is open in } Y \text{ and } W \cap X \in \mathcal{U}\}$. It is easy to verify that $\mathcal{F}$ is an open filter on Y. Let T be an open set

in Y such that $T \notin \mathcal{F}$. So, $T \cap X \notin \mathcal{U}$. Let $S = X \backslash c\ell_X(T \cap X)$. Then by 2.3(e)(2), $S \in \mathcal{U}$. Since $c\ell_Y(T \cap X) \cap S = c\ell_X(T \cap X) \cap S = \emptyset$, it follows that $S \subseteq Y \backslash c\ell_Y(T \cap X) = Y \backslash c\ell_Y T$. Hence, $Y \backslash c\ell_Y T \in \mathcal{F}$. By 2.3(d), this shows $\mathcal{F}$ is an open ultrafilter on Y. As Y is H-closed, $\mathcal{F}$ converges to a unique point $y_{\mathcal{U}} \in Y$. Define $f : \kappa X \to Y$ by $f(x) = x$ for $x \in X$ and $f(\mathcal{U}) = y_{\mathcal{U}}$ for $\mathcal{U} \in \kappa X \backslash X$. Since $f|X = id_X$, we only need to show that f is continuous. Since X is open in κX and $f|X : X \to Y$ is continuous, it follows that f is continuous at points of X. Let $\mathcal{U} \in \kappa X \backslash X$ and W be an open neighborhood of $y_{\mathcal{U}}$ in Y. Since $\mathcal{F}$ converges to $y_{\mathcal{U}}$, it follows that $W \in \mathcal{F}$. So, $W \cap X \in \mathcal{U}$. Thus, $\{\mathcal{U}\} \cup (W \cap X)$ is an open neighborhood of $\mathcal{U}$ in κX and $f[\{\mathcal{U}\} \cup (W \cap X)] = \{y_{\mathcal{U}}\} \cup (W \cap X) \subseteq W$. Hence, f is continuous. By 4.1(b), f is unique.

(3) By (2), $\kappa X \geqslant Z$. By hypothesis, $Z \geqslant \kappa X$. Thus, $\kappa X \equiv_X Z$. By (2) and (3), we have that $\kappa X = \bigvee \mathcal{H}(X)$.

■

(o) <u>Definition</u>. For a space X, the H-closed extension κX is called the **Katětov extension** of X. For $Y \in \mathcal{H}(X)$, the unique continuous function $f : \kappa X \to Y$ such that $f|X = id_X$ is called the **Katětov function** of Y.

Since κX is the projective maximum of $\mathcal{H}(X)$ and since $\mathcal{H}$ is productive but not closed-hereditary, it follows that the converse of 4.1(k) is false.

Here are some additional properties of the Katětov extension.

(p) <u>Proposition</u>. Let X be a space. Then:

(1) $\kappa X \backslash X$ is a closed, discrete subspace of κX,

(2) if $U \subseteq X$ is open, then $c\ell_{\kappa X} U = c\ell_X U \cup \{\mathcal{U} \in \kappa X \backslash X : U \in \mathcal{U}\}$,

(3) if U, $V \subseteq X$ are open in X and $U \cap V = \emptyset$, then $(c\ell_{\kappa X} U \cap c\ell_{\kappa X} V) \backslash X = \emptyset$,

(4) if U and V are open in X and $c\ell_X U \cap c\ell_X V = \emptyset$, then $c\ell_{\kappa X} U \cap c\ell_{\kappa X} V = \emptyset$; in particular, if U is clopen in X, then $c\ell_{\kappa X} U$ is clopen in κX,

(5) X is C^*-embedded in κX, and

(6) if A is a closed nowhere dense subset of X, then A is closed in κX.

Proof. The proof of (1) is immediate from the proof of 4.8(n)(1).

(2) Since $(c\ell_{\kappa X} U) \cap X = c\ell_X U$, it suffices to show that $c\ell_{\kappa X} U \backslash X = \{\mathcal{U} \in \kappa X \backslash X : U \in \mathcal{U}\}$. Now, $\mathcal{U} \in c\ell_{\kappa X} U \backslash X$ iff $U \cap V \neq \emptyset$ for all $V \in \mathcal{U}$. By 2.3(d), $U \cap V \neq \emptyset$ for all $V \in \mathcal{U}$ iff $U \in \mathcal{U}$.

(3) By (2), $(c\ell_{\kappa X} U \cap c\ell_{\kappa X} V) \backslash X = \{\mathcal{U} \in \kappa X : U \in \mathcal{U}\} \cap \{\mathcal{U} \in \kappa X : V \in \mathcal{U}\} = \{\mathcal{U} \in \kappa X : U \cap V \in \mathcal{U}\} = \emptyset$ as $U \cap V = \emptyset$.

(4) Since $c\ell_{\kappa X} U \cap c\ell_{\kappa X} V = [(c\ell_{\kappa X} U \cap c\ell_{\kappa X} V) \backslash X] \cup [(c\ell_{\kappa X} U \cap c\ell_{\kappa X} V) \cap X]$ and by (3), $(c\ell_{\kappa X} U \cap c\ell_{\kappa X} V) \backslash X = \emptyset$, then $c\ell_{\kappa X} U \cap c\ell_{\kappa X} V = (c\ell_{\kappa X} U \cap c\ell_{\kappa X} V) \cap X = c\ell_X U \cap c\ell_X V = \emptyset$. If U is clopen in X, then $c\ell_X U \cap c\ell_X (X \backslash U) = \emptyset$. So, $c\ell_{\kappa X} U \cap c\ell_{\kappa X} (X \backslash U) = \emptyset$. But $\kappa X = c\ell_{\kappa X} (U \cup (X \backslash U)) = c\ell_{\kappa X} U \cup c\ell_{\kappa X} (X \backslash U)$. Thus, $c\ell_{\kappa X} U = \kappa X \backslash c\ell_{\kappa X} (X \backslash U)$ and

$c\ell_{\kappa X}U$ is clopen.

(5) Let $f \in C(X,K)$ where K is a compact space. To find a continuous function $F \in C(\kappa X,K)$ such that $F|X = f$, it suffices by 4.1(m) to show for disjoint closed sets $A, B \subseteq K$, $c\ell_{\kappa X}f^{\leftarrow}[A] \cap c\ell_{\kappa X}f^{\leftarrow}[B] = \emptyset$. Since K is compact, there are open sets $U, V \subseteq K$ such that $A \subseteq U$, $B \subseteq V$, and $c\ell_K U \cap c\ell_K V = \emptyset$. Since f is continuous, then $f^{\leftarrow}[U]$ and $f^{\leftarrow}[V]$ are open and $f^{\leftarrow}[A] \subseteq f^{\leftarrow}[U] \subseteq c\ell_X f^{\leftarrow}[U] \subseteq f^{\leftarrow}[c\ell_K U]$ and $f^{\leftarrow}[B] \subseteq f^{\leftarrow}[V] \subseteq c\ell_X f^{\leftarrow}[V] \subseteq f^{\leftarrow}[c\ell_K V]$. Since $f^{\leftarrow}[c\ell_K U] \cap f^{\leftarrow}[c\ell_K V] = \emptyset$, then $c\ell_X f^{\leftarrow}[U] \cap c\ell_X f^{\leftarrow}[V] = \emptyset$. By (4), $c\ell_{\kappa X}f^{\leftarrow}[U] \cap c\ell_{\kappa X}f^{\leftarrow}[V] = \emptyset$. Thus, $c\ell_{\kappa X}f^{\leftarrow}[A] \cap c\ell_{\kappa X}f^{\leftarrow}[B] = \emptyset$.

(6) Suppose A is closed and nowhere dense in X. Then $X\setminus A$ is open and dense in X. If $\mathcal{U}$ is an open ultrafilter on X, then $X\setminus A$ meets $\mathcal{U}$. By 2.3(d), $X\setminus A \in \mathcal{U}$. Hence, $(X\setminus A) \cup (\kappa X\setminus X)$ is open in κX. So, $\kappa X\setminus[(X\setminus A) \cup (\kappa X\setminus X)] = A$ is closed in κX.

■

Now, we compare $\kappa\mathbb{N}$ and $\beta\mathbb{N}$.

(q) **Examples**

(1) By 4.8(n), $\kappa\mathbb{N}$ is an H-closed extension of $\mathbb{N}$. Since we can find a partition $\{A_n : N \in \mathbb{N}\}$ of $\mathbb{N}$ such that each A_n is infinite, and since each A_n is contained in some free ultrafilter $\mathcal{U}_n$ on $\mathbb{N}$, then $|\kappa\mathbb{N}\setminus\mathbb{N}| \geq \aleph_0$. Thus, by 4.8(p)(1), $\kappa\mathbb{N}\setminus\mathbb{N}$ is a closed, discrete and infinite subspace of $\kappa\mathbb{N}$. Hence, $\kappa\mathbb{N}$ is not compact and $\kappa\mathbb{N} \not\equiv_{\mathbb{N}} \beta\mathbb{N}$. Since $\beta\mathbb{N} \in H(\mathbb{N})$, then $\kappa\mathbb{N} \geq \beta\mathbb{N}$. So, there are at least two distinct H-closed extensions of $\mathbb{N}$ in which $\mathbb{N}$ is C^*-embedded. If U and V are disjoint open sets in $\kappa\mathbb{N}$, then $U \cap \mathbb{N}$

and $V \cap \mathbb{N}$ are disjoint clopen sets in $\mathbb{N}$. So, $c\ell_{\mathbb{N}}(U \cap \mathbb{N}) \cap c\ell_{\mathbb{N}}(V \cap \mathbb{N}) = \emptyset$. By 4.8(p)(4), $c\ell_{\kappa\mathbb{N}}(U \cap \mathbb{N}) \cap c\ell_{\kappa\mathbb{N}}(V \cap \mathbb{N}) = \emptyset$. But $c\ell_{\kappa\mathbb{N}}(U \cap \mathbb{N}) = c\ell_{\kappa\mathbb{N}}U$ and $c\ell_{\kappa\mathbb{N}}(V \cap \mathbb{N}) = c\ell_{\kappa\mathbb{N}}V$. So, $c\ell_{\kappa\mathbb{N}}U \cap c\ell_{\kappa\mathbb{N}}V = \emptyset$. Since $\kappa\mathbb{N}$ is Hausdorff, it follows that $\kappa\mathbb{N}$ is Urysohn. So, $\kappa\mathbb{N}$ is an example of an H-closed, Urysohn space which is not compact.

(2) Consider the space X of 4.8(d). Since $\{(1/n,1/m) : n \in \mathbb{N}, |m| \in \mathbb{N}\}$ is a countable, dense set of isolated points, X can be considered as a H-closed extension of $\mathbb{N}$. In particular, $|X\backslash\mathbb{N}| = \aleph_0$ and $X\backslash\mathbb{N}$ is a closed, infinite, discrete subspace of X. So, X is not compact. Hence $\beta\mathbb{N} \not\geq X$. It is trivial to verify that X is semiregular. So X is an example of an H-closed, semiregular space which is not compact.

(3) Let $Z = \{(1/n,1/m) : n, m \in \mathbb{N}\} \cup \{(1/n,0) : n \in \mathbb{N}\} \cup \{p^+\}$ with the subspace topology inherited from the space X described in 4.8(d). Since Z is a regular closed subset of X, by 4.8(e) Z is H-closed. Since $\{(1/n,1/m) : n, m \in \mathbb{N}\}$ is a countable, dense set of isolated points, it follows that Z can be considered as an H-closed extension of $\mathbb{N}$. Z is not semiregular since the regular open sets containing p^+ do not form a neighborhood base of p^+. On the other hand, it is easy to verify that Z is Urysohn.

(r) **Definition**. Let X be a space and $A \subseteq X$.

(1) Let $c\ell_\Theta A$ denote $\{x \in X : \text{for every open neighborhood } V \text{ of } x, (c\ell_X V) \cap A \neq \emptyset\}$. Then $c\ell_\Theta A$ is called the **Θ-closure** of A.

(2) If $\mathcal{F}$ is a filter on X, then $a_\Theta(\mathcal{F})$ is used to denote $\cap\{c\ell_\Theta F : F \in \mathcal{F}\}$.

In general, "$c\ell_\Theta$" is **not** a closure operator as it is not necessarily true that $c\ell_\Theta(c\ell_\Theta A) = c\ell_\Theta A$ in a non-regular space. For example, let X be the space in 4.8(d) and $A = \{(1/n,1/m) : n, m \in \mathbb{N}\}$. Then, $c\ell_\Theta(c\ell_\Theta A)\setminus c\ell_\Theta A = \{p^-\}$.

(s) **Proposition**. Let X and Y be spaces, $f \in F(X,Y)$, $A \subseteq X$, and U an open subset of X. Then:

(1) $c\ell_\Theta A$ is closed,

(2) $c\ell_\Theta U = c\ell_X U$,

(3) if f is Θ-continuous, then $f[c\ell_X U] \subseteq c\ell_\Theta f[U]$,

(4) X is H-closed iff for every filter $\mathcal{F}$ on X, $a_\Theta(\mathcal{F}) \neq \emptyset$, and

(5) f is a Θ-homeomorphism iff $f : X(s) \to Y(s)$ is a homeomorphism.

Proof

(1) Let $x \notin c\ell_\Theta A$. Then there is an open set V such that $x \in V$ and $(c\ell_X V) \cap A = \emptyset$. Thus, $V \cap c\ell_\Theta A = \emptyset$. So, $c\ell_\Theta A$ is closed.

(2) If V is an open set, then $V \cap U = \emptyset$ iff $(c\ell_X V) \cap U = \emptyset$. Thus, it is immediate that $c\ell_\Theta U = c\ell_X U$.

(3) Let $x \in c\ell_X U$ and W be an open neighborhood of f(x) in Y. So, there is an open neighborhood V of x such that $f[c\ell_X V] \subseteq c\ell_Y W$. Since $V \cap U \neq \emptyset$, it follows that $\emptyset \neq f[U \cap V] \subseteq f[U] \cap f[V] \subseteq f[U] \cap c\ell_Y W$. Thus $f(x) \in c\ell_\Theta f[U]$.

(4) Suppose $\mathcal{F}$ is a filter on X such that $a_\Theta \mathcal{F} = \emptyset$. Let $\mathcal{G} = \{V \in \mathcal{F} : V \text{ is open}\}$. Then $\mathcal{G}$ is an open filter on X. If $x \in X$, there is an open neighborhood W of x and some $F \in \mathcal{F}$ such that

$(c\ell_X W) \cap F = \emptyset$. Thus, $V = X \setminus c\ell_X W \in \mathcal{F}$. Hence, $V \in \mathcal{G}$ and $W \cap V = \emptyset$ which implies $x \notin c\ell_X V$ and $x \notin a_X(\mathcal{G})$. This shows $a_X(\mathcal{G}) = \emptyset$ and X is not H-closed. Conversely, suppose X is not H-closed. Then, there is an open filter $\mathcal{G}$ on X such that $a_X(\mathcal{G}) = \emptyset$. By (2), $a_\Theta(\mathcal{G}) = a_X(\mathcal{G})$. Let $\mathcal{F} = \{F \subseteq X : F \supseteq U$ for some $U \in \mathcal{G}\}$. Then $\mathcal{F}$ is a filter and $\mathcal{F} \supseteq \mathcal{G}$. So, $a_\Theta(\mathcal{F}) \subseteq a_\Theta(\mathcal{G}) = \emptyset$.

(5) Suppose $f : X(s) \to Y(s)$ is a homeomorphism. Since $id_X : X \to X(s)$ and $id_Y : Y \to Y(s)$ are Θ-homeomorphisms by 4.8(h)(7) it follows that $f = id_Y^{\leftarrow} \circ f \circ id_X : X \to Y$ is a Θ-homeomorphism by 4.8(h)(1). Conversely, suppose $f : X \to Y$ is a Θ-homeomorphism. Since $f = id_Y \circ f \circ id_X^{\leftarrow} : X(s) \to Y(s)$ is a Θ-homeomorphism by 4.8(h)(1,7), it suffices to assume that X and Y are semiregular (i.e., $X = X(s)$ and $Y = Y(s)$). So, we need to show that $f : X \to Y$ is closed; then by symmetry, $f^{\leftarrow} : Y \to X$ will be closed. Since f is a bijection and $\{c\ell_X U : U$ open in $X\}$ is a closed base for X, we only need to show $f[c\ell_X U]$ is closed in Y for each open set $U \subseteq X$. Let $y \in Y \setminus f[c\ell_X U]$, $z = f^{\leftarrow}(y)$, and $W = X \setminus c\ell_X U$. Then $z \in W$. Since $f^{\leftarrow}$ is Θ-continuous, there is an open neighborhood V of y such that $f^{\leftarrow}[c\ell_Y V] \subseteq c\ell_X W$. Since $(c\ell_X W) \cap U = \emptyset$, it follows that $f^{\leftarrow}[c\ell_Y V] \cap U = \emptyset$. So, $c\ell_Y V \cap f[U] = \emptyset$. Thus, $y \notin c\ell_\Theta f[U]$. This shows $c\ell_\Theta f[U] \subseteq f[c\ell_X U]$. By (2) and (3), we have that $f[c\ell_X U] = f[c\ell_\theta U] \subseteq c\ell_\Theta f[U]$. Combining the two set inclusions, we have that $f[c\ell_X U] = c\ell_\Theta f[U]$. By (1), we have that $f[c\ell_X U]$ is closed.

■

(t) **Example**. In 4.8(q) it is shown that $\kappa\mathbb{N}$ and $\beta\mathbb{N}$ are not equivalent extensions of $\mathbb{N}$, and that $\kappa\mathbb{N}$ is projectively larger than $\beta\mathbb{N}$ (i.e., $\kappa\mathbb{N} > \beta\mathbb{N}$). Recall that $\kappa\mathbb{N} \setminus \mathbb{N}$ consists of the set of free open ultrafilters on $\mathbb{N}$ which, because $\mathbb{N}$ is discrete, coincides with the set of free z-ultrafilters on $\mathbb{N}$. Hence the continuous function $j : \kappa\mathbb{N} \to \beta\mathbb{N}$ that witnesses the fact that $\kappa\mathbb{N} \geq \beta\mathbb{N}$ is just the identity function on the common underlying set of $\kappa\mathbb{N}$ and $\beta\mathbb{N}$. Thus j is a continuous bijection but not a homeomorphism. In 4L the reader is invited to verify that $(\kappa\mathbb{N})(s) = \beta\mathbb{N}$ and that $j : (\kappa\mathbb{N})(s) \to \beta\mathbb{N}$ is a homeomorphism. Thus by 4.8(s)(5), $j : \kappa\mathbb{N} \to \beta\mathbb{N}$ is a continuous θ-homeomorphism that is not a homeomorphism.

Chapter 4 — Problems

4A. Locally compact spaces and one-point compactifications

(1) Prove 4.3(a).

(2) Let X be a locally compact, but not compact space; find a Wallman base L such that $w_L X$ is the one-point compactification of X.

(3) Let α be an ordinal and ω_α the woset defined after 2.6(j). If $cf(\omega_\alpha) > \aleph_0$, prove $\beta\omega_\alpha \equiv_{\omega_\alpha} \omega_\alpha + 1$. (Hint: Use 2.6(q)(6).)

(4) Let X be a Tychonoff space. Prove $\beta X \equiv_X \hat{X}$ ($\hat{X}$ is the one-point compactification of X) iff whenever $Z_1, Z_2 \in Z(X)$ and $Z_1 \cap Z_2 = \emptyset$, then either Z_1 or Z_2 is compact.

4B. Wallman base properties

(1) Prove 4.4(c).

Let X be a Tychonoff space and let $L = \{X \backslash U : U \text{ open in } X\}$.

(2) Prove L is a Wallman base for X iff X is normal.

(3) If L is a Wallman base for X, prove $w_L X \equiv_X \beta X$.

4C. Binomial expansion. Show that $(1 - t^2)^{1/2}$ has a binomial expansion $\Sigma_{n=0}^{\infty} c_n t^n$ which converges uniformly to $(1-t^2)^{1/2}$ in the interval $[0,1]$.

4D. Ordered Archimedian fields. Prove that an ordered field F is Archimedian iff F is order isomorphic to a subfield of $\mathbb{R}$. (Hint: First prove $\mathbb{Q}$ is ordered and ring isomorphic to a subfield Q of F. Show Q is order dense in F and that each element of F is uniquely determined by a Dedekind cut of Q.)

4E. Regular subrings

(1) Prove 4.5(a) and 4.5(i)(1,2,3).

(2) Let T be a compactification of a space X. Prove that $\{f|X : f \in C^*(T)\}$ is a regular subring $\mathcal{Q}_T$ of $C^*(X)$. (See 4.5(o)(1).)

(3) Let $\mathcal{Q}$ be a regular subring of $C^*(X)$ where X is a Tychonoff space and $E = C(X,[0,1]) \cap \mathcal{Q}$. Define $e : X \to [0,1]^E$ by $e(x)(f) = f(x)$ for $f \in E$ and $x \in X$, and let $Y = [0,1]^E$.

(a) Prove that e is an embedding function, and show that $c\ell_Y e[X]$ is a compactification of X.

(b) Prove $m_{\mathcal{Q}} X \equiv_X c\ell_Y e[X]$.

Note that if T is a compactification of a space X, then by (2) $\mathcal{Q}_T = \{f|X : f \in C^*(T)\}$ is a regular subring of $C^*(X)$ and by 4.5(o), $T \equiv_X m_{\mathcal{Q}_T}X$. By (3) $m_{\mathcal{Q}_T}X \equiv c\ell_Y e[X]$; hence $T \equiv_X c\ell_Y e[X]$. In particular, the product construction in (3) is another method of constructing T once $\circ_T$ is known.

(4) Let T be a compactification of a Tychonoff space X and $\mathcal{Q} = \{f|X : f \in C^*(T)\}$. By (2), $\mathcal{Q}$ is a subring of $C^*(X)$. For $f \in C^*(T)$, define $\phi(f) = f|X$. So, $\phi \in F(C^*(T),\mathcal{Q})$. Prove ϕ is a norm-preserving ring isomorphism from $C^*(T)$ onto $\mathcal{Q}$. ("Norm-preserving" means that $\|f\|' = \|\phi(f)\|''$ where $\|f\|'$ is the norm of f in $C^*(T)$ and $\|\phi(f)\|''$ is the norm of $\phi(f)$ in $C^*(X)$.)

4F. Tychonoff compactifications. Let X be an infinite Tychonoff space and $\mathcal{B}$ be an open base for X. Let $(\mathcal{B},\mathcal{B}) = \{(U,V) \in \mathcal{B} \times \mathcal{B} :$ there is some $f \in C(X,[0,1])$ such that $f[U] \subseteq \{0\}$ and $f[X\setminus V] \subseteq \{1\}\}$. If $(U,V) \in (\mathcal{B},\mathcal{B})$, pick some $f_{U,V} \in C(X,[0,1])$ such that $f_{U,V}[U] \subseteq \{0\}$ and $f_{U,V}[X\setminus V] \subseteq \{1\}$. Let $\mathcal{C} = \{f_{U,V} : (U,V) \in (\mathcal{B},\mathcal{B})\}$. Define $e \in F(X,[0,1]^{\mathcal{C}})$ by $e(x)(f_{U,V}) = f_{U,V}(x)$ for $x \in X$. Let $Y = [0,1]^{\mathcal{C}}$.

(1) Show that $c\ell_Y e[X]$ is a compactification of X and $w(c\ell_Y e[X]) \leq |\mathcal{B}|$.

(2) Prove that if X is second countable, then X has a metrizable compactification.

4G. Wallman vs. Gelfand techniques. Let X be a Tychonoff space and let $\mathfrak{a}$ be a regular subring of $C^*(X)$. Show:

(1) $L = Z[\mathfrak{a}]$ is a Wallman base,

(2) $w_L X \geq m_{\mathfrak{a}} X$;

(3) $w_L X \equiv_X m_{\mathfrak{a}} X$ iff $c\ell_Y Z(f) \cap c\ell_Y Z(g) = \emptyset$ whenever $f, g \in \mathfrak{a}$, and $Z(f) \cap Z(g) = \emptyset$, where $Y = m_{\mathfrak{a}} X$ (Hint: Use 4.2(h)); and

(4) if $\hat{\mathbb{R}}$ is the one-point compactification of $\mathbb{R}$ and $L = \{Z(f) : f \in \mathfrak{a}\}$, where $\mathfrak{a} = \mathfrak{a}_{\hat{\mathbb{R}}}$, then $w_L \mathbb{R} \equiv_{\mathbb{R}} \beta\mathbb{R}$ and $m_{\mathfrak{a}} \mathbb{R} \equiv_{\mathbb{R}} \hat{\mathbb{R}}$.

4H. Isomorphism between $C^*(X)$ and $C^*(\beta X)$. Prove that the function $\Phi : C^*(X) \to C^*(\beta X)$ defined by $\Phi(f) = f^\beta$ is a ring isomorphism.

4I. Generating $K_0(X)$. Let X be a zero-dimensional space. A family $E \subseteq B(X)$ is called a **Boolean base** for X if E is a Boolean subalgebra of $B(X)$ and E is a base for X. Let $BB(X) = \{E \subseteq B(X) : E$ is a Boolean base for X$\}$. Define $\Psi : BB(X) \to K_0(X)$ by $\Psi(E) = c_E X$ (defined after 4.7(a)). Now $BB(X)$ is partially ordered by inclusion.

(1) Prove that Ψ is an order isomorphism.

(2) Let X be locally compact and $E = \{C \in B(X) : C$ or

$X \setminus C$ is compact}. Prove that $c_E X$ is the one-point compactification of X.

4J. Extensions of functions to $\beta_0 X$.

(1) Formulate statements parallel to 4.6(k) and 4.6(l) in the setting of zero-dimensional spaces, i.e., for $\beta_0 X$ where X is zero-dimensional. Now prove the statements.

(2) For a zero-dimensional space X, prove that the function Ψ defined by $\Psi(A) = cl_{\beta_0 X} A$ is a Boolean algebra isomorphism from $B(X)$ onto $B(\beta_0 X)$.

4K. Urysohn spaces

(1) Prove that the space described in 4.8(d) is not Urysohn.

(2) Show that the subspace $\{p^+\} \cup \{(\frac{1}{n},0) : n \in \mathbb{N}\} \cup \{(\frac{1}{n},\frac{1}{m}) : n, m \in \mathbb{N}\}$ of the space described in 4.8(d) is Urysohn but not regular.

(3) Prove that a subspace of an Urysohn space is Urysohn.

(4) Let $\{X_i : i \in I\}$ be a family of nonempty spaces. Prove that $\Pi\{X_i : i \in I\}$ is Urysohn iff X_i is Urysohn for each $i \in I$.

(5) Let $f, g \in \Theta C(X,Y)$ where X and Y are spaces and Y is Urysohn. If $\{x \in X : f(x) = g(x)\}$ is dense in X, prove that $f = g$. Compare this with 1.6(d).

(6) Give examples of spaces X and Y and functions $f, g \in$

$\theta C(X,Y)$ such that $\{x \in X : f(x) = g(x)\}$ is dense in X but $f \neq g$. (Hint: Use (1) above.)

(7) Prove that a space X is Urysohn iff X(s) is Urysohn. (Hint: Use 2.2(f).)

(8) If Y is an H-closed, Urysohn extension of a space X, use 4.8(k) to deduce that Y(s) is compact and use 2.2(i)(2) to deduce that Y(s) is a compactification of X(s). This shows that X(s) is Tychonoff.

(9) Prove that a space X is H-closed and Urysohn (equivalent to X(s) being compact) iff for each open cover $\mathcal{C}$, $\{int_X(c\ell_X U) : U \in \mathcal{C}\}$ has a finite subcover.

4L. $\beta\mathbb{N}$ vs. $\kappa\mathbb{N}$. This problem continues the examination of the spaces $\beta\mathbb{N}$ and $\kappa\mathbb{N}$ begun in 4.8(t).

(1) Use 4.8(p)(4) to show that $\kappa\mathbb{N}$ is Urysohn. By 4K(8), deduce that $\kappa\mathbb{N}(s)$ is a compactification of $\mathbb{N}(s)$ $(=\mathbb{N})$.

(2) Use 4.8(n) to show that $\kappa\mathbb{N} \geq \beta\mathbb{N}$. Let $f : \kappa\mathbb{N} \to \beta\mathbb{N}$ be the continuous function such that $f(n) = n$ for all $n \in \mathbb{N}$. Use 2.2(g) to show that $\kappa\mathbb{N}(s) \geq \beta\mathbb{N}$.

(3) Show that $f_s : \kappa\mathbb{N}(s) \to \beta\mathbb{N}$ is a homeomorphism (f is defined in (2) and f_s is defined in 2.2(g)).

(4) Let $g = f^{\leftarrow}$ (f is defined in (2)). Show that g is θ-continuous. (Hint: Use 4.8(h)(7).)

(5) Show that $g|(\beta\mathbb{N}\setminus\mathbb{N}) : \beta\mathbb{N}\setminus\mathbb{N} \to \kappa\mathbb{N}$ is θ-continuous (g is defined in (4)) but $g|(\beta\mathbb{N}\setminus\mathbb{N}) : \beta\mathbb{N}\setminus\mathbb{N} \to \kappa\mathbb{N}\setminus\mathbb{N}$ is not θ-continuous.

(6) For an infinite discrete space D, prove that $(\kappa D)(s) \equiv_D \beta D$.

(7) Let $h : \mathbb{N} \to \kappa\mathbb{N}$ be the continuous function defined by: $h(n) = n$ for each $n \in \mathbb{N}$. Show that for each $p \in \beta\mathbb{N}$, $\{h[S] : S \in O^p\}$ converges to $g(p)$, where O^p is as defined in 4.1(l) and g is as defined in (4) above.

(8) Show that if h has a continuous extension, then it must be g. Use (5) to conclude that h does not have a continuous extension. This illustrates that 4.1(l) does not remain true if the assumption that the range space is regular is replaced by the assumption that the range space is Urysohn.

(9) Suppose that X and Z are arbitrary Hausdorff spaces, $f \in C(X,Z)$, and Y is an extension of X. Show that if condition 4.1(l)(2) is satisfied, then f has a θ-continuous extension $F : Y \to Z$.

4M. <u>C^*-embedding in H-closed extensions</u>. Let X be a Tychonoff space and $Y \in \mathcal{H}(X)$. Prove that X is C^*-embedded in Y iff $Y \geq \beta X$.

4N. <u>H-sets</u>. A subspace X of a space Y is an **H-set** if for each cover $\mathcal{C}$ of X by open subsets of Y, there is a finite subfamily $\mathcal{S} \subseteq \mathcal{C}$ such that $X \subseteq \bigcup\{c\ell_Y C : C \in \mathcal{S}\}$.

(1) If X is an H-set of a space Y, prove that X is closed in Y.

(2) If X is an H-set of a space Y and Y is a subspace of Z, prove that X is an H-set of Z.

(3) If X is an H-closed subspace of a space Y, prove that X is an H-set of Y.

(4) Find a subspace Z of the H-closed space X described in 4.8(d) such that Z is an H-set but not H-closed.

(5) Let X be a subspace of a space Y. Prove that X is an H-set of Y iff for every open filter $\mathcal{F}$ on Y such that $F \cap X \neq \emptyset$ for all $F \in \mathcal{F}$, $a_Y(\mathcal{F}) \cap X \neq \emptyset$.

(6) Let $f \in \Theta C(X,Y)$ where X, Y are spaces. Let A be an H-set of X. Prove that f[A] is an H-set of Y.

(7) Prove that if Y is H-closed and X is a subspace of Y, then $c\ell_\Theta X$ is an H-set.

(8) Find an H-closed space X and a subspace Z such that $c\ell_\Theta Z$ is not H-closed. (Hint: Let X be the H-closed space described in 4.8(d).)

(9) If A is an H-set of a space X, $B \in R(X)$, and $B \subseteq A$, prove that B is an H-closed subspace of X.

(10) Prove that an H-set of a regular space is compact.

(11) Let A be a subspace of an H-closed, Urysohn space X. Prove that A is an H-set of X iff A is a compact subspace of X(s). (Hint: Use (6), (10), and 4.8(k).)

40. <u>Properties of Θ-continuous functions</u>.

(1) Let $\{X_i : i \in I\}$ and $\{Y_i : i \in I\}$ be families of spaces

and let $f_i \in \Theta C(X_i,Y_i)$ for each $i \in I$. Prove that $\Pi_i f_i : \Pi\{X_i : i \in I\} \to \Pi\{Y_i : i \in I\}$ is Θ-continuous.

(2) Let $f \in \Theta C(X,Y)$ where X, Y are spaces and let K be a compact subset of Y. Prove that $f^{\leftarrow}[K]$ is a closed set in X.

(3) Use (2) and 4N(6) to show that if $f \in \Theta C(X,Y)$ where X is compact and Y is a space, then f is perfect. Note that this result slightly improves 1.8(b).

4P. <u>The supremum of a set of extensions as an inverse limit.</u>
Let X be a space.

(1) Let $Y, Z \in E(X)$. Prove that $Y \vee Z \equiv_X c\ell_{Y\times Z}\{(x,x) \in Y \times Z : x \in X\}$. (Hint: First, show that $f \in F(X,Y\times Z)$, defined by $f(x) = (x,x)$ for $x \in X$, is an embedding.)

(2) Let $\emptyset \neq S \subseteq E(X)$ and $U = \{\vee\{Y : Y \in F\} : F$ is a finite subfamily of $S\}$. Show that under the usual ordering on $E(X)$, $(U,\leq)$ is a directed set and that $\vee U \equiv_X \vee S$.

(3) If $Y, Z \in U$ and $Y \geq Z$, let $f_{Y,Z} \in F(Y,Z)$ be the unique continuous function such that $f_{Y,Z}(x) = x$ for $x \in X$. Let Y_∞ denote the inverse limit of $(U,f_{Y,Z})$ (see 2U). Prove that $Y_\infty \equiv_X \vee U$.

4Q. Upper semicontinuous decompositions of compact spaces. If X and Y are spaces and $f \in C(X,Y)$, then $\{f^{\leftarrow}(y) : y \in f[X]\}$ is a partition of X into closed sets. An interesting topological question is to determine when a partition of X into closed sets is induced by a continuous function. Let P be a partition of X into closed sets. Define $f_P \in F(X,P)$ by $f_P(x) = A$ where A is the unique set in P containing x. The usual method is to give P the quotient topology induced by f_P, i.e., the topology generated by the open base $\{U \subseteq P : f_P^{\leftarrow}[U]$ is open in $X\}$. One obstacle is that this topology on P may not be Hausdorff even though singletons of P are closed. A partition of X into closed sets is called an **upper semicontinuous decomposition** if for each $A \in P$ and open set $U \subseteq X$ such that $A \subseteq U$, there is an open set $W \subseteq X$ such that $A \subseteq W = f_P^{\leftarrow}[f_P[W]] \subseteq U$.

(1) If $f \in C(R,S)$ is a surjection where R, S are spaces and f is closed, prove that f is a quotient function.

(2) Let $Y = \{(x,y) : 0 \leqslant x \leqslant 1,\ 0 \leqslant y \leqslant 1\}$, $A_n = \{(\frac{1}{n},y) : 0 \leqslant y \leqslant 1\}$ for $n \in \mathbb{N}$, $X = Y \setminus \bigcup\{A_n : n \in \mathbb{N}\}$, and $P = \{A_n : n \in \mathbb{N}\} \cup \{\{x\} : x \in X\}$. Prove that P with the quotient topology induced by f_P is not Hausdorff.

(3) Let X be compact and let P be a partition of X into closed sets (equipped with the quotient topology). Prove that the following are equivalent:

(a) P is a Hausdorff space,

(b) f_P is closed, and

(c) P is an upper semicontinuous decomposition of X.

Let X be a Tychonoff space, $Y \in K(X)$, and $D_Y = \{P :$ P is an upper semicontinuous decomposition of Y such that $\{x\} \in$ P for each $x \in X\}$.

(4) For each $P \in D_Y$, show that $f_P | X \in F(X,P)$ is an embedding and P is a compactification of $f_P[X]$.

By (4), it follows that P is a compactification of X which we denote, henceforth, as Y_p. Note that D_Y is partially ordered by refinement, i.e., if $P, Q \in D_Y$, then $P \leqslant Q$ if for each $A \in P$, there is some $B \in Q$ such that $A \subseteq B$; cf. 2.1(f)(2).

(5) Prove that the function Φ from D_Y to $\{Z \in K(X) : Z \leqslant Y\}$ defined by $\Phi(P) = Y_p$ is a bijection that is order reversing, i.e., if $P \leqslant Q$, then $\Phi(P) \geqslant \Phi(Q)$.

4R. <u>More on Σ-products</u>. This problem is a continuation of 1X and the definitions and notations developed in 1X will be used in this problem. Let $\{X_a : a \in A\}$ be an infinite family of nonempty compact spaces, $Y = \Pi\{X_a : \in A\}$, and let $b \in Y$.

(1) Suppose each X_a is separable.

(a) Prove that $\beta\Sigma(b) \equiv_{\Sigma(b)} Y$. (Hint: Use 1X(6).)

(b) If $|A| > \aleph_0$, prove that $\beta Z \equiv_Z \beta\Sigma(b)$ where $Z = \beta\Sigma(b)\setminus\Sigma(b)$. (Hint: Let $c \in Y\setminus\Sigma(b)$. Note that $\Sigma(c) \subseteq Y\setminus\Sigma(b) \subseteq Y$ and use (a).)

Now, we will establish (1) without the separability restriction. For the remainder of this problem, let $f : \Sigma(b) \to \mathbb{R}$ be a continuous function and $\Sigma'(b) = \{x \in \Sigma(b) : |\{a \in A : x_a \neq b_a\}| < \aleph_0\}$. Recall from 1X, that for $x \in Y$ and $T \subseteq A$, $S_T(x)$ denotes $\{y \in Y : x_a = y_a$ for all $a \in T\}$. In particular, note that as a product of compact spaces, $S_T(x)$ is compact and that $S_{A\setminus T}(b) \subseteq \Sigma(b)$ whenever T is countable. For $x \in \Sigma(b)$, let $A(x) = \{a \in A : x_a \neq b_a\}$. For $n \in \mathbb{N}$, let $B_n = \{a \in A :$ there are points $x^a, y^a \in \Sigma'(b)$ such that $|f(x^a) - f(y^a)| > 1/n$ and $(x^a)_c = (y^a)_c$ for $c \in A\setminus\{a\}\}$. The first step is to establish that $|B_n| \leq \aleph_0$ for each $n \in \mathbb{N}$. For each $a \in B_n$, choose x^a and y^a in $\Sigma'(b)$ such that $|f(x^a) - f(y^a)| > 1/n$, $(x^a)_a \neq b_a$, and $(x^a)_c = (y^a)_c$ for $c \in A\setminus\{a\}$.

(2) If B_n is infinite, show that $|B_n| \leq |\{x^a : a \in B_n\}|$. (Hint: Use that $A(x^a)$ is finite for each $a \in B_n$.)

(3) Assume B_n is uncountable for some $n \in \mathbb{N}$.

(a) Show there is an uncountable subset $C \subseteq B_n$ and a finite subset $F \subset A$ such that $A(x^a) \cap A(x^c) = F$ for $a, c \in C$ and $a \neq c$. (Hint: Use the Δ-system lemma of 3R(3).)

Let Δ be a countably infinite subset of C and $T = \cup\{A(x^a) : a \in \Delta\}$.

(b) Show that $\{x^a : a \in \Delta\}$ has an accumulation point, say x, in $S_{A\setminus T}(b)$, that (x,x) is an accumulation point of $\{(x^a,y^a) : a \in \Delta\}$, and that $(f(x),f(x))$ is an accumulation point of $\{f(x^a),f(y^a)) : a \in \Delta\}$. (Hint: Note that $A(y^a) \subseteq A(x^a)$.)

(c) Use the fact that $|f(x^a) - f(y^a)| \geqslant 1/n$ for all $a \in \Delta$ to obtain a contradiction. Now conclude that B_n is countable for all $n \in \mathbb{N}$.

Let $B = \cup\{B_n : n \in \mathbb{N}\}$. Note that by (3), B is countable.

(4) Let $x, y \in \Sigma'(b)$ such that $\{a \in A : x_a \neq y_a\} \cap B = \emptyset$.

(a) If $|\{a \in A : x_a \neq y_a\}| = 2$, show that $f(x) = f(y)$. (Hint: Let $c \in A$ such that $x_c \neq y_c$ and define $z \in \Sigma'(b)$ by $z_c = y_c$ and $z_a = x_a$ for all $a \neq c$. Use that $c \notin B$ to show that $f(x) = f(z)$ and $f(z) = f(y)$.)

(b) Show that $f(x) = f(y)$. (Hint: Use induction.)

For each $x \in \Sigma(b)$ and $T \subseteq A$, let x_T denote the unique element in $S_{A \setminus T}(b)$ such that $x_a = (x_T)_a$ for $a \in T$.

(5) For $x \in \Sigma'(b)$, prove that $f(x) = f(x_B)$. (Hint: Use (4)(b).)

(6) Let $x \in \Sigma(b)$ and U and V be basic open sets in Y such that $x \in U$ and $x_B \in V$.

(a) Prove there are finite sets $F, G \subset A$ such that $x_F \in U$ and $(x_B)_G \in V$.

(b) Let $H = F \cup G$. Prove $x_H \in U$, $(x_B)_H \in V$, and $f(x_H) = f((x_B)_H)$. (Hint: To prove $f(x_H) = f(x_B)_H$, first prove that $(x_B)_H = (x_H)_B$ and apply (5).)

(c) Show that $f(x) = f(x_B)$. (Hint: Assume that $f(x) \neq f(x_B)$ and use that $\mathbb{R}$ is Hausdorff to obtain a contradiction to (b).)

Define $g : \Pi_B[Y] \to \mathbb{R}$ by $g(x) = f(z)$ where z is any element of $\Sigma(b)$ such that $\Pi_B(z) = x$. By (6), g is well-defined.

(7) Show that g is continuous and $(g \circ \Pi_B) | \Sigma(b) = f$.

(8) Show that f has a continuous extension $g \circ \Pi_B : Y \to \mathbb{R}$. (Hint: Use (7).)

(9) Show that $\Sigma(b)$ is C-embedded in Y, that $\beta\Sigma(b)$

$\equiv_{\Sigma(b)} Y$, and that $\Sigma(b)$ is pseudocompact.

4S. <u>H-closed subspaces</u>. Let X be an H-closed space and $\mathcal{H}$ a chain of nonempty H-closed subspaces of X.

(1) For each $H \in \mathcal{H}$ and $p \notin H$, show there exists an open set $U(p,H)$ such that $p \notin c\ell_X U(p,H)$ and $H \subseteq c\ell_X(U(p,H) \cap H)$.

(2) Let $n \in \mathbb{N}$ and $H_i \in \mathcal{H}$ for $1 \leqslant i \leqslant n$. Suppose $p_i \in X \setminus H_i$ for $1 \leqslant i \leqslant n$ and $H_1 \subseteq H_2 \subseteq \ldots \subseteq H_n$. Use induction to show that $H_i \cap U(p_1,H_1) \cap U(p_2,H_2) \cap \ldots \cap U(p_i,H_i) \neq \varnothing$ for each $i \leqslant n$.

(3) Let $\mathcal{F}$ be the open filter generated by finite intersections of elements of $\{U(p,H) : H \in \mathcal{H}, p \notin H\}$. Prove that $a_X(\mathcal{F}) = \cap\mathcal{H}$ and since X is H-closed, conclude that $\cap\mathcal{H} \neq \varnothing$.

(4) Prove that in a space (not necessarily H-closed), the intersection of a chain of nonempty H-closed subspaces is nonempty.

(5) (Alexander's Subbase Theorem). Prove that a space is compact iff the intersection of any chain of nonempty closed sets is nonempty. (Hint: Use Zorn's Lemma.)

(6) Show that if every closed subspace of a space Z is H-closed, then Z is compact.

(7) Find a family $\mathcal{S}$ of nonempty H-closed subspaces of the space X described in 4.8(d) such that $\mathcal{S}$ has the finite

intersection property and $\bigcap S = \emptyset$.

4T. T_1 extensions. In this problem, the assumption that all spaces are Hausdorff is dropped. Recall that a space is T_1 iff every singleton is closed; in particular, Hausdorff spaces are T_1. Let $X = (0,1)$ (the open unit interval) and κ be a cardinal. Let $e_\kappa X$ denote $X \cup \kappa$ with the topology generated by the open base: $\{U \subseteq X \cup \kappa$: $U \cap X$ is open in X and if $U \cap \kappa \neq \emptyset$, then for some $n \in \mathbb{N}$, $(0,\frac{1}{n}) \subseteq U\}$.

(1) Prove that $e_\kappa X$ is a T_1 extension of X.

(2) If $\kappa \geqslant 2$, prove that $e_\kappa X$ is not Hausdorff.

(3) Prove that there is no T_1 extension Y of X such that $Y \geqslant Z$ for every T_1 extension Z of X.

4U. The Stone-Čech compactification of a discrete space

(1) Let D be an infinite discrete space. Prove that $|\beta D| = 2^m$ where $m = 2^{|D|}$. (Hint: By 3O(4), the compact space C_m has density $|D|$. So, there is a function $f \in F(D,C_m)$ such that $f[D]$ is dense in C_m. Use 4.2(b) to deduce that $|\beta D| \geqslant |C_m| = 2^m$. To prove the reverse inequality, note that βD is a collection of ultrafilters on D, i.e., $\beta D \subseteq \mathbb{P}(\mathbb{P}(D))$.)

(2) Let X be a separable Tychonoff space. Prove that

$|\beta X| \leq 2^{(2^{\aleph_0})}$. (Hint: Note that $d(\beta X) \leq d(X)$ and apply 2N(2)(a).)

(3) Let μ be an infinite cardinal and D a discrete space of cardinality μ. Prove that $S(\mathbb{P}(\mu))$ is homeomorphic to βD where $S(\mathbb{P}(\mu))$ is the Stone space of $\mathbb{P}(\mu)$. (Hint: Use the fact that βD is the collection of ultrafilters on D.)

(4) Let μ be an infinite cardinal and D a discrete space such that $|D| = \mu$. Prove that $S(\mathbb{P}(\mu)/FN)$ is homeomorphic to $\beta D \backslash D$. (Hint: Use 3L(4).)

4V. Strongly zero-dimensional vs. zero-dimensional. Let $J = \mathbb{R} \backslash \mathbb{Q}$. For $x \in J$, let $J_x = \{x + r : r \in \mathbb{Q}\}$ and $\mathcal{J} = \{J_x : x \in J\}$.

(1) Prove that $J_x \cap J_y \neq \emptyset$ iff $J_x = J_y$.

(2) Prove that J_x is dense in $\mathbb{R}$ and $\mathbb{R} \backslash J_x$ is dense in $\mathbb{R}$ for each $x \in J$.

(3) Prove $|\mathcal{J}| = 2^{\aleph_0}$. (Hint: Note that $|J_x| = \aleph_0$ and $\cup \mathcal{J} = J$.)

Re-index $\mathcal{J}$ by $\mathcal{J} = \{J_\alpha : \alpha < 2^{\aleph_0}\}$ so that $J_\alpha \cap J_\beta = \emptyset$ whenever $\alpha \neq \beta$. For $\alpha < \omega_1$, let $U_\alpha = \mathbb{R} \backslash \cup\{J_\beta : \alpha < \beta < \omega_1\}$, $X = \cup\{\{\alpha\} \times U_\alpha : \alpha < \omega_1\}$, and $Y = X \cup (\{\omega_1\} \times \mathbb{R})$. Give X and Y the subspace topology from $(\omega_1 + 1) \times \mathbb{R}$; so, X and Y are Tychonoff spaces. Let $f \in C(X)$.

(4) Show that X is zero-dimensional and dense in Y,

(5) Let $r \in \mathbb{R}$. Show that for some $\alpha < \omega_1$, $(\omega_1 \setminus \alpha) \times \{r\} \subseteq X$.

(6) For each $r \in \mathbb{R}$, show that there is some $\beta < \omega_1$ such that f is constant on $(\omega_1 \setminus \beta) \times \{r\}$. (Hint: See 2.6(q)(5) and its proof.) Denote this constant value on $(\omega_1 \setminus \beta) \times \{r\}$ by k_r. Define $F \in F(Y,\mathbb{R})$ by setting $F(\omega_1,r) = k_r$ and $F|X = f$.

(7) Show that there exists $\alpha_0 < \omega_1$ such that for each $q \in \mathbb{Q}$, F is constant on $((\omega_1+1)\setminus\alpha_0) \times \{q\}$.

(8) If $r \in \mathbb{R}$, $\{r_n : n \in \mathbb{N}\} \subseteq \mathbb{Q}$ such that $(r_n) \to r$, and $\alpha_0 < \beta < \delta < \omega_1$ where $(\beta,r) \in X$, then show that $f(\beta,r) = f(\delta,r) = F(\omega_1,r)$.

(9) Let $(\beta,r) \in X$ where $\beta > \alpha_0$ and $r \in \mathbb{R}$. Let U be an open neighborhood of $F(\omega_1,r)$ in $\mathbb{R}$ and let V be an open neighborhood of $F(\omega_1,r)$ such that $c\ell_{\mathbb{R}}V \subseteq U$. Since f is continuous at (β,r), there is some $\epsilon > 0$ and $\gamma \in [\alpha_0,\beta]$ such that $f[[\gamma,\beta] \times (r-\epsilon,r+\epsilon) \cap X] \subseteq V$. Show that $F[[\alpha_0,\omega_1] \times (r-\epsilon,r+\epsilon) \cap Y] \subseteq U$. In particular, conclude that F is continuous and that X is C-embedded in Y.

(10) Prove that X is not strongly zero-dimensional. (Hint: Use 4V(9) and 4.5(p)(3) to obtain that $X \subseteq Y \subseteq \beta Y \equiv_X \beta X$. As Y contains a copy of $\mathbb{R}$, show that βX is not zero-dimensional. Apply 4.7(g,h) to deduce that X is not strongly zero-dimensional.)

In this problem, we have constructed a zero-dimensional

space X which is not strongly zero-dimensional. Thus, by 4.7(g), $\beta X > \beta_0 X$.

4W. <u>Fan-Gottesman</u> <u>compactifications</u>. Let X be a regular space. A FG-base $\mathcal{B}$ is a base of open sets satisfying:

(FG1) $\emptyset, X \in \mathcal{B}$,

(FG2) if $U, V \in \mathcal{B}$, then $U \cap V \in \mathcal{B}$,

(FG3) if $U \in \mathcal{B}$, then $X \backslash c\ell_X U \in \mathcal{B}$, and

(FG4) for each open set U in X and $V \in \mathcal{B}$ such that $c\ell_X V \subseteq U$, there is a set $W \in \mathcal{B}$ such that $c\ell_X V \subseteq W \subseteq c\ell_X W \subseteq U$.

(1) If $U, V \in \mathcal{B}$, W is an open set, and $c\ell_X U \cap c\ell_X V \subseteq W$, prove that there is some $R \in \mathcal{B}$ such that $c\ell_X U \cap c\ell_X V \subseteq R \subseteq c\ell_X R \subseteq W$.

A nonempty family $\mathcal{F} \subseteq \mathcal{B}$ is a **binding family** if for $F_1, \ldots, F_n \subseteq \mathcal{F}$, $c\ell_X F_1 \cap \ldots \cap c\ell_X F_n \neq \emptyset$. Let $b_{\mathcal{B}} X = \{\mathcal{F} \subseteq \mathcal{B} : \mathcal{F}$ is a maximal binding family$\}$. For $U \in \mathcal{B}$, let $S(U) = \{\mathcal{F} \in b_{\mathcal{B}} X :$ for some $V \in \mathcal{F}$, $c\ell_X V \subseteq U\}$.

(2) Prove that $S(\emptyset) = \emptyset$, $S(X) = b_{\mathcal{B}} X$, and $S(U \cap V) = S(U) \cap S(V)$ for $U, V \in \mathcal{B}$.

(3) If $\mathcal{U} \in b_{\mathcal{B}} X$, $U \in \mathcal{B}$, and $U \notin \mathcal{U}$, then show that there is some $V \in \mathcal{U}$ such that $c\ell_X V \cap c\ell_X U = \emptyset$; in particular, $\mathcal{U} \in S(X \backslash c\ell_X U)$. (Hint: Use (1) plus induction.)

By (2), $\{S(U) : U \in \mathcal{B}\}$ is a base for a topology on $b_{\mathcal{B}}X$.

(4) Prove that $b_{\mathcal{B}}X$ is Hausdorff.

(5) If $U \in \mathcal{B}$ and $\mathcal{U} \in b_{\mathcal{B}}X$, prove that $U \in \mathcal{U}$ iff $\mathcal{U} \in c\ell S(U)$ (where closure is in $b_{\mathcal{B}}X$). Hence conclude that $b_{\mathcal{B}}X \setminus c\ell_X S(U) = S(X \setminus c\ell_X U)$.

(6) If $U_1, \ldots, U_n \in \mathcal{B}$ and $c\ell S(U_1) \cap \ldots \cap c\ell S(U_n) \neq \varnothing$, then prove that $c\ell_X U_1 \cap \ldots \cap c\ell_X U_n \neq \varnothing$.

(7) Prove that $b_{\mathcal{B}}X$ is compact. (Hint: Use (6) and (5).)

(8) For $x \in X$, prove that $\mathcal{U}_x = \{U \in \mathcal{B} : x \in c\ell_X U\}$ is a maximal binding family.

(9) Define $e \in F(X, b_{\mathcal{B}}X)$ by $e(x) = \mathcal{U}_x$ for $x \in X$. Prove that e is a dense embedding. (Hint: Show that e is one-to-one, e[X] is dense in $b_{\mathcal{B}}X$, and that $e^{\leftarrow}[S(U)] = U$ for $U \in \mathcal{B}$.)

One easy consequence of (9) is that a regular space with a base satisfying FG1, FG2, FG3, FG4 is Tychonoff. A compactification $Y \in \mathcal{K}(X)$ is called a **Fan-Gottesman compactification** if there is some FG-base $\mathcal{B}$ such that $Y \equiv_X b_{\mathcal{B}}X$.

(10) Let $\mathcal{B}$ and $\mathcal{C}$ be FG-bases for a space X such that $\mathcal{B} \subseteq \mathcal{C}$. Prove $b_{\mathcal{C}}X \geq b_{\mathcal{B}}X$. (Hint: If $\mathcal{V} \in b_{\mathcal{C}}X$, show that $\mathcal{V} \cap \mathcal{B}$ is a $\mathcal{B}$-maximal binding family. Define $f \in F(b_{\mathcal{C}}X, b_{\mathcal{B}}X)$ by $f(\mathcal{V}) = \mathcal{V} \cap \mathcal{B}$ for $\mathcal{V} \in b_{\mathcal{C}}X$. If $x \in$

X show that $f(e_{\mathcal{C}}(X)) = e_{\mathcal{B}}(X)$, where "$e_{\mathcal{C}}$" (respectively, "$e_{\mathcal{B}}$") is the "e" defined in (9) relative to $\mathcal{C}$ (respectively, $\mathcal{B}$). If $U \in \mathcal{B}$ and $\mathcal{V} \in b_{\mathcal{C}}X$ such that $f(\mathcal{V}) \in S_{\mathcal{B}}(U)$ ($S_{\mathcal{B}}(U)$ is $S(U)$ relative to $\mathcal{B}$), show there is some $V \in \mathcal{V} \cap \mathcal{B}$ such that $c\ell_X V \subseteq U$. By (FG4), there is some $W \in \mathcal{B}$ such that $c\ell_X V \subseteq W \subseteq c\ell_X W \subseteq U$. Show that f is continuous by proving that $\mathcal{V} \in S_{\mathcal{C}}(W)$ and $f[S_{\mathcal{C}}(W)] \subseteq S_{\mathcal{B}}(U)$).

(11) Let X be a locally compact, noncompact space and $\mathcal{B} = \{U \subseteq X : U$ is open and either $c\ell_X U$ or $X \backslash U$ is compact$\}$. Prove that $\mathcal{B}$ is a FG-base for X and $b_{\mathcal{B}}X$ is the one-point compactification of X.

4X. <u>The Freudenthal compactification</u>. Let X be a Tychonoff space. Some of the notation and definitions used in this problem are contained in 4W.

(1) If $Y \in \mathcal{K}(X)$ and $W \subseteq Y$ is an open set such that $bd_Y W \subseteq X$, then prove that $bd_X(W \cap X)$ is compact.

An extension Y of a space X has a **relatively zero-dimensional remainder** if $\{W \subseteq Y : W$ is open and $bd_Y W \subseteq X\}$ is a base for Y. A space X is **rimcompact** if X has a base with compact boundaries.

(2) If $Y \in \mathcal{K}(X)$ and Y has a relatively zero-dimensional remainder, prove that X is rimcompact.

(3) Suppose $Y \in \mathcal{K}(X)$ has a relatively zero-dimensional remainder and $\mathcal{B} = \{W \cap X : W \text{ is open in } Y \text{ and } bd_Y W \subseteq X\}$. Prove that $\mathcal{B}$ is a FG-base and that $Y \equiv_X b_{\mathcal{B}}X$. (Hint: For $y \in Y$, let $\mathcal{W}_y = \{W \cap X : W \text{ open in } Y, bd_Y W \subseteq X, \text{ and } y \in c\ell_Y W\}$. Show that $\mathcal{W}_y$ is a $\mathcal{B}$-maximal binding family and if $\mathcal{U}$ is a $\mathcal{B}$-maximal binding family, then $\mathcal{U} = \mathcal{W}_y$ for some $y \in Y$.)

(4) If X is rimcompact and $\mathcal{B} = \{U : U \text{ is open and } bd_X U \text{ is compact}\}$, prove that $\mathcal{B}$ is a FG-base for X and $b_{\mathcal{B}}X$ has a relatively zero-dimensional remainder. (Hint: Show that $bd_X S(U) \subseteq X$ for each $U \in \mathcal{B}$.)

If $\mathcal{B}$ is defined as in (4), then the FG-compactification $b_{\mathcal{B}}X$ is called the **Freudenthal compactification** of X and is denoted as FX.

(5) If X is rimcompact, prove that FX is the projective maximum of the set of compactifications with relatively zero-dimensional remainders. (Hint: Use (3), (4) and 4W(10).)

4Y. <u>The Stone-Čech compactification via CR-filters.</u> Let X be a Tychonoff space. For $A, B \subseteq X$, let $A << B$ denote the statement that there is a function $f \in C^*(X)$ such that $f[c\ell_X A] \subseteq \{0\}$, $f[c\ell_X(X \setminus B)] \subseteq \{1\}$, and $f[X] \subseteq [0,1]$. Note that by 1.10(d) and (f), $A << B$ iff there is a family of open sets $\{U_r : r \in \mathbb{Q}\}$ such that $c\ell_X A \subseteq U_r \subseteq c\ell_X U_r \subseteq U_s \subseteq$

$c\ell_X U_s \subseteq \text{int}_X B$ whenever $r < s$. An open filter F on X is a **completely regular filter** (abbreviated as **CR-filter**) if for each $U \in F$, there is an open set $V \in F$ such that $V \ll U$. Let $Y = \{\mathcal{U} : \mathcal{U}$ is a maximal CR-filter on X$\}$. For each open set U in X, let $c(U) = \{\mathcal{U} \in Y : U \in \mathcal{U}\}$. Let U, V be open sets of X.

(1) If $V \ll U$, prove that $X \backslash c\ell_X U \ll X \backslash c\ell_X V$ and that there is an open set W of X such that $V \ll W$ and $W \ll U$.

(2) If $V \ll U$ and $R \ll S$ where R, S are open sets in X, prove that $V \cap R \ll U \cap S$.

(3) If $V \ll U$ and $\{W_r : r \in \mathbb{Q}\}$ is a family of open sets such that $c\ell_X V \subseteq W_r \subseteq c\ell_X W_r \subseteq W_s \subseteq c\ell_X W_s \subseteq X \backslash U$ whenever $r < s$, then prove that $W_r \ll W_s$ whenever $r < s$.

(4) Prove that $\{c(U) : U$ open in X$\}$ forms a base for a topology on Y.

(5) Prove that a CR-filter F is a maximal CR-filter iff for each pair of open sets U and V such that $V \ll U$, either $U \in F$ or there exists $F \in F$ such that $V \cap F = \emptyset$. (Hint: If F is maximal, and if $V \cap F \neq \emptyset$ for each $F \in F$, use (1) and (2) to show that if G is an open filter generated by $\{W \cap F : W$ is open, $V \ll W$, and $F \in F\}$, then G is a CR-filter. Conclude that $F \subseteq G$ and use the maximality of F to deduce that $U \in F$.)

Note that Y is Hausdorff, and that if $U \ll V$ then $c\ell_Y c(U) \subseteq c(V)$.)

(6) For $x \in X$, let $N(x)$ be the set of open neighborhoods of x. Prove that $N(x)$ is a maximal CR-filter.

(7) Define $e \in F(X,Y)$ by $e(x) = N(x)$ for $x \in X$. Prove that e is a dense embedding. (Hint: First, prove that e is one-to-one and e[X] is dense. Then prove that $e^{\leftarrow}[c(U)] = U$ for each open set $U \subseteq X$.)

(8) Prove that Y is compact. (Hint: Assume that F is a closed filter on Y such that $\bigcap F = \varnothing$. Using the remark at the end of (5), show that for each $U \in Y$ there exists $U_U \in U$ and $F_U \in F$ such that $F_U \cap c\ell_Y c(U_U) = \varnothing$. Choose $W_U \in U$ such that $W_U \ll U_U$. Let $G = \{V \in \tau(X) :$ there exists $F \in F$ and $R \in \tau(X)$ such that $F \subseteq c(R)$ and $R \ll V\}$. Use (2) and (3) to prove that G is a CR-filter. Thus G is contained in some maximal CR-filter W. Find $R \in \tau(X)$ such that $F_W \subseteq c(R)$ and $R \cap U_W = \varnothing$. Infer that both W_W and $X \setminus c\ell_X W_W$ belong to W, which is a contradiction.)

(9) Prove that $Y \equiv_X \beta X$. (Hint: Use 4.6(g)(5). If Z_1 and Z_2 are disjoint zero-sets of X, find open sets U, V, W such that $Z_1 \ll U \ll V \ll W \ll X \setminus Z_2$. Show that $e[Z_1] \subseteq c(U)$, $e[Z_2] \subseteq c(X \setminus c\ell_X W)$, and $c\ell_Y c(U) \cap c\ell_Y c(X \setminus c\ell_X W) = \varnothing$.)

4Z. Proximities. The reader is advised to solve 4Y before attempting this problem. Let X be a set, and let $\ll$ be a binary relation on $\mathbb{P}(X)$ which satisfies, for $A, B, C \subseteq X$, the following:

(P1) $\varnothing \ll \varnothing$,

(P2) $A \ll B$ implies $A \subseteq B$,

(P3) $A \ll B$ implies $X \setminus B \ll X \setminus A$,

(P4) $A \ll (B \cap C)$ iff $A \ll B$ and $A \ll C$,

(P5) $A \ll B$ implies for some $D \subseteq X$, $A \ll D$ and $D \ll B$, and

(P6) if $x, y \in X$ and $x \neq y$, then $\{x\} \ll X \setminus \{y\}$.

A binary relation on X satisfying (P1) – (P6) is called a **proximity** on X and $(X, \ll)$ is called a **proximity space**.

(The usual definition of a proximity on a set X is that it is a binary relation δ on $\mathbb{P}(X)$ which satisfies, for $A, B, C \subseteq X$, the following axioms.

(Q1) $\varnothing \not\delta A$

(Q2) $\{a\} \delta \{a\}$ for each $a \in X$

(Q3) $A \delta B$ implies $B \delta A$

(Q4) $A \delta (B \cup C)$ iff $A \delta B$ or $A \delta C$

(Q5) $A \not\delta B$ implies there exist subsets S, T of X such that $S \cap T = \varnothing$, $A \not\delta (X \setminus S)$, and $B \not\delta (X \setminus T)$.

(Q6) $\{a\} \delta \{b\}$ implies $a = b$.

If we define a binary relation δ on $\mathbb{P}(X)$ be saying that A $\ll$ B iff A δ $(X\setminus B)$, then $\ll$ satisfies (P1) — (P6) iff δ satisfies (Q1) — (Q6). Thus our definition of a proximity is equivalent to the usual one. We have chosen our definition because it leads more naturally to the definition of "p-filters". Such filters are used in constructing "proximal compactifications" of a proximity space (see 4AA).)

(1) If $A \subseteq B \ll C \subseteq D$, prove that $A \ll D$.

(2) Prove that $\emptyset \ll A$ for all $A \subseteq X$, and hence $X \ll X$.

(3) If $A_i \ll B_i$ for $i = 1, \ldots, n$, prove that $\cap\{A_i : i = 1, \ldots, n\} \ll \cap\{B_i : i = 1, \ldots, n\}$ and $\cup\{A_i : i = 1, \ldots, n\} \ll \cup\{B_i : i = 1, \ldots, n\}$.

Let $\tau(\ll) = \{U \subseteq X : x \in U \text{ implies } \{x\} \ll U\}$.

(4) Prove that $\tau(\ll)$ is a topology on X. Then show that if $F \subseteq X$, $\mathrm{int}_X F = \{x \in X : \{x\} \ll F\}$.

The topology $\tau(\ll)$ on X is called a **proximity topology** and the **topology generated** by $\ll$.

(5) Let $Y \subseteq X$. For $A, B \subseteq Y$ define $A \ll_Y B$ to mean that $A \ll X\setminus(Y\setminus B)$. Show that the binary relation $\ll_Y$ defined on $\mathbb{P}(Y)$ satisfies (P1) — (P6), and that the subspace topology induced on Y by $\tau(\ll)$ is the same as the topology $\tau(\ll_Y)$.

(6) Prove that $A << B$ iff $c\ell_X A << \text{int}_X B$. (Hint: One direction follows from (1). If $A << B$, use (P5) to obtain $D, E \subseteq X$ such that $A << D$, $D << E$, and $E << B$. Now show that $c\ell_X A \subseteq D$, $E \subseteq \text{int}_X B$, and apply (1).)

(7) If $A << B$, show that A and $X \backslash B$ are completely separated in X. (Hint: Use (6) and (P5) to obtain a separating chain; now, apply 1.10(f).)

(8) Prove that X is a Tychonoff space.

(9) Let Z be a compact space. Define a binary relation $<<$ on $\mathbb{P}(Z)$ as follows: $A << B$ if A and $Z \backslash B$ are completely separated in Z. Prove that $<<$ is a proximity on Z and that $\tau(Z) = \tau(<<)$.

(10) Prove there is a unique proximity $<<$ on Z such that $\tau(Z) = \tau(<<)$. (Hint: By (9) there is at least one such proximity. Let $[[$ be another proximity on Z for which $\tau(Z) = \tau([[)$. To show $[[= <<$ (defined in (9)), by (7) it suffices to show that if A and $Z \backslash B$ can be completely separated in Z, then $A \ [[\ B$. Show that if A and $Z \backslash B$ are completely separated in Z, then $c\ell_Z A \subseteq \text{int}_Z B$. As $\text{int}_Z B \in \tau([[)$, conclude that for each $p \in c\ell_Z A$, $\{p\} \ [[\ \text{int}_Z B \subseteq B$.

Find $U_p \in \tau(Z)$ such that $\{p\} \ [[\ U_p \ [[\ B$. Use (3) to obtain $U \in \tau(Z)$ such that $c\ell_Z A \subseteq U \ [[\ \text{int}_Z B$. Use (1) to conclude that $A \ [[\ B$.)

Note: Let Z be a compactification of a Tychonoff space Y. Let $<<^Z$ be the unique proximity on Z such that $\tau(Z) = \tau(<<^Z)$. By (5) the subspace proximity $(<<^Z)_Y$, denoted as $<<_Y^Z$, has the property that $\tau(Y) = \tau(<<_Y^Z)$. In particular, note that if A, B $\subseteq$ Y and A and Y $\setminus$ B are completely separated in Z, then A $<<_Y^Z$B. Also, note that A $<<_Y^{\beta Y}$B iff A and Y $\setminus$ B are completely separated in Y (see 1.9(f) and 4.6(g)).

(11) If S and T are compactifications of Y, prove that S $\geqslant$ T iff $<<_T \subseteq <<_S$ ($<<_T \subseteq <<_S$ means that A $<<_T$B implies A $<<_S$B). (Hint: Use 4.2(h).)

4AA. More proximities. Much of the notation used in this problem was introduced in 4Z. Let X be a Tychonoff space, and let $<<$ be a proximity on X such that $\tau(X) = \tau(<<)$. A filter F on X is called a **p-filter** if for each A $\in$ F, there is some B $\in$ F such that B $<<$ A. Maximal p-filters are called **p-ultrafilters**.

(1) Let F be a p-filter. Show that F is a p-ultrafilter iff A $<<$ B implies X$\setminus$A $\in$ F or B $\in$ F (see 4Y(5)).

(2) For x $\in$ X, let $N(x)$ denote the set of neighborhoods of x. Show that $N(x)$ is a p-ultrafilter.

If G is a filter, let $\tilde{G}$ = {A $\subseteq$ X : A $>>$ G for some G $\in$ G}.

(3) Prove that $\tilde{G}$ is a p-filter.

(4) If U is an ultrafilter on X, prove that U contains an unique p-ultrafilter, namely, $\tilde{U}$.

(5) Prove that each p-filter is contained in some

p-ultrafilter.

Let cX be the set of all p-ultrafilters on X. For $A \subseteq X$, let $o(A) = \{\mathcal{U} \in cX : A \in \mathcal{U}\}$. Define $\lhd$ on cX by $\mathcal{A} \lhd \mathcal{B}$ iff there are sets A, $B \subseteq X$ such that $A << B$, $\mathcal{A} \subseteq o(A)$, and $cX \setminus \mathcal{B} \subseteq o(X \setminus B)$.

(6) Prove that $\lhd$ is a proximity on cX.

(7) Define $\lambda : X \to cX$ by $\lambda(x) = \mathcal{N}(x)$. Prove that λ is a dense embedding (here cX is given the topology $\tau(\lhd)$; see 4Z(4)).

(8) If A, $B \subseteq X$ prove that $A << B$ iff $\lambda(A) \lhd_{\lambda[X]} \lambda(B)$ (see 4Z(5) for the definition of the subspace proximity).

(9) Prove that $\tau(\lhd)$ is a compact topology on cX. (Hint: If $\mathcal{U}$ is an open ultrafilter on cX, prove that the filter $\mathcal{V}$ generated by $\{\lambda^{\leftarrow}[U] : U \in \mathcal{U}\}$ is a p-filter on X and hence contained in some p-ultrafilter $\mathcal{W}$. Show that $\mathcal{U}$ converges to $\mathcal{W}$. Conclude that cX is H-closed. Use 4Z(8) and 4.8(c) to show that cX is compact.)

(10) If Z is a compactification of X, $<<^Z$ is the proximity introduced after 4Z(10), and cX is the compactification of X constructed in (6) — (9) above using the proximity $<<_X^Z$ (see 4Z(5)), prove that $Z \equiv_X cX$ (Use 4Z(11)). Thus conclude that there is a one-to-one correspondence between $\mathcal{K}(X)$ and the set $\rho(X) = \{<< : <<$ is a proximity on X such that $\tau(<<) = \tau(X)\}$.

(11) Prove that the function $\Psi : \mathcal{K}(X) \to \rho(X)$ defined by

$\Psi(Z) = <<_X^Z$ is a bijection.

(12) If X is infinite, prove that $|K(X)| \leq 2^{(2^{|X|})}$.

(13) If D is an infinite, discrete space, prove that $|K(D)| = 2^{(2^{|D|})}$. (Hint: By 4U, $|\beta D \setminus D| = 2^{(2^{|D|})}$. Fix $p \in \beta D \setminus D$ and for each $q \in \beta D \setminus (D \cup \{p\})$, let $P_q = \{p,q\} \cup \{\{r\} : r \in \beta D \setminus \{p,q\}\}$. Show that $P_q \in \mathcal{D}_X$ ($\mathcal{D}_X$ is defined after 4Q(3)). Let Y_q denote the compactification of X induced by P_q. Use 4Q(5) to deduce that $\{Y_q : q \in \beta D \setminus (D \cup \{p\})\}$ is a family of non-equivalent compactifications of X. Thus $|K(X)| \geq |\beta D \setminus (D \cup \{p\})| = 2^{(2^{|D|})}$. Apply (12) to obtain the reverse inequality.)

4AB. Non-equivalent extensions may be homeomorphic. Let C denote the Cantor space (cf. 3.3 and 3K) and $X = C \times \mathbb{N}$. Let $\hat{X}$ and $\hat{\mathbb{N}}$ be the one-point compactifications of X and $\mathbb{N}$, respectively. Prove that $\hat{X} \not\equiv_X C \times \hat{\mathbb{N}}$ but that $\hat{X}$ and $C \times \hat{\mathbb{N}}$ are homeomorphic. (Hint: To prove the latter statement, use 3.3(e).)

4AC. The maximal ideal space of C(X). Let X be a Tychonoff space.

(1) If M is a maximal ideal in C(X), prove that $Z(M)$ (= $\{Z(f) : f \in M\}$) is a z-ultrafilter on X.

(2) If $\mathcal{U}$ is a z-ultrafilter on X, prove that $Z^{\leftarrow}[\mathcal{U}]$ (= $\{f \in C(X) : Z(f) \in \mathcal{U}\}$) is a maximal ideal in C(X).

Let $M(C(X))$ denote the set of all maximal ideals in $C(X)$. Define a function $\Psi : M(C(X)) \to w_{Z(X)}X$ by $\Psi(M) = Z(M)$.

(3) Prove that Ψ is a bijection.

(4) If $x \in X$, prove that $\Psi^{\leftarrow}(A^x) = \{f \in C(X) : f(x) = 0\}$.

For $p \in \beta X$, let M^p denote $Z^{\leftarrow}[A^p]$. So, by (1), (2), and (3), $M(C(X)) = \{M^p : p \in \beta X\}$. For $f \in C(X)$, let $S(f) = \{M^p : f \in M^p\}$.

(5) If $f \in C(X)$, prove that $\Psi[S(f)] = S(Z(f))$. ($S(Z(f))$ is defined in 4.4(a).)

By (5) and 4.4(c), $\{S(f) : f \in C(X)\}$ is a closed base for the closed sets of a topology on $M(C(X))$. Define $\lambda : X \to M(C(X))$ by $\lambda(x) = M^x$.

(6) Prove that λ is a dense embedding and $M(C(X))$ is a compact space. (Hint: Use the fact that Ψ is a homeomorphism.)

(7) Prove that $M(C(X)) \equiv_X \beta X$.

4AD. <u>Countably compact separable spaces and feebly compact spaces with countable π-weight</u>.

(1) $[P(\kappa^+)]$. Let X be a countably compact separable space. Show that each open cover of X of cardinality κ has a

finite subfamily whose union is dense in X. (Hint: Let $D = \{x_n : n \in \omega\}$ be a dense subset of X. Assume that $\{U_\alpha : \alpha < \kappa\}$ is an open cover of X without a finite subfamily whose union is dense in X. For each $\alpha < \kappa$, let $A(\alpha) = \{n \in \omega : x_n \in U_\alpha\}$, and for $F \subseteq \kappa$, let $A(F) = \bigcup\{A(\alpha) : \alpha \in F\}$. Apply $P(\kappa^+)$ to $\{\omega \setminus A(F) : F$ is a finite subset of $\kappa\}$ to obtain an infinite subset $S \subseteq \omega$ such that $S \cap A(\alpha)$ is finite for each $\alpha < \kappa$. Use countable compactness to obtain a point $p \in c\ell_X\{x_n : n \in S\}\setminus\{x_n : n \in S\}$. If $p \in U_\alpha$, show that $S \cap A(\alpha)$ is infinite, a contradiction.)

(2) $[P(\kappa^+)]$. Show that each countably compact, separable space of cardinality κ is H-closed.

(3) $[P(2^{\aleph_0})]$. Show that each regular, countably compact, separable space of cardinality less than $2^{\aleph_0}$ is compact.

(4) $[P(\kappa^+)]$. Show that each feebly compact space of cardinality κ with countable π-weight is H-closed.

(5) $[P(2^{\aleph_0})]$. Show that each regular, feebly compact space of cardinality less than $2^{\aleph_0}$ with countable π-weight is compact.

(6) $[P(2^{\aleph_0})]$. Show that a pseudocompact, Tychonoff space of cardinality less than $2^{\aleph_0}$ with countable π-weight is compact. Infer than any maximal almost disjoint family of subsets of $\mathbb{N}$ must have cardinality $2^{\aleph_0}$ (see 1N).

4AE. Martin's axiom and countably compact, perfect spaces. In this problem we show that if we assume MA($\aleph_1$), then every countably compact, perfect, regular space is compact. The techniques used generalize those used to prove 3.5(l). One can obtain slightly stronger results than those presented here; see the Notes.

(1) MA(κ). Let X be a space with *ccc* and let κ be a regular uncountable cardinal. If $\mathcal{U}$ is a family of κ open subsets of X, show that there is a subfamily $\mathcal{G}$ of $\mathcal{U}$ such that $|\mathcal{G}| = \kappa$ and $\mathcal{G}$ has the finite intersection property. (Hint: Let $\mathcal{U} = (U_\alpha)_{\alpha<\kappa}$ and let $V_\alpha = \cup\{\mathrm{int}_X c\ell_X U_\gamma : \alpha \leqslant \gamma < \kappa\}$. Arguing as in the proof of 3.5(j), show there exists $\alpha_0 < \kappa$ such that $\alpha_0 \leqslant \alpha < \kappa$ implies $c\ell_X V_\alpha = c\ell_X V_{\alpha_0}$. Let $Y = c\ell_X V_{\alpha_0}$ and form the Stone space $S(RO(Y))$. Show that since X is *ccc*, it follows that $S(RO(Y))$ is *ccc*. For each $\alpha \in (\alpha_0,\kappa)$ let $V_\alpha^* = \cup\{\lambda(\mathrm{int}_Y c\ell_Y U_\delta) : \delta \in [\alpha,\kappa)\}$. Show that V_α^* is a dense open subset of $S(RO(Y))$. Apply MA(κ) to find an $RO(Y)$-ultrafilter $\mathcal{V} \in \cap\{V_\alpha^* : \alpha_0 \leqslant \alpha < \kappa\}$. Use the fact that $\mathcal{V}$ is closed under finite intersections to show that there is a subfamily of $\{\mathrm{int}_Y c\ell_Y U_\alpha : \alpha_0 \leqslant \alpha < \kappa\}$ of cardinality κ that has the finite intersection property. Now consider the corresponding subfamily of $\mathcal{U}$.)

(2) Show that a countably compact perfect space contains no uncountable discrete subspaces. (Hint: If $D = \{x_\alpha$

$: \alpha < \omega_1\}$ were discrete, for each $\alpha < \omega_1$ find an open set U_α such that $U_\alpha \cap D = \{x_\alpha\}$. Write $\bigcup\{U_\alpha : \alpha < \omega_1\}$ as a union of countably many closed sets (possible as X is perfect) and show that one of them contains an uncountable closed discrete subset of X, contradicting the countable compactness of X.)

(3) $[MA(\aleph_1)]$. Prove that a countably compact, regular, perfect space is hereditarily separable. (Hint: Let X be countably compact, regular, and perfect and assume that X is not hereditarily separable. Show that there exists $\{y_\alpha : \alpha < \omega_1\} \subseteq X$ such that $y_\alpha \notin c\ell_X\{y_\delta : \delta < \alpha\} = Y_\alpha$ for each $\alpha < \omega_1$. Let $Y = \bigcup\{c\ell_X\{y_\delta : \delta < \alpha\} : \alpha < \omega_1\}$. Use 3.5(l)(1) to show that Y is closed in X, and hence countably compact, regular, and perfect. For each $\alpha < \omega_1$ find an open subset U_α of Y such that $y_\alpha \in U_\alpha$ and $Y_\alpha \cap c\ell_Y U_\alpha = \emptyset$. Use 3.5(l)(2) to show that Y is *ccc*. Apply (1) to find an uncountable subset S of ω_1 such that $\{U_\alpha : \alpha \in S\}$ has the finite intersection property. If $\beta \in S$, let $F_\beta = \bigcap\{c\ell_Y U_\alpha : \alpha \leqslant \beta \text{ and } \alpha \in S\}$. Show that $F_\beta \neq \emptyset$. Inductively choose $\{x_\mu : \mu < \omega_1\} \subseteq Y$ and $\{\beta_\mu : \mu < \omega_1\} \subseteq S$ such that $\{x_\mu : \mu < \delta\} \subseteq Y_{\beta_\delta}$ and $x_\delta \in F_{\beta_\delta}$ for each $\delta < \omega_1$. Prove that $\{x_\mu : \mu < \omega_1\}$ is discrete, which contradicts (2).)

(4) $(MA[\aleph_1])$. Prove that a countably compact, regular, perfect space is compact. (Hint: It suffices to prove that X is Lindelöf. If X is not Lindelöf, find $\{y_\alpha : \alpha$

$< \omega_1\} = Y \subseteq X$ such that for each $\beta < \omega_1$, $\{y_\alpha : \alpha < \beta\} = Y_\beta$ is open in Y. For each $\beta < \omega_1$, find open subsets U_β and V_β of X such that $Y \cap U_\beta = Y_{\beta+1}$ and $y_\beta \in V_\beta \subseteq c\ell_X V_\beta \subseteq U_\beta$. By writing $V = \cup\{V_\beta : \beta < \omega_1\}$ as a countable union of closed sets of X, find a closed subset F of X such that $F \subseteq V$ and $F \cap Y$ is uncountable. Let $H = c\ell_X(Y \cap F)$. By (3) H is separable; use 4AD(1) to obtain a contradiction.)

Remark: 4AE(4) can be proved using only the axiom $P(\aleph_2)$ rather than the strictly stronger assumption $MA(\aleph_1)$. See the Notes.

4AF. Pseudocompact products. Let X and Y be spaces. A function $f : X \to Y$ is called **z-closed** if for every zero set $Z \subseteq X$, $f[Z]$ is closed. Recall that $\Pi_X : X \times Y \to X$ is used to denote the projection function (defined in 1.1(f)).

(1) Let Z be a zero set in $X \times Y$ and $p \in X \backslash \Pi_X[Z]$.

(a) Show that there is a function $h \in C^*(X \times Y)$ such that $\{p\} \times Y \subseteq h^{\leftarrow}(1)$ and $Z = Z(h)$. (Hint: Since $Z = Z(f)$ for some $f \in C^*(X \times Y)$, consider $h(x,y) = (f(p,y))^{-1}(f(x,y))$ where $(x,y) \in X \times Y$.)

(b) Assume $\Pi_X[Z]$ is not closed in X, and let $p \in c\ell_X \Pi_X[Z] \backslash \Pi_X[Z]$. Using the h in (1)(a), show that there are open sets U_n, V_n in X and W_n in Y for $n \in \mathbb{N}$ such that for $m \in \mathbb{N}$, (i) $p \in U_m$, (ii) $(V_m \times W_m) \cap Z \neq \emptyset$, (iii) $h[V_m \times W_m] \subseteq [0,1/3)$, (iv) $h[U_m \times W_m] \subseteq (2/3,1]$, and (v) $U_{m+1} \cup V_{m+1} \subseteq U_m$. (Hint: First, pick $(x_1,y_1) \in Z$ and open neighborhoods U_1 of p, V_1 of x_1, and W_1 of y_1 such that $h[V_1 \times W_1] \subseteq [0,1/3)$ and $h[U_1 \times W_1] \subseteq (2/3,1]$. Since $U_1 \cap \Pi_X[Z] \neq \emptyset$, there is some $(x_2,y_2) \in Z$ such that $x_2 \in U_1$. Find open neighborhoods U_2 of p, V_2 of x_2, and W_2 of y_2 such that $h[V_2 \times W_2] \subseteq [0,1/3)$, $h[U_2 \times W_2] \subseteq (2/3,1]$, and $U_2 \cup V_2 \subseteq U_1$. Continue by induction.)

(2) If $X \times Y$ is feebly compact, prove that

(a) X and Y are feebly compact and

(b) Π_X is z-closed.

(Hint: Assume Π_X is not z-closed and obtain the sets Z(h), U_n, V_n in X, and W_n in Y as in (1) above. Show that $\{V_n \times W_n : n \in \mathbb{N}\}$ is a family of nonempty pairwise disjoint open sets. Use 1.11(b) to obtain a point $(q,r) \in X \times Y$ with the property that for every neighborhood $R \times T$ of (q,r), $A = \{n \in \mathbb{N} : (V_n \times W_n) \cap (R \times T) \neq \emptyset\}$ is infinite. Show that $h(q,r) \leq 1/3$. If n and $n + k \in A$ where $n, k \in \mathbb{N}$, then show that V_{n+k}

$\subseteq U_{n+k-1} \subseteq \ldots \subseteq U_n$. Since $(R \times T) \cap (V_{n+k} \times W_{n+k}) \neq \emptyset$, show that $(R \times T) \cap (U_n \times W_n) \neq \emptyset$. Conclude that $h(q,r) \geq 2/3$, which is a contradiction.)

(3) If X is feebly compact and $f \in C(X)$, prove that $f[X]$ is compact.

(4) If X is feebly compact and Y is H-closed, prove that $X \times Y$ is feebly compact.

(5) Prove that the following are equivalent for a space X:

(a) X is pseudocompact,

(b) for each $f \in C(X)$, f is bounded above,

(c) for each $f \in C(X)$, f is bounded below, and

(d) for each $f \in C(X)$ such that $f > \underline{0}$, $f \geq \underline{\epsilon}$ for some positive real number ϵ.

(6) Suppose X and Y are feebly compact and Π_X is z-closed. Show that $X \times Y$ is pseudocompact. (Hint: Let $f \in C(X \times Y)$. Use the pseudocompactness of Y to show that $g(x) = \sup\{f(x,y) : y \in Y\}$ is defined for each $x \in X$; in fact using (3), show that for each $x \in X$, $g(x) = f(x,y)$ for some $y \in Y$. To show g is continuous, let $p \in X$ and $\epsilon > 0$. Note that $Z = \{(x,y) : |f(x,y) - f(p,y)| \geq \epsilon\}$ is a zero-set; so, $U = X \backslash \Pi_X[Z]$ is an open neighborhood of p. Let $x \in U$ and $y, q \in Y$ such that $g(x) = f(x,y)$ and $g(p) = f(p,q)$. Note that $g(x) + \epsilon = f(x,y) + \epsilon \geq f(x,q) + \epsilon > f(p,q) = g(p) \geq f(p,y) > f(x,y) - \epsilon = g(x) - \epsilon$. Conclude that g is continuous and that f is bounded above since X is

pseudocompact. Now apply (5).)

(7) (Tamano's Theorem). Let X and Y be nonempty Tychonoff spaces. Prove that $X \times Y$ is pseudocompact iff X and Y are pseudocompact and Π_X is z-closed. (Hint: Use the above results plus the results in 1.11.)

(8) Prove that a product of nonempty spaces is feebly compact iff each countable product is feebly compact. (Hint: Use 1.11(b).)

4AG. Product of two Stone-Čech compactifications. The reader will need to know the results of 4AF before solving this problem. Let X and Y be Tychonoff spaces, $W = X \times \beta Y$, and $\Pi_X : X \times Y \to X$ the projection function defined in 1.1(f).

(1) If X or Y is finite, prove that $X \times Y$ is C^*-embedded in $\beta X \times \beta Y$.

(2) If Π_X is z-closed, Z is a zero set in $X \times Y$, and $(x,p) \in c\ell_W Z$, prove that $(x,p) \in c\ell_W(Z \cap (\{x\} \times Y))$. (Hint: Assume that $(x,p) \notin c\ell_W(Z \cap (\{x\} \times Y))$. There is a function $f \in C(W)$ and a neighborhood U of (x,p) such that $f[c\ell_W(Z \cap (\{x\} \times Y))] = \{1\}$ and $f[U] = \{0\}$. Obtain a contradiction by showing that $Z(f) \cap Z \cap (\{x\} \times Y) = \varnothing$ and $x \in c\ell_X \Pi_X[Z \cap Z(f)] \setminus \Pi_X[Z \cap Z(f)]$.)

(3) If Π_X is z-closed, prove that $X \times Y$ is C^*-embedded in $X \times \beta Y$. (Hint: By 4.6(h), it suffices to show that if Z_1 and Z_2 are disjoint zero sets of $X \times Y$, then $c\ell_W Z_1 \cap$

$c\ell_W Z_2 = \emptyset$. Assume there is some point $(x,p) \in c\ell_W Z_1 \cap c\ell_W Z_2$ where $x \in X$ and $p \in \beta Y \setminus Y$. Let $T = \{x\} \times \beta Y$. Use (2) to show that $(x,p) \in c\ell_T(Z_1 \cap (\{x\} \times Y)) \cap c\ell_T(Z_2 \cap (\{x\} \times Y))$. Note that $Z_1 \cap (\{x\} \times Y)$ and $Z_2 \cap (\{x\} \times Y)$ are disjoint zero sets in $\{x\} \times Y$. Use 4.6(g,h) to deduce that $c\ell_T(Z_1 \cap (\{x\} \times Y)) \cap c\ell_T(Z_2 \cap (\{x\} \times Y)) = \emptyset$ as $\{x\} \times Y$ is C^*-embedded in T.)

(4) If $X \times Y$ is C^*-embedded in $X \times \beta Y$, prove that Π_X is z-closed. (Hint: Assume there is a zero set Z and a point $p \in c\ell_X \Pi_X[Z] \setminus \Pi_X[Z]$. By 4AF(1), there is a function $h \in C^*(X \times Y)$ such that $\{p\} \times Y \subseteq h^{\leftarrow}(1)$ and $Z = Z(h)$. Let H be a continuous extension of h to $X \times \beta Y$. Show that $\{p\} \times \beta Y \subseteq H^{\leftarrow}(1)$, $Z \subseteq Z(H)$, and there are disjoint open neighborhoods U of $\{p\} \times \beta Y$ and V of Z(H). Using compactness, show there is an open set S such that $\{p\} \times \beta Y \subseteq S \times \beta Y \subseteq U$. Now, show that $p \in S$ and $S \cap \Pi_X[Z(H)] = \emptyset$, a contradiction.)

Recall that a space X is a **P-space** if every G_δ set of X is open. (See 1W.)

(5) If Π_X is z-closed, prove that X is a P-space or Y is pseudocompact. (Hint: Assume X is not a P-space and Y is not pseudocompact. Use 1W(2) to find $f \in C(X)$ for which Z(f) is not open; using 4AF(5), show that there is some $g \in C(Y)$ such that $g > \underline{0}$ but $\inf\{g(y) : y \in Y\}$

$= 0$. For $(x,y) \in X \times Y$, define $h(x,y) = |f(x)|/g(y)$. Show that h is continuous and $h[Z(f) \times Y] = \{0\}$. Let $A = h^{\leftarrow}[[1,\infty)]$. Note that A is a zero set by 1.4(j), and show that $\Pi_X[A] = X \backslash Z(f)$. Conclude that Π_X is not z-closed.)

(6) (a) Prove that countable subsets of a Tychonoff P-space with disjoint closures are contained in disjoint cozero sets. (Hint: Let $A = \{a_n : n \in \mathbb{N}\}$ and $B = \{b_n : n \in \mathbb{N}\}$ have disjoint closures in a P-space R. Choose a cozero neighborhood U_1 of a_1 such that $c\ell_{\mathbb{R}}U_1 \cap c\ell_{\mathbb{R}}B = \emptyset$. Choose a cozero neighborhood V_1 of b_1 such that $(c\ell_R U_1 \cup c\ell_R A) \cap c\ell_R V_1 = \emptyset$. Choose a cozero neighborhood U_2 of a_2 such that $(c\ell_R U_1 \cup c\ell_R U_2) \cap (c\ell_R V_1 \cup c\ell_R B) = \emptyset$. Continue by induction. Let $U = \cup\{U_n : n \in \mathbb{N}\}$ and $V = \cup\{V_n : n \in \mathbb{N}\}$. Show that $A \subseteq U$, $B \subseteq V$, and U and V are disjoint cozero sets.)

(b) Show that every countable subset of a regular P-space is both closed and discrete. (Hint: If A is a countable subset of R, note that $R \backslash A$ is a G_δ-set.)

(c) Prove that countable subsets of Tychonoff P-spaces are C^*-embedded. (Hint: Use 1.9(h), 1W(2), and 6(a) and (b).)

(d) Prove that countable subsets of a Tychonoff P-space are C-embedded. (Hint: Use 1.9(j),

1W(2), and 6(c).)

(e) Prove that a pseudocompact, Tychonoff P-space is finite.

(7) (Glicksburg Theorem). Prove that the product $X \times Y$ is C^*-embedded in $\beta X \times \beta Y$ iff either one of the spaces X or Y is finite, or $X \times Y$ is pseudocompact. (Hint: By (1), one can assume $X \times Y$ is pseudocompact. Use 4AF(7) and (3) to conclude that $X \times Y$ is C^*-embedded in $X \times \beta Y$. Show that $X \times \beta Y$ is pseudocompact, and, again, using 4AF(7) and (3), conclude that $X \times \beta Y$ is C^*-embedded in $\beta X \times \beta Y$. To show the converse, apply (4), (5), (6)(e) and 4AF(7).)

(8) If K is an infinite compact space, prove that $\beta(X \times K) \equiv_{X \times K} \beta X \times K$ iff X is pseudocompact, i.e., $X \times K$ is pseudocompact iff X is pseudocompact. (Hint: Use 4AF(4) for one direction and (7) with 4AF(7) for the other direction.)

4AH. Remote points.

(1) Let Y be a normal space and $\mathcal{F}$ a filter base of closed sets on Y. Let $\mathcal{O}(\mathcal{F}) = \{U \subseteq Y : U \text{ is open and } U \supseteq F \text{ for some } F \in \mathcal{F}\}$.

(a) Prove that $\mathcal{O}(\mathcal{F})$ is a CR-filter on Y. (See 4Y for the definition of a CR-filter.)

(b) If $\cap F = \emptyset$, prove that $a_Y O(F) = \emptyset$.

(c) If H is a filter base of closed sets on Y and H misses F ("misses" is defined in 2.3(c)(5)), show that $O(H)$ misses $O(F)$.

For a Tychonoff space Y, a point $p \in \beta Y \setminus Y$ is a **remote point** of Y if $p \notin c\ell_{\beta Y} D$ for every nowhere dense set D in Y. Let D be the open filter of all open, dense subsets of Y.

(2) Let Y be a Tychonoff space and consider βY as constructed in 4Y where $p \in \beta Y \setminus Y$ iff p is a free, maximal CR-filter. Prove that $p \in \beta Y \setminus Y$ is a remote point of Y iff $D \subseteq p$.

Let X be a space and $n \in \mathbb{N}$. A nonempty family F of subsets of X has the **n-intersection property** (abbreviated as n-i.p.) if $A_1, A_2, \ldots, A_n \in F$ implies $A_1 \cap A_2 \cap \ldots \cap A_n \neq \emptyset$. In particular, F has the finite intersection property (f.i.p.) iff F has m-i.p. for each $m \in \mathbb{N}$. The space X is a **$\bar{G}$-space** if for each nonempty open set U of X and each $n \in \mathbb{N}$, there is a family G_n of open sets with n-i.p. such that for each $D \in D$, $c\ell_X G \subseteq U \cap D$ for some $G \in G_n$.

(3) Let $\{U_n : n \in \mathbb{N}\}$ be a locally finite family of pairwise disjoint, nonempty open sets in a $\bar{G}$-space X. For each $n \in \mathbb{N}$, let G_n be a family of open sets with n-i.p. such

for each $D \in \mathcal{D}$, $c\ell_X G \subseteq U_n \cap D$ for some $G \in \mathcal{G}_n$. Let $\mathcal{G} = \{\cup\{f(n) : n \in \mathbb{N}\} : f \in \Pi\{\mathcal{G}_n : n \in \mathbb{N}\}\}$.

(a) Prove that $\mathcal{H} = \{G \setminus c\ell_X(U_1 \cup \ldots \cup U_m) : G \in \mathcal{G}, m \in \mathbb{N}$ has f.i.p. and $a_X\mathcal{H} = \emptyset\}$.

(b) If $\mathcal{O}(\mathcal{H})$ is the open filter generated by $\{c\ell_X H : H \in \mathcal{H}\}$ (see (1)), prove that $\mathcal{D} \subseteq \mathcal{O}(\mathcal{H})$.

(4) Let Y be a normal, non-feebly compact $\bar{G}$ space.

(a) Prove that Y has a remote point. (Hint: Apply 1.11(b) to obtain a family $\{U_n : n \in \mathbb{N}\}$ of pairwise disjoint nonempty open sets which is locally finite. Apply (3), (1), and (2) to obtain a remote point.)

(b) Prove that Y has at least $2^{(2^{\aleph_0})}$ remote points. (Hint: Start with the family $\{U_n : n \in \mathbb{N}\}$ as in (a). For each $\mathcal{U} \in \beta\mathbb{N}\setminus\mathbb{N}$, let $\mathcal{G}(\mathcal{U}) = \{\cup\{f(n) : n \in A\} : f \in \Pi\{\mathcal{G}_n : n \in \mathbb{N}\}$ and $A \in \mathcal{U}\}$ and $\mathcal{H}(\mathcal{U}) = \{G \setminus c\ell_Y(U_1 \cup \ldots \cup U_m) : G \in \mathcal{G}(\mathcal{U})$ and $m \in \mathbb{N}\}$. Use (1) to show that $\mathcal{O}(\mathcal{H}(\mathcal{U}))$, the open filter generated by $\{c\ell_Y H : H \in \mathcal{H}(\mathcal{U})\}$, is a free CR-filter, $\mathcal{D} \subseteq \mathcal{O}(\mathcal{H}(\mathcal{U}))$, and if $\mathcal{V} \in \beta\mathbb{N}\setminus\mathbb{N}$ and $\mathcal{U} \neq \mathcal{V}$, then $\mathcal{O}(\mathcal{H}(\mathcal{U}))$ misses $\mathcal{O}(\mathcal{H}(\mathcal{V}))$. Complete the proof by using (2) and the fact that

$|\beta\mathbb{N}\setminus\mathbb{N}| = 2^{(2^{\aleph_0})}$ by 4U.)

A family $\mathcal{B}$ of nonempty open subsets of a space X is called a **$\bar{\pi}$-base** for X (see 2N(5) for related concepts) if for each nonempty open set $U \subseteq X$, there is some $B \in \mathcal{B}$ such that $c\ell_X B \subseteq U$.

(5) Let X be a space with a countable $\bar{\pi}$-base $\mathcal{B} = \{B_n : n \in \mathbb{N}\}$. For each nonempty open set U, let $\#(U)$ be the smallest integer m such that $c\ell_X B_m \subseteq U$. Let $r = \#(U)$, $M = \{k \in \mathbb{N} : k \leqslant r \text{ and } B_k \cap U \neq \emptyset\}$, and $G_1(U) = B_r$. If $G_n(V)$ is defined for every nonempty set V, inductively define $G_{n+1}(U)$ by

$$G_{n+1}(U) = \cup\{G_n(U \cap B_k) : k \in M\} \cup G_n(U).$$

For each $n \in \mathbb{N}$, let $\mathcal{G}_n(U) = \{G_n(U \cap D) : D \in \mathcal{D}\}$.

(a) Show that $G_1(U) \subseteq G_2(U) \subseteq \ldots \subseteq G_n(U)$ for $n \in \mathbb{N}$.

(b) Show that $c\ell_X G_n(U) \subseteq U$ for $n \in \mathbb{N}$. (Hint: Prove by induction.)

(c) Show that $\mathcal{G}_n(U)$ has n-i.p. whenever $n = 1$.

(d) Fix $n \in \mathbb{N}$. Suppose $\mathcal{G}_n(U)$ has n-i.p. for every nonempty open set U. Let $D_1,\ldots,D_n,D_{n+1}$ be dense open sets indexed so that

$$s = \#(D_1 \cap U) \leqslant \#(D_2 \cap U) \leqslant \ldots \leqslant \#(D_{n+1} \cap U).$$

(i) Show that $B_s \subseteq G_1(U \cap D_k)$ for $1 \leqslant k \leqslant n+1$.

(ii) Show that $B_s \subseteq G_k(U \cap D_k)$ for $1 \leqslant k \leqslant n+1$. (Hint: Use (i) and (a).)

(iii) Let $L = G_n(U \cap B_s \cap D_2) \cap \ldots \cap G_n(U \cap B_s \cap D_{n+1})$. Show that $L \neq \emptyset$. (Hint: Use n-i.p. and the fact that $\emptyset \neq B_s \subseteq U \cap D_1 \subseteq U$.)

(iv) Show that $G_n(U \cap B_s \cap D_i) \subseteq G_{n+1}(U \cap D_i)$ for $2 \leqslant i \leqslant n+1$. (Hint: Use the definition of $G_{n+1}(U \cap D_i)$ and note that $B_s \cap U \cap D_i \neq \emptyset$ and $s \leqslant \#(U \cap D_i)$ for $2 \leqslant i \leqslant n+1$.)

(v) Show that $G_{n+1}(U)$ has (n+1)-i.p. (Hint: First note that $L \subseteq G_n(U \cap B_s \cap D_2) \subseteq B_s$ by (a). Next, note that $B_s = G_1(U \cap D_1) \subseteq G_{n+1}(U \cap D_1)$ by (a). Finally, apply (iv) to obtain that $\emptyset \neq L \subseteq G_{n+1}(U \cap D_1) \cap \ldots \cap G_{n+1}(U \cap D_{n+1})$.)

(e) Conclude that a space with a countable $\bar{\pi}$-base is a $\bar{G}$-space.

(6) Show that a normal, non-feebly compact space with countable π-weight has at least $2^{(2^{\aleph_0})}$ remote points.

(Hint: Show that a regular space with a countable $\bar{\pi}$-base has countable π-weight. Now use (5)(e) and (4)(b).)

(7) Let Y be a Tychonoff space. Show that Y is not pseudocompact iff there is a nonempty G_δ-set G in βY such that $G \subseteq \beta Y \backslash Y$. (Hint: If Y is not pseudocompact, show that there is an unbounded $f \in C(Y)$ such that $f \geq \underline{1}$. Now $g = \frac{1}{f} \in C^*(Y)$ and has an extension $h \in C(\beta Y)$. Show that Z(h) is a nonempty G_δ-set in βY such that $Z(h) \subseteq \beta Y \backslash Y$. If G is an nonempty G_δ-set of βY such that $G \subseteq \beta Y \backslash Y$, show that $G = Z(f)$ for some $f \in C(\beta Y)$. Let $g = f|Y$. Now $Z(g) = \emptyset$; so, $\frac{1}{g} \in C(Y)$. Show that $\frac{1}{g}$ is unbounded.)

(8) If Z is a compact space and G is a G_δ-subset of Z, prove that $Z \backslash G$ is σ-compact and regular (and hence normal).

(9) If $Y \subseteq Z \subseteq \beta Y$, prove that remote points of Z are also remote points of Y.

(10) If Y is a Tychonoff, non-pseudocompact space with countable π-weight, prove that Y has $2^{(2^{\aleph_0})}$ remote points. (Hint: Use (6), (7), (8), (9) and 1.11(d).)

CHAPTER 5

Maximum *P*—Extensions

5.1 Introductory Remarks

In Chapter 4 we constructed the Stone-Čech compactification βX of a Tychonoff space X. One of the many characterizations of βX that we obtained is the following: if X is Tychonoff, K is compact, and $f \in C(X,K)$ then there exists $\beta f \in C(\beta X,K)$ such that $\beta f | X = f$ (see 4.6(g)). Put informally, this says that every continuous function from X to K has a continuous extension to βX.

Proceeding in the same manner, we later constructed the maximum zero-dimensional compactification $\beta_0 X$ of a zero-dimensional space X. We characterized $\beta_0 X$ in the following way: if X is zero-dimensional, K is compact and zero-dimensional, and $f \in C(X,K)$, then there exists $\beta_0 f \in C(\beta_0 X,K)$ such that $\beta_0 f | X = f$ (this follows from 4.7(d)(2)). Thus every continuous function from X to K has a continuous extension to $\beta_0 X$.

One might ask what it is about compactness and "compact and zero-dimensional" that gives rise to the constructions βX and $\beta_0 X$ with their parallel properties. One might also ask whether there are any other topological properties P such that each "suitable" space X has a P-extension $\gamma_P X$ with the following property: if Y has P and $f \in C(X,Y)$, then f has a continuous extension to $\gamma_P X$. This chapter is devoted to answering these questions. We begin our investigation by making a precise definition of the space $\gamma_P X$

mentioned above (see 4.1(j) for notation and terminology used below).

We remind ourselves that a "topological property" can be thought of as a class P of topological spaces with this attribute: if X is in P and X is homeomorphic to Y, then Y is in P.

(a) **Definition**. Let P be a topological property. A space T is a **maximum P-extension** of a space X if

(1) T is a P-extension of X (i.e., $T \in P(X)$) and

(2) if Y has P and if $f \in C(X,Y)$, then there exists $Pf \in C(T,Y)$ such that $Pf|X = f$.

We note immediately that by 1.6(d), Pf is unique; i.e., if $g \in C(T,Y)$ and $g|X = f$, then $g = Pf$. Some other immediate consequences of the definition are recorded below.

(b) **Proposition**. Let X be a space. Then

(1) a maximum P-extension of X is a projective maximum of $P(X)$, and

(2) if T_1 and T_2 are maximum P-extensions of X, then $T_1 \equiv_X T_2$.

Proof. (1) is obvious, and (2) follows from (1) and the remarks following 4.1(h).

(c) **Definition**. The unique (up to equivalence) maximum P-extension of a space X (if it exists) is denoted by $\gamma_P X$. (Note that $\gamma_P X = X$ iff X has P).

Note that there exist spaces X and topological properties for which the set of P-extensions of X has a projective maximum, but X has no maximum P-extension. For example, the set of H-closed extensions of $\mathbb{N}$ has a projective maximum, i.e., $\kappa\mathbb{N}$ (see 4.8(n)), but $\mathbb{N}$ has no maximum H-closed extension (this is a consequence of 5.8(c) below). As implicitly noted above, the Stone-Čech compactification of Tychonoff space is a typical example of a maximum P-extension. If P is compactness and X is a Tychonoff space, then βX is the maximum P-extension of X.

In this chapter we investigate the following questions: for which spaces X, and for which topological properties P, does a maximum P-extension exist? If X has a maximum P-extension, what does it look like? The answers to these questions essentially appear in 5.3(b), 5.6(b), 5.7(b), and 5.8(c). We also study four extension properties, namely compactness, compactness and zero-dimensionality, realcompactness, and m-boundedness; realcompactness is discussed in great detail. A wealth of other extension properties are introduced in the problems.

It is obvious that if X has a maximum P-extension, then X must be embeddable as a dense subspace of a space with P (for example, a connected space with more than one point has no maximum compact zero-dimensional extension). More generally, if P is a topological property, then the spaces that are the candidates for the honour of having a maximum P-extension turn out to be those spaces which are homeomorphic to a subspace of some product of spaces with P. It is to these spaces that we now turn our attention.

5.2 P-regular and P-compact spaces

(a) **Definition**. Let P be a topological property. A space X is called **P-regular** (respectively, **P-compact**) if it is homeomorphic to a subspace (respectively, a closed subspace) of a product of spaces each of which has P. The class of P-regular (respectively, P-compact) spaces is denoted by $\mathrm{Reg}(P)$ (respectively, $K(P)$).

Let us consider some examples. If P is the property of being homeomorphic to $[0,1]$, then $\mathrm{Reg}(P)$ is the class of Tychonoff spaces and $K(P)$ is the class of compact spaces (see 1.10(i)). If P is the property of being compact and zero-dimensional, then $\mathrm{Reg}(P)$ is the class of zero-dimensional spaces and $K(P)$ is the class of compact zero-dimensional spaces (1F(3) and 4.7(b)).

The following theorem gives a criterion for deciding whether a space is P-regular.

(b) **Theorem**. Let P be a topological property and X a space. Then:

(1) Suppose $\{X_\alpha : \alpha \in I\}$ is a set of spaces each of which is a product of finitely many P-spaces, and suppose $\cup\{C(X,X_\alpha) : \alpha \in I\}$ separates points and closed sets of X. For each $\alpha \in I$ and $f \in C(X,X_\alpha)$, let X_f be homeomorphic to X_α. Then the evaluation map $e : X \to \Pi\{X_f : f \in \cup\{C(X,X_\alpha), \alpha \in I\}$ given by

$$e(x) = \langle f(x)\rangle_{f \in \cup\{C(X,X_\alpha) : \alpha \in I\}} \qquad (\text{for } x \in X)$$

is a homeomorphism from X onto $e[X]$; thus X is P-regular.

(2) Let X be P-regular. Then there exists a set $\{X_\alpha : \alpha \in I\}$ of spaces, each a product of finitely many P-spaces, such that $\cup\{C(X,X_\alpha) : \alpha \in I\}$ separates points and closed sets of X.

Proof

(1) This is a direct consequence of the Embedding Theorem 1.7(j).

(2) Let $\{Y_\beta : \beta \in \Sigma\}$ be a set of P-spaces and suppose that X is a subspace of $\Pi\{Y_\beta : \beta \in \Sigma\} = Y$. Let I denote the set of nonempty finite subsets of Σ. Let A be a closed subset of X and let $p = \langle p_\beta \rangle_{\beta \in \Sigma}$ be a point of $X \setminus A$. Find $F \in I$ and open subsets V_β of Y_β, for each $\beta \in F$, such that $p \in \cap\{\Pi_\beta^{\leftarrow}[V_\beta] : \beta \in F\}$ and $\cap\{\Pi_\beta^{\leftarrow}[V_\beta] : \beta \in F\} \cap A = \emptyset$. (As usual Π_β denotes the βth projection map on Y.) Let $f_F = \Pi_F | X$, where $\Pi_F : Y \to \Pi\{Y_\beta : \beta \in F\} = X_F$ is the projection map. Then $f_F \in C(X,X_F)$ and $f_F(p) \in \Pi\{V_\beta : \beta \in F\}$, which is open in X_F and disjoint from $f_F[A]$. Thus $f_F(p) \notin c\ell_{X_F} f_F[A]$. Hence $\cup\{C(X,X_F) : F \in I\}$ separates points and closed subsets of X, and so $\{X_F : F \in I\}$ is the required set of spaces.

∎

The proof of the following is obvious, but it is sufficiently important to record formally.

(c) **Proposition**. Let P be a topological property. Then $P \subseteq K(P) \subseteq Reg(P)$, $Reg(K(P)) = Reg(P)$, $K(K(P)) = K(P)$, $Reg(Reg(P)) =$

Reg(P), and K(Reg(P)) = Reg(P).

As we shall see in the next section, the question of whether a space X has a maximum P-extension can be answered succinctly as follows. If X has a maximum P-extension, then X is P-regular; a necessary and sufficient condition for each P-regular space to have a maximum P-extension is that K(P) = P.

5.3 Characterizations of extension properties

Let P be a topological property. If a space X is to have a maximum P-extension — in fact, if it is to have any P-extensions — then it must be P-regular. So if we seek topological properties P such that "as many spaces as possible" have maximum P-extensions, the best we can do is to require that each P-regular space have a maximum P-extension. This motivates the following definition.

(a) **Definition**. A topological property P is called an **extension property** if each P-regular space has a maximum P-extension.

In this section we will characterize those topological properties that are extension properties. Recall from Chapter 1 that a topological property P is productive if the topological product of a set of spaces with P has P, and closed-hereditary if each closed subspace of a space with P has P.

(b) **Lemma**. Extension properties are closed-hereditary and productive.

Proof. Let P be an extension property. If X is a P-space and A is a closed subspace of X, let $j : A \to X$ denote the inclusion map. As A is P-regular $\gamma_P A$ exists, and there is a continuous function $Pj : \gamma_P A \to X$ such that $Pj | A = j$. Now

$$Pj[\gamma_P A] = Pj[c\ell_{\gamma_P A} A] \subseteq c\ell_X j[A] = A,$$

the last equality following because A is closed in X. If $\gamma_P A \setminus A \neq \emptyset$, this would give a contradiction to 1.8(i); hence $\gamma_P A = A$ and so A is a P-space. Thus P is closed-hereditary.

If $\{X_\alpha : \alpha \in I\}$ is a set of P-spaces, let $X = \Pi\{X_\alpha : \alpha \in I\}$. Recall Π_α denotes the αth projection from X onto X_α. Evidently X is P-regular so $\gamma_P X$ exists. Hence for each $\alpha \in I$ there is a continuous function $P\Pi_\alpha : \gamma_P X \to X_\alpha$ such that $P\Pi_\alpha | X = \Pi_\alpha$. Define $f : \gamma_P X \to X$ by requiring that $\Pi_\alpha \circ f = P\Pi_\alpha$ for each $\alpha \in I$. By 1.7(d)(2) $f \in C(\gamma_P X, X)$ and $f | X = id_X$. Hence by 1.8(i), $\gamma_P X \setminus X = \emptyset$, and so $\gamma_P X = X$. Thus X has P and so P is productive.

■

The reader may find it interesting to compare the next theorem (which characterizes extension properties) to 4.1(k), which gives a sufficient (but not necessary) condition for the existence of projective maxima.

(c) **Theorem**. Let P be a topological property. The following are equivalent:

(1) P is closed-hereditary and productive.

(2) P is an extension property.

(3) $P = K(P)$.

Proof

(1) $\Longleftrightarrow$ (3) Obviously $K(P)$ is closed-hereditary and productive, and is the smallest property containing P and possessing these properties.

(2) $\Rightarrow$ (1) This is a restatement of 5.3(b).

(1) $\Rightarrow$ (2) Let $X \in \mathrm{Reg}(P)$ and let S be a set of cardinality $2^{2^{|X|}}$. Let F be the set of topological spaces with P whose underlying set is a subset of S. Then F is a set (rather than a proper class) of spaces. For each $Y \in F$ and each $f \in C(X,Y)$ let Y_f denote a homeomorphic copy of Y. Put $M = \Pi\{Y_f : Y \in F$ and $f \in C(X,Y)\}$.

By 5.2(b) there is a set $\{X_\alpha : \alpha \in \Sigma\}$ of spaces and a set $\{f_\alpha : \alpha \in \Sigma\}$ of functions such that $f_\alpha \in C(X,X_\alpha)$, each X_α is a product of finitely many P-spaces, and $\{f_\alpha : \alpha \in \Sigma\}$ separates points and closed sets of X. By hypothesis, $c\ell_{X_\alpha} f_\alpha[X]$ is a P-space; denote it by T_α. As each X_α is Hausdorff it follows by 4.1(c) that $|T_\alpha| \leqslant 2^{2^{|f_\alpha[X]|}} \leqslant 2^{2^{|X|}}$, so each T_α is homeomorphic to some member of F. It follows that $\cup\{C(X,Y) : Y \in F\}$ separates points and closed sets of X. Hence by 1.7(j) the evaluation map $e : X \to M$ given by

$$e(x) = \langle f(x) \rangle_{f \in C(X,Y),\, Y \in F} \qquad \text{(for } x \in X)$$

is an embedding of X in M. We identify X with e[X] in this manner, i.e., by identifying each $x \in X$ with e(x). Thus if $Y \in F$ and $f \in C(X,Y)$ then $\Pi_f | e[X]$ is regarded as a function on X and $\Pi_f(e(x)) = f(x)$.

Suppose Z is a P-space and $g \in C(X,Z)$. As P is closed-hereditary $c\ell_Z g[X]$ has P. As above $|c\ell_Z[(g \circ e)[X]]| \leqslant 2^{2^{|X|}}$, so there is a space $W \in F$ and a homeomorphism $h : c\ell_Z[g[X]] \to W$. Thus $h \circ g \in C(X,W) = C(X,W_{h \circ g})$. Thus $h^{\leftarrow} \circ \Pi_{h \circ g} \in C(M,Z)$ where $\Pi_{h \circ g}$ is the projection map from M onto $W_{h \circ g}$. Let $\bar{g} = h^{\leftarrow} \circ \Pi_{h \circ g} | c\ell_M e[X]$; then $\bar{g} \in C(c\ell_M e[X],Z)$ and if $x \in X$, $\bar{g}(e(x)) = h^{\leftarrow} \circ \Pi_{h \circ g}(x) = g(x)$. Thus $\bar{g}$ is a continuous extension of g to $c\ell_M e[X]$. As P is closed-hereditary and productive, $c\ell_M e[X]$ has P, so $c\ell_M e[X]$ is the maximum P-extension of e[X], i.e., of X. The result follows.

■

We note two obvious but important points. First, suppose that P is an extension property, X and Y are P-regular, and $f \in C(X,Y)$. Then there exists $Pf \in C(\gamma_P X, \gamma_P Y)$ such that $Pf|X = f$. Second, for each topological property P there exists a smallest extension property Q that contains P and for which $\text{Reg}(P) = \text{Reg}(Q)$, namely $K(P)$; this follows from 5.2(c) and 5.3(c).

We now give a criterion for deciding whether a given P-compact extension of a P-regular space X is $\gamma_{K(P)}X$.

(d) **Theorem**. Let P be a topological property, X a P-regular space, and T a P-compact extension of X. The following are equivalent:

(1) $T \equiv_X \gamma_{K(P)}X$ (up to equivalence) and

(2) if Y is a P-space and $f \in C(X,Y)$, then f extends continuously to $f^* : T \to Y$.

Proof

(1) ⇒ (2) Since by 5.2(c) each P-space has the property $K(P)$, this follows from the definition of $\gamma_{K(P)}X$.

(2) ⇒ (1) Let Z be P-compact and let $g \in C(X,Z)$. If we can find $g^* \in C(T,Z)$ such that $g^*|T = g$, then by definition of $\gamma_{K(P)}X$, we must have $T = \gamma_{K(P)}X$ (up to equivalence). As Z is P-compact there exists a set $\{P_i : i \in I\}$ of P-spaces and an embedding $h : Z \to \Pi\{P_i : i \in I\} = P$ such that $h[Z]$ is closed in P. For each $i \in I$, $\Pi_i \circ h \circ g \in C(X,P_i)$ (where $\Pi_i : P \to P_i$ is the ith projection map), so by hypothesis there exists $k_i \in C(T,P_i)$ such that $k_i|X = \Pi_i \circ h \circ g$. Define $k : T \to P$ by requiring that $\Pi_i \circ k = k_i$ for each $i \in I$. Then $k \in C(T,P)$ by 1.7(d)(2) and $k|X = h \circ g$, so $k[T] = k[c\ell_T X] \subseteq c\ell_P h[Z] = h[Z]$ since $h[Z]$ is closed in P. Now $h^{\leftarrow} : h[Z] \to Z$ is a homeomorphism so we may define $g^* \in C(T,Z)$ by $g^* = h^{\leftarrow} \circ k$. Then $g^*|X = h^{\leftarrow} \circ h \circ g = g$ and g^* is the required continuous extension of g.

■

(e) **Definition**. Let P be a topological property. A subspace X of a space Y is **P-embedded in** Y if, whenever Z has P and $f \in C(X,Z)$, there exists $F \in C(Y,Z)$ such that $F|X = f$.

We have already encountered special cases of P-embedding; if P is "homeomorphic to $\mathbb{R}$", then P-embedding is C-embedding, and if P is "homeomorphic to [0,1]", then P-embedding is C^*-embedding.

We can rephrase 5.3(d) above in terms of P-embedding as follows.

(f) **Theorem**. Let P be a topological property. Then the maximum P-compact extension of the P-regular space X is the unique (up to equivalence) P-compact extension of X in which X is P-embedded.

An extremely useful characterization of P-compactness can be deduced from 5.3(d).

(g) **Corollary**. Let P be a topological property and X a P-regular space. The following are equivalent:

(1) X is P-compact and

(2) if T is a P-regular extension of X and $p \in T \setminus X$, then there exists a P-space Y and $f \in C(X,Y)$ such that f cannot be continuously extended to $f^* \in C(X \cup \{p\},Y)$.

Proof

(1) $\Rightarrow$ (2) If (2) fails, find a P-regular extension T of X and a point $p \in T \setminus X$ such that for each P-space Y and each $f \in C(X,Y)$, f extends continuously to $f^* \in C(X \cup \{p\},Y)$. Then $\gamma_{K(P)}(X \cup \{p\}) = S$ is a proper P-compact extension of X such that if $f \in C(X,Y)$ and Y is a P-space, then f extends continuously to $f' \in$

C(S,Y). By 5.3(d) above $S \equiv_X \gamma_{K(P)}X$. Thus X is properly contained in $\gamma_{K(P)}X$, so (1) fails.

(2) ⇒ (1) If (1) fails, then $\gamma_{K(P)}X$ is a proper P-extension of X, contradicting the truth of (2).

■

5.4 E—compact spaces

Suppose we are given a (Hausdorff) space E. How do we go about finding a topological property E such that E is E-compact? If E is such a property, then every E-regular space has a maximum E-compact extension. Since E must be E-regular, each subspace of a product of copies of E must be E-regular; call such spaces E-completely regular. Evidently the class of closed subspaces of products of copies of E is an extension property, and is the smallest extension property to which E belongs. Call this extension property E-compactness; then every E-completely regular space has a maximum E-compact extension.

Let us formalize these notions.

(a) <u>**Definition**</u>. Let E be a space containing more than one point and let P_E be the property of being homeomorphic to E. A space X is called **E-completely regular** if X is P_E-regular, and **E-compact** if X is P_E-compact. An extension property P that is P_E-compactness for some space E is called a **simply generated** extension property.

Thus a space is E-completely regular (respectively, E-compact) iff it is homeomorphic to a subspace (respectively, closed subspace) of some topological power of E. Although the term "E-regular" would be more in accordance with our previous usage that "E-completely regular", we use the latter term because of its widespread use in the literature. By 5.3(c) $K(P_E)$ is an extension property and each E-completely regular space has a maximum E-compact extension, which we denote by $\gamma_E X$ (rather than $\gamma_{K(P_E)} X$, which would have been in accordance with our previous notation). This provides us with a rich source of examples of extension properties.

(b) **Definition**. Let X and E be spaces. A subset of X is called E-**open** if it is of the form $f^{\leftarrow}[V]$ where V is an open subset of some finite power E^n and $f \in C(X,E^n)$. The following theorem provides two useful characterizations of E-completely regular spaces.

(c) **Theorem**. Let X and E be spaces. The following are equivalent:

(1) X is E-completely regular,

(2) for each closed subset A of X and each $p \in X \setminus A$ there is a positive integer n and $f \in C(X,E^n)$ such that $f(p) \notin cl_{E^n} f[A]$; i.e., $\cup\{C(X,E^n) : n \in \mathbb{N}\}$ separates points and closed sets of X, and

(3) the E-open subsets of X form a base for the open subsets of X.

Proof. That (1) and (2) are equivalent is a direct consequence of 5.2(b).

(2) ⇒ (3) Let W be open in X and $p \in W$. Find a positive integer n and $f \in C(X,E^n)$ such that $f(p) \notin c\ell_{E^n} f[X \setminus W]$. Put $V = f^{\leftarrow}[E^n \setminus c\ell_{E^n} f[X \setminus W]]$. Then $p \in V \subseteq W$ and V is E-open. Thus (3) follows.

(3) ⇒ (2) Let p and A be as in (2). Find $f \in C(X,E^n)$ and an open subset V of E^n such that $p \in f^{\leftarrow}[V] \subseteq X \setminus A$. Then $f(p) \in V \subseteq E^n \setminus f[A]$ and so $f(p) \notin c\ell_{E^n} f[A]$.

■

In general we cannot replace "$\cup\{C(X,E^n) : n \in \mathbb{N}\}$" by "C(X,E)" in (2) above, although examples illustrating this are difficult to describe; see 5AE for a reference to one. However, in the special cases which will be of most interest to us we can make this replacement, as we will see in 5.5(a).

The characterization of the maximum P-compact extension of a P-regular space given in 5.3(d), and the characterization of P-compactness which follows from it (see 5.3(g)), can be specialized in the following manner to the case where P-compactness is E-compactness for some space E.

(d) **Theorem**. Let X and E be spaces and let X be E-completely regular. Then:

(1) An E-compact extension T of X is (equivalent to) $\gamma_E X$ iff for each $f \in C(X,E)$, there exists $f^* \in C(T,E)$ such that $f^* | X = f$.

(2) X is E-compact iff for each E-completely regular extension T of X, and each $p \in T \setminus X$, there exists $f \in C(X,E)$ such that f cannot be continuously extended to $f^* \in C(X \cup \{p\},E)$.

Proof. See 5.3(d) and 5.3(g).

■

(e) **Definition**. A subspace X of a space Y is E-**embedded** in Y if each $f \in C(X,E)$ extends continuously to $f^* \in C(Y,E)$.

The preceding theorem shows that $\gamma_E X$ is the unique (up to equivalence) E-compact extension of X in which X is E-embedded. Obviously X is E-embedded in Y iff X is P_E-embedded in Y.

5.5 Examples of E-compactness

We now consider E-complete regularity and E-compactness when E is one of $\mathbb{R}$, [0,1], $\mathbb{N}$, and $\mathbb{2}$. We begin by showing that in each case, "functions to E suffice" in (2) of Theorem 5.4(c). Explicitly:

(a) **Theorem**. Let E be one of $\mathbb{R}$, [0,1], $\mathbb{N}$, and $\mathbb{2}$. The following are equivalent for a space X:

(1) X is E-completely regular,

(2) C(X,E) separates the points and closed subsets of X, and

(3) the family $\{f^{\leftarrow}[V] : V \text{ open in } E \text{ and } f \in C(X,E)\}$ forms a base for the open sets of X.

Proof

(2) ⇒ (1) This follows immediately from 5.4(c).

(1) ⇒ (2) If E is $\mathbb{N}$, each finite power of E is

homeomorphic to E, so the implication follows from 5.4(c) in this case. If $E = \mathbb{N}$ we leave the proof to the reader.

If E is $\mathbb{R}$ or $[0,1]$ and n is a positive integer, let d be a metric on E^n compatible with the topology of E^n such that $d(x,y) \leq 1$ for each pair of points x and y of E^n. Suppose A is closed in X and $p \in X \backslash A$. By (1), and its equivalence to 5.4(c)(2), there exists a positive integer n and $f \in C(X,E^n)$ such that $f(p) \notin c\ell_{E^n} f[A]$. We can define $g \in C(E^n,E)$ by letting $g(y)$ be the d-distance from y to $c\ell_{E^n} f[A]$ for each $y \in E^n$. Then $g \circ f \in C(X,E)$, $c\ell_E[(g \circ f)[A]] = \{0\}$, and $(g \circ f)(p) \neq 0$. Thus $C(X,E)$ separates points and closed subsets of X.

Finally it is an easy exercise to prove that (2) and (3) are equivalent.

■

It follows from the above theorem that a space X is $\mathbb{R}$-completely regular or $[0,1]$-completely regular iff the cozero-sets of X form a base for the open sets of X. According to 1.4(e), this is equivalent to X being completely regular, i.e., Tychonoff. Hence:

(b) **Theorem**. The following are equivalent for a space X:

(1) X is $\mathbb{R}$-completely regular,

(2) X is $[0,1]$-completely regular, and

(3) X is Tychonoff.

(c) <u>**Definition**</u>

(1) $\mathbb{R}$-compact spaces are also called **realcompact spaces.** (Thus realcompact spaces are spaces homeomorphic to a closed subspace of a power of $\mathbb{R}$.)

(2) The maximum realcompact extension of the Tychonoff space X is called the **Hewitt realcompactification** of X and is denoted by υX. (To be consistent with the notation introduced after 5.4(a) we should denote this by $\gamma_{\mathbb{R}}X$, but υX is the universally used notation.)

An extended discussion of realcompact spaces and the extension υX appears in 5.11. However, [0,1]-compact spaces are already familiar to us as we now see.

(d) <u>**Theorem**</u>

(1) The [0,1]-compact spaces are precisely the compact spaces.

(2) The maximum [0,1]-compact extension of the Tychonoff space X is its Stone-Čech compactification βX.

<u>Proof</u>

(1) As compactness is a closed-hereditary, productive property and [0,1] is compact, each [0,1]-compact space is compact. Conversely, every compact space K is Tychonoff and thus [0,1]-completely regular, and hence has a maximum [0,1]-compact extension $\gamma_{[0,1]}K$. But K is closed in each Hausdorff space containing it, so $\gamma_{[0,1]}K = K$. Thus K is [0,1]-compact.

(2) The characterization of βX given in 4.6(g)(3) shows that βX is up to equivalence the maximum compact extension (i.e., compactification) of X.

■

We now examine the situation for $\mathbb{N}$ and $\mathbb{2}$. The reader should recall the construction of $\beta_0 X$ described in 4.7.

(e) **Theorem**

(1) Let E be a space with more than one point. Then E-complete regularity is zero-dimensionality iff E is zero-dimensional.

(2) $\mathbb{2}$-complete regularity and $\mathbb{N}$-complete regularity are zero-dimensionality.

(3) $\mathbb{2}$-compactness is the property of being compact and zero-dimensional.

(4) $\beta_0 X$ is the maximum $\mathbb{2}$-compact extension of the zero-dimensional space X.

Proof

(1) One direction is obvious. For the non-trivial direction, suppose E is zero-dimensional. If X is E-completely regular, then X is a subspace of a power of E. Zero-dimensionality is productive and hereditary (see 1.5(c) and 1F(3)): thus X is zero-dimensional. Conversely, suppose X is zero-dimensional. It is easy to see that $C(X,\mathbb{2})$ separates the points and closed sets of X, so by 5.5(a) X is $\mathbb{2}$-completely regular. As $\mathbb{2}$ is homeomorphic to a subspace of E (since $|E| > 1$), X is E-completely regular.

(2) This follows from (1).

(3) As compactness and zero-dimensionality are closed-hereditary productive properties enjoyed by $\mathbb{2}$, $\mathbb{2}$-compact spaces are compact and zero-dimensional. As noted above, a zero-dimensional space X is $\mathbb{2}$-completely regular. As a compact zero-dimensional space is closed in any power of $\mathbb{2}$ in which it is embedded, it is 2-compact.

(4) This follows from the characterization of $\beta_0 X$ given in 4.7(d)(2).

■

In 5E we will give a characterization of the maximum $\mathbb{N}$-compact extension $\gamma_{\mathbb{N}} X$ of the zero-dimensional space X as a subspace of $\beta_0 X$; this will closely resemble the description of υX as a subspace of βX given in 5.11(b) below.

5.6 Tychonoff extension properties

(a) **Definition**. A **Tychonoff extension property** P is an extension property for which Reg(P) is the class of Tychonoff spaces.

We have already seen two Tychonoff extension properties, namely compactness and realcompactness. We now give a characterization of Tychonoff extension properties.

(b) **Theorem**. Let P be a topological property such that each P-regular space is Tychonoff. The following are equivalent:

(1) P is a Tychonoff extension property, and

(2) P is a closed-hereditary, productive property possessed by

all compact spaces.

Proof

(1) ⇒ (2) By 5.3(c) P is closed-hereditary and productive. If K is compact, then K is P-regular. Hence $\gamma_P K$ exists and contains K as a dense subspace. But K is closed in $\gamma_P K$, so K = $\gamma_P K$; hence K has P.

(2) ⇒ (1) This follows from 5.3(c).

■

As another example of Tychonoff extension property we consider m-boundedness.

(c) **Definition**. Let m be an infinite cardinal. A topological space X is said to be m-**bounded** if X is Tychonoff and if each subset of X of cardinality no greater than m has a compact closure in X.

(d) **Theorem**. m-boundedness is a Tychonoff extension property.

Proof. Obviously all compact spaces are m-bounded. If S is a closed subspace of the m-bounded space X, if $A \subseteq S$, and if $|A| \leq$ m then $c\ell_S A$ is compact since $c\ell_S A = c\ell_X A$. Thus m-boundedness is closed-hereditary. Let $\{X_i : i \in I\}$ be a set of m-bounded spaces and denote $\Pi\{X_i : i \in I\}$ by X. If $A \subseteq X$ and $|A| \leq m$, then $|\Pi_i[A]| \leq m$ and so $c\ell_{X_i}\Pi_i[A]$ is compact for each $i \in I$. Hence $\Pi\{c\ell_{X_i}\Pi_i[A] : i \in I\}$ is a compact subset of X and evidently contains A; thus $c\ell_X A$ is compact. It follows that m-boundedness is productive and thus by the preceding theorem is a Tychonoff extension property.

■

In 5B we invite the reader to show that m-boundedness is not E-compactness for any space E. Thus not all extension properties are simply generated. A more complicated example appears in 5P.

5.7 Zero-dimensional extension properties

(a) **Definition**. A topological property P is called a **zero-dimensional extension property** if P is an extension property and Reg(P) is the class of zero-dimensional spaces.

The proof of the following analogue of 5.6(b) is left to the reader.

(b) **Theorem**. Let P be a topological property such that each space with P is zero-dimensional. The following are equivalent:

(1) P is a zero-dimensional extension property and

(2) P is closed-hereditary, productive, and possessed by all compact zero-dimensional spaces.

"Compact zero-dimensional" and $\mathbb{N}$-compactness are two zero-dimensional extension properties which we have already considered. There are many other examples; in fact, if we are given a Tychonoff extension property P we can manufacture a zero-dimensional extension property P_0 from it as follows: a space X is defined to have P_0 if X has P and is zero-dimensional. We can also go in the opposite direction; if P is a zero-dimensional extension property, then the topological property $[0,1] \times P$ is defined as follows: X has $[0,1] \times P$ if X is a closed subspace of a product of a P-space with a power of $[0,1]$. One can check that $[0,1] \times P$ is a Tychonoff extension property and that $([0,1] \times P)_0 = P$. The reader is invited to work out the proof of these assertions in 5D.

Since realcompactness is a Tychonoff extension property, it follows that "realcompact and zero-dimensional" is a zero-dimensional extension property. It is tempting to conjecture that a space will be realcompact and zero-dimensional iff it is $\mathbb{N}$-compact, but this is untrue. The $\mathbb{N}$-compact spaces are all realcompact and zero-dimensional (since $\mathbb{N}$ is realcompact), but the opposite inclusion does not hold (see 5G).

5.8 Hausdorff extension properties

Having defined Tychonoff and zero-dimensional extension properties, it is natural for us to define Hausdorff extension properties as follows.

(a) <u>**Definition**</u>. A topological property P is called a **Hausdorff extension property** if P is an extension property and Reg(P) is the class of Hausdorff spaces.

This concept is not fruitful, however, for when we try to characterize Hausdorff extension properties we discover a surprising fact; the only Hausdorff extension property is "Hausdorff". We prove this below.

(b) **Lemma**. Each space can be embedded as a closed nowhere dense subspace of an H-closed space.

Proof. Let X be a space and let Y be any non-discrete space. Let p be a non-isolated point of Y. Then $X \times \{p\}$ is homeomorphic to X, and is a closed nowhere dense subspace of $X \times Y$. By 4.8(p)(6) we see that $X \times \{p\}$ is closed in $\kappa(X \times Y)$.

■

(c) **Theorem**. If P is a Hausdorff extension property, then every Hausdorff space has P. Thus "Hausdorff" is the only Hausdorff extension property.

Proof. Let P be a Hausdorff extension property and let K be an H-closed space. Then K is P-regular and thus has a maximum P-extension $\gamma_P K$. But K is H-closed and hence closed in $\gamma_P K$ by definition. Thus $\gamma_P K = K$ and so K has P.

If X is a space, by the preceding lemma there is an H-closed space K such that X is (homeomorphic to) a closed subspace of K. By 5.3(c) P is closed-hereditary; thus, since K has P, X also has P.

■

It follows from the above result that "H-closed" is not a Hausdorff extension property. Thus there exist spaces X and Y, and $f \in C(X,Y)$, such that f does not extend continuously to $\kappa f \in C(\kappa X, \kappa Y)$. This is perhaps surprising in light of the fact that in many ways H-closed spaces are to Hausdorff spaces as compact spaces are to Tychonoff spaces, and in light of the similarities between the Stone-Čech compactification and the Katětov H-closed extension noted in Chapter 4. However, it gives us an indication of how to develop for Hausdorff spaces a "theory of extensions" that is parallel to the theories of Tychonoff, and zero-dimensional, extension properties developed in the remainder of this chapter and its problems. Instead of considering the set C(X,Y) (for two given spaces X and Y) we should consider all functions from X to Y that can be extended to the "same sort" of function from κX to κY. We will characterize such functions in 7.6(b).

5.9 More on Tychonoff and zero-dimensional extension properties

We now develop the basic properties of Tychonoff and zero-dimensional extension properties. This development culminates in another characterization of such properties, and the construction of the maximum P-extension as a subspace of βX or $\beta_0 X$.

(a) __Lemma__. Let X and Y be two spaces and suppose that $f \in C(X,Y)$. Let A and B be subspaces of X and Y respectively such that $A = f^{\leftarrow}[B]$. Then:

(1) A is homeomorphic to a closed subspace of $X \times B$ and

(2) if P is an extension property and X and B have P, then A has P.

__Proof__

(1) The mapping $h : A \to X \times B$ defined by $h(x) = (x,f(x))$ is

easily verified to be an embedding of A in $X \times B$. The function $g : X \times B \to Y \times B$ defined by $g(x,y) = (f(x),y)$ is obviously continuous. The set $D = \{(y,y) : y \in B\}$ is closed in $Y \times B$ (as all spaces are Hausdorff), so $g^{\leftarrow}[D]$ is closed in $X \times B$. But $g^{\leftarrow}[D] = h[A]$ so A is embedded as a closed subspace of $X \times B$.

(2) This follows from 5.3(b) and (1) above.

■

(b) **Lemma**. Let P be a closed-hereditary topological property of Tychonoff (respectively zero-dimensional) spaces. The following are equivalent:

(1) if X and Y are Tychonoff (respectively, zero-dimensional) and $f : X \to Y$ is a perfect continuous surjection and Y has P, then X has P and

(2) if X has P and K is a compact (respectively, compact zero-dimensional) space then $X \times K$ has P.

Proof. We prove the Tychonoff case; the zero-dimensional case is the same.

(2) $\Rightarrow$ (1) Let X and Y be Tychonoff spaces and let $f : X \to Y$ be a perfect continuous surjection. By 4.2(c) there is a continuous extension $\beta f : \beta X \to \beta Y$ and since f is a perfect continuous surjection, by 4.2(f) $(\beta f)^{\leftarrow}[Y] = X$. Hence by the preceding lemma X is homeomorphic to a closed subspace of $\beta X \times Y$. By hypothesis such a subspace has P, so X has P.

(1) $\Rightarrow$ (2) By problem 1M the projection Π_X from $X \times K$ onto X is a perfect continuous surjection.

■

If P-regularity is the property of being Tychonoff or zero-dimensional, the following is an immediate consequence of (b) above.

(c) **Lemma**. Let P be an extension property, let Y be a space with P, let X be a P-regular space, and let $f : X \to Y$ be a perfect continuous surjection. Then X has P.

Proof. As P is an extension property, f can be continuously extended to $Pf : \gamma_P X \to Y$. By 1.8(i) $Pf[\gamma_P X \setminus X] \subseteq Y \setminus Pf[X] = Y \setminus f[X] = \emptyset$. Thus $\gamma_P X \setminus X = \emptyset$ and so X has P.

■

Of course, a perfect continuous pre-image of a P-regular space need not be P-regular; examples abound of compact non-zero-dimensional spaces being mapped onto compact zero-dimensional spaces. See 8H for a much more involved example of the phenomenon.

We next show that extension properties are closed under the formation of arbitrary intersections. This result, together with (c) above, will allow us to formulate a new characterization of extension properties.

(d) **Lemma**. Let P be a closed-hereditary productive topological property. If $\{X_\alpha : \alpha \in A\}$ is a set of subspaces of a space Z, and if each X_α has P, then $\cap\{X_\alpha : \alpha \in A\}$ has P.

Proof. Let $X = \Pi\{X_\alpha : \alpha \in A\}$ and let $Y = \cap\{X_\alpha : \alpha \in$

$A\}$. Define $h \in F(Y,X)$ by $h(y)(\alpha) = y$ for each $\alpha \in A$ and $y \in Y$. Let $\Delta = h[Y]$. It is easily checked that h is one-to-one. If $\Pi_\alpha : X \to X_\alpha$ is the αth projection map, then $\Pi_\alpha \circ h = id_Y$ so h is continuous. If V is open in Y, find W open in Z such that $W \cap Y = V$. It is easily checked that for each $\alpha \in A$, $\Delta \cap \Pi_\alpha^{\leftarrow}[W \cap X_\alpha] = h[V]$; therefore, h is an open map onto Δ. Thus Y and Δ are homeomorphic.

If $z \in X \setminus \Delta$ there exist distinct indices α and β in A such that $\Pi_\alpha(z) \neq \Pi_\beta(z)$. Find disjoint open sets U and V of Z such that $\Pi_\alpha(z) \in U$ and $\Pi_\beta(z) \in V$. Then $z \in \Pi_\alpha^{\leftarrow}[U \cap X_\alpha] \cap \Pi_\beta^{\leftarrow}[V \cap X_\beta] \subseteq X \setminus \Delta$; therefore, Δ is closed in X. Since P is closed-hereditary and productive, Δ (and hence Y) has P.

■

(e) **Theorem**. Let P be a topological property such that the P-regular spaces either are precisely the Tychonoff spaces or are precisely the zero-dimensional spaces. The following are equivalent:

(1) P is an extension property and

(2) (i) all compact P-regular spaces have P, (ii) if Y has P, X is P-regular and $f : X \to Y$ is a perfect continuous surjection, then X has P, and (iii) if $\{X_\alpha : \alpha \in A\}$ is a set of subspaces of a space X, and if each X_α has P, then $\cap\{X_\alpha : \alpha \in A\}$ has P.

Proof

(1) $\Rightarrow$ (2) Statement (i) follows from 5.6(b) and 5.7(b); statements (ii) and (iii) follow from (c) and (d) above.

(2) $\Rightarrow$ (1) We consider the Tychonoff case; the zero-dimensional case is similar (with "$\beta_0 X$" replacing "βX"). Let X

be Tychonoff and let $T = \bigcap\{S : X \subseteq S \subseteq \beta X$ and S has $P\}$. Note that βX has P by (i), so T is a well-defined extension of X. We will show that (up to equivalence) T is the maximum P-extension of X, and so P must be an extension property.

Let Y be a P-space and let $f \in C(X,Y)$. Let $J = \bigcap\{M : f[X] \subseteq M \subseteq Y$ and M has $P\}$. As Y has P, J is a well-defined subspace of Y, and by (iii) J has P. We claim that f[X] is dense in J. If it were not, choose $p \in J \setminus c\ell_Y f[X]$. Now $c\ell_{\beta Y} f[X]$ has P by (i), and $p \notin c\ell_{\beta Y} f[X]$. As $c\ell_{\beta Y} f[X]$ and J are subspaces of βY with P, it follows from (iii) that their intersection is a subspace of Y with P that contains f[X] but not p. This contradicts the definition of J and choice of p, and our claim follows. By 4.6(g)(3), f extends continuously to a surjection $\beta f \in C(\beta X, \beta J)$, and evidently $X \subseteq (\beta f)^{\leftarrow}[J] \subseteq \beta X$. As f[X] is dense in J, it follows from 1.8(f) that $\beta f | (\beta f)^{\leftarrow}[J]$ is a perfect map from $(\beta f)^{\leftarrow}[J]$ onto J; therefore, by (ii) $(\beta f)^{\leftarrow}[J]$ has P. Thus $T \subseteq (\beta f)^{\leftarrow}[J]$ and so $\beta f | T$ is a continuous extension of f mapping T into J and hence into Y. By (iii) T has P, so T is a maximum P-extension of X.

∎

Evidently (1) implies (2) in the above theorem for any extension property (not just Tychonoff or zero-dimensional ones). However, if one tries to prove that (2) implies (1) for an arbitrary extension property, one is stymied by the lack of spaces to play the role of βX and βY in the argument above.

(f) **Corollary**. Let X be a Tychonoff (respectively, zero-dimensional) space. If P is a Tychonoff (respectively, zero-dimensional) extension property, then (up to equivalence) $\gamma_P X$ is a dense subspace of βX (respectively $\beta_0 X$) that contains X; explicitly,

$$\gamma_P X = \cap\{T : X \subseteq T \subseteq \beta X \text{ and } T \text{ has } P\}$$

$$(\text{respectively, } \gamma_P X = \cap\{T : X \subseteq T \subseteq \beta_0 X \text{ and } T \text{ has } P\}).$$

Proof. See the proof that (2) ⇒ (1) above.

■

Hence, if P is a Tychonoff (respectively, zero-dimensional) extension property, one way of describing $\gamma_P X$ is to specify which free z-ultrafilters (respectively, clopen ultrafilters) of X we must adjoin to X in order to construct $\gamma_P X$. We will describe the Hewitt realcompactification υX, the maximum $\mathbb{N}$-compact extension $\gamma_{\mathbb{N}} X$, and the maximum m-bounded extension in this manner. Also, if P and Q are two Tychonoff (respectively, zero-dimensional) extension properties, $\gamma_P X$ and $\gamma_Q X$ will both be subspaces of βX (respectively, $\beta_0 X$) and it makes sense to talk about the relationship between these extensions. This relationship is investigated in 5.9(h) below and in problems 5T, 5U, 5W, and 5X.

(g) **Definition**

(1) Two topological properties P and Q are called **co-regular** if Reg P = Reg Q.

(2) Let P be an extension property. Then $\beta_P X$ denotes

βX (respectively, $\beta_0 X$) if P is a Tychonoff (respectively, zero-dimensional) extension property.

(h) **Theorem**. Let P and Q be co-regular extension properties and let P-regularity be either the Tychonoff property or zero-dimensionality. If $P \subseteq Q$ then $\gamma_Q X \subseteq \gamma_P X$ and $\gamma_Q X = \bigcap\{T : X \subseteq T \subseteq \gamma_P X \text{ and } T \text{ has } Q\}$.

Proof. As $\gamma_P X$ has Q, by 5.9(f) $X \subseteq \gamma_Q X \subseteq \gamma_P X \subseteq \beta_P X$. If $X \subseteq T \subseteq \beta_P X$ and T has Q, then by 5.9(d) $T \cap \gamma_P X$ has Q. Thus the second assertion follows from the first.

■

5.10 Two examples of maximum P-extensions

In this section we construct the maximum P-extension of the Tychonoff space X as a subspace of βX for two properties P, namely realcompactness and m-boundedness. A similar construction of $\gamma_P X$ when P is $\mathbb{N}$-compactness is outlined in 5E.

First we construct the maximum realcompact extension of X, i.e., its Hewitt realcompactification υX. Recall from 4.6(b) that $\beta X \setminus X$ may be regarded as the set of free z-ultrafilters on X. Then υX can be characterized as a subspace of βX in the following manner.

(a) **Definition**. A z-ultrafilter $\mathcal{a}$ on a Tychonoff space is said to have the **countable intersection property** (C.I.P.), cf. 2P, if each countable subfamily of $\mathcal{Q}$ has nonempty intersection.

(b) **Theorem**. Let X be a Tychonoff space. Then:

(1) Up to equivalence υX is the unique realcompact extension of X in which X is C-embedded.

(2) $\upsilon X = X \cup \{\mathcal{Q} \in \beta X \setminus X : \mathcal{Q}$ has C.I.P.$\}$.

Proof

(1) This follows from 5.4(e) and the comments preceding it.

(2) Let $T = X \cup \{\mathcal{Q} \in \beta X \setminus X : \mathcal{Q}$ has C.I.P.$\}$. By (1) it suffices to verify two things: (i) if $f \in C(X)$, then f extends continuously to $\upsilon f \in C(T)$, and (ii) T is realcompact.

To verify (i), let $\alpha\mathbb{R} = \mathbb{R} \cup \{\omega\}$ denote the one-point compactification of the locally compact space $\mathbb{R}$. By 4.2(e) and 4.6(k) f extends continuously to $\alpha f : \beta X \to \alpha\mathbb{R}$ and if $\mathcal{Q} \in \beta X \setminus X$ then $\alpha f(\mathcal{Q}) = \bigcap\{Z \in Z(\alpha\mathbb{R}) : f^{\leftarrow}[Z] \in \mathcal{Q}\}$. Suppose $\mathcal{Q} \in \beta X \setminus X$ and $\alpha f(\mathcal{Q}) = \omega$. Then for each positive integer n, $\omega \notin [-n,n] \in Z(\mathbb{R})$ so $f^{\leftarrow}[-n,n] \notin \mathcal{Q}$. By 1.4(j) and 2.3(d)(3) it follows that $f^{\leftarrow}[\alpha\mathbb{R}\setminus(-n,n)] \in \mathcal{Q}$. (Note that $\alpha\mathbb{R}$ is a metric space so all closed subsets of it are zero-sets of it.) But $f^{\leftarrow}[\alpha\mathbb{R}\setminus(-n,n)] = f^{\leftarrow}[\mathbb{R}\setminus(-n,n)]$ as $\omega \notin f[X]$. Since $\bigcap\{f^{\leftarrow}[\mathbb{R}\setminus(-n,n)] : n \in \mathbb{N}\} = \emptyset$, $\mathcal{Q}$ does not have C.I.P. It follows from this that $\alpha f[T] \subseteq \mathbb{R}$, and so $\alpha f|T$ is the required continuous extension of f to T.

To verify (ii), suppose $\mathcal{Q} \in \beta X \setminus T$. There exists a countable subfamily $\{Z_n : n \in \mathbb{N}\}$ of $\mathcal{Q}$ such that $\bigcap\{Z_n : n \in \mathbb{N}\} = \emptyset$. Without loss of generality we may assume that $Z_{n+1} \subseteq Z_n$ for each $n \in \mathbb{N}$. By 1.4(f) for each $n \in \mathbb{N}$ we can find $f_n \in C(X)$

such that $f_n[X] \subseteq [0,1]$ and $Z_n = Z(f_n)$. Define a function $g : X \to \mathbb{R}$ by $g(x) = \sum_{n=1}^{\infty} 2^{-n} f_n(x)$. By 1B6 (a generalization of the Weierstrass M-test) and the proof of 1.4(i), g is a well-defined continuous function; clearly, $Z(g) = \cap\{Z(f_n) : n \in \mathbb{N}\} = \emptyset$. Thus we can define $h \in C(X)$ by putting $h(x) = 1/g(x)$ for each $x \in X$. As h maps X into $\alpha\mathbb{R}$, by 4.2(e) there exists $\alpha h \in C(\beta X, \alpha\mathbb{R})$ such that $\alpha h | X = h$.

We now claim that $\alpha h(\mathcal{Q}) = \omega$. To see this, suppose $r \in \mathbb{R}$ and $\alpha h(\mathcal{Q}) = r$. As $X = h^{\leftarrow}[[r-1,r+1]] \cup h^{\leftarrow}[\mathbb{R} \setminus (r-1,r+1)]$, by 4.6(k) we see that $h^{\leftarrow}[[r-1,r+1]] \in \mathcal{Q}$. Choose $m \in N$ such that $\sum_{n>m} 2^{-n} < (r+1)^{-1}$. Now $Z_m \cap h^{\leftarrow}[[r-1,r+1]] \neq \emptyset$ as $\mathcal{Q}$ is a filter containing Z_m and $h^{\leftarrow}[[r-1,r+1]]$; choose $p \in Z_m \cap h^{\leftarrow}[[r-1,r+1]]$. Then $p \in \bigcap_{n \leq m} Z_n$ so

$$0 < g(p) = \sum_{n>m} 2^{-n} f_n(p) \leq \sum_{n>m} 2^{-n} < (r+1)^{-1}.$$

Thus $h(p) = 1/g(p) > r + 1$, which is a contradiction. Thus $\alpha h(\mathcal{Q}) = \omega$ as claimed and $(\alpha h)^{\leftarrow}[\mathbb{R}] \subseteq \beta X \setminus \{\mathcal{Q}\}$.

From the verification of (i) above it is evident that $T \subseteq (\alpha h)^{\leftarrow}[\mathbb{R}]$, so by combining our results we see that $T = \cap\{(\alpha f)^{\leftarrow}[\mathbb{R}] : f \in C(X, \alpha\mathbb{R})\}$. By 5.9(c) each $(\alpha f)^{\leftarrow}[\mathbb{R}]$ is realcompact, so by 5.9(d) T is realcompact. Thus (ii) is verified and $T = \upsilon X$.

■

(c) **Corollary**. A Tychonoff space is realcompact iff each z-ultrafilter on X with C.I.P. is fixed.

Proof. From the preceding theorem, each z-ultrafilter on X with C.I.P. is fixed iff $\upsilon X \backslash X = \emptyset$ iff $X = \upsilon X$ iff X is realcompact. ■

A good source of information about realcompactness is the classic text by Gillman and Jerison [GJ], where the interested reader can find a more algebraic approach to the construction and characterization of υX. The monograph by Weir [We], devoted exclusively to realcompact spaces (which he calls "Hewitt-Nachbin spaces"), provides an encyclopaedic collection of detailed information on the subject. We will prove some of the more important results about realcompactness in the next section. In particular, 5.11(m) provides us with a large class of realcompact spaces.

If X is Tychonoff and m is an infinite cardinal, let mX denote the maximum m-bounded extension X (see 5.6(c)). We characterize mX as a subspace of βX.

(d) **Theorem**. $mX = \bigcup\{cl_{\beta X} S : S \subseteq X \text{ and } |S| \leq m\}$.

Proof. Let $\bigcup\{cl_{\beta X} S : S \subseteq X \text{ and } |S| \leq m\} = T$. Let $A \subseteq T$ and suppose $|A| \leq m$. If $a \in A$, there exists a subset $S(a)$ of X such that $a \in cl_{\beta X} S(a)$ and $|S(a)| \leq m$. Thus:

$$cl_T A \subseteq cl_T[\bigcup\{cl_{\beta X} S(a) : a \in A\}].$$
$$\subseteq cl_T[cl_{\beta X}[\bigcup\{S(a) : a \in A\}]].$$

Now $|\bigcup\{S(a) : a \in A\}| \leq m \cdot m = m$, so $cl_{\beta X}[\bigcup\{S(a) : a \in A] \subseteq$

T. Thus $c\ell_T A = c\ell_{\beta X} A$ and so $c\ell_T A$ is compact. Thus T is m-bounded.

Let Y be m-bounded and let $f \in C(X,Y)$. Recall that $\beta f \in C(\beta X, \beta Y)$ extends f to βX. If $\beta f[T] \subseteq Y$, then $\beta f|T$ will be a continuous extension of f to T that maps T into Y; thus T will be the maximum m-bounded extension mX of X.

If $p \in T$ find $S \subseteq X$ such that $|S| \leqslant m$ and $p \in c\ell_{\beta X} S$. Thus $\beta f(p) \in \beta f[c\ell_{\beta X} S] = c\ell_{\beta Y} f[S]$. But $|f[S]| \leqslant m$ and since Y is m-bounded, it follows that $c\ell_{\beta Y} f[S] \subseteq Y$. Thus $\beta f(p) \in Y$ and so $T = mX$.

■

The reader should not infer from our characterizations of υX and mX that an explicit description of $\gamma_P X$ as a subspace of $\beta_P X$ is always easy to obtain. On the contrary, there are many extension properties *P* for which an explicit description of the points of $\beta_P X \backslash X$ that belong to $\gamma_P X$ is unknown at the time this is written. (For example, define *P* by saying that a space X has *P* iff X is Tychonoff and all pseudocompact closed subsets of X are compact. Then *P* is easily shown to be a Tychonoff extension property, but an explicit description of $\gamma_P X$, analogous to 5.10(b)(2) or 5.10(d), eludes us. See 5X for more detail.) Nonetheless, the fact that $\gamma_P X$ can be viewed as "living between" X and $\beta_P X$ is a powerful tool that we will exploit repeatedly in forthcoming chapters, particularly in 8.3.

5.11 Realcompact spaces and extensions

In this section we develop in more detail the theory of realcompact spaces. We focus on two themes, namely the relationship between X and υX, and the structure of $\beta X \backslash X$ when X is realcompact.

(a) **Lemma**. Every Lindelöf Tychonoff space is realcompact.

Proof. Let $\mathcal{A}$ be a free z-ultrafilter on the Lindelöf space X. Then $\{X \backslash Z : Z \in \mathcal{A}\}$ is an open cover of X, so there exists a countable subfamily $\{Z_n : n \in \mathbb{N}\}$ of $\mathcal{A}$ such that $\cup\{X \backslash Z_n : n \in \mathbb{N}\} = X$. Then $\cap\{Z_n : n \in \mathbb{N}\} = \emptyset$ so $\mathcal{A}$ does not have C.I.P. It follows from 5.10(c) that X is realcompact.

■

(b) **Theorem**. Let X be a Tychonoff space and let $p \in \beta X$. The following are equivalent:

(1) $p \in \beta X \backslash \upsilon X$,

(2) there exists $S \in Z(\beta X)$ such that $p \in S$ and $S \cap X = \emptyset$,

and

(3) there exists a G_δ-set G of βX such that $p \in G$ and $G \cap X = \emptyset$.

Proof

(1) ⇒ (2) Let $\mathcal{A}^p = \{Z \in Z(X) : p \in c\ell_{\beta X} Z\}$; thus

$\mathcal{A}^p$ is the z-ultrafilter on X associated with p (see 4.6(b)). Then by 5.10(b) $\mathcal{A}^p$ does not have C.I.P. Find $\{Z_n : n \in \mathbb{N}\} \subseteq \mathcal{A}^p$ such that $\cap\{Z_n : n \in \mathbb{N}\} = \varnothing$. Because X is z-embedded in βX (see 1.9(m)), for each $n \in \mathbb{N}$ there exists $S_n \in Z(\beta X)$ such that $S_n \cap X = Z_n$. Evidently $cl_{\beta X}Z_n \subseteq S_n$, so $p \in \cap\{S_n : n \in \mathbb{N}\} = S$. By 1.4(i), $S \in Z(\beta X)$, and evidently $S \cap X = \varnothing$.

(2) ⇒ (1) Let S be as given in (2). Find $f \in C(\beta X)$ such that $S = Z(f)$. Then $\beta X \backslash S = f^{\leftarrow}[\mathbb{R}\backslash\{0\}]$. Since $\mathbb{R}\backslash\{0\}$ is Lindelöf, then by 5.11(a) it is realcompact. It follows from 5.9(a)(2) that $\beta X \backslash S$ is realcompact. As $S \cap X = \varnothing$ it follows that $X \subseteq \beta X \backslash S \subseteq \beta X$. Hence by 5.9(f), $\upsilon X \subseteq \beta X \backslash S$. It follows that $p \in \beta X \backslash \upsilon X$.

Finally, note that since zero-sets are G_δ-sets, and every G_δ-set is the union of zero-sets, (2) and (3) are equivalent.

■

The following characterization of realcompactness is an immediate consequence of the above theorem.

(c) **Corollary**. The following are equivalent for a Tychonoff space X:

(1) X is realcompact and

(2) if $p \in \beta X \backslash X$ there exists $S \in Z(\beta X)$ such that $p \in S$ and $S \cap X = \varnothing$.

(d) **Corollary**. An F_σ-subset of a realcompact space is realcompact.

Proof. Let X be realcompact and let $F = \cup\{A_n : n \in \mathbb{N}\}$, where each A_n is a closed subspace of X. Let $j : \beta F \to c\ell_{\beta X}F$ be the Stone extension of the function that embeds F in $c\ell_{\beta X}F$.

Let $p \in \beta F\backslash F$. By 1.8(i), $j(p) \in c\ell_{\beta X}F\backslash F$. We will find $Z \in Z(c\ell_{\beta X}F)$ such that $j(p) \in Z$ and $Z \cap F = \emptyset$. By 1.4(h), $j^{\leftarrow}[Z]$ will be a zero-set of βF containing p and disjoint from F. As p was arbitrary, it will follow that F is realcompact.

Suppose $j(p) \in c\ell_{\beta X}F\backslash X$. By 5.11(c) there exists $S \in Z(\beta X)$ such that $j(p) \in S$ and $S \cap X = \emptyset$. Then $S \cap c\ell_{\beta X}F$ is the required Z described above.

Suppose $j(p) \in (c\ell_{\beta X}F) \cap X$; then $j(p) \in c\ell_X F\backslash F$. For each $n \in \mathbb{N}$ there exists $Z_n \in Z(X)$ such that $j(p) \in Z_n$ and $Z_n \cap A_n = \emptyset$. As X is z-embedded in βX and $\cap\{Z_n : n \in \mathbb{N}\} \in Z(X)$ (see 1.4(i)), there exists $S \in Z(\beta X)$ such that $S \cap X = \cap\{Z_n : n \in \mathbb{N}\}$. Thus $j(p) \in S \cap c\ell_{\beta X}F \in Z(c\ell_{\beta X}F)$ and $(S \cap c\ell_{\beta X}F) \cap F = \emptyset$. Hence $S \cap c\ell_{\beta X}F$ is the required Z described above.

■

(e) **Corollary**. Every cozero-set of a realcompact space is realcompact.

We now establish a correspondence between the zero-sets of X and those of υX. First we establish a lemma.

(f) **Lemma**. If $\emptyset \neq Z \in Z(\upsilon X)$ then $Z \cap X \neq \emptyset$.

Proof. As $X \subseteq \upsilon X \subseteq \beta X$, υX is z-embedded in βX. Thus there exists $S \in Z(\beta X)$ such that $S \cap \upsilon X = Z$. Since $S \cap \upsilon X \neq$

$\emptyset$, it follows from 5.11(b) that $S \cap X \neq \emptyset$. Thus $Z \cap X \neq \emptyset$.

■

(g) **Theorem**. The function $Z \to c\ell_{\upsilon X}Z$ is a lattice isomorphism from $Z(X)$ onto $Z(\upsilon X)$. Thus $c\ell_{\upsilon X}(Z \cap X) = Z$ for each $Z \in Z(\upsilon X)$; if $\mathcal{a} \in \upsilon X \setminus X$ and $Z \in Z(\upsilon X)$, then $\mathcal{a} \in Z$ iff $Z \cap X \in \mathcal{a}$. (Here $\mathcal{a}$ is a Z-ultrafilter on X.)

Proof. We first prove that if $Z \in Z(X)$, then $c\ell_{\upsilon X}Z \in Z(\upsilon X)$. More explicitly, we show that if $f \in C(X)$, then $c\ell_{\upsilon X}Z(f) = Z(\upsilon f)$, where $\upsilon f \in C(\upsilon X)$ and $\upsilon f | X = f$. (υf exists as X is C-embedded in υX.) Obviously $c\ell_{\upsilon X}Z(f) \subseteq Z(\upsilon f)$. Conversely, suppose $p \in \upsilon X \setminus c\ell_{\upsilon X}Z(f)$. Find $g \in C(\upsilon X)$ such that $g(p) = 0$ and $g[c\ell_{\upsilon X}Z(f)] = \{1\}$. If $\upsilon f(p) = 0$, then $p \in Z(g) \cap Z(\upsilon f)$. But $Z(g) \cap Z(\upsilon f) \cap X = Z(g) \cap Z(f) = \emptyset$, which contradicts the preceding lemma. Thus $c\ell_{\upsilon X}Z(f) = Z(\upsilon f)$.

It follows that $Z \to c\ell_{\upsilon X}Z$ is a well-defined function from $Z(X)$ into $Z(\upsilon X)$. It is easily seen to be one-to-one. If $Z \in Z(\upsilon X)$, let $Z = Z(f)$. Using the results of the previous paragraph, we see that $c\ell_{\upsilon X}(Z \cap X) = c\ell_{\upsilon X}(Z(f|X)) = Z(\upsilon(f|X)) = Z(f)$ (see 1.6(d)). Thus our function is onto. It is obvious that if $Z_1, Z_2 \in Z(X)$, then $Z_1 \subseteq Z_2$ iff $c\ell_{\upsilon X}Z_1 \subseteq c\ell_{\upsilon X}Z_2$. Thus by 2.4(c)(2) our function is a lattice isomorphism.

■

We now prove two theorems about the structure of $\beta X \setminus X$ when X is locally compact and realcompact.

(h) **Theorem**. If X is a locally compact, realcompact space, then $Z(\beta X\setminus X) \subseteq R(\beta X\setminus X)$.

Proof. It suffices to show that if $Z \in Z(\beta X\setminus X)$ then $\text{int}_{\beta X\setminus X}Z \neq \emptyset$. To see that this is indeed sufficient, suppose that $p \in Z\setminus c\ell_{\beta X\setminus X}(\text{int}_{\beta X\setminus X}Z)$ for some $Z \in Z(\beta X\setminus X)$. Find $f \in C^*(\beta X\setminus X)$ such that $f(p) = 0$ and $f[c\ell_{\beta X\setminus X}\text{int}_{\beta X\setminus X}Z] = \{1\}$. Then $\emptyset \neq Z \cap Z(f) \in Z(\beta X\setminus X)$ and $\text{int}_{\beta X\setminus X}(Z \cap Z(f)) = \emptyset$.

Since X is locally compact, $\beta X\setminus X$ is compact (see 4.3(e)) and thus C^*-embedded in βX (see 1.9(i)). Thus $\beta X\setminus X$ is z-embedded in βX (see 1.9(m)). Let $Z \in Z(\beta X\setminus X)$, and choose $f \in C^*(\beta X)$ so that $Z = Z(f)\setminus X$. Let $p \in Z$. By 5.11(c) there exists $g \in C^*(\beta X)$ such that $p \in Z(g)$ and $Z(g) \cap X = \emptyset$. Let $h = |f| + |g|$. Then $p \in Z(h) \subseteq Z$. Since $Z(h) \neq \emptyset$ and $Z(h) \cap X = \emptyset$, we can inductively construct a sequence $\{x_n : n \in \mathbb{N}\} \subseteq X$ such that $0 < h(x_{n+1}) < \min\{\frac{1}{n+1}, h(x_n)\}$ for each $n \in \mathbb{N}$. As $Z(h) \cap X = \emptyset$, $\{x_n : n \in \mathbb{N}\}$ is a closed discrete subset of X. Since X is locally compact, we can inductively construct a pairwise disjoint sequence $\{V_n : n \in \mathbb{N}\}$ of open subsets of X such that $x_n \in V_n \subseteq h^{\leftarrow}[[0,\frac{1}{n})]$ and $c\ell_X V_n$ is compact. Since $Z(h) \cap X = \emptyset$ it follows that $\{c\ell_X V_n : n \in \mathbb{N}\} \cup \{X \setminus \cup\{V_n : n \in \mathbb{N}\}\}$ is a locally finite family of closed subsets of X. For each $n \in \mathbb{N}$ define $k_n \in C^*(X)$ so that $k_n(x_n) = 1$, $k_n[X\setminus V_n] = \{0\}$, and $\underline{0} \leq k_n \leq \underline{1}$. Define $k \in F(X,\mathbb{R})$ to be $\Sigma\{k_n : n \in \mathbb{N}\}$. By 1.6(c), $k \in C^*(X)$. Now $\{x_n : n \in \mathbb{N}\}$ is a closed noncompact subset of X; therefore, $\emptyset \neq c\ell_{\beta X}\{x_n : n \in \mathbb{N}\} \setminus X = S$. Since $k[\{x_n : n \in \mathbb{N}\}] = \{1\}$, it follows that $(\beta k)[S] = \{1\}$. Thus $(\beta X\setminus X)\setminus Z(\beta k)$ is an nonempty open subset of $\beta X\setminus X$.

We claim that $(\beta X\setminus X)\setminus Z(\beta k) \subseteq Z(h)$. For suppose

$q \in (\beta X\backslash X)\backslash Z(\beta k)$. Then $(\beta k)^{\leftarrow}[(\beta k)(q)/2, 3(\beta k)(q)/2)]$ is a βX-neighborhood of q disjoint from $X\backslash\cup\{V_n : n \in \mathbb{N}\}$; therefore, $q \in cl_{\beta X}\cup\{V_n : n \in \mathbb{N}\}$. If $h(q) \neq 0$, then find $m \in \mathbb{N}$ such that $h(q) > \frac{1}{m}$. Evidently this implies that $q \notin cl_{\beta X}\cup\{V_n : n > m\}$ and so $q \in cl_{\beta X}\cup\{V_n : 1 \leq n \leq m\}$. As each $cl_X V_n$ is compact, this implies that $q \in cl_X\cup\{V_n : 1 \leq n \leq m\} \subseteq X$, which is a contradiction. Thus $\emptyset \neq (\beta X\backslash X)\backslash Z(\beta k) \subseteq Z(h) \subseteq Z$, so $int_{\beta X\backslash X}Z \neq \emptyset$.

■

One immediate consequence of the above theorem is the following:

(i) **Theorem**. Let X be a realcompact space, let $S \subseteq \beta X\backslash X$, and let $Y = X \cup S$. Then

(1) If Y is pseudocompact then S is dense in $\beta X\backslash X$ and

(2) if X is locally compact and S is dense in $\beta X\backslash X$, then Y is pseudocompact.

Proof.

(1) Suppose S is not dense in $\beta X\backslash X$. Then there is an open subset V of βX such that $\emptyset \neq V\backslash X \subseteq (\beta X\backslash X)\backslash S$. Let $p \in V\backslash X$. As βX is Tychonoff, there exists $Z_0 \in Z(\beta X)$ such that $p \in Z_0 \subseteq V$. As X is realcompact, by 5.11(c) there exists $Z_1 \in Z(\beta X)$ such that $p \in Z_1$ and $Z_1 \cap X = \emptyset$. Thus $p \in Z_0 \cap Z_1 \subseteq \beta X\backslash Y$. Now $Z_0 \cap Z_1 \in Z(\beta X)$, so find $f \in C(\beta X)$ such that $Z(f) = Z_0 \cap Z_1$. Define $g \in F(Y,\mathbb{R})$ by $g(y) = 1/f(y)$ for each $y \in Y$. By 1B(5) g is a well-defined member of C(Y). Since $Z(f) \neq \emptyset$, it follows that $g \notin C^*(Y)$; thus Y is not pseudocompact (see 1.11(c)).

(2) Suppose Y is not pseudocompact. Choose $f \in C(Y) \setminus C^*(Y)$. By 4.6(g), f can be continuously extended to $f^* : \beta Y \to \alpha\mathbb{R}$, where $\alpha\mathbb{R}$ denotes the one-point compactification $\mathbb{R} \cup \{\omega\}$ of $\mathbb{R}$. Now $f^*[\beta Y]$ is compact, so if $f^*[\beta Y] \subseteq \mathbb{R}$, then $f^*[\beta Y]$ would be bounded. Thus f[Y] would be a bounded subset of $\mathbb{R}$, contradicting our choice of f. It follows that $(f^*)^{\leftarrow}(\omega) \neq \emptyset$. Since $\alpha\mathbb{R}$ is metrizable, by 1.4(g,h), $(f^*)^{\leftarrow}(\omega)$ is a nonempty zero-set of βY disjoint from Y. Now $\beta Y \equiv_X \beta X$ by 4.5(p), so by 5.11(h), $\mathrm{int}_{\beta X \setminus X}(f^*)^{\leftarrow}(\omega) \neq \emptyset$. But $[\mathrm{int}_{\beta X \setminus X}(f^*)^{\leftarrow}(\omega)] \cap S = \emptyset$, so S is not dense in $\beta X \setminus X$.

■

Our final task is to prove that the realcompact spaces constitute a reasonably broad class of Tychonoff spaces. Frequently, when proving that a space is realcompact, one is led to the problem of deciding whether its discrete subspaces are realcompact. Unfortunately the answer to this question is rather complex. In 5I the reader is invited to prove that the discrete space D(m) is realcompact iff m is not a Ulam-measurable cardinal (see 2P). Since the consistency of ZFC implies the consistency of the nonexistence of measurable cardinals, this usually does not pose a problem. However, it means that many theorems asserting the realcompactness of a space X of a certain type must have this form: if X has no discrete subspaces of measurable cardinality (i.e., if all discrete subspaces of X are realcompact), and if such-and-such is true, then X is realcompact.

In 5.11(l) below we prove that a normal θ-refinable space with no discrete subspaces of measurable cardinality is realcompact. Since θ-refinability is one of the weakest conditions asserting the existence of open covers with certain local finiteness properties, this theorem is rather powerful (and it proof is commensurately complicated). In particular, every metric space is paracompact (see [Wi]), every paracompact space is metacompact (see [Wi]), and every metacompact space is θ-refinable. Thus each metric space, paracompact space, and normal metacompact space is, in the absence of measurable cardinals, realcompact.

(j) <u>Definition</u>. A space X is **θ-refinable** if given an open cover $\mathcal{C}$ of X, there exists a sequence $\{\mathcal{C}(i) : i \in \mathbb{N}\}$ of open covers of X, each refining $\mathcal{C}$, such that for each $x \in X$ there exists $i(x) \in \mathbb{N}$ such that $|\{C \in \mathcal{C}(i(x)) : x \in C\}| < \aleph_0$.

(k) <u>Lemma</u>. Let $\mathcal{F}$ be a collection of closed subsets of a space X with the property that if $\mathcal{F}_0 \subseteq \mathcal{F}$, then $\cup\mathcal{F}_0$ and $\cup(\mathcal{F}\setminus\mathcal{F}_0)$ are completely separated in X. If $|\mathcal{F}|$ is not Ulam-measurable and if $p \in \beta X\setminus\cup\{c\ell_{\beta X}F : F \in \mathcal{F}\}$, then there exists $f \in C(\beta X)$ such that $f(p) = 0$, and $f[\cup\mathcal{F}] \subseteq (0,1]$, and $\underline{0} \leq f \leq \underline{1}$.

<u>Proof</u>. If $p \notin c\ell_{\beta X}(\cup\mathcal{F})$ then we can produce f because βX is Tychonoff. So, assume $p \in c\ell_{\beta X}(\cup\mathcal{F})$.

Give the set $\mathcal{F}$ (whose elements are closed subsets of X) the discrete topology. As $|\mathcal{F}|$ is not Ulam-measurable, this topology is realcompact. Define $\phi : \cup\mathcal{F} \to \mathcal{F}$ as follows: if $x \in \cup\mathcal{F}$, then $\phi(x)$ is the unique member of $\mathcal{F}$ to which x belongs. If B and C are disjoint closed subsets of $\beta\mathcal{F}$, then $\phi^{\leftarrow}[B] = \cup\{F : F \in B \cap \mathcal{F}\}$ and $\phi^{\leftarrow}[C] = \cup\{F : F \in C \cap \mathcal{F}\}$; therefore, by hypothesis $\phi^{\leftarrow}[B]$ and

$\phi^{\leftarrow}[C]$ are completely separated in X. Thus $c\ell_{\beta X}\phi^{\leftarrow}[B] \cap c\ell_{\beta X}\phi^{\leftarrow}[C] = \emptyset$ by 4.6(g)(6). It follows from 4.1(m) that there is a continuous function $k : c\ell_{\beta X}(\cup F) \to \beta F$ such that $k|\cup F = \phi$. Let $F \in F$. As $\{F\}$ is open in βF, it follows that

$$k^{\leftarrow}[\beta F\backslash\{F\}] \supseteq c\ell_{\beta X}[(\cup F)\backslash F] = c\ell_{\beta X}(\cup F)\backslash c\ell_{\beta X}F \ ;$$

the latter equality follows from our hypotheses on F and from 4.6(g)(6). Hence $k^{\leftarrow}(F) \subseteq c\ell_{\beta X}F$. It follows that $k(p) \in \beta F\backslash F$ (since $p \notin \cup\{c\ell_{\beta X}F : F \in F\}$).

As F is realcompact, there exists $g \in C(\beta F)$ such that $g(F) > 0$ for each $F \in F$, and $g(k(p)) = 0$; to see this, choose S as in 5.11(c)(2) and choose $g \in C(\beta F)$ so that $\underline{1} \geq g \geq \underline{0}$ and $S = Z(g)$. Thus $g \circ k \in C(c\ell_{\beta X}(\cup F))$, and $g \circ k[\cup F] \subseteq (0,1]$ while $(g \circ k)(p) = 0$. As $c\ell_{\beta X}(\cup F)$ is closed in X and hence C-embedded, there exists $f \in C(\beta X)$ such that $f|c\ell_{\beta X}(\cup F) = g \circ k$ and $\underline{0} \leq f \leq \underline{1}$. Evidently f has the required properties.

■

(l) **Theorem**. Let X be a normal θ-refinable space. If X has no closed discrete subsets of measurable cardinality, then X is realcompact.

Proof. Let $p \in \beta X\backslash X$. We will produce an $f \in C(\beta X)$ for which $p \in Z(f)$ and $Z(f) \subseteq \beta X\backslash X$. It will follow from 5.11(c)(2) that X is realcompact.

Since βX is Hausdorff there exists an open cover U of X such that $p \in \beta X\backslash\cup\{c\ell_{\beta X}U : U \in U\}$. As X is θ-refinable, there exists a collection $\{\mathcal{V}_i : i \in \mathbb{N}\}$ of open refinements of $\mathcal{U}$ such that for each $x \in X$ there exists $i(x) \in \mathbb{N}$ for which $\{V \in \mathcal{V}_{i(x)} : x \in V\}$ is finite.

Fix $i \in \mathbb{N}$. We will inductively construct functions $\{f_{i,j} : j \in \mathbb{N}\} \subseteq C(\beta X)$ for which the following conditions hold: (i) $f_{i,j}(p) = 0$ for each $j \in \mathbb{N}$ and (ii) if $k \in \mathbb{N}$, $x \in X$, and $|\{V \in \mathcal{V}_i : x \in V\}| = k$ then there exists $s \in \{1,\ldots,k\}$ such that $f_{i,s}(x) > 0$.

Let $f_{i,1} = \underline{0}$. Now let $n \in \mathbb{N}$ and inductively assume that we have constructed functions $\{f_{i,j} : j \leqslant n\} \subseteq C(\beta X)$ such that

(1) $f_{i,j}(p) = 0$ for each $j \leqslant n$ and $\underline{0} \leqslant f_{i,j} \leqslant \underline{1}$, and

(2) if $k \leqslant n$, $x \in X$, and $|\{V \in \mathcal{V}_i : x \in V\}| = k$ then there exists $s \in \{1,\ldots,k\}$ such that $f_{i,s}(x) > 0$.

Let $[\mathcal{V}_i]^{n+1} = \{\mathcal{A} \subseteq \mathcal{V}_i : |\mathcal{A}| = n + 1\}$. For each $\mathcal{A} \in [\mathcal{V}_i]^{n+1}$ let $F(\mathcal{A}) = \cap\{Z(f_{i,j}|X) : 1 \leqslant j \leqslant n\} \backslash \cup\{V \in \mathcal{V}_i : V \notin \mathcal{A}\}$.

We claim that $\{F(\mathcal{A}) : \mathcal{A} \in [\mathcal{V}_i]^{n+1}\}$ is a locally finite family of closed subsets of X. To show this, we will produce, for each $x \in X$, a neighborhood $N(x)$ of x such that $N(x) \cap F(\mathcal{A}) \neq \varnothing$ for at most one $\mathcal{A} \in [\mathcal{V}_i]^{n+1}$. (Thus, in particular, if $\mathcal{A}_1 \neq \mathcal{A}_2$ then $F(\mathcal{A}_1) \cap F(\mathcal{A}_2) = \varnothing$.) First suppose that $|\{V \in \mathcal{V}_i : x \in V\}| > n$. Choose $\{V_s : s = 1 \text{ to } n + 1\} = \mathcal{A}_0 \subseteq \mathcal{V}_i$ such that $x \in \cap\mathcal{A}_0 = W$. If $W \cap F(\mathcal{A}) \neq \varnothing$ for some $\mathcal{A} \in [\mathcal{V}_i]^{n+1}$ then $\mathcal{A}_0 \subseteq \mathcal{A}$, so $\mathcal{A}_0 = \mathcal{A}$ as $|\mathcal{A}_0| = n + 1 = |\mathcal{A}|$. Thus x has a neighborhood meeting at most one $F(\mathcal{A})$.

If $|\{V \in \mathcal{V}_i : x \in V\}| \leqslant n$ then by our induction hypothesis (2) there exists $s \leqslant n$ such that $f_{i,s}(x) > 0$. But $F(\mathfrak{a}) \subseteq Z(f_{i,s} | X)$ for each $\mathfrak{a} \in [\mathcal{V}_i]^{n+1}$, so $f_{i,s}^{\leftarrow}[(0,+\infty)]$ is a neighborhood of x missing each $F(\mathfrak{a})$. Hence $\{F(\mathfrak{a}) : \mathfrak{a} \in [\mathcal{V}_i]^{n+1}\} = \mathcal{F}$ is a locally finite collection of closed sets as claimed.

It follows that if $\mathcal{F}_0 \subseteq \mathcal{F}$, then $\bigcup\mathcal{F}_0$ and $\bigcup(\mathcal{F}\backslash\mathcal{F}_0)$ are disjoint closed subsets of X. Since X is normal, these sets are completely separated in X. By 5.11(k) above there exists $f_{i,n+1} \in C(\beta X)$ such that $f_{i,n+1}(p) = 0$ and $f_{i,n+1}(x) > 0$ if $x \in \bigcup\{F(\mathfrak{a}) : \mathfrak{a} \in [\mathcal{V}]^{n+1}\}$. Thus (1) is satisfied if n is replaced by $n + 1$. To see that (2) is satisfied when n is replaced by $n + 1$, suppose $x \in X$ and that $|\{V \in \mathcal{V}_i : x \in V\}| \leqslant k$, where $k \leqslant n + 1$. If $k \leqslant n$, then our induction hypotheses show that there exists $s \leqslant n$ such that $f_{i,s}(x) > 0$. If $k = n + 1$, let $\{V \in \mathcal{V}_i : x \in V\} = \mathfrak{a}_0$. Then $\mathfrak{a}_0 \in [\mathcal{V}_i]^{n+1}$. If there exists $s \leqslant n$ such that $f_{i,s}(x) > 0$ then (2) is satisfied for our x when n is replaced by $n + 1$. If there is no such s, then $x \in F(\mathfrak{a}_0)$ and $f_{i,n+1}(x) > 0$. Thus (2) is always satisfied when n is replaced by $n + 1$.

By our choice of $\{\mathcal{V}_i : i \in \mathbb{N}\}$, if $x \in X$ there will exist $(n,m) \in \mathbb{N} \times \mathbb{N}$ such that $|\{V \in \mathcal{V}_n : x \in V\}| \leqslant m$, and hence $f_{n,m}(x) > 0$. Let $f(x) = \Sigma\{2^{-(i+j)}f_{i,j}(x) : i \in \mathbb{N} \text{ and } j \in \mathbb{N}\}$. Then $f(p) = 0$, $f(x) > 0$ if $x \in X$, and $f \in C(\beta X)$ (see 1B(6)). It follows from 5.11(c) that X is realcompact.

■

(m) **Corollary**. A paracompact, metric, or normal metacompact space with no discrete subspaces of Ulam-measurable cardinality (equivalently, with no discrete subspaces that fail to be realcompact) is realcompact.

Chapter 5 — Problems

5A. Projective maxima vs. maximum P-extensions. In this problem we outline a proof that there is no maximum H-closed extension of the discrete space $\mathbb{N}$.

(1) Suppose that $h\mathbb{N}$ were a maximum H-closed extension of $\mathbb{N}$. Show that $h\mathbb{N} \equiv_{\mathbb{N}} \kappa\mathbb{N}$.

(2) Let Y be a countably infinite space, let $j : \mathbb{N} \to Y$ be a bijection, and suppose there exists $f \in C(\kappa\mathbb{N}, \kappa Y)$ such that $f|\mathbb{N} = j$. Show that each closed nowhere dense subset F of Y is H-closed. (Hint: Consider $f[c\ell_{\kappa\mathbb{N}} j^{\leftarrow}[F]]$.)

(3) Let j be a bijection from $\mathbb{N}$ onto the rationals $\mathbb{Q}$. Show that j cannot be extended to a continuous function $f \in C(\kappa\mathbb{N}, \kappa\mathbb{Q})$.

(4) Show that there is no maximum H-closed extension of $\mathbb{N}$.

5B. An extension property that is not simply generated. Let m be an infinite cardinal and let $P(m)$ be the property of being Tychonoff and m-bounded. We saw in 5.6(d) that $P(m)$ is a Tychonoff extension property.

Let E be a space for which E-complete regularity is the

Tychonoff property. Let ω_α be an ordinal such that $cf(\omega_\alpha) > \max\{m, |E|\}$, and give ω_α the order topology. By 2.5(n), ω_α is normal.

(1) Show that ω_α is m-bounded and $|\beta\omega_\alpha \setminus \omega_\alpha| = 1$.

(2) If $f \in C(\omega_\alpha, E)$ show that there exists exactly one $x_0 \in E$ such that $f^{\leftarrow}(x_0)$ is an unbounded subset of ω_α.

(3) If $f \in C(\omega_\alpha, E)$, show that f can be continuously extended to $f^* \in C(\beta\omega_\alpha, E)$.

(4) Show that $P(m)$ is not E-compactness for any space E.

5C. <u>The Hausdorff property vs. E-complete regularity</u>. Show that there is no space E such that the class of E-completely regular spaces is the class of Hausdorff spaces. (Hint: If such an E exists, show that every Hausdorff space would have to be E-compact. Now argue as in 5B.)

5D. <u>Tychonoff vs. zero-dimensional extension properties</u>.

(1) Prove 5.7(b).

(2) If P is a Tychonoff extension property show that P_0 (defined in 5.7) is a zero-dimensional extension property.

(3) If P is a zero-dimensional extension property show that $[0,1] \times P$ (defined in 5.7) is a Tychonoff extension property and $([0,1] \times P)_0 = P$.

(4) If P is a zero-dimensional extension property, show that

a Tychonoff space X is $([0,1] \times P)$-compact iff there is a perfect continuous surjection from X onto a space with P. (Hint: Use 1M, 5.9(a), and 4.2(f).)

(5) Let P be a zero-dimensional extension property. If X is a zero-dimensional space, let $f : \beta X \to \beta_0 X$ extend the identity map. Show that $\gamma_{[0,1]\times P}X = f^{\leftarrow}[\gamma_P X]$. (Hint: For one direction, let Y be closed in K $\times$ P where K is compact and P has P. If $g \in C(X,Y)$ consider $\beta(\Pi_K \circ g)$ and $P(\Pi_P \circ g) \circ f$ on $f^{\leftarrow}[\gamma_P X]$ and construct a product map into a product; then use 5.3(d).)

5E. The construction of $\gamma_{\mathbb{N}}X$. Recall that $\gamma_{\mathbb{N}}X$ denotes the maximum $\mathbb{N}$-compact extension of the zero-dimensional space X (see the remarks after 5.4(a)). Let X be a zero-dimensional space, and let $\alpha\mathbb{N} = \mathbb{N} \cup \{\omega\}$ be the one-point compactification of $\mathbb{N}$. Define the countable intersection property (C.I.P.) for clopen ultrafilters on X is the obvious way. Put $T_0 = X \cup \{\mathcal{Q} \in \beta_0 X \setminus X : \mathcal{Q}$ has C.I.P.$\}$. Show the following:

(1) If $f \in C(X,\mathbb{N})$, there is a unique $g \in C(\beta_0 X, \alpha\mathbb{N})$ such that $g(x) = f(x)$ for each $x \in X$. Denote g by αf.

(2) Let $f \in C(X,\mathbb{N})$. Then $\alpha f[T_0] \subseteq \mathbb{N}$.

(3) Let $\mathcal{Q} \in \beta_0 X \setminus T_0$. Then there exists $f \in C(X,\mathbb{N})$ such that $\alpha f(\mathcal{Q}) = \omega$.

(4) $T_0 = \bigcap\{(\alpha f)^{\leftarrow}[\mathbb{N}] : f \in C(\beta_0 X, \alpha\mathbb{N}) \text{ and } f[X] \subseteq \mathbb{N}\}$.

(5) $T_0 \equiv_X \gamma_{\mathbb{N}} X$.

(6) X is $\mathbb{N}$-compact iff each clopen ultrafilter on X with C.I.P. is fixed.

5F. <u>More properties of the Hewitt realcompactification</u>. Let X be a Tychonoff space.

(1) Let $\mathcal{A}$ be a z-ultrafilter on X. Show that $\mathcal{A}$ has C.I.P. iff the intersection of each countable subfamily of $\mathcal{A}$ belongs to $\mathcal{A}$.

(2) Show that if $\{Z_n : n \in \mathbb{N}\}$ is a countable subfamily of $Z(X)$, then $c\ell_{\upsilon X}[\bigcap\{Z_n : n \in \mathbb{N}\}] = \bigcap\{c\ell_{\upsilon X} Z_n : n \in \mathbb{N}\}$.

(3) Show that if T is a realcompact extension of X for which $c\ell_T[\bigcap\{Z_n : n \in \mathbb{N}\}] = \bigcap\{c\ell_T Z_n : n \in \mathbb{N}\}$ for each countable subset $\{Z_n : n \in \mathbb{N}\}$ of $Z(X)$, then $T \equiv_X \upsilon X$. (Hint: Use 5.10(b), 4.6(h), and 1.9(j).)

(4) Show that the following are equivalent:

(a) $\upsilon X = \beta X$,

(b) X is pseudocompact, and

(c) each nonempty zero-set of βX intersects X.

(5) Show that a space is compact iff it is pseudocompact and realcompact.

(6) Show that βX is not first countable at any point of

$\beta X \setminus X$. (Hint: If $p \in \beta X \setminus X$ and βX is first countable at p, let $T = \beta X \setminus \{p\}$. Infer that $\beta X = \beta T$ (see 4.5(p)(3)) and that $\{p\} \in Z(\beta T)$. Use 1Q(7), together with what you know about the cardinality of $\beta\mathbb{N} \setminus \mathbb{N}$, to show that T is pseudocompact. Show that this yields a contradiction.)

(7) Show that $\upsilon(X_\delta) = (\upsilon X)_\delta$, in the sense that $j^{\leftarrow}[X]$ is homeomorphic to X_δ and is dense and C-embedded in the realcompact space $(\upsilon X)_\delta$. (The space X_δ is defined in 1W(6), and $j : (\upsilon X)_\delta \to \upsilon X$ is the identity function on the underlying set of υX.)

(8) Show that if $X \subseteq T_n \subseteq \beta X$ and T_n is realcompact for each $n \in \mathbb{N}$, then $\cup\{T_n : n \in \mathbb{N}\}$ is realcompact. (Hint: Use 5.11(b).)

5G. <u>Realcompact zero-dimensional spaces vs. $\mathbb{N}$-compact spaces.</u> See 5E for notation and useful results.

(1) Let X be a zero-dimensional space. Show that if $\{B_n : n \in \mathbb{N}\}$ is a countable subfamily of $B(X)$, then $c\ell_{\gamma_{\mathbb{N}}X}[\cap\{B_n : n \in \mathbb{N}\}] = \cap\{c\ell_{\gamma_{\mathbb{N}}X}B_n : n \in \mathbb{N}\}$.

(2) If X is realcompact, and if each zero-set of X is the intersection of countably many clopen subsets of X, show that X is $\mathbb{N}$-compact.

(3) If X is strongly zero-dimensional, show that $\gamma_{\mathbb{N}}X \equiv_X \upsilon X$. (Hint: Use 4.7(j).)

(4) Show that the following conditions on a zero-dimensional

space X are equivalent:

(a) $\gamma_{\mathbb{N}}X = \beta_0 X$,

(b) there is no continuous surjection from X onto $\mathbb{N}$, and

(c) X is pseudocompact.

(Hint: (c) $\Rightarrow$ (b) is trivial. To show that (b) $\Rightarrow$ (a), assume (a) fails and use 5E(5) to select a decreasing sequence of nonempty clopen subsets of X with empty intersection. Use this sequence to define the desired surjection. To show (a) $\Rightarrow$ (c), assume (c) fails. Find $(x_n)_{n\in\mathbb{N}}$ and $f \in C(X)$ such that $|f(x_{n+1})| \geq |f(x_n)| + 1$ for each $n \in \mathbb{N}$. Find $A_n \in B(X)$ such that $\emptyset \neq A_n \subseteq f^{\leftarrow}[(|f(x_n)| - 1/3, |f(x_n)| + 1/3)]$ for each n. Now use 5E.) (Note: Not all realcompact zero-dimensional spaces are $\mathbb{N}$-compact; see 5P.)

5H. <u>Weakly homogeneous spaces and $\mathbb{N}$-compactness</u>. A Tychonoff space X is **weakly homogeneous** if, for all $p \in \beta X\backslash X$, the set $\{\beta f(p) : f \in C(X,X)\}$ is dense in $\beta X\backslash X$.

(1) Show that an $\mathbb{N}$-compact space is weakly homogeneous. (Hint: Let X be $\mathbb{N}$-compact. If V is a nonempty open subset of $\beta X\backslash X$, find U open in βX such that $\emptyset \neq c\ell_{\beta X}U\backslash X \subseteq V$. Apply 1Q(6) to find a countable closed

discrete subspace A of $c\ell_X(U \cap X)$. If $p \in \beta X \setminus X$, find $h \in C(X,\mathbb{N})$ such that $(\beta h)(p) \in \beta\mathbb{N}\setminus\mathbb{N}$. Map $\mathbb{N}$ onto A and compose.)

(2) If X is locally compact, realcompact, zero-dimensional and weakly homogeneous, show that X is $\mathbb{N}$-compact. (Hint: If not, find $r \in \beta_0 X \setminus X$ such that $\beta_0 g(r) \in \mathbb{N}$ for each $g \in C(X,\mathbb{N})$. Let $k : \beta X \to \beta_0 X$ be the Stone extension of the function embedding X into $\beta_0 X$, and choose $p \in k^{\leftarrow}(r)$. Let $T = X \cup \{(\beta f)(p) : f \in C(X,X)\}$. Show that $\mathbb{N}$ is a continuous image of T and invoke 5.11(i).)

5I. Realcompactness and $\mathbb{N}$-compactness for discrete spaces. Let m be an infinite cardinal.

(1) Show that $\gamma_{\mathbb{N}} D(m) = \upsilon D(m) = D(m) \cup \{\mathcal{U} \in \beta D(m)\setminus D(m) : \mathcal{U}$ has C.I.P.$\}$.

(2) Show that the following conditions on m are equivalent:

(a) m is not Ulam-measurable,

(b) D(m) is realcompact, and

(c) D(m) is $\mathbb{N}$-compact.

5J. Extension properties and ordinal spaces.

(1) Let E be the property of being homeomorphic to some

countably compact ordinal space. Let $X \in \text{Reg}(E)$. Show that $X \in K(E)$ iff for each free clopen ultrafilter A on X there exists an ordinal $\delta(A)$ of uncountable cofinality, and a subset $\{U_\alpha : \alpha < \delta(A)\}$ of A, such that $\bigcap\{U_\alpha : \alpha < \delta(A)\} = \varnothing$ and for each $\alpha_0 < \delta(A)$, $U_{\alpha_0} \subseteq \bigcap\{U_\alpha : \alpha < \alpha_0\}$.

(2) Let E be as in (1). Construct $\gamma_{K(E)}X$ as a subspace of $\mathcal{B}_0(X)$ if $X \in \text{Reg}(E)$.

5K. Sups and infs of extension properties.

(1) Let $\{P_\alpha : \alpha \in I\}$ be a set of Tychonoff extension properties. Then there is a smallest Tychonoff extension property P such that $P_\alpha \subseteq P$ for each $\alpha \in I$ (denote P by $\bigvee_\alpha P_\alpha$). Find an explicit description of P in terms of the P_α's.

There is also a largest extension property Q such that $Q \subseteq P_\alpha$ for each $\alpha \in I$ (denote Q by $\bigwedge_\alpha P_\alpha$). Find an explicit description of Q in terms of the P_α's.

(2) Let X be a Tychonoff space. Show that $\gamma_{\bigvee_\alpha P_\alpha} X = \bigcap\{\gamma_{P_\alpha} X : \alpha \in I\}$.

(3) Find Tychonoff extension properties P and Q and a Tychonoff space X, such that $\gamma_{P \wedge Q} X \neq \gamma_P X \cup \gamma_Q X$. (Hint: If you choose P and Q properly, you can let X be $D(\aleph_1)$.)

(4) Statements similar to the above are valid for

zero-dimensional extension properties; find them and prove them.

(5) If E and F are two spaces, show that $K(P_E) \vee K(P_F) = K(P_{E \times F})$.

5L. Extension properties and compact subspaces.

(1) Let P be an extension property, let X be P-regular, and let A be a P-embedded subspace of X. Show that $c\ell_{\gamma_P X} A \equiv_A \gamma_P A$. Conclude that if A has P then $c\ell_{\gamma_P X} A = A$.

(2) Let P be a zero-dimensional extension property, and let Y be a zero-dimensional space that has the form $X \cup K$, where X has P and K is compact. Prove that Y has P. (Hint: If $p \in \gamma_P Y \setminus Y$, separate p from K by a clopen set and use (1).)

(3) Let C denote the Cantor set constructed by deleting "open middle thirds" from [0,1]. Let E denote the set of endpoints of these deleted open intervals. Let S denote the following subspace of $\mathbb{R}^2$:

$$S = \cup\{M(x) : x \in C\} \text{ where}$$

L(x) denotes (for each $x \in C$) the straight line segment joining (x,0) to $(\frac{1}{2},\frac{1}{2})$, and

$$M(x) = \begin{cases} \{(r,y) \in L(x) : y \text{ is rational}\} & \text{if } x \in E \\ \{(r,y) \in L(x) : y \text{ is irrational}\} & \text{if } x \in C \setminus E. \end{cases}$$

Prove that S is connected and that $S \setminus \{(\frac{1}{2},\frac{1}{2})\}$ is totally disconnected.

(4) Let P be the following topological property: X has P if X is Tychonoff and the connected components of X are compact. Show that P is a Tychonoff extension property.

(5) Prove that (2) above fails if "zero-dimensional" is replaced by "Tychonoff" by using the property P and the space S.

(6) Prove that (2) remains true if "a zero-dimensional extension property" is replaced by "realcompactness" and if Y is assumed to be Tychonoff (rather than zero-dimensional). (Hint: Let $\mathfrak{a}$ be a free z-ultrafilter on Y. Find $Z_0 \in \mathfrak{a}$ such that $Z_0 \cap K = \emptyset$. Now find a free z-ultrafilter on X that contains $\{Z \in \mathfrak{a} : Z \subseteq Z_0\}$.)

5M. Hereditarily P-spaces. Let P be an extension property and let Y be a P-regular space. Consider the following set of conditions on Y.

(a) If X is P-regular, $f \in C(X,Y)$, and $f^{\leftarrow}(y)$ is compact for each $y \in Y$, then X has P.

(b) If X is P-regular and $f : X \to Y$ is a continuous

bijection then X has P.

(c) Each open subspace of Y has P.

(d) Each subspace of Y has P.

(e) If $y \in Y$ then $Y \setminus \{y\}$ has P.

(1) Show that (a) $\Rightarrow$ (b) $\Rightarrow$ (c) $\Longleftrightarrow$ (d) $\Longleftrightarrow$ (e). (Hint: For (b) $\Rightarrow$ (c), let $S \subseteq Y$ and let $X = S \oplus (Y \setminus S)$.)

(2) Suppose that P has the following property: if X is P-regular and $X = S \cup K$, where S has P and K is compact, then X has P. Show that conditions (a) — (e) are equivalent. (Hint: Let $Pf : \Upsilon_P X \to Y$ extend f. Note that $X \subseteq (Pf)^{\leftarrow}[Y \setminus \{y\}] \cup f^{\leftarrow}(y) \subseteq \Upsilon_P X$, and apply 5.9(a).)

(3) Show that if P is a zero-dimensional extension property then conditions (a) — (e) on the zero-dimensional space Y are equivalent. (See 5L.)

(4) Show that if P is realcompactness and if Y is Tychonoff, then conditions (a) — (e) on Y are equivalent.

5N. m-compact and zero-dimensionally m-compact spaces. Let U be a clopen ultrafilter (respectively, z-ultrafilter) on the zero-dimensional (respectively, Tychonoff) space X and let m be an infinite cardinal. Then U is said to have the **m-intersection property** if every intersection of fewer than m members of U is nonempty. The zero-dimensional (respectively, Tychonoff) space X is called **zero-dimensionally m-compact** (respectively, **m-compact**) if every clopen ultrafilter (respectively, z-ultrafilter) on X with the m-intersection property is fixed.

(Note that what we called the countable intersection property in 5.10(a) would, in this terminology, be called the $\aleph_1$-intersection property.)

(1) Show that a zero-dimensional (respectively, Tychonoff) space X is zero-dimensionally $\aleph_0$-compact (respectively, $\aleph_0$-compact) iff it is compact, and is zero-dimensionally $\aleph_1$-compact (respectively $\aleph_1$-compact) iff it is $\mathbb{N}$-compact (respectively, realcompact). (See 5E(6).)

(2) If X is Tychonoff define $\upsilon_m X$ to be $X \cup \{\mathcal{U} \in \beta X \setminus X : \mathcal{U}$ has the m-intersection property$\}$. (Here we view points of $\beta X \setminus X$ as free z-ultrafilters on X; see 4.6(b).) Show that $\upsilon_m X$ is m-compact. (Hint: If $m \geqslant \aleph_1$, note that $\upsilon_m X \subseteq \upsilon X$ and use 5.11(g).)

(3) Let X be Tychonoff, let Y be m-compact, and let $f \in C(X,Y)$. Prove that there exists $F \in C(\upsilon_m X,Y)$ such that $F|X = f$. (Hint: Note that $\beta f[\upsilon_m X] \subseteq \upsilon Y$ and use 5.11(g) to show that if $\mathcal{U} \in \upsilon_m X$ and $Z \in \beta f(\mathcal{U})$, then $f^{\leftarrow}[Z] \in \mathcal{U}$.)

(4) Let $P(m)$ denote the property of being m-compact. Prove that $P(m)$ is a Tychonoff extension property and that $\gamma_{P(m)} X = \upsilon_m X$ for each Tychonoff space X.

(5) If X is zero-dimensional, define $\upsilon_{m,o} X$ to be $X \cup \{\mathcal{U} \in \beta_0 X \setminus X : \mathcal{U}$ has the m-intersection property$\}$. Define $P(m,o)$ to be the property of being zero-dimensionally m-compact. By arguing as in (2) — (5) above, show that

$P(m,o)$ is a zero-dimensional extension property and that $\gamma_{P(m,o)}X = \upsilon_{m,o}X$ for each zero-dimensional space X.

Warning: Being zero-dimensional and m-compact is **not** equivalent to being zero-dimensionally m-compact. See 5P(19).

50. E-compactness vs. m-compactness and zero-dimensional m-compactness. In this problem we show that for each infinite cardinal m, m-compactness and zero-dimensional m-compactness are E-compactness for suitably chosen spaces E.

(1) Let k be an infinite cardinal and let X be a Tychonoff space that can be written as $X = \cup\{K_i : i \in I\}$, where each K_i is compact and $|I| \leq k$. Show that X is k^+-compact.

(2) Let k be an infinite cardinal and define $P(k^+)$ to be $[0,1]^k \setminus \{\underline{0}\}$. Show that $P(k^+)$ can be written as the union of k compact subsets and therefore is k^+-compact.

(3) Show that every $P(k^+)$-compact space is k^+-compact.

(4) Show that every k^+-compact space is $P(k^+)$-compact. (Hint: Suppose X were a k^+-compact space that is not $P(k^+)$-compact. Choose $\mathcal{U} \in \gamma_{P(k^+)}X \setminus X$. (We regard $\mathcal{U}$ as a free z-ultrafilter on X.) Find $\{Z_i : i \in I\} \subseteq \mathcal{U}$ such that $|I| = k$ and $\cap\{Z_i : i \in I\} = \emptyset$. For each $i \in I$ find $f_i \in C(X,[0,1])$ such that $Z_i = Z(f_i)$. Construct $f \in C(X,P(k^+))$ using these f_i's so that f cannot be continuously extended to $X \cup \{\mathcal{U}\}$. Show this yields a contradiction.) Thereby conclude that

k^+-compactness is $P(k^+)$-compactness for each infinite cardinal k.)

(5) Let m be an uncountable cardinal that is not a successor (i.e., there is no k such that $m = k^+$). Define $P(m)$ to be $\Pi\{P(k^+) : k \text{ is a cardinal less than } m\}$. Prove that m-compactness is $P(m)$-compactness. Thus conclude that for every infinite cardinal k there exists a space $E(k)$ such that k-compactness is $E(k)$-compactness.

(6) Using $\{0,1\}^k \setminus \{\underline{0}\}$ in place of $[0,1]^k \setminus \{\underline{0}\}$, show that for each infinite cardinal k there is a space $F(k)$ such that zero-dimensional k-compactness is $F(k)$-compactness.

5P. Realcompact zero-dimensional spaces vs. E-compact spaces. In this lengthy problem we show that if it is assumed that Ulam-measurable cardinals do not exist, then there is no space E such that a space is realcompact and zero-dimensional iff it is E-compact. (In other words, the property of being realcompact and zero-dimensional is not simply generated.) The approach is to find a proper class C of cardinal numbers such that for each $m \in C$, there exists a realcompact zero-dimensional space $X(m)$ that is not zero-dimensionally m^+-compact. The process of constructing the spaces $X(m)$ is rather involved, and utilizes the results of several previous problems.

Let m be an infinite cardinal.

Let X be a metric space of weight m and cardinality 2^m. (We will produce such an X in (14) below.) Let λ denote the initial ordinal of cardinality 2^m. Write X as $\{x_\alpha : \alpha < \lambda\}$ (where $\alpha \neq \gamma$ implies $x_\alpha \neq x_\gamma$), and for each $\alpha < \lambda$, let $X(\alpha) = \{x_\gamma : \gamma < \alpha\}$. Put $S = \{(F,G) : F, G \in \mathbb{P}(X), |F| \leqslant m, |G| \leqslant m, \text{ and } |c\ell_X F \cap c\ell_X G| = 2^m\}$.

(1) Show that $|S| = 2^m$. Then index S as $\{(F_\alpha,G_\alpha) : \alpha < \lambda\}$ so that $(F_\alpha,G_\alpha) = (F_\beta,G_\beta)$ iff $\alpha = \beta$.

(2) Show that there exists a one-to-one function $\phi : \lambda \to \lambda$ such that $\phi(0) = \min\{\alpha < \lambda : F_0 \cup G_0 \subseteq X(\alpha)$ and $x_\alpha \in c\ell_X F_0 \cap c\ell_X G_0\}$ and $\phi(\gamma) = \min(\{\alpha < \lambda : F_\gamma \cup G_\gamma \subseteq X(\alpha)$ and $x_\alpha \in c\ell_X F_\gamma \cap c\ell_X G_\gamma\}\setminus\{\phi(\beta) : \beta < \gamma\})$. (Use a transfinite inductive construction.) Note that $0 \notin \phi[\lambda]$.

(3) Let $\gamma < \lambda$. Show that there exists a sequence $s_\gamma : \mathbb{N} \to X(\phi(\gamma))$ such that for each $i \in \mathbb{N}$, $d(s_\gamma(i),x_{\phi(\gamma)}) \leqslant 1/i$, $s_\gamma(2i-1) \in F_\gamma$, and $s_\gamma(2i) \in G_\gamma$ (d denotes the metric on X). Observe that if $\gamma < \lambda$ and $i \in \mathbb{N}$ then $s_\gamma(i) = x_\delta$ for some $\delta < \phi(\gamma)$.

(4) Show that for each $j \in \mathbb{N}$ and $\alpha < \lambda$, there exists a set $L(x_\alpha,j)$ defined as follows:

$$L(x_\alpha,j) = \{x_\alpha\} \text{ if } \alpha \notin \phi[\lambda]$$
$$L(x_\alpha,j) = \{x_\alpha\} \cup \bigcup\{L(s_\gamma(i),i) : i \geqslant 2j\} \text{ if } \alpha = \phi(\gamma)$$

(Define the sets using transfinite recursion.)

(5) Show that if $x, y \in X$, $j \in \mathbb{N}$, and $x \in L(y,j)$, then $L(x,i) \subseteq L(y,j)$ for some $i \in \mathbb{N}$. (Use transfinite

induction.)

(6) Show that if for each $x \in X$ we define $\{L(x,j) : j \in \mathbb{N}\}$ to be an open neighborhood base at x, this gives a valid definition of a topology $\hat{\tau}$ on the underlying set of the metric space X (use (5)). Denote this new space by $\hat{X}$.

(7) Using transfinite induction, prove that if $x \in L(y,j)$, then $d(x,y) < 1/j$. Hence infer that the topology of $\hat{X}$ contains that of X, and so $\hat{X}$ is Hausdorff.

(8) Prove that each $L(x,j)$ is a countable, compact subspace of $\hat{X}$. (Use transfinite induction, and invoke (5) to prove compactness.) Hence show that $\hat{X}$ is a zero-dimensional, locally compact, locally countable space.

(9) If $A, B \subseteq \hat{X}$ and $|c\ell_X A \cap c\ell_X B| = 2^m$, prove that $c\ell_{\hat{X}} A \cap c\ell_{\hat{X}} B \neq \emptyset$. (Hint: Find $\gamma < \lambda$ such that $(F_\gamma, G_\gamma) \in S$, F_γ is X-dense in A, and G_γ is X-dense in B. Then show that $x_{\phi(\gamma)} \in c\ell_{\hat{X}} A \cap c\ell_{\hat{X}} B$).

(10) Show that if m is not Ulam-measurable then X is hereditarily realcompact, and so $\hat{X}$ is realcompact (see 5M(4)).

(11) Show that if $A \in B(\hat{X})$ then $|c\ell_X A \cap c\ell_X(X \setminus A)| < 2^m$ (see (9)).

(12) Suppose, in addition to the hypothesis originally postulated about X, we assume that if $V \in RO(X)$ and $\emptyset \neq V \neq X$, then $|bd_X V| = 2^m$. Let $U = \{A \in B(\hat{X}) : |\hat{X} \setminus A| < 2^m\}$. Prove that U is a free clopen ultrafilter on $\hat{X}$ with the m^+-intersection property; hence

infer that $\hat{X}$ is not zero-dimensionally m^+-compact. (Hint: The hypothesis about $bd_X V$ enables us to use (11) to show that if $A \in B(\hat{X})$, then either $|A| < 2^m$ and $X \setminus A$ is dense in X, or else $|X \setminus A| < 2^m$ and A is dense in X.)

In the next three parts of this problem we produce a metric space satisfying the hypotheses imposed on X in (12) above.

(13) Prove that there is a proper class of cardinal numbers m for which $2^m = m^{\aleph_0}$ (if k is a cardinal let $m = 2^k + 2^{2^k} + 2^{2^{2^k}} + \ldots$).

(14) Let m be an infinite cardinal with the property that $2^m = m^{\aleph_0}$. Let H(m) be a Hilbert space (see $[W_i]$ for definition of Hilbert space) with an orthonormal basis of cardinality m. Prove that H(m) is a metric space of cardinality 2^m and weight m.

(15) Prove that if m and H(m) are as in (13) and (14), and if $V \in RO(X)$ and $\emptyset \neq V \neq X$, then $|bd_{H(m)}V| = 2^m$. (Hint: Choose $\underline{x} \in V$, $\underline{y} \in H(m) \setminus c\ell_{H(m)}V$, and $\delta > 0$ such that the open ball of radius δ and center $\underline{x}$ (respectively, $\underline{y}$) is contained in V (respectively, $H(m) \setminus c\ell_{H(m)}V$). Let $G = \{\underline{z} \in H(m) : \|z\| = \delta/2\}$. If $z \in G$, let $I(\underline{z}) = \{\underline{x} + \underline{z} + \theta(\underline{y} - \underline{x}) : 0 \leqslant \theta \leqslant 1\}$. Prove that $I(\underline{z})$ is homeomorphic to [0,1] and hence infer that $I(\underline{z})$ is connected. If we define $\equiv$ on G by $\underline{z}_1 \equiv \underline{z}_2$

if $I(\underline{z}_1) \cap I(\underline{z}_2) \neq \emptyset$, show that $\equiv$ is an equivalence relation on G and that each equivalence class has at most 2 members. Prove that $I(\underline{z}) \cap \mathrm{bd}_{H(m)}V \neq \emptyset$ for each $\underline{z} \in G$, and thus deduce that $|\mathrm{bd}_{H(m)}V| = 2^m$.)

(16) Prove that if $2^m = m^{\aleph_0}$ and if measurable cardinals do not exist, then $\widehat{H(m)}$ is a realcompact zero-dimensional space that is not zero-dimensionally m^+-compact.

(17) Prove that if the property of being realcompact and zero-dimensional were E-compactness for some space E, then there would exist a cardinal number k such that every realcompact zero-dimensional space is zero-dimensionally k-compact.

(18) Prove that if measurable cardinals do not exist, then there is no space E such that a space X is realcompact and zero-dimensional iff X is E-compact. (Combine (13), (15) and (17).)

(19) If $X = \mathbb{R}^2$, show that the space $\hat{X}$ defined in (6) is realcompact, zero-dimensional, but not $\mathbb{N}$-compact.

5Q. <u>Metacompact δ-normally separated spaces need not be realcompact.</u> Recall (see 1R) that a Tychonoff space is δ-normally separated if each zero-set is completely separated from each closed set disjoint from it. In 5.11(l) we showed that (barring measurable cardinals) every normal θ-refinable space is realcompact. In this problem we show that "normal" cannot be weakened to "δ-normally separated", even if

"θ-refinable" is strengthened to "metacompact".

(1) Let $X = \omega_2 + 1$, topologized as follows: if $\alpha < \omega_2$ then α is isolated in X, and if $\omega_2 \in U \subseteq X$ then U is open iff $|X \setminus U| \leq \omega_1$. Let $Y = \omega_1 + 1$, topologized in the analogous manner (all points but ω_1 are isolated and the neighborhoods of ω_1 are co-countable). Let $T = X \times Y \setminus \{(\omega_2,\omega_1)\}$. Prove that T is a Tychonoff P-space and thus is δ-normally separated (see 1W).

(2) Prove that $\upsilon T = X \times Y$, and conclude that T is not realcompact. (Argue as in 2R.)

(3) Prove directly that T is metacompact.

5R. <u>$\mathcal{U}$-compactness and countable compactness</u>. Let $\mathcal{U}$ be a free ultrafilter on $\mathbb{N}$, let X be a space, and let $\{x_n : n \in \mathbb{N}\}$ be a sequence in X. A point $y \in X$ is called a <u>$\mathcal{U}$-limit point</u> of $\{x_n : n \in \mathbb{N}\}$ if for each neighborhood V of y, the set $\{n \in \mathbb{N} : x_n \in V\}$ belong to $\mathcal{U}$.

(1) If p and q are both $\mathcal{U}$-limit points of a sequence $\{x_n : n \in \mathbb{N}\}$, show that $p = q$. Thus if a sequence $\{x_n : n \in \mathbb{N}\}$ has a $\mathcal{U}$-limit point it is unique; we denote it by $\mathcal{U}\text{-lim}\{x_n : n \in \mathbb{N}\}$.

(2) Find an example of a faithfully indexed sequence $\{x_n : n \in \mathbb{N}\}$ in a space X, and a point $p \in X$, such that $p = \mathcal{U}\text{-lim}\{x_n : n \in \mathbb{N}\}$ for every $\mathcal{U} \in \beta\mathbb{N}\setminus\mathbb{N}$. What

can be said about the topology of the subspace $\{x_n : n \in \mathbb{N}\} \cup \{p\}$ in this situation?

We define a Tychonoff space X to be **U-compact** if every sequence in X has a U-limit point.

(3) Show that every U-compact space is countably compact.

(4) Show that a Tychonoff space is $\aleph_0$-bounded iff it is U-compact for every $U \in \beta\mathbb{N}\setminus\mathbb{N}$.

(5) Prove that a Tychonoff space X is U-compact iff each $f \in C(\mathbb{N},X)$ extends continuously to $f^U \in C(\mathbb{N} \cup \{U\},X)$, where $\mathbb{N} \cup \{U\}$ is regarded as a subspace of $\beta\mathbb{N}$.

(6) Show that U-compactness is a Tychonoff extension property.

(7) Let X be a Tychonoff space. If $X \subseteq T \subseteq \beta X$ define UT to be

$T \cup \{p \in \beta X\setminus T : p$ is the U-limit point of some sequence in $T\}$.

Define by transfinite induction an ω_1-sequence $\{X(\alpha) : \alpha < \omega_1 + 1\}$ of subspaces of βX as follows:

(a) $X(0) = X$

(b) $X(\alpha+1) = UX(\alpha)$

(c) $X(\lambda) = U[\cup\{X(\alpha) : \alpha < \lambda\}]$ if λ is a countable limit ordinal

(d) $X(\omega_1) = \cup\{X(\alpha) : \alpha < \omega_1\}$.

If $[U]$ denotes the property of being U-compact, prove that $\gamma_{[U]}X = X(\omega_1)$.

(8) Determine the cardinality of $\Upsilon_{[U]}\mathbb{R}$, where $U \in \beta\mathbb{N}\setminus\mathbb{N}$. Is $\Upsilon_{[U]}\mathbb{R}$ $\aleph_0$-bounded?

(9) Prove that the following are equivalent for a Tychonoff space X:

(a) every power of X is countably compact,

(b) X^{2^c} is countably compact, and

(c) X is U-compact for some $U \in \beta\mathbb{N}\setminus\mathbb{N}$.

(10) Prove that if P is a Tychonoff extension property such that every space with P is countably compact, then there exists $U \in \beta\mathbb{N}\setminus\mathbb{N}$ such that every space with P is U-compact.

(11) Prove that if P is a Tychonoff extension property, then either $\mathbb{N}$ has P or there exists $U \in \beta\mathbb{N}\setminus\mathbb{N}$ such that every space with P is U-compact. Conclude that if $\mathbb{N}$ does not have P, then every space with P is countably compact.

(12) Prove that the continuous Tychonoff image of a U-compact space is U-compact.

(13) Let P be a Tychonoff extension property that is preserved by continuous surjection onto Tychonoff spaces. Show that either every Tychonoff space whose cardinality is not Ulam-measurable has P, or else there exists $U \in \beta\mathbb{N}\setminus\mathbb{N}$ such that every space with P is U-compact. (Hint: See 5I.)

5S. P-pseudocompactness. Let P be either a Tychonoff or a zero-dimensional extension property. Let $\beta_P X$ denote either βX (if P is a Tychonoff extension property) or $\beta_0 X$ (if P is a zero-dimensional extension property). A P-regular space X is called **P-pseudocompact** if $\gamma_P X$ is compact. The class of P-pseudocompact spaces is denoted by P'

(1) Prove that X is P-pseudocompact iff $\gamma_P X \equiv_X \beta_P X$.

(2) If P is realcompactness, prove that X is P-pseudocompact iff X is pseudocompact. (This is the origin of the term "P-pseudocompact".)

(3) If P is either $\mathbb{N}$-compactness, or the property of being realcompact and zero-dimensional, prove that X is P-pseudocompact iff X is pseudocompact and zero-dimensional (see 5G).

(4) Prove that a P-regular extension of a P-pseudocompact space is P-pseudocompact.

(5) Prove that a P-regular continuous image of a P-pseudocompact space is P-pseudocompact.

(6) Show that a P-regular space X is P-pseudocompact iff whenever S has P and $f \in C(X,S)$, then $c\ell_S f[X]$ is compact.

(7) Show that a clopen subspace of a P-pseudocompact space is P-pseudocompact.

(8) If a P-regular space X is the union of a finite set of P-pseudocompact subspaces, show that X is P-pseudocompact.

(9) Show that if X is P-regular and $B \in B(X)$, then B is P-pseudocompact iff $c\ell_{\gamma_P X} B$ is compact (see 5L(1)).

(10) If P is $\aleph_0$-boundedness, show that a Tychonoff space X is P-pseudocompact iff each point of $\beta X \setminus X$ is in the βX-closure of a countable subset of X.

5T. <u>Co-regular extension properties</u>. Let P and Q be co-regular extension properties (see 5.9(g)), and let P-regularity be either the Tychonoff property or zero-dimensionality. Note that by 5.9(f) $\gamma_P X$ and $\gamma_Q X$ can both be regarded as subspaces of $\beta_P X (= \beta_Q X)$.

(1) Show that if $P \subseteq Q$, then $Q' \subseteq P'$ (see 5S for notation).

(2) Show that $Q \subseteq P'$ iff for every P-regular space X, $\gamma_Q X \setminus \gamma_P X$ is dense in $\beta_P X \setminus \gamma_P X$. (Hint: Use regular closed neighborhoods.)

(3) Show that if X is any Tychonoff space, then $\aleph_0 X \setminus \upsilon X$ is dense in $\beta X \setminus \upsilon X$ (see 5.10(d) for the definition of $\aleph_0 X$).

(4) Show that if U is a free ultrafilter on $\mathbb{N}$, then $\gamma_{[U]} \mathbb{R} \setminus \mathbb{R}$ is a dense subset of $\beta\mathbb{R} \setminus \mathbb{R}$ (see 5R).

5U. <u>More on co-regular extension properties</u>. A Tychonoff space is called **almost compact** if $|\beta X \setminus X| \leqslant 1$.

(1) Prove that every almost compact space is

pseudocompact. (Hint: Use 1Q(6).)

(2) Let P be a Tychonoff extension property. Show that if X is almost compact, then either $X \in P$ or $X \in P'$. Show that X is compact iff $X \in P \cap P'$.

(3) Let X and Y be almost compact, noncompact spaces. Let $\beta X \setminus X = \{x_0\}$ and $\beta Y \setminus Y = \{y_0\}$. Let $S = \beta X \times \beta Y \setminus \{(x_0,y_0)\}$. Prove that S is almost compact. (Hint: Use pseudocompact $\times$ compact = pseudocompact and $X \times Y$ pseudocompact $\Rightarrow \beta(X \times Y) = \beta X \times \beta Y$; see 4AF(4) and 4AG(8).)

Henceforth let P and Q denote two Tychonoff extension properties.

(4) If a Tychonoff extension property P is possessed by every almost compact space, show that every Tychonoff space has P. (Hint: Use 5.9(d).)

(5) Suppose $P' \subseteq Q$, and let X, Y, and S be as in (3). Prove that either $S \in P$ or $S \in Q$. Then show that if there exists an almost compact space that does not belong to Q, then every Tychonoff space belongs to P, and so P is the property of being Tychonoff. The same is true if P and Q are interchanged. (Hint: S contains copies of X and Y.)

(6) Show that the following are equivalent:

(a) $P' \subseteq Q$ and

(b) either P or Q is the property of being

Tychonoff.

5V. Pseudocompact regular closed subsets. Let X be a Tychonoff space.

(1) Let A be a non-pseudocompact member of $R(X)$. Show that there exists a countably infinite pairwise disjoint family $B = \{B_n : n \in \mathbb{N}\} \subseteq R(X)$ such that B is a locally finite family of subsets of X and $\cup B \subseteq \mathrm{int}_X A$.

(2) Let A and B be as in (1). Show that there exists $g \in C(X)$ such that $g^{\leftarrow}(n) \cap B_n \neq \emptyset$ for each $n \in \mathbb{N}$. (Hint: Use 1.6(c).)

(3) Let A be a non-pseudocompact member of $R(X)$. Prove that $c\ell_{\upsilon X}A$ is not compact. (Consider $\upsilon g \mid c\ell_{\upsilon X}A$, where $\upsilon g \in C(\upsilon X)$ and $\upsilon g \mid X = g$.)

(4) If $A \in R(X)$, show that A is pseudocompact iff $c\ell_{\upsilon X}A$ is compact.

5W. Copseudocompact extension properties. Let P and Q be co-regular extension properties. Then P and Q are called **copseudocompact** if $P' = Q'$.

(1) Let P and Q be zero-dimensional extension properties. Show that $P' = Q'$ iff for each zero-dimensional space X, $\beta_0 X \setminus \gamma_P X$ and $\beta_0 X \setminus \gamma_Q X$ are each dense in

$\beta_0 X \backslash (\gamma_P X \cap \gamma_Q X)$. (Hint: See the remarks following 5.9(f), and use 5S(9).)

(2) Show that if X is a realcompact zero-dimensional space, then $\beta_0 X \backslash \upsilon_{\mathbb{N}} X$ is dense in $\beta_0 X \backslash X$.

(3) Let P be a Tychonoff extension property. Show that P' is pseudocompact iff for each Tychonoff space X, $\beta X \backslash \upsilon X$ is dense in $\beta X \backslash (\gamma_P X \cap \upsilon X)$. (Hint: Proceed as in (1), using (2).)

See 6AB for an application of these results.

5X. Maximum extension properties

(1) Let S be a topological property such that Reg(S) is the class of Tychonoff spaces. Suppose that each Tychonoff continuous image of a space with S has S. Define $\hat{S}$ to be the following topological property: X has $\hat{S}$ if X is Tychonoff and each subspace of X with S has compact X-closure. Prove that $\hat{S}$ is a Tychonoff extension property.

(2) Give a succinct description of $\hat{S}$ if S is (a) complete regularity, (b) compactness, (c) separable and Tychonoff, and (d) connected and Tychonoff.

(3) If P is a Tychonoff extension property, show that $\widehat{P'}$ is a Tychonoff extension property that is copseudocompact with P (see 5W).

(4) Show that if P and Q are copseudocompact Tychonoff extension properties, then $Q \subseteq \widehat{P'}$. Thus conclude

that Tychonoff extension properties can be partitioned into "copseudocompactness classes", each of which has a largest member.

(5) Do (1) — (4) substituting "zero-dimensional" for "Tychonoff".

5Y. Maximum $K(P)$-extensions and stable ultrafilters

(1) Let P be a topological property for which $\mathrm{Reg}(P)$ is the class of Tychonoff spaces. Show that up to equivalence

$$\gamma_{K(P)}X = X \cup \{p \in \beta X \setminus X : \text{for all } Y \in P \text{ and all } f \in C(X,Y),\ \beta f(p) \in Y\}.$$

(Hint: Use 5.3(g).)

(2) If P is as in (1), we define a z-ultrafilter $\mathcal{A}$ on X to be **P-stable** if, whenever Y is a space with P and $f \in C(X,Y)$, there exists $Z \in \mathcal{A}$ such that $c\ell_Y f[Z]$ is compact. Let $T = X \cup \{\mathcal{A} \in \beta X \setminus X : \mathcal{A}$ is P-stable$\}$. Prove that $T \subseteq \gamma_{K(P)}X$.

(3) Suppose that P is as in (1), and in addition suppose all spaces with P are locally compact. Prove that $T = \gamma_{K(P)}X$.

(4) Formulate and prove "zero-dimensional versions" of (1) — (3).

5Z. $\beta\underline{X}\setminus\upsilon\underline{X}$ has no G_δ-points

(1) Let X be a Tychonoff space, let $\emptyset \neq Z \in Z(\beta X)$, and suppose that $Z \cap X = \emptyset$. Prove that Z contains a copy of $\beta\mathbb{N}\setminus\mathbb{N}$ (and hence $|Z| \geq 2^{c}$). (Hint: Use 4.5(p), 4.6(j), 1Q(6), and 5F(4).)

(2) Let X be a Tychonoff space. Show that if $p \in \beta X\setminus\upsilon X$ then $\{p\}$ is not a G_δ-set of $\beta X\setminus\upsilon X$. (Hint: If not, use 5.11(b) to construct a small zero-set that will contradict (1).) In particular, $\beta X\setminus\upsilon X$ has no isolated points.

5AA. Ultrarealcompact spaces. A space X is called **ultrarealcompact** if it is $[0,1] \times \mathbb{N}$—compact. Prove that the following conditions on the Tychonoff space X are equivalent.

(1) X is ultrarealcompact,

(2) every free z-ultrafilter on X contains a countable decreasing sequence of clopen sets with empty intersection, and

(3) for each $p \in \beta X\setminus X$ there is a countable disjoint open cover $\mathcal{C}$ of X such that $p \notin \cup\{cl_{\beta X}C : C \in \mathcal{C}\}$.

(Hint: For (1) $\Rightarrow$ (2), use 5.4(d)(2) and extend $f \in C(X,[0,1] \times \mathbb{N})$ to $\beta f \in C(\beta X,([0,1] \times \mathbb{N}) \cup\{\infty\})$.)

5AB. Extension properties and discrete spaces

(1) Let P be a Tychonoff or zero-dimensional extension property. Prove that the following are equivalent for each infinite cardinal m:

(a) D(m) has P and

(b) the topological sum of m spaces with P has P.

(Hint: If $\{X(i) : i < m\}$ is a set of spaces with P, map $X = \oplus\{X(i) : i < M\}$ onto D(m). Note that $X \subseteq \oplus\{\beta X(i) : i < m\} \subseteq \beta X$ in effect, and use 4.2(f) and 5.9(f) to conclude that if $x_0 \in \Upsilon_P X \setminus X$ then $x_0 \in \beta X(i_0) \setminus X(i_0)$ for some $i_0 < m$. If P has P and $f \in C(X(i_0),P)$ find a related $F \in C(X,P)$ and consider $F | X(i_0) \cup \{x_0\}$. Then note 5.3(g).)

(2) Let P be as in (1) and assume D(m) has P. Show that if $\{X(i) : i \in I\}$ is a collection of no more than m P-regular spaces then $\Upsilon_P X \equiv_X \oplus\{\Upsilon_P X(i) : i \in I\}$, where X denotes $\oplus\{X(i) : i \in I\}$. (Hint: Use 5L(1), 5.9(d), and 5.9(f).)

5AC. (p,S)-compactness. Let S be a noncompact Tychonoff space. Fix $p \in \beta S \setminus S$. A space X is called **(p,S)-compact** if for each $f \in C(S,X)$, f can be continuously extended to $f^{\#} : S \cup \{p\} \to X$.

(1) Prove that (p,S)-compactness is a Tychonoff extension

property.

(2) Let U be a free ultrafilter on $\mathbb{N}$ (and thus a point of $\beta\mathbb{N}\setminus\mathbb{N}$). Prove that $(U,\mathbb{N})$-compactness is the same as the U-compactness defined in 5R.

(3) Let E be a space for which E-complete regularity is the Tychonoff property. Prove that the following are equivalent:

(a) E is not (p,S)-compact for any $p \in \beta S\setminus S$ and

(b) S is E-compact.

(Hint: Use 5.4(d).)

(4) Let P be a Tychonoff extension property. Let S be a P-pseudocompact space (see 5S) and let $p \in \beta S\setminus S$. Show that P is contained in (p,S)-compactness.)

5AD. <u>Hewitt realcompactifications of products</u>. Throughout this problem all hypothesized spaces are assumed to be Tychonoff.

(1) Prove that if the product space $X \times Y$ is C^*-embedded in $\upsilon X \times \upsilon Y$, then it is C-embedded in $\upsilon X \times \upsilon Y$ (see 1.9(j) and 5.11(b)).

(2) Let X be a space and K a compact space. Suppose that $f \in C(X\times K)$. Fix $x \in X$ and define $f_x : K \to \mathbb{R}$ by $f_x(y) = f(x,y)$ for each $y \in K$. Prove that $f_x \in C(K)$.

(3) Let X and K be as in (2) and let $f \in C(X\times K)$ be fixed. Give C(K) the metric described in the paragraph before

4.5(a) and define $\phi : X \to C(K)$ as follows:

$$\phi(x) = f_x.$$

Prove that ϕ is continuous. (Hint: Let $x_0 \in X$ and suppose $\epsilon > 0$. For each $y \in K$ find $U(y)$ and $V(y)$, open in X and K respectively, such that

$$|f(x_0,y) - f(p,q)| < \epsilon/2 \quad \text{whenever} \quad (p,q) \in U(y) \times V(y)$$

Extract a finite subcover from $\{V(y) : y \in K\}$ and let U be the intersection of the corresponding $U(y)$'s. Show that $\phi[U]$ is a subset of the open ball of radius ϵ centered at f_{x_0}.)

(4) Assume that the cardinality of K is not Ulam-measurable (see 2P). Show that there exists $\phi^* : \upsilon X \to C(K)$ such that $\phi^* | X = \phi$ (see 5.11(m)).

(5) Define $f^\# : \upsilon X \times K \to \mathbb{R}$ as follows:

$$f^\#(x,y) = (\phi^*(x))(y)$$

Show that $f^\# | X \times K = f$.

(6) Prove that $f^\#$ is continuous. (Hint: Let $(p,y) \in \upsilon X \times K$ and let $\epsilon > 0$ be given. Find an open subset U of υX such that $p \in U$ and $\phi^*[U]$ is contained in the open ball in $C(K)$ centered at $\phi^*(p)$ and of radius $\epsilon/2$. As $\phi^*(p) \in C(K)$, find an open subset V of K such that $y \in$

V and $|(\phi^*(p))(y) - (\phi^*(p))(z)| < \epsilon/2$ whenever $z \in V$. Now show that $|f^\#(p,y) - f^\#(q,z)| < \epsilon$ whenever $(q,z) \in U \times V$.)

(7) Prove that if X is a space and K is compact, then $\upsilon(X\times K) \equiv_{X\times K} (\upsilon X) \times K$.

(8) Prove that if Y is locally compact and the cardinality of Y is not Ulam-measurable, then $X \times Y$ is C-embedded in $\upsilon X \times Y$ for each space X. (Hint: Let $f \in C(X\times Y)$, and suppose $y \in Y$. Let $K(y)$ be a compact Y-neighborhood of y. Use (7) to find $g_y \in C(\upsilon X \times K(y))$ such that $g_y | X \times K(y) = f | X \times K(y)$. Prove that if $x \in \upsilon X$ and $z \in K(y_1) \cap K(y_2)$, then $g_{y_1}(x,z) = g_{y_2}(x,z)$. Define $g : \upsilon X \times Y \to \mathbb{R}$ by $g | \upsilon X \times K(y) = g_y$ for each $y \in Y$. Prove that g is a well-defined continuous extension of f. (See 1.6(a).)

(9) Prove that if Y is a locally compact realcompact space whose cardinality is not Ulam-measurable and if X is any space, then $\upsilon(X\times Y) \equiv_{X\times Y} (\upsilon X) \times Y$.

More general results than the above have been proved; see the Notes.

5AE. <u>$C(X,E)$ need not separate points and closed sets of X if X is E-completely regular</u>. Let E be a space containing more than one point with this property: the only continuous functions from E to E are constant functions and the identity function, i.e., $C(E,E) = \{id_E\} \cup \{f \in F(E,E) : |f[E]| = 1\}$. (In Corollary 2, page 87 of [dG] it is asserted that there is a subspace of $\mathbb{R}^2$ with this property.)

(1) Let p and q be distinct points of E and let U and V be disjoint open subsets of E containing p and q respectively. Show that there does not exist a $G \in C(E^2,E)$ such that $G(p,q) \notin c\ell_E G[E^2 \setminus (U \times V)]$. (Hint: Suppose there were such a G. Define $g : E \to E$ as follows: $g(x) = G(p,x)$. Prove that $g = id_E$ and infer that $G(p,q) = q$. "Turn the argument through 90°" and infer that $G(p,q) = p$, a contradiction.)

(2) Prove that although E^2 is E-completely regular, $C(E^2,E)$ does not separate points and closed sets of E^2 (see 1.7(i)). Conclude that "$\vee\{C(X,E^n) : n \in \mathbb{N}\}$" cannot be replaced by $C(X,E)$ in 5.4(c).

CHAPTER 6

Extremally Disconnected Spaces and Absolutes

6.1 Introduction

One of the best behaved classes of functions encountered in general topology is the class of perfect functions, which was discussed in 1.8. As we have already seen, two topological spaces, one of which is the perfect continuous image of the other, will have many topological properties in common. (Examples of a number of such properties are given in 1J.) Perfect continuous surjections also play an important role in compactification theory (see 4.2(f,g) and in the study of extension properties (see 5.9(c)). Roughly speaking, if you cannot construct a homeomorphism between X and Y, constructing a perfect continuous surjection between them is the next best goal.

Because perfect functions are so well-behaved, two natural questions to ask, when confronted with a new topological property *P*, are the following. If X has *P*, what kinds of spaces will be perfect continuous images of X? What spaces can be mapped onto X be a perfect continuous surjection? Perhaps if we can find a "nice", easy-to-handle space that is linked to X in one of these ways, we can study the properties of that space and infer from them the properties of X.

If X is regular, such a "nice" space does indeed exist. It is called the **Iliadis absolute** of X and is denoted by EX. It can be mapped onto X by a perfect continuous surjection k_X which has the additional desirable feature of being "irreducible"; this means that if A is a proper closed subset of EX (i.e., $EX\setminus A\neq\emptyset$), then $k_X[A]$ is a proper closed subset of X. The space EX is zero-dimensional (which is nice), and (what is nicer) it is "extremally disconnected", which means that its open sets have open closures. Also, in a sense made precise in 6.7(a), EX is unique with respect to these properties.

Suppose X is not regular. By 1J no regular (and thus no zero-dimensional) space can be mapped onto X be a perfect continuous surjection, and no perfect continuous image of X is regular. But two nice things still happen. First, the zero-dimensional extremally disconnected space EX and the function $k_X : EX \to X$ can still be constructed, and it is still true that k_X is a perfect surjection. Although k_X is no longer continuous, it is Θ-continuous (see 4.8(g)(2)), which for many purposes is almost as good as continuity. Second, we can still construct an (essentially unique) extremally disconnected space PX (called the **Banaschewski absolute** of X) and a perfect irreducible continuous surjection $\Pi_X : PX \to X$. The space PX fails to be regular (as predicted above), and hence fails to be zero-dimensional. But PX is still "nice" in that it is extremally disconnected. If X is regular, PX and EX coincide, as do k_X and Π_X. The unmodified term "absolute" will mean "Iliadis absolute".

The spaces PX and X not only share those topological properties preserved by perfect continuous surjections; they also share many other topological properties by virtue of the irreducibility of Π_X. Examples of such properties are cellularity, π-weight, density character (see 2N) and feeble compactness (see 1.11).

Absolutes arise naturally when we try to solve the following problem: what Tychonoff spaces are perfect images of realcompact spaces? Must all such images be realcompact? This latter question easily reduces to the following problem: if EX is realcompact, must X be realcompact? The answer turns out to be "no"; a characterization of perfect Tychonoff images of realcompact spaces is presented in 6U.

The absolute EX of a space X also arises naturally in the following situation. Suppose we can associate with each space X some algebraic object A(X) (the categorically minded reader should view A as being a functor from a category whose objects are topological spaces to a category whose objects are rings, Boolean algebras, or some other algebraic structure). Suppose we form some canonical algebraic "completion" $\widehat{A(X)}$ of A(X), and then ask whether $\widehat{A(X)}$ can be represented as A(X*) for some suitably chosen space X*. The answer to such questions is often as follows: If X is "sufficiently nice", then $\widehat{A(X)}$ is isomorphic to A(EX). To clarify these ideas, we give three examples below.

Let X be a compact zero-dimensional space. We associate with it the Boolean algebra B(X), form its minimal completion $\widehat{B(X)}$ (see 3.4(h,j), and then ask if there is a space X* such that B(X*) = $\widehat{B(X)}$. The answer is "yes"; B(EX) is isomorphic to $\widehat{B(X)}$.

Let X be a Tychonoff space. We can form the lattice-ordered ring C(X), and construct its Dedekind-MacNeille completion $\widehat{C(X)}$ (see 2.5). It turns out that $\widehat{C(X)}$ is also a lattice-ordered ring, and we then ask whether there exists a Tychonoff space $\hat{X}$ such that C($\hat{X}$) is isomorphic (as a lattice-ordered ring) to $\widehat{C(X)}$. The answer is not always "yes", but for a large class of spaces (including the normal countably paracompact spaces) the answer will be "yes"; C(EX) is isomorphic to $\widehat{C(X)}$. See 8.5 for details.

Let R be a commutative ring. One can embed R in its

"complete ring of quotients" Q(R); this construction is a generalization of the well-known procedure for embedding an integral domain in a field (see [La] for details). When one applies this construction to the ring $C^*(X)$ (where X is a Tychonoff space), one discovers that the maximal ideal space of $Q(C^*(X))$ is homeomorphic to $E(\beta X)$. A detailed discussion of this is beyond the scope of this book; the interested reader is referred to [FGL] for an extended treatment.

In addition to the above algebraic applications of the absolute, we mention an application from category theory and one from topological measure theory. The reader unfamiliar with category theory will find the technical vocabulary used below discussed in detail at the beginning of Chapter 9.

Let $\mathcal{C}$ be a category. An object X of $\mathcal{C}$ is said to be **projective** in $\mathcal{C}$ if, whenever Y and Z are objects of $\mathcal{C}$, $g : Y \to Z$ is a morphism of $\mathcal{C}$, and $f : Y \to Z$ is an epimorphism in $\mathcal{C}$ there is a morphism $h : X \to Y$ of $\mathcal{C}$ such that $g = f \circ h$. (The motivation for such a definition arises partially from the study of projective modules in ring theory.) The original impetus behind the study of absolutes was the problem of characterizing the projective objects in the category of compact spaces and continuous functions. In 1958, Gleason [Gl] solved this problem by showing that the projective objects of this category are precisely the compact extremally disconnected spaces. He also constructed EX (when X is compact) as part of the solution of this problem. Other authors interpreted absolutes in a similar category-theoretic manner for other categories whose objects are topological spaces.

Finally, Wheeler [Wh] has applies the theory of absolutes to topological measure theory. Roughly speaking, he relates various spaces of measures on X to corresponding spaces of measures on EX, and then uses the fact that Baire sets of extremally disconnected spaces have a relatively simple structure to analyze these spaces of measures on EX. The results obtained are "mapped back" to X to obtain results concerning the structure of spaces of measures on X.

We hope that we have convinced the reader that absolutes have considerable usefulness outside this field of general topology. Thus motivated, we concentrate in the chapter on constructing the absolute of a space and deriving its basic properties. Discussion of some of its applications is postponed to a later chapter.

Recall our blanket assumption that all hypothesized topological spaces are Hausdorff.

6.2 Characterizations of extremally disconnected spaces

(a) Definition. A space X is said to be **extremally disconnected** if the closure of each open subset of X is also open.

Obviously every discrete space is extremally disconnected, but the reader unfamiliar with extremally disconnected spaces might be hard put to think of other examples. However, as a consequence of the following theorems we will have several techniques available for constructing such spaces. Recall that $B(X)$, $R(X)$, and $RO(X)$ denote respectively the Boolean algebras of clopen, regular closed, and regular open subsets of X.

(b) **Theorem**. The following are equivalent for a space X:

(1) X is extremally disconnected,

(2) each dense subset of X is extremally disconnected,

(3) each open subset of X is extremally disconnected,

(4) if U and V are open subsets of X, then $c\ell_X(U \cap V) = c\ell_X U \cap c\ell_X V$,

(5) if U and V are disjoint open subsets of X, then $c\ell_X U \cap c\ell_X V = \varnothing$,

(6) $B(X) = R(X) = RO(X)$, and

(7) κX is extremally disconnected.

Proof

(1) ⇒ (2) Let S be a dense subset of X and let V be an open subset of S. There exists an open subset W of X such that $V = S \cap W$. By hypothesis $c\ell_X W$ is open in X. But

$$S \cap c\ell_X W = S \cap c\ell_X V = c\ell_S V \quad \text{(see 1A)}$$

therefore $c\ell_S V$ is open in S. Thus S is extremally disconnected.

(2) ⇒ (3) Let V be an open subset of X. Then $V \cup (X \setminus c\ell_X V) = S$ is dense in X and thus extremally disconnected. If W is an open subset of V, then W is also open in S. But $c\ell_S W = c\ell_V W$; it follows that $c\ell_V W$ is open in S, and therefore in V. Hence V is extremally disconnected.

(3) ⇒ (4) By hypothesis X is extremally disconnected. Thus $c\ell_X U$ and $c\ell_X V$ are clopen in X. Thus,

$c\ell_X(U \cap V) = c\ell_X(U \cap c\ell_X V) = c\ell_X(c\ell_X U \cap c\ell_X V) = c\ell_X U \cap c\ell_X V$.

(4) ⇒ (5) This is obvious.

(5) ⇒ (1) Let U be open in X. Since $U \cap (X \setminus c\ell_X U) = \varnothing$, by (5) $c\ell_X U = X \setminus c\ell_X(X \setminus c\ell_X U)$, and so $c\ell_X U$ is open in X. Thus X is extremally disconnected.

We have now shown that conditions (1) — (5) are equivalent.

(1) ⇒ (6) Obviously $B(X) \subseteq R(X)$ for any space X. If $A \in R(X)$ then $A = c\ell_X(\mathrm{int}_X A)$; so as X is extremally disconnected, A is open. Hence $A \in B(X)$ and $B(X) = R(X)$. Then $RO(X) = \{X \setminus A : A \in R(X)\} = \{X \setminus A : A \in B(X)\} = B(X)$.

(6) ⇒ (7) Let V be open in κX. Then $c\ell_X(V \cap X) \in R(X)$, and by hypothesis $c\ell_X(V \cap X) \in B(X)$. Hence by 4.8(p)(4), $c\ell_{\kappa X}(c\ell_X(V \cap X)) \in B(\kappa X)$. But by 1A

$$c\ell_{\kappa X}(c\ell_X(V \cap X)) = c\ell_{\kappa X}(V \cap X) = c\ell_{\kappa X}V;$$

therefore, $c\ell_{\kappa X}V$ is open in κX. Hence κX is extremally disconnected.

(7) ⇒ (1) The proof of (1) ⇒ (2) shows that a dense subspace of an extremally disconnected space is extremally disconnected.

■

There are several other important characterizations of extremally disconnected Tychonoff spaces.

(c) **Theorem**. The following are equivalent for the Tychonoff space X:

(1) X is extremally disconnected,

(2) each dense subspace of X is C^*-embedded in X,

(3) each open subspace of X is C^*-embedded in X, and

(4) βX is extremally disconnected.

Proof

(1) ⇒ (2) Let S be a dense subspace of X and let Z_1 and Z_2 be disjoint zero-sets of S. Find disjoint open sets V_1 and V_2 of S such that $Z_1 \subseteq V_1$ and $Z_2 \subseteq V_2$. Let $V_i = W_i \cap S$ where W_i is an open subset of X (i = 1,2). Then $W_1 \cap W_2 = \emptyset$ since S is dense in X. By the previous theorem $c\ell_X W_1 \in B(X)$ and $c\ell_X W_1 \cap c\ell_X W_2 = \emptyset$. Thus $\chi_{c\ell_X W_1} \in C^*(X)$ and completely separates Z_1 and Z_2 in X. Hence by 1.9(h) S is C^*-embedded in X.

(2) ⇒ (3) Let V be open in X and put $S = V \cup (X \setminus c\ell_X V)$. Then $V \in B(S)$ so V is C^*-embedded in S. As S is dense in X, by hypothesis S is C^*-embedded in X. Hence V is C^*-embedded in X.

(3) ⇒ (4) Let V be open in βX and put $S = (V \cap X) \cup (X \setminus c\ell_X(V \cap X))$; then $\chi_{V \cap X} \in C^*(S)$. As S is open in X by hypothesis there exists $f \in C^*(X)$ such that $f|S = \chi_{V \cap X}$. For the function f^β defined in 4.5(r), observe that

$$f^{\beta}[c\ell_{\beta X}V] = f^{\beta}[c\ell_{\beta X}(V \cap X)]$$
$$= c\ell_{\beta X}f[V \cap X]$$
$$= \{1\}$$

Similarly $f^{\beta}[c\ell_{\beta X}(X \setminus c\ell_{\beta X}V)] = \{0\}$. It follows that

$$c\ell_{\beta X}V \cap c\ell_{\beta X}(X \setminus c\ell_{\beta X}V) = \emptyset.$$

Since $\beta X = c\ell_{\beta X}V \cup c\ell_{\beta X}(X \setminus c\ell_{\beta X}V)$, it follows that $c\ell_{\beta X}V$ is clopen.

(4) ⇒ (1) This is a consequence of the implication (1) ⇒ (2) of the previous theorem.

■

Related to the equivalence of (1) and (4) in 6.2(b) is the following fact: compact extremally disconnected spaces are the "Stone dual" of complete Boolean algebras.

(d) **Theorem**. Let B be a Boolean algebra. The following are equivalent:

(1) B is complete and

(2) S(B) is extremally disconnected.

Proof

(1) ⇒ (2) Let U be open in S(B). There is a subset $\{b_i : i \in I\}$ of B such that $U = \bigcup\{\lambda(b_i) : i \in I\}$ (see 3.2(a) for notation). Let $b = \bigvee\{b_i : i \in I\}$ (b exists as B is complete). We claim that $\lambda(b) =$

$c\ell_{S(B)}U$.

Obviously $b_i \leqslant b$ so $\lambda(b_i) \subseteq \lambda(b)$ for each $i \in I$. Thus $U \subseteq \lambda(b)$ and so $c\ell_{S(B)}U \subseteq \lambda(b)$. Conversely, suppose that $\lambda(b) \setminus c\ell_{S(B)}U \neq \emptyset$. Find $a \in B$ such that $0 \neq \lambda(a) \subseteq \lambda(b) \setminus c\ell_{S(B)}U$. Then $c\ell_{S(B)}U \subseteq \lambda(b) \cap (S(B) \setminus \lambda(a)) = \lambda(b \wedge a')$, and so $b_i \leqslant b \wedge a' < b$ for each $i \in I$. This contradicts the definition of b. Thus $c\ell_{S(B)}U = \lambda(b)$, which is open, and so S(B) is extremally disconnected.

[A quicker proof of (1) ⇒ (2), which leans more heavily on the theory of completions of Boolean algebras developed in Chapter 3, goes as follows. By 3.2(d), $B(S(B))$ is isomorphic to B; therefore, $B(S(B))$ is complete. But $R(S(B))$ is the minimal completion of its dense subalgebra $B(S(B))$; since $B(S(B))$ is complete, then $B(S(B)) = R(S(B))$. By 6.2(b), S(B) is extremally disconnected.]

(2) ⇒ (1) Since S(B) is extremally disconnected, by 6.2(b) $B(S(B)) = R(S(B))$. By 2.2(c), $R(S(B))$ is complete, and by 3.2(d), $B(S(B))$ is isomorphic to B. Thus B is complete.

■

(e) **Corollary**

(1) Every compact extremally disconnected space is homeomorphic to the Stone space of some complete Boolean algebra.

(2) Every complete Boolean algebra is isomorphic to the Boolean algebra of clopen sets of some compact extremally disconnected space.

Proof

(1) Let X be compact and extremally disconnected. Then by

6.2(b), $B(X) = R(X)$, and so by 2.2(c), $B(X)$ is complete. By 3.2(h) X is homeomorphic to $S(B(X))$.

(2) Let B be a complete Boolean algebra. By 6.2(d), $S(B)$ is compact and extremally disconnected. By 3.2(d), B is isomorphic to $B(S(B))$.

■

6.3 Examples of extremally disconnected spaces

We can now add to our stock of extremally disconnected spaces. If D is a discrete space, then by 6.2(b) and 6.2(c) βD and κD are extremally disconnected. Furthermore, if $D \subseteq X \subseteq \beta D$ or $D \subseteq Y \subseteq \kappa D$, then X, Y and any open subspaces of X and Y will be extremally disconnected. This provides us with a varied collection of extremally disconnected spaces. However, any space constructed by these methods will have a dense set of isolated points.

To find examples of extremally disconnected spaces without isolated points, we can proceed as follows. Let X be a space without isolated points; then by 3.2(m) $R(X)$ has no atoms and so $S(R(X))$ has no isolated points. But $R(X)$ is a complete Boolean algebra (see 2.2(c)) so $S(R(X))$ is a compact extremally disconnected space without isolated points. Dense subspaces of $S(R(X))$ will be extremally disconnected (by 6.2(b)) and have no isolated points.

In fact, every extremally disconnected Tychonoff space is a dense subspace of the Stone space of some complete Boolean algebra, as we will see.

6.4 Extremally disconnected spaces and zero-dimensionality

It is evident that an extremally disconnected space has many clopen subsets. If it also has enough separation, it will be zero-dimensional. Explicitly, the situation is as follows:

Theorem. The following are equivalent for an extremally disconnected space X:

(1) X is strongly zero-dimensional (defined in 4.7(h)),

(2) X is zero-dimensional,

(3) X is regular, and

(4) X is semiregular.

Proof. Obviously (1) $\Rightarrow$ (2) $\Rightarrow$ (3) $\Rightarrow$ (4). To show that (4) implies (1), let V be an open subset of the extremally disconnected semiregular space X and let $p \in V$. Find $U \in RO(X)$ such that $p \in U \subseteq V$. By 6.2(c), $B(X) = RO(X)$ and so $U \in B(X)$. Thus X is zero-dimensional and therefore Tychonoff. By 6.2(c) βX is extremally disconnected; as it is semiregular, the above argument shows that βX is zero-dimensional. Thus X is strongly zero-dimensional.

■

The standard example of an extremally disconnected space that is not semiregular is κD, where D is an infinite discrete space. As κD is H-closed but not compact (the latter fact follows from the fact that $\kappa D \backslash D$ is infinite (see 4.8(q)(1)) and 4.8(p)(1)), κD is not regular (see 4.8(c)) and hence, by the preceding theorem, not semiregular. Note that if $\mathcal{U} \in \kappa D \backslash D$, then $\kappa D \backslash (D \cup \{\mathcal{U}\})$ is closed in κD, but $\mathcal{U}$

and $\kappa D \backslash (D \cup \{U\})$ cannot be put inside disjoint open subsets of κD.

6.5 Irreducible functions

(a) **Definition**. Let X and Y be spaces and let f be a closed surjection from X onto Y. Then f is called **irreducible** if, whenever A is a proper closed subset of X, $f[A] \neq Y$.

Note that irreducible functions are closed but are not required to be continuous. Examples of irreducible functions are easy to find. Any closed bijection is irreducible; if X is Tychonoff and αX is a compactification of X, then the Stone extension of the inclusion map is irreducible. Our interest in irreducible functions arises primarily from the fact that for each space X, the absolute of X is mapped onto X by a perfect irreducible Θ-continuous function. (See 4.8 for a discussion of Θ-continuous functions.)

We list some of the basic properties of irreducible functions in the following lemma.

(b) **Lemma**. Let $f \in F(X,Y)$ be irreducible. Then:

(1) if $g \in F(Y,Z)$ is irreducible, then $g \circ f : X \to Z$ is irreducible;

(2) if T is a space, $h : X \to T$ is a closed surjection, and there is a surjection $k : T \to Y$ such that $f = k \circ h$, then h is irreducible;

(3) if U is a nonempty open subset of X, then $\text{int}_Y f[U] \neq \varnothing$;

(4) if S is a dense subset of Y, then $f^{\leftarrow}[S]$ is a dense subset of X and $f|f^{\leftarrow}[S]$ is an irreducible function from $f^{\leftarrow}[S]$ onto S; and

(5) if T is a space, $k : T \to X$ is a closed function, and $h : T \to Y$ is a surjection such that $h = f \circ k$, then k is a surjection.

Proof

(1) As f and g are closed surjections, so is $g \circ f$. If A is a proper closed subset of X, then $f[A] \neq Y$, so $g[f[A]] = (g \circ f)[A] \neq Z$. Thus $g \circ f$ is irreducible.

(2) If h were not irreducible, there would be a proper closed subset A of X such that $h[A] = T$. Thus $f[A] = k[h[A]] = k[T] = Y$, which is a contradiction.

(3) By hypothesis $Y \setminus f[X \setminus U]$ is a nonempty open subset of Y, and is obviously contained in $f[U]$.

(4) As f is closed, $f[c\ell_X f^{\leftarrow}[S]] \supseteq c\ell_Y S = Y$; thus $c\ell_X f^{\leftarrow}[S] = X$ as f is irreducible. If A is closed in X, then $f[A \cap f^{\leftarrow}[S]] = f[A] \cap S$ which is closed in S. If $f^{\leftarrow}[S] \setminus A \neq \emptyset$, then $A \neq X$ and $f[A] \neq Y$; as $f[A]$ is closed in Y and S is dense, $S \cap f[A] \neq S$ and so $f|f^{\leftarrow}[S]$ is irreducible.

(5) Since $Y = h[T] = f[k[T]]$, $k[T]$ is a closed subset of X, and as f is irreducible, it follows that $k[T] = X$.

■

We now show that every perfect surjection has a restriction that is irreducible.

(c) **Theorem**. Let f be a perfect surjection (not necessarily continuous) from a space X onto a space Y. Then there exists a closed subset C of X such that $f[C] = Y$ and $f|C$ is an irreducible perfect surjection from C onto Y.

<u>Proof</u>. Let $\mathcal{F} = \{F : F$ is a closed subset of X and $f[F] = Y\}$. Obviously $X \in \mathcal{F}$ so $\mathcal{F} \neq \varnothing$. Partially order $\mathcal{F}$ by reverse inclusion; i.e., put $F_1 \leqslant F_2$ if $F_2 \supseteq F_1$. By Zorn's lemma there is a maximal chain $\mathcal{C}$ in $\mathcal{F}$. Let $C = \cap\{F : F \in \mathcal{C}\}$. Evidently C is a closed subset of X; we will show that C has the required properties.

If $y \in Y$ then $\{F \cap f^{\leftarrow}(y) : F \in \mathcal{C}\}$ is a family of closed subsets of the compact space $f^{\leftarrow}(y)$. As $f[F] = Y$ for each $F \in \mathcal{C}$ and $\mathcal{C}$ is a chain, this family has the finite intersection property. Hence by compactness, $\cap\{F \cap f^{\leftarrow}(y) : F \in \mathcal{C}\} \neq \varnothing$, i.e., $C \cap f^{\leftarrow}(y) \neq \varnothing$. As y was arbitrarily chosen it follows that $f[C] = Y$. As f is perfect and C is closed in X, $f|C$ is a perfect map from C onto Y (see 1.8(f)).

Let A be a proper closed subset of C. Then A is closed in X and by the maximality of $\mathcal{C}$, $A \notin \mathcal{F}$. Thus $f[A] \neq Y$ and so $f|C$ is irreducible.

■

We conclude this section by showing that irreducible Θ-continuous surjections preserve regular closed sets and preserve the property of being nowhere dense.

(d) **<u>Theorem</u>**. Let f be an irreducible Θ-continuous function from X onto Y. Then:

(1) if A and B are closed subsets of X such that $\text{int}_X(A \cap B)$

$= \emptyset$, then $int_Y(f[A] \cap f[B]) = \emptyset$;

(2) if C is a closed nowhere dense subset of X, then f[C] is a closed nowhere dense subset of Y;

(3) the map $A \to f[A]$ is a Boolean algebra isomorphism from $R(X)$ onto $R(Y)$; and

(4) if Y is extremally disconnected then f is one-to-one; if f is also continuous then f is a homeomorphism.

Proof

(1) If $int_Y(f[A] \cap f[B]) \neq \emptyset$, by the Θ-continuity of the surjection f there exists a nonempty open subset V of X such that $f[c\ell_X V] \subseteq c\ell_Y int_Y(f[A] \cap f[B]) \subseteq f[A] \cap f[B]$ (see 4.8(g)). By hypothesis $V\setminus(A \cap B) \neq \emptyset$, and so $(X\setminus V) \cup (A \cap B) \neq X$. Thus either $(X\setminus V) \cup A \neq X$ or $(X\setminus V) \cup B \neq X$. But $Y = f[X\setminus V] \cup f[V] \subseteq f[X\setminus V] \cup f[A] = f[(X\setminus V) \cup A]$; therefore, by the irreducibility of f, $(X\setminus V) \cup A = X$. A similar argument shows that $(X\setminus V) \cup B = X$, which gives a contradiction. Thus $int_Y(f[A] \cap f[B]) = \emptyset$.

(2) This is a special case of (1) where C is used in place of both A and B.

(3) First we show that if $A \in R(X)$, then $f[A] \in R(Y)$. As f[A] is closed in Y it suffices to show that $f[A]\setminus c\ell_Y int_Y f[A] = \emptyset$.

Suppose $a \in A$ and $f(a) \notin c\ell_Y int_Y f[A]$. As f is Θ-continuous there is an open subset V of X such that $a \in V$ and $f[c\ell_X V] \subseteq c\ell_Y[Y\setminus c\ell_Y int_Y f[A]]$. Thus $f[c\ell_X V] \cap int_Y f[A] = \emptyset$. As $A \in R(X)$, $V \cap int_X A \neq \emptyset$; thus by 6.5(b)(3) $int_Y f[V \cap int_X A] \neq \emptyset$. But $int_Y f[V \cap int_X A] \subseteq f[c\ell_X V] \cap int_Y f[A]$, which gives a contradiction.

Thus $f[A] = c\ell_Y \mathrm{int}_Y f[A]$ and so $f[A] \in R(X)$. Thus we can define a function $\phi : R(X) \to R(Y)$ by letting $\phi(A) = f[A]$.

Next we show that ϕ maps $R(X)$ onto $R(Y)$. If $B \in R(Y)$, let $p \in f^{\leftarrow}[\mathrm{int}_Y B]$. As f is Θ-continuous there is an open subset $V(p)$ of X such that $p \in V(p)$ and $f[c\ell_X V(p)] \subseteq c\ell_Y \mathrm{int}_Y B = B$. If $V(p) \setminus c\ell_X f^{\leftarrow}[\mathrm{int}_Y B] \neq \varnothing$, then by 6.6(b)(3) $\mathrm{int}_Y f[V(p) \setminus c\ell_X f^{\leftarrow}[\mathrm{int}_Y B]]$ is a nonempty open subset of Y contained in B and disjoint from $\mathrm{int}_Y B$, which is a contradiction; thus $V(p) \subseteq c\ell_X f^{\leftarrow}[\mathrm{int}_Y B]$. It follows that $c\ell_X f^{\leftarrow}[\mathrm{int}_Y B] = c\ell_X[\cup\{V(p) : p \in f^{\leftarrow}[\mathrm{int}_Y B]\}]$ and so $c\ell_X f^{\leftarrow}[\mathrm{int}_Y B] \in R(X)$.

We claim that $f[c\ell_X f^{\leftarrow}[\mathrm{int}_Y B]] = B$. As f is closed, $f[c\ell_X f^{\leftarrow}[\mathrm{int}_Y B]] \supseteq c\ell_Y(\mathrm{int}_Y B) = B$. Conversely, if $p \in X$ and $f(p) \notin B$, by Θ-continuity there is an open subset W of X such that $p \in W$ and $f[c\ell_X W] \subseteq c\ell_Y(Y \setminus B) = Y \setminus \mathrm{int}_Y B$. Thus $W \cap f^{\leftarrow}[\mathrm{int}_Y B] = \varnothing$ and so $p \notin c\ell_X f^{\leftarrow}[\mathrm{int}_Y B]$. Thus $f[c\ell_X f^{\leftarrow}[\mathrm{int}_Y B]] = B$ as claimed, and ϕ maps $R(X)$ onto $R(Y)$.

Next, suppose that A_1 and A_2 are distinct members of $R(X)$. Without loss of generality assume that $(\mathrm{int}_X A_1) \setminus A_2 \neq \varnothing$. As f is irreducible there exists $p \in Y \setminus f[(X \setminus \mathrm{int}_X A_1) \cup A_2]$. As f is surjective there exists $a \in X$ such that $f(a) = p$, and evidently $a \in \mathrm{int}_X A_1$. Since $p \notin f[A_2]$ it follows that $f[A_1] \neq f[A_2]$, and so ϕ is one-to-one.

Finally, let A_1 and A_2 belong to $R(X)$. Then $A_1 \leq A_2$ iff $A_2 = A_1 \vee A_2$ iff $f[A_2] = f[A_1 \vee A_2]$ (as ϕ is one-to-one) iff $f[A_2] = f[A_1] \vee f[A_2]$ iff $\phi(A_1) \leq \phi(A_2)$. Thus ϕ is an order isomorphism from $R(X)$ onto $R(Y)$. As a Boolean algebra is determined by its order structure (see 3.1(n)(1)), ϕ is a Boolean algebra isomorphism.

(4) Let Y be extremally disconnected and let x_1 and x_2 be

distinct points of X. Find an open subset U of X such that $x_1 \in U$ and $x_2 \in X \setminus c\ell_X U$. Then $f(x_1) \in \phi(c\ell_X U)$ and $f(x_2) \in \phi(c\ell_X(X \setminus c\ell_X U))$. Since Y is extremally disconnected, by 6.2(b) $B(Y) = R(Y)$ and so by (3) above $\phi(c\ell_X U) \cap \phi(c\ell_X(X \setminus c\ell_X U)) = \phi(c\ell_X U) \wedge \phi(c\ell_X(X \setminus c\ell_X U))$. Thus

$$\begin{aligned}\phi(c\ell_X U) \cap \phi(c\ell_X(X \setminus c\ell_X U)) &= \phi(c\ell_X U \wedge c\ell_X(X \setminus c\ell_X U)) \\ &= \phi(\emptyset) \quad \text{as } \phi \text{ is a Boolean algebra homomorphism} \\ &= \emptyset\end{aligned}$$

so $f(x_1) \neq f(x_2)$. Hence f is one-to-one and thus is a closed bijection from X onto Y. If f were also continuous it would be a homeomorphism.

■

We note that a θ-continuous closed bijection onto an extremally disconnected space need not be a homeomorphism. The function $\overleftarrow{j}$: $\beta N \to \kappa N$ described in 4.8(t) is such a function.

Other useful results on irreducible θ-continuous functions can be found in 8.4(i).

6.6 The construction of the Iliadis absolute

In this section we construct the absolute EX of a space X and show that it is unique (up to homeomorphism). Our construction involves the Stone space of $\mathcal{R}(X)$, and the reader unfamiliar with the material discussed in 2.2 and 3.2 is advised to review it before proceeding. In 6.8 we will discuss a second widely-used method for constructing EX.

(a) **Definition**. Let X be a space. The **Gleason space** of X is defined to be the Stone space of the Boolean algebra $\mathcal{R}(X)$, and is denoted by θX.

Thus the elements of θX are ultrafilters on $\mathcal{R}(X)$, and $\mathcal{B}(\theta X) = \{\lambda(A) : A \in \mathcal{R}(X)\}$ (recall $\lambda(A) = \{\mathcal{U} \in \theta X : A \in \mathcal{U}\}$). The space θX gets its name from A. M. Gleason, who described in [Gℓ] the construction and properties of EX in the case where X is compact (in this case, as we shall see, $EX = \theta X$).

(b) **Definition**. Let X be a space. The space $\{\mathcal{U} \in \theta X : \cap\mathcal{U} \neq \varnothing\}$, equipped with the subspace topology inherited from θX, is called the **Iliadis absolute** of X and denoted by EX. (Sometimes the term "Iliadis absolute" will be used to refer to the pair (EX,k_X), where k_X is as defined below, rather than just to the space EX.)

As noted earlier, the unmodified term "absolute" will mean "Iliadis absolute" (another absolute is introduced in 6.11).

Thus EX consists of the fixed ultrafilters on $\mathcal{R}(X)$, viewed as a subspace of $S(\mathcal{R}(X))$.

(c) **Definition**. Let X be a space and let $x \in X$. We denote by $\mathcal{F}(x)$ the family of regular closed neighborhoods of x, i.e., $\mathcal{F}(x) =$

$\{A \in R(X) : x \in \mathrm{int}_X A\}$. Evidently $F(x)$ is a filter on $R(X)$.

(d) **Lemma**. Let X be a space.

(1) If $U \in EX$, then $\cap U$ contains exactly one point.

(2) If $x \in X$, there exists $U \in EX$ such that $\cap U = \{x\}$.

Proof

(1) Let p and q be distinct points of X. Choose $A \in R(X)$ such that $p \in \mathrm{int}_X A$ and $q \notin A$. If $p \in \cap U$, then for each $B \in U$, $(\mathrm{int}_X A) \cap B \neq \emptyset$. Thus $A \wedge B \neq \emptyset$, and by the maximality of U it follows that $A \in U$ (see 2.3(d)). Thus $q \notin \cap U$ and so $\cap U$ contains at most (and thus exactly) one point.

(2) By 2.3(d)(2) there is an ultrafilter $U(x)$ on $R(X)$ such that $F(x) \subseteq U(x)$. If $A \in U(x)$ then $x \in A$; if not, then $x \in X \setminus A = \mathrm{int}_X(c\ell_X(X \setminus A))$, and so $c\ell_X(X \setminus A) \in F(x)$. Thus $\emptyset = A \wedge c\ell_X(X \setminus A) \in U(x)$, which is impossible. It follows that $x \in \cap U(x)$.

■

If $U \in EX$, denote by $k_X(U)$ the unique point of X belonging to $\cap U$. By the preceding lemma k_X is a well-defined surjection from EX onto X. We now derive the properties of the pair (EX, k_X) that will characterize EX as the absolute of X.

(e) **Theorem**. Let X be a space. Then:

(1) EX is a dense extremally disconnected zero-dimensional subspace of θX, and $\theta X \equiv_{EX} \beta(EX)$ (see 4.1(d)).

(2) Let $U \in \theta X$ and $x \in X$. Then $U \in EX$ and $k_X(U) = x$ iff $F(x) \subseteq U$.

(3) If $A \in R(X)$, then $k_X[EX \cap \lambda(A)] = A$.

(4) Let $x \in X$ and $B \in R(X)$. Then $k_X^{\leftarrow}(x) \subseteq \lambda(B)$ iff $x \in \text{int}_X B$.

(5) k_X is a perfect irreducible Θ-continuous surjection from EX onto X.

(6) k_X is continuous iff X is regular.

(7) $B(EX) = R(EX) = \{\lambda(A) \cap EX : A \in R(X)\}$.

(8) The function $EX \cap \lambda(A) \to k_X[EX \cap \lambda(A)]$ is a Boolean algebra isomorphism from $B(EX)$ onto $R(X)$.

(9) If $\{A_n : n \in \mathbb{N}\}$ is a decreasing sequence of elements of $R(X)$, then $k_X[\cap\{EX \cap \lambda(A_n) : n \in \mathbb{N}\}] = \cap\{A_n : n \in \mathbb{N}\}$.

Proof

(1) By 6.2(d), θX is a compact extremally disconnected zero-dimensional space. To show that EX is dense in θX, it suffices to show that if $A \in R(X) \setminus \{\emptyset\}$ then $\lambda(A) \cap EX \neq \emptyset$. Choose $x \in \text{int}_X A$. Then $A \in F(x)$ and as in the preceding lemma there exists $U \in EX$ such that $F(x) \subseteq U$ and $\cap U = \{x\}$. Thus $U \in \lambda(A) \cap EX$ and EX is dense in θX. By 6.2(b) EX is extremally disconnected. As θX is zero-dimensional, so is EX. By 6.2(c) EX is C^*-embedded in θX, and so by 4.6(g) $\theta X \equiv_{EX} \beta(EX)$.

(2) Suppose $F(x) \subseteq U$. If $A \in R(X)$ and $x \notin A$, then $x \in \text{int}_X A'$ (recall $A' = c\ell_X(X \setminus A)$, the Boolean-algebraic complement of A in $R(X)$; see 3.1(e)(4)). Thus $A' \in F(x)$ and so $A' \in U$. Hence $A \notin U$, and it follows that $x \in \cap U$. Thus $U \in EX$ and $k_X(U) = x$. If $F(x) \nsubseteq U$, find $A \in F(x) \setminus U$. By the maximality of U, $A' \in U$ (see 2.3(e)(3)). Since $x \in \text{int}_X A = X \setminus A'$, it follows that $x \notin$

$\cap \mathcal{U}$. Thus even if $\mathcal{U} \in EX$, we would have $k_X(\mathcal{U}) \neq x$.

(3) Let $A \in R(X)$ and let $\mathcal{U} \in EX \cap \lambda(A)$. Then $A \in \mathcal{U}$ so the unique point in $\cap \mathcal{U}$ belongs to A. Thus $k_X[EX \cap \lambda(A)] \subseteq A$. Conversely, suppose $x \in A$, and consider the collection $S = F(x) \cup \{A\}$. We claim that any finite subcollection drawn from S has a nonempty infimum in $R(X)$. To see this, let $F_1, \ldots, F_n \in F(x)$. Then $x \in \cap\{int_X F_i : i = 1 \text{ to } n\} = int_X(\wedge\{F_i : i = 1 \text{ to } n\})$ (see 2.2(c)). This implies that $int_X(\wedge\{F_i : 1 \leqslant i \leqslant n\}) \cap int_X A \neq \varnothing$; thus $A \wedge (\wedge\{F_i : 1 \leqslant i \leqslant n\}) \neq \varnothing$ and out claim is justified. By 2.3(d) there exists $\mathcal{U} \in \theta X$ such that $F(x) \cup \{A\} \subseteq \mathcal{U}$. By (2) above $\mathcal{U} \in EX$ and $k_X(\mathcal{U}) = x$. Also $\mathcal{U} \in \lambda(A)$ and $x \in k_X[EX \cap \lambda(A)]$. Thus $k_X[EX \cap \lambda(A)] \supseteq A$.

(4) Suppose $x \in int_X B$ and $\mathcal{U} \in k_X^{\leftarrow}(x)$. By (2), $F(x) \subseteq \mathcal{U}$. Since $B \in F(x)$, it follows that $B \in \mathcal{U}$ and $\mathcal{U} \in \lambda(B)$. Thus $k_X^{\leftarrow}(x) \subseteq \lambda(B)$. If $x \notin int_X B$ then $x \in X \setminus int_X B = B'$. By (3) above $x \in k_X[EX \cap \lambda(B')]$, and $\varnothing \neq k_X^{\leftarrow}(x) \cap \lambda(B') = k_X^{\leftarrow}(x) \setminus \lambda(B)$.

(5) We begin by showing that k_X is a compact map. It suffices to show that $k_X^{\leftarrow}(x)$ is closed in θX for $x \in X$. Suppose $\mathcal{U} \notin k_X^{\leftarrow}(x)$. Then $x \notin \cap \mathcal{U}$, which implies there is an $A \in \mathcal{U}$ such that $x \notin A$. So, $x \in int_X A'$. By (4), $k_X^{\leftarrow}(x) \subseteq \lambda(A')$. Since $A \wedge A' = \varnothing$, then $\lambda(A) \cap \lambda(A') = \varnothing$. Thus $\mathcal{U} \in \lambda(A)$ and $\lambda(A) \cap k_X^{\leftarrow}(x) = \varnothing$. This shows that $k_X^{\leftarrow}(x)$ is closed in θX.

Next we show that k_X is a closed map. Let F be a closed subset of EX and suppose $x \in X \setminus k_X[F]$. Then $k_X^{\leftarrow}(x) \cap F = \varnothing$ so for each $\mathcal{U} \in k_X^{\leftarrow}(x)$ we can choose $A(\mathcal{U}) \in R(X)$ such that $\mathcal{U} \in \lambda(A(\mathcal{U})) \subseteq EX \setminus F$. By the compactness of $k_X^{\leftarrow}(x)$ find $\mathcal{U}_1, \ldots, \mathcal{U}_n$

such that $k_X^{\leftarrow}(x) \subseteq \cup\{\lambda(A(U_i)) : 1 \leqslant i \leqslant n\} = \lambda(A)$, where $A = \cup\{A(U_i) : 1 \leqslant i \leqslant n\}$. Evidently, $\lambda(A) \cap F = \emptyset$, and by (4) above $x \in \text{int}_X A$. Thus $F \subseteq EX \setminus \lambda(A) = \lambda(A')$, and by (3) $k_X[F] \subseteq A' = X \setminus \text{int}_X A$. Therefore each point of $X \setminus k_X[F]$ has a neighborhood disjoint from $k_X[F]$, and k_X is therefore closed. Hence k_X is a perfect surjection.

To prove that k_X is irreducible, let F be a proper closed subset of EX. Then there exists $A \in R(X)$ such that $\emptyset \neq \lambda(A) \cap EX \subseteq EX \setminus F$. Thus $F \subseteq \lambda(A') \cap EX$ and by (3) $k_X[F] \subseteq A'$. As $\emptyset \neq \lambda(A)$, $X \setminus A' \neq \emptyset$. Thus k_X is irreducible.

To prove that k_X is Θ-continuous let $U \in EX$, let V be open in X, and let $k_X(U) \in V$. Put $A = c\ell_X V$; by (4) $U \in \lambda(A) \cap EX$. By (3) $k_X[EX \cap \lambda(A)] = c\ell_X V$. Thus k_X is Θ-continuous at the arbitrarily chosen point U, and therefore Θ-continuous.

(6) By (1) EX is zero-dimensional and hence regular. By 1.8(h) a perfect continuous image of a regular space is regular; so if k_X is continuous then X is regular. Conversely, by 4.8(h)(3) if X is regular then k_X, being Θ-continuous, must be continuous.

(7) By 3.2(d)(2), $B(\theta X) = \{\lambda(A) : A \in R(X)\}$; therefore, $\{\lambda(A) \cap EX : A \in R(X)\} \subseteq B(EX)$. Conversely if $C \in B(EX)$, by (1) above and 4.6(g)(5) $c\ell_{\theta X} C = c\ell_{\beta(EX)} C \in B(\beta(EX)) = B(\theta X)$, so there exists $A \in R(X)$ such that $c\ell_{\theta X} C = \lambda(A)$. Thus $C = EX \cap c\ell_{\theta X} C = \lambda(A) \cap EX$. Also, by 6.3(b), $B(EX) = R(EX)$.

(8) This follows from (5) above and 6.5(d)(3).

(9) Obviously $k_X[\cap\{EX \cap \lambda(A_n) : n \in \mathbb{N}\}] \subseteq \cap\{A_n : n \in \mathbb{N}\}$ by (3). Conversely, if $x \in \cap\{A_n : n \in \mathbb{N}\}$, then $F(x) \cup \{A_n : n \in \mathbb{N}\}$ is easily seen to be a filter base on $R(X)$ and hence is contained in an ultrafilter α on $R(X)$ which converges to x since $F(x) \subseteq \alpha$.

Thus $\alpha \in \bigcap\{EX \cap \lambda(A_n) : n \in \mathbb{N}\}$ and $k_X(\alpha) = x$.

6.7 The uniqueness of the absolute

In 6.6 we saw that associated with each space X there is a pair (EX,k_X) consisting of an extremally disconnected zero-dimensional space EX and a perfect irreducible θ-continuous surjection k_X from EX onto X. We wish to show that this pair (EX,k_X) is unique (up to homeomorphism) in the sense made precise by 6.7(a) below.

(a) **Theorem**. Let X be a space and let (Y,f) be a pair consisting of an extremally disconnected zero-dimensional space Y and a perfect irreducible θ-continuous surjection f from Y onto X. Then there exists a homeomorphism h from EX onto Y such that $f \circ h = k_X$.

Proof. As Y is extremally disconnected, by 6.2(b), $B(Y) = R(Y)$. Thus by 6.5(d) the map $B \rightarrow f[B]$ is a Boolean algebra isomorphism ϕ from $B(Y)$ onto $R(X)$.

If $\mathcal{U} \in EX$, let $\mathcal{U}' = \phi^{\leftarrow}[\mathcal{U}] = \{B \in B(Y) : f[B] \in \mathcal{U}\}$. Since $\mathcal{U}$ is an ultrafilter on $R(X)$, $\mathcal{U}'$ is an ultrafilter on $B(Y)$. As Y is zero-dimensional, this implies that there is at most one point in $\bigcap\mathcal{U}'$. Evidently $\{B \cap f^{\leftarrow}(k_X(\mathcal{U})) : B \in \mathcal{U}'\}$ has the finite intersection property; for if $B_1, \ldots, B_n \in \mathcal{U}'$, then $\wedge\{f[B_i] : 1 \leqslant i \leqslant n\} \in \mathcal{U}$ so $k_X(\mathcal{U}) \in \wedge\{f[B_i] : 1 \leqslant i \leqslant n\} = f[\bigcap\{B_i : 1 \leqslant i \leqslant n\}]$ (as ϕ preserves infima). Since $f^{\leftarrow}(k_X(\mathcal{U}))$ is compact and each $B \in \mathcal{U}'$ is closed in Y, this implies that $\bigcap\mathcal{U}' \neq \varnothing$. Thus for each

$\mathcal{U} \in EX$ there is exactly one point in $\cap\mathcal{U}'$; we will denote this point by $h(\mathcal{U})$. This defines a function $h : EX \to Y$. Evidently $f \circ h(\mathcal{U}) = \cap\{f[B] : B \in \mathcal{B}(Y) \text{ and } f[B] \in \mathcal{U}\} = \cap\mathcal{U} = k_X(\mathcal{U})$; so $f \circ h = k_X$.

If $y \in Y$, let $\mathcal{U}(y) = \{f[B] : B \in \mathcal{B}(Y) \text{ and } y \in B\}$. As $\{B \in \mathcal{B}(Y) : y \in B\}$ is an ultrafilter on $\mathcal{B}(Y)$, $\mathcal{U}(y)$ is an ultrafilter on $\mathcal{R}(X)$. Clearly $f(y) \in \cap\mathcal{U}(y)$ and by 6.6(d), $\{f(y)\} = \cap\mathcal{U}(y)$. Evidently $h(\mathcal{U}(y)) = y$, so h maps EX onto Y.

If $\mathcal{U}$ and $\mathcal{W}$ are distinct members of EX, find $A \in \mathcal{R}(X)$ such that $A \in \mathcal{U}$ and $A' \in \mathcal{W}$ (see 2.3(e)(3)). As ϕ is an isomorphism there exists a unique $B \in \mathcal{B}(Y)$ such that $\phi(B) = A$; thus $\phi(Y \backslash B) = A'$. Thus $B \in \mathcal{U}'$ and $Y \backslash B \in \mathcal{W}'$; therefore, $h(\mathcal{U}) \neq h(\mathcal{W})$. Thus h is one-to-one.

If F is closed in EX, there is a family $\{A_i : i \in I\} \subseteq \mathcal{R}(X)$ such that $F = \cap\{\lambda(A_i) \cap EX : i \in I\}$. For each $i \in I$ there is a unique $B_i \in \mathcal{B}(Y)$ such that $f[B_i] = A_i$. A straightforward calculation shows that $h[F] = \cap\{B_i : i \in I\}$, and so h is a closed mapping.

Finally, suppose $\mathcal{U} \in EX$, $B \in \mathcal{B}(Y)$, and $h(\mathcal{U}) \in B$. Then $f[B] \in \mathcal{U}$ and so $\mathcal{U} \in \lambda(f[B])$. But if $\mathcal{W} \in \lambda(f[B])$ then $f[B] \in \mathcal{W}$; this implies that $B \in \mathcal{W}'$, and so $h(\mathcal{W}) \in B$. Thus $h[\lambda(f[B])] \subseteq B$. It follows that h is continuous at the arbitrarily chosen point $\mathcal{U}$, and thus is continuous. Therefore h, being a closed continuous bijection, is a homeomorphism.

■

(b) **Corollary**. Let X and Y be spaces and let g be a perfect Θ-continuous surjection from Y onto X. Suppose that $\phi : S \to X$ is a perfect irreducible Θ-continuous surjection from the extremally disconnected space S onto X. Then there exists a perfect irreducible Θ-continuous surjection f from S onto a closed subspace of Y such that $g \circ f = \phi$.

Proof. By 6.5(c) there is a closed subspace C of Y such that $g|C : C \to X$ is a perfect, irreducible and onto. By 4.8(h)(4), $g|C \in \Theta C(C,X)$. By 4.8(h), 6.5(b), and 1.8(e) $(g|C) \circ k_C$ is a perfect irreducible Θ-continuous surjection from EC onto X. By the previous theorem there are homeomorphisms $h_1 : EX \to S$ and $h_2 : EX \to EC$ such that $\phi \circ h_1 = k_X$ and $(g|C) \circ k_C \circ h_2 = k_X$. Then $k_C \circ h_2 \circ h_1^{\leftarrow}$ is the required mapping f.

■

A note of caution and a note of optimism should be sounded here. First, the above corollary does not remain true if we omit the work "irreducible" from the two places in which it occurs in the statement of the corollary. This will be demonstrated in 9.8(n). However, if in the corollary we replace the two occurrences of the term "Θ-continuous" by "continuous", we can then drop the two occurrences of the word "irreducible" in the statement of the corollary and be left with a true statement. This will be proved in 6.11(d).

Let us review what we have accomplished so far. For each space X we have constructed an extremally disconnected zero-dimensional space EX and a perfect irreducible Θ-continuous surjection k_X from EX onto X. We have also shown that if Y is an extremally disconnected zero-dimensional space and f is a perfect irreducible Θ-continuous surjection from Y onto X, then there is a homeomorphism h from EX onto Y such that $f \circ h = k_X$. Thus such a pair (Y,f), no matter how it is constructed, is "the same as" the pair (EX,k_X). We formalize this situation by making the following definition.

(c) **Definition**. Let X and Y be spaces and let f be a function from Y onto X. Then the expression "$(EX,k_X) \sim (Y,f)$" is an abbreviation of the following statement: Y is extremally disconnected and zero-dimensional, and f is a perfect irreducible Θ-continuous surjection from Y onto X. In this situation we will say that "(Y,f) is the absolute of X (up to equivalence)."

Note that if Y is a space and $f : Y \to X$ is a function, then $(EX,k_X) \sim (Y,f)$ iff there is a homeomorphism $h : EX \to Y$ such that $f \circ h = k_X$. Of course, in such cases the underlying set of Y will not, in general, be the set of convergent filters on $R(X)$. Nonetheless there will usually be an obvious way to associate the points of Y with the convergent ultrafilters on $R(X)$. The first example of such a representation of EX is given in the next section.

6.8 The construction of EX as a space of open ultrafilters

In our construction of EX in 6.6 we took as the underlying set of EX the set of convergent ultrafilters on $R(X)$. Since $R(X)$ and $RO(X)$ are isomorphic Boolean algebras (see 3B(2)), we could equally well have constructed EX by using the convergent ultrafilters on $RO(X)$ as our underlying set. The reader is invited to work out the details of this approach to the construction of EX in problem 6E.

However, instead of just using the regular open sets of X as the underlying lattice from which we build EX, we can use the lattice $\tau(X)$ of all open subsets of X instead. This approach to the construction of EX was popularized by Iliadis and Fomin (see [Il_2] and [IF]) and has been used ever since by a variety of authors. We will build an extremally disconnected zero-dimensional space (which we will denote temporarily by E′X) whose underlying set is the set of convergent ultrafilters on $\tau(X)$, and construct a perfect irreducible Θ-continuous map k_X' from E′X onto X. By 6.7(a) it follows that $(EX,k_X) \sim (E'X,k_X')$. We will describe an explicit homeomorphism from EX onto E′X. Subsequently we will abandon the "E′X" notation and identify E′X with EX and k_X' with k_X.

Recall that an open ultrafilter $\mathcal{U}$ on X converges iff $\cap\{c\ell_X U : U \in \mathcal{U}\} = a(\mathcal{U}) \neq \emptyset$ (see 2.3(c)(3) and 2.3(f)). We formalize some of the above notions.

(a) Definition. Let X be a space. The set of all convergent open ultrafilters on X will be denoted by E′X. Recall that if $x \in X$, then $\{U \in \tau(X) : x \in U\}$ is denoted by $\mathcal{N}(x)$ (see 1.2(d)(6)).

(b) Lemma. Let X be a space, let $x \in X$, and let $\mathcal{U} \in E'X$. Then:

(1) $\mathcal{N}(x) \subseteq \mathcal{U}$ iff $x \in a(\mathcal{U})$;

(2) $a(\mathcal{U})$ contains exactly one point; and

(3) there exists $\mathcal{U}(x) \in E'X$ such that $a(\mathcal{U}(x)) = \{x\}$.

Proof

(1) If $x \in a(\mathcal{U})$, $U \in \mathcal{N}(x)$, and $V \in \mathcal{U}$, then $x \in U \cap$

$c\ell_X V$. Thus $U \cap V \neq \emptyset$ for each $V \in \mathcal{U}$, so by 2.3(d)(3) $U \in \mathcal{U}$. If $x \notin a(\mathcal{U})$, there exists $V \in \mathcal{U}$ such that $x \notin c\ell_X V$. Then $X \backslash c\ell_X V \in \mathcal{N}(x) \backslash \mathcal{U}$.

(2) This follows from 2.3(h).

(3) By 2.3(d)(2) there exists an ultrafilter $\mathcal{U}(x)$ on $\tau(X)$ such that $\mathcal{N}(x) \subseteq \mathcal{U}(x)$. Now apply (1) and (2).

■

Recall that when we constructed EX, we first constructed a compact extremally disconnected space θX (the Stone space of $R(X)$) and then defined EX to be a certain dense subspace of θX. We proceed in an analogous fashion when we build $E'X$.

(c) **Definition**. Let X be a space, and let $\theta' X$ denote the set of all open ultrafilters on X. If $U \in \tau(X)$, let $O(U) = \{\mathcal{U} \in \theta' X : U \in \mathcal{U}\}$. Therefore, $E'X = \{\mathcal{U} \in \theta' X : a(\mathcal{U}) \neq \emptyset\}$.

(d) **Lemma**. Let $U, V \in \tau(X)$. Then:

(1) $O(U) = \emptyset$ iff $U = \emptyset$;

(2) $O(U \cap V) = O(U) \cap O(V)$;

(3) $\theta' X \backslash O(U) = O(X \backslash c\ell_X U)$ and in particular $O(U) = O(\mathrm{int}_X c\ell_X U)$;

(4) $O(U) = \theta' X$ iff U is dense in X; in particular $O(X) = \theta' X$; and

(5) $\{O(U) : U \in \tau(X)\}$ is a base for a Hausdorff topology on $\theta' X$.

Proof

(1) $0(\varnothing) = \{\mathcal{U} \in \theta'X : \varnothing \in \mathcal{U}\} = \varnothing$. If $U \in \tau(X)\backslash\{\varnothing\}$, choose $x \in U$. Then $U \in \mathcal{N}(x)$; so as indicated in 6.8(b)(3), there exists $\mathcal{U}(x) \in E'X$ such that $U \in \mathcal{U}(x)$. Then $\mathcal{U}(x) \in 0(U)$ so $0(U) \neq \varnothing$.

(2) $\mathcal{U} \in 0(U \cap V)$ iff $U \cap V \in \mathcal{U}$ iff $U \in \mathcal{U}$ and $V \in \mathcal{U}$ iff $\mathcal{U} \in 0(U)$ and $\mathcal{U} \in 0(V)$ iff $\mathcal{U} \in 0(U) \cap 0(V)$.

(3) By (1) and (2) $0(U) \cap 0(X\backslash c\ell_X U) = \varnothing$. Conversely if $\mathcal{U} \in \theta'X\backslash 0(U)$ then $U \notin \mathcal{U}$. So by 2.3(e)(2) $X\backslash c\ell_X U \in \mathcal{U}$, and thus $\mathcal{U} \in 0(X\backslash c\ell_X U)$. The result follows.

(4) If $U \in \tau(X)$ then $X\backslash c\ell_X U = \varnothing$ iff U is dense in X. The result now follows from (1) and (3).

(5) That $\{0(U) : U \in \tau(X)\}$ is a base for a topology follows from (1), (2) and (4). By 2.3(d) distinct points of $\theta'X$ contain disjoint sets, so by (1) and (2) above $\theta'X$ is Hausdorff.

■

Henceforth we will regard $\theta'X$ as a topological space equipped with the topology described in 6.8(d)(5). We now can topologize $E'X$ and find an irreducible Θ-continuous surjection from $E'X$ onto X.

(e) **Definition**. Let X be a space.

(1) The set $E'X$ is topologized by giving it the subspace topology inherited from the space $\theta'X$ described above. (Explicitly, $\{0(U) \cap E'X : U \in \tau(X)\}$ is an open base for the topology of $E'X$.) Henceforth $E'X$ is regarded as a topological space.

(2) If $\mathcal{U} \in E'X$, let $k_X'(\mathcal{U})$ denote the unique point in $a(\mathcal{U})$.

This defines a function from $E'X$ onto X (see 6.8(b)).

(f) **Theorem**. Let X be a space. Then:

(1) if $\{U_i : i \in I\} \subseteq \tau(X)$, then $c\ell_{\theta'X} \vee\{O(U_i) : i \in I\} = O(\vee\{U_i : i \in I\})$;

(2) $B(\theta'X) = \{O(U) : U \in \tau(X)\}$ and $\theta'X$ is zero-dimensional;

(3) $\theta'X$ is a compact extremally disconnected space;

(4) $E'X$ is a dense extremally disconnected zero-dimensional subspace of $\theta'X$;

(5) $k_X'[O(U) \cap E'X] = c\ell_X U$ for each $U \in \tau(X)$;

(6) $k_X'^{\leftarrow}(x) \subseteq O(U)$ iff $x \in int_X c\ell_X U$ for $x \in X$ and $U \in \tau(X)$; and

(7) k_X' is a perfect irreducible Θ-continuous function from $E'X$ onto X.

Proof

(1) By (2) of the preceding lemma if $U, V \in \tau(X)$ and $U \subseteq V$ then $O(U) \subseteq O(V)$. Thus $\vee\{O(U_i) : i \in I\} \subseteq O(\vee\{U_i : i \in I\})$. By (3) of the preceding lemma $O(\vee\{U_i : i \in I\}) \in B(\theta'X)$ so $c\ell_{\theta'X} \vee\{O(U_i) : i \in I\} \subseteq O(\vee\{U_i : i \in I\})$. Now suppose that $\mathcal{U} \in O(\vee\{U_i : i \in I\})$ and $O(V)$ is a basic neighborhood of $\mathcal{U}$ in $\theta'X$ (for some $V \in \tau(X)$). Then $\emptyset \neq O(\vee\{U_i : i \in I\}) \cap O(V) = O([\vee\{U_i : i \in I\}] \cap V)$; so there exists $j \in I$ such that $U_j \cap V \neq \emptyset$. Thus $O(U_j) \cap O(V) \neq \emptyset$, and it follows that $\mathcal{U} \in c\ell_{\theta'X} \vee\{O(U_i) : i \in I\}$. Hence (1) follows.

(2) From (3) of the previous lemma $\{O(U) : U \in \tau(X)\} \subseteq B(\theta'X)$. Conversely, if $B \in B(\theta'X)$, then $B = \vee\{O(U_i) : i \in I\}$ for

some $\{U_i : i \in I\} \subseteq \tau(X)$. By (1) above $B = c\ell_{\theta'X}B = O(U)$ where $U = \vee\{U_i : i \in I\}$. It follows that $\theta'X$ is zero-dimensional.

(3) Any open subset V of $\theta'X$ is of the form $\cup\{O(U_i) : i \in I\}$ for some $\{U_i : i \in I\} \subseteq \tau(X)$, so by (1) $c\ell_{\theta'X}V$ is open in $\theta'X$. Thus $\theta'X$ is extremally disconnected.

Suppose that $\{U_i : i \in I\} \subseteq \tau(X)$ and that $\theta'X = \cup\{O(U_i) : i \in I\}$. Suppose further that $\theta'X\backslash\cup\{O(U_i) : i \in F\} \neq \emptyset$ for each finite subset F of I. By 6.8(d)(3), and (1) above, $\theta'X\backslash\cup\{O(U_i) : i \in F\} = O(X\backslash c\ell_X\vee\{U_i : i \in F\})$. Thus $\{X\backslash c\ell_X\vee\{U_i : i \in F\} : F$ is a finite subset of $I\}$ is a collection of nonempty open subsets of X with the finite intersection property. By 2.3(d) there is an open ultrafilter $\mathcal{U}$ on X containing this family. Obviously $\{U_i : i \in I\} \cap \mathcal{U} = \emptyset$; so $\mathcal{U} \in \theta'X\backslash\cup\{O(U_i) : i \in I\}$, which is a contradiction. This shows that any cover of $\theta'X$ by basic open sets has a finite subcover, and so $\theta'X$ is compact.

(4) Let $U \in \tau(X)\backslash\{\emptyset\}$ and choose $x \in U$. By 6.8(b)(3) find $\mathcal{U}(x) \in E'X$ such that $a(\mathcal{U}(x)) = \{x\}$. It follows that $U \in \mathcal{U}(x)$. Thus $\mathcal{U}(x) \in O(U) \cap E'X$, and so $E'X$ is dense in $\theta'X$. By 6.2(b) $E'X$ is extremally disconnected, and it is zero-dimensional because $\theta'X$ is.

(5) If $\mathcal{U} \in O(U)$, then $U \in \mathcal{U}$, so $k_X'(\mathcal{U}) \in a(\mathcal{U}) \subseteq c\ell_XU$. Conversely, if $x \in c\ell_XU$, then $\mathcal{N}(x) \cup \{U\}$ has the finite intersection property and thus by 2.3(d) is contained in an ultrafilter $\mathcal{U}$ on $\tau(X)$. Then $\mathcal{U} \in O(U)$ and as $\mathcal{N}(x) \subseteq \mathcal{U}$, $a(\mathcal{U}) = \{x\}$. Thus $\mathcal{U} \in E'X$ and $x \in k_X'[O(U) \cap E'X]$.

(6) If $x \notin int_Xc\ell_XU$, then $x \in c\ell_X(X\backslash c\ell_XU)$. Then by 2.3(d) there is an open ultrafilter $\mathcal{U}$ on X such that $\mathcal{N}(x) \cup \{X\backslash c\ell_XU\} \subseteq \mathcal{U}$. By 6.8(b)(1) $k_X'(\mathcal{U}) = x$, and $\mathcal{U} \in O(X\backslash c\ell_XU)$

$= \theta' X \setminus O(U)$ (see 6.8(d)(3)). Thus $k_X'^{\leftarrow}(x) \not\subseteq O(U)$. Conversely, if $x \in \mathrm{int}_X c\ell_X U$ and $\mathcal{U} \in k_X'^{\leftarrow}(x)$, then $(\mathrm{int}_X c\ell_X U) \cap c\ell_X V \neq \emptyset$ for each $V \in \mathcal{U}$. Thus $X \setminus c\ell_X U \notin \mathcal{U}$, and so $\mathcal{U} \notin O(X \setminus c\ell_X U)$ $= \theta' X \setminus O(U)$. Thus $\mathcal{U} \in O(U)$.

(7) We begin by showing that if $x \in X$, then $(k_X')^{\leftarrow}(x)$ is a compact subset of $E'X$. It suffices to show that $(k_X')^{\leftarrow}(x)$ is closed in $\theta' X$. If $\mathcal{U} \in \theta' X \setminus (k_X')^{\leftarrow}(x)$, by 6.8(b)(1) there exists an open subset V of X such that $x \in V$ and $V \notin \mathcal{U}$. By (6), $(k_X')^{\leftarrow}(x) \subseteq O(V)$, while $\mathcal{U} \in \theta' X \setminus O(V) = O(X \setminus c\ell_X V)$. Thus $\mathcal{U}$ has a $\theta' X$-neighborhood disjoint from $k_X'^{\leftarrow}(x)$, and so $k_X'^{\leftarrow}(x)$ is closed in $\theta' X$.

Next we show that k_X' is a closed map. Let F be a closed subset of $E'X$ and let $x \in X \setminus k_X'[F]$. Then $(k_X')^{\leftarrow}(x) \cap F = \emptyset$, and as $(k_X')^{\leftarrow}(x)$ is compact, we can argue as in the proof of 6.6(e)(5) and find $U_1, \ldots, U_n \in \tau(X)$ such that $(k_X')^{\leftarrow}(x) \subseteq [\cup\{O(U_i) : 1 \leqslant i \leqslant n\}] \cap E'X \subseteq E'X \setminus F$. By (1) and (2), $\cup\{O(U_i) : 1 \leqslant i \leqslant n\}] \cap E'X = O(U) \cap E'X$ where $U = \cup\{U_i : 1 \leqslant i \leqslant n\}$. Thus $F \subseteq E'X \setminus O(U) = O(X \setminus c\ell_X U) \cap E'X$ so by (5) $k_X'[F] \subseteq c\ell_X(X \setminus c\ell_X U) = X \setminus \mathrm{int}_X c\ell_X U$. By (6), $x \in \mathrm{int}_X c\ell_X U$, which is open and disjoint from $k_X'[F]$. Thus $k_X'[F]$ is closed in X.

We now have shown that k_X' is a perfect map. It is irreducible, for if F is a proper closed subset of $E'X$, then there exists $U \in \tau(X) \setminus \{\emptyset\}$ such that $F \subseteq E'X \setminus O(U)$. Then $k_X'[F] \subseteq X \setminus U$ (as shown above), so k_X' is irreducible.

If $\mathcal{U} \in E'X$ and $k_X'(\mathcal{U}) \in U \in \tau(X)$, then $U \in \mathcal{U}$ by the maximality of $\mathcal{U}$ (see 2.3(d)) and so $\mathcal{U} \in O(U)$. By (5), $k_X'[O(U) \cap E'X] = c\ell_X U$; therefore, k_X' is Θ-continuous.

■

It is a consequence of 6.7(a) and the preceding theorem that there is a homeomorphism $h : EX \to E'X$ such that $k_X = k_X' \circ h$. Thus $(EX, k_X) \sim (E'X, k_X')$. If $\mathcal{U} \in EX$, it is evident that $h(\mathcal{U})$ must be $\{U \in \tau(X) : c\ell_X U \in \mathcal{U}\}$. In 6E(2) we invite the reader to show that h is a well-defined homeomorphism with the desired properties.

We have now developed the two most widely used constructions of the absolute, and we shall use each of them frequently henceforth. If X is a given space, then k_X will always denote the perfect irreducible θ-continuous surjection from EX onto X that we described in 6.6.

6.9 Elementary properties of EX

In this section we develop enough properties of the absolute to allow us in 6.10 to identify the absolutes of certain specific spaces.

(a) **Proposition**. Let X be a dense subspace of a space T. Then $(EX, k_X) \sim (k_T^{\leftarrow}[X], k_T | k_T^{\leftarrow}[X])$.

Proof. As k_T is irreducible and X is dense in T, by 6.5(b)(4) $k_T^{\leftarrow}[X]$ is dense in ET. Thus by 6.2(b) $k_T^{\leftarrow}[X]$ is extremally disconnected. By 1.8(f)(2) $k_T | k_T^{\leftarrow}[X]$ is a perfect surjection from $k_T^{\leftarrow}[X]$ onto X, by 6.5(b)(4) it is irreducible, and by 4.8(h)(5) it is

Θ-continuous. As ET is zero-dimensional so is $k_T^{\leftarrow}[X]$; so by 6.7(c) $(EX, k_X) \sim (k_T^{\leftarrow}[X], k_T^{\leftarrow} | k_T^{\leftarrow}[X])$.

■

The above proposition allows us to identify $k_T^{\leftarrow}[X]$ with EX whenever T is an extension of X. Henceforth we will frequently do this without explicitly saying so.

(b) **Theorem**. Let X be a space. Then:

(1) EX is compact iff X is H-closed iff $\theta X = EX$,

(2) if hX is an H-closed extension of X then E(hX) and $\beta(EX)$ are equivalent extensions of EX,

(3) if X is Tychonoff then $E(\beta X)$ and $\beta(EX)$ are equivalent extensions of EX (i.e., $E(\beta X) \equiv_{EX} \beta(EX)$),

(4) if X is Tychonoff then $(\beta(EX), \beta k_X)$ is the absolute of βX (up to equivalence) (i.e., $(E(\beta X), k_{\beta X})) \sim (\beta(EX), \beta k_X)$ (see 4.2(d) for notation), and

(5) for $\gamma_1 X, \gamma_2 X \in \mathcal{H}(X)$, there exists a homeomorphism h: $E(\gamma_1 X) \to E(\gamma_2 X)$ such that $h[k_{\gamma_1 X}^{\leftarrow}[X]] = k_{\gamma_2 X}^{\leftarrow}[X]$ and

$(k_{\gamma_2 X} | k_{\gamma_2 X}^{\leftarrow}[X]) \circ (h | k_{\gamma_1 X}^{\leftarrow}[X]) = k_{\gamma_1 X} | k_{\gamma_1 X}^{\leftarrow}[X]$.

Proof

(1) If EX is compact then it is H-closed (see 4.8(c)); since k_X is a Θ-continuous surjection, by 4.8(h)(6) it follows that X is H-closed. Conversely, if X is H-closed then by 4.8(b) $E'X = \theta' X$ (see 6.8(c)). As $\theta' X$ is compact (6.8(f)), $E'X$ must be compact. But $E'X$ and EX are homeomorphic, so EX is compact. By 6.6(e) EX is

dense in the compact space θX, so $EX = \theta X$ iff EX is compact.

(2) Using 6.9(a), we identify (EX, k_X) with $(k_{hX}^{\leftarrow}[X], k_{hX} | k_{hX}^{\leftarrow}[X])$. As $k_{hX}^{\leftarrow}[X]$ is dense and C^*-embedded in E(hX) (see 6.2(c)and 6.9(a)), and as E(hX) is compact by (1), it follows from 4.6(g) that E(hX) and $\beta(k_{hX}^{\leftarrow}[X])$ are equivalent extensions of $k_{hX}^{\leftarrow}[X]$, i.e., equivalent extensions of EX.

(3) This is a special case of (2).

(4) Since $k_X : EX \to X$ is continuous (see 6.6(e)(6)), it has a continuous extension $\beta k_X : \beta(EX) \to \beta X$. By 6.2(c) $\beta(EX)$ is extremally disconnected. As $\beta(EX)$ is compact, βk_X is perfect; as k_X is an irreducible surjection, so is βk_X. Thus $(E(\beta X), k_{\beta X}) \sim (\beta(EX), \beta k_X)$.

(5) By 6.9(a) there is a homeomorphism $h_i : EX \to k_{\Upsilon_i X}^{\leftarrow}[X]$ such that $(k_{\Upsilon_i X} | k_{\Upsilon_i X}^{\leftarrow}[X]) \circ h_i = k_X$ $(i = 1,2)$. Thus $h_2 \circ h_1^{\leftarrow}$ is a homeomorphism $h' : k_{\Upsilon_1 X}^{\leftarrow}[X] \to k_{\Upsilon_2 X}^{\leftarrow}[X]$ satisfying $(k_{\Upsilon_2 X} | k_{\Upsilon_2 X}^{\leftarrow}[X]) \circ h' = k_{\Upsilon_1 X} | k_{\Upsilon_1 X}^{\leftarrow}[X]$. Now by the proof of (2) above, $\beta(k_{\Upsilon_i X}^{\leftarrow}[X])$ and $E(\Upsilon_i X)$ $(i = 1,2)$ are equivalent extensions of $k_{\Upsilon_i X}^{\leftarrow}[X]$; by the above there is a homeomorphism $h : E(\Upsilon_1 X) \to E(\Upsilon_2 X)$ such that $h | k_{\Upsilon_1 X}^{\leftarrow}[X] = h'$. Then h has the required properties.

■

Each of the statements 6.9(b)(3) and 6.9(b)(4) is often

abbreviated by writing the ambiguous equation "$\beta(EX) = E(\beta X)$". This equation can be interpreted in two ways: first, that there exists k such that $(\beta(EX),k)$ is the absolute of βX (up to equivalence), or second, that $k_{\beta X}^{\leftarrow}[X]$ is a dense subspace S of $E(\beta X)$ homeomorphic to EX, and $E(\beta X)$, viewed as an extension of S, is equivalent to βS. As we saw above, each of these interpretations is correct.

In 6.9(a) we were given a dense subspace X of a space T, and we showed how to represent the absolute of X in terms of the absolute of T. Now we do the opposite; we show how the absolute of T can be represented as a subset of $\theta' X$ that lies between $E'X$ and $\theta' X$. In doing this it will be useful to introduce the concept of the adherence in T of a filter $\mathcal{U}$ on X. We define this to be the set $\cap\{cl_T F : F \in \mathcal{U}\}$, and denote it by $a_T(\mathcal{U})$. This is an obvious generalization of definition 2.3(c)(3).

(c) __Lemma__. Let X be a dense subspace of T. Then:

(1) if $\mathcal{U} \in \theta' X$, then $|a_T(\mathcal{U})| \leq 1$ and

(2) if $p \in T$, there exists $\mathcal{U} \in \theta' X$ such that $a_T\mathcal{U} = \{p\}$.

__Proof__

(1) If x and y were distinct points in $a_T(\mathcal{U})$, there would be disjoint open subsets U and V of T with $x \in U$ and $y \in V$. Then $(U \cap X) \cap W \neq \varnothing \neq (V \cap X) \cap W$ for each $W \in \mathcal{U}$. By 2.3(d)(3) $U \cap X$ and $V \cap X$ belong to $\mathcal{U}$. which is a contradiction as they are disjoint.

(2) By 2.3(d)(2) there exists $\mathcal{U} \in \theta'X$ that contains $\{U \cap X : p \in U \in \tau(T)\}$. One easily verifies that $a_T\mathcal{U} = \{p\}$.

■

We now represent ET as a subspace of $\theta'X$.

(d) **Theorem**. Let X be a dense subspace of a space T. Let $E''T = \{\mathcal{U} \in \theta'X : a_T(\mathcal{U}) \neq \emptyset\}$. Define $f : E''T \to T$ by letting $f(\mathcal{U})$ be the unique point in $a_T(\mathcal{U})$. Then:

(1) $(ET, k_T) \sim (E''T, f)$, and

(2) $E''T = \theta'X$ iff T is H-closed.

Proof

(1) Since $E'X \subseteq E''T \subseteq \theta'X$, then $E''T$ is dense in $\theta'X$ and hence is extremally disconnected by 6.2(b)(2). By the preceding lemma f is a well-defined surjection. The proof that it is perfect, irreducible, and Θ-continuous is essentially the same as the proof in 6.8(f)(7) that k_X' has those properties. We do not include the details, but invite the reader in 6E to supply them.

(2) By 6.9(b)(1) ET is compact iff T is H-closed. By (1) ET is compact iff $E''T$ is compact. As $E'X \subseteq E''T \subseteq \theta'X$, $E''T$ is dense in $\theta'X$ and thus is compact iff it is all of $\theta'X$.

■

Of course, we can also represent ET as a subspace of θX that contains EX. The procedure for doing so is completely analogous to that described in 6.9(d) above.

Next we show that the restriction of k_X to the set of isolated points of EX is a homeomorphism onto the set of isolated points of X. We use the construction of EX developed in 6.6. We denote the set of isolated points of X by I(X).

(e) **Proposition**. Let X be a space. Then $k_X | I(EX)$ is a bijection from I(EX) onto I(X) and $k_X^{\leftarrow}[I(X)] = I(EX)$.

Proof. Let $\mathcal{U} \in I(EX)$. Then $EX \setminus \{\mathcal{U}\}$ is a proper closed subset of EX so $k_X[EX \setminus \{\mathcal{U}\}] \neq X$. Thus $X \setminus \{k_X(\mathcal{U})\} = k_X[EX \setminus \{\mathcal{U}\}]$ so as k_X is a closed map, $k_X(\mathcal{U}) \in I(X)$. It also follows that $|k_X^{\leftarrow}[k_X(\mathcal{U})]| = 1$ so $k_X | I(EX)$ is one-to-one.

If $x \in I(X)$ and $\mathcal{U} \in k_X^{\leftarrow}(x)$, then $\{x\} \in \mathcal{U}$ and so $\mathcal{U} \in \lambda(\{x\})$. If $\mathcal{W} \in \lambda(\{x\})$, then $\{x\} \in \mathcal{W}$; so if $A \in \mathcal{U}$ and $B \in \mathcal{W}$, then $\varnothing \neq \{x\} \subseteq A \wedge B$. It follows from the maximality of $\mathcal{U}$ and $\mathcal{W}$ that $\mathcal{U} = \mathcal{W}$, and so $\lambda(\{x\}) = \{\mathcal{U}\}$. Thus $k_X^{\leftarrow}(x) \subseteq I(EX)$ and k_X maps I(EX) onto I(X).

■

(f) **Proposition**. If X and Y are spaces, and if there exists a perfect irreducible Θ-continuous surjection f from X onto Y, then $(EX, f \circ k_X)$ is the absolute of Y (up to equivalence).

Proof. Let f be a perfect irreducible Θ-continuous surjection from X onto Y. Then $f \circ k_X$ is a perfect irreducible θ-continuous surjection from EX onto Y; so, by 6.7(a) $(EY, k_Y) \sim (EX, f \circ k_X)$.

■

We should warn the reader that the converse of the preceding proposition is not true. There exist spaces X and Y such that EX and EY are homeomorphic and such that neither X nor Y can be mapped onto the other by a perfect irreducible Θ-continuous surjection (for example, $\mathbb{R}$ and $\mathbb{R} \times \mathbb{N}$).

6.10 Examples of absolutes

Although we have given two detailed descriptions of the construction of the absolute of a space X, until now we have not given any examples to show what the absolute looks like in specific cases. We will now show that the absolutes of certain spaces are spaces with which we are already familiar in other contexts. In addition we prove a theorem which will imply that if K and L are two compact metric spaces without isolated points, then EK and EL are homeomorphic.

(a) **Theorem**

(1) Let X be an extremally disconnected zero-dimensional space and let αX be a compactification of X. Then $(E(\alpha X),k_{\alpha X}) \sim (\beta X,j)$ where j denotes the Stone extension $\beta(id_X) : \beta X \rightarrow \alpha X$ of id_X.

(2) If D is a discrete space and αD is a compactification of D, then $(E(\alpha D),k_{\alpha D}) \sim (\beta D,j)$.

(3) Let m be an infinite cardinal and let κ_m be the smallest ordinal of cardinality m^+, equipped with the order topology. Then $E(\kappa_m)$ is homeomorphic to $mD(m^+)$ (see 5.10(d)).

Proof

(1) Evidently j is irreducible because $j[\beta X \setminus X] = \alpha X \setminus X$ and $j|X = id_X$. Obviously j is a perfect continuous surjection. By 6.2(c) βX is extremally disconnected so by 6.7(a) $(E(\alpha X), k_{\alpha X}) \sim (\beta X, j)$.

(2) This is a special case of (1).

(3) By 2.6(q)(6) and 4.6(g) $\beta\kappa_m = \kappa_m + 1$. The non-limit ordinals of κ_m form a dense discrete subspace of κ_m of cardinality m^+, and each is isolated in κ_m, so we can regard $\beta\kappa_m$ as a compactification of $D(m^+)$. Let $j : \beta D(m^+) \to \beta\kappa_m$ be the Stone extension of the embedding of $D(m^+)$ in $\beta\kappa_m$. By (2) above $\beta D(m^+) = E(\beta\kappa_m)$ and by 6.9(a), $E(\kappa_m) = j^{\leftarrow}[\kappa_m]$. We now show that $j^{\leftarrow}[\kappa_m] = mD(m^+)$ (see 5.10(d) for a description of $mD(m^+)$).

If $\alpha \in mD(m^+)$, there is a subset S of $D(m^+)$ such that $|S| \leq m$ and $\alpha \in c\ell_{\beta D(m^+)} S$. Thus $j(\alpha) \in c\ell_{\beta\kappa_m} S \subseteq \kappa_m$ (since S is a subset of κ_m of cardinality less than m^+, and $cf(\kappa_m) = m^+$). Thus $mD(m^+) \subseteq j^{\leftarrow}[\kappa_m]$. Conversely, if $\alpha \in \kappa_m$ then the set S of all non-limit ordinals of κ_m that precede $\alpha+1$ is a subset of κ_m of cardinality less than m^+, and $\alpha \in c\ell_{\kappa_m} S = c\ell_{\beta\kappa_m} S$. Thus $j^{\leftarrow}(\alpha) \subseteq j^{\leftarrow}[c\ell_{\beta\kappa_m} S] = c\ell_{\beta D(m^+)} S$ (see 6.5(d)(3)). But $c\ell_{\beta D(m^+)} S \subseteq mD(m^+)$, so $j^{\leftarrow}[\kappa_m] \subseteq mD(m^+)$ and $(E\kappa_m, k_{\kappa_m}) \sim (mD(m^+), j|mD(m^+))$.

■

Recall (see 3.4(d)(4)) that a subset S of a Boolean algebra B is said to be **dense** in B if for each $b \in B \setminus \{0\}$ there exists $s \in S$ such that $0 < s \leqslant b$. Dense subalgebras of Boolean algebras are related to absolutes, as we see in the next few results.

(b) <u>Lemma</u>. Let j be an embedding of a Boolean algebra A into a Boolean algebra B. The following are equivalent:

(1) $j[A]$ is dense in B and

(2) the dual map $\lambda(j) : S(B) \to S(A)$ is a perfect irreducible continuous surjection.

<u>Proof</u>. By 3.2(f)(3) $\lambda(j)$ is perfect and continuous. As j is one-to-one by 3.2(f)(4) $\lambda(j)$ maps $S(B)$ onto $S(A)$.

(1) $\Rightarrow$ (2) Let F be a proper closed subset of $S(B)$. Find $b \in B$ such that $0 \neq \lambda(b) \subseteq S(B) \setminus F$. By hypothesis there exists $a \in A$ such that $0 \neq j(a) \leqslant b$. If $\mathcal{W} \in F$, then $\mathcal{W} \in S(B) \setminus \lambda(b)$ and $b \notin \mathcal{W}$. Thus $j(a) \notin \mathcal{W}$ so $j(a') = j(a)' \in \mathcal{W}$. It follows from the definition of the dual map $\lambda(j)$ (see 3.2(e)) that if $\lambda(j)(\mathcal{W}) = \mathcal{U}$, then $a' \in \mathcal{U}$ and so $\mathcal{U} \in \lambda(a') = S(A) \setminus \lambda(a)$. It follows that $\lambda(j)[F] \subseteq S(A) \setminus \lambda(a)$ and so $\lambda(j)$ is irreducible.

(2) $\Rightarrow$ (1) Suppose $\lambda(j)$ is irreducible. If $b \in B \setminus \{0\}$, then $\varnothing \neq S(A) \setminus \lambda(j)[S(B) \setminus \lambda(b)]$; thus there exists $a \in A \setminus \{0\}$ such that

$$(*) \qquad \lambda(a) \subseteq S(A) \setminus \lambda(j)[S(B) \setminus \lambda(b)] .$$

Now suppose $\mathcal{W} \in \lambda(j(a))$. Then $j(a) \in \mathcal{W}$ and so $a \in \{x \in A : j(x) \in \mathcal{W}\}$. But $\{x \in A : j(x) \in \mathcal{W}\} = \lambda(j)(\mathcal{W})$ (see 3.2(e)) and so

$a \in \lambda(j)(\mathcal{W})$. Thus $\lambda(j)(\mathcal{W}) \in \lambda(a)$. By (*) this means that $\mathcal{W} \in \lambda(b)$. It follows that $\lambda(j(a)) \subseteq \lambda(b)$, and so $j(a) \leqslant b$. Hence $j[A]$ is dense in B.

■

(c) <u>Lemma</u>. Let A be a dense subalgebra of a Boolean algebra B. Then b is an atom of B iff $b \in A$ and b is an atom of A.

<u>Proof</u>. Let $b \in B$. As A is dense in B, there exists $a \in A$ such that $0 < a \leqslant b$. If b is an atom of B, it follows that $a = b$ and so $b \in A$. Obviously b is an atom of A. Conversely if b is an atom of A, suppose there exists $c \in B$ such that $0 < c \leqslant b$. As A is dense in B there exists $a \in A$ such that $0 < a \leqslant c$. Thus $0 \leqslant a \leqslant b$, so as b is an atom of A, $a = b$. Thus $a = c = b$ and b is an atom of B.

■

Recall from 2N(5) that a π-base of an infinite topological space X is a collection $\mathcal{S}$ of non-empty open subsets of X with the property that if V is a nonempty open subset of X, then there exists $S \in \mathcal{S}$ such that $S \subseteq V$. The π-weight of X (abbreviated $\pi w(X)$) is defined to be $\min\{m : X \text{ has a } \pi\text{-base of cardinality } m\}$. If $\pi w(X) = \aleph_0$ we say that X has **countable π-weight**.

(d) <u>Theorem</u>. If X is a space without isolated points and if X has countable π-weight, then θX is homeomorphic to $E(\mathbf{2}^{\aleph_0})$ ($\mathbf{2}^{\aleph_0}$ is the Cantor space described in 3K(1) and 3.3(e)).

Proof. Let $\mathcal{S}$ be a countable π-base of X and let $\mathcal{Q}$ be the subalgebra of $\mathcal{R}(X)$ generated by $\{c\ell_X S : S \in \mathcal{S}\}$. As $\mathcal{S}$ is countable it follows from 3.1(l)(2) that so is $\mathcal{Q}$. If $H \in \mathcal{R}(X)\backslash\{\emptyset\}$, find $V \in \mathcal{S}$ such that $V \subseteq int_X H$; thus $\emptyset \neq c\ell_X V \subseteq H$ and so $\mathcal{Q}$ is dense in $\mathcal{R}(X)$. As X has no isolated points $\mathcal{R}(X)$ has no atoms (see 3.2(m)); so by the preceding lemma neither has $\mathcal{Q}$. By 3.3(g) $\mathcal{Q}$ is isomorphic to $\mathcal{B}(\mathfrak{Z}^{\aleph_0})$; therefore, there is an embedding $j : \mathcal{B}(\mathfrak{Z}^{\aleph_0}) \to \mathcal{R}(X)$ such that $j[\mathcal{B}(\mathfrak{Z}^{\aleph_0})]$ $(=\mathcal{Q})$ is dense in $\mathcal{R}(X)$. By 6.10(b) the Stone dual $\lambda(j)$ is a perfect irreducible continuous surjection from $S(\mathcal{R}(X))$ onto $S(\mathcal{B}(\mathfrak{Z}^{\aleph_0}))$. Now $S(\mathcal{R}(X)) = \theta X$ and by 3.2(h) $S(\mathcal{B}(\mathfrak{Z}^{\aleph_0}))$ is homeomorphic to $\mathfrak{Z}^{\aleph_0}$, so there is a perfect irreducible Θ-continuous surjection f from the extremally disconnected space θX onto $\mathfrak{Z}^{\aleph_0}$. By 6.7(a) $(E(\mathfrak{Z}^{\aleph_0}), k_{\mathfrak{Z}^{\aleph_0}}) \sim (\theta X, f)$.

■

(e) Corollary. Let X be an H-closed space without isolated points and let X have countable π-weight. Then EX is homeomorphic to $E(\mathfrak{Z}^{\aleph_0})$.

Proof. By 6.9(b) $EX = \theta X$ and by the preceding theorem θX is homeomorphic to $E(\mathfrak{Z}^{\aleph_0})$.

■

(f) Corollary. Let K and L be two compact metric spaces without isolated points. Then EK and EL are homeomorphic.

Proof. K and L are both second countable and hence have

countable π-weight. Thus EK and EL are each homeomorphic to $E(2^{\aleph_0})$.

■

6.11 The Banaschewski absolute

One disadvantage of EX is that the perfect irreducible map k_X from EX onto X is continuous only if X is regular. We now construct for each Hausdorff space X, an extremally disconnected space PX and a perfect irreducible continuous surjection $\Pi_X : PX \to X$ such that the pair (PX, Π_X) is unique in the same way that the pair (EX, k_X) was shown in 6.7 to be unique. However, in gaining the continuity of Π_X, we also lose something; the space PX is not zero-dimensional except when X is regular, and in that case PX is homeomorphic to EX.

(a) **Lemma**. Let X be a space. Define S to be the following collection of subsets of EX:

$$S = \{\lambda(A) \cap k_X^{\leftarrow}[V] : V \in \tau(X) \text{ and } A \in R(X)\}.$$

Then S is an open base for a Hausdorff topology τ_S on the underlying set of EX, and τ_S contains the topology of EX.

Proof. S is closed under finite intersections, contains $\emptyset$ and EX, and contains $\{\lambda(A) \cap EX : A \in R(X)\}$, which is a base for the open sets of the Hausdorff space EX.

■

(b) **Definition**. We denote the underlying set of EX, equipped with the topology τ_S, by PX. The identity map on the underlying set of EX, viewed as a bijection from the space PX onto the space EX, will be denoted by j_{PX}. Denote $k_X \circ j_{PX}$ by Π_X; then Π_X is a function from PX onto X.

Evidently the functions Π_X and k_X are identical as functions, and we shall often use this fact without comment. It is clear that τ_S is the weakest topology on the underlying set of EX that both contains the topology of EX and "makes k_X continuous."

(c) **Theorem**. Let X be a space. Then

(1) PX is extremally disconnected and EX = (PX)(s) (see 2.2(e))

and

(2) Π_X is a perfect irreducible continuous surjection from PX onto X.

Proof

(1) Since the topology of PX contains that of EX, it is evident that $B(EX) \subseteq B(PX)$. We will show that if $U \in \tau_S$, then $c\ell_{PX}U \in B(EX)$. This will show both that PX is extremally disconnected and that $B(EX) = B(PX)$. From these facts it follows that $RO(PX) = B(EX)$, and so $(PX)(s) = EX$.

Let U be open in PX. Then U is of the form $\cup\{\lambda(A_i) \cap k_X^{\leftarrow}[V_i] : i \in I\}$, where $\{A_i : i \in I\} \subseteq R(X)$ and $\{V_i : i \in I\} \subseteq \tau(X)$. Let $W = \cup\{(\text{int}_X A_i) \cap V_i : i \subseteq I\}$. We will show that

$$(*) \qquad c\ell_{PX}U = \lambda(c\ell_X W) \cap EX .$$

Let $\mathcal{U} \in \lambda(c\ell_X W) \cap EX$. To show that $\mathcal{U} \in c\ell_{PX}U$, it suffices to show that if $\mathcal{U} \in \lambda(B) \cap k_X^{\leftarrow}[T]$, where $B \in R(X)$ and $T \in \tau(X)$, then $\lambda(B) \cap k_X^{\leftarrow}[T] \cap U \neq \varnothing$. Under our assumptions

$$\mathcal{U} \in \lambda(c\ell_X W) \cap \lambda(B) \cap k_X^{\leftarrow}[T] = \lambda(B \wedge c\ell_X W) \cap k_X^{\leftarrow}[T];$$

so

$$\begin{aligned} k_X(\mathcal{U}) \in T \cap k_X[EX \cap \lambda(B \wedge c\ell_X W)] \\ &= T \cap (B \wedge c\ell_X W) \quad \text{(see 6.6(e)(3))} \\ &= T \cap c\ell_X(\text{int}_X B \cap \text{int}_X c\ell_X W). \end{aligned}$$

Thus $T \cap \text{int}_X B \cap \text{int}_X c\ell_X W \neq \varnothing$, from which it follows that $T \cap \text{int}_X B \cap W \neq \varnothing$. Thus there exists $i_0 \in I$ such that $T \cap \text{int}_X B \cap \text{int}_X A_{i_0} \cap V_{i_0} \neq \varnothing$. If $x \in T \cap \text{int}_X B \cap \text{int}_X A_{i_0} \cap V_{i_0}$, by 6.6(e)(4), $\varnothing \neq k_X^{\leftarrow}(x) \subseteq k_X^{\leftarrow}[T] \cap \lambda(B) \cap \lambda(A_{i_0}) \cap k_X^{\leftarrow}[V_{i_0}]$ and so $\lambda(B) \cap k_X^{\leftarrow}[T] \cap U \neq \varnothing$.

Now let $\mathcal{U} \in c\ell_{PX}U$. If $\mathcal{U} \notin \lambda(c\ell_X W)$, then $\mathcal{U} \in EX \backslash \lambda(c\ell_X W) = EX \cap \lambda(c\ell_X(X \backslash c\ell_X W))$. Thus $\lambda(c\ell_X(X \backslash c\ell_X W)) \cap U \neq \varnothing$. Thus there exists $i_0 \in I$ such that $\lambda(c\ell_X(X \backslash c\ell_X W)) \cap \lambda(A_{i_0}) \cap k_X^{\leftarrow}[V_{i_0}] \neq \varnothing$. By 6.6(e)(3) this implies that

$$\varnothing \neq V_{i_0} \cap c\ell_X[\text{int}_X c\ell_X(X \backslash c\ell_X W) \cap \text{int}_X A_{i_0}],$$

and so $\varnothing \neq V_{i_0} \cap (X \backslash c\ell_X W) \cap \text{int}_X A_{i_0}$. This contradicts the definition of W. Thus $U \in \lambda(c\ell_X W) \cap EX$ and (*) holds. It follows that $R(PX) = B(EX)$, and our other claims follow.

(2) As j_{PX} and k_X are surjections, Π_X is also a surjection. If V is open in X, then $\Pi_X^{\leftarrow}[V] = k_X^{\leftarrow}[V] \in \tau_S$ and Π_X is continuous.

To prove that Π_X is a compact function, suppose $x \in X$ and $\Pi_X^{\leftarrow}(x) \subseteq \bigvee\{\lambda(A_i) \cap k_X^{\leftarrow}[V_i] : i \in I\}$, where $\{A_i : i \in I\} \subseteq R(X)$ and $\{V_i : i \in I\} \subseteq \tau(X)$. As k_X and Π_X are the same function, we note that if $\Pi_X^{\leftarrow}(x) \cap k_X^{\leftarrow}[V_i] \neq \varnothing$, then $\Pi_X^{\leftarrow}(x) \subseteq k_X^{\leftarrow}[V_i]$. Let $I' = \{i \in I : \Pi_X^{\leftarrow}(x) \subseteq k_X^{\leftarrow}[V_i]\}$. Then $\Pi_X^{\leftarrow}(x) \subseteq \bigvee\{\lambda(A_i) : i \in I'\}$. As k_X is compact, there exists a finite subset F of I' such that $\Pi_X^{\leftarrow}(x) \subseteq \bigvee\{\lambda(A_i) : i \in F\}$. Then $\Pi_X^{\leftarrow}(x) \subseteq \bigvee\{\lambda(A_i) \cap k_X^{\leftarrow}[V_i] : i \in F\}$ and so Π_X is a compact function.

Let H be a closed subset of PX. If $x_0 \in X \setminus \Pi_X[H]$, then $\Pi_X^{\leftarrow}(x_0)$, which is a compact subset of PX, is disjoint from H. By a standard compactness argument there exist finite collections $\{A_i : i \in F\} \subseteq R(X)$ and $\{V_i : i \in F\} \subseteq \tau(X)$ such that $\Pi_X^{\leftarrow}(x_0) \subseteq \bigvee\{\lambda(A_i) \cap k_X^{\leftarrow}[V_i] : i \in F\} \subseteq PX \setminus H$. Without loss of generality assume that $\Pi_X^{\leftarrow}(x_0) \cap \lambda(A_i) \cap k_X^{\leftarrow}[V_i] \neq \varnothing$ for each $i \in F$; then $x_0 \in \bigcap\{V_i\ i \in F\}$. Now $\bigvee\{\lambda(A_i) : i \in F\} = \lambda(\bigvee\{A_i : i \in F\})$; therefore, by 6.6(e)(4) $x_0 \in [\mathrm{int}_X \bigvee\{A_i : i \in F\}] \cap [\bigcap\{V_i : i \in F\}] = T$, which is open in X. If $x \in T$, by 6.6(e)(4) $\Pi_X^{\leftarrow}(x) \subseteq \lambda(\bigvee\{A_i : i \in F\}) \cap [\bigcap\{k_X^{\leftarrow}[V_i] : i \in F\}] \subseteq \bigvee\{\lambda(A_i) \cap k_X^{\leftarrow}[V_i] : i \in F\} \subseteq PX \setminus H$, and so $T \cap \Pi_X[H] = \varnothing$. It follows that Π_X is a closed function.

Finally, since j_{PX} is a bijection, k_X is an irreducible map, and $\Pi_X = k_X \circ j_{PX}$, then obviously Π_X is also irreducible. The proof is now complete.

■

We want to show that the pair (PX,Π_X) is unique in the same sense as the pair (EX,k_X) was shown to be unique in 6.7(a). To do this we first prove a more general result, which will have considerable importance in Chapter 9 when the category-theoretic aspects of absolutes are discussed.

(d) **Theorem**. Let E be an extremally disconnected space, let X and Y be spaces, let $g : E \to X$ be perfect and continuous, and let $f : Y \to X$ be a perfect continuous surjection. Then:

(1) There is a perfect continuous function $h : E \to Y$ such that $f \circ h = g$.

(2) If, in addition, g is an irreducible surjection and f is irreducible, then h is a perfect irreducible continuous surjection.

Proof

(1) Let $S = \{(e,y) \in E \times Y : g(e) = f(y)\}$ and let $j = \Pi_E | S$ (Π_E is the projection function from $E \times Y$ onto E). Note that S is a closed subspace of $E \times Y$. We will show that $j : S \to E$ is a perfect continuous surjection.

Obviously j is continuous, and j is surjective as f is. If $e \in E$, then $j^{\leftarrow}(e) = \{e\} \times f^{\leftarrow}(g(e))$; thus j is a compact map as f is. If F is a closed subset of S and $e \in E \backslash j[F]$, then F and $\{e\} \times f^{\leftarrow}(g(e))$ are disjoint closed subsets of S (and thus of $E \times Y$), one of which is compact. Using a standard compactness argument one finds open sets $D_1, \ldots, D_n$ of E and $W_1, \ldots, W_n$ of Y such that $e \in \cap\{D_i : 1 \leqslant i \leqslant n\}$ $(=D)$ and $\{e\} \times f^{\leftarrow}(g(e)) \subseteq \cup\{D_i \times W_i : i \leqslant 1 \leqslant n\} \subseteq (E \times$

$Y)\setminus F$. Let $W = \cup\{W_i : 1 \leqslant i \leqslant n\}$. Then $f^{\leftarrow}(g(e)) \subseteq W$ and $g(e) \in X\setminus f[Y\setminus W]$, which is open in X as f is a closed function. Let $U = D \cap g^{\leftarrow}[X\setminus f[Y\setminus W]]$; then U is open in E, $e \in U$, and $U \cap j[F] = \emptyset$. Thus j is a closed map, and thus is a perfect continuous surjection from S onto E. Similarly one proves that $\Pi_Y | S$ is a perfect continuous map from S into (but perhaps not onto) Y.

By 6.5(c) there is a closed subset A of S such that $j|A$ is an irreducible perfect continuous function onto E. By 6.5(d)(4), $j|A$ is a homeomorphism. Let $h = (\Pi_Y | S) \circ (j|A)^{\leftarrow}$. Then h is a perfect continuous map from E to Y and $g = f \circ h$.

(2) This follows immediately from (1), 6.5(b)(5), and 6.5(b)(2).

■

(e) **Theorem**. Let X be a space and let (Y,f) be a pair consisting of an extremally disconnected space Y and a perfect continuous irreducible surjection $f : Y \to X$. Then there exists a homeomorphism $h : PX \to Y$ such that $f \circ h = \Pi_X$.

Proof. By (d) above there exists a perfect irreducible continuous function $h : PX \to Y$ such that $f \circ h = \Pi_X$. Hence by 6.5(d)(4) h is a homeomorphism.

■

We shall call the pair (PX, Π_X) the **Banaschewski absolute** of the space X. In view of the preceding theorem, any pair (Y,f) consisting of an extremally disconnected space Y and a perfect irreducible continuous surjection $f : Y \to X$ will be identified with (and called) the Banaschewski absolute of X. Thus we have the following definition, which is an analogue of 6.7(c).

(f) **Definition**. Let X and Y be spaces, and let f be a function from Y onto X. Then the expression "$(PX,\Pi_X) \sim (Y,f)$" is an abbreviation of the following statement: Y is extremally disconnected and f is a perfect irreducible continuous surjection from Y onto X. In this situation we will say that "(Y,f) is the Banaschewski absolute of X (up to equivalence)."

(g) **Theorem**. Let T be an extension of X. Then $\Pi_T^{\leftarrow}[X]$ is dense in PT, and $(PT,\Pi_T) \sim (\Pi_T^{\leftarrow}[X], \Pi_T | \Pi_T^{\leftarrow}[X])$.

Proof. The proof is essentially the same as that of 6.9(a). ■

(h) **Proposition**. The following are equivalent for a space X:

(1) $(EX,k_X) \sim (PX,\Pi_X)$ and

(2) X is regular.

Proof

(1) ⇒ (2)

EX, being zero-dimensional, is always regular; so if EX is homeomorphic to PX, then PX is regular. By 1.8(h) a perfect continuous image of a regular space is regular, so X is regular.

(2) ⇒ (1)

By 6.6(e)(6) k_X is continuous, and from the definition of the topology on PX, it immediately follows that PX and EX have the same topology. As $k_X = \Pi_X$ as functions, (1) follows.

■

In 6.9(b)(3,4), we prove that if X is Tychonoff, then (i) EX is homeomorphic to a dense subspace of $E(\beta X)$ and $E(\beta X) \equiv_{EX} \beta(EX)$ and (ii) $(E(\beta X), k_{\beta X}) \sim (\beta(EX), \beta k_X)$. We now show that "P is to κ as E is to β" (see (j) below). First we need a lemma.

(i) **Lemma**. A space X is H-closed iff PX is H-closed.

Proof. By 6.9(b) X is H-closed iff EX is compact. By 4.8(c) EX is compact iff EX is H-closed. By 4.8(h)(8) Y is H-closed iff Y(s) is H-closed. By 6.11(c)(1) (PX)(s) = EX. Combining these facts gives us the lemma.

■

(j) **Theorem**. Let X be any space. Then:

(1) PX is homeomorphic to the dense subspace $\Pi_{\kappa X}^{\leftarrow}[X]$ of $P(\kappa X)$,

(2) $P(\kappa X) \equiv_{PX} \kappa(PX)$, and

(3) $(P(\kappa X), \Pi_{\kappa X}) \sim (\kappa(PX), \Pi_{\kappa X} \circ j)$ where $j : \kappa(PX) \to P(\kappa X)$ is the continuous function such that $j|PX = id_{PX}$.

Proof. Consider the perfect irreducible continuous surjection $\Pi_{\kappa X} : P(\kappa X) \to \kappa X$ defined in 6.11(b). By 6.11(g) we can identify $\Pi_{\kappa X}^{\leftarrow}[X]$ with PX; also, $\Pi_{\kappa X}|\Pi_{\kappa X}^{\leftarrow}[X]$ is a perfect irreducible continuous surjection from the extremally disconnected space $\Pi_{\kappa X}^{\leftarrow}[X]$ onto X. We will denote $\Pi_{\kappa X}^{\leftarrow}[X]$ by PX, and

$\Pi_{\kappa X} | \Pi_{\kappa X}^{\leftarrow}[X] = \Pi_{\kappa X} | PX$ by Π_X. By 6.11(i) above $P(\kappa X)$ is an H-closed extension of PX, and as $\kappa(PX)$ is the projective maximum of $H(PX)$, there exists a continuous surjection $j : \kappa(PX) \to P(\kappa X)$ such that $j | PX = id_{PX}$. The situation is summarized in Figure 1.

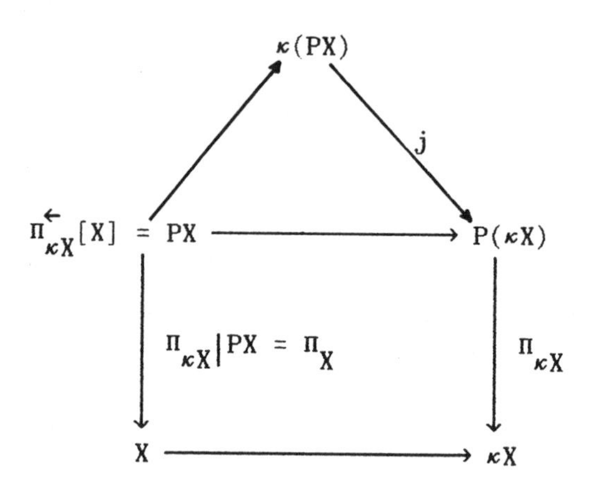

Figure 1

There are two ways of interpreting the assertion that $P(\kappa X)$ and $\kappa(PX)$ are the same. One interpretation of this equality is the assertion that $\kappa(PX)$ and $P(\kappa X)$ are equivalent H-closed extensions of PX (which has been identified as $\Pi_{\kappa X}^{\leftarrow}[X]$). Another interpretation of this equality is that there is a perfect irreducible continuous surjection from $\kappa(PX)$ onto κX (see the remarks preceding 6.11(f) and note that by 6.2(b) $\kappa(PX)$ is extremally disconnected). We shall prove that j is a homeomorphism from $\kappa(PX)$ onto $P(\kappa X)$. This will verify that $\kappa(PX)$ and $P(\kappa X)$ are equivalent extensions of PX, and it will also imply that $j \circ \Pi_{\kappa X} : \kappa(PX) \to \kappa X$ is a perfect irreducible continuous surjection.

Obviously $\Pi_{\kappa X}[P(\kappa X)\setminus PX] = \kappa X\setminus X$ and $j[\kappa(PX)\setminus PX] = P(\kappa X)\setminus PX$ (the second equality follows from 1.8(i)). We now assert that $\Pi_{\kappa X}|P(\kappa X)\setminus PX$ and $j|\kappa(PX)\setminus PX$ are one-to-one functions, and thus bijections onto $\kappa X\setminus X$ and $P(\kappa X)\setminus PX$ respectively.

Let $\mathcal{U}_1, \mathcal{U}_2 \in P(\kappa X)\setminus PX$ such that $\mathcal{U}_1 \neq \mathcal{U}_2$. Then $\mathcal{U}_1$ and $\mathcal{U}_2$ are open ultrafilters on κX and each converges to a point of $\kappa X\setminus X$. Let $V_1 \in \mathcal{U}_1$, $V_2 \in \mathcal{U}_2$ such that $V_1 \cap V_2 = \emptyset$. Then $(c\ell_{\kappa X}V_1 \cap c\ell_{\kappa X}V_2)\setminus X = (c\ell_{\kappa X}(V_1 \cap X) \cap c\ell_{\kappa X}(V_2 \cap X))\setminus X = \emptyset$ (see 4.8(p)(3)). But from the definition of $\Pi_{\kappa X}$ (see 6.11(b)) $\Pi_{\kappa X}(\mathcal{U}_i) \in c\ell_{\kappa X}V_i$ (i=1,2). Thus $\Pi_{\kappa X}(\mathcal{U}_1) \neq \Pi_{\kappa X}(\mathcal{U}_2)$ and $\Pi_{\kappa X}|P(\kappa X)\setminus PX$ is one-to-one.

Now let $\mathcal{W}_1$ and $\mathcal{W}_2$ be distinct members of $\kappa(PX)\setminus PX$, i.e., distinct free open ultrafilters on PX. Then there are open subsets W_1 and W_2 of PX such that $W_1 \in \mathcal{W}_1$, $W_2 \in \mathcal{W}_2$, and $W_1 \cap W_2 = \emptyset$. Thus $\mathcal{W}_i \in c\ell_{\kappa(PX)}W_i$ (i=1,2). As $P(\kappa X)$ is extremally disconnected, it follows from 6.2(b) that $c\ell_{P(\kappa X)}W_1 \cap c\ell_{P(\kappa X)}W_2 = \emptyset$. As j is continuous, $j(\mathcal{W}_i) \in c\ell_{P(\kappa X)}W_i$, so $j(\mathcal{W}_1) \neq j(\mathcal{W}_2)$. Therefore, $j|\kappa(PX)\setminus PX$ is one-to-one.

Thus j is a continuous bijection from $\kappa(PX)$ onto $P(\kappa X)$. To show that j is a homeomorphism, it suffices to show that j is closed. We prove two preliminary statements.

First, we claim that if M is a closed nowhere dense subset of PX, then M is closed in $P(\kappa X)$. To see this, note that by 6.5(d)(2) $\Pi_X[M]$ is a closed nowhere dense subset of X. Thus by 4.8(p)(6) $\Pi_X[M]$ is closed in κX. Thus $\Pi_{\kappa X}^{\leftarrow}[\Pi_X[M]]$ is closed in $P(\kappa X)$ and contained in PX. As it contains M as a closed subset, M is closed in $P(\kappa X)$.

Next, we claim that $P(\kappa X)\backslash PX$ is a closed discrete subset of $P(\kappa X)$. It is closed because $PX = \Pi_{\kappa X}^{\leftarrow}[X]$ and X is open in κX. To see it is discrete, note that we proved above that $\Pi_{\kappa X}|P(\kappa X)\backslash PX$ is a continuous bijection from $P(\kappa X)\backslash PX$ onto the discrete space $\kappa X\backslash X$.

Now we can prove that j is a closed function. If S is closed in $\kappa(PX)$, write S as $(c\ell_{\kappa(PX)}\mathrm{int}_{\kappa(PX)}S) \cup (PX \cap \mathrm{bd}_{\kappa(PX)}S) \cup ((\mathrm{bd}_{\kappa(PX)}S)\backslash PX)$. By 4.8(e) $c\ell_{\kappa(PX)}\mathrm{int}_{\kappa(PX)}S$ is H-closed, so by 4.8(h)(6) $j[c\ell_{\kappa(PX)}\mathrm{int}_{\kappa(PX)}S]$ is H-closed. Since $(\mathrm{bd}_{\kappa(PX)}S) \cap PX$ is a closed nowhere dense subset of PX, it is closed in $P(\kappa X)$ by the second last paragraph. Finally, $j[\mathrm{bd}_{\kappa(PX)}S\backslash PX] \subseteq P(\kappa X)\backslash PX$, and by the previous paragraph every subset of $P(\kappa X)\backslash PX$ is closed in $P(\kappa X)$. Thus $j[S]$ is the union of three subsets of $P(\kappa X)$, each of which is closed in $P(\kappa X)$; thus $j[S]$ is closed in $P(\kappa X)$.

It follows that j is a homeomorphism, and so $\kappa(PX)$ and $P(\kappa X)$ are equivalent extensions of PX. In this sense $P(\kappa X) \equiv_{PX} \kappa(PX)$. Also, $\Pi_{\kappa X}\circ j$ is a continuous closed surjection from $\kappa(PX)$ onto κX. It is irreducible as j and $\Pi_{\kappa X}$ both are (see 6.5(b)(1)). It is compact, for if $x \in X$ then $(\Pi_{\kappa X}\circ j)^{\leftarrow}(x) = \Pi_{\kappa X}^{\leftarrow}(x)$, which is a compact subset of PX and thus of $\kappa(PX)$, while if $\alpha \in \kappa X\backslash X$ then $(\Pi_{\kappa X}\circ j)^{\leftarrow}(\alpha)$ contains only one element and thus is compact. Thus $\Pi_{\kappa X}\circ j$ is a perfect irreducible continuous surjection from $\kappa(PX)$ onto κX, and so in this sense $(P(\kappa X),\Pi_{\kappa X}) \sim (\kappa(PX),\Pi_{\kappa X}\circ j)$.

■

Chapter 6 - Problems

6A. Absolutes of sums and products

(1) Let $\{X_i : i \in I\}$ and $\{Y_i : i \in I\}$ be two sets of spaces and for each $i \in I$, let f_i be a perfect irreducible Θ-continuous surjection from X_i onto Y_i. Show that

$$\Pi_i f_i : \Pi\{X_i : i \in I\} \to \Pi\{Y_i : i \in I\}$$

and $\oplus f_i : \oplus\{X_i : i \in I\} \to \oplus\{Y_i : i \in I\}$ are both perfect irreducible Θ-continuous surjections.

(2) If $\{X_i : i \in I\}$ is a set of extremally disconnected spaces, show that $\oplus\{X_i : i \in I\}$ is extremally disconnected. (But, as we will see in problem 6P, the product of two extremally disconnected spaces, neither of which is discrete, will be extremally disconnected only under exceptional circumstances.)

(3) If $\{X_i : i \in I\}$ is a set of spaces show that

$$(E(\oplus\{X_i : i \in I\}), k_{\oplus\{X_i : i \in I\}}) \sim (\oplus\{EX_i : i \in I\}, \oplus k_i)$$

and

$$(E(\Pi\{X_i : i \in I\}), k_{\Pi\{X_i : i \in I\}}) \sim (E(\Pi\{EX_i : i \in I\}), (\Pi_i k_{X_i}) \circ k_{\Pi\{EX_i : i \in I\}}).$$

(4) Show that

$$(E(\beta\mathbb{N} \times \beta\mathbb{N}), k_{\beta\mathbb{N} \times \beta\mathbb{N}}) \sim (\beta\doteqdot, f)$$

for a suitably chosen f.

(5) Prove the "P, Π_X" analogue of (3).

(6) State and prove the "P, κ" analogue to (4).

6B. Topological properties shared by X and EX

(1) Prove that X has P iff PX has P, where P is any one of the following topological properties: σ-compactness, countable compactness, the Lindelöf property (see 1J). (Recall EX = PX iff X is regular.)

(2) Prove that the following are equivalent for a Tychonoff space X:

(a) if Y can be mapped onto X by a perfect continuous surjection, then Y is pseudocompact,

(b) X is countably compact, and

(c) every closed subset of X is pseudocompact.

(3) Prove that there exist spaces X and Y, and perfect continuous surjections $f : X \to Y$, such that Y has P and X lacks P if P is one of the following topological properties: density character m (see 2N(2)), π-weight m (see 2N(5)), feeble compactness (see 1.11(a) and (2) above), cellularity m (see 2N(6)).

(4) Prove that X has P iff EX has P for each property P described in (3) above.

6C. Countable extremally disconnected spaces. Find examples of two countable extremally disconnected zero-dimensional spaces without isolated points, one with countable π-weight and one of π-weight $2^{\aleph_0}$. (Hint: Consider E$\mathbb{Q}$ and E($\mathfrak{P}^c$) and use 6B(3).)

6D. Absolutes of spaces of countable π-weight. Show the following:

(1) Let X be a space and let $A \in R(X)$. Show that EX is homeomorphic to EA $\oplus$ E(A$'$).

(2) Let X be an infinite H-closed space of countable π-weight. Then EX is homeomorphic to precisely one of E($\mathfrak{P}^\omega$), $\beta\mathbb{N}$, E($\mathfrak{P}^\omega$) $\oplus$ $\beta\mathbb{N}$, or E($\mathfrak{P}^\omega$) $\oplus$ F where F is a finite discrete space.

(3) There exists an infinite separable H-closed space H such that EH is not homeomorphic to any of E($\mathfrak{P}^\omega$), $\beta\mathbb{N}$, E($\mathfrak{P}^\omega$) $\oplus$ $\beta\mathbb{N}$, or E($\mathfrak{P}^\omega$) $\oplus$ F where F is a finite discrete space. (Hint: See 6B(4) and 6C.)

(4) Every infinite second countable space (in particular, every infinite separable metric space) has as its absolute a dense subspace of one of E($\mathfrak{P}^\omega$), $\beta\mathbb{N}$, E($\mathfrak{P}^\omega$) $\oplus$ $\beta\mathbb{N}$, or E($\mathfrak{P}^\omega$) $\oplus$ F where F is a finite discrete space.

6E. Other descriptions of EX

(1) Construct EX and k_X by using a subspace of $S(R0(X))$ as the underlying set for EX. Mimic the proof of 6.6(d,e) to obtain the properties of EX and k_X under this construction.

(2) Prove that the function h described after the proof of 6.8(f) is a well-defined homeomorphism from EX onto $E'X$ such that $k_X' \circ h = k_X$.

(3) Let X be dense in T. Find an explicit homeomorphism $k : EX \to k_T^{\leftarrow}[X]$ such that $k_T | k_T^{\leftarrow}[X] \circ h = k_X$.

(4) Prove that for any space X, $(E(EX), k_{EX}) \sim (EX, id_{EX})$. Find and prove a similar result for PX.

(5) Complete the proof of 6.9(d).

6F. Embedding EX in βD

(1) Let X be an infinite Tychonoff (respectively, zero-dimensional) space. Show that X can be embedded in $[0,1]^{w(X)}$ (respectively, $\{\mathbf{2}\}^{w(X)}$). (Hint: Use 1.7(j), 2N(3)(b), and either cozero or clopen sets.)

Let κ be an infinite cardinal.

(2) Show that $d([0,1]^{2^\kappa}) \leq \kappa$ and $d(\mathbf{2}^{2^\kappa}) \leq \kappa$ (see 3O(4)).

(3) Prove that $[0,1]^{2^\kappa}$ and $\mathbf{2}^{2^\kappa}$ are continuous images of $\beta D(\kappa)$.

(4) Prove that $\beta D(\kappa)$ is the continuous image of $\beta D(\kappa)\backslash D(\kappa)$. (Hint: Since $|\kappa \times \omega| = \kappa$, there is a surjection $f : D(\kappa) \to D(\kappa)$ such that $|f^{\leftarrow}(\alpha)| = \aleph_0$ for each $\alpha \in D(\kappa)$. Show that $\beta f[\beta D(\kappa)\backslash D(\kappa)] = \beta D(\kappa)$.)

(5) Let G be an extremally disconnected Tychonoff space and suppose that $w(G) \leq 2^{\kappa}$. Show that G can be embedded in $\beta D(\kappa)\backslash D(\kappa)$. (Hint: Use (1), (3), (4), 1.8(f), and 6.5(c,d).)

(6) Let $d(X) = \kappa$. Prove that $w(EX) \leq 2^{\kappa}$. (Hint: See 6B(4), choose an appropriate $Y \subseteq EX$, and compare $B(Y)$ and $B(\beta(EX))$.)

(7) Let $d(X) = \kappa$. Prove that EX can be embedded in $\beta D(\kappa)$. (This follows from (6), and can also be proved directly utilizing the techniques used in (3) and (5).)

(8) Prove that if $2^{\kappa} = 2^{\aleph_0}$ then $\beta D(\kappa)$ can be embedded in $\beta\mathbb{N}$.

(9) Prove that if $2^{\aleph_1} = 2^{\aleph_0}$ then $\beta\mathbb{N}\backslash\mathbb{N}$ contains a C^*-embedded discrete subspace of cardinality $\aleph_1$. Thus in particular, MA + $\neg$CH implies that $\beta\mathbb{N}\backslash\mathbb{N}$ contains such a subspace (see 3.5(t)).

(10) Prove that if $2^{\aleph_0} < 2^{\aleph_1}$, then $\beta\mathbb{N}\backslash\mathbb{N}$ cannot contain a C^*-embedded discrete subspace of cardinality $\aleph_1$. In particular, the continuum hypothesis implies that $\beta\mathbb{N}\backslash\mathbb{N}$ contains no such subspace.

(11) A space X is called **collectionwise Hausdorff** if, whenever $\{d(\alpha) : \alpha \in A\}$ is a closed discrete subspace of X, there exists a pairwise disjoint collection $\{V(\alpha) : \alpha$

$\in A\}$ of open subsets of X such that $d(\alpha) \in V(\alpha)$ for each $\alpha \in A$. Suppose $2^{\aleph_1} = 2^{\aleph_0}$ and choose D as described in (9). Prove that $\mathbb{N} \cup D$ is a normal space of character at most $2^{\aleph_0}$ (see 2N(4)) that is not collectionwise Hausdorff. (It is consistent with the usual axioms of set theory that no such spaces exist; see the Notes.)

6G. Extending functions to the Banaschewski absolute

(1) Let X and Y be spaces, and let E be an extremally disconnected space. Let $f : X \to Y$ be a perfect continuous surjection, and let $g \in C(E,Y)$. Show there is a continuous function $\bar{g} : E \to X$ such that $f \circ \bar{g} = g$. (Hint: Modify the proof of 6.11(d).)

(2) Let X and Y be spaces, and let $f \in C(X,Y)$. Show there exists $\bar{f} \in C(PX,PY)$ such that $\Pi_Y \circ \bar{f} = f \circ \Pi_X$.

(3) If $f \in C(X,Y)$, call f a **c-map** if $f^{\leftarrow}[Y \setminus bd_Y A]$ is dense in X for each $A \in R(Y)$. Prove that $\bar{f}$ is unique iff f is a c-map. (Hint: If there exists $g \in C(PX,PY)$ such that $\Pi_Y \circ g = f \circ \Pi_X$ and such that $g \neq \bar{f}$, find $\alpha \in PX$ such that $\bar{f}(\alpha) \neq g(\alpha)$. Consider $\bar{f}^{\leftarrow}[\lambda(c\ell_X U)] \cap g^{\leftarrow}[PY \setminus \lambda(c\ell_X U)]$ where U is a "correctly chosen" open subset of Y.

Conversely, if $\emptyset \neq U = int_X f^{\leftarrow}[bd_Y A]$ for some $A \in R(Y)$ consider $f \circ \Pi_X | \lambda(c\ell_X U)$:

$\lambda(c\ell_X U) \to A$, $\Pi_Y | \lambda(A) : \lambda(A) \to A$, and apply (1). Do the same with $c\ell_Y(Y \setminus A)$ in place of A. This yields two different ways of defining a function on $\lambda(c\ell_X U)$; together with $\bar{f} | PX \setminus \lambda(c\ell_X U)$, these yield our functions.)

(4) Show that if $f \in C(X,Y)$ and f is either open or irreducible, then f is a c-map.

(5) Show that the composition of two c-maps need not be a c-map. (Hint: Consider $k_{[0,1]} : E[0,1] \to [0,1]$ and $g : D \to E[0,1]$, where D is discrete and g is a bijection.)

(6) If $f : X \to Y$ is a perfect irreducible continuous surjection and $\bar{f} \in C(PX,PY)$ is the unique function such that $\Pi_Y \circ \bar{f} = f \circ \Pi_X$, prove that $\bar{f}$ is a homeomorphism and that there is a perfect irreducible continuous surjection $g : PY \to X$ such that $f \circ g = \Pi_Y$.

6H. Basic subalgebras. In this problem we generalize the construction of EX given in 6.6.

Let X be a space. A Boolean subalgebra $\mathfrak{a}$ of $R(X)$ is called a **basic subalgebra** of $R(X)$ if $\mathfrak{a}$ is a base for the closed subsets of X. Note that this is a topological, not a Boolean-algebraic, concept.

Let $\mathfrak{a}$ be a (not necessarily basic) Boolean subalgebra of $R(X)$. An ultrafilter α on $\mathfrak{a}$ is called **fixed** if $\cap\{A : A \in \alpha\}$ (henceforth denoted $\cap\alpha$) is nonempty. Let $E_{\mathfrak{a}}X = \{\alpha \in S(\mathfrak{a}) : \cap\alpha \neq \varnothing\}$ and give $E_{\mathfrak{a}}X$ the subspace topology inherited from the Stone space $S(\mathfrak{a})$.

(1) Show that $E_{\mathcal{R}(X)}X = EX$.

Henceforth $\mathcal{Q}$ denotes a basic subalgebra of $\mathcal{R}(X(s))$ (see 2.2(e)).

(2) Show that:

(a) if $\alpha \in E_{\mathcal{Q}}X$ then $|\cap\alpha| = 1$ and

(b) if $x \in X$ there exists $\alpha \in E_{\mathcal{Q}}X$ such that $\cap\alpha = \{x\}$.

(3) If $\alpha \in E_{\mathcal{Q}}X$ let $k_{\mathcal{Q}}(\alpha)$ denote the unique point of X in $\cap\alpha$. Show that $k_{\mathcal{Q}}$ is a well-defined perfect irreducible θ-continuous surjection from $E_{\mathcal{Q}}X$ onto X.

(4) Show that $E_{\mathcal{Q}}X$ is compact iff X is H-closed iff $\mathcal{B}(E_{\mathcal{Q}}X) = \{\lambda(A) : A \in \mathcal{Q}\}$.

(5) Show that $k_{\mathcal{Q}}$ is continuous iff X is regular.

(6) Let $H(\mathcal{Q}) = X \backslash \cup\{bd_X A : A \in \mathcal{Q}\}$. If $x \in X$, show that $|k_{\mathcal{Q}}^{\leftarrow}(x)| = 1$ iff $x \in H(\mathcal{Q})$.

(7) Show that if $H(\mathcal{Q})$ is dense in X, then $H(\mathcal{Q})$ is θ-homeomorphic to $k_{\mathcal{Q}}^{\leftarrow}[H(\mathcal{Q})]$. If X is regular then $H(\mathcal{Q})$ (whether dense in X or not) is homeomorphic to $k_{\mathcal{Q}}^{\leftarrow}[H(\mathcal{Q})]$.

(8) Let X be zero-dimensional, let $A \in \mathcal{R}(X)$, and let $\mathcal{B}_A(X)$ denote the subalgebra of $\mathcal{R}(X)$ generated by $\mathcal{B}(X)$ together with A. Show that $\mathcal{B}_A(X)$ is a basic

subalgebra of $\mathcal{R}(X)$ and $E_{\mathcal{B}_A(X)}X = A \oplus A'$ (up to homeomorphism).

6I. Basic subalgebras and compact metric spaces

(1) Let K be a compact metric space without isolated points. Show that there is a basic subalgebra $\mathfrak{a}$ of $\mathcal{R}(K)$ such that $E_{\mathfrak{a}}K$ is homeomorphic to $\mathfrak{k}^{\omega}$. (Hint: Use 3.3(g).)

(2) Show that every compact metric space without isolated points is a perfect irreducible continuous image of $\mathfrak{k}^{\omega}$.

(3) Let J denote the irrational points of [0,1]. Show there exists a perfect irreducible continuous surjection $f : \mathfrak{k}^{\omega} \to [0,1]$ such that $f|f^{\leftarrow}[J]$ is a homeomorphism from $f^{\leftarrow}[J]$ onto J. Hence conclude that $\mathfrak{k}^{\omega}$ contains a dense copy of the irrationals. (Hint: Use 6H(6), and consider closed intervals with rational endpoints.)

(4) If f is a perfect irreducible continuous surjection from $\mathfrak{k}^{\omega}$ onto a compact metric space K, show there is a dense subspace T of $\mathfrak{k}^{\omega}$ such that $f|T$ is a homeomorphism from T onto f[T]. (Hint: Use the Baire category theorem and consider $\{\mathrm{bd}_K f[B] : B \in \mathcal{B}(\mathfrak{k}^{\omega})\}$.)

(5) Verify that in (3) (but not necessarily in (4)), $|f^{\leftarrow}(x)| \leq 2$ for each $x \in [0,1]$.

6J. Compact subsets of absolutes

(1) Let A be a subset of an H-closed, Urysohn space X. Prove that $k_X^{\leftarrow}[A]$ is a compact subset of EX iff A is an H-set of X. (Hint: H-sets are defined in 4N. For one direction use 4N(6). For the other, use 4N(11), 6.6(e)(5), and 1.8(d).)

(2) Show that $\kappa\mathbb{N}\setminus\mathbb{N}$ is an H-set of $\kappa\mathbb{N}$ and verify directly that $k_{\kappa\mathbb{N}}^{\leftarrow}[\kappa\mathbb{N}\setminus\mathbb{N}]$ is compact.

6K. Basically disconnected spaces. A Tychonoff space is said to be **basically disconnected** if each of its cozero sets has an open closure. Obviously each extremally disconnected Tychonoff space is basically disconnected. Prove that the following statements are equivalent for a Tychonoff space X:

(1) X is basically disconnected,

(2) if $C \in \text{coz } X$, V is open in X, and $C \cap V = \varnothing$, then $c\ell_X C \cap c\ell_X V = \varnothing$,

(3) βX is basically disconnected, and

(4) X is zero-dimensional and $B(X)$ is a σ-complete Boolean algebra (i.e., each countable subset of $B(X)$ has a supremum and infimum in $B(X)$).

6L. <u>F-spaces</u>. A Tychonoff space X is called an **F-space** if each cozero-set of X is C^*-embedded in X. Show that each of the following is true.

(1) X is an F-space iff βX is an F-space.

(2) If X is locally compact and σ-compact, then $\beta X \setminus X$ is an F-space. (Hint: If $C \in \text{coz}(\beta X \setminus X)$, show that $X \cup C$ is normal and C is a closed subspace of it.)

(3) Every regular P-space (see 1W) is basically disconnected, and every basically disconnected space is an F-space.

(4) C^*-embedded subspaces of F-spaces are F-spaces.

(5) If X is a Tychonoff space and if A and B are two countable subsets of X, neither of which meets the closure of the other, then A and B are contained in disjoint cozero-sets of X. (Define inductively a countable sequence of cozero-sets of X which covers $A \cup B$.)

(6) Countable subsets of F-spaces are C^*-embedded. (Use (5).)

(7) Separable subspaces of F-spaces are C^*-embedded F-spaces.

(8) If a Tychonoff space has countable cellularity, then it is an F-space iff it is extremally disconnected. (Hint: Consider maximal families of pairwise disjoint cozero-sets contained in a given open set.)

(9) Every convergent sequence in an F-space is eventually constant.

(10) A point of first countability of an F-space is isolated (and hence metrizable F-spaces are discrete).

6M. Cardinalities of absolutes

(1) Let Y be a locally compact F-space. Show that each open subset of Y without isolated points contains a copy of $\beta\mathbb{N}$.

(2) Show that $|EX| \leq 2^{|R(X)|}$ for any space X. Find spaces Y and Z without isolated points for which $|EY| = 2^{|R(Y)|}$ and $|EZ| < 2^{|R(Z)|}$.

(3) If X is a locally compact separable space without isolated points, show that every nonempty open subset of EX has cardinality 2^{c} (see 2N(2)(a) and 6B(3)).

6N. Pseudocompact metacompact Tychonoff spaces are compact

(1) Let X be a pseudocompact Tychonoff space and let $\mathcal{U}$ be a point-finite open cover of X. If $U \in \mathcal{U}$, let U' be an open subset of βX such that $U' \cap X = U$. Put $\mathcal{U}' = \{U' : U \in \mathcal{U}\}$ and let $Y = \cup\mathcal{U}'$. Show that $\mathcal{U}'$ is a point-finite open cover of Y. (Hint: Use 5F(4).)

(2) If $y \in Y$, let $(\mathcal{U}')_y = \{U' \in \mathcal{U}' : y \in U'\}$. If $n \in \mathbb{N}$, let $F_n = \{y \in Y : |(\mathcal{U}')_y| \leq n\}$. Show

that $\{F_n : n \in \mathbb{N}\}$ is a closed cover of Y.

(3) Let $W = \cup\{\text{int}_Y F_n \setminus F_{n-1} : n \in \mathbb{N}\}$ where $F_0 = \emptyset$. Prove that W is a dense open subset of Y. (Hint: First use the Baire category theorem to verify that $\cup\{\text{int}_Y F_n : n \in \mathbb{N}\}$ is dense in Y.)

(4) Prove that there exists a finite subfamily of $\mathcal{U}'$ with dense union in Y. (Hint: Assume not. Inductively choose $\{y(n) : n \in \mathbb{N}\} \subseteq Y$ so that $y(n+1) \in W \setminus \cup\{\cup(\mathcal{U}')_{y(k)} : k \leqslant n\}$. For each $n \in \mathbb{N}$ find $m(n) \in \mathbb{N}$ such that $y(n) \in \text{int}_Y F_{m(n)} \setminus F_{m(n)-1}$ and let $V_n = X \cap \text{int}_Y F_{m(n)} \cap [\cap(\mathcal{U}')_{y(n)}]$. Prove that $\{V_n : n \in \mathbb{N}\}$ is an infinite, locally finite family of open subsets of X, thereby contradicting the pseudocompactness of X.)

(5) Prove that there exists a finite subfamily of $\mathcal{U}$ with dense union in X.

(6) Prove that a pseudocompact metacompact Tychonoff space is compact.

(7) Prove that a feebly compact metacompact regular space is compact. (Hint: Consider (EX, k_X).)

60. P-points of extremally disconnected spaces. Let p be a non-isolated point of an extremally disconnected space X.

(1) Show that there exists a collection $\mathcal{S}$ of pairwise disjoint open subsets of X such that $\cup\mathcal{S}$ is dense in X and $p \in X \setminus \cup\{c\ell_X S : S \in \mathcal{S}\}$. (Hint: Consider a maximal family.)

(2) Let $\mathcal{U} = \{\mathcal{F} \subseteq \mathcal{S} : p \in c\ell_X[\bigcup \mathcal{F}]\}$. Show that $\mathcal{U}$ is a free ultrafilter on the set $\mathcal{S}$.

(3) Assume that $|X|$ is not Ulam-measurable. Show that there exists a countable collection $\{\mathcal{F}_n : n \in \mathbb{N}\} \subseteq \mathcal{U}$ such that if $S \in \mathcal{S}$, there exists $n(S) \in \mathbb{N}$ such that $S \notin \mathcal{F}_{n(S)}$.

(4) Let $G = \bigcap\{c\ell_X[\bigcup \mathcal{F}_n] : n \in \mathbb{N}\}$. Show that G is a nowhere dense G_δ-set of X that contains p, and conclude that p is not a P-point of X (see 1W).

(5) Show that if X is an extremally disconnected space whose cardinality is not Ulam-measurable, then its P-points are all isolated.

(6) Let m be an Ulam-measurable cardinal, and let $\mathcal{W}$ be a free ultrafilter on D(m) that is closed under countable intersection. Let $X = D(m) \cup \{\mathcal{W}\}$, regarded as a subspace of $\beta D(m)$. Show that X is an extremally disconnected P-space.

6P. Products of extremally disconnected spaces

(1) Let X and Y be Tychonoff spaces. If $X \times Y$ is an F-space, show that either X or Y is a P-space (see 1W). (Hint: Assume that neither X nor Y is a P-space. Define $h \in C(X \times Y)$ by the formula

$$h(x,y) = |f(x)| - |g(y)|,$$

where f and g have been chosen after meditating on 1W(2). Now consider $h^{\leftarrow}[(-\infty,0)]$ and $h^{\leftarrow}[(0,+\infty)]$.)

(2) Let X and Y be Tychonoff spaces whose cardinalities are not Ulam-measurable. Show that $X \times Y$ is extremally disconnected iff one factor is discrete and the other factor is extremally disconnected. (Hint: Use 6O(5).)

(3) Show that a space is extremally disconnected iff its semiregularization is extremally disconnected.

(4) Let X and Y be any two (Hausdorff) spaces whose cardinalities are not Ulam-measurable. Show that $X \times Y$ is extremally disconnected iff one factor is extremally disconnected and the other factor is discrete. (Hint: Consider $(X \times Y)(s)$.)

6Q. <u>Separable extremally disconnected Tychonoff spaces</u>

(1) Show that every separable subspace of $\beta\mathbb{N}$ is extremally disconnected. (Hint: Use 6L(7,8).)

(2) Let X be a separable Tychonoff space. Show that X is extremally disconnected iff it is homeomorphic to a subspace of $\beta\mathbb{N}$. (Use (1) and 6F.) See 6C for examples.

6R. <u>Open maps and extremally disconnected spaces</u>. Show that the following conditions on a Tychonoff space E are equivalent:

(1) E is extremally disconnected,

(2) if X is any Tychonoff space and $f : X \to E$ is open and continuous, then $\beta f : \beta X \to \beta E$ is an open map, and

(3) if X is any Tychonoff space and $f : X \to E$ is a continuous open surjection, then $\beta f : \beta X \to \beta E$ is an open function.

(Hint: For (3) ⇒ (1), let U be open in E and let $X = E \oplus U$.)

6S. Another generalization of the absolute. Let S be a sublattice of the lattice $\tau(X)$ of open subsets of the space X. If S is an open base for X, we will call S a **basic sublattice** of $\tau(X)$. Let $\Omega'(X,S)$ denote the set of ultrafilters on S. If $S \in S$, let $O(S,S) = \{\alpha \in \Omega'(X,S) : S \in \alpha\}$. Henceforth assume S is a basic sublattice of $\tau(X)$.

(1) Show that $\{O(S,S) : S \in S\}$ is closed under finite intersection and forms an open base for a Hausdorff topology on $\Omega'(X,S)$.

We denote $\Omega'(X,S)$, so topologized, by $\theta'(X,S)$.

(2) Prove that if $S \in S$ then $O(S,S) \in B(\theta'(X,S))$. Hence conclude that $\theta'(X,S)$ is zero-dimensional.

(3) Let $E'(X,S) = \{\alpha \in \theta'(X,S) : \alpha$ converges to a point of $X\}$. Show that $E'(X,S)$ is a dense subspace of $\theta'(X,S)$.

(4) If $\alpha \in E'(X,S)$, denote by $k_{X,S}(\alpha)$ the unique point of X to which α converges. Prove that $k_{X,S}$ is a well-defined Θ-continuous surjection from $E'(X,S)$ onto X.

(5) If $S = \tau(X)$, show that $\theta'(X,S) = \theta'X$, $E'(X,S) = E'X$, and $k_{X,S} = k_X'$.

A basic sublattice S of $\tau(X)$ is said to be **complemented** if given $S \in S$, there exists $T \in S$ such that $S \cap T = \emptyset$ and $S \cup T$ is dense in X.

(6) If S is a complemented basic sublattice of $\tau(X)$, show that:

(a) $\theta'(X,S)$ is compact,

(b) $k_{X,S} : E'(X,S) \to X$ is a perfect θ-continuous irreducible surjection,

(c) $E'(X,S)$ is compact iff X is H-closed, and

(d) $k_{X,S}$ is continuous iff X is regular.

(7) Let $f : Y \to X$ be a perfect irreducible continuous surjection from Y onto the regular space X. Assume that Y is zero-dimensional. Show that there exists a complemented basic sublattice S of $\tau(X)$, and a homeomorphism $h : E'(X,S) \to Y$, such that $k_{X,S} =$

$f \circ h$. (Hint: Let $S = \{X \setminus f[Y \setminus U] : U \in B(Y)\}$.)

(8) If the basic sublattice S is not complemented, show that $\theta'(X,S)$ need not be compact and $k_{X,S}$ need not be perfect. (Hint: Let X be a connected compact F-space (such as $\beta H \setminus H$, where $H = [0,+\infty)$), and let $S = \mathrm{coz}(X)$. Show that $k_{X,S}$ is a continuous bijection but is not a homeomorphism.)

6T. "Almost P" spaces. Let P be a topological property. A space X is called "almost-P" if there exists a space Y with P and a perfect continuous surjection $f : Y \to X$. Let $\mathcal{a}P$ (respectively, $\mathcal{a}PT$) denote the class of almost-P (respectively, almost-P and Tychonoff) spaces.

(1) Prove that if P is a Tychonoff extension property (see 5.6), then so is $\mathcal{a}PT$.

Henceforth assume P is a given Tychonoff extension property.

(2) Prove that a space X has $\mathcal{a}P$ iff $EX \in P$. (Thus in particular if $X \in P$ then $EX \in P$.)

(3) Let X be a Tychonoff space and let A and B be closed subspaces of X with $\mathcal{a}PT$. Prove that $A \cup B \in \mathcal{a}PT$. (It is **not** true that the union of two closed realcompact subspaces of a Tychonoff space need be realcompact. Thus the obvious generalization of this

result to arbitrary Tychonoff extension properties fails.)

(4) Prove that P and APT need not be copseudocompact extension properties. (Hint: Let S be "connected and Tychonoff" and consider $\hat{S}$; see 5X.)

6U. Almost realcompact spaces and extensions. According to the previous problem, a space is **almost realcompact** if it is the perfect continuous image of a realcompact space. In particular, note by 6T(2) that a space X is almost realcompact iff EX is realcompact.

(1) Prove that the following are equivalent for a regular space X:

(a) X is almost realcompact, and

(b) if $\mathcal{U}$ is an ultrafilter on $R(X)$ with the countable intersection property, then $\cap\mathcal{U} \neq \emptyset$.

(Hint: See 5E(6) and 5G(3).)

(2) A Tychonoff space X is called **weak c.b** if, given a decreasing sequence $\{F_n : n \in \mathbb{N}\}$ of regular closed subsets of X with empty intersection, there exists a sequence $\{Z_n : n \in \mathbb{N}\} \subseteq Z(X)$ such that $F_n \subseteq Z_n$ for each $n \in \mathbb{N}$ and $\cap\{Z_n : n \in \mathbb{N}\} = \emptyset$. Prove that an almost realcompact weak c.b. space is realcompact. (See 8K for other sufficient conditions for almost realcompact spaces to be realcompact.)

(3) Prove that an almost realcompact space is H-closed iff it is feebly compact (see 1.11(a)). Thus show that "almost realcompact Tychonoff" is copseudocompact with realcompactness. (See 5W.)

(4) Let X be a Tychonoff space and define a_1X to be $X \cup \{p \in \beta X \setminus X :$ there exists an ultrafilter $\mathcal{U}$ on $\mathcal{R}(X)$ with the countable intersection property such that $\{p\} = \bigcap\{cl_{\beta X}A : A \in \mathcal{U}\}$. Let P denote the property of being almost realcompact and Tychonoff. Then P is an extension property (see 6T(1)); denote $\gamma_P X$ by aX. Prove that $a_1X \subseteq aX$.

(5) If $n > 1$ define a_nX inductively to be $a_1k_{\beta X}[\upsilon E(a_{n-1}X)]$. Show that $\bigcup\{E(a_nX) : n \in \mathbb{N}\}$ is a realcompact subspace of $\beta(EX)$. (Hint: Use 5F(8).)

(6) Prove that $aX = \bigcup\{a_nX : n \in \mathbb{N}\}$.

6V. <u>An almost realcompact, non-realcompact Tychonoff space</u>

(1) Let $S = D(\omega_1) \cup \{p\}$, topologized as follows. Each point of $D(\omega_1)$ is isolated, and a subset X of S that contains p is open in S iff $|S \setminus X| \leq \aleph_0$. Show that this is a valid definition of a zero-dimensional Hausdorff topology on S.

(2) Let $\mathbb{N}^* = \mathbb{N} \cup \{\omega\}$ denote the one-point compactification of $\mathbb{N}$, and let $A = S \times \mathbb{N}^* \setminus \{(p,\omega)\}$. (The space A is sometimes called the **Dieudonné plank**). Show that $A = G \cup L$, where G is a dense realcompact subspace of A

and L is a Lindelöf subspace of A.

(3) Let Y be realcompact and let L be a Lindelöf subspace of βY. Prove that $Y \cup L$ is realcompact. (Hint: Use 5.11(c).)

(4) Prove that if $X = G \cup L$ where X is Tychonoff, G is a dense realcompact subspace of X, and L is a Lindelöf subspace of X, then X is almost realcompact. (Hint: Consider $f^{\leftarrow}[X]$, where $f : \beta G \to \beta X$ is the Stone extension of the embedding map $j : G \to X$, and use 1J.) Hence conclude that A is almost realcompact.

(5) Prove that $\upsilon A \equiv_A S \times \mathbb{N}^*$ and hence conclude that A is not realcompact. (Hint: If $f \in C(A)$, show that there exists a neighborhood U of p in S such that for each $n \in \mathbb{N}$, f is constant on $U \times \{n\}$. Consider the behavior of f on $U \times \{\omega\} \setminus \{(p,\omega)\}$ and use this to extend f to (p,ω).)

6W. Non-normal absolutes

(1) If X is a pseudocompact, non-countably compact Tychonoff space (e.g., the space Ψ of 1N) and if Y is a space for which EX and EY are homeomorphic, show that Y is not δ-normally separated (see 1R). In particular, Y is not normal.

Let $P = (\omega_1+1) \times (\omega_1+1) \setminus \{(\omega_1,\omega_1)\}$, where $\omega_1 + 1$ has the order topology. In 2T we showed that P is a pseudocompact non-normal space. It follows from

2T(6) that $\beta P \equiv_P (\omega_1+1) \times (\omega_1+1)$.

(2) Define an equivalence relation $\sim$ on P as follows. For each limit ordinal $\alpha < \omega_1$, $\{(\alpha,\omega_1), (\alpha,\alpha), (\omega_1,\alpha)\}$ is an equivalence class, and all other equivalence classes are singletons. Let T be the quotient space of P associated with this equivalence relation, and let $f : P \to T$ be the quotient map. Prove that f is a perfect irreducible continuous surjection, and that T is a regular space.

(3) Prove that if A and B are disjoint closed subsets of T, then at least one of them is compact.

(4) Prove that T is normal but that ET is not normal.

6X. Minimal extremally disconnected spaces. A space (X,τ) is a **minimal extremally disconnected space** if it is extremally disconnected and there is no other extremally disconnected Hausdorff topology on X that is strictly contained in τ.

(1) Show that a minimal extremally disconnected space is zero-dimensional (consider the semiregularization; see 6P(3)).

(2) Let X and Y be zero-dimensional extremally disconnected spaces and let $j : X \to Y$ be a continuous bijection that is not a homeomorphism. Show that there exists a nonempty clopen subset A of X such that $\mathrm{int}_Y j[A] = \emptyset$.

(3) Prove that the following are equivalent for an extremally disconnected space X:

(a) X is a minimal extremally disconnected space, and

(b) X is zero-dimensional and has no nonempty clopen subset A such that $c\ell_{\beta X}(X\setminus A)\setminus X$ contains a continuous one-to-one image of A.

(Hint: To show (a) $\Rightarrow$ (b), let $\emptyset \neq A \in B(X)$, let $H \subseteq c\ell_{\beta X}(X\setminus A)\setminus X$, and let H be a continuous one-to-one image of A. Now consider $(X\setminus A) \cup H$. To show (b) $\Rightarrow$ (a), suppose (a) fails. Then produce j and A as in (2) above and consider the Stone extension of j.)

(4) Show that if X is a minimal extremally disconnected space and $X \subseteq T \subseteq \beta X$, then T is a minimal extremally disconnected space.

(5) Show that the following are equivalent for a locally compact extremally disconnected space.

(a) X is a minimal extremally disconnected space.

(b) If $A \in B(X)\setminus\{\emptyset\}$ and A is compact, then A is not homeomorphic to any subspace of $\beta X\setminus X$.

(6) Let B be a subset of E([0,1]) of cardinality less than 2^{c}. Prove that E([0,1])\B is a noncompact zero-dimensional minimal extremally disconnected space. (Hint: Use 6M.)

6Y. <u>Co-absolutes of $\beta X\setminus X$.</u> Two spaces X and Y are called **co-absolute** if EX and EY are homeomorphic. In 6.10(f) we proved that any two compact metric spaces without isolated points are co-absolute. In this problem we show that if the pseudocompact closed subspaces of the Tychonoff space X are compact, then $k_{\beta X} \mid \beta(EX) \setminus EX)$ is a perfect irreducible continuous surjection onto $\beta X \setminus X$, and hence $\beta(EX) \setminus EX$ and $\beta X \setminus X$ are co-absolute. (Here, as in 6.9(b)(3), we identify $E(\beta X)$ and $\beta(EX)$.)

Our approach is to show that $\{c\ell_{\beta X} A \setminus X : A \in R(X)\}$ is a basic subalgebra (denoted $R^*(X)$) of $R(\beta X \setminus X)$, and that $E_{R^*(X)}(\beta X \setminus X)$ is homeomorphic to $\beta(EX) \setminus EX$ (see 6H).

If A is a closed subspace of X, we denote $c\ell_{\beta X} A \setminus X$ by A^*.

(1) Let $A \in R(X)$ and let B be closed in X. Show that if $A^* \subseteq B^*$ then $c\ell_X(A \setminus B)$ is pseudocompact. (Hint: Let $S = c\ell_X(A \setminus B)$ and $V = (\text{int}_X A) \setminus B$. Show that $S = c\ell_X V$. Now assume S is not pseudocompact, and apply 1Q(6) to find a countably infinite discrete subspace D of V that is closed and C-embedded in S. Prove that D and B are completely separated in X by first showing that D and $\text{bd}_X V$ are completely separated in S.)

(2) Let F be a family of closed subsets of X satisfying these conditions: (i) F is closed under finite unions, and (ii) $\{F^* : F \in F\}$ is a base for the closed subsets of $\beta X \setminus X$. Prove that the following statments are equivalent:

(a) $c\ell_{X^*}(X^* \setminus F^*) = [c\ell_X(X \setminus F)]^*$ for each $F \in F$,

(b) if $F \in \mathcal{F}$ then $X^* = F^*$ implies that $c\ell_X(X\backslash F)$ is compact for each $F \in \mathcal{F}$,

(c) if $F \in \mathcal{F}$ then $X^* \subseteq c\ell_{\beta X}F$ implies $X^* \subseteq int_{\beta X}c\ell_{\beta X}F$, and

(d) if $F \in \mathcal{F}$ then $int_{X^*}F^* = (int_{\beta X}c\ell_{\beta X}F)\backslash X$.

(3) Prove that the following conditions on a space X are equivalent:

(a) if $A \in R(X)$ then $A^* \in R(X^*)$,

(b) $R(X)$ satisfies the (equivalent) conditions (a) — (d) of (2), and

(c) $A \mapsto A^*$ is a Boolean algebra homomorphism from $R(X)$ into $R(X^*)$.

(4) Prove that if X is a space in which pseudocompact closed subspaces are compact, then $A \mapsto A^*$ is a Boolean algebra homomorphism from $R(X)$ onto a basic subalgebra (see 6H) of $R(\beta X\backslash X)$.

(5) Let X be any Tychonoff space. Prove that $K(X)$, the set of compact members of $R(X)$, is an ideal of $R(X)$ and $S(R(X)/K(X))$ is homeomorphic to $c\ell_{\beta EX}(\beta(EX)\backslash EX)$ (see 3L).

(6) Prove that if X is a space in which pseudocompact closed subspaces are compact, then $\{A^* : A \in R(X)\}$ is isomorphic to $R(X)/K(X)$ (see 3L), and so $S(\{A^* : A \in R(X)\})$ is homeomorphic to $c\ell_{\beta EX}(\beta(EX)\backslash EX)$.

(7) If X is any Tychonoff space, prove that $k_{\beta X} | \beta(EX)\setminus EX$ is a perfect continuous surjection from $\beta(EX)\setminus EX$ onto $\beta X\setminus X$; prove that if pseudocompact closed subspaces of X are compact, then this map is also irreducible, and so $\beta(EX)\setminus EX$ and $\beta X\setminus X$ are co-absolute.

(8) By considering the ordinal space ω_1, and the space $\omega_1 \times \omega_1$, show that the following phenomona can occur:

(a) $A \mapsto A^*$ can map $R(X)$ into $R(\beta X\setminus X)$ but not be a Boolean algebra homomorphism, and

(b) A^* need not belong to $R(\beta X\setminus X)$, even if A belongs to $R(X)$ (see 4AG(8) for help in identifying $\beta(\omega_1 \times \omega_1)$).

(9) Let X be any Tychonoff space. Let PS(X) denote the set of pseudocompact members of $R(X)$. Prove that:

(a) PS(X) is an ideal of $R(X)$,

(b) $A \mapsto c\ell_{\beta X}A\setminus \upsilon X$ is a Boolean algebra homomorphism from $R(X)$ into $R(\beta X\setminus \upsilon X)$ (see 5V).

(c) $\beta(EX)\setminus E(\upsilon X)$ is co-absolute with $\beta X\setminus \upsilon X$, and

(d) $S(R(X)/PS(X))$ is homeomorphic to $c\ell_{\beta EX}(\beta(EX)\setminus E(\upsilon X))$.

(10) Prove that the converse of (4) fails by considering a pseudocompact, nowhere locally compact space.

6Z. PX as an inverse limit. In this problem we construct PX as an inverse limit of a system of spaces each of which can be mapped onto X by a closed irreducible finite-to-one continuous function. See 2U for background material on inverse limits.

(1) Let X be a space and let S denote the set of all nonempty finite subsets of $R(X)$, partially ordered by inclusion. Show that S is a directed set.

(2) Suppose $H \subseteq F \in S$. Define $M(H,F)$ to be

$$[\wedge\{A : A \in H\}] \wedge [\wedge\{A' : A \in F \setminus H\}]$$

(We define $\wedge\varnothing$ to be X). Prove each of the following statements:

(a) $\vee\{M(H,F) : H \in \mathbb{P}(F)\} = X$ for each $F \in S$.

(Hint: Choose $F \in S$. If $x \in X$ let U be an ultrafilter on $R(X)$ converging to x and let $H = F \cap U$.)

(b) If $H_1, H_2 \in \mathbb{P}(F)$ and $H_1 \neq H_2$ then $M(H_1,F) \wedge M(H_2,F) = \varnothing$.

(3) If $F \in S$ define $X(F)$ to be the direct sum

$\oplus\{M(H,F) : H \in \mathbb{P}(F)$ and $M(H,F) \neq \emptyset\}$. (Thus, points of $X(F)$ are of the form (x,H), where $H \in \mathbb{P}(F)$ and $x \in M(H,F)$; see 1.2(h).) Define $k(F) : X(F) \to X$ as follows: $k(F)((x,H)) = x$. Prove that $k(F)$ is a closed continuous irreducible finite-to-one function. (Use (2) above.)

(4) Let $\emptyset \neq F \subseteq G \in S$. If $H \in \mathbb{P}(F)$, show that $M(H,F) = \cup\{M(H',G) : H \subseteq H'$ and $H' \setminus F \subseteq G \setminus F\}$.

(5) Let F, G be as in (4). Define $k(G,F) : X(G) \to X(F)$ as follows: $k(G,F)((x,H)) = (x,H \cap F)$ for each $H \in \mathbb{P}(G)$ and $x \in M(H,G)$. Show that $k(G,F)$ is a well-defined closed continuous finite-to-one surjection from $X(G)$ onto $X(F)$.

(6) Let $H \in S$ and $\emptyset \neq F \subseteq G \subseteq H$.

(a) Show that $k(F) \circ k(G,F) = k(G)$ and hence conclude that $k(G,F)$ is irreducible.

(b) Show that $k(G,F) \circ k(H,G) = k(H,F)$.

(7) Show that $\langle\{X(F)\}, k(G,F), S\rangle$ is an inverse system in which the "bonding maps" are closed, finite-to-one (and thus perfect), irreducible, and continuous (see 2U).

(8) Let $X(\infty)$ be the inverse limit of the above system (see 2U). As in 2U, let $f(F) = \Pi_{X(F)} | X(\infty)$ for each $F \in S$.

(a) If $\emptyset \neq F \subseteq G \in S$ show that $f(F) =$

$k(G,F) \circ f(G)$

(b) Show that if F, $G \in S$ then $k(F) \circ f(F) = k(G) \circ f(G)$.

Denote $k(F) \circ f(F)$ by k; then k is a function from $X(\infty)$ to X.

(9) Prove that k is perfect. (Hint: By 1.8(g), $\Pi\{f(F) : F \in S\} : \Pi\{X(F) : F \in S\} \to \Pi\{X : F \in S\}$ is perfect. Check that its range is a subset of

$$\Delta_X = y \in \Pi X : F \in S\} : \text{for all } F,\ G \in S,\ \Pi_F(y) = \Pi_G(y)\},$$

and that Δ_X is homeomorphic to X. Now restrict the product map to $X(\infty)$ and relate it to k.)

(10) Prove that $k[X(\infty)] = X$. (Hint: Let $x_0 \in X$ and let U be an ultrafilter on $R(X)$ that converges to x_0. For each $F \in S$ show that there is precisely one member of $\mathbb{P}(F)$ — call it $H(F,x_0)$ — such that $M(H(F,x_0),F) \in U$. (Use (2).) Use these to construct a point in $k^{\leftarrow}(x_0)$.)

(11) Prove that $k : X(\infty) \to X$ is irreducible (2U(1) is useful here).

(12) Prove that $X(\infty)$ is extremally disconnected. (Hint: Let $V \in RO(X(\infty))$ and $A = c\ell_{X(\infty)}V$. Then $k[A] \in R(X)$ (see 6.5(d)), so $\{k[A]\} \in S$. Set $F(V) = \{k[A]\}$. Define $g = f(F(V)) \circ \chi_{k[A]}$; then $g \in C(X(\infty),\not{2})$ and use this to argue that A is open in $X(\infty)$.)

(13) Conclude that $(PX,\Pi_X) \sim (X(\infty),k)$.

6AA. $\underline{\beta[0,\infty)\backslash[0,\infty)}$ __is an indecomposable continuum__. A **continuum** is a compact connected space. (Some authors insist that continua be metrizable; we do not.) A continuum is **indecomposable** if it cannot be written as the union of two proper subcontinua. It is **hereditarily indecomposable** if all its subcontinua are indecomposable. In this problem we prove that $\beta[0,\infty)\backslash[0,\infty)$ is an indecomposable (but not hereditarily indecomposable) continuum, and that $\beta\mathbb{R}^n\backslash\mathbb{R}^n$ is a decomposable continuum if $n \geqslant 2$. We shall use results from 6Y; the reader is advised to work through that problem before attempting this one.

Denote $[0,\infty)$ by H, and $\beta[0,\infty)\backslash[0,\infty)$ by H*. (This latter notation is consistent with that used in 6Y.)

(1) Prove that the intersection of a chain of subcontinua of a space X is a subcontinuum of X.

(2) Show that H*, and $\beta\mathbb{R}^n\backslash\mathbb{R}^n$ for $n > 1$, are continua.

(3) Prove that a continuum is indecomposable iff every proper subcontinuum of it has empty interior.

(4) Suppose that E is a subcontinuum of H* and $\varnothing \neq E \neq$ H*. Show that there exists $Z \in Z(H)$ such that both Z and $H\backslash Z$ are unbounded, and E is contained in a connected component of Z*. (See 6Y for notation.)

(5) If $\text{int}_{H^*}E \neq \varnothing$, show that there exists $B \in R(H)$ such

that $\emptyset \neq B^* \subseteq E$ (see 6Y(4)).

(6) Show that there exists $n_0 \in \mathbb{N}$ such that $B \cap [n_0,\infty) \subseteq Z \cap [n_0,\infty)$.

(7) Show there exists an increasing sequence $\{\lambda_n : n \in \mathbb{N}\} \subseteq H$ with these properties: (a) $\lambda_1 = 0$ and $\lambda_{n+1} \geq \lambda_n + 1$ for each $n \in \mathbb{N}$; (b) $\{\lambda_n : n \in \mathbb{N}\} \cap Z = \emptyset$; and (c) $Z \cap [\lambda_n,\lambda_{n+1}] \neq \emptyset$ for each $n \in \mathbb{N}$.

(8) Show that there exists an increasing sequence $\{\alpha_n : n \in \mathbb{N}\} \subseteq H$ such that $\alpha_1 = 0$ and $\alpha_n = \min\{\lambda_i : \lambda_i > \alpha_{n-1}$ and $[\alpha_{n-1},\lambda_i] \cap B \neq \emptyset\}$.

(9) Let $F = Z \cap \bigcup\{[\alpha_{2n-1},\alpha_{2n}] : n \in \mathbb{N}\}$ and $G = Z \cap \bigcup\{[\alpha_{2n},\alpha_{2n+1}] : n \in \mathbb{N}\}$. Prove that $\{E \cap F^*, E \cap G^*\}$ is a decomposition of E into disjoint nonempty closed sets. Conclude from this that H^* is indecomposable.

(10) Prove that if G is closed in H^* and if $A \in \mathcal{B}(G)$, then there exist closed subsets S and T of H such that $A = S^* \cap G$ and $G \backslash A = T^* \cap G$ (see 4.6(g)(4)).

(11) Let $\{\lambda_n : n \in \mathbb{N}\}$ be an infinite closed discrete subspace of H such that $\lambda_{n+1} > \lambda_n$ for each $n \in \mathbb{N}$. Let $\mathcal{U} \in \beta\mathbb{N}\backslash\mathbb{N}$, and let $K = \bigcap\{\bigcup\{[\lambda_{2n-1},\lambda_{2n}] : n \in U\}^* : U \in \mathcal{U}\}$. Prove that K is a subcontinuum of H^*. (Hint: If not, find S and T as in (10). Use compactness to find $V \in \mathcal{U}$ such that $S^* \cap T^* \cap [\bigcup\{[\lambda_{2n-1},\lambda_{2n}] : n \in V\}]^* = \emptyset$. Then find $W \in \mathcal{U}$ such that $S \cap T \cap [\bigcup\{[\lambda_{2n-1},\lambda_{2n}] : n \in W\}] = \emptyset$. Use the characterization of ultrafilters on $\mathbb{N}$, together with the fact that $S^* \cap K \neq \emptyset \neq T^* \cap K$, to find infinitely many $n \in W$ such that $S \cap [\lambda_{2n-1},\lambda_{2n}] \neq \emptyset \neq T$

$\cap\ [\lambda_{2n-1},\lambda_{2n}]$.)

(12) Produce a decomposable subcontinuum of H*. (Hint: Juxtapose a couple of K's from (11).)

(13) If $n > 1$, prove that $\beta\mathbb{R}^n\setminus\mathbb{R}^n$ is decomposable. (Consider $\{(x_1,\ldots,x_n) \in \mathbb{R}^n : x_1 \geq 0\}$, etc.)

6AB. $\underline{\beta X\setminus \upsilon X}$ vs. $\underline{\beta X\setminus X}$

(1) Suppose that X is a Tychonoff space all of whose pseudocompact closed subspaces are compact. Show that $\beta X\setminus \upsilon X$ is dense in $\beta X\setminus X$. (Hint: Use 5W.)

(2) Let X be a Tychonoff space, and assume that X is a P-space (1W), or almost realcompact (6U), or metacompact. Prove that $\beta X\setminus \upsilon X$ is dense in $\beta X\setminus X$.

(3) Assume MA + $\neg$CH and let X be a perfectly normal space (see 3.5(k)). Prove that $\beta X\setminus \upsilon X$ is dense in $\beta X\setminus X$. (Hint: See 4AE and 1R(2).)

6AC. P-points and $\underline{\beta X\setminus X}$ under CH

(1) Let X be a compact space of weight no greater than $\aleph_1$. Prove that if each nonempty G_δ-set of X has a nonempty interior, then X has a dense set of P-points (see 1W). (Hint: Let U be an open subset of X and let $\{B(\alpha) : \alpha < \omega_1\}$ be an open base for X. Inductively

define a collection $\{V(\alpha) : \alpha < \omega_1\}$ of open subsets of X as follows. Let $V(0) = U$. Suppose that $\alpha_0 < \omega_1$ and that $\{V(\alpha) : \alpha < \alpha_0\}$ has been defined so that:

(a) $\alpha < \delta < \alpha_0$ implies $\emptyset \neq c\ell_X V(\delta) \subseteq V(\alpha)$ and

(b) if $\alpha < \alpha_0$ then either $V(\alpha) \cap B(\alpha) = \emptyset$ or else $V(\alpha) \subseteq B(\alpha)$.

Now consider $\bigcap\{V(\alpha) : \alpha < \omega_1\}$.

(2) Let X be as above, and in addition assume that X has no isolated points. Prove that X has a dense set of $2^{\aleph_1}$ P-points. (Each $V(\alpha)$ can be chosen in two disjoint ways.)

(3) Show that if we weaken the assumptions in (1) and (2) from "X is compact" to "X is locally compact", the conclusion remains true.

(4) Assume the continuum hypothesis. Suppose that X is locally compact, realcompact, and noncompact, and suppose that $|C^*(X)| = c$. Prove that $\beta X \setminus X$ has a dense set of 2^c P-points (see 5.11(h) and 5Z(2)). Thus in particular, $\beta\mathbb{N}\setminus\mathbb{N}$ has a dense set of 2^c P-points.

6AD. <u>P-points of $\beta X \setminus X$ under Martin's axiom</u>. In this problem we show that if X is a "sufficiently small" locally compact, non-pseudocompact space then $\beta X \setminus X$ will have a dense set of P-points. To prove this, we use an (apparently) slightly strengthened version of P(c) (see 3.5(q)), which we derive directly from Martin's axiom. This problem generalizes 6AC.

(1) If X is a space prove that $|R(X)| \leq 2^{\pi w(X)}$ (see 2N(5)). Thus conclude that if X is semiregular, then $w(X) \leq 2^{\pi w(X)}$. Show that $w(\kappa\mathbb{N}) > 2^{\pi w(\kappa\mathbb{N})}$.

Now let X be a locally compact σ-compact noncompact space of countable π-weight. Let $X = \bigcup\{K(n) : n \in \mathbb{N}\}$ where each K(n) is compact.

(2) Prove that $w(\beta X\setminus X) = 2^{\aleph_0}$ (use (1) and 5Z(1)).

(3) Prove that X has a countable π-base $\mathcal{B} = \{B(n) : n \in \mathbb{N}\}$ that is closed under finite unions and whose members have compact closures in X.

Now let $\aleph_0 \leq \kappa < c$ and let $\{V(\alpha) : \alpha < \kappa\}$ be a filter base on $RO(\beta X)$ such that $\bigcap\{V(\alpha)\setminus X : \alpha < \kappa\} \neq \varnothing$. Let $P = \{\langle A,F\rangle : A \in \mathcal{B}$ and F is a finite subset of $\kappa\}$. If $\langle A,F\rangle, \langle B,G\rangle \in P$ define $\langle A,F\rangle \leq \langle B,G\rangle$ to mean that $B \subseteq A$, $G \subseteq F$, and $A\setminus V(\alpha) \subseteq B$ for each $\alpha \in G$.

(4) Show that $\leq$ as defined above is partial order on P that satisfies the countable chain condition.

(5) If $\alpha \in \kappa$, let $D(\alpha) = \{\langle A,F\rangle \in P : \alpha \in F\}$. Show that $D(\alpha)$ is dense in P.

(6) If $n \in \mathbb{N}$ let $E(n) = \{\langle A,F\rangle \in P : A\setminus K(n) \neq \varnothing\}$.

Show that $E(n)$ is dense in P.

(7) Assume Martin's axiom. Show that there exists $C \in R(\beta X)$ such that $\emptyset \neq C \setminus X \subseteq \bigcap \{c\ell_{\beta X} V(\alpha) \setminus X : \alpha < \kappa\}$. (Take the closure of the union of the first components of members of a filter on P meeting each member of $\{D(\alpha) : \alpha < \kappa\} \cup \{E(n) : n \in \mathbb{N}\}$.) Hence conclude that $\text{int}_{\beta X \setminus X} \bigcap \{c\ell_{\beta X} V(\alpha) \setminus X : \alpha < \kappa\} \neq \emptyset$ (see 6Y(4)).

(8) Prove that if Martin's axiom holds and if X is as above, then $\beta X \setminus X$ has a dense set of 2^{c} P-points. (Adapt the argument of 6AC(1), using (2) and (7) above.)

(9) Check that if $X = \mathbb{N}$, the above argument reduces to a direct application of P(c).

Now we relax the requirement that X be σ-compact.

For the remainder of this problem assume that X is locally compact, not pseudocompact, and of countable π-weight.

(10) Using 5F(4) choose $Z \in Z(\beta X)$ such that $\emptyset \neq Z \subseteq \beta X \setminus X$. Prove that $\text{int}_{\beta X \setminus X} Z \neq \emptyset$ (generalize the proof of 5.11(h)).

(11) Let $Y = \beta X \setminus Z$. Show that $\beta Y \setminus Y = Z$ and infer that if Martin's axiom holds then $\beta Y \setminus Y$ has a dense set of 2^{c} P-points.

(12) By considering $P(\beta Y \setminus Y) \cap \text{int}_{\beta X \setminus X} Z$ show that $P(\beta X \setminus X) \cap Z$ is dense in Z if Martin's axiom holds. Recall from 1W that P(T) denotes the set of P-points of

the space T.

(13) Conclude that if Martin's axiom holds than $\beta X \setminus X$ has a dense set of 2^{c} P-points.

CHAPTER 7

H-CLOSED EXTENSIONS

In this chapter we begin a detailed investigation of the set $H(X)$ of all H-closed extensions of a space X. We begin by considering strict and simple extensions of a space. We then construct and study the Fomin extension σX of an arbitrary space X, the Banaschewski-Fomin-Šanin extension μX of a semiregular space X, and one-point H-closed extensions of locally H-closed spaces. Next we consider the interrelationships among certain partitions of $\sigma X \backslash X$ and the poset structure of $H(X)$. We characterize and study those $f \in C(X,Y)$ that can be extended to a function $\kappa f \in C(\kappa X, \kappa Y)$. The chapter concludes with the study of Θ-equivalent H-closed extensions.

7.1 Strict and simple extensions

Let Y be an extension of X. In this section we investigate ways of enlarging and reducing the topology on Y so that the new space thus obtained remains an extension of X and has the same neighborhood filter trace on X as Y had. We will also show that the set of such topologies on Y has a largest and smallest member.

(a) **Definition**. Let Y be an extension of X and let $p \in Y$.

(1) We denote $\{U \cap X : U \text{ open in } Y \text{ and } p \in U\}$ by 0^p (see proof of 4.1(c)). Sometimes 0^p is denoted by $0_Y{}^p$ when more than one extension of X is under discussion and there is a possibility of

confusion. We call $\{0^p : p \in Y\}$ the **neighborhood filter trace** of Y on X. (Note that 0^p is an open filter on X, and if $x \in X$ then 0^x is the filter of open neighborhoods of x in X.)

(2) If U is an open subset of X, let $oU = \{p \in Y : U \in 0^p\}$. Again, the notation "$o_Y U$" will be used if there is ambiguity about the identity of Y.

(b) Lemma. Let $Y \in \mathcal{H}(X)$ and $f : \kappa X \to Y$ be the unique continuous function such that $f|X = id_X$ (see 4.8(n)(2)). If $\mathcal{U} \in \kappa X \setminus X$ and $p \in Y$, then $f(\mathcal{U}) = p$ iff $0^p \subseteq \mathcal{U}$.

Proof. Obvious. ∎

(c) Proposition. If Y is an extension of X and U and V are open subsets of X, then:

(1) $oX = Y$, $o(\varnothing) = \varnothing$,

(2) $(oU) \cap X = U$,

(3) $oU \cap oV = o(U \cap V)$,

(4) $oU = \bigcup\{W : W \text{ is open in } Y \text{ and } W \cap X \subseteq U\}$,

(5) $\{oU : U \in \tau(Y)\}$ is an open base for a Hausdorff topology $\tau^{\#}$ on Y that is contained in the original topology of Y,

(6) $(Y, \tau^{\#})$ is an extension of X, and

(7) $oU = Y \setminus c\ell_Y(X \setminus U)$.

Proof

(1) Since X belongs to every open filter on X and $\varnothing$ belongs to no open filter on X, $oX = Y$ and $o(\varnothing) = \varnothing$.

(2) If $x \in U$, then $U \in 0^x$; so $x \in (oU) \cap X$. Hence $U \subseteq (oU) \cap X$. Conversely, if $p \in (oU) \cap X$, then $U \in 0^p$. So, there is an open set W in Y such that $p \in W$ and $W \cap X = U$. Since $p \in X$, it follows that $p \in W \cap X = U$.

(3) Let $p \in Y$. Then $p \in oU \cap oV$ iff $U \in 0^p$ and $V \in 0^p$ iff $U \cap V \in 0^p$ iff $p \in o(U \cap V)$.

(4) Let $p \in oU$. Since $U \in 0^p$, there is an open set W in Y such that $p \in W$ and $W \cap X = U$. If $q \in W$, then $q \in o(W \cap X) = oU$; in particular, $W \subseteq oU$. So, $oU = \bigcup\{W : W \text{ open in } Y \text{ and } W \cap X \subseteq U\}$.

(5) That $\{oU : U \in \tau(Y)\}$ is a base for a topology on Y follows from (1) and (3). That this topology is contained in the original topology of Y follows from (4). To show it is Hausdorff, let p and q be distinct points of Y. Find disjoint members of U and V of $\tau(Y)$ such that $p \in U$ and $q \in V$. Thus $(U \cap X) \cap (V \cap X) = \emptyset$. By (4) it follows that $p \in U \subseteq o(U \cap X)$ and $q \in V \subseteq o(V \cap X)$; therefore the topology is Hausdorff.

(6) It follows from (2) above that X inherits the same subspace topology from $(Y,\tau(Y))$ as it does from $(Y,\tau^{\#})$, and that X is dense in $(Y,\tau^{\#})$.

(7) Since $(Y \backslash c\ell_Y(X \backslash U)) \cap X = X \backslash c\ell_X(X \backslash U)$, then by (4), $Y \backslash c\ell_Y(X \backslash U) \subseteq oU$. Since $oU \cap (X \backslash U) = \emptyset$, then $oU \cap c\ell_Y(X \backslash U) = \emptyset$. Hence $oU \subseteq Y \backslash c\ell_Y(X \backslash U)$.

■

(d) **Definition**. The space $(Y,\tau^{\#})$ described in the preceding proposition is denoted by $Y^{\#}$. Its topology will henceforth be denoted by $\tau(Y^{\#})$. We say that Y is a **strict extension** of X if $Y = Y^{\#}$ (i.e., if

$\tau(Y) = \tau(Y^{\#})$.)

Note that the topology of $Y^{\#}$ depends not only on the topology of Y, but also on the topology of X. Thus, if Y were an extension both of X and T, then $Y^{\#}$ defined with reference to X might be different from $Y^{\#}$ defined with reference to T. Thus, strictly speaking (sorry about that!), we should write $Y^{\#,X}$, or some such, to reflect this dependence of $Y^{\#}$ on X. In practice, however, there will never be any ambiguity about the identity of X, and we will use the notation defined above.

We derive more properties of oU.

(e) **Proposition**. Let Y be an extension of a space X and let U be open in Y. Then:

(1) $X \cap o(U \cap X) = U \cap X$,

(2) $U \subseteq o(U \cap X) \subseteq c\ell_Y(U \cap X) = c\ell_Y U$,

(3) $int_Y c\ell_Y U = int_Y c\ell_Y(U \cap X) = o(int_X c\ell_X(U \cap X))$, and

(4) if Y is semiregular, then Y is a strict extension.

Proof

(1) This follows from 7.1(c)(2).

(2) Since $o(U \cap X)$ and U are open in Y and X is dense in Y, it follows that $c\ell_Y(o(U \cap X)) = c\ell_Y(o(U \cap X) \cap X)$ and $c\ell_Y U = c\ell_Y(U \cap X)$. Hence, by (1), $c\ell_Y(o(U \cap X)) = c\ell_Y U$. If $p \in U$ then $U \cap X \in \mathcal{O}^p$ and $p \in o(U \cap X)$.

(3) Since $c\ell_Y U = c\ell_Y(U \cap X)$, it follows that $int_Y c\ell_Y U = int_Y c\ell_Y(U \cap X)$ which gives the first equality. Let $R = int_X c\ell_X(U \cap$

$X)$ and $W = \text{int}_Y c\ell_Y(U \cap X)$. Evidently $W = Y \backslash c\ell_Y(X \backslash R)$, which is oR by 7.1(c)(7).

(4) If Y is semiregular, then the topology of Y is generated by $\{\text{int}_Y c\ell_Y U : U \text{ is open in } Y\}$. Thus, by (3), the topology of Y is as coarse as the topology of $Y^{\#}$. It is always true by 7.1(c)(5) that the topology of $Y^{\#}$ is as coarse as the topology of Y. So, $Y = Y^{\#}$ and Y is a strict extension of X.

■

The converse of 7.1(e)(4) is not true. Note that the non-semiregular space Y described in 2.2(h)(2) is an extension of $Y \backslash \{p+\}$ and since $Y \backslash (Y \backslash \{p+\})$ is a singleton, it easily follows Y is a strict extension of $Y \backslash \{p+\}$.

We next consider another way of modifying the topology of an extension Y of X to obtain a new extension of X.

(f) **Proposition**. Let Y be an extension of a space X. Let $B = \{U \cup \{p\} : U \in O^p \text{ and } p \in Y\}$. Then:

(1) B is an open base for a topology τ^+ on Y that contains $\tau(Y)$.

(2) (Y, τ^+) is an extension of X.

(3) $Y \backslash X$ is a closed discrete subspace of (Y, τ^+).

Proof

(1) Let $U \cup \{p\}$, $V \cup \{q\} \in B$ where $U \in O^p$ and $V \in O^q$. If $p = q$ then $(U \cup \{p\}) \cap (V \cup \{q\}) = (U \cap V) \cup \{p\} \in B$. If $p \neq q$ and $r \in (U \cup \{p\}) \cap (V \cup \{q\})$ then $(U \cup \{p\}) \cap (V \cup \{q\}) = U \cap V$

and $r \in X$. Thus $U \cap V = (U \cap V) \cup \{r\} \in \mathcal{B}$. It follows that $\mathcal{B}$ is a base for a topology τ^+ on Y. If $V \in \tau(Y)$ then $V = \bigcup\{(V \cap X) \cup \{r\} : r \in V\}$, and so $V \in \tau^+$. Thus (Y,τ^+) is Hausdorff.

(2) Since $(U \cup \{p\}) \cap X = U$ for each $U \in 0^p$, the subspace topology X inherits from $(Y,\tau(Y))$ is the same as that inherited from (Y,τ^+). Thus (Y,τ^+) is an extension of X.

(3) If $p \in X$ then $X \in 0^p$ and $X \cup \{p\} = X \in \mathcal{B}$. Thus X is open in (Y,τ^+). If $r \in Y\backslash X$ then $X \cup \{r\} \in \mathcal{B}$ and $\{r\} = (Y\backslash X) \cap (X \cup \{r\})$. Thus $Y\backslash X$ is a discrete subspace of (Y,τ^+).

■

(g) **Definition**. The space (Y,τ^+) described in the preceding proposition is denoted by Y^+. Its topology will henceforth be denoted by $\tau(Y^+)$. We say that Y is a **simple extension** of Y if $Y = Y^+$ (i.e., if $\tau(Y) = \tau(Y^+)$).

We have already encountered an example of a simple extension of a space, namely the Katětov H-closed extension.

We next consider the relationship among Y^+, $Y^\#$, and other extensions of X with the same underlying set as Y.

(h) **Proposition**. Let Y be an extension of a space X. Then:

(1) $\tau(Y^\#) \subseteq \tau(Y^+)$ and $\{0_{Y^\#}{}^p : p \in Y^\#\} = \{0_Y{}^p : p \in Y\} = \{0_{Y^+}{}^p : p \in Y^+\}$.

(2) Let Z be an extension of X such that Z and Y have the same underlying set. Then $\tau(Y^\#) \subseteq \tau(Z) \subseteq \tau(Y^+)$ iff $\{0_Y{}^p : p \in$

$Y\} = \{0_Z{}^p : p \in Z\}$. In this case $\tau(Z^+) = \tau(Y^+)$ and $\tau(Z^\#) = \tau(Y^\#)$.

(3) $\tau(Y^+) = \tau((Y^+)^+)$ and $\tau(Y^\#) = \tau((Y^\#)^\#)$.

Proof

(1) The first part of (1) follows from 7.1(c) and (f). Since $\tau(Y^\#) \subseteq \tau(Y) \subseteq \tau(Y^+)$, it follows that $0_{Y^\#}{}^p \subseteq 0_Y{}^p \subseteq 0_{Y^+}{}^p$ for each $p \in Y$. If $U \in 0_{Y^+}{}^p$, then $U = Y \cap W$ where $p \in W$ and W is open in Y^+. Thus $W = \cup\{V_i \cup \{r_i\} : i \in I\}$ where $r_i \in Y$ and $V_i \in 0_Y{}^{r_i}$ for each $i \in I$. There exists $j \in I$ such that $p \in V_j \cup \{r_j\}$. Evidently $V_j \subseteq U$ and $p \in oV_j$. Thus $p \in oU$ and $U \in 0_{Y^\#}{}^p$. Therefore, $0_{Y^+}{}^p = 0_{Y^\#}{}^p$. Hence, $\{0_{Y^\#}{}^p : p \in Y^\#\} = \{0_Y{}^p : p \in Y\} = \{0_{Y^+}{}^p : p \in Y^+\}$.

(2) If $\tau(Y^\#) \subseteq \tau(Z) \subseteq \tau(Y^+)$, then for $p \in Y$, $0_{Y^\#}{}^p \subseteq 0_Z{}^p \subseteq 0_{Y^+}{}^p$. Thus, by (1), we have that $\{0_Z{}^p : p \in Z\} = \{0_Y{}^p : p \in Y\}$. Conversely, suppose $\{0_Z{}^p : p \in Z\} = \{0_Y{}^p : p \in Y\}$. Then by a proof similar to the proof of (1), we have that $\tau(Z^+) = \tau(Y^+)$ and $\tau(Z^\#) = \tau(Y^\#)$. Thus, $\tau(Y^\#) = \tau(Z^\#) \subseteq \tau(Z) \subseteq \tau(Z^+) = \tau(Y^+)$.

(3) This follows immediately from (2).

■

Recall from 2.3(c)(5) that if F and G are filters on a set X, then $\mathcal{F}$ **meets** if $F \cap G \neq \emptyset$ for all $F \in \mathcal{F}$ and $G \in \mathcal{G}$.

(i) **Proposition**. Let Y be an extension of a space X. The following are equivalent:

(1) Y is H-closed,

(2) each open filter on X meets 0^p for some $p \in Y$, and

(3) each open ultrafilter on X contains 0^p for some $p \in Y$.

Proof

(1) ⇒ (2) Let $\mathcal{F}$ be an open filter on X. Then $\mathcal{G} = \{W : W$ open in Y, $W \cap X \in \mathcal{F}\}$ is an open filter on Y. Since Y is H-closed, by 4.8(b), there is a point $p \in a_Y(\mathcal{G})$. It easily follows that 0^p meets $\mathcal{F}$.

(2) ⇒ (3) If $\mathcal{U}$ is an open ultrafilter on X, then $\mathcal{U}$ meets 0^p for some $p \in Y$ by the hypothesis. Thus, $\mathcal{U} \wedge 0^p$ $(=\{U \cap V : U \in \mathcal{U}, V \in 0^p\})$ is an open filter base on X and $\mathcal{U} \wedge 0^p \supseteq \mathcal{U}$. Since $\mathcal{U}$ is maximal among open filters on X, it follows that $\mathcal{U} = \mathcal{U} \wedge 0^p$ and so $\mathcal{U} \supseteq 0^p$.

(3) ⇒ (1) Let $\mathcal{G}$ be an open filter on Y and $\mathcal{F} = \{U \cap X : U \in \mathcal{G}\}$. Then $\mathcal{F}$ is an open filter on X. By 2.3(d)(2), $\mathcal{U} \supseteq \mathcal{F}$ for some open ultrafilter $\mathcal{U}$ on X. Now, by hypothesis, $\mathcal{U} \supseteq 0^p$ for some $p \in Y$. Thus $p \in ad_X(\mathcal{U}) \subseteq ad_X(\mathcal{F}) \subseteq ad_Y(\mathcal{G})$. By 4.8(b), Y is H-closed.

■

(j) **Corollary**. Let Y be an extension of a space X. The following are equivalent:

(1) Y is H-closed,

(2) Y^+ is H-closed, and

(3) $Y^{\#}$ is H-closed.

Proof. This follows from 7.1(i) and 7.1(h)(1).

■

7.2 The Fomin extension

There are two well-known H-closed extensions of a space; the Katětov extension (introduced in 4.8) and the Fomin extension defined in this section. The Katětov extension κX of a space X is a simple extension, and $0_{\kappa X}\mathcal{U} = \mathcal{U}$ for each $\mathcal{U} \in \kappa X \setminus X$. The strict extension $(\kappa X)^{\#}$ of X is H-closed by 7.1(j); it is called the **Fomin extension** of X and is denoted by σX. Thus $\{o_{\kappa X}U : U \text{ is open in } X\}$ is a base for the open sets of σX.

(a) **Proposition.** Let X be a space. Then:

(1) if $p \in \kappa X = \sigma X$ then $0_{\kappa X}p = 0_{\sigma X}p$, and

(2) if $\mathcal{U} \in \kappa X \setminus X = \sigma X \setminus X$, then $0_{\kappa X}\mathcal{U} = 0_{\sigma X}\mathcal{U} = \mathcal{U}$.

Proof

(1) This follows from 7.1(h)(1) and the definition of σX.

(2) From the definition of κX (prior to 4.8(n)) it follows immediately that $0_{\kappa X}\mathcal{U} = \mathcal{U}$ (recall that points of $\kappa X \setminus X$ are free open ultrafilters on X).

■

The properties developed in 7.1(e) can be sharpened for the extensions κX and σX.

(b) **Proposition**. Let X be a space, let Y be σX or κX, and let $U \in \tau(X)$. Then:

(1) $c\ell_Y U = (c\ell_X U) \cup oU$,

(2) $o(\text{int}_X c\ell_X U) = (\text{int}_X c\ell_X U) \cup oU$, (note by 7.2(a)(1) there is no ambiguity in the "oU" notation),

(3) there is a continuous Θ-homeomorphism j from κX onto σX,

(4) closed nowhere dense subsets of X are closed in Y, and

(5) $(Y \backslash X) \backslash oU = o(X \backslash c\ell_X U) \backslash X$

Proof

(1) Note by 7.2(a)(2) that $0_{\kappa X}{}^{\mathcal{U}} = 0_{\sigma X}{}^{\mathcal{U}} = \mathcal{U}$. Thus $\mathcal{U} \in c\ell_Y U \backslash X$ iff $V \cap U \neq \varnothing$ for each $V \in \mathcal{U}$ iff $U \in \mathcal{U}$ (as $\mathcal{U}$ is an open ultrafilter). By 7.2(a)(2) $U \in \mathcal{U}$ iff $U \in 0_Y{}^{\mathcal{U}}$, i.e., iff $\mathcal{U} \in oU$.

(2) This follows from the fact that if $\mathcal{U}$ is an open ultrafilter on X, then $U \in \mathcal{U}$ iff $\text{int}_X c\ell_X U \in \mathcal{U}$.

(3) Let $j : \kappa X \to \sigma X$ be the identity function on the underlying set of κX. If $p \in \kappa X$ and $U \in 0^p$, then $j[U \cup \{p\}] \subseteq oU$ (see 7.1(a)(2)); so, j is a continuous bijection. Conversely, if $p \in \kappa X$ and $U \in 0^p$, then $p \in oU$ and by 7.2(a)(2), $j^{\leftarrow}[c\ell_{\sigma X} oU] = c\ell_X U \cup oU = c\ell_{\kappa X} U$. Thus j is Θ-continuous and it follows that j is a Θ-homeomorphism.

(4) If F is a closed nowhere dense subset of X, it follows from (1) that $Y \backslash X \subseteq o(X \backslash F)$. Thus by 7.1(c)(2), $F = Y \backslash o(X \backslash F)$.

(5) Note that $Y \setminus X = [c\ell_Y U \cup c\ell_Y(X \setminus c\ell_X U)] \setminus X =$ $(oU \setminus X) \cup [o(X \setminus c\ell_X U) \setminus X]$ by (1). But $oU \cap o(X \setminus c\ell_X U) = \emptyset$ by 7.1(c)(3).

■

As we saw above, κX and σX are closely related in many ways. However, there are two major differences between κX and σX; first, $\sigma X \setminus X$ is not necessarily discrete, and second, X is not necessarily open in σX.

From 7.2(b)(5) it is clear that $oU \setminus X \in B(\sigma X \setminus X)$, and thus $\sigma X \setminus X$ is a zero-dimensional space (and thus Tychonoff). Actually, more is true about the remainder $\sigma X \setminus X$; it is a subspace of an extremally disconnected space, as we see below.

By 6.9(d) there is a perfect irreducible Θ-continuous surjection $k_{\sigma X}' : \theta' X \to \sigma X$ defined as follows: $k_{\sigma X}'(\mathcal{U})$ is the unique point in $\bigcap\{c\ell_{\sigma X} U : U \in \mathcal{U}\}$ for each $\mathcal{U} \in \theta' X$. Let $S = (k_{\sigma X}')^{\leftarrow}[X]$. By 6.9(a) $(EX, k_X) \sim (S, k_{\sigma X}' \mid S)$ and by 6.9(b), $\theta' X \equiv_S \beta S$. We will prove that $k_{\sigma X}' \mid \theta' X \setminus S$ is a homeomorphism from $\theta' X \setminus S$ onto $\sigma X \setminus X$, thereby proving that there is an extremally disconnected space S which in fact is homeomorphic to EX such that $\sigma X \setminus X$ is homeomorphic to $\beta S \setminus S$. We formalize these ideas below.

(c) **Theorem**. Let X be a space. Then:

(1) $\sigma X \setminus X$ is homeomorphic to $\beta(EX) \setminus EX$, and

(2) if $A \subseteq \sigma X \setminus X$ then A is compact iff A is closed in σX.

Proof

(1) Note that $\theta'X\backslash S$ and $\sigma X\backslash X$ have the same underlying set, namely $\{\mathcal{U} : \mathcal{U}$ is a free open ultrafilter on $X\}$. (See 6.8 for a discussion of $\theta'X$.) Denote $k_{\sigma X}'|\theta'X\backslash S$ by h. From the definition of h it follows that if $\mathcal{U} \in \theta'X\backslash S$, then $h(\mathcal{U}) = \bigcap\{c\ell_{\sigma X}U : U \in \mathcal{U}\} = \bigcap\{c\ell_{\sigma X}U\backslash X : U \in \mathcal{U}\}$ (as $\mathcal{U}$ is free) $= \bigcap\{oU : U \in \mathcal{U}\}$ (see 7.2(b)) $= \mathcal{U}$; therefore, $h = id_{\theta'X\backslash S}$. If $U \in \tau(X)$ then $0U\backslash S$ is a basic open set of $\theta'X\backslash S$; evidently $\mathcal{U} \in 0U\backslash S$ iff $U \in \mathcal{U}$ and $\mathcal{U} \in \sigma X\backslash X$ iff $\mathcal{U} \in o_{\sigma X}U\backslash X$. Thus $h[0U\backslash S] = o_{\sigma X}U\backslash X$; so, h is a homeomorphism.

(2) If A is a compact subset of $\sigma X\backslash X$ it is obviously closed in σX. Conversely, if A is a subset of $\sigma X\backslash X$ that is closed in σX, then $k_{\sigma X}'^{\leftarrow}[A] = h^{\leftarrow}[A]$ is closed in $\theta'X\backslash S$ by (1). If $\mathcal{U} \in S$ and $k_{\sigma X}'(\mathcal{U}) = x_0 \in X$, there exists $U \in \tau(X)$ such that $x_0 \in U$ and $(o_{\sigma X}U) \cap A = \varnothing$. Thus $\mathcal{U} \in 0U$, and $(0U) \cap h^{\leftarrow}[A] = (0U\backslash S) \cap h^{\leftarrow}[A]$, which is empty as h is a bijection. Thus $\mathcal{U} \notin c\ell_{\theta'X}h^{\leftarrow}[A]$; therefore, $h^{\leftarrow}[A]$ is closed in $\theta'X$ and hence is compact. As h is a homeomorphism, it follows that A is compact.

■

Note that if E is extremally disconnected and zero-dimensional, then $\sigma E\backslash E$ and $\beta E\backslash E$ are homeomorphic, but σE and βE are not homeomorphic in general (see 7.7(g)). Evidently $\sigma E\backslash E$ and $\beta E\backslash E$ are "attached to E is different ways".

7.3 One-point H-closed extensions

In this section we construct one-point H-closed extensions of certain spaces, and show how the existence of such extensions is related to the poset structure of $H(X)$.

(a) **Definition**. A space is **locally H-closed** if every point has a neighborhood which is H-closed. An extension Y of X is a **one-point H-closed extension** of X if Y is H-closed and $Y\setminus X$ is a singleton.

The proofs of the following two results are similar to the proofs of the corresponding results for locally compact spaces, and are left an an exercise (7D) for the reader.

(b) **Proposition**. Let X be a space.

(1) X is locally H-closed iff X is open in σX iff X is open in every extension of X.

(2) X is locally H-closed and not H-closed iff X has an one-point H-closed extension.

(c) **Corollary**. If X is locally H-closed then $\sigma X\setminus X$ is compact.

Proof. This follows from 7.3(b)(1) and 7.2(c)(2).

■

A locally compact space has a unique one-point compactification, but there may be many non-equivalent one-point

H-closed extensions of a locally H-closed space. The following example illustrates this.

(d) **Example**. Let $X = \{(1/n,1/m) : n, m \in \mathbb{N}\} \cup \{(1/n,0) : n \in \mathbb{N}\}$ with the subspace topology induced by $\mathbb{R}^2$. Evidently X is locally compact and therefore is locally H-closed. Let $Y = X \cup \{(0,0)\}$ with this topology: a subset U of Y is open in Y iff (1) $U \cap X$ is open in X, and (2) if $(0,0) \in U$ then there exists $k \in \mathbb{N}$ such that $\{(1/n,1/m) : m \in \mathbb{N}, n \geq k\} \subseteq U$. It is easy to verify that with this topology, Y is an H-closed extension of X but is not semiregular. Also, Y(s) is the one-point compactification of X. Hence Y and Y(s) are two non-equivalent H-closed extensions of X.

By 4.8(n)(3) κX is a projective maximum in $H(X)$. The existence of a projective minimum in $H(X)$ requires additional structure on X, as the next result shows.

(e) **Proposition**. Let X be a space. The following are equivalent:

(1) X is locally H-closed,

(2) $H(X)$ has a projective minimum, and

(3) $E(X)$ has a projective minimum.

Proof. Evidently X is H-closed iff $H(X) = E(X) = \{X\}$. Therefore assume X is not H-closed.

(1) ⇒ (2) and (3) Suppose X is locally H-closed. Let X^b be $X \cup \{\infty\}$, where $\infty \notin X$. Give X^b the following topology: a subset U of X^b is open in X^b iff (1) $U \cap X$ is open in X, and (2) if $\infty \in U$ there exists a subset V of U such that $X \setminus V$ is H-closed. It is easy to

verify that this defines a topology on X^b, and that with this topology X^b is a one-point H-closed extension of X (see 7D). Let Y be an extension of X and define $f : Y \to X^b$ as follows: $f|X = id_X$ and $f[Y\setminus X] = \{\infty\}$ if $Y\setminus X \neq \emptyset$. Since X is open in Y by 7.3(b), f is continuous at each point of X. Suppose $y \in Y\setminus X$ and $f(y) \in U$ where U is open in X^b. For some $V \subseteq U$, $X\setminus V$ is H-closed. Thus $V \cup \{\infty\}$ is an open neighborhood of ∞ in X^b and $X\setminus V$ is closed in Y. Thus $f^{\leftarrow}[V \cup \{\infty\}] = Y\setminus(X\setminus V)$ is open in Y, contains y, and is contained in $f^{\leftarrow}[U]$; hence f is continuous. It follows that Y is projectively larger than X^b.

(3) ⇒ (2) If Y is a projective minimum in $E(X)$, then Y is a continuous image of κX, and hence is H-closed.

(2) ⇒ (1) Suppose Y is a projective minimum in $H(X)$. Since X is not H-closed, then $Y\setminus X \neq \emptyset$. By 7.3(b), it suffices to show that $Y\setminus X$ is a singleton. Let p and q be distinct points of $Y\setminus X$. Let Z be Y with p and q identified and give Z the induced quotient topology. It is straightforward to show that Z is a Hausdorff extension of X. Since Z is the continuous image of Y, it follows that Z is H-closed. Thus Y is projectively larger than Z. Since Y is a projective minimum, this is a contradiction. Hence $Y\setminus X$ is a singleton.

■

By 4.1(e,j), if X is a space then $H(X)$ and $E(X)$ are sets. The cardinalities of $H(X)$ and $E(X)$ can be related to the cardinalities of certain power-set reiterations of X, as we see in the next result.

(f) **Proposition**. For an infinite space X, $|H(X)| \leq$

$|E(X)| \leqslant |\mathbb{P}(\mathbb{P}(\mathbb{P}(\mathbb{P}(X))))|$.

<u>Proof</u>. Obviously $|H(X)| \leqslant |E(X)|$. Let $Y \in E(X)$. Then $p \to 0^p$ is a one-to-one function from Y into $\mathbb{P}(\mathbb{P}(X))$. Hence the elements of $E(X)$ correspond to subsets of $\mathbb{P}(\mathbb{P}(X))$ with different topologies; note that if distinct extensions correspond to the same set, then the extensions generate distinct topologies on the subset. If $A \subseteq \mathbb{P}(\mathbb{P}(X))$, then the number of topologies on A is no more than $|\mathbb{P}(\mathbb{P}(\mathbb{P}(\mathbb{P}(X))))|$. Thus,

$$|E(X)| \leqslant |\mathbb{P}(\mathbb{P}(\mathbb{P}(X)))| \cdot |\mathbb{P}(\mathbb{P}(\mathbb{P}(\mathbb{P}(X))))| = |\mathbb{P}(\mathbb{P}(\mathbb{P}(\mathbb{P}(X))))|.$$

■

In 7F an example of a space X is given such that $|H(X)| = |\mathbb{P}(\mathbb{P}(\mathbb{P}(\mathbb{P}(X))))|$.

Recall from Chapter 2 that a partially ordered set with joint (respectively, meets) is called an **upper semilattice** (respectively, **lower semilattice**). An upper (respectively, lower) semilattice with arbitrary joint (respectively, meets) is called **complete**. Thus, a complete lattice is a lattice with arbitrary joint and meets.

(g) <u>**Proposition**</u>. Let X be a space. Then

(1) $H(X)$ is a complete upper semilattice, and

(2) $H(X)$ (respectively, $E(X)$) is a complete lattice iff X is locally H-closed.

<u>Proof</u>

(1) Let $\mathcal{Q}$ be a nonempty subset of $H(X)$. By 4.1(g), $\mathcal{Q}$ has a

supremum, say Z, in $E(X)$. Since $H(X) \subseteq E(X)$, Z will be a supremum of $\mathfrak{a}$ in $H(X)$ if $Z \in H(X)$. Since $\kappa X \geq Y$ for all $Y \in \mathfrak{a}$ and $\kappa X \in E(X)$, it follows that $\kappa X \geq Z$. Thus Z is H-closed.

(2) By 2.1(e), $H(X)$ (respectively, $E(X)$) is complete iff $H(X)$ (respectively, $E(X)$) has a smallest element. By 7.3(e), this is equivalent to X being locally H-closed.

■

7.4 Partitions of $\sigma X \setminus X$

In this section we show that each H-closed extension of a space X induces a partition of $\sigma X \setminus X$ into compact sets, and that each partition of $\sigma X \setminus X$ into compact sets is induced by some H-closed extension of X. Already we have noted that κX and σX have the same underlying sets and that $\kappa X \setminus X$ is discrete while $\sigma X \setminus X$ is a zero-dimensional Tychonoff space. Thus, if Y is an H-closed extension of X, $f : \kappa X \to Y$ is the continuous function that fixes the points of X, and $p \in Y \setminus X$, then $f^{\leftarrow}(p)$ can be considered as a subset of $\sigma X \setminus X$. Denote the partition $\{f^{\leftarrow}(p) : p \in Y \setminus X\}$ of $\sigma X \setminus X$ by $P(Y)$. (Although this is the same notation as used for the set of P-points of Y, the context will always make clear which meaning is intended.)

(a) **Theorem**. Let X be a space. Then:

(1) if $Y \in H(X)$ then $P(Y)$ is a partition of $\sigma X \setminus X$ into compact sets, and

(2) if P is a partition of $\sigma X \setminus X$ into compact sets, then

there is a largest H-closed extension $Y(P)$ of X such that $P(Y(P)) = P$.

<u>Proof</u>

(1) Let $p \in Y \setminus X$ and $A = f^{\leftarrow}(p)$. By 7.2(c), it suffices to show that A is closed in σX. Suppose $y \in \sigma X \setminus A$. Then $f(y)$ and p are distinct points and contained in disjoint open sets. By the continuity of f, there are disjoint open sets U and V in κX such that $A \subseteq U$ and $y \in V$. Evidently $A \cap \text{int}_{\kappa X} c\ell_{\kappa X} V = \emptyset$. By 7.1(e)(3), $\text{int}_{\kappa X}(c\ell_{\kappa X} V)$ is open in σX. So, A is closed in σX.

(2) Let P be a partition of $\sigma X \setminus X$ into compact sets. Then $P \cup \{\{x\} : x \in X\}$ is a partition of κX that induces an equivalence relation R. Let $Y(P)$ be the quotient space $\kappa X/R$ and $f : \kappa X \to Y(P)$ the induced quotient function. Since $f^{\leftarrow}[f[A]] = A$ for $A \subseteq X$, $f[X]$ is open in $Y(P)$ and $f|X : X \to Y(P)$ is an embedding. Now, we show that $Y(P)$ is Hausdorff. Let $p, q \in Y(P)$ such that $p \neq q$. Now, $f^{\leftarrow}(p)$ and $f^{\leftarrow}(q)$ are disjoint compact subsets in σX and are therefore contained respectively in disjoint open subsets U and V of σX. Let $U' = f^{\leftarrow}(p) \cup (U \cap X)$ and $V' = f^{\leftarrow}(q) \cup (V \cap X)$. Then U' is open in κX; to see this let $\mathcal{U} \in f^{\leftarrow}(p)$. If $W \in \mathcal{U}$ then $\mathcal{U} \in c\ell_{\kappa X} W$ and $U \cap c\ell_{\kappa X} W \neq \emptyset$. Thus $(U \cap X) \cap W \neq \emptyset$; therefore, $U \cap X \in \mathcal{U}$ by the maximality of $\mathcal{U}$. Hence $\{\mathcal{U}\} \cup (U \cap X)$ is a κX-neighborhood of $\mathcal{U}$ contained in U', and so U' is open in κX. Similarly V' is open in κX. As $U' = f^{\leftarrow}[f[U']]$ and $V' = f^{\leftarrow}[f[V']]$, it follows that $f[U']$ and $f[V']$ are disjoint open sets in $Y(P)$ containing p and q respectively. Thus, $Y(P)$ is Hausdorff. Since $Y(P)$ is the continuous image of κX, $Y(P)$ is H-closed. So, $Y(P)$ is an H-closed extension of X. Now f is the unique continuous function from κX to $Y(P)$ that fixes points of X (here we identify $x \in X$ with

$\{x\} \in f[X]$); so, $P = P(Y(P))$. Finally, we need to show that if Z is an H-closed extension of X and $P(Z) = P(Y(P))$, then $Y(P) \geq Z$. If $g : \kappa X \to Z$ is the unique continuous function that fixes points of X, then define the function $h : Y(P) \to Z$ by $h(f(y)) = g(y)$ for each $Y \in \kappa X$. Thus $h \circ f = g$ and since f is a quotient function, h is continuous. Clearly, h fixes the points of Z; so, $Y(P) \geq Z$.

■

Using 7.4(a) we can make an interesting comparison between the construction of compactifications of Tychonoff spaces and the construction of H-closed extensions of Hausdorff spaces. In 4Q(5) we saw that if X is Tychonoff and αX is a compactification of X, then there is a one-to-one correspondence between $\{\delta X \in K(X) : \delta X \geq \alpha X\}$ and the collection of all upper semicontinuous decompositions D of βX into compact subsets with the property that if $x \in X$ then $\{x\} \in D$. In particular the collection of all such upper semicontinuous decompositions of βX is in one-to-one correspondence with $K(X)$. By contrast, if X is Hausdorff there is a many-to-one correspondence $Y \to P(Y)$ from $H(X)$ into the set of all decompositions D of σX into compact sets with the property that if $x \in X$ then $\{x\} \in D$. Upper semicontinuity is not necessary in this latter case. In the case where X is Tychonoff, this gives us some intuitive measure of the difference between $K(X)$ and $H(X)$.

We now give a corollary to 7.4(a).

(b) **Corollary**. Let X be a space. Then:

(1) If $Y_1, Y_2 \in H(X)$ and $P(Y_1) \neq P(Y_2)$, then Y_1 and Y_2 are not equivalent extensions of X.

(2) $|H(X)| \geq |\{P : P$ is a partition of $\sigma X \setminus X$ into compact subsets$\}|$.

(3) If X is an infinite discrete space, then $|H(X)| \geq |\mathbb{P}(\mathbb{P}(\mathbb{P}(X)))|$.

Proof. Both (1) and (2) are immediate from 7.4(a).

(3) First we show $|\kappa X| = |\mathbb{P}(\mathbb{P}(X))|$. Since X is discrete, then $\kappa X \setminus X = \{\mathcal{U} : \mathcal{U}$ is a free ultrafilter on $X\} = \beta X \setminus X$. By 4U $|\beta X \setminus X| = |\mathbb{P}(\mathbb{P}(X))|$. For an infinite set Y, the number of partitions of Y into finite subsets is precisely $|\mathbb{P}(Y)|$. Thus, $|H(X)| \geq |\mathbb{P}(\mathbb{P}(\mathbb{P}(X)))|$.

■

We can think of a subset of $\sigma X \setminus X$ as a set of free open ultrafilters on X. This viewpoint yields a characterization of compact subsets of $\sigma X \setminus X$.

(c) **Proposition**. Let X be a space and let $A \subseteq \sigma X \setminus X$. Then A is compact iff $\cap A$ is a free open filter on X and $A = \{\mathcal{U} \in \sigma X \setminus X : \cap A \subseteq \mathcal{U}\}$.

Proof. Let $S = \{\mathcal{U} \in \sigma X \setminus X : \cap A \subseteq \mathcal{U}\}$. Obviously $A \subseteq S$ always. Suppose A is compact and let $\mathcal{F} = \cap A$. Clearly, $\mathcal{F}$ is an open filter on X. Let $x \in X$. There are disjoint open sets U and V of σX with $A \subseteq U$ and $x \in V$. If $\mathcal{V} \in A$, then $U \cap X \in \mathcal{V}$ and so $U \cap X \in \mathcal{F}$. Now $x \in V \cap X$ and $(V \cap X) \cap (U \cap X) = \varnothing$. So, $\mathcal{F}$ is free. If $\mathcal{U} \in (\sigma X \setminus X) \setminus A$, as A is compact there are disjoint open subsets U and V of σX containing A and $\mathcal{U}$ respectively. Now $U \cap$

$X \in \mathcal{F}$ and $V \cap X \in \mathcal{U}$. But $(U \cap X) \cap (V \cap X) = \emptyset$ which implies that $\mathcal{F} \nsubseteq \mathcal{U}$. Thus $S \subseteq A$. Conversely, suppose $\cap A$ is a free open filter and $S \subseteq A$. Now, if $x \in X$, there is an open neighborhood U of x and $V \in \mathcal{F}$ such that $U \cap V = \emptyset$. Thus, $oU \cap oV = \emptyset$. Since $V \in \mathcal{F} \subseteq \mathcal{V}$ for all $\mathcal{V} \in A$, it follows that $A \subseteq oV$. Thus $x \notin c\ell_{\sigma X}A$. Likewise, if $\mathcal{U} \in \sigma X \setminus (X \cup A)$, then $\cap A \nsubseteq \mathcal{U}$. Hence there are open sets $U \in \mathcal{U}$ and $V \in \cap A$ such that $U \cap V = \emptyset$. So, $oU \cap oV = \emptyset$. Since $\mathcal{U} \in oU$ and $A \subseteq oV$, it follows that $\mathcal{U} \notin c\ell_{\sigma X}A$. Thus, A is closed in σX and by 7.2(c)(2), A is compact.

■

If X is a space, P is a partition of $\sigma X \setminus X$ into compact subsets, $A \in P$, and Y(P) is as described in 7.4(a)(2), then there is a point $y \in Y(P) \setminus X$ such that $\cap A = 0_{Y(P)}{}^{y}$. This proof of this fact is left to the reader; see 7E.

7.5 Minimal Hausdorff spaces

A space X is called **minimal Hausdorff** if X has no strictly coarser Hausdorff topology. In this section we characterize those spaces with minimal Hausdorff extensions, and construct such extensions.

(a) **Proposition**. Let X be a space. The following are equivalent:

(1) X is minimal Hausdorff,

(2) X is semiregular and H-closed, and

(3) every open filter with an unique adherent point converges.

Proof

(1) ⇒ (2) Suppose X is minimal Hausdorff. By 2.2(f)(1) X(s) is also Hausdorff. Hence X = X(s) and X is semiregular. Let F be a free open filter on X and x ∈ X. It is straightforward to check that $\{U \subseteq X : U \text{ open in } X \text{ and } x \notin U\} \cup F$ is a strictly coarser Hausdorff topology on X; this is a contradiction as X is minimal Hausdorff. So, X is H-closed by 4.8(b).

(2) ⇒ (1) Suppose X is semiregular and H-closed. Let X′ be X with a coarser Hausdorff topology. Then id : X → X′ is continuous. If U is open in X, then by 4.8(e), $c\ell_X U$ is H-closed. So, $id(c\ell_X U) = c\ell_X U$ is H-closed as a subspace of X′ by 4.8(h)(6); thus, $c\ell_X U$ is closed in X′. Since X is semiregular, $\{c\ell_X U : U \text{ open in } X\}$ is a closed base for X. Hence $id^{\leftarrow}$ is continuous, and so id is a homeomorphism. Thus X is minimal Hausdorff.

(1) ⇒ (3) Let (X,τ) be minimal Hausdorff. Let F be an open filter such that $a(F) = \{x\}$. Let $\{U \subseteq X : U \text{ open in } X \text{ and } x \notin U\} \cup F = \tau'$; then τ' is a Hausdorff topology on X contained in τ. Thus F converges (in(X,τ)) to x; if not τ' would be strictly contained in τ.

(3) ⇒ (1) Suppose X′ is X with a coarser Hausdorff topology. Let x ∈ X′, let F be the open neighborhood filter of x in X′, and let G be the open filter generated by F in X. Then $a_X(G) = \{x\}$. Thus G converges to x in X. Since this is true for each x ∈ X, X and X′ must have the same topology. Thus X is minimal Hausdorff.

■

Recall (see 4.8(j)) that a space is **Urysohn** if distinct points have disjoint closed neighborhoods.

(b) **Corollary**

(1) A space is compact iff it is minimal Hausdorff and Urysohn.

(2) The product of nonempty spaces is minimal Hausdorff iff each coordinate space is minimal Hausdorff.

(3) A space X is H-closed iff X(s) is minimal Hausdorff.

(4) If Y ∈ H(X) then Y(s) is a minimal Hausdorff extension of X(s).

Proof. (1) follows from 4.8(k) and (2) follows from the corresponding results about H-closed spaces (4.8(l)) and semiregular spaces (2.2(j)). If X is H-closed, then X(s) is H-closed by 4.8(h)(8) and is semiregular by 2.2(f)(5,6); therefore, by 7.5(a) X(s) is minimal Hausdorff. Conversely, if X(s) is minimal Hausdorff, then X(s) is H-closed by 7.5(a). So by 4.8(h)(8), X is H-closed. This establishes (3). Now (4) follows from (3) and 2.2(i).

■

By 7.5(b)(3) an H-closed space has a coarser minimal Hausdorff topology. In contrast, we now give an example of a Hausdorff space which has no coarser minimal Hausdorff topology.

(c) **Example**. Let $\mathbb{Q} = \{x_n : n \in \mathbb{N}\}$ be the rationals with the usual topology. We will show that no coarser topology is minimal Hausdorff. Let $\mathbb{Q}'$ be $\mathbb{Q}$ with a coarser Hausdorff topology. There is a nonempty open subset U_1 in $\mathbb{Q}'$ such that $x_1 \notin c\ell_{\mathbb{Q}'}U_1$. Since U_1 is also open in $\mathbb{Q}$, it is an infinite set; thus there is a nonempty open subset U_2 of $\mathbb{Q}'$ such that $U_2 \subseteq U_1$ and $x_2 \notin c\ell_{\mathbb{Q}'}U_2$. By induction, there is a nonempty open set U_n in $\mathbb{Q}'$ such that $x_n \notin c\ell_{\mathbb{Q}'}U_n$ and $U_1 \supseteq U_2 \supseteq \dots \supseteq U_n$. Since $\bigcap\{c\ell_{\mathbb{Q}'}U_n : n \in \mathbb{N}\} = \emptyset$, $\mathbb{Q}$ is not H-closed and hence, not minimal Hausdorff.

When X is H-closed, there is a coarser minimal Hausdorff topology on X, namely the topology of X(s). The next result reveals that more is true: the topology of X(s) is in fact the minimum Hausdorff topology contained in the topology of X.

(d) **Proposition**. If X is an H-closed space, then the topology on X(s) is the smallest of the Hausdorff topologies coarser than the topology of X.

Proof. If X' is X with a coarser Hausdorff topology, then the identity function $id : X \to X'$ is continuous. Arguing as in the proof that (2) implies (1) in 7.5(a), we see that $id^{\leftarrow} : X' \to X(s)$ is continuous.

■

A natural question to ask is which Hausdorff spaces can be embedded or densely embedded in a minimal Hausdorff space. By 2.2(i) semiregularity is hereditary on dense subspaces, so if a space X has a minimal Hausdorff extension, then X must be semiregular. The converse is also true; every semiregular space has a minimal Hausdorff extension.

(e) **Proposition.**

(1) A space can be densely embedded in a minimal Hausdorff space iff it is semiregular.

(2) Any space can be embedded as a closed nowhere dense subspace of a minimal Hausdorff space.

Proof

(1) As noted above, only one direction needs to be proved. Suppose X is semiregular. Then by 7.5(b)(4) $(\kappa X)(s)$ is a minimal Hausdorff extension of X(s). But X(s) = X.

(2) By 2G any space X can be embedded as a closed nowhere dense subspace of a semiregular space Y and κY can be embedded as a closed nowhere dense subspace of a semiregular space Z. But by (1), Z can be densely embedded in a minimal Hausdorff space W. Clearly $\text{int}_W X = \emptyset$. Since X is closed and nowhere dense in Y, it follows that $Y \setminus X$ is open and dense in Y; so, $\kappa Y \setminus X$ is open in κY. Thus, X is closed in κY. But κY is H-closed and, hence, closed in W. It follows that X is closed in W.

■

(f) **Definition.** Let X be a semiregular space. Then its minimal Hausdorff extension $(\kappa X)(s)$ is denoted by μX and is called the **Banaschewski-Fomin-Šanin extension** of X. The set of minimal Hausdorff extensions of X is denoted by $\mathcal{M}(X)$.

Note that in general, the extensions κX, σX, and μX of the semiregular space X are three pairwise non-equivalent extensions (see, for example, 7.7(g)). Unfortunately $M(X)$ may not have a projective maximum even though μX is an obvious candidate; however if $M(X)$ does have a projective maximum, then it is μX (this is shown in 7.7(g)). Now we show that $\mu\mathbb{N}$ is not a projective maximum in $M(\mathbb{N})$.

(g) **Example**. Let X be the noncompact H-closed space described in 4.8(d). It is straightforward to show that X is semiregular; so, X is minimal Hausdorff. Also, X has a countable discrete dense subspace and hence $X \in M(\mathbb{N})$. By 4L(6), $\mu\mathbb{N} \equiv_{\mathbb{N}} \beta\mathbb{N}$; so, $\mu\mathbb{N}$ is compact. If $\mu\mathbb{N} \geq X$, then X would be compact, which it is not.

(h) **Proposition**. Let X be a semiregular space. For each open set U in X, let oU be as in 7.2(b). Then:

(1) $\{oU : U \in RO(X)\}$ is a base for μX,

(2) $c\ell_{\mu X} oU = oU \cup (c\ell_X U)$ for $U \in RO(X)$,

(3) $c\ell_{\mu X}(bd_X U) = bd_X(U)$ for $U \in RO(X)$,

(4) $\mu X \setminus X = \sigma X \setminus X$, and

(5) $(\mu X \setminus X) \setminus oU = o(X \setminus c\ell_X U) \setminus X$ if $U \in RO(X)$.

Proof. The proof of (1) follows from 7.1(e)(3) and (2) follows from 2.2(f)(2) and 7.2(b)(1). By (2), $\mu X \setminus X \subseteq oU \cup o(X \setminus c\ell_X U)$ if $U \in RO(X)$; so, (3) follows. Since $\mu X = (\kappa X)(s)$, μX and σX have the same underlying set. As $\{oU \setminus X : U \in \tau(X)\}$ is a base for $\sigma X \setminus X$ and $\{oU \setminus X : U \in RO(X)\}$ is a base for $\mu X \setminus X$, it follows by 7.2(b)(2) that $\sigma X \setminus X = \mu X \setminus X$. Finally, (5) follows immediately from (2).

■

Note that by 7.5(h)(4) and 7.2(c)(1), it follows that $\beta(EX)\setminus EX$ and $\mu X\setminus X$ are homeomorphic.

7.6 p-maps

In Chapter 4 we established that if X and Y are Tychonoff spaces and $f \in C(X,Y)$, then f can be continuously extended to a function $\beta f \in C(\beta X, \beta Y)$ such that $\beta f|X = f$. Unfortunately the analogous result is not true for the Katětov extension; there exist spaces X and Y, and $f \in C(X,Y)$, for which there does not exist an $F \in C(\kappa X, \kappa Y)$ such that $F|X = f$ (see 5A for a specific example, and 7.6(k) for a class of examples). Our goal in this section is, for given spaces X and Y, to identify those functions in C(X,Y) which can be extended to a continuous function between the corresponding Katětov extensions. Also, we will identify the H-closed subspaces of an H-closed space.

(a) <u>**Definition**</u>

(1) An open cover of a space is a **p-cover** if there is a finite subfamily of the cover whose union is dense.

(2) A continuous function $f : X \to Y$ is a **p-map** if for each p-cover $\mathcal{Q}$ of Y, $f^{\leftarrow}(\mathcal{Q}) = \{f^{\leftarrow}[A] : A \in \mathcal{Q}\}$ is a p-cover of X.

Note that a continuous function with an H-closed domain is a p-map.

(b) **Theorem**. A continuous function $f : X \to Y$ has a continuous extension $F : \kappa X \to \kappa Y$ iff f is a p-map.

Proof. Suppose f has a continuous extension $F : \kappa X \to \kappa Y$ and $\mathcal{Q}$ is a p-cover of Y. Assume $f^{\leftarrow}(\mathcal{Q})$ is not a p-cover. Then $\{X \backslash cl_X f^{\leftarrow}(A) : A \in \mathcal{Q}\}$ has the finite intersection property and is contained in a free open ultrafilter $\mathcal{U}$ on X. If $F(\mathcal{U}) \in Y$, then there is $A \in \mathcal{Q}$ such that $F(\mathcal{U}) \in A$. Thus, $\mathcal{U} \in F^{\leftarrow}(A)$ and $X \cap F^{\leftarrow}(A) \in \mathcal{U}$. But $X \cap F^{\leftarrow}(A) = f^{\leftarrow}(A)$, and this is a contradiction as $X \backslash cl_X f^{\leftarrow}(A) \in \mathcal{U}$. So, $F(\mathcal{U}) = \mathcal{V} \in \kappa Y \backslash Y$. There is a finite family $\mathcal{F} \subseteq \mathcal{Q}$ such that $\cup\mathcal{F}$ is dense in Y. Since $\mathcal{V}$ is maximal, there exists $A \in \mathcal{F}$ such that $A \in \mathcal{V}$. So, $F^{\leftarrow}(\{\mathcal{V}\} \cup A)$ is open and contains $\mathcal{U}$. Hence, $f^{\leftarrow}(A) = F^{\leftarrow}(\{\mathcal{V}\} \cup A) \cap X \in \mathcal{U}$, and this too is a contradiction, as $X \backslash cl_X f^{\leftarrow}(A) \in \mathcal{U}$. This shows that $f^{\leftarrow}[\mathcal{Q}]$ is a p-cover.

Conversely, suppose f is a p-map. Let $\mathcal{U}$ be a free open ultrafilter on X, and let $\mathcal{G} = \{V \subseteq Y : V \text{ is open in } Y \text{ and } V \supseteq f(U) \text{ for some } U \in \mathcal{U}\}$. Clearly, $\mathcal{G}$ is an open filter on Y.

Case 1. $\mathcal{G}$ is a free open filter. Suppose $V \subseteq Y$ is open. Then for each $y \in bd_Y V$, there is an open neighborhood $W(y)$ of y such that $W(y) \notin \mathcal{G}$. Let $\mathcal{Q} = \{V, Y \backslash cl_Y V\} \cup \{W(y) : y \in bd_Y V\}$; then $\mathcal{Q}$ is a p-cover of Y. So by hypothesis $f^{\leftarrow}[\mathcal{Q}]$ is a p-cover of X. There is a finite subfamily $\mathcal{F}$ of $\mathcal{Q}$ such that $\cup f^{\leftarrow}[\mathcal{F}]$ is dense in X. Hence, there is an element $A \in \mathcal{F}$ such that $f^{\leftarrow}(A) \in \mathcal{U}$ by the maximality of $\mathcal{U}$. So, $A \in \mathcal{G}$. But, $A = V$ or $A = Y \backslash cl_Y V$. This shows that $\mathcal{G}$ is an open ultrafilter. Define $F(\mathcal{U}) = \mathcal{G}$.

Case 2. $\mathcal{G}$ is a fixed open filter. There is a point $z \in \bigcap\{c\ell_Y G : F \in \mathcal{G}\}$. Let V be an open neighborhood of z. For each $y \in bd_Y V$, let W(y) be an open neighborhood of y such that $z \notin c\ell_X W(y)$. Now $\mathcal{A} = \{V, Y \setminus c\ell_Y V\} \cup \{W(y) : y \in bd_Y V\}$ is a p-cover of Y. As in Case 1, there is $A \in \mathcal{A}$ such that $f^{\leftarrow}(A) \in \mathcal{U}$. Since $z \notin c\ell_Y W(y)$ and $z \notin c\ell_Y(Y \setminus c\ell_Y V)$, then $A = V$. By definition of $\mathcal{G}$, $V \in \mathcal{G}$. Hence $\mathcal{G}$ converges to z. Define $F(\mathcal{U}) = z$.

If $x \in X$, define $F(x) = f(x)$; so, $F : \kappa X \to \kappa Y$ is an extension of f. Since X (respectively, Y) is open in κX (respectively, κY), it follows that F is continuous at points of X. Let $\mathcal{U} \in \kappa X \setminus X$. If $F(\mathcal{U}) = \mathcal{G} \in \kappa Y \setminus Y$, let $G \in \mathcal{G}$. By the definition of F there exists $U \in \mathcal{U}$ such that $G \supseteq f[U]$. Thus $U \cup \{\mathcal{U}\}$ is a neighborhood of $\mathcal{U}$ in κX which is mapped into $G \cup \{\mathcal{G}\}$ by F, and so F is continuous at $\mathcal{U}$. If $F(\mathcal{U}) = z \in Y$, and V is a neighborhood in Y (and thus a basic neighborhood in κY) of z, then as noted above, $V \in \mathcal{G}$ so there exists $U \in \mathcal{U}$ such that $V \supseteq f[U]$. Thus $U \cup \{\mathcal{U}\}$ is a neighborhood of $\mathcal{U}$ in κX which is mapped into V by F, and again F is continuous at $\mathcal{U}$. Hence F is continuous.

■

(c) **Notation**. If $f : X \to Y$ is a p-map, its continuous extension $F : \kappa X \to \kappa Y$ (which is unique by 1.6(d)) is denoted by κf.

(d) **Proposition**. If $f \in C(X,Y)$ is either an open function or a dense embedding, then f is p-map.

Proof. Both types of functions have the property that if U is open and dense in Y, then $f^{\leftarrow}[U]$ is dense in X. So, if $\mathcal{A}$ is a p-cover of Y and F is a finite subfamily of $\mathcal{Q}$ such that $\bigcup F$ is dense, then $\bigcup f^{\leftarrow}(F)$ is dense; hence, $f^{\leftarrow}(\mathcal{Q})$ is a p-cover.

■

The compact subspaces of a compact space are easily identified; they are precisely its closed subspaces. However, the H-closed subspaces of an H-closed space are not precisely the closed subspaces of an H-closed space, as closed sets may not be H-closed; e.g., the closed subspace $\{(1/n,0) : n \in \mathbb{N}\}$ in the H-closed space X in 7.5(g) is not H-closed. Thus, the set of H-closed subspaces of an H-closed space lies somewhere between the set of closed subspaces and the set of regular closed subspaces. We now develop a characterization of H-closed subspaces using p-maps. First we prove two lemmas:

(e) __Lemma__. Let $f : X \to Y$ be a p-map and let $A \in R(X)$. If $\mathrm{int}_Y c\ell_Y f[A] = \emptyset$, then $c\ell_Y f[A]$ is H-closed.

__Proof__. Let C be an open cover of $c\ell_Y f[A]$. Write C as $\{U_i \cap c\ell_Y f[A] : i \in I\}$ where each U_i is an open subset of Y. As $\mathrm{int}_Y c\ell_Y f[A] = \emptyset$, $\{Y \setminus c\ell_Y f[A]\} \cup \{U_i : i \in I\}$ is a p-cover of Y. Thus by hypothesis $\{f^{\leftarrow}[Y \setminus c\ell_Y f[A]]\} \cup \{f^{\leftarrow}[U_i] : i \in I\}$ is a p-cover of X. Hence there is a finite subset F of I such that $f^{\leftarrow}[Y \setminus c\ell_Y f[A]] \cup \{f^{\leftarrow}[U_i] : i \in F\}$ is dense in X. Since $f^{\leftarrow}[Y \setminus c\ell_Y f[A]] \cap \mathrm{int}_X A = \emptyset$, it follows that $(\mathrm{int}_X A) \cap \bigcup\{f^{\leftarrow}[U_i] : i \in F\}$ is dense in $\mathrm{int}_X A$. Thus $f[\mathrm{int}_X A] \cap \bigcup\{U_i : i \in F\}$ is dense in $f[\mathrm{int}_X A]$. As $A \in R(X)$, $\mathrm{int}_X A$ is dense in A and it follows that $\bigcup\{U_i : i \in F\} \cap c\ell_Y f[A]$ is dense in $c\ell_Y f[A]$. Thus by 4.8(b) $c\ell_Y f[A]$ is H-closed.

■

(f) **Lemma**. Let $f : X \to Y$ be a p-map. If $A \in R(X)$, then $c\ell_Y f[A] = B \cup H$, where $B \in R(Y)$ and H is an H-closed nowhere dense subset of Y.

Proof. Let $B = c\ell_Y \mathrm{int}_Y c\ell_Y f[A]$; then $B \in R(Y)$. Let $H = c\ell_Y [c\ell_Y f[A] \backslash B]$. Then H is a closed nowhere dense subset of Y and $c\ell_Y f[A] = B \cup H$. It remains to show that H is H-closed. Let $c\ell_X [(\mathrm{int}_X A) \cap f^{\leftarrow}[Y \backslash B]] = E$. Then $E \in R(X)$ and it is straightforward to verify that $c\ell_Y f[E] = H$. It follows from that 7.6(e) that H is H-closed.

■

(g) **Theorem**. Let $A \subseteq X$ and $i : A \to X$ and $j : c\ell_X A \to X$ be the inclusion functions. Then:

(1) i is a p-map iff j is a p-map,

(2) i is a p-map iff $c\ell_X A = B \cup H$ where $B \in R(X)$ and H is an H-closed, nowhere dense subset of X,

(3) if X is H-closed, then i is a p-map iff $c\ell_X A$ is H-closed, and

(4) a subspace of an H-closed space Y is H-closed iff it is of the form $B \cup H$, where $B \in R(Y)$ and H is an H-closed nowhere dense subset of Y.

Proof

(1) Let $\mathcal{A}$ be p-cover of X. Let $\mathcal{F}$ be finite subfamily of $\mathcal{A}$. If $\cup\{U \cap A : U \in \mathcal{F}\}$ is dense in A, then it is dense in $c\ell_X A$. Conversely, suppose $D = \cup\{U \cap c\ell_X A : U \in \mathcal{F}\}$ is dense in $c\ell_X A$; since D is open in $c\ell_X A$, it follows that $D \cap A$ is dense in $c\ell_X A$. Hence, $\cup\{U \cap A : U \in \mathcal{F}\}$ is dense in A.

(2) By 7.6(f) above, one direction is immediate. For the other, suppose $A = B \cup H$ where $B \in R(X)$ and H is an H-closed nowhere dense subset of X. Let $\mathcal{Q}$ be a p-cover of X. There is a finite family $\mathcal{F} \subseteq \mathcal{Q}$ such that $\cup\mathcal{F}$ is dense in X. So, $\mathrm{int}_X B \cap (\cup\mathcal{F})$ is dense in $\mathrm{int}_X B$ which implies that $\mathrm{int}_X B \cap (\cup\mathcal{F})$ is dense in $c\ell_X(\mathrm{int}_X B) = B$. So $\cup\{U \cap A : U \in \mathcal{F}\}$ is dense in B. Also, $\{U \cap H : U \in \mathcal{Q}\}$ has a finite subfamily $\mathcal{G} \subseteq \mathcal{Q}$ such that $\cup\{U \cap H : U \in \mathcal{G}\}$ is dense in H. Thus, $\cup\{U \cap A : U \in \mathcal{F} \cup \mathcal{G}\}$ is dense in A. This implies that $i^{\leftarrow}(\mathcal{Q})$ is a p-cover.

(3) If i is a p-map, then by (2), $c\ell_X A$ is H-closed since regular closed subspaces are H-closed and finite unions of H-closed subspaces are H-closed. Conversely if $c\ell_X A$ is H-closed, then j is p-map since the domain of j is H-closed. By (1), i is p-map.

(4) This follows immediately from (2) and (3).

■

The problem of determining which of the closed subspaces of an H-closed space are H-closed is thus reduced to determining which of the closed, nowhere dense subspaces are H-closed. It also follows from (g) above that if every closed nowhere dense subspace of an H-closed space X is H-closed, then every closed subspace of X is H-closed; by 4S(6) it also follows that X is compact.

(h) **Corollary**. If A is a closed subspace of X, then the inclusion function $i : A \to X$ is a p-map iff $c\ell_{\kappa X}A$ and κA are equivalent extensions of A.

Proof. If i is a p-map, then i has a continuous extension $\kappa i : \kappa A \to \kappa X$. Thus, $\kappa i(\kappa A) = c\ell_{\kappa X}A$ is H-closed. Let $p \in c\ell_{\kappa X}A \setminus A$ and 0^p be the neighborhood trace of p on A relative to the extension $c\ell_{\kappa X}A$ of A. Now, p is an open ultrafilter on X and $0^p = \{U \cap A : U \in p\}$. Let $\mathcal{U}$ be an open ultrafilter on A such that $\mathcal{U} \supseteq 0^p$. Now A is open in $c\ell_{\kappa X}A$ and $c\ell_{\kappa X}A$ has the simple extension topology; thus to show that $c\ell_{\kappa X}A \equiv_A \kappa A$ it suffices to show that $0^p = \mathcal{U}$. Now, $A = c\ell_X(\mathrm{int}_X A) \cup H$ where H is an H-closed, nowhere dense subspace. Thus, each member of 0^p and $\mathcal{U}$ intersects $\mathrm{int}_X A$ and $\{V \cap \mathrm{int}_X A : V \in p\} = \{V \cap \mathrm{int}_X A : V \in 0^p\} \subseteq \{U \cap \mathrm{int}_X A : U \in \mathcal{U}\}$. But $\{V \cap \mathrm{int}_X A : V \in p\}$ and $\{U \cap \mathrm{int}_X A : U \in \mathcal{U}\}$ are both open ultrafilters on $\mathrm{int}_X A$, and hence they are equal. As $\mathrm{int}_X A \in \mathcal{U}$, it follows that $0^p = \mathcal{U}$. Conversely, suppose $f : \kappa A \to c\ell_{\kappa X}A$ is a homeomorphism such that $f(x) = x$ for $x \in A$. Thus, $f|A = i$. Hence, $f : \kappa A \to \kappa X$ is a continuous extension of i. By 7.6(b), i is p-map.

■

We conclude this section by finding conditions on a space X that are equivalent to the bijection $id_X : D(X) \to X$ being a p-map (here D(X) denotes the underlying set of X equipped with the discrete topology). Since id_X is obviously continuous, this will lead us to examples of continuous functions that are not p-maps. We need several preliminary lemmas.

(i) Lemma. The following are equivalent for a space X:

(1) X is compact and

(2) if λ is an ordinal and if $\{C(\alpha) : \alpha < \lambda\}$ is a family of nonempty closed subsets of X such that $\alpha_1 < \alpha_2 < \lambda$ implies that $C(\alpha_1) \supseteq C(\alpha_2)$, then $\cap\{C(\alpha) : \alpha < \lambda\} \neq \emptyset$.

Proof

(1) $\Rightarrow$ (2) If F is a finite subset of λ then $\cap\{C(\alpha) : \alpha \in F\} = C(\max F) \neq \emptyset$. Thus $\{C(\alpha) : \alpha < \lambda\}$ has the finite intersection property, and so $\cap\{C(\alpha) : \alpha < \lambda\} \neq \emptyset$.

(2) $\Rightarrow$ (1) Let $\mathcal{C} = \{\mathcal{A} \subseteq \mathbb{P}(X) : \mathcal{A}$ is a collection of closed subsets of X with F.I.P., and $\cap\mathcal{A} = \emptyset\}$. If (1) fails, then $\mathcal{C} \neq \emptyset$ so we can let $\lambda = \min\{\delta : \delta$ is a cardinal and there exists $\mathcal{A} \in \mathcal{C}$ such that $|\mathcal{A}| = \delta\}$. Choose $\mathcal{A} \in \mathcal{C}$ such that $|\mathcal{A}| = \lambda$ and write $\mathcal{A} = \{A_\alpha : \alpha < \lambda\}$. For each $\alpha < \lambda$ let $C_\alpha = \cap\{A_\beta : \beta < \alpha\}$. Then $\cap\{C_\alpha : \alpha < \lambda\} = \cap\mathcal{A} = \emptyset$. By (2) there exists $\sigma < \lambda$ such that $C_\sigma = \emptyset$. Then $\{A_\beta : \beta < \sigma\} \in \mathcal{C}$, which violates the minimality of λ. Thus (1) cannot fail if (2) holds.

■

(j) Lemma. The following are equivalent for a space X:

(1) each closed nowhere dense subset of X is compact, and

(2) the set of non-isolated points of X is compact.

Proof

(2) $\Rightarrow$ (1) Each closed nowhere dense subset of X is a

subset of the set of non-isolated points of X. If this latter set is compact, closed nowhere dense sets will also be compact.

(1) $\Rightarrow$ (2) Let $\{F(\alpha) : \alpha < \lambda\}$ be a family of closed nonempty subsets of the set S of non-isolated points of X such that $\alpha_1 < \alpha_2 < \lambda$ implies $F(\alpha_1) \supseteq F(\alpha_2)$. By the previous lemma, to show S is compact it suffices to show that $\cap\{F(\alpha) : \alpha < \lambda\} \neq \varnothing$.

If there exists $\alpha_0 < \lambda$ such that $F(\alpha) \cap (F(\alpha_0)\setminus int_X F(\alpha_0)) \neq \varnothing$ for each $\alpha < \lambda$ with $\alpha_0 < \alpha$, then $\{F(\alpha) \cap (F(\alpha_0)\setminus int_X F(\alpha_0)) : \alpha < \lambda\}$ is a collection of closed subsets of the closed nowhere dense (and therefore compact) space $F(\alpha_0)\setminus int_X F(\alpha_0)$ with the finite intersection property. Thus $\cap\{F(\alpha) \cap (F(\alpha_0)\setminus int_X F(\alpha_0)) : \alpha < \lambda\} \neq \varnothing$ and we are done. Hence for each $\alpha < \lambda$ we may assume there exists $\delta(\alpha)$ such that $\alpha < \delta(\alpha) < \lambda$ and $F(\delta(\alpha)) \subset int_X F(\alpha)$ (for if $F(\mu) = int_X F(\alpha)$ for each μ satisfying $\delta(\alpha) \leqslant \mu < \lambda$, then $\cap\{F(\alpha) : \alpha < \lambda\} = int_X F(\alpha) \neq \varnothing$ and we are done). Thus we can find a cofinal subset $\{\mu(\alpha) : \alpha < \lambda\}$ of λ such that $\alpha_1 < \alpha_2 < \lambda$ implies $\mu(\alpha_1) < \mu(\alpha_2)$ and $F(\mu(\alpha+1)) \subset int_X F(\mu(\alpha))$ for each $\alpha < \lambda$.

For each $\alpha < \lambda$ choose $a(\alpha) \in int_X F(\mu(\alpha))\setminus F(\mu(\alpha+1))$ and let $A = \{a(\alpha) : \alpha < \lambda\}$. If $\alpha < \lambda$ then $\{a(\alpha)\} = A \cap (int_X F(\mu(\alpha))\setminus F(\mu(\alpha+1)))$; therefore, A is a discrete subset of X. Since $A \subseteq S$, no point of A is isolated in X. It easily follows that $c\ell_X A$ is closed and nowhere dense in X. Thus $c\ell_X A$ is compact. But $\{F(\alpha) \cap c\ell_X A : \alpha < \lambda\}$ is a collection of closed subsets of $c\ell_X A$ with the finite intersection property (since $\{\mu(\alpha) : \alpha < \lambda\}$ is cofinal in λ). Thus $\cap\{F(\alpha) \cap c\ell_X A : \alpha < \lambda\} \neq \varnothing$ and we are done.

■

(k) <u>**Theorem**</u>. Let X be a space. The following are equivalent:

(1) $id_X : D(X) \to X$ is a p-map,

(2) each p-cover of X has a finite subcover,

(3) the set of non-isolated points of X is compact, and

(4) each closed nowhere dense subset of X is compact.

<u>Proof</u>

(1) ⇒ (4) First suppose that $id_X : D(X) \to X$ is a p-map, and let S be a closed nowhere dense subset of X. Then $f^{\leftarrow}[S] \in R(D(X))$, and by 7.6(f) $S = B \cup H$ where $B \in R(X)$ and H is H-closed and nowhere dense. As $int_X S = \varnothing$, $B = \varnothing$ and hence S is H-closed. Every closed subset of S is closed nowhere dense in X so by the above argument every closed subset of S is H-closed. It follows from 4S(6) that S is compact.

(4) ⇒ (3) This is the hard direction of 7.6(j).

(3) ⇒ (2) Let I(X) denote the set of isolated points of X and let C be a p-cover of X. Let F_1 be a finite subfamily of C such that $\cup F_1$ is dense in X. Then $I(X) \subseteq \cup F_1$. As $X \backslash I(X)$ is compact, there is a finite subfamily F_2 of C that covers $X \backslash I(X)$. Thus $F_1 \cup F_2$ is a finite subcover of C.

(2) ⇒ (1) This is obvious.

■

7.7 An equivalence relation on $H(X)$

In this section we use some of the tools developed in 7.5 to investigate further the partitions of the remainder of the Fomin extension introduced in 7.4. In particular we develop a necessary and sufficient condition for two H-closed extensions of a space X to induce the same partition of $\sigma X \setminus X$. We begin by defining an equivalence relation on $H(X)$ induced by the partitions of $\sigma X \setminus X$ into compact subsets.

(a) **Definition**. Two H-closed extensions Y_1 and Y_2 of a space X are said to be **Θ-equivalent** if $P(Y_1) = P(Y_2)$. Let $H^\theta(X) = \{Y(P(Z)) : Z \in H(X)\}$. (See 7.4(a) and the paragraph before 7.4(a) for notation.)

Obviously Θ-equivalence is an equivalence relation on $H(X)$. Evidently $H^\theta(X)$ is a set of H-closed extensions with one representative (the largest, in fact) from each Θ-equivalence class of H-closed extensions of X. Before we establish a theorem characterizing Θ-equivalent H-closed extensions, a couple of lemmas are needed. Let us recall that if $Y \in H(X)$, then by 7.5(b)(4) Y(s) is a minimal Hausdorff extension of X(s).

(b) **Lemma**. Let $Z \in H(X)$ for a space X. Then there is a continuous bijection $h : Y(P(Z)) \to Z$ that leaves X pointwise fixed; also, h is a Θ-homeomorphism and $(Y(P(Z)))(s)$ and Z(s) are equivalent extensions of X(s).

Proof. Let $f : \kappa X \to Y(P(Z)))$ and $g : \kappa X \to Z$ be the Katětov functions of Y(P(Z)) and Z, respectively. As in the proof of 7.4(a)(2), there is a continuous $h : Y(P(Z)) \to Z$ that leaves X pointwise fixed and satisfies $h \circ f = g$. Let $y \in Z \setminus X$; then $g^{\leftarrow}(y) \in P(Z) = P(Y(P(Z))) = \{f^{\leftarrow}(z) : z \in Y(P(Z)) \setminus X\}$. There is unique $z \in Y(P(Z)) \setminus X$ such that $g^{\leftarrow}(y) = f^{\leftarrow}(z)$. Hence, $h^{\leftarrow}(y) = \{z\}$ and h is a bijection. Since $\{h^{\leftarrow}(U) : U \text{ open in } Z\}$ is a coarser Hausdorff topology on the H-closed space $Y(P(Z))$, then by 7.5(d), $h^{\leftarrow} : Z \to Y(P(Z))(s)$ is continuous. Thus, by 4.8(h)(1,7), $h^{\leftarrow} : Z \to Y(P(Z))$ is Θ-continuous. As h is continuous, h is a Θ-homeomorphism. Evidently, h is a homeomorphism from $Y(P(Z))(s)$ onto $Z(s)$.

■

(c) __Lemma__. Let X be a space. Then:

(1) there is a continuous bijection $t : \kappa X \to \kappa(X(s))$ that leaves X pointwise fixed and

(2) if $Y \in \mathcal{H}(X)$, and $f : \kappa X \to Y$ and $g : \kappa(X(s)) \to Y(s)$ are the Katětov functions (see 4.8(o)) for Y and $Y(s)$ respectively, and $s : Y \to Y(s)$ is the identity function, then $s \circ f = g \circ t$.

__Proof__

(1) Let $j : X \to X(s)$ be the identity map on the underlying set of X; it is continuous. It is easy to see that if V is a dense open subset of $X(s)$ then V is dense in X, and from this it follows that j is a p-map. Hence by 7.6(b) there is a continuous function $\kappa j : \kappa X \to \kappa(X(s))$ such that $\kappa j | X = j$. As $j[X] = X(s)$, and $\kappa j[\kappa X]$ is an H-closed subset of $\kappa(X(s))$ containing X, it follows that κj maps κX onto $\kappa(X(s))$. It remains to show that κj is one-to-one.

If $\mathcal{U} \in \kappa X \setminus X$, let $\mathcal{G}_{\mathcal{U}} = \{V \subseteq X(s) : V \text{ is open in } X(s) \text{ and } U \subseteq V \text{ for some } U \in \mathcal{U}\}$. As noted in the proof of 7.6(b), $\mathcal{G}_{\mathcal{U}}$ is an open ultrafilter on $X(s)$, and because $\mathcal{U}$ is free and $R(X) = R(X(s))$

it is evident that $G_{\mathcal{U}}$ is also free. Thus $\kappa j(\mathcal{U}) = G_{\mathcal{U}}$ (see 7.6(b)). If $\mathcal{U}_1 \neq \mathcal{U}_2$ find $U_i \in \mathcal{U}_i$ $(i = 1,2)$ such that $U_1 \cap U_2 = \emptyset$. Then $int_X cl_X U_i \in G_i$ $(i = 1,2)$ and $int_X cl_X U_1 \cap int_X cl_X U_2 = \emptyset$; therefore, $G_1 \neq G_2$. Thus κj is the desired bijection t.

(2) Note that $s \circ f$ and $g \circ t$ both are continuous functions from κX to Y(s). Let h denote the function embedding X(s) in Y(s). Also $s \circ f | X = h \circ j$ because X, with the subspace topology inherited from Y(s), is just X(s) (see 2.2(i)). Meanwhile, $g \circ t | X = h \circ j$. By 1.6(d) $s \circ f = g \circ t$.

■

(d) **Theorem**. Let X be a space and $Y_1, Y_2 \in H(X)$. The following are equivalent:

(1) Y_1 and Y_2 are θ-equivalent,

(2) there is a θ-homeomorphism $m : Y_1 \to Y_2$ that leaves X pointwise fixed,

(3) $Y_1(s)$ and $Y_2(s)$ are equivalent extensions of X(s), and

(4) there is a continuous bijection from $Y(P(Y_1))$ onto Y_2 that leaves X pointwise fixed.

Proof

(1) ⇒ (2) Since $P(Y_1) = P(Y_2)$, it follows that $Y(P(Y_1)) \equiv_X Y(P(Y_2))$. By 7.7(b), there are θ-homeomorphisms $f_i : Y(P(Y_i)) \to Y_i$, $i = 1,2$, that leave X pointwise fixed. Thus, $f_2 \circ f_1^{\leftarrow} : Y_1 \to Y_2$ is a θ-homeomorphism that leaves X pointwise fixed.

(2) ⇒ (3) By 4.8(s)(5), $m : Y_1(s) \to Y_2(s)$ is a homeomorphism. Since $m(x) = x$ for $x \in X(s)$, then $Y_1(s)$ and $Y_2(s)$ are equivalent extensions of X(s).

(3) ⇒ (4) By 7.7(b) and our hypothesis, there is a homeomorphism h between $(Y(P(Y_1)))(s)$ and $Y_2(s)$ that leaves X(s) pointwise fixed. Let $f : \kappa X \to Y(P(Y_1))$, $f' : \kappa(X(s)) \to (Y(P(Y_1)))(s)$, $g : \kappa X \to Y_2$, and $g' : \kappa(X(s)) \to Y_2(s)$ be the Katětov functions for $Y(P(Y_1))$, $(Y(P(Y_1)))(s)$, Y_2, and $Y_2(s)$ respectively (see diagram). Let t be the continuous bijection defined in 7.7(c) and let $s_1 : Y(P(Y_1)) \to (Y(P(Y_1)))(s)$ and $s_2 : Y_2 \to Y_2(s)$ be the identity functions. Since $h \circ f' | X(s) = g' | X(s)$, then $h \circ f' = g'$. By 7.7(c) (see the following diagram), $h \circ s_1 \circ f = s_2 \circ g$. Thus, if $k = s_2^{\leftarrow} \circ h \circ s_1 : Y(P(Y_1)) \to Y_2$, then $k \circ f = g$. Since $Y(P(Y_1))$ has the quotient topology induced by f, then $k : Y(P(Y_1)) \to Y_2$ is continuous. This completes the proof of (4).

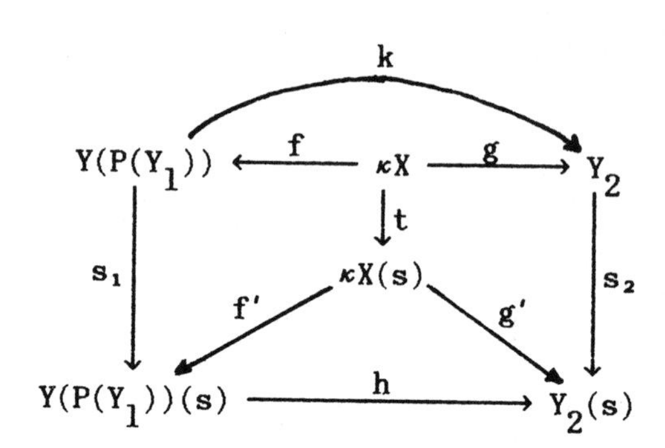

(4) ⇒ (1) Let $k : Y(P(Y_1)) \to Y_2$ be the continuous bijection that leaves X pointwise fixed. Let $f : \kappa X \to Y(P(Y_1))$ and $g : \kappa X \to Y_2$ be the Katetov functions of $Y(P(Y_1))$ and Y_2, respectively (see 4.8(o)). Thus, $k \circ f = g$. Since k is a bijection, it follows that $P(Y(P(Y_1))) = P(Y_2)$. But $P(Y(P(Y_1))) = P(Y_1)$ by 7.4(a).

So, Y_1 and Y_2 are Θ-equivalent.

■

If P_1 and P_2 are partitions of a set T, recall from 1.2(j) that P_1 **refines** P_2 (denoted by $P_1 \leqslant P_2$) if for each $A \in P_1$ there exists $B \in P_2$ such that $A \subseteq B$.

(e) **Proposition**. Let X be a space and $Y_1, Y_2 \in H(X)$. Then:

(1) if $Y_1 \geqslant Y_2$, then $P(Y_1) \leqslant P(Y_2)$ (here, and in (2), "$\leqslant$" refers to partition refinement and "$\geqslant$" refers to the order on $H(X)$),

(2) if $P(Y_1) \leqslant P(Y_2)$ then $Y(P(Y_1)) \geqslant Y(P(Y_2))$,

(3) $H^\theta(X)$ is a complete upper semilattice,

(4) $M(X(s)) = \{Y(s) : Y \in H(X)\}$,

(5) the function from $H^\theta(X)$ to $M(X(s))$ that sends Y to Y(s) is a bijection, and

(6) if X is infinite and discrete, then $|M(X)| = |H^\theta(X)| = |\mathbb{P}(\mathbb{P}(\mathbb{P}(X)))|$.

Proof

(1) Let $h : Y_1 \to Y_2$ be the continuous extension of the identity function on X. Let $f_i : \kappa X \to Y_i$ be the Katětov function of Y_i for $i = 1,2$. Since $h \circ f_1 | X = f_2 | X$, it follows that $h \circ f_1 = f_2$. If $y \in Y_1 \backslash X$, then $f_1^{\leftarrow}(y) \subseteq f_1^{\leftarrow}[h^{\leftarrow}(h(y))] = f_2^{\leftarrow}(h(y))$. So, $P(Y_1) \leqslant P(Y_2)$.

(2) If $P(Y_1) \leqslant P(Y_2)$, there is a function $g \in F[Y(P(Y_1)), Y(P(Y_2))]$ such that $g(x) = x$ for $x \in X$ and if $f_i : \kappa X \to Y_i$ is the

Katetov function for Y_i, $i = 1,2$, then $g \circ f_1 = f_2$. Since $Y(P(Y_1))$ has the quotient topology induced by f_1, it follows that g is continuous. So, $Y(P(Y_1)) \geq Y(P(Y_2))$.

(3) Let $\emptyset \neq \mathcal{Q} \subseteq H^\theta(X)$. By 7.3(g), $\mathcal{Q}$ has a supremum in $H(X)$, say T. Let $Y = Y(P(T))$. Thus, $Y \in H^\theta(X)$. Since $Y \geq T$, then $Y \geq Z$ for all $Z \in \mathcal{Q}$. Suppose $S \in H^\theta(X)$ and $S \geq Z$ for all $Z \in \mathcal{Q}$. Then $S \geq T$. So, $Y(P(S)) \geq Y(P(T)) = Y$ by (2). But $S \in H^\theta(X)$ implies $Y(P(S)) = S$ by 7.4(a). This shows that Y is the supremum of $\mathcal{Q}$ in $H^\theta(X)$.

(4) By 7.5(b)(4), $M(X(s)) \supseteq \{Y(s) : Y \in H(X)\}$. Let $Z \in M(X(s))$; we will produce $Y \in H(X)$ such that $Y(s) = Z$. Let τ denote the topology of Z. Define τ' as follows: $\tau' = \{U \subseteq Z : U \cap X$ is open in X and if $p \in U \setminus X(s)$ then there exists $V \in \tau$ such that $p \in V$ and $V \cap X \subseteq U \cap X\}$. It is straightforward to check that τ' is a topology on Z that contains τ and induces the topology of X on X(s). Thus (Z,τ') is an extension of X. We will show that $(Z,\tau')(s) = (Z,\tau)$. Then by 7.5(b) $(Z,\tau') \in H(X)$ and hence $(Z,\tau) \in \{Y(s) : Y \in H(X)\}$. Denote (Z,τ') by T. To show $T(s) = Z$ it will suffice to show that $R(Z) = R(T)$.

Let $A \in R(Z)$. Then there exists an open subset V of X(s) that $A = c\ell_Z V$. Since $\tau \subseteq \tau'$, it follows that $c\ell_T V \subseteq c\ell_Z V$. Conversely, let $p \in c\ell_Z V$ and let U be a τ'-neighborhood of p. Then there exists $W \in \tau$ such that $p \in W$ and $W \cap X \subseteq U \cap X$. As $p \in c\ell_Z V$, it follows that $W \cap V \neq \emptyset$; therefore, $W \cap V \cap X \neq \emptyset$. As $W \cap X \subseteq U \cap X$, $U \cap V \neq \emptyset$; thus $p \in c\ell_T V$. Hence $A = c\ell_Z V = c\ell_T V \in R(T)$. Therefore, $R(Z) \subseteq R(T)$. Conversely, if $B \in R(T)$ then $B = c\ell_T V$ where V is open in X. Let W =

$\mathrm{int}_X c\ell_X V$, which is open in X(s). Clearly B $= c\ell_T W$; as the argument above shows that $c\ell_T(c\ell_{X(s)} W) = c\ell_Z W$, it follows that B $\in R(Z)$. Thus $R(T) = R(Z)$ and as noted above, this completes the proof.

(5) This is immediate from 7.7(d).

(6) By the proof of 7.4(b)(3), $|H^\theta(X)| \geq |\mathbb{P}(\mathbb{P}(\mathbb{P}(X)))|$. Now $|H^\theta(X)| = |\{P : P$ is a partition of $\sigma X \setminus X$ into compact subsets$\}|$. As noted in the proof of 7.4(b)(3), $|\sigma X \setminus X| = |\beta X \setminus X| = |\mathbb{P}(\mathbb{P}(X))|$. But the number of partitions of an infinite set A is $\leq |\mathbb{P}(A \times A)| = |\mathbb{P}(A)|$. So, $|H^\theta(X)| = |\mathbb{P}(\mathbb{P}(\mathbb{P}(X)))|$.

■

Let X be a semiregular space. By 7.7(e)(5), $M(X) = \{Y(s) : Y \in H^\theta(X)\}$. However, $M(X)$ with the usual ordering $\leq$ inherited from $E(X)$ is not necessarily a complete upper semilattice as there is no largest element when X $= \mathbb{N}$ (see 7.5(g) and the following 7.7(g)). Since there is a bijection between $H^\theta(X)$ and $M(X)$ by 7.7(e), and $H^\theta(X)$ is a complete upper semilattice, the order induced on $M(X)$ from $H^\theta(X)$ will be a complete upper semilattice (see 7G). These facts together imply that the bijection defined in 7.7(e)(5) is **not** necessarily an order isomorphism from $H^\theta(X)$ (with the order inherited from $H(X)$) onto $M(X)$ (with the order inherited from $E(X)$).

(f) **Proposition**. Let X be a semiregular space. Then:

(1) if $(M(X),\leq)$ has a largest element, it is μX, and

(2) for Y $\in H(X)$, Y(s) is the least element in the equivalence class of H-closed extensions Θ-equivalent to Y.

<u>Proof</u>

(1) If $Y \in \mathcal{M}(X)$, then $\kappa X \geq Y$. By 7.7(e)(1), $P(\kappa X) \leq P(Y)$. But $P(\mu X) = P(\kappa X)$; so $P(\mu X) \leq P(Y)$ for all $Y \in \mathcal{M}(X)$. If $Z \in \mathcal{M}(X)$ and $Z \geq \mu X$, then by 7.7(e), $P(Z) \leq P(\mu X)$. So, $P(Z) = P(\mu X)$. By 7.7(d) $Z(s)$ $(=Z)$ and $\mu X(s)$ $(=\mu X)$ are equivalent extensions of $X(s)$ $(=X)$.

(2) This is immediate from 7.7(d) and 7.5(d).

■

We conclude this section with an example of a space X such that $\mu X \not\equiv_X \sigma X$.

(g) **<u>Example</u>**. Let $X = E\mathbb{R}$. Since X is extremally disconnected, then $c\ell_X U \cap c\ell_X V = \emptyset$ whenever U and V are disjoint open sets of X. Thus by 7B, $\mu X \equiv_X \beta X$. But if $k_{\mathbb{R}} : E\mathbb{R} \to \mathbb{R}$ is the perfect irreducible continuous surjection defined in 6.6(e), then $A = k_{\mathbb{R}}^{\leftarrow}(\mathbb{N})$ is a noncompact closed nowhere dense subspace of X. Consequently $X \setminus A \in \mathcal{U}$ for each $\mathcal{U} \in \sigma X \setminus X$ and so $(X \setminus A) \cup (\sigma X \setminus X)$ is an open set of σX that is disjoint from A. Hence A is closed in σX. Since A is noncompact, σX is not compact, so $\sigma X \not\equiv_X \mu X$.

Chapter 7 - Problems

7A. Strict extensions. Let Y be an extension of X.

(1) Show that Y is a strict extension of X iff $\{c\ell_Y A : A \subseteq X\}$ is a closed base for Y.

(2) If Y is a strict extension of X, prove that $\pi w(Y) = \pi w(X)$ (see 2N for definition).

(3) Give an example of an extension Y of a space X such that $w(Y) \neq w(X)$ (see 2N for definition).

7B. The equivalence of the Stone-Čech compactification with H-closed extensions.

(1) If X is Tychonoff, show that $\kappa X \geqslant \sigma X \geqslant \mu X \geqslant \beta X$. (Hint: 4.8(n), 7.2(b)(1), 7.5(d), and 2.2(g)(2).)

(2) Let X be Tychonoff and $f : \kappa X \to \beta X$ the unique continuous function such that $f(x) = x$ for $x \in X$. Prove the following are equivalent:

(a) f is one-to-one,

(b) κX is Urysohn, and

(c) $(\kappa X)(s) \equiv_X \beta X$.

(Hint: See 4.8(k).)

A closed set A in a space X is said to be **regularly**

nowhere dense if there are disjoint open sets U, V in X such that $A \subseteq c\ell_X U \cap c\ell_X V$.

(3) Prove that X is Tychonoff and $\mu X \equiv_X \beta X$ iff X is semiregular and every regularly nowhere dense set in X is compact. (Hint: In proving one direction, it suffices, by (2), to show that κX is Urysohn. It might be helpful to use 7.2(a)(1) and 2.2(f)(2).)

(4) If $bd_X U$ is compact for every open set $U \subseteq X$, prove that X is normal. (Hint: First, show that if $F \subseteq X$ is closed, then $bd_X F$ is compact.)

(5) Prove the following are equivalent:

(a) the space X is Tychonoff and $\sigma X \equiv_X \beta X$,

(b) σX is regular, and

(c) for each open set $U \subseteq X$, $bd_X U$ is compact.

Note that by 7.6(j), (c) is equivalent to the set of non-isolated points of X being compact.

(6) If X is Tychonoff and $\kappa X \equiv_X \beta X$, prove that X is locally compact and $\kappa X \backslash X$ is finite.

(7) If X is locally compact and the set of non-isolated points of X is compact, prove that $X = X_1 \cup X_2$ where X_1 and X_2 are disjoint closed sets in X (i.e., $X = X_1 \oplus X_2$), X_1 is compact and X_2 is discrete.

(8) Prove that if X is Tychonoff, then $\kappa X \equiv_X \beta X$ iff X is compact. (Hint: If $\kappa X \equiv_X \beta X$, use the previous

problems and 7.6(j) to deduce that $X = X_1 \oplus X_2$ where X_1 is compact and X_2 is discrete. Show that $\kappa X \equiv_X X_1 \oplus \kappa X_2$. If X_2 is infinite, then $|\kappa X \setminus X| = |\kappa X_2 \setminus X_2|$, which is infinite since X_2 is infinite and discrete. This contradicts (6). Now, deduce that X is compact.)

7C. S-equivalent H-closed extensions. Let X be a space and Y, Z $\in$ $H(X)$. Define Y and Z to be **S-equivalent** if $Y^\# \equiv_X Z^\#$.

(1) Prove the following are equivalent:

(a) Y is S-equivalent to Z,

(b) $Y^+ \equiv_X Z^+$,

(c) $\{0^p : p \in Y\} = \{0^p : p \in Z\}$, and

(d) $Y^+ \geqslant Z \geqslant Y^\#$.

For each space X, let $H^+(X) = \{Y^+ : Y \in H(X)\}$ and let $H^\#(X) = \{Y^\# : Y \in H(X)\}$.

(2) If Y and Z are S-equivalent H-closed extensions of X, show that Y and Z are Θ-equivalent (see 7.7). (Hint: Use (1)(d).)

(3) Show that $H^\theta(X) \subseteq H^+(X)$, and if X is semiregular, show that $M(X) \subseteq H^\#(X)$. (Hint: Use (2) to show that Y(P(Y)) is a simple extension of X for each Y $\in$ $H(X)$.)

(4) If D is an infinite discrete space, prove that $|H^+(D)|$

$= |H^{\#}(D)| = |\mathbb{P}(\mathbb{P}(\mathbb{P}(D)))|$. (Hint: Use (1)(c), 7.7(e)(6), and (3).)

7D. <u>One-point H-closed extensions and locally H-closed spaces.</u> Let X be locally H-closed and not H-closed. Let $\hat{X} = X \cup \{\infty\}$ where $\infty \notin X$, and let G be the intersection of all free open filters on X. Let $X^{\#}$ denote $\hat{X}$ with the topology $\{U : U$ open in $X\} \cup \{\{\infty\} \cup U : U \in G\}$.

(1) Prove that G is a free open filter on X and that $X^{\#}$ is a one-point H-closed extension of X. (Hint: Use local H-closedness to prove that G is free (and hence $X^{\#}$ is Hausdorff), and the open filter characterization appearing in 4.8(b)(3) to prove that $X^{\#}$ is H-closed.)

Let H be the intersection of all free open ultrafilters on X. Let X^{+} denote $\hat{X}$ with the topology $\{U : U$ open in $X\}$ $\cup \{\{\infty\} \cup U : U \in H\}$.

(2) Prove that H is a free open filter on X and that X^{+} is a one-point H-closed extension of X. (Hint: See the hint for (1).)

(3) (a) Prove that $H = \{U \subseteq X : U$ is open in X and $X \setminus \mathrm{int}_X(c\ell_X U)$ is H-closed$\}$; this is true for an arbitrary space X. (Hint: If U is an open subset of X and $c\ell_X U$ is not H-closed, show that U is contained in some free open ultrafilter on X.)

(b) Prove that $\mathcal{H} \cap RO(X)$ is a free open filter base on X. (Hint: Use that $\mathcal{H}$ is free and the fact that for an open set $U \subseteq X$, $c\ell_X(\text{int}_X(c\ell_X U)) = c\ell_X U$.)

(4) (a) Prove that $\mathcal{G}$ is the open filter generated by $\mathcal{H} \cap RO(X)$. (Hint: Use 2.3(k) to prove that $\mathcal{H} \cap RO(X) \subseteq \mathcal{G}$.)

(b) Prove that $\mathcal{G}$ is the open filter generated by the open filter base $\{U \subseteq X : U \text{ is open and } X \backslash U \text{ is H-closed}\}$.

(5) Let Y denote a one-point H-closed extension of X with $p \in Y \backslash X$. Prove $\mathcal{G} \subseteq \mathcal{O}^p \subseteq \mathcal{H}$. In particular, prove X^+ (respectively, $X^\#$) is the projective maximum (respectively, minimum) of the set of one-point H-closed extensions of X. Thus we have $X^+ \geqslant Y \geqslant X^\#$. (Hint: In proving $\mathcal{O}^p \subseteq \mathcal{H}$, show that if $\mathcal{U}$ is an open ultrafilter on X, then $\{W \subseteq X^+ : W \text{ is open in } X^+ \text{ and } W \cap X \in \mathcal{U}\}$ is an open ultrafilter on X^+.)

(6) Prove that X has a unique one-point H-closed extension iff every nowhere dense closed set is contained in a H-closed subspace of X. (Hint: By (5) it suffices to show that $\mathcal{G} = \mathcal{H}$ iff every nowhere dense closed set of X is contained in some H-closed subspace of X.)

(7) Find a non-H-closed space in which every closed, nowhere dense set is H-closed.

(8) Prove 7.3(b). (Hint: In proving that X being open in σX implies that X is locally H-closed, it is helpful to

use 7.2(b)(1) and 4.8(e).)

Let T be a space and let $Y \in H(T)$ (respectively, $E(T)$). The Y is a **projective minimal** element of $H(T)$ (respectively, $E(T)$) if $Z \in H(T)$ (respectively, $E(T)$) and $Y \geqslant Z$ imply $Y \equiv_T Z$.

(9) For a space T, prove that $H(T)$ (respectively, $E(T)$) has a projective minimum iff $H(T)$ (respectively, $E(T)$) has a projective minimal element, (cf. 7.3(e)).

(10) Let $\{X_a : a \in I\}$ be a family of nonempty spaces. Prove that $\Pi\{X_a : a \in I\}$ is locally H-closed iff all but a finite number of the spaces $\{X_a : a \in I\}$ are H-closed and X_a is locally H-closed for each $a \in I$.

7E. Partitions of the remainder of the Fomin extension. Let X be a space and P be a partition of $\sigma X \setminus X$ by compact subsets of $\sigma X \setminus X$.

(1) If $A \in P$, prove there is some $y \in Y(P) \setminus X$ such that $\cap A = 0^y$.

(2) If $y \in Y(P) \setminus X$, prove there is $A \in P$ such that $\cap A = 0^y$.

7F. Cardinality of the lattice of H-closed extensions of a discrete space. Let D be an infinite discrete space. Recall, by 4U, that $|\beta D| = |\mathbb{P}(\mathbb{P}(D))|$. Let $\mathcal{F}$ be an ultrafilter on $\beta D \setminus D$. Since $\beta D \setminus D$ is compact, $\mathcal{F}$ converges to a unique point $x_{\mathcal{F}} \in \beta D \setminus D$. By 7B(2,3), we can assume that κD and βD have the same underlying set.

(1) Prove that $\{U \subseteq \kappa D : U$ open in κD and $x_{\mathcal{F}} \in U$ implies $F \subseteq U$ for some $F \in \mathcal{F}\}$ is a Hausdorff topology on κD. Let $\kappa_{\mathcal{F}} D$ denote κD with this new topology.

(2) Prove that $\tau(\kappa D) \supseteq \tau(\kappa_{\mathcal{F}} D) \supseteq \tau(\sigma D)$ and that $\kappa_{\mathcal{F}} D$ is an H-closed extension of D. (Hint: Use the fact that $\sigma D \equiv_D \beta D$ by 7B(5) and that $\sigma D = (\kappa D)^{\#}$ by definition of σD.)

(3) Prove that every point of $\kappa_{\mathcal{F}} D \setminus D$ is isolated in $\kappa_{\mathcal{F}} D \setminus D$ except for $x_{\mathcal{F}}$ and the neighborhood system of $x_{\mathcal{F}}$ in $\kappa_{\mathcal{F}} D \setminus D$ is $\mathcal{F}$.

(4) If $\mathcal{F}$ and $\mathcal{G}$ are ultrafilters on $\beta D \setminus D$, prove that $\kappa_{\mathcal{F}} D \equiv_D \kappa_{\mathcal{G}} D$ iff $\mathcal{F} = \mathcal{G}$.

(5) Prove that $|\mathcal{H}(D)| = |\mathbb{P}(\mathbb{P}(\mathbb{P}(\mathbb{P}(D))))|$. (Hint: To prove one direction, prove that $|\mathcal{H}(D)| \geqslant |\{\mathcal{F} : \mathcal{F}$ is an ultrafilter on $\beta D \setminus D\}|$ and use 4U to note that $|\{\mathcal{F} : \mathcal{F}$ is an ultrafilter on $\beta D \setminus D\}| = |\mathbb{P}(\mathbb{P}(\beta D \setminus D))|$.)

<u>Note</u>. For an infinite discrete space D, we have established in this problem that $|\mathcal{H}(D)| = |\mathbb{P}(\mathbb{P}(\mathbb{P}(\mathbb{P}(D))))|$. By 7C and 7.7(e), it follows that $|\mathcal{H}^{+}(D)| = |\mathcal{H}^{\#}(D)| = |\mathcal{H}^{\theta}(D)| = |\mathcal{M}(D)| = |\mathbb{P}(\mathbb{P}(\mathbb{P}(D)))|$ and

by 4AA(13), $|K(D)| = |\mathbb{P}(\mathbb{P}(D))|$. Also, we know that $H(D) \supseteq H^+(D) \supseteq H^\theta(D)$ and $H(D) \supseteq H^\#(D) \supseteq M(D) \supseteq K(D)$.

7G. Another order on $M(X)$. Let X be a semiregular space. For Y_1, $Y_2 \in M(X)$, define $Y_1 \trianglelefteq Y_2$ if $Y(P(Y_1)) \leqslant Y(P(Y_2))$.

(1) Show that $(M(X), \trianglelefteq)$ is a complete upper semilattice.

(2) Show that $Y(P(\mu X)) \equiv_X \kappa X$.

(3) Show that $\mu X \trianglerighteq Y$ for all $Y \in M(X)$.

This last fact asserts that in $M(X)$ with this new partial ordering, μX is the projective maximum.

7H. Compact, irreducible surjections with compact or H-closed domain.

(1) Prove that if $f : X \to Y$ is a perfect irreducible surjection and if Y is H-closed, then X is H-closed.

(2) Let X be a regular space, Y a set, and $f \in F(X,Y)$ a compact, irreducible surjection. Prove that $\{f[A] : A$ closed in $X\}$ is a closed base for a semiregular Hausdorff topology on Y and $f : X \to Y$ is a perfect, irreducible, Θ-continuous surjection. In particular, if X is compact, show that Y is minimal Hausdorff. (Hint: First use the regularity of X to prove that f is Θ-continuous. Then use 6.5(d) to show that Y (which is easily shown to be Hausdorff) is also semiregular.)

(3) Let X be an H-closed space. Show that the set $A \subseteq X$ is H-closed iff there is a compact subspace B of EX such that $k_X|B \in \Theta C(B,A)$ and $k_X[B] = A$. (Hint: Find $B \subseteq EX$ so that $k_X|B$ is perfect and irreducible (see 1.8(f)and 6.5(c)). Then use (2) and (1) to conclude that $k_X|B$ is Θ-continuous and that B is compact.)

(4) Use (3) and 7.6(i) to prove that a space X is compact if every closed set is H-closed. (Hint: Let $\{A_\alpha : \alpha < \lambda\}$ be a chain of nonempty closed sets in X for some ordinal λ. Use (3) to find a compact subset B_0 of EX such that $k_X|B_0 \in \Theta C(B_0,A_0)$ and $k_X[B_0] = A_0$. Suppose $\{B_\alpha : \alpha < \gamma\}$ is defined for some $\gamma < \lambda$ where $\{B_\alpha : \alpha < \gamma\}$ is a decreasing chain of compact sets such that $k_X|B_\alpha \in \Theta C(B_\alpha,A_\alpha)$ and $k_X[B_\alpha] = A_\alpha$ for $\alpha < \gamma$. If $\gamma = \beta + 1$, let $B = B_\beta$ and $A = A_\beta$; if γ is a limit ordinal, let $B = \cap\{B_\alpha : \alpha < \gamma\}$ and $A = \cap\{A_\alpha : \alpha < \gamma\}$. Now B is compact and $k_X[B] = A \supseteq A_\gamma$. Apply (2) and (3) to obtain a compact subset $B_\gamma \subseteq B$ such that $k_X[B_\gamma] = A_\gamma$ and $k_X|B_\gamma \in \Theta C(B_\gamma,A_\alpha)$.)

7I. Fully disconnected H-closed extensions. A space X is **fully disconnected** if $\{p\} = \cap\{B \in \mathcal{B}(X) : p \in B\}$ for each point $p \in X$.

(1) Prove that a space X is fully disconnected iff X(s) is fully disconnected.

(2) Prove that a fully disconnected, minimal Hausdorff space is compact and zero-dimensional.

(3) Using (2), conclude that a space X is fully disconnected and H-closed iff X(s) is zero-dimensional and compact.

(4) Show that the space Z described in 4.8(q)(3) is fully disconnected and H-closed but is not zero-dimensional.

(5) Use the technique of 2G(3) to embed the space Z of (4) in a semiregular space Y. Show that Y is fully disconnected and semiregular but is not zero-dimensional.

(6) If a space X has a fully disconnected, H-closed extension, prove that X(s) is zero-dimensional.

(7) Give an example of a fully disconnected, Tychonoff space which has no fully disconnected, H-closed extension.

For the remainder of this problem, X will denote a space such that X(s) is zero-dimensional. Let $H_0(X) = \{Y \in H(X) : Y \text{ is fully disconnected}\}$. Let $\kappa_0 X$ denote the set $X \cup \{\mathcal{U} : \mathcal{U}$ is a free maximal $B(X)$-filter$\}$.

(8) Prove that $\{U \subseteq \kappa_0 X : U \cap X$ is open in X and if $\mathcal{U} \in U \setminus X$, then for some $A \in \mathcal{U}$, $A \subseteq c\ell_X(U \cap X)\}$ is a topology on $\kappa_0 X$.

(9) Let $C \in B(X)$. Prove that $c\ell_{\kappa_0 X} C \cap c\ell_{\kappa_0 X}(X \setminus C) = \varnothing$; in particular, $c\ell_{\kappa_0 X} C \in B(\kappa_0 X)$.

(10) Prove that $\kappa_0 X$ is the projective maximum of $H_0(X)$ and

that $(\kappa_0 X)(s) \equiv_{X(s)} \beta_0(X(s))$.

(11) Prove that $\kappa X \equiv_X \kappa_0 X$ iff each regular open subset of $X(s)$ has a compact boundary in $X(s)$.

An open cover C of a space T is a **po-cover** if there is a finite family $C_1, C_2, \ldots, C_n$ of C and clopen sets $B_1, B_2, \ldots, B_n$ such that $B_i \subseteq c\ell_T C_i$ for $1 \leqslant i \leqslant n$ and $T = \cup\{B_i : 1 \leqslant i \leqslant n\}$. A function $f \in C(T,S)$, for spaces T and S, is a **po-map** if for each po-cover C of S, $\{f^{\leftarrow}[C] : C \in C\}$ is a po-cover of T.

(12) Let Y be a space such that Y(s) is zero-dimensional and let $f \in C(X,Y)$. Prove that there is a continuous function $F \in C(\kappa_0 X, \kappa_0 Y)$ such that $F|X = f$ iff f is a po-map. (Hint: See the proof of 7.6(b).)

7J. <u>A generalization of 5.8(b)</u>. Let P be a topological property such that there is some $T \in P$ for which $|T| \geqslant 2$. A space Y is defined to be **P-closed** if $Y \in \text{Reg}(P)$ and whenever Y is a subspace of a space $Z \in P$, then Y is closed in Z. Throughout this problem, we will assume that P has this property: if $X \in \text{Reg}(P)$, then there is an extension Y of X such that Y is P-closed and $X \subseteq Y^+ \subseteq \kappa X$ (Y^+ is the simple extension of X corresponding to Y).

(1) Show that the property "Urysohn" satisfies this condition. (Hint: First note that $P = \text{Reg}(P)$ in this

case (see 4K). Let X be a Urysohn space and let $\mathcal{M} \subseteq \kappa X \setminus X$ be a maximal subset with respect to this property: if $\mathcal{U}, \mathcal{V} \in \mathcal{M}$ and $\mathcal{U} \neq \mathcal{V}$, there exist $U \in \mathcal{U}$ and $V \in \mathcal{V}$ such that $c\ell_X U \cap c\ell_X V = \emptyset$. Let $Y = X \cup \mathcal{M}$, viewed as a subspace of κX. Show that Y is Urysohn and "Urysohn-closed".)

(2) Prove that each space in Reg(P) is a closed subspace of some P-closed space. (Hint: First note that the Cantor space (see 3.3) is P-regular. Conclude that the one-point compactification of $\mathbb{N}$ is P-regular. Now argue as in 5.8(b).)

(3) Prove that if P is an extension property, then P = Reg(P).

7K. <u>Tychonoff's embedding of a Hausdorff space in an H-closed space</u>. Let ~ be an equivalence relation on the space X whose equivalence classes are closed subsets of X. Let X(~) denote the set of equivalence classes induced by ~. Define $p : X \to X(\sim)$ by $p(x) = [x]$ for each $x \in X$, where $[x]$ denotes the (unique) equivalence class containing x. If $A \subseteq X$, let $p^{\#}(A) = \{y \in X(\sim) : p^{\leftarrow}(y) \subseteq A\}$.

(1) Show that $\{p^{\#}(U) : U \in \tau(X)\}$ is a base for a topology (not necessarily Hausdorff) on X(~). Note that this topology need not be the quotient topology induced on X(~) by p, and p need not be continuous with respect to it.

(2) If X is normal, show that $X(\sim)$ is Hausdorff.

Let $S = \{x \in X : \{x\} \in X(\sim)\}$.

(3) Show that $p[S \cap A] \subseteq p^{\#}(A)$ for $A \subseteq X$.

(4) If S is dense in X and U is open in X, prove that $p^{\leftarrow}[c\ell_{X(\sim)}p^{\#}[U]] \supseteq c\ell_X U$; in particular, p is Θ-continuous.

(5) If X is H-closed and S is dense in X, prove that every open cover of $X(\sim)$ has a finite subfamily whose union is dense.

Let $\{U_a : a \in A\}$ be a base for X, $I_a = [0,1]$ with the usual topology for each $a \in A$, and $Y = \Pi\{I_a : a \in A\}$. For $x \in X$, define $X(x) \subseteq Y$ by

$$X(x)(a) = \begin{cases} \{0\} & \text{if } x \in U_a \\ \{1\} & \text{if } x \in X \backslash c\ell_X U_a \\ I_a & \text{otherwise} \end{cases}$$

and $X(x) = \Pi\{X(x)(a) : a \in A\}$. Note that for $a \in A$, $\Pi_a(X(x)) = X(x)(a)$.

(6) If x and y are distinct points of X, show that $X(x) \cap X(y) = \emptyset$. Let $S = Y \backslash \cup\{X(x) : x \in X\}$ and $F = \{X(x) : x \in X\} \cup \{\{y\} : y \in S\}$.

(7) Show that S is dense in Y.

(8) Show that F is a partition of Y into closed sets.

Let $\sim$ be the equivalence relation on Y whose set of equivalence classes is F.

(9) Prove that $Y(\sim)$ is H-closed (remember to verify that it is Hausdorff).

Define $e : X \to Y(\sim)$ by $e(x) = p(X(x))$ where $p : Y \to Y(\sim)$ is as defined at the beginning of the problem.

(10) For each $a \in A$, prove that $e[U_a] = p^{\#}[\Pi\{Z_b : b \in A\}] \cap e[X]$ where $Z_b = I_b$ for $b \neq a$ and $Z_a = [0,1/4)$.

(11) Prove that e is an embedding of X into $Y(\sim)$.

This problem establishes that an arbitrary space can be embedded in an H-closed space; however, the embedding is not necessarily dense. Since a closed subspace of an H-closed space is not necessarily H-closed, the closure of e[X] is not necessarily H-closed and thus this embedding does not necessarily give rise to a dense embedding of X in an H-closed space.

7L. A generalization of semiregularity. Let A and M be subsets of a space X. A is said to be **regular open relative to** M if $A = \text{int}_X(A \cup (M \cap c\ell_X A))$. Let $\alpha_M(A)$ denote $\text{int}_X(A \cup (M \cap c\ell_X A))$. Note that $\alpha_X(A) = \text{int}_X c\ell_X(A)$ and $\alpha_M(A) \in \tau(X)$.

(1) Prove that $\alpha_M(A) = \alpha_X(A) \cap \text{int}_X(A \cup M)$.

(2) Let $A \subseteq B \subseteq X$. Prove that $\alpha_M(A) \subseteq \alpha_M(B)$.

(3) Prove that $\alpha_M(\alpha_M(A)) = \alpha_M(A)$.

(4) If A is closed, prove that $\alpha_M(A) = \alpha_X(A)$.

(5) If A and $B \in \tau(X)$, prove that $\alpha_M(A \cap B) = \alpha_M(A) \cap \alpha_M(B)$.

(6) Prove that $\alpha_M(A) = \alpha_X(A)$ iff $\alpha_X(A)\backslash A \subseteq M$.

(7) If M is regular open and $A \subseteq M$, prove that $\alpha_M(A) = \alpha_X(A)$.

By (5), $\{\alpha_M(U) : U \in \tau(X)\}$ is a basis for a topology on X; this space is denoted by X_M. Note that if $M \in RO(X)$, $\tau(X(s)) = \tau(X_X) \subseteq \tau(X_M) \subseteq \tau(X_\emptyset) = \tau(X)$. The space X is said to be **semiregular relative to** M if $\tau(X) = \tau(X_M)$.

Let Y be an extension of X.

(8) Let W be an open set in Y. Prove that $\alpha_{Y\backslash X}(W) = o(W \cap X)$.

(9) Prove that Y is a strict extension of X iff Y is semiregular relative to $Y\backslash X$.

Note that (9) and 7A(1) give two characterizations of strict extensions.

7M. <u>Submaximal and maximal H-closed spaces</u>. Let X be a set, let τ be a topology on X, and let $\tau(s)$ denote the semiregularization of τ. Let $\mathcal{D} = \{D \subseteq X : D \text{ is dense in } (X,\tau(s))\}$ and let

$E(\tau) = \{\sigma : \sigma$ is a topology on X and $\sigma(s) = \tau(s)\}$. If $\sigma \in E(\tau)$ let $M(\sigma) = D \cap \sigma$.

(1) Prove that if $\sigma \in E(\tau)$, then $M(\sigma)$ is a filter base on $(D,\subseteq)$ that is closed under finite intersection.

(2) If $\sigma \in E(\tau)$, let $\lambda(\sigma)$ denote the filter on $(D,\subseteq)$ generated by $M(\sigma)$. Prove that the topology on X generated by $\lambda(\sigma) \cup \tau(s)$ is σ.

(3) Let F be a filter on $(D,\subseteq)$ and let $\phi(F)$ be the topology on X generated by $\tau(s) \cup F$. Prove that $\phi(F) \in E(\tau)$.

(4) Let $\mathbb{F}$ denote the collection of filters on $(D,\subseteq)$. Prove that $\lambda : E(\tau) \to \mathbb{F}$ (as defined in (2)) is one-to-one and if $\sigma_1, \sigma_2 \in E(\tau)$ with $\sigma_1 \subseteq \sigma_2$, then $\lambda(\sigma_1) \subseteq \lambda(\sigma_2)$.

(5) Prove that if $\sigma \in E(\tau)$ then $\phi(\lambda(\sigma)) = \sigma$. Prove that if $F \in \mathbb{F}$ then $\lambda(\phi(F)) \supseteq F$.

(6) Prove that if $\sigma \in E(\tau)$ and $V \in \sigma \cup \tau$, then $\text{int}_\sigma c\ell_\sigma V = \text{int}_\tau c\ell_\tau V$.

(7) Prove that each element of the poset $(E(\tau),\subseteq)$ is contained in a maximal element of $(E(\tau),\subseteq)$. (Hint: Use (6) to show that nonempty chains in $(E(\tau),\subseteq)$ have upper bounds, and apply Zorn's lemma.)

A space (X,τ) is said to be **submaximal** if τ is a maximal member of $E(\tau)$.

(8) Prove that if F is a maximal filter on D then $(X,\phi(F))$

is a submaximal space.

(9) Let Y be a space. Prove that Y is a submaximal space iff every dense set of Y is open. (Hint: For one direction, use (6).)

(10) Prove that a subspace of a submaximal space is submaximal.

(11) Prove that each subspace of a submaximal space is the intersection of a closed set and an open set, i.e., is locally closed.

(12) Let K be an infinite discrete space and Y an extension of K. Prove that Y is a submaximal space iff Y is a simple extension of K.

(13) Give an example of a connected, submaximal space. (Hint: Start with a connected space Y and let F be a maximal filter on D. Let σ be the topology generated by $\tau(Y) \cup F$. Show that (Y,σ) is a connected, submaximal space.)

An H-closed space Y is said to be **maximal H-closed** if, whenever σ is a topology on Y such that (Y,σ) is H-closed and $\sigma \supseteq \tau(Y)$, then $\sigma = \tau(Y)$.

(14) Prove that a space is maximal H-closed iff it is H-closed and every dense set is open. (Hint: By 4.8(h)(8), it follows that a maximal H-closed space is submaximal. Conversely, suppose that Y is H-closed and submaximal, and σ is an H-closed topology on Y such that $\sigma \supseteq \tau(Y)$. If $U \in RO(Y,\sigma)$ and $V = Y\backslash c\ell_{\tau(Y)}U$, show

that the σ-closure and τ-closure of U are the same and hence infer that $U = Y \backslash cl_Y V$. Note that id : $(Y,\sigma) \to (Y,\tau(Y))$ is a continuous bijection, and infer from 4.8(e) and the above that $RO(Y,\sigma) = RO(Y,\tau(Y))$. Conclude that $\sigma(s) = \tau(Y)(s)$.)

7N. H-closed extensions with relatively zero-dimensional remainder. Let X be a space. An open base B for X is called a **b-basis** for X if: (i) $X \in B$; (ii) if $U, V \in B$, then $U \cap V \in B$; and (iii) if $U \in B$, then $X \backslash cl_X U \in B$. So a FG-base (see 4W) is a b-base. Let $B_0 = B \cap RO(X)$.

(1) Prove that B_0 is a Boolean subslgebra of $RO(X)$.

(2) If W is a free open filter, prove that $W \cap B$ is a free B-filter. Then show that a B-filter F is a maximal B-filter iff for each $B \in B$, either $B \in F$ or $X \backslash cl_X B \in F$.

Let $M = \{F : F \text{ is a free maximal } B\text{-filter}\}$ and $X(M) = X \cup M$. If $U \in B$, let $G(U) = U \cup \{F \in M : U \in F\}$.

(3) Prove that $\tau(X) \cup \{U \cup \{F\} : U \in F, F \in M\}$ is a basis for a Hausdorff topology on $X(M)$.

Let $\kappa_B X$ denote $X(M)$ with the topology generated by $\tau(X)$

$\cup \{U \cup \{F\} : U \in F, F \in M\}$.

(4) Prove that $\kappa_B X$ is an H-closed extension of X. (Hint: Use (2).)

(5) If U, V $\in$ B, prove that $G(U \cap V) = G(U) \cap G(V)$.

Let $\sigma_B X$ denote $X(M)$ with the topology generated by $\{G(U) : U \in B\}$.

(6) Prove that $\sigma_B X$ is an H-closed extension of X and $w(\sigma_B X) \leqslant |B|$ (see 2N for the definition of weight).

(7) Let U $\in$ B. Prove that $bd_Y G(U) \subseteq X$ where $Y = \sigma_B X$. (Hint: Show that $c\ell_Y G(U) = c\ell_X U \cup G(U)$.)

Recall from 4X that an extension Z of X is said to have **relatively zero-dimensional remainder** if $\{W \subseteq Z : W \text{ open and } bd_Z W \subseteq X\}$ is a basis for Z. Note that in this case $Z \backslash X$ is a zero-dimensional subspace.

(8) Let Z be an extension of X with relatively zero-dimensional remainder. Prove that Z is a strict extension of X. (Hint: Let $W \subseteq Z$ be open such that $bd_Z W \subseteq X$. Prove that $W = o(W \cap X)$.)

(9) Let Z be an H-closed extension of X with relatively zero-dimensional remainder. Let $\mathcal{Q} = \{U \in \tau(Z) : bd_Z U \subseteq X\}$ and $C = \{U \cap X : U \in \mathcal{Q}\}$. Prove that

(a) $\mathcal{Q}$ is a b-basis for Z,

(b) $\mathcal{C}$ is a b-basis for X, and

(c) $Z \equiv_X \sigma_{\mathcal{C}} X$.

(10) Prove that $\kappa_{\mathcal{B}} X \equiv_X \sigma_{\mathcal{B}} X$ iff $\mathcal{M}$ is finite. (Hint: One direction is easy. If $\kappa_{\mathcal{B}} X \equiv_X \sigma_{\mathcal{B}} X$, then for each $\mathcal{F} \in \mathcal{M}$, there is an open set $U_{\mathcal{F}} \in \mathcal{B}$ such that $\mathcal{F} \in G(U_{\mathcal{F}}) = U_{\mathcal{F}} \cup \{\mathcal{F}\}$. For $x \in X$, there is an open set $U_x \in \mathcal{B}$ such that $x \in G(U_x) \subseteq X$. Consider the open cover $\{U_{\mathcal{F}} \cup \{\mathcal{F}\} : \mathcal{F} \in \mathcal{M}\} \cup \{U_x : x \in X\}$ of the H-closed space $\kappa_{\mathcal{B}} X$.)

(11) Let $\tau = \tau(X)$. Prove that τ is a b-basis, $\sigma_\tau X = \sigma X$, and $\kappa_\tau X = \kappa X$.

(12) If X is semiregular, prove that $\mu X \equiv_X \sigma_{RO(X)} X$. (Hint: Use 7.1(e)(3).)

(13) Let D be an infinite discrete space and $CF(D) = \{F \subseteq X : F \text{ is finite or } D \backslash F \text{ is finite}\}$ (see 3.1(e)(5)). Prove that $CF(D)$ is a b-basis for D, that $\sigma_{CF(D)} D$ is a one-point extension of D, and $\kappa_{CF(D)} D \equiv_D \sigma_{CF(D)} D$.

70. <u>Characterizations of the Katětov and Fomin H-closed extensions.</u> Let Y be an extension of X. The space X is said to be **hypercombinatorially embedded in** Y if whenever F and H are closed sets in X and $F \cap H$ is nowhere dense in X, then $c\ell_Y F \cap c\ell_Y H = F \cap H$.

(1) Suppose that X is hypercombinatorially embedded in Y.

Prove that 0^p is an open ultrafilter for each $p \in Y\setminus X$. (Hint: Let $U \in \tau(X)$. To show that either $U \in 0^p$ or $X\setminus c\ell_X U \in 0^p$, let $F = c\ell_X U$ and $H = c\ell_X(X\setminus c\ell_X U)$. Show that $p \notin c\ell_Y F$ or $p \notin c\ell_Y H$. If $p \notin c\ell_Y F$, show that $X\setminus c\ell_X U \in 0^p$.)

(2) If $\mathcal{U}$ is an open ultrafilter on X, $A \subseteq X$ is closed, and $\mathcal{U}$ meets A (i.e., $U \cap A \neq \emptyset$ for each $U \in \mathcal{U}$), show that $\mathcal{U}$ meets $\text{int}_X A$. (Hint: Note that $X\setminus \text{bd}_X A \in \mathcal{U}$ as $\text{bd}_X A$ is nowhere dense.)

(3) Suppose that X is hypercombinatorially embedded in Y. Let $F_1, \ldots, F_n$ be closed sets in X for $n \in \mathbb{N}$ such that $\cap\{F_i : 1 \leqslant i \leqslant n\}$ is nowhere dense. Prove that $\cap\{c\ell_Y F_i : 1 \leqslant i \leqslant n\} = \cap\{F_i : 1 \leqslant i \leqslant n\}$. (Hint: Use (1) and (2).)

(4) If X is hypercombinatorially embedded in Y, prove that $\kappa X \geqslant Y \geqslant \sigma X$. (Hint: Use (1) and 7.1(h).)

(5) Prove that Y is a simple extension of X iff $Y\setminus X$ is a closed discrete subspace of Y.

(6) Let Y be H-closed. Prove that $Y \equiv_X \kappa X$ iff $Y\setminus X$ is a closed, discrete subspace of Y and X is hypercombinatorially embedded in Y. (Hint: Use (4) and (5).)

(7) Let Y be H-closed. Prove that $Y \equiv_X \sigma X$ iff $\{c\ell_Y A : A \subseteq X\}$ is a closed base for Y (i.e., Y is a strict extension of X) and X is hypercombinatorially embedded in Y. (Hint: Use (4) and 7A(1).)

7P. Almost H-closed spaces. A space X is **almost H-closed** if $\kappa X \setminus X$ is a singleton or empty.

(1) Prove that a space X is almost H-closed iff for each pair of disjoint open sets U and V, $c\ell_X U$ or $c\ell_X V$ is H-closed.

(2) If X is not feebly compact, show that $|\kappa X \setminus X| \geq 2^{(2^{\aleph_0})}$. (Hint: Let $\{U_n : n \in \mathbb{N}\}$ be a locally finite family of pairwise disjoint nonempty open sets. For each $p \in \beta\mathbb{N} \setminus \mathbb{N}$, let $F(p) = \{\cup\{U_n : n \in A\} : A \in p\}$. Show that $F(p)$ is a free open filter base on X. If $q \in \beta\mathbb{N} \setminus (\mathbb{N} \cup \{p\})$, show that there are open sets $U \in F(p)$ and $V \in F(q)$ such that $U \cap V = \emptyset$.)

(3) Prove that an almost H-closed space is locally H-closed and feebly compact. (Hint: If $\sigma X \setminus X$ is empty or a singleton, then X is open in σX. Use 7.3(b).)

(4) Let p be a non-isolated point of an H-closed extremally disconnected space T. Show that $T \setminus \{p\}$ is almost H-closed.

(5) Consider the space ω_1 defined in 2.6(q). By 4A(3), $|\beta\omega_1 \setminus \omega_1| = 1$. Show that ω_1 is not almost H-closed and hence that $|\kappa\omega_1 \setminus \omega_1| > 1$.

(6) If $A \subseteq X$ and A is almost H-closed, prove that $c\ell_X A \setminus A$ contains at most one point.

(7) If $U \subseteq X$ is open and $oU \setminus U$ contains at most one point (oU is relative to σX), prove that $c\ell_X U$ is almost H-closed.

(8) Prove that the following are equivalent:

(a) $\kappa X \equiv_X \sigma X$,

(b) $\kappa X \setminus X$ is finite, and

(c) X is the finite union of almost H-closed subspaces.

(Hint: Use 7N(10) to prove (a) $\Longleftrightarrow$ (b). Suppose $\kappa X \setminus X = \{\mathcal{U}_1, \ldots, \mathcal{U}_n\}$ for some $n \in \mathbb{N}$. Show that there are open sets $U_i \in \mathcal{U}_i$ for $1 \leqslant i \leqslant n$ such that $U_i \cap U_j = \emptyset$ for $1 \leqslant i < j \leqslant n$. Let $U_0 = X \setminus \bigcup\{c\ell_X U_i : 1 \leqslant i \leqslant n\}$. Show that $\{c\ell_X U_0, \ldots, c\ell_X U_n\}$ is a cover of X by almost H-closed spaces.)

7Q. <u>Θ-closed subsets and Martin's Axiom.</u> A subset A of a space X is **Θ-closed** in X if for each $p \in X \setminus A$, there is an open set U such that $p \in U \subseteq c\ell_X U \subseteq X \setminus A$; in terms of the notation introduced in 4.8(r), A is Θ-closed iff $c\ell_\Theta A = A$.

(1) Let X be a space and $A \subseteq X$. Prove the following are equivalent:

(a) A is Θ-closed in κX,

(b) A is Θ-closed in every extension of X,

(c) A is Θ-closed in σX,

(d) A is Θ-closed in some H-closed extension of X, and

(e) $k_X^{\leftarrow}(A)$ is compact ($k_X : EX \to X$ is defined in 6.6).

(Hint: Use 7.2(b)(1) to show that (a) and (c) are equivalent and 4.8(n)(2) to show (d) implies (a). To show (c) implies (e), first use 6.9(b)(2) to obtain that $E(\sigma X)$ is a compact extension of EX and then show, for each $p \in \sigma X \setminus A$, that $k_{\sigma X}^{\leftarrow}(p)$ and $k_{\sigma X}^{\leftarrow}(A)$ are contained in disjoint clopen sets in $E(\sigma X)$. To show (e) implies (b), let T be an extension of X. Note that ET is an extension of EX (see 6.9(a)); find disjoint open sets U and V of T such that $0U$ and $0V$ respectively contain the disjoint compact spaces $k_T^{\leftarrow}(p)$ ($p \in T \setminus A$) and $k_T^{\leftarrow}(A)$ ($= k_X^{\leftarrow}(A)$). Use 6.8(f)(6) to conclude that $p \in \text{int}_T c\ell_T U$ and $A \subseteq \text{int}_T c\ell_T V$, and use 2.2(a)(4) to finish the proof.)

(2) Let Z be an H-closed, Urysohn space and $A \subseteq Z$.

(a) Prove that A is θ-closed in Z iff A is an H-set in Z. (Hint: Use 4N.)

(b) Let $f : X \to Z$ be a θ-continuous function and Y an extension of the space X. Prove that f has a θ-continuous extension from Y to Z iff for each pair of disjoint θ-closed subsets B and C of Z, $c\ell_Y f^{\leftarrow}[B] \cap c\ell_Y f^{\leftarrow}[C] = \varnothing$. (This result extends 4.1(m).)

(3) Let κ be an infinite cardinal such that $\aleph_0 \leqslant \kappa < 2^{\aleph_0}$. Prove MA($\kappa$) is equivalent to this statement: (*) no H-closed space with *ccc* is the union of less than κ Θ-closed nowhere dense subsets. (Hint: To show that MA(κ) implies (*), let X be an H-closed space with *ccc* which is the union of less than κ Θ-closed nowhere dense subsets. Obtain a contradiction by showing EX is a compact space with *ccc* which is the union of less that κ closed, nowhere dense subsets of EX.)

7R. <u>Examples of p-maps</u>.

(a) Show that irreducible continuous surjections are p-maps.

(b) Let $f \in C(X,Y)$ and let Y be either Urysohn or Lindelöf. Prove that f is a p-map iff, for each $A \in R(X)$, $f[A] = B \cup H$ where $B \in R(Y)$ and H is an H-closed subset of Y. (Hint: For the Lindelöf case, use an inductive argument together with an application of the countable case of 4S(4).)

7S. <u>Examples of H-closed spaces</u>. Let D be a dense subset of a space X. For each $n \in \mathbb{N}$ let $X(D^n) = (D \times \{0,1,2,\ldots,n-1\}) \cup (X\setminus D)$. If $A \subseteq X$, let $A^n = (A\setminus D) \cup ((A \cap D) \times \{0,1,\ldots,n-1\})$. A set $U \subseteq X(D^n)$ is defined to be open if $U \cap (D\times\{i\})$ is open in $D \times \{i\}$ (with the product topology) for $i = 0,1,\ldots,n-1$, and for $x \in U \cap (X\setminus D)$, there is an open set V in X such that $x \in V^n \subseteq U$. Note that the space X(D), (i.e., $n = 1$), is homeomorphic to X with the topology generated by $\tau(X) \cup \{D\}$.

(1) If $n, m \in \mathbb{N}$ and $n \leq m$, prove that $X(D^n)$ is a regular closed subspace of $X(D^m)$.

(2) If X is H-closed, prove $X(D^n)$ is H-closed for $n \in \mathbb{N}$. In particular, the property of being H-closed is preserved when the topology of an H-closed space is enlarged by making any dense set open.

(3) If $n \geq 2$ and X is semiregular, prove $X(D^n)$ is semiregular.

In particular, note that if X is minimal Hausdorff and $n \geq 2$, then by (2) and (3) and 7.5(a), $X(D^n)$ is minimal Hausdorff.

(4) Consider the compact subspace $K = \{(0,0)\} \cup \{(1/n,0) : n \in \mathbb{N}\} \cup \{(1/n,1/m) : n, m \in \mathbb{N} \text{ and } m \geq n\}$ of $\mathbb{R}^2$ and $D = K \setminus \{(1/n,0) : n \in \mathbb{N}\}$. It is clear that D is dense in K. By (2) and (3), $K(D^2)$ is a minimal Hausdorff space and by (1), $K(D)$ is an H-closed subspace of $K(D^2)$. Prove that $K(D^2)$ is homeomorphic to the space X described in 4.8(d).

(5) Let $Z = (\mathbb{N} \times \omega) \cup \{\infty\}$ and $\{\mathbb{N}_i : i \in \mathbb{N}\}$ be a family of pairwise disjoint infinite subsets of $\mathbb{N}$ such that $\mathbb{N} = \bigcup\{\mathbb{N}_i : i \in \mathbb{N}\}$. A set $U \subseteq Z$ is open if $(n,0) \in U$ implies that for some $m \geq n$, $\{(n,p) : p \geq m\} \cup \{(k,\ell) : k \geq m, \ell \in \mathbb{N}_n\} \subseteq U$ and if $\infty \in U$ implies that for some $m \in \mathbb{N}$, $\{(k,\ell) : k \geq m, \ell \in \mathbb{N} \setminus \bigcup\{\mathbb{N}_i : 1 \leq i \leq m\}\} \subseteq U$. Prove that Z is minimal Hausdorff,

not compact, and every closed subset of Z is an H-set (see 4N). Note that by 4S, a space in which every closed subset is H-closed is compact.

7T. Quasiregular H-closed space, Baire spaces, and MA. A space X is **quasiregular** if for every nonempty open set $U \subseteq X$, there is a nonempty open set V such that $c\ell_X V \subseteq U$. Any extension of a discrete space is quasiregular.

(1) Show that the space described in 4.8(d) and used in 7.5(g) is quasiregular, minimal Hausdorff but is not regular.

A space is **Baire** if the countable intersection of open dense subsets is dense.

(2) Prove that a quasiregular, feebly compact space is Baire.

(3) If X is quasiregular, prove that σX is quasiregular; in particular, by (2), this shows that σX is a Baire space.

(4) Let I be the compact subspace [0,1] of $\mathbb{R}$ with the usual topology and $D = \mathbb{Q} \cap [0,1]$. By 7S, $I(D^2)$ is a minimal Hausdorff space. Show that $I(D^2)$ is not a Baire space.

Note that the space $I(D^2)$ defined in (4) has no Baire extension.

(5) Let I and D be as defined in (4) and let $X = ((I\setminus\mathbb{Q}) \times \{0\}) \cup (\cup\{D \times \{n\} : |n| \in \mathbb{N}\})$. A set $U \subseteq X$ is open if $(x,k) \in U$ implies there is an open neighborhood V of x in I such that: (i) for $k = 0$, $((V\setminus\mathbb{Q}) \times \{0\}) \cup (\cup\{D \times \{n\} : |n| \in \mathbb{N}\}) \subseteq U$; (ii) for $k = 1$ or -1, there is some $n \in \mathbb{N}$ such that $((V \cap \mathbb{Q}) \times \{k\}) \cup (\cup\{(V \cap \mathbb{Q}) \times \{j\} : k \cdot j \geq n\}) \subseteq U$; and (iii) for $|k| > 1$, $(V \cap \mathbb{Q}) \times \{k\} \subseteq U$. Show that X is minimal Hausdorff and the countable union of nowhere dense, minimal Hausdorff subspaces. (Hint: Show that X is H-closed and semiregular and that $Y = ((I\setminus\mathbb{Q}) \times \{0\}) \cup (D \times \{-1,+1\})$ is a nowhere dense, minimal Hausdorff subspace. Now use the fact that $X\setminus Y$ is countable.)

(6) Let κ be a cardinal such that $\aleph_0 \leq \kappa < 2^{\aleph_0}$. Prove that MA($\kappa$) is equivalent to this statement: for every quasiregular, H-closed space with *ccc*, if $\{U_\alpha : \alpha < \kappa\}$ is a family of open dense subsets of X, then $\cap\{U_\alpha : \alpha < \kappa\} \neq \emptyset$. (Hint: The proof of one direction is easy. To prove that MA(κ) implies the statement, carefully trace through the proof that (4) implies (1) in 3.5(i) and show that compactness can be replaced by "H-closed and quasiregular.")

7U. <u>Accumulation points of nowhere dense sets</u>. In this problem we investigate spaces without isolated points which have points that are not accumulation points of any nowhere dense subset of the space.

(1) If X is a first countable space without isolated points, prove that every point in X is an accumulation point of some nowhere dense subset of X.

(2) Let p be a remote point of $\beta\mathbb{Q}$ (see 4AH). Prove that $\mathbb{Q} \cup \{p\}$ is a perfectly normal space and p is not an accumulation point of any nowhere dense subset of $\mathbb{Q} \cup \{p\}$.

(3) Let I be the unit interval and $\mathcal{F}$ a maximal filter on $\{D \subseteq I : D \text{ is dense in } I\}$. Let σ be the topology generated by $\tau(I) \cup \mathcal{F}$ (see 7M). Prove:

(a) (I,σ) is an H-closed space without isolated points,

(b) every nowhere dense subset of (I,σ) is closed, and

(c) no point of (I,σ) is an accumulation point of any nowhere dense subset of (I,σ).

(4) Let X be a regular space without isolated points. For $x \in X$, let $\mathcal{R}_x = \{R \in \mathcal{R}(X) : x \in R\}$. Let $\mathcal{C}$ be a maximal chain in $\mathcal{R}_x$ and $D = \bigcap\mathcal{C}$.

(a) Show that $x \notin c\ell_X \mathrm{int}_X D$. (Hint: Assume $x \in c\ell_X \mathrm{int}_X D$ and find an open set $U \subset \mathrm{int}_X D$ such that $x \in c\ell_X U \subset c\ell_X \mathrm{int}_X D \subseteq D \subseteq C$ for all $C \in \mathcal{C}$ to obtain a contradiction of the maximality of $\mathcal{C}$.)

(b) Consider the closed, nowhere dense set E =

$D \setminus \text{int}_X D$. Suppose $x \notin c\ell_X(E \setminus \{x\})$. Use the regularity of X to obtain an open neighborhood U of x such that $E \cap c\ell_X U = \{x\}$. Show that $\mathcal{D} = \{c\ell_X(U \cap \text{int}_X C) : C \in \mathcal{C}\} \subseteq \mathcal{R}_X$ and $\{x\} = \cap \mathcal{D}$.

(5) Prove that each point of a compact space X without isolated points is an accumulation point of a nowhere dense set of X. (Hint: Use (4) to find a chain $\mathcal{C} \subseteq \mathcal{R}_X$ such that $\{x\} = \cap \mathcal{C}$. Show there is a well-ordered subchain $\mathcal{E} = \{R_\alpha : \alpha < \lambda\}$, of $\mathcal{C}$ (where λ is a limit ordinal), such that $\mathcal{E}$ is cofinal in $\mathcal{C}$. In particular, show that $\{x\} = \cap \mathcal{E}$, and for each $\alpha \in \lambda$, $(\text{int}_X R_\alpha) \setminus R_{\alpha+1} \neq \varnothing$. Let $p_\alpha \in (\text{int}_X R_\alpha) \setminus R_{\alpha+1}$ for $\alpha \in \lambda$. Show that $N = \{p_\alpha : \alpha < \lambda\}$ is a nowhere dense set such that $x \in c\ell_X N \setminus N$.)

(6) Let X be a space and $p \in X$. Let $Y_p = EX/k^{\leftarrow}(p)$, i.e., the quotient space of EX with $k^{\leftarrow}(p)$ identified to a point, denoted as $\bar{p}$. Let $f : EX \to Y_p$ be the induced quotient map, i.e., $f(y) = y$ for $y \in EX \setminus k^{\leftarrow}(p)$ and $f(y) = \bar{p}$ for $y \in k^{\leftarrow}(p)$. Define $g : Y_p \to X$ by $g(y) = k(y)$ for $y \neq \bar{p}$ and $g(\bar{p}) = p$. Suppose X is H-closed.

(a) Prove that Y_p is compact (and Hausdorff) and zero-dimensional.

(b) Prove that $f : EX \to Y_p$ is perfect, irreducible,

continuous and onto and that $f|(EX\setminus k^{\leftarrow}(p))$ is a homeomorphism onto $Y_p\setminus\{\bar{p}\}$.

(c) Prove that $g : Y_p \to X$ is perfect, irreducible, Θ-continuous and onto.

(d) If p is a regular point of X, i.e., if for each neighborhood U of p there is a neighborhood V of p such that $c\ell_X V \subseteq U$, prove that g is continuous at $\bar{p}$.

(7) Let p be a regular point of an H-closed space X without isolated points. Prove p is an accumulation point of some nowhere dense subset of X. (Hint: Apply (5) to Y_p (defined in (6)) to find a nowhere dense set $A \subseteq Y_p$ such that $\bar{p}$ is an accumulation point of g[A] and that g[A] is nowhere dense in X; use (6)(c).)

(8) If p is a regular point of a space X, show that p is a regular point of μX. (Hint: Use 7.5(h).)

(9) Let $X = \mathbb{Q} \cup \{p\}$ where p is a remote point of $\beta\mathbb{Q}$ (see 4AH).

(a) Show that p is an accumulation point of some nowhere dense subset of μX. (Hint: Use (8).)

(b) Let $D = \mathbb{Q} \cup \{p\}$. Clearly, D is dense in $\beta\mathbb{Q}$. Show that $(\beta\mathbb{Q})(D^2)$ (defined in 7S) is a minimal Hausdorff space in which (p,0) is not the accumulation point of any nowhere dense subset of $(\beta\mathbb{Q})(D^2)$.

7V. Generalized Remote Points. Before attempting this problem, the reader will find a review of 4AH helpful. As in 6.8(c), the symbol ΘX will denote the set of all open ultrafilters on the space X.

For a space X, let $\mathcal{D} = \{D \subseteq X : D \text{ is dense and open}\}$. An open filter $\mathcal{F}$ on X is **saturated** if $\mathcal{F} = \{W \subseteq X : W \text{ is open and } \mathrm{int}_X c\ell_X W \in \mathcal{F}\}$.

(1) Let $\mathcal{F}$ be an open filter on a space X. Prove that the following are equivalent:

(a) $\mathcal{F}$ is saturated,

(b) $\mathcal{D} \subseteq \mathcal{F}$, and

(c) $\mathcal{F} = \cap\{\mathcal{U} \in \Theta X : \mathcal{F} \subseteq \mathcal{U}\}$.

Let Y be an extension of a space X. A point $p \in Y \backslash X$ is a **Y-remote point** of X if $p \notin c\ell_Y D$ for each nowhere dense set D in X. As in 4AH, a βX-remote point of X is simply called a remote point of X.

(2) If X is a space and $Y = \kappa X$ or σX, prove that each point $p \in Y \backslash X$ is a Y-remote point of X. (Hint: Use 7.2(b)(4).)

(3) Let Y be an extension of a space X and $p \in Y \backslash X$. Prove that p is a Y-remote point of X iff 0^p is a saturated open filter.

Let F be an open filter on a space X. Define F^+ to be $\cap\{U \in \Theta X : F \subseteq U\}$ and F_s to be $\{U \in \tau(X) : U \supseteq \text{int}_X c\ell_X W$ for some $W \in F\}$. Clearly, $F^+ \supseteq F \supseteq F_s$; by 2.3(k), $F^+ = \{U \in \tau(X) : \text{int}_X c\ell_X U \in F\}$.

(4) Let F and G be open filters on a space X and Y, Z $\in H(X)$.

(a) Prove the following are equivalent:

(i) $\{U \in \theta X : F \subseteq U\} =$
$\{U \in \theta X : G \subseteq U\}$,

(ii) $F^+ \supseteq G \supseteq F_s$,

(iii) $F^+ = G^+$, and

(iv) $F_s = G_s$.

(b) For $p \in Y$, prove that $(0_Y{}^p)_s$ is the $\tau(X)$-open filter generated by $0_{Y(s)}{}^p$.

(c) Prove that Y and Z are Θ-equivalent iff $\{(0_Y{}^p)_s : p \in Y\} = \{(0_Z{}^p)_s : p \in Z\}$ (cf. 7C).

(5) Let Y be an extension of a space X and $p \in Y \setminus X$.

(a) Prove that 0^p is saturated iff $(0^p)^+ = 0^p$.

(b) If Y is semiregular, prove that $(0^p)_s = 0^p$.

Note: If X is a Tychonoff space and $p \in \beta X \setminus X$ is a βX-remote point of X, then 0^p is a free, maximal CR-filter (see 4Y) and $(0^p)^+ = 0^p = (0^p)_s$.

The next goal is to examine when a semiregular space X has μX-remote points.

(6) (a) Let Y and Z be extensions of a space X and let $q \in Y \setminus X$ and $p \in Z \setminus X$. Let $f \in C(Y,Z)$ such that $f(x) = x$ for $x \in X$ and $f(q) = p$. Prove that $0_Z{}^p \subseteq 0_Y{}^q$ and that if p is a Z-remote point of X, then q is a Y-remote point of X. (Hint: Use continuity to prove the first part and (1) to prove the second part.)

(b) If X is a Tychonoff, non-feebly compact space with countable π-weight, prove that X has $2^{(2^{\aleph_0})}$ μX-remote points. (Hint: Use (a), 7B(1), and 4AH.)

By re-examining the proof of 4AH, it is possible to improve 6(b).

(7) Let X be a semiregular, non-feebly compact space with a countable $\bar{\pi}$-base ($\bar{\pi}$-base is defined in 4AH). Prove that X has $2^{(2^{\aleph_0})}$ μX-remote points. (Hint: In 4AH(3)(a), let $\mathcal{K}$ be the open filter generated by $\mathcal{H}$.

First, show that $D \subseteq K_s$ and that K_s is free. Next, show that any maximal $RO(X)$-filter U containing K_s is also a maximal open filter; hence, $U \in \mu X \setminus X$. Using 7.5(h), show that $0_{\mu X}{}^{U} = U$ (cf. 7N(12)). Conclude that U is a μX-remote point of X.)

Let (X,d) be a metric space. Let $\epsilon > 0$. A subset $A \subseteq X$ is **ϵ-discrete** if for every $x, y \in A$, $x \neq y$, $d(x,y) \geqslant \epsilon$. For $x \in X$, let $S(x,\epsilon) = \{y \in X : d(x,y) < \epsilon\}$.

(8) Let (X,d) be a metric space without isolated points, F a closed nowhere dense subset, and $\epsilon > 0$.

(a) If A is a nonempty ϵ-discrete subset of X, $0 < \delta < \epsilon/2$, and $U = \cup\{S(x,\delta) : x \in A\}$, prove that $c\ell_X U = \cup\{c\ell_X S(x,\delta) : x \in A\}$.

(b) For $n \in \mathbb{N}$, let $G_n = \{x \in X : d(x,F) < 1/n\}$, A_n be a maximal $1/n$-discrete subset of $G_n \setminus G_{n+1}$ (which exists by Zorn's lemma), and $A = \cup\{A_n : n \in \mathbb{N}\}$.

(i) Show that $F = \cap\{G_n : n \in \mathbb{N}\}$, $c\ell_X G_{n+1} \subseteq G_n$ for $n \in \mathbb{N}$, and $c\ell_X A \setminus A = F$.

(ii) Show that A is a discrete subset of X.

(iii) If $V = X \setminus c\ell_X A$, prove that $F \subseteq \mathrm{bd}_X V$.

(iv) For each $n \in \mathbb{N}$, let B_n be a maximal $1/n$-discrete subset of $(G_n \setminus G_{n+1}) \cap V$ and $B = \bigcup\{B_n : n \in \mathbb{N}\}$. Prove that $F = c\ell_X B \setminus B$, B is a discrete subset of X, $A \cap c\ell_X B = \emptyset$, and $B \cap c\ell_X A = \emptyset$.

(v) Let $p \in A_n$ for some $n \in \mathbb{N}$. Prove there is some $\delta(p) > 0$ such that $\delta(p) < 1/n$, $S(p,\delta(p)) \subseteq X \setminus c\ell_X B$, and $S(p,\delta(p)) \cap A = \{p\}$.

(vi) For each $p \in A$, fix some $\delta(p) > 0$, provided by (v), and let $\epsilon(p) = \delta(p)/3$ and $W = \bigcup\{S(p,\epsilon(p)) : p \in A\}$. For $p \in B$, similarly define $\epsilon(p)$ and $U = \bigcup\{S(p,\epsilon(p)) : p \in B\}$. Prove that $U \cap V = \emptyset$ and $c\ell_X W \cap c\ell_X U = F$.

(vii) Using (vi), conclude that each closed nowhere dense set of X is regularly nowhere dense (defined after 7B(2)).

(c) Prove that every point of $\mu X \setminus X$ is a μX-remote point. (Hint: Use 7.5(h)(3,5).)

(9) Let X be a noncompact, metric space without isolated points and $f : \mu X \to \beta X$ the continuous function such that $f(x) = x$ for $x \in X$ (which exists by 7B(1)). Prove there is a point $p \in \beta X \setminus X$ which is not a remote

point of X but, yet, each point $q \in f^{\leftarrow}(p)$ is a μX-remote point of X. (Hint: By 8(c), it suffices to show that $\beta X \setminus X$ contains a point which is not a remote point. Assume not and obtain a contradiction to 7.6(j).)

Let X and Y be spaces and $y \in Y$. A function $f : X \to Y$ is **closed at y** if for each open set $U \supseteq f^{\leftarrow}(y)$, there is an open set W such that $U \supseteq f^{\leftarrow}[W] \supseteq f^{\leftarrow}(y)$.

(10) Let X and Y be spaces. Prove that a function $f : X \to Y$ is closed iff f is closed at each point $y \in Y$.

(11) Let Y and Z be strict extensions of a space X, $p \in Z \setminus X$, $f : Y \to Z$ a continuous function such that $f(x) = x$ for $x \in X$ and $f^{\leftarrow}(p)$ is compact. Prove that f is closed at p iff $0_Z{}^p = \cap\{0_Y{}^p : q \in f^{\leftarrow}(p)\}$. (Hint: Use these two facts which follow from 6(a): (1) $0_Z{}^p \subseteq \cap\{0_Y{}^q : q \in f^{\leftarrow}(p)\}$ and (2) $f^{\leftarrow}[o_Z V] \subseteq o_Y V$ whenever V is an open subset of X.)

(12) Let X be a Tychonoff space and $f : \sigma X \to \beta X$ be the continuous function such that $f(x) = x$ for $x \in X$ (which exists by 7B(1)). Let $p \in \beta X \setminus X$. Prove that f is closed at p iff p is a remote point of X. (Hint: Use 7.1(e)(4), paragraph before 7.4(a), and 7.4(a)(1) to show that (11) applies. So, f is closed at p iff $0_{\beta X}{}^p = \cap\{0_{\sigma X}{}^q : q \in f^{\leftarrow}(p)$. Apply (1) and (3).)

CHAPTER 8

Further Properties and Generalizations of Absolutes

8.1 Introduction

In Chapter 6 we constructed the Iliadis absolute (EX,k_X) and the Banaschewski absolute (PX,Π_X) of a space X and developed their basic properties. In this chapter we examine in greater detail the interaction of absolutes with other topological constructions.

In 8.2 we investigate when the various H-closed extensions we have constructed "commute with" one or the other of the two absolutes we have constructed. In 8.3 we look at the relationship between $E(\gamma_P X)$ and $\gamma_P(EX)$ where X is a Tychonoff (respectively, zero-dimensional) space and P is a Tychonoff (respectively, zero-dimensional) extension property. In 8.4 we study the family of all pairs (Y,f) consisting of a space Y and a perfect irreducible continuous surjection $f : Y \to X$; such pairs are called "covers" of X, and the family of covers of X has (PX,Π_X) as its "largest" member. We finish the chapter by investigating the relationship between C(X) and C(EX) in 8.5.

8.2 Absolutes and H-closed extensions

Let hX be some canonical H-closed extension of the space X (such as the Katětov extension, the Fomin extension, the

Banaschewski-Fomin-Šanin extension if X is semiregular, etc.). As noted in 6.9(a), we can identify $k_{hX}^{\leftarrow}[X]$ with EX and $k_{hX} | k_{hX}^{\leftarrow}[X]$ with k_X. Thus as $k_{hX}^{\leftarrow}[X]$ is dense in E(hX) by 6.4(b), and E(hX) is compact by 6.9(b)(1), it follows that $E(hX) \in \mathcal{H}(EX)$. Thus it makes sense to ask if E(hX) is "the same" H-closed extension of EX as h(EX) is.

There are two ways of interpreting what "the same" means in the preceding sentence. Since $E(hX) \in \mathcal{H}(EX)$, one interpretation is that E(hX) and h(EX) are equivalent extensions of EX. Explicitly, this means that $E(hX) \equiv_{k_{hX}^{\leftarrow}[X]} h(k_{hX}^{\leftarrow}[X])$. The second interpretation is that h(EX) is extremally disconnected and zero-dimensional and there exists a perfect irreducible θ-continuous surjection f from h(EX) onto hX. Explicitly, this means that $(E(hX), k_{hX}) \sim (h(k_{hX}^{\leftarrow}[X]), f)$. We will show that these two interpretations are equivalent.

(a) **Lemma**. Let X be a space. Then

$$E(hX) \equiv_{k_{hX}^{\leftarrow}[X]} h(k_{hX}^{\leftarrow}[X]) \text{ iff}$$

$$(E(hX), k_{hX}) \sim (h(k_{hX}^{\leftarrow}[X]), f) \text{ for some } f.$$

Proof. As noted above, we can (and will) identify $k_{hX}^{\leftarrow}[X]$ with EX. Suppose that $E(hX) \equiv_{EX} h(EX)$. Then there exists a homeomorphism $j : h(EX) \to E(hX)$ that fixes EX pointwise. Hence h(EX) is zero-dimensional and extremally disconnected, and it is evident that $k_{hX} \circ j$ is the required f — i.e., $(E(hX), k_{hX}) \sim (h(EX), k_{hX} \circ j)$. Conversely, suppose that $(E(hX), k_{hX}) \sim (h(EX), f)$. Then h(EX) is extremally disconnected and zero-dimensional, and hence compact. Then by 6.2(c) it follows that $h(EX) \equiv_{EX} \beta(EX)$. But $E(hX) \equiv_{EX} \beta(EX)$ by 6.9(b)(2), so we conclude that $E(hX) \equiv_{EX} h(EX)$.

■

(b) **Definition**. Let hX be a "canonical" H-closed extension of a space X. The statement "E(hX) = h(EX)" abbreviates the (equivalent) statements in 8.2(a).

The following corollary to 8.2(a) follows easily from 6.9(b)(2).

(c) **Corollary**. Let hX be a canonical H-closed extension of X. Then E(hX) = h(EX) iff EX is C^*-embedded in h(EX) and h(EX) is regular.

We now give necessary and sufficient conditions for E to commute with some of the H-closed extensions that we have previously discussed. We defer to problem 8A our consideration of when E commutes with the one-point H-closed extensions $X^\#$ and X^+ of the locally H-closed space X that were constructed in 7D. Of course, we already know that if X is Tychonoff then $E(\beta X) = \beta(EX)$; see 6.9(b)(3).

(d) **Theorem**. Let X be any space. Then:

(1) $E(\kappa X) = \kappa(EX)$ iff X is H-closed,

(2) $E(\sigma X) = \sigma(EX)$ iff the set of non-isolated points of X is Θ-closed in κX (see 7Q), and

(3) if X is semiregular then $E(\mu X) = \mu(EX)$.

Proof

(1) By 6.9(b)(2) $E(\kappa X) = \kappa(EX)$ iff $\kappa(EX)$ is the Stone-Čech compactification of EX. By 7B(8) this happens iff EX is compact, which by 6.9(b)(1) is equivalent to X being H-closed.

(2) By 6.9(b)(2) $E(\sigma X) = \sigma(EX)$ iff $\sigma(EX) = \beta(EX)$. By 7B(5) $\sigma(EX) = \beta(EX)$ iff the set of non-isolated points of EX is compact. By 6.9(e) and 7Q(1), this is equivalent to the set of non-isolated points of X being Θ-closed in κX.

(3) By 6.9(b)(2), $E(\mu X) = \mu(EX)$ iff $\mu(EX) = \beta(EX)$. By 7B(3) this occurs iff every regularly nowhere dense subset of EX is compact. But EX is extremally disconnected; so, its regularly nowhere dense sets are empty. Hence the condition is always fulfilled and $E(\mu X) = \mu(EX)$ for every semiregular space X.

■

When we attempt to study the commutativity of P with the various H-closed extensions, we are hampered by the fact that we have no "PX-analogue" to the very useful 6.9(b) that we used above when investigating the commutativity of E. However, the remarks concerning E that we made preceding 8.2(a) apply to P as well. By 6.11(g) we know that if hX is a canonical H-closed extension of a space X then $(PX, \Pi_X) \sim (\Pi_{hX}^{\leftarrow}[X], \Pi_{hX} | \Pi_{hX}^{\leftarrow}[X])$ so we can identify PX with $\Pi_{hX}^{\leftarrow}[X]$. Hence P(hX) and h(PX) are both H-closed extensions of PX, and it makes sense to ask if they are equivalent extensions of PX. Secondly, we can ask if h(PX) is extremally disconnected and if there exists a perfect irreducible continuous surjection from h(PX) onto hX. In contrast to 8.2(a), these two interpretations are not always equivalent; there are cases in which the second does not imply the first (see 8B). However, the first interpretation implies the second. Hence 8.2(e) below is an analogue to one direction of 8.2(a).

(e) **Lemma**. Let X be a space. If

$$P(hX) \equiv_{\Pi_{hX}^{\leftarrow}[X]} h(\Pi_{hX}^{\leftarrow}[X]) \text{ then}$$

$$(P(hX), \Pi_{hX}) \sim (h(\Pi_{hX}^{\leftarrow}[X]), f) \text{ for some } f.$$

Proof. As noted above, we can (and will) identify $\Pi_{hX}^{\leftarrow}[X]$ with PX. Suppose that $P(hX) \equiv_{PX} h(PX)$. Then there exists a homeomorphism $j : h(PX) \to P(hX)$ that fixes PX pointwise. Thus h(PX) is extremally disconnected and it is evident that $\Pi_{hX} \circ j$ is the required f.

■

(f) **Definition**. Let h be a "canonical" H-closed extension of a space X. The statement "P(hX) = h(PX)" abbreviates the statement "$P(hX) \equiv_{\Pi_{hX}^{\leftarrow}[X]} h(\Pi_{hX}^{\leftarrow}[X])$". Furthermore, when considering the equation "P(hX) = h(PX)", PX is to be interpreted as $\Pi_{hX}^{\leftarrow}[X]$.

We now give necessary and sufficient conditions for P to commute with some of the H-closed extensions that we have constructed. Recall that we have already derived one such result in 6.11(j), namely that $P(\kappa X) = \kappa(PX)$ for any space X. In 8C we indicate a different way of proving this result.

We begin by showing that $P(\sigma X) = \sigma(PX)$ for every space X. To do this we will use the characterization of σX given in 7O(7), and the reader is advised to work through this problem before continuing with the next two results. Recall that if T is an extension of X, then X is hypercombinatorially embedded in T if whenever F and H are closed sets of X and $F \cap H$ is nowhere dense in X, then $c\ell_T F \cap c\ell_T H = F \cap H$.

(g) **Lemma**. Let T be an extension of a space X. Suppose f is a perfect irreducible continuous surjection from a space S onto T. If X is hypercombinatorially embedded in T, then $f^{\leftarrow}[X]$ is hypercombinatorially embedded in S.

Proof. Let A and B be closed in $f^{\leftarrow}[X]$ and suppose $A \cap B$ is nowhere dense in $f^{\leftarrow}[X]$. By 6.5(d)(1) $f[A] \cap f[B]$ is nowhere dense in X, and $f[A]$ and $f[B]$ are closed in X. Thus by hypothesis $c\ell_T f[A] \cap c\ell_T f[B] = f[A] \cap f[B]$. Hence $c\ell_S A \cap c\ell_S B \subseteq f^{\leftarrow}[c\ell_T f[A] \cap c\ell_T f[B]] \subseteq f^{\leftarrow}[f[A] \cap f[B]] \subseteq f^{\leftarrow}[X]$. But $(c\ell_S A) \cap f^{\leftarrow}[X] = A$ and $(c\ell_S B) \cap f^{\leftarrow}[X] = B$, so $c\ell_S A \cap c\ell_S B = A \cap B$. The lemma follows. ∎

(h) **Theorem**. For any space X, $P(\sigma X) = \sigma(PX)$.

Proof. By 7O(7), X is hypercombinatorially embedded in σX; so, by 8.2(g) PX ($= \Pi_{\sigma X}^{\leftarrow}[X]$) is hypercombinatorially embedded in $P(\sigma X)$. Since $P(\sigma X)$ is H-closed by 6.11(i), if we can show that $P(\sigma X)$ is a strict extension of PX then by 7O(7) we can conclude that $P(\sigma X)$

$\equiv_{PX}\sigma(PX)$. Since $\{0U \cap \Pi_{\sigma X}^{\leftarrow}[V] : U, V \in \tau(\sigma X)\}$ is an open base for $P(\sigma X)$ (see 6.8 and 6.11(b)) it will suffice to verify the following:

(a) if $U \in \tau(\sigma X)$ then $0U = o_{P(\sigma X)}(0U \cap \Pi_{\sigma X}^{\leftarrow}[X])$ and

(b) if $\{U_i : i \in I\} \subseteq \tau(X)$ then $\Pi_{\sigma X}^{\leftarrow}[\bigvee\{oU_i : i \in I\}] = \bigvee\{o_{P(\sigma X)}\Pi_{\sigma X}^{\leftarrow}[U_i] : i \in I\}$.

(Note that when we apply the $o_{P(\sigma X)}$ operator, we are regarding $P(\sigma X)$ as an extension of $\Pi_{\sigma X}^{\leftarrow}[X]$; see 7.1(a). Also note that if $V \in \tau(\sigma X)$, then V has the form $\bigvee\{oU_i : i \in I\}$, where $U_i \in \tau(X)$; see 7.1(c)(5) and 7.2(a).)

To verify (a), note that

$$\begin{aligned} o_{P(\sigma X)}(0U \cap \Pi_{\sigma X}^{\leftarrow}[X]) &= P(\sigma X)\setminus c\ell_{P(\sigma X)}(\Pi_{\sigma X}^{\leftarrow}[X]\setminus 0U) \text{ (see 7.1(c)(7))} \\ &= P(\sigma X)\setminus c\ell_{P(\sigma X)}(\Pi_{\sigma X}^{\leftarrow}[X] \cap 0(\sigma X\setminus c\ell_{\sigma X}U)) \\ &\qquad \text{(see 6.8(d)(3))} \\ &= P(\sigma X)\setminus 0(\sigma X\setminus c\ell_{\sigma X}U) \\ &= 0U \text{ (see 6.8(d)(3)).} \end{aligned}$$

To verify (b) it suffices to show that if U is open in X, then $\Pi_{\sigma X}^{\leftarrow}[oU] = o_{P(\sigma X)}(\Pi_{\sigma X}^{\leftarrow}[U])$. Now $\Pi_{\sigma X}^{\leftarrow}[oU]$ is open in $P(\sigma X)$ and $\Pi_{\sigma X}^{\leftarrow}[oU] \cap \Pi_{\sigma X}^{\leftarrow}[X] = \Pi_{\sigma X}^{\leftarrow}[oU \cap X] = \Pi_{\sigma X}^{\leftarrow}[U]$ (see 7.1(c)(2)); so, by definition of $o_{P(\sigma X)}(\Pi_{\sigma X}^{\leftarrow}[U])$ it follows that $\Pi_{\sigma X}^{\leftarrow}[oU] \subseteq o_{P(\sigma X)}(\Pi_{\sigma X}^{\leftarrow}[U])$. Also, $o_{P(\sigma X)}(\Pi_{\sigma X}^{\leftarrow}[U]) \cap \Pi_{\sigma X}^{\leftarrow}[X] = \Pi_{\sigma X}^{\leftarrow}[U]$ (see 7.1(c)(2)); so, $\Pi_{\sigma X}^{\leftarrow}[oU] \cap \Pi_{\sigma X}^{\leftarrow}[X] = o_{P(\sigma X)}(\Pi_{\sigma X}^{\leftarrow}[U]) \cap \Pi_{\sigma X}^{\leftarrow}[X]$. If $\alpha \in (P(\sigma X)\setminus\Pi_{\sigma X}^{\leftarrow}[X])\setminus\Pi_{\sigma X}^{\leftarrow}[oU]$, then by 7.2(b)(5), $\alpha \in (P(\sigma X)\setminus\Pi_{\sigma X}^{\leftarrow}[X]) \cap \Pi_{\sigma X}^{\leftarrow}[o(X\setminus c\ell_X U)]$. Thus if $\alpha \in W \in \tau(P(\sigma X))$, then $W \cap \Pi_{\sigma X}^{\leftarrow}[o(X\setminus c\ell_X U)] \neq \emptyset$; thus

$$\emptyset \neq W \cap \Pi_{\sigma X}^{\leftarrow}[o(X \backslash c\ell_X U)] \cap \Pi_{\sigma X}^{\leftarrow}[X]$$
$$= W \cap \Pi_{\sigma X}^{\leftarrow}[o(X \backslash c\ell_X U) \cap X]$$
$$= W \cap \Pi_{\sigma X}^{\leftarrow}[X \backslash c\ell_X U] \quad \text{(see 7.1(c)(2))}.$$

Thus $\alpha \notin o_{P(\sigma X)}(\Pi_{\sigma X}^{\leftarrow}[U])$ and we have shown that $\Pi_{\sigma X}^{\leftarrow}[oU] = o_{P(\sigma X)}(\Pi_{\sigma X}^{\leftarrow}[U])$, thereby verifying (b). Thus $P(\sigma X)$ is a strict extension of $\Pi_{\sigma X}^{\leftarrow}[X]$. The theorem follows.

■

In 8D we sketch a computationally complicated, but more direct, proof that $P(\sigma X) = \sigma(PX)$; this proof is similar to the proof of 6.11(j).

Our next goal is to determine when $P(\mu X) = \mu(PX)$. In doing this we must remember that the extension μY exists iff Y is semiregular.

(i) __Theorem__. Let X be a space. Then:

(1) PX is semiregular iff X is regular and

(2) if X is regular then $\mu(PX) = P(\mu X)$ iff every regularly nowhere dense subset of X is compact (see 7B(2)).

__Proof__

(1) Since PX is extremally disconnected, by 6.4 PX is semiregular iff PX is regular. By 6.11(h) PX is regular iff X is regular.

(2) Suppose X is regular. By 6.11(h) $PX = EX$, so by 8.2(d)(3) $\mu(PX) = \mu(EX) = E(\mu X)$. Hence $\mu(PX) = P(\mu X)$ iff $P(\mu X) =$

$E(\mu X)$. By 6.11(h) this occurs iff μX is regular; so, by 4.8(k) it occurs iff μX is compact. But by 7B(1) $\mu X \geqslant \beta X$, so by definition of βX, μX is compact iff $\mu X \equiv_X \beta X$. By 7B(3) this occurs iff every regularly nowhere dense subset of X is compact.

■

8.3 Absolutes and extension properties

In this section we wish to investigate the relationship between $E(\gamma_P X)$ and $\gamma_P(EX)$ if P is a Tychonoff or zero-dimensional extension property. We will be able to say relatively little about the situation in the most general setting, but for specific extension properties - in particular realcompactness - we will be able to characterize those X for which $E(\gamma_P X) = \gamma_P(EX)$.

Suppose that P is a Tychonoff extension property. Then by 6.9(b) $E(\beta X) = \beta(EX)$, and we can interpret EX as being $k_{\beta X}^{\leftarrow}[X]$ and k_X as being $k_{\beta X} | EX$. Since $X \subseteq \gamma_P X \subseteq \beta X$ by 5.9(f), we see that $EX \subseteq k_{\beta X}^{\leftarrow}[\gamma_P X] \subseteq \beta(EX)$. By 6.9(a) $(E(\gamma_P X), k_{\gamma_P X}) \sim (k_{\beta X}^{\leftarrow}[\gamma_P X], k_{\beta X} | k_{\beta X}^{\leftarrow}[\gamma_P X])$ so we can interpret $k_{\beta X}^{\leftarrow}[\gamma_P X]$ as being $E(\gamma_P X)$ and $k_{\beta X} | k_{\beta X}^{\leftarrow}[\gamma_P X]$ as being $k_{\gamma_P X}$. Thus we have $EX \subseteq E(\gamma_P X) \subseteq \beta(EX)$. By 5.9(a) $E(\gamma_P X)$ has P, so by 5.9(f) we have $\gamma_P(EX) \subseteq E(\gamma_P X)$. The situation is summarized in the commutative diagram below.

$$\begin{array}{ccccccc}
EX = k_{\beta X}^{\leftarrow}[X] & \hookrightarrow & \Upsilon_P(EX) & \hookrightarrow & E(\Upsilon_P X) = k_{\beta X}^{\leftarrow}[\Upsilon_P X] & \hookrightarrow & \beta(EX) \\
\downarrow k_X & & & & \downarrow k_{\Upsilon_P X} & & \downarrow k_{\beta X} = \beta(k_X) \\
X & & \hookrightarrow & & \Upsilon_P X & \hookrightarrow & \beta X
\end{array}$$

Evidently the same arguments hold if P is a zero-dimensional extension property; we merely replace βX by $\beta_0 X$ is the diagram. With the above interpretation of the symbols involved, we can state the following result, whose proof is given above.

(a) **Theorem**. If P is a Tychonoff (respectively, zero-dimensional) extension property and X is a Tychonoff (respectively, zero-dimensional) space, then $EX \subseteq \Upsilon_P(EX) \subseteq E(\Upsilon_P X) \subseteq \beta(EX) = E(\beta X)$ (respectively, $= E(\beta_0 X)$).

An obvious question to ask is this: under what conditions, both on P and on X, is $\Upsilon_P(EX) = E(\Upsilon_P X)$? (I.e., when is the inclusion $\Upsilon_P(EX) \subseteq E(\Upsilon_P X)$ actually an equality?) We have already seen in 6.9(b) that if P is compactness, and X is any Tychonoff space, then $\Upsilon_P(EX) = E(\Upsilon_P X)$. The rest of this section will be devoted to providing some partial answers to this question. Our best results will be obtained for extension properties contained in almost realcompactness (see 6U), but we begin by considering more general cases.

(b) **Theorem**. Let X be a Tychonoff (respectively, zero-dimensional) space and P a Tychonoff (respectively, zero-dimensional) extension property. The following conditions are equivalent:

(1) $\Upsilon_P(EX) = E(\Upsilon_P X)$,

(2) the extension Pk_X of the map $k_X : EX \to X$ is a perfect continuous surjection from $\Upsilon_P(EX)$ onto $\Upsilon_P X$ (see 5.1(a) for notation),

(3) if T is any P-regular space and $f : T \to X$ is a perfect irreducible continuous surjection, then $Pf : \Upsilon_P T \to \Upsilon_P X$ is a perfect continuous surjection, and

(4) if T is any P-regular space and $f : T \to X$ is a perfect continuous surjection, there exists a closed subset T_0 of $\Upsilon_P T$ such that $Pf | T_0 : T_0 \to \Upsilon_P X$ is a perfect continuous surjection.

Proof. We will prove the theorem for Tychonoff extension properties; the proof for zero-dimensional extension properties is similar.

(1) $\Rightarrow$ (2) From 8.3(a) and the diagram preceding it we see that $Pk_X = k_{\Upsilon_P X} | \Upsilon_P(EX)$. Thus if $\Upsilon_P(EX) = E(\Upsilon_P X)$, it follows that $Pk_X = k_{\Upsilon_P X}$, which is a perfect continuous surjection from $E(\Upsilon_P X)$ onto $\Upsilon_P X$.

(2) $\Rightarrow$ (3) Let f be as hypothesized in (3). By 6.5(b)(1) and 1.8(e) $f \circ k_T$ is a perfect continuous irreducible surjection, so by 6.7(a) there is a homeomorphism $h : ET \to EX$ such that $k_X \circ h = f \circ k_T$. Applying 1.6(d) several times we see that $Pk_X \circ Ph = Pf \circ Pk_T$ (each side of the equation has domain $\Upsilon_P(ET)$) and that

$k_{\beta X} \circ \beta h = \beta f \circ k_{\beta T}$ (each side of the equation has domain $\beta(ET) = E(\beta T)$). Now βf is the extension of Pf to βT; so, by 4.2(g) Pf will be a perfect map from $\gamma_P T$ onto $\gamma_P X$ iff $\beta f[\beta T \setminus \gamma_P T] \subseteq \beta X \setminus \gamma_P X$ (since $Pf[\gamma_P T] \subseteq \gamma_P X$).

If $t \in \beta T \setminus \gamma_P T$ then $k_{\beta T}^{\leftarrow}(t) \subseteq \beta(ET) \setminus k_{\beta T}^{\leftarrow}[\gamma_P T] = \beta(ET) \setminus E(\gamma_P T) \subseteq \beta(ET) \setminus \gamma_P(ET)$ by 8.3(a). As h is a homeomorphism, Ph is a homeomorphism from $\gamma_P(ET)$ onto $\gamma_P(EX)$; so, by 4.2(g) $\beta h(s) \in \beta(EX) \setminus \gamma_P(EX)$ for each $s \in k_{\beta T}^{\leftarrow}(t)$. By hypothesis Pk_X is perfect, and by 4.2(g) $k_{\beta X}(\beta h(s)) \in \beta X \setminus \gamma_P X$. Thus $\beta f(k_{\beta T}(s)) \in \beta X \setminus \gamma_P X$, i.e., $\beta f(t) \in \beta X \setminus \gamma_P X$ as required.

(3) ⇒ (4) Let f be as hypothesized in (4). By 6.5(c) there is a closed subset F of T such that $f|F$ is a perfect irreducible continuous surjection from F onto X. By hypothesis $P(f|F) : \gamma_P F \to \gamma_P X$ is a perfect continuous surjection. Let $T_0 = cl_{\gamma_P T} F$. Then T_0 has P by 5.3(b), and by the definition of $\gamma_P F$ there exists $g \in C(\gamma_P F, T_0)$ such that $g|F = id_F$. Now $(Pf)|T_0 \in C(T_0, \gamma_P X)$ and so $(Pf)|T_0 \circ g \in C(\gamma_P F, \gamma_P X)$ (see diagram).

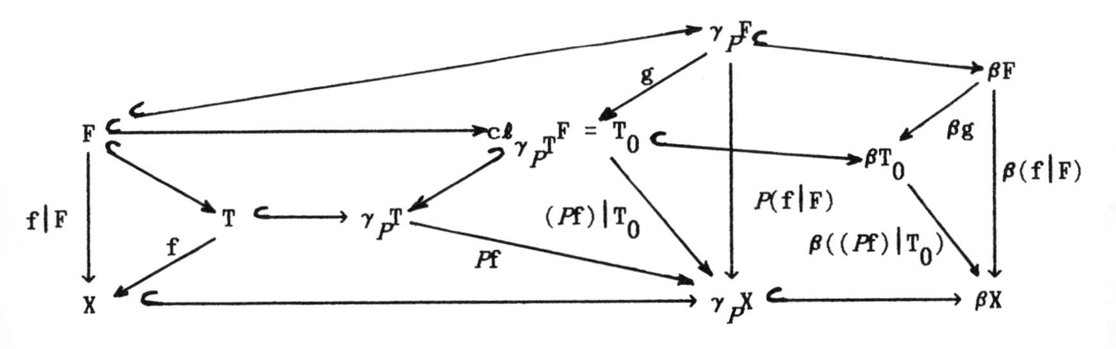

It is routine to show that $((Pf)|T_0 \circ g)|F = f|F$; so, by 1.6(d) we have $(Pf)|T_0 \circ g = P(f|F)$. Applying 1.6(d) again, we have $\beta(f|F) = \beta(Pf|T_0) \circ \beta g$. Since $P(f|F)$ is perfect and $Pf[T_0] \subseteq \Upsilon_P X$, by 4.2(g) $Pf|T_0$ will be a perfect surjection from T_0 onto $\Upsilon_P X$ iff $\beta(Pf|T_0)[\beta T_0 \setminus T_0] \subseteq \beta X \setminus \Upsilon_P X$. Let $t \in \beta T_0 \setminus T_0$ and $x \in (\beta g)^{\leftarrow}(t)$. Then $x \in \beta F \setminus \Upsilon_P F$, and as $P(f|F)$ is perfect, by 4.2(g) $\beta(f|F)(x) \in \beta X \setminus \Upsilon_P X$. But $\beta(f|F)(x) = (\beta(Pf|T_0) \circ \beta g)(x) = \beta(Pf|T_0)(t)$, and our result follows.

(4) ⇒ (1) By (4) there is a closed subset T_0 of $\Upsilon_P(EX)$ such that $Pk_X|T_0$ is a perfect map from T_0 onto $\Upsilon_P X$. Thus $k_{\Upsilon_P X}$ maps the closed set $c\ell_{E(\Upsilon_P X)} T_0$ of $E(\Upsilon_P X)$ onto $\Upsilon_P X$ (see the diagram preceding 8.3(a)). As $k_{\Upsilon_P X}$ is irreducible it follows that T_0 is dense in $E(\Upsilon_P X)$ and thus $T_0 = \Upsilon_P(EX)$. Hence Pk_X is a perfect map from $\Upsilon_P(EX)$ onto $\Upsilon_P X$. By 4.2(g) and the diagram preceding 8.3(a), it follows that $k_{\Upsilon_P X}[E(\Upsilon_P X) \setminus \Upsilon_P(EX)] \subseteq \beta X \setminus \Upsilon_P X$, and hence $\Upsilon_P(EX) = E(\Upsilon_P X)$.

■

We will prove in 8.3(c) below that another set of conditions on an extension property P are equivalent. After doing this we will relate the conditions in 8.3(b) above to those in 8.3(c).

(c) **Theorem**. Let P be a Tychonoff (respectively, zero-dimensional) extension property and let X be a Tychonoff (respectively, zero-dimensional) space. The following conditions are equivalent. (We label them (5) to (8) to put them in sequence with those in 8.3(b)):

(5) every countable locally finite family of open subsets of X is locally finite in $\Upsilon_P X$,

(6) if F is a locally finite family of open subsets of X, and if $|F|$ is not Ulam-measurable (see 2P), then F is locally finite in $\Upsilon_P X$,

(7) if M is a first countable Tychonoff (zero-dimensional, if P-regularity is zero-dimensionality) space then $X \times M$ is P-embedded in $\Upsilon_P X \times M$ (see 5.3(e)), and

(8) if M is a strongly zero-dimensional metric space, then $X \times M$ is P-embedded in $\Upsilon_P X \times M$.

Proof

(5) $\Rightarrow$ (6) Let F be as hypothesized in (6), and suppose that (6) fails. Then F is infinite and there exists $x_0 \in \Upsilon_P X \setminus X$ such that F is not locally finite at x_0. If there were a countable infinite subset G of F such that $x_0 \in \bigcap\{cl_{\Upsilon_P X} G : G \in G\}$, then as G is locally finite (since F is) we would contradict (5). Hence the family $H = \{F \in F : x_0 \in cl_{\Upsilon_P X} F\}$ is finite, and for the family $F_0 = F \setminus H$, $|F_0| = |F|$. Let N be a neighborhood base at x_0 in $\Upsilon_P X$. If $U \in N$ let $F_0(U) = \{F \in F_0 : U \cap F \neq \emptyset\}$. Evidently $\bigcap\{F_0(U) : U \in U\} \supseteq F_0(\bigcap U)$ if U is a finite subfamily of N; so, $\{F_0(U) : U \in N\}$ is a filter base on the lattice $(\mathbb{P}(F_0),\subseteq)$ and hence is contained in some ultrafilter α on $(\mathbb{P}(F_0),\subseteq)$. Since $F_0 \cap H = \emptyset$, it follows that $\bigcap\{F_0(U) : U \in N\} = \emptyset$; so, α is a free ultrafilter on $(\mathbb{P}(F_0),\subseteq)$. Since $|F_0|$ is not Ulam-measurable, neither is $|\mathbb{P}(F_0)|$ (see 2P(5)); by 2P(2) α does not have the countable intersection property. We will now show that in fact α must have the countable intersection property. Thus the assumption that (6) fails will have led to a contradiction and we will be done.

Let $\{S_n : n \in \mathbb{N}\}$ be a decreasing sequence of members of α, and for each $n \in \mathbb{N}$ let $V_n = \cup S_n$ (recall each S_n is a subset of F_0, and hence is a collection of open subsets of X). Thus $\{V_n : n \in \mathbb{N}\}$ is a decreasing sequence of open subsets of X. Note that $x_0 \in c\ell_{\Upsilon_P X} V_n$ for each $n \in \mathbb{N}$; for if $x_0 \notin c\ell_{\Upsilon_P X} V_n$ there exists $U \in N$ such that $U \cap (\cup S_n) = \emptyset$. But then $F_0(U) \cap S_n = \emptyset$, which contradicts the fact that $F_0(U)$ and S_n both are in α. Thus $\{V_n : n \in \mathbb{N}\}$ is not locally finite in $\Upsilon_P X$; so, by (5) $\{V_n : n \in \mathbb{N}\}$ is not locally finite in X. So choose $x_1 \in \cap\{c\ell_X V_n : n \in \mathbb{N}\}$. Since F_0 is locally finite in X, for each $n \in \mathbb{N}$ we can find $S_n \in S_n$ such that $x_1 \in c\ell_X S_n$. As $\{S_n : n \in \mathbb{N}\}$ is locally finite (being a subfamily of F_0), it follows that it is in fact finite. Thus $\cap S_n \neq \emptyset$ (as $S_n : n \in \mathbb{N}\}$ is decreasing). Hence α has the countable intersection property, and we have achieved the previously advertised contradiction.

(6) $\Rightarrow$ (7) We will assume that X is a Tychonoff space and that P is a Tychonoff extension property, and indicate in parentheses the modifications needed for the zero-dimensional case.

We begin by showing that $X \times M$ is C^*-embedded in $\Upsilon_P X \times M$. By 4.1(m) it suffices to show that if E_1 and E_2 are disjoint closed subsets of $[0,1]$ ($\{0,1\}$ in the zero-dimensional case) and $f \in C(X \times M, [0,1])$ then $c\ell_Y f^{\leftarrow}[E_1] \cap c\ell_Y f^{\leftarrow}[E_2] = \emptyset$ (where Y denotes $\Upsilon_P X \times M$). To this end, let $(y_0, t_0) \in Y \setminus (X \times M)$. Define $h \in C(X,[0,1])$ by $h(x) = f(x,t_0)$ for each $x \in X$. Since $X \subseteq \Upsilon_P X \subseteq \beta X$ by 5.9(f), there exists $g \in C(\Upsilon_P X,[0,1])$ such that $g|X = h$ (replace βX by $\beta_0 X$ in the zero-dimensional case). Now $y_0 \in \Upsilon_P X \setminus X$; since $E_1 \cap E_2 = \emptyset$ we can assume that $g(y_0) \notin E_2$. Choose an open subset U of $[0,1]$ such that $E_1 \cup \{g(y_0)\} \subseteq U \subseteq c\ell_{[0,1]} U \subseteq$

$[0,1]\setminus E_2$, and let $\{V_n : n \in \mathbb{N}\}$ be a neighborhood base at t_o in M with $V_{n+1} \subseteq V_n$ for each n. Set

$$H_n = \bigcup\{H : H \text{ is open in } X \text{ and } H \times V_n \subseteq f^{\leftarrow}[U]\} \quad \text{for each } n \in \mathbb{N}\}.$$

Then

$$\begin{aligned} f[(c\ell_X H_n \times V_n) \cap f^{\leftarrow}[E_2]] &= f[c\ell_X H_n \times V_n] \cap E_2 \\ &\subseteq (c\ell_{[0,1]} f[H_n \times V_n]) \cap E_2 \\ &\subseteq (c\ell_{[0,1]} U) \cap E_2 \\ &= \emptyset. \end{aligned}$$

So $(c\ell_X H_n \times V_n) \cap f^{\leftarrow}[E_2] = \emptyset$. Choose an open set T of [0,1] so that $E_1 \cup \{g(y_o)\} \subseteq T \subseteq c\ell_{[0,1]}T \subseteq U$. Now put

$$G_n = (g^{\leftarrow}[T] \cap X)\setminus c\ell_X H_n$$

Evidently $H_{n+1} \supseteq H_n$ for each $n \in \mathbb{N}$; so, $\{G_n : n \in \mathbb{N}\}$ is a decreasing sequence of open sets of X. It is routine to verify that $g^{\leftarrow}[U] \cap X = \bigcup\{H_n : n \in \mathbb{N}\}$, and from this it follows that $\bigcap\{c\ell_X G_n : n \in \mathbb{N}\} = \emptyset$; in other words, $\{G_n : n \in \mathbb{N}\}$ is a locally finite collection of open subsets of X. By hypothesis it is a locally finite family in $\Upsilon_p X$; hence there exists $m \in \mathbb{N}$ such that
$y_o \notin c\ell_{\Upsilon_p X} G_m$. Let $W = g^{\leftarrow}[U]\setminus c\ell_{\Upsilon_p X} G_m$. Then $W \times V_m$ is a Y-neighborhood of (y_o,t_o). A straightforward calculation shows that $W \cap X \subseteq c\ell_X H_m$; since $c\ell_X H_m \times V_m \subseteq f^{\leftarrow}[c\ell_{[0,1]}U] \subseteq (X \times$

$M)\backslash f^{\leftarrow}[E_2]$, it follows that $(W \times V_m) \cap f^{\leftarrow}[E_2] = \varnothing$. Thus $(y_o,t_o) \notin c\ell_Y f^{\leftarrow}[E_2]$. Thus our assumption that $g(y_o) \notin E_2$ has led to the conclusion that $(y_o,t_o) \notin c\ell_Y f^{\leftarrow}[E_2]$. But if $(y_o,t_o) \in c\ell_Y f^{\leftarrow}[E_1]$ then $g(y_o) = f(y_o,t_o) \in f[c\ell_Y f^{\leftarrow}[E_1]] \subseteq c\ell_{[0,1]}E_1 = E_1 \subseteq [0,1]\backslash E_2$. Hence $c\ell_Y f^{\leftarrow}[E_1] \cap c\ell_Y f^{\leftarrow}[E_2] = \varnothing$. It follows (as noted above) that $X \times M$ is C^*-embedded in $\Upsilon_P X \times M$; so, $\Upsilon_P X \times M \subseteq \beta(X \times M)$.

If J has P and $f \in C(X \times M,J)$ then f extends continuously to $F \in C(\Upsilon_P(X \times M), J)$. Now $\Upsilon_P(X \times M)$ is a subspace of $\beta(X \times M)$, so to show that $X \times M$ is P-embedded in $\Upsilon_P X \times M$ it suffices to show that $\Upsilon_P X \times M \subseteq \Upsilon_P(X \times M)$. Suppose not; choose $(y_1,t_1) \in (\Upsilon_P X \times M)\backslash\Upsilon_P(X \times M)$ and define Z as follows:

$$Z = \{z \in \Upsilon_P X : (z,t_1) \in (\Upsilon_P X \times M) \cap \Upsilon_P(X \times M)\}.$$

Evidently Z is homeomorphic to $\Upsilon_P(X \times M) \cap (\Upsilon_P X \times \{t_1\})$, and so by 5.9(e) Z has P. But $X \subseteq Z \subseteq \Upsilon_P X$ and $y_1 \in \Upsilon_P X\backslash Z$, which contradicts 5.9(f). Hence $X \times M$ is P-embedded in $\Upsilon_P X \times M$.

(7) ⇒ (8) Obvious.

(8) ⇒ (5) Let $\{G_n : n \in \mathbb{N}\}$ be a countable locally finite family of open sets of X that is not locally finite at $x_o \in \Upsilon_P X\backslash X$; in other words, suppose (5) fails. Let $T = (X \times \mathbb{N}) \cup \{\infty\}$ and topologize T as follows: each point of $X \times \mathbb{N}$ is isolated, and $\{W_n : n \in \mathbb{N}\}$ is a neighborhood base at ∞, where $W_n = (X \times \{i \in \mathbb{N} : i > n\}) \cup \{\infty\}$. If we define $d : T \times T \to [0,+\infty)$ by $d((x_1,n_1),$

$(x_2,n_2)) = 1$ if $(x_1,n_1) \neq (x_2,n_2)$ and $d((x,n),\infty) = 1/n$, then d is a metric compatible with the topology of T. It is straightforward to verify that if F and H are disjoint closed subsets of T then there is a clopen subset A of T containing F and disjoint from H; hence by 4.7(g) T is a strongly zero-dimensional metric space.

For each $n \in \mathbb{N}$ and each $y \in G_n$ there exists $f_{ny} \in C(X,[0,1])$ such that $f_{ny}(y) = 0$ and $f_{ny}[X\backslash G_n] \subseteq \{1\}$. (Use $\{0,1\}$ in place of $[0,1]$ if P-regularity is zero-dimensionality). Now define $f \in F(X \times T,[0,1])$ as follows:

$$\begin{aligned} f(z,t) &= f_{ny}(z) \text{ if } t = (y,n) \in G_n \times \{n\} \\ &= 1 \text{ otherwise.} \end{aligned}$$

We now verify that f is continuous. Let $p_0 = (x,t) \in X \times T$. If $t \in G_n \times \{n\}$ for some $n \in \mathbb{N}$ and $f(x,t) \in V$ where V is open in $[0,1]$, then $(x,t) \in f_{ny}^{\leftarrow}[V] \times \{t\} \subseteq f^{\leftarrow}[V]$; so, f is continuous at p_0 as $f_{ny}^{\leftarrow}[V] \times \{t\}$ is open in $X \times T$. If $t \in T\backslash(\{\infty\} \cup \cup\{G_n \times \{n\} : n \in \mathbb{N}\})$, then $(x,t) \in X \times \{t\} \subseteq f^{\leftarrow}(f(x,t))$ and so f is continuous at p_0. It remains to verify that f is continuous at p_0 when $t = \infty$. In this case $f(p_0) = 1$. As $\{G_n : n \in \mathbb{N}\}$ is locally finite in X, there exists $j \in \mathbb{N}$ and an open set U of X such that $x \in U$ and $U \cap G_n = \varnothing$ if $n > j$. Now $(U \times W_j) \cap (G_n \times (G_n \times \{n\})) = \varnothing$ for each $n \in \mathbb{N}$; so, it follows that $f[U \times W_j] = \{1\}$ and f is continuous at p_0. Thus $f = C(X \times T,[0,1])$.

We now show that f cannot be extended continuously to $\gamma_P X \times T$; specifically, we show that f cannot be extended continuously to $(X \times T) \cup \{(x_0,\infty)\}$. For let $V \times W_k$ be a basic neighborhood of (x_0,∞) in $\Upsilon_p X \times T$. By definition of x_0, V meets infinitely many G_n so there exists $m > k$ for which we can find $y \in V \cap G_m$. Then both $p_1 = (y,(y,m))$ and $p_2 = (y,\infty)$ belong to $V \times W_k$ and $f(p_1) = f_{my}(y) = 0$ while $f(p_2) = 1$. Thus any $\Upsilon_p X \times T$-neighborhood of (x_0,∞) is mapped by any extension of f to a set that contains both 0 and 1. Hence f cannot be extended continuously to $\Upsilon_p X \times T$.

■

Now we link the equivalent conditions given in 8.3(b) to the equivalent conditions given in 8.3(c). First we need a definition and a lemma.

(d) <u>Definition</u>. Let X and Y be spaces and let $f \in C(X,Y)$. Then f is called **(countably) biquotient** if, whenever $y \in Y$ and $\mathcal{C}$ is a (countable) cover of $f^{\leftarrow}(y)$ by open sets of X, there exists a neighborhood V of y in Y, and a finite subcollection $\mathcal{F}$ of $\mathcal{C}$, such that $V \subseteq \cup\{f[U] : U \in \mathcal{F}\}$.

(e) <u>Lemma</u>. Let X and Y be spaces and let $f \in C(X,Y)$. Then:

(1) if f is either open or perfect, then f is biquotient, and

(2) if f is a countably biquotient surjection, then f is a quotient map.

<u>Proof</u>

(1) If f is perfect and $y \in Y$, let $\mathcal{C}$ be a collection of open sets of X that covers $f^{\leftarrow}(y)$. As $f^{\leftarrow}(y)$ is compact there exists a

finite subcollection $\mathcal{F}$ of $\mathcal{C}$ such that $f^{\leftarrow}(y) \subseteq \bigcup\mathcal{F}$. Then $\bigcup\{Y \setminus f[X \setminus F] : F \in \mathcal{F}\}$ is a Y-neighborhood V of y by 1.8(c) and evidently $V \subseteq \bigcup\{f[F] : F \in \mathcal{F}\}$. If f is open then choose $C \in \mathcal{C}$ such that $f^{\leftarrow}(y) \cap C \neq \varnothing$; then f[C] is the required V.

(2) We must show that if $W \subseteq Y$ and $f^{\leftarrow}[W]$ is open in X, then W is open in Y. But $\{f^{\leftarrow}[W]\}$ is an open cover of $f^{\leftarrow}(y)$ for each $y \in W$, so by hypothesis there exists a Y-neighborhood V(y) of y such that $V(y) \subseteq f[f^{\leftarrow}[W]] = W$. Thus $W = \bigcup\{V(y) : y \in W\}$ and hence is open in Y.

■

(f) **Lemma**. Consider the following conditions on a Tychonoff or zero-dimensional extension property P and a P-regular space X.

(9) The extension $Pk_X : \Upsilon_P(EX) \to \Upsilon_P X$ is a biquotient surjection.

(10) The extension $Pk_X : \Upsilon_P(EX) \to \Upsilon_P X$ is a countably biquotient surjection.

Then (9) implies (10) and each of the equivalent conditions of 8.3(b) implies (9).

Proof. Biquotient maps are countably biquotient; so, (9) implies (10). Since $k_X : EX \to X$ is a perfect irreducible continuous surjection, if 8.3(b)(3) holds then $Pk_X : \Upsilon_P(EX) \to \Upsilon_P X$ is a perfect surjection. By 8.3(e) above Pk_X would then be biquotient.

■

Now we can relate the conditions of 8.3(b) to those of 8.3(c). See 8E for another condition equivalent to those given below.

(g) **Theorem**. Let P be a Tychonoff or zero-dimensional extension property. The following are equivalent:

(1) if the P-regular space X satisfies (10) of 8.3(f) then it satisfies (5) of 8.3(c),

(2) $\mathbb{N}$ has P (equivalently, it is not true that every space with P is countably compact; see 5R(11)), and

(3) every extremally disconnected realcompact space has P.

Proof.

(1) $\Rightarrow$ (2) As $\mathbb{N}$ is P-regular and extremally disconnected we know that $E\mathbb{N} = \mathbb{N}$ and thus $Pk_{\mathbb{N}}$ is a homeomorphism and thus certainly countably biquotient. Hence (10) of 8.3(f) is satisfied; so, by hypothesis (5) of 8.3(c) is satisfied. If $x_0 \in \gamma_P\mathbb{N}\setminus\mathbb{N}$, then $\{\{n\} : n \in \mathbb{N}\}$ would be a countable family of open sets that is locally finite in $\mathbb{N}$, yet fails to be locally finite in $\gamma_P\mathbb{N}$ as each neighborhood of x_0 would intersect infinitely many members of the family. Hence (5) would be violated. We conclude that $\gamma_P\mathbb{N} = \mathbb{N}$, i.e., that $\mathbb{N}$ has P.

(2) $\Rightarrow$ (3) By (2) every $\mathbb{N}$-compact space has P. By 6.4 every extremally disconnected realcompact space is strongly zero-dimensional, and hence by 5G(3) is $\mathbb{N}$-compact. The result follows.

(3) $\Rightarrow$ (1) Suppose that X is a P-regular space satisfying (10) of 8.3(f). We must show that X satisfies (5) of 8.3(c). So, let $\{G_n : n \in \mathbb{N}\}$ be a countable family of open subsets of X that is locally finite in X and let $y \in \gamma_P X\setminus X$. We may assume $G_1 = X$.

For each $n \in \mathbb{N}$ let $H_n = c\ell_{EX} k_X^{\leftarrow}[G_n]$. Then H_n is clopen in EX as EX is extremally disconnected. Let $D_n = H_n \setminus \bigcup\{H_i : i > n\}$ for each $n \in \mathbb{N}$. Because $\{G_n : n \in \mathbb{N}\}$ is locally finite in X it follows easily that $\{H_n : n \in \mathbb{N}\}$ is locally finite in EX. Thus $\bigcup\{H_i : i > n\}$ is closed in EX and so D_n is clopen in EX. As $G_1 = X$ it quickly follows that $\bigcup\{D_n : n \in \mathbb{N}\} = EX$; so, $EX = \oplus\{D_n : n \in \mathbb{N}\}$ (see 1.2(h)). As $\mathbb{N}$ has P, it follows from 5AB(2) that $\gamma_P(EX) = \oplus\{\gamma_P D_n : n \in \mathbb{N}\}$. Because Pk_X is countably biquotient onto $\gamma_P X$ by hypothesis, and $\{\gamma_{\mathbb{P}} D_n : n \in \mathbb{N}\}$ is a countable open cover of $(Pk_X)^{\leftarrow}(y)$, there exists a neighborhood U of y in $\gamma_P X$, and $m \in \mathbb{N}$ such that $U \subseteq \bigcup\{Pk_X[\gamma_P D_n] : n \leq m\}$. Since

$$\begin{aligned}\varnothing &= \bigcup\{\gamma_P D(n) : n \leq m\} \cap [\bigcup\{H_i : i > m\}] \\ &\supseteq \bigcup\{\gamma_P D(n) : n \leq m\} \cap (Pk_X)^{\leftarrow}[\bigcup\{G_i : i > m\}],\end{aligned}$$

it follows that $U \cap (\bigcup\{G_i : i > m\}) = \varnothing$. As y was arbitrarily chosen in $\gamma_P X \setminus X$, it follows that $\{G_n : n \in \mathbb{N}\}$ is locally finite in $\gamma_P X$. Thus (5) of 8.2(c) holds as claimed.

■

We have now shown, in particular, that if P is a Tychonoff or zero-dimensional extension property possessed by $\mathbb{N}$ and if X is a P-regular space for which $\gamma_P(EX) = E(\gamma_P X)$, then any countable family (equivalently, any family whose cardinality is not Ulam-measurable) of open sets of X that is locally finite in X must also be locally finite in $\gamma_P X$. The final result in the theory that we are developing will be to prove the converse of this implication if every space with P is almost realcompact.

(h) **Theorem**. Let P be a Tychonoff or zero-dimensional extension property. Suppose every space with P is almost realcompact. If X is a P-regular space satisfying condition (6) of 8.3(c), then X satisfies condition (2) of 8.3(b).

Proof. Suppose X does not satisfy condition (2) of 8.3(b). By 4.2(f) there is a point $p \in \beta(EX)\setminus\gamma_P(EX)$ such that $k_{\beta X}(p) \in \gamma_P X$ (see the diagram preceding 8.3(a)). Since P is contained in almost realcompactness, it follows that every extremally disconnected space with P is realcompact (see 6U) and so $\upsilon(EX) \subseteq \gamma_P(EX)$ (see 5.9(h)). Thus by 5.11(c) there exists $h \in C(\beta(EX),[0,1])$ such that $h(p) = 0$ and $h(y) > 0$ if $y \in EX$. For each $n \in \mathbb{N}$ let $H_n = EX \cap \text{int}_{EX} h^{\leftarrow}[[0,1/n]]$. Evidently $\{H_n : n \in \mathbb{N}\}$ is a locally finite (in EX) collection of regular open sets of EX. For each $n \in \mathbb{N}$ let $G_n = X\setminus k_X[EX\setminus H_n]$. Now $EX\setminus H_n = A_n \in R(EX)$ and by 6.5(d)(3) $c\ell_X G_n = (k_X[A_n])' = k_X[A_n{}'] = k_X[c\ell_{EX} H_n]$. As k_X is a closed function each G_n is open in X. We now show that $\{G_n : n \in \mathbb{N}\}$ is locally finite in X. Since $\{H_n : n \in \mathbb{N}\}$ is locally finite in EX, it follows that $k_X^{\leftarrow}(r) \cap \bigcap\{c\ell_{EX} H_n : n \in A\} = \varnothing$ for each $r \in X$ and each infinite subset A of $\mathbb{N}$. As $k_X^{\leftarrow}(r)$ is compact for each $r \in X$, and $\{H_n : n \in \mathbb{N}\}$ is decreasing, it follows that there exists some $j \in \mathbb{N}$ such that $k_X^{\leftarrow}(r) \cap c\ell_{EX} H_j = \varnothing$. Thus $r \notin k_X[c\ell_{EX} H_j] = c\ell_X G_j$. But $\{G_n : n \in \mathbb{N}\}$ is decreasing as $\{H_n : n \in \mathbb{N}\}$ is; so, $r \notin c\ell_X G_n$ if $n \geq j$. Hence $\{G_n : n \in \mathbb{N}\}$ is locally finite. However, it is evident that $p \in c\ell_{\beta(EX)} H_n$ for each n; thus, $k_{\beta X}(p) \in \bigcap\{c\ell_{\beta X} G_n : n \in \mathbb{N}\}$. As $k_{\beta X}(p) \in \gamma_P X$ it follows that $\{G_n : n \in$

$\mathbb{N}\}$ is not locally finite in $\gamma_P X$. Hence X does not satisfy condition (6) of 8.3(c).

■

We collect our results into one major theorem.

(i) **Theorem**. Let P be a Tychonoff or zero-dimensional extension property and let X be P-regular. If every space with P is almost realcompact, then conditions (1) to (10) listed in 8.3(b), 8.3(c), and 8.3(f) are equivalent conditions on X.

Proof. If $\mathbb{N}$ has P this follows from 8.3(b), 8.3(c), 8.3(f), 8.3(g) and 8.3(h). If $\mathbb{N}$ does not have P then every space with P is countably compact (see 5R(11)), and so P is the property of being compact and P-regular (see 6U(3) and 4.8(c)). The equivalences are now immediate.

■

It is worth noting that 8.3(h) is the strongest possible result in the following sense: if P is a Tychonoff or zero-dimensional extension property that is not just Reg(P), and if there is a space with P that is not almost realcompact and whose cardinality is not Ulam-measurable, then one can produce a P-regular space X satisfying (6) of 8.3(c) but not (2) of 8.3(b). The reader is invited to prove this assertion in 8F.

Our next goal is to use 8.3(i) to obtain an internal condition on a Tychonoff space X that is equivalent to $E(\upsilon X) = \upsilon(EX)$. To do this it will be useful to introduce the notion of a stable ultrafilter on

$R(X)$. This notion is a slight variation on the notion of a P-stable z-ultrafilter that was discussed in 5Y.

(j) **Definition**. An ultrafilter $\mathcal{Q}$ on $R(X)$ is called **stable** if, for each $f \in C(X)$, there exists $\mathcal{Q}(f) \in \mathcal{Q}$ such that $cl_{\mathbb{R}} f[\mathcal{Q}(f)]$ is compact.

The central lemma linking stable ultrafilters on $R(X)$ to the equality $E(\upsilon X) = \upsilon(EX)$ is the following.

(k) **Lemma**. The following are equivalent for an ultrafilter $\mathcal{Q}$ on $R(X)$:

(1) $\mathcal{Q}$ is stable and

(2) the unique point of $\cap\{cl_{\beta X} A : A \in \mathcal{Q}\}$ belongs to υX.

Proof

(1) $\Rightarrow$ (2) By 3B(4) $\{cl_{\beta X} A : A \in \mathcal{Q}\}$ is an ultrafilter on $R(\beta X)$ and as we saw when we constructed the absolute in 6.6, there is a unique point p in $\cap\{cl_{\beta X} A : A \in \mathcal{Q}\}$. Suppose that (2) fails and that $p \in \beta X \backslash \upsilon X$. By 5.11(c) there exists a function $f \in C^*(\beta X)$ such that $f(p) = 0$ and $f(x) > 0$ for each $x \in X$. For each $n \in \mathbb{N}$, $f^{\leftarrow}[[0,1/n)]$ is a neighborhood of p and so $(int_{\beta X} cl_{\beta X} A) \cap f^{\leftarrow}[[0,1/n)] \neq \varnothing$ for each $A \in \mathcal{Q}$. It follows that $A \cap f^{\leftarrow}[[0,1/n)] \neq \varnothing$ for each $A \in \mathcal{Q}$ and $n \in \mathbb{N}$. Since $f(x) > 0$ for each $x \in X$ we can define $f \in C(X)$ by letting $g(x) = 1/f(x)$. Let $A \in \mathcal{Q}$; for each $n \in \mathbb{N}$ we can choose $a_n \in A \cap f^{\leftarrow}[[0,1/n)]$. Then $g(a_n) > n$ and g is unbounded on A. Thus $\mathcal{Q}$ is not stable and (1) fails.

(2) ⇒ (1) Suppose $\cap\{c\ell_{\beta X}A : A \in \mathcal{Q}\} = \{p\}$ and let $f \in C(X)$. Then by 5.10(b) there exists $\upsilon f \in C(\upsilon X)$ such that $\upsilon f | X = f$. Thus $\upsilon f(p) = r \in \mathbb{R}$. Let $\delta > 0$ and put $B = c\ell_{\upsilon X}[(\upsilon f)^{\leftarrow}[(r-\delta, r+\delta)]]$. Then $B \in R(\upsilon X)$ and B is an υX-neighborhood of p; so, it follows that $\mathrm{int}_{\upsilon X}c\ell_{\upsilon X}B \cap \mathrm{int}_{\upsilon X}c\ell_{\upsilon X}A \neq \emptyset$ for each $A \in \mathcal{Q}$. By 3B(4) $C = B \cap X \in R(X)$ and it follows from the above that $C \wedge A \neq \emptyset$ for each $A \in \mathcal{Q}$. Thus by 2.3(d)(3) $C \in \mathcal{Q}$. Evidently $f[C] \subseteq [r-\delta, r+\delta]$ and so f is bounded on some member of $\mathcal{Q}$. As f was arbitrarily chosen, $\mathcal{Q}$ is stable.

■

Now we can given an internal characterization of those Tychonoff spaces X for which $E(\upsilon X) = \upsilon(EX)$.

(l) **Theorem.** The following are equivalent for any Tychonoff space X:

(1) $E(\upsilon X) = \upsilon(EX)$,

(2) if $\{A_n : n \in \mathbb{N}\}$ is a decreasing sequence of members of $R(X)$ and $\cap\{A_n : n \in \mathbb{N}\} = \emptyset$ then $\cap\{c\ell_{\upsilon X}A_n : n \in \mathbb{N}\} = \emptyset$,

(3) if $\{A_n : n \in \mathbb{N}\}$ is a decreasing sequence of members of $R(X)$ then $c\ell_{\upsilon X}[\cap\{A_n : n \in \mathbb{N}\}] = \cap\{c\ell_{\upsilon X}A_n : n \in \mathbb{N}\}$, and

(4) every stable ultrafilter on $R(X)$ has the countable intersection property.

Proof

(1) ⇒ (2) If $\{A_n : n \in \mathbb{N}\}$ is a decreasing family of members of $R(X)$ and $\bigcap\{A_n : n \in \mathbb{N}\} = \emptyset$, then $\{int_X A_n : n \in \mathbb{N}\}$ is a locally finite family of open sets of X. Since $E(\upsilon X) = \upsilon(EX)$, by 8.3(i) it follows that $\{int_X A_n : n \in \mathbb{N}\}$ is a locally finite family in υX. Thus $\bigcap\{c\ell_{\upsilon X}(int_X A_n) : n \in \mathbb{N}\} = \emptyset$, i.e., $\bigcap\{c\ell_{\upsilon X} A_n : n \in \mathbb{N}\} = \emptyset$.

(2) ⇒ (3) Obviously $c\ell_{\upsilon X}[\bigcap\{A_n : n \in \mathbb{N}\}] \subseteq \bigcap\{c\ell_{\upsilon X} A_n : n \in \mathbb{N}\}$. To prove the opposite inclusion, suppose that $p \notin c\ell_{\upsilon X}[\bigcap\{A_n : n \in \mathbb{N}\}]$. Find $B \in R(\upsilon X)$ such that $p \in int_{\upsilon X} B$ and $B \cap [\bigcap\{A_n : n \in \mathbb{N}\}] = \emptyset$. By 3B(4) $C = B \cap X \in R(X)$ and $\{C \wedge A_n : n \in \mathbb{N}\}$ is a decreasing sequence in $R(X)$ with empty intersection. By (2) we deduce that $\bigcap\{c\ell_{\upsilon X}(C \wedge A_n) : n \in \mathbb{N}\} = \emptyset$. If $p \in \bigcap\{c\ell_{\upsilon X} A_n : n \in \mathbb{N}\}$ and W is an υX-neighborhood of p, it follows from 3B(4) that $W \cap int_{\upsilon X} B \cap int_{\upsilon X} c\ell_{\upsilon X} A_n \neq \emptyset$ each n, and so $W \cap (C \wedge A_n) \neq \emptyset$. Hence $p \in \bigcap\{c\ell_{\upsilon X}(C \wedge A_n) : n \in \mathbb{N}\}$, which from the above cannot happen. Thus $p \notin \bigcap\{c\ell_{\upsilon X} A_n : n \in \mathbb{N}\}$ and the opposite inclusion is proved.

(3) ⇒ (4) Let $\mathcal{Q}$ be a stable ultrafilter on $R(X)$ and let $\{B_n : n \in \mathbb{N}\}$ be a countable subfamily of it. Let $A_n = \wedge\{B_i : 1 \leq i \leq n\}$. Then $\{A_n : n \in \mathbb{N}\}$ is a decreasing sequence of members of $\mathcal{Q}$. Since $\mathcal{Q}$ is stable it follows from 8.3(k) that there exists $p \in \upsilon X$ such that $p \in \bigcap\{c\ell_{\upsilon X} A_i : i \in \mathbb{N}\}$. By (3) it follows that $\bigcap\{A_i : i \in \mathbb{N}\} \neq \emptyset$ and so $\bigcap\{B_i : i \in \mathbb{N}\} \neq \emptyset$. Thus $\mathcal{Q}$ has the countable intersection property.

(4) ⇒ (1) By 8.3(i) it suffices to show that if $\{G_n : n \in \mathbb{N}\}$ is a locally finite family of open subsets of X, then $\{G_n : n \in \mathbb{N}\}$ is locally finite in υX. For each $n \in \mathbb{N}$ let

$$V_n = \cup\{G_i : i \in \mathbb{N}\}\backslash c\ell_X[\cup\{G_i : 1 \leq i \leq n\}].$$

Then $\{c\ell_X V_n : n \in \mathbb{N}\}$ is a decreasing sequence in $R(X)$. Suppose that $p \in \cap\{c\ell_{\upsilon X} V_n : n \in \mathbb{N}\}$. It is straightforward to verify that $\{c\ell_X V_n : n \in \mathbb{N}\} \cup \{B \cap X : B \in R(\upsilon X)$ and $p \in \text{int}_{\upsilon X} B\}$ is a filter subbase on $R(X)$ and hence is contained in some ultrafilter $\mathcal{Q}$ on $R(X)$ (see 2.3(d)). As $B \cap X \in \mathcal{Q}$ for all regular closed υX-neighborhoods B of p, it follows that $\cap\{c\ell_{\beta X} A : A \in \mathcal{Q}\} = \{p\} \subseteq \upsilon X$. Thus by 8.3(k) $\mathcal{Q}$ is stable, and by (4) $\mathcal{Q}$ has the countable intersection property. It follows that $\cap\{c\ell_X V_n : n \in \mathbb{N}\} \neq \emptyset$. If $q \in \cap\{c\ell_X V_n : n \in \mathbb{N}\}$, then for each $n \in \mathbb{N}$ and each neighborhood U of q, $U \cap V_n \neq \emptyset$ so there exists $j_n \geq n$ such that $U \cap G_{j_n} \neq \emptyset$. It follows that $\{G_{j_n} : n \in \mathbb{N}\}$ is not locally finite at q, in contradiction to the definition. We thus conclude that $\cap\{c\ell_{\upsilon X} V_n : n \in \mathbb{N}\} = \emptyset$. From this it quickly follows (arguing as above) that $\{G_n : n \in \mathbb{N}\}$ is locally finite in υX as required.

■

8.4 Covers of topological spaces

In Chapter 4 we introduced the notion of an extension of a space X, and we defined a partial order on the set of all (equivalence classes of) extensions of X. We later studied, for particular topological properties P, the "largest" P-extension of a P-regular space. In fact Chapter 5 was devoted entirely to studying those topological properties P for which each P-regular space X has a maximum P-extension.

In this section we go part way towards developing a theory that is in some ways "dual" to the theory of maximum P-extensions. Instead of studying extensions of a space X (i.e., spaces in which X is densely embedded) we will study **covers** of a space (i.e., pairs (Y,f), where Y is a space that is mapped onto X by a perfect irreducible continuous surjection f). Just as we defined what it meant for two extensions to be equivalent, and then identified equivalent extensions, we will define what it means for two covers to be equivalent, and then identify equivalent covers. We then will define a partial order on the "set" of covers of a space X in a manner similar to that in which we defined a partial order on the "set" $E(X)$ of extensions of X on 4.1. Then we will define "covering properties" in a way similar to that used to define "extension properties" in Chapter 5. Finally, we construct "minimum P-covers" that are analogous to the "maximum P-extensions" discussed in Chapter 5, and look at some specific examples. A category-theoretic interpretation of what we are doing appears in 9.7 and in 9M.

We being by formalizing the definition of "cover". The Iliadis and Banaschewski absolutes introduced in Chapter 6 are "projective covers" in certain categories (see Chapter 9), and our use of the word "cover" is intended to be a generalization of this terminology. Other authors (e.g., [MR_1]) have used the word "resolution".

(a) <u>**Definition**</u>

(1) Let X be a space. A pair (Y,f) is called a **cover** of X if Y is a space and f is a perfect irreducible continuous surjection from Y

onto X.

(2) A perfect irreducible continuous surjection will be called a **covering map**.

The reader should note that the term "covering map" is given a different meaning by other authors. Note that by 1J if (Y,f) is a cover of a regular space X, then Y must be regular. As noted above, the Banaschewski absolute (PX,Π_X) is a cover of X, and we can view covers as being a generalization of absolutes.

Instead of proceeding directly to a consideration of the set of covers of a given space, we first define the poset of spaces "covered by" a given space.

(b) **Definition**. Let E be a space. Then AB(E) will denote a collection of pairs (c_X,X) such that: (i) X is a space and $c_X : E \to X$ is a covering map; (ii) if (c_X,X) and (c_Y,Y) are distinct members of AB(E) then there is no homeomorphism $h : X \to Y$ such that $c_Y = h \circ c_X$; (iii) if Y is a space and there is a covering map $f \in C(E,Y)$, then there is a pair $(c_Z,Z) \in AB(E)$ and a homeomorphism $h : Y \to Z$ such that $h \circ f = c_Z$.

Suppose $g \in C(E,Y)$ and $f \in C(E,X)$ are covering maps. We consider the pairs (g,Y) and (f,X) to be **equivalent** if there is a homeomorphism $h : Y \to X$ such that $f = h \circ g$. In constructing AB(E) we are choosing exactly one representative from each "equivalence class" defined by the above relation. This is essentially the same process that we used in constructing $E(X)$ in 4.1. In fact, if we had defined an extension of a space X to be a pair (i, eX) consisting of a dense embedding map $i : X \to eX$ and a space eX, then our definition of equivalence in 4.1(d) would have been rephrased as follows: (i_1, e_1X) is equivalent to (i_2, e_2X) if there is a homeomorphism $h : e_1X \to e_2X$ such that $h \circ i_1 = i_2$. When we studied extensions we thought of a space X as being a dense subspace of an "extension space" eX, and we thought of the dense embedding map $i : X \to eX$ as being the inclusion map. Hence we seldom mentioned i explicitly. In considering analogies between extensions and covers in what is to follow, it is useful to think of an extension of X as a **pair** (i,eX) as described above.

It should be noted that there may be distinct members of $AB(E)$ whose second components are homeomorphic - i.e., there may be pairs (c_X,X) and (c_Y,Y) such that X and Y are homeomorphic but there is no homeomorphism $h : X \to Y$ such that $c_Y = h \circ c_X$. The reader should compare this to 4AB, where exactly the same phenomenon occurs for extensions.

Finally, note that $\{X : (c_X,X) \in AB(E)\}$ is a set (rather than a proper class).

(c) <u>Definition</u>. Let E be a space and let (c_X,X) and (c_Y,Y) be in $AB(E)$. Define (c_X,X) to be **projectively smaller** than (c_Y,Y), written as $(c_X,X) \leq (c_Y,Y)$, if there exists $f \in C(Y,X)$ such that $f \circ c_Y = c_X$. (Note that since c_X is onto, then so is f).

(d) <u>Lemma</u>. Let X, Y, Z be spaces, $f \in C(X,Y)$, $g \in C(Y,Z)$, and f be surjective. Then:

(1) if $g \circ f$ is perfect, then both f and g are perfect and

(2) if g is surjective and $g \circ f$ is a covering map, then f and g

are covering maps.

Proof. Let $h = g \circ f$.

(1) If $y \in Y$, then one easily checks that $f^{\leftarrow}(y)$ is a closed subset of the compact set $h^{\leftarrow}(g(y))$; thus, f is a compact function. Similarly, if $z \in Z$, then $g^{\leftarrow}(z)$ is a closed subset of $f[h^{\leftarrow}(z)]$ which is the continuous image of the compact space $h^{\leftarrow}(z)$ and hence is compact. Thus, g is a compact function. If A is a closed subset of Y, then it is easy to verify that $g[A] = h[f^{\leftarrow}[A]]$; so, as f is continuous and h is closed, g[A] is closed. Thus, g is closed and compact, i.e., perfect. Suppose that A is a closed subset of X and that f[A] is not closed in Y. Choose $p \in c\ell_Y f[A] \setminus f[A]$. Then $g(p) \in c\ell_Z g[f[A]] = c\ell_Z h[A] = h[A]$ (as g is continuous and h is closed). It follows that $A \cap h^{\leftarrow}(g(p))$ is a nonempty compact subset of X, and so $f[A \cap h^{\leftarrow}(g(p))] = f[A] \cap g^{\leftarrow}[g(p)]$ is a compact subset of Y. Hence, we can choose disjoint open subsets U and V of Y such that $p \in U$ and $f[A] \cap g^{\leftarrow}(g(p)) \subseteq V$. Thus, $A \setminus f^{\leftarrow}[V]$ is a closed subset of X disjoint from $h^{\leftarrow}(g(p))$; so, as h is closed it follows that $Z \setminus h[A \setminus f^{\leftarrow}[V]]$ is a neighborhood of g(p) in Z. But $p \in c\ell_Y(U \cap f[A])$; so, $g(p) \in c\ell_Z g[U \cap f[A]]$. Thus, $g[U \cap f[A]] \cap (Z \setminus h[A \setminus f^{\leftarrow}[V]]) \neq \varnothing$; so, there exists some $a \in f^{\leftarrow}[U] \cap A$ such that $g(f(a)) \notin h[A \setminus f^{\leftarrow}[V]]$. Thus $a \in f^{\leftarrow}[V]$, and so $f^{\leftarrow}[U] \cap f^{\leftarrow}[V] \neq \varnothing$. This is a contradiction, and so, f is a closed function.

(2) By (1), we need to show that f and g are irreducible. Let B be a proper closed subset of Y. Then as f is onto, $f^{\leftarrow}[B]$ is a proper closed subset of X. Thus, $Z \neq h[f^{\leftarrow}[B]] = g[B]$ as h is irreducible. Hence, g is irreducible. By 6.5(b)(2), f is irreducible.

■

(e) **Theorem**. Let E be a space. The relation $\leqslant$ defined in $AB(E)$ in 8.4(c) is a partial order.

Proof. It is obvious that $\leqslant$ is reflexive and transitive; so all we need to prove is that $\leqslant$ is antisymmetric. Suppose (c_X,X) and (c_Y,Y) are in $AB(E)$ with $(c_X,X) \leqslant (c_Y,Y)$ and $(c_Y,Y) \leqslant (c_X,X)$. There are surjections $f \in C(X,Y)$ and $g \in C(Y,X)$ such that $f \circ c_X = c_Y$ and $g \circ c_Y = c_X$. Since $id_Y \circ c_Y = c_Y = f \circ c_X = f \circ g \circ c_Y$ and c_Y is onto, it follows that $id_Y = f \circ g$. A similar argument shows that $g \circ f = id_X$. Thus f is a homeomorphism and by (ii) of the definition of $AB(E)$, $(c_Y,Y) = (c_X,X)$.

■

(f) **Theorem**. Let E be a space. Then the poset $(AB(E),\leqslant)$ is a complete upper semilattice, and if $\varnothing \neq S \subseteq AB(E)$ and S has a lower bound, then $\wedge S$ exists in $AB(E)$.

Proof. Let $\varnothing \neq S \subseteq AB(E)$; we will construct the supremum of S in $(AB(E),\leqslant)$. Let $Z = \Pi\{X : (c_X,X) \in S\}$, and let Π_X denote the projection function from Z onto X. Define $g : E \to Z$ as follows: $\Pi_X(g(p)) = c_X(p)$ for each $(c_X,X) \in S$. By 1.7(d)(2) g is continuous; note that $\Pi_X \circ g = c_X$ if $(c_X,X) \in S$. Let $T = g[E]$, and $\Pi_X' = \Pi_X|T$ if $(c_X,X) \in S$. Define $g' : E \to T$ by $g'(p) = g(p)$ if $p \in E$. Since $c_X = \Pi_X' \circ g'$ and g' and Π_X' are surjective, it follows from 8.4(d) that g' and Π_X' are covering maps for each $(c_X,X) \in S$. By (iii) of the definition of $AB(E)$, there is a pair $(c_R,R) \in AB(E)$

and a homeomorphism $h : R \to T$ such that $h \circ c_R = g'$. Since $(\Pi' \circ h) \circ c_R = \Pi' \circ g' = c_X$ it follows that $(c_R, R) \geq (c_X, X)$ for each $(c_X, X) \in S$. Now suppose that $(c_U, U) \in AB(E)$ such that $(c_U, U) \geq (c_X, X)$ for each $(c_X, X) \in S$. Thus there is a continuous surjection $m_X : U \to X$ such that $m_X \circ c_U = c_X$. For each $u \in U$ there exists some $y \in E$ such that $c_U(y) = u$; define $f : U \to T$ by $f(u) = g'(y)$. To show that f is well-defined, suppose $z \in E$ and $c_U(z) = u = c_U(y)$. Now $\Pi_X' \circ g'(z) = c_X(z) = m_X \circ c_U(z) = m_X(u) = m_X \circ c_U(y) = c_X(y) = \Pi_{X'} \circ g'(y)$ whenever $(c_X, X) \in S$; so, $g'(z) = g'(y)$. Thus f is well-defined, and $f \circ c_U = g'$. Since g' maps E onto T, f maps U onto T. As g' is continuous and c_U is a quotient function, it follows that f is continuous. Since $(h^{\leftarrow} \circ f) \circ c_U = c_R$, it follows that $(c_U, U) \geq (c_R, R)$. Thus $(c_R, R) = \bigvee S$. This completes the proof that $(AB(E), \leq)$ is a complete upper semilattice. The argument used in the proof of 2.1(e) shows that if $\varnothing \neq S \subseteq AB(E)$ and S has a lower bound, then $\bigwedge S$ exists in $(AB(E), \leq)$.

■

Let X be a space. By 8.4(b)(iii) we can assume, without loss of generality, that $(c_X, X) \in AB(PX)$ and that $c_X = \Pi_X$. (Here PX and Π_X are as defined in 6.11.) In fact the notation "AB(E)" was chosen because when E is extremally disconnected $(c_X, X) \in AB(E)$ iff E is homeomorphic to the Banaschewski absolute of X.

(g) **Definition**. Let X be a space. Define $D(X)$ to be the set $\{(c_Y, Y) \in AB(PX) : (c_Y, Y) \geq (c_X, X)\}$, and give $D(X)$ the order inherited from $(AB(PX), \leq)$.

(h) **Theorem**. Let X be a space. Then $(D(X),\leq)$ is a complete lattice in which the infimum (respectively, supremum) of a nonempty subset S of $D(X)$ in $(D(X),\leq)$ is the same as the infimum (respectively, supremum) of S in $(AB(PX),\leq)$.

Proof. Let $\emptyset \neq S \subseteq D(X)$ and let $(c_T,T) = \bigvee S$ in $(AB(PX),\leq)$. Let $(c_S,S) \in S$. Since $(c_T,T) \geq (c_S,S)$ and $(c_S,S) \geq (c_X,X)$ it follows that $(c_T,T) \geq (c_X,X)$ and so $(c_T,T) \in D(X)$ and is an upper bound of S in $(D(X),\leq)$. Let (c_R,R) be an upper bound of S in $(D(X),\leq)$. Then $(c_R,R) \in AB(PX)$ and (c_R,R) is an upper bound of S in $(AB(PX),\leq)$. This shows that $\bigvee S$ (in $(D(X),\leq)$ = $\bigvee S$ (in $(AB(PX),\leq)$). Since $S \subseteq D(X)$, (c_X,X) is a lower bound of S in $AB(PX)$. By 8.4(f) $\bigwedge S$ exists in $(AB(PX),\leq)$. An argument similar to that given above shows that $\bigwedge S$ (in $(D(X),\leq)$) = $\bigwedge S$ (in $(AB(PX),\leq)$).

■

Now we return to our consideration of covers of X. First we need a lemma.

(i) **Lemma**. Let X, Y, and Z be spaces. Let $f : X \to Y$, $g : X \to Y$, and $h : Y \to Z$ be irreducible Θ-continuous surjections. If $h \circ f = h \circ g$ then $f = g$. (In particular, this holds if f, g, and h are covering maps.)

Proof. Suppose not; choose $x \in X$ such that $f(x) \neq g(x)$. Let U and V be disjoint regular open subsets of Y containing $f(x)$ and $g(x)$ respectively. As f and g are Θ-continuous there exists $W \in$

$RO(X)$ such that $x \in W$, $f[c\ell_X W] \subseteq c\ell_Y U$, and $g[c\ell_X W] \subseteq c\ell_Y V$. Thus $\text{int}_Y c\ell_Y f[c\ell_X W] \subseteq \text{int}_Y c\ell_Y U = U$. By 6.5(d)(1,3) $\emptyset \neq \text{int}_Y f[c\ell_X W] = Y \setminus f[X \setminus W]$. As $U \cap c\ell_Y V = \emptyset$ it follows that $(Y \setminus f[X \setminus W]) \cap c\ell_Y V = \emptyset$, and so $(Y \setminus f[X \setminus W]) \cap g[c\ell_X W] = \emptyset$. As h is irreducible there exists $z \in Z \setminus h[f[X \setminus W]]$. Thus $h^{\leftarrow}(z) \subseteq Y \setminus f[X \setminus W] \subseteq Y \setminus g[c\ell_X W]$. Thus $f^{\leftarrow}(h^{\leftarrow}(z)) \subseteq W$ while $g^{\leftarrow}(h^{\leftarrow}(z)) \cap c\ell_X W = \emptyset$. It follows that $f^{\leftarrow}(h^{\leftarrow}(z)) \cap g^{\leftarrow}(h^{\leftarrow}(z)) = \emptyset$. As f and h are surjective there exists $a \in f^{\leftarrow}(h^{\leftarrow}(z))$. Evidently $h \circ f(a) \neq h \circ g(a)$, in contradiction to hypothesis. Thus $f = g$ as claimed.

■

(j) Definition. Two covers (Y,f) and (Z,g) of a space X are **equivalent** if there is a homeomorphism $h : Y \to Z$ such that $g \circ h = f$.

The set $D(X)$ defined in 8.4(g) is not a set of covers of X, for if $(c_Y, Y) \in D(X)$ then c_Y is a function from PX to Y, rather than from Y to X. However, we can use the poset $(D(X), \leqslant)$ to define a set of covers $C(X)$, and an ordering on $C(X)$, such that $D(X)$ and $C(X)$ are order-isomorphic.

(k) Lemma. Let $(c_Y, Y) \in D(X)$. Then there is a unique covering map $g_Y : Y \to X$ such that $c_X = g_Y \circ c_Y$.

Proof. Evidently there exists $g_Y \in C(Y,X)$ such that $c_X = g_Y \circ c_Y$; by 8.4(d) g_Y is a covering map. If $f : Y \to X$ and $c_X = f \circ c_Y$, then $f \circ c_Y = g_Y \circ c_Y$, and as c_Y is surjective it follows that $f = g_Y$.

■

(l) **Definition**. Let $\mathcal{C}(X) = \{(Y, g_Y) : (c_Y, Y) \in \mathcal{D}(X)\}$ where g_Y is the map defined in (k) above.

(m) **Lemma**. Let (c_Y, Y) and (c_Z, Z) be in $\mathcal{D}(X)$. Let $f \in C(Y,Z)$. If f is onto, then $c_Z = f \circ c_Y$ iff $g_Y = g_Z \circ f$.

Proof. If $c_Z = f \circ c_Y$, then $g_Z \circ c_Z = c_X = g_Y \circ c_Y$; so, $g_Z \circ f \circ c_Y = g_Y \circ c_Y$. As c_Y is surjective, it follows that $g_Z \circ f = g_Y$. Conversely, if $g_Y = g_Z \circ f$, then $g_Z \circ c_Z = c_X = g_Y \circ c_Y = g_Z \circ f \circ c_Y$. By 8.4(d,i) it follows that $c_Z = f \circ c_Y$.

■

(n) **Theorem**. Let X be a space.

(1) If (Y, g_Y) and (Z, g_Z) are distinct members of $\mathcal{C}(X)$ then there is no homeomorphism $h : Y \to Z$ such that $g_Z \circ h = g_Y$.

(2) If W is a space and $f : W \to X$ is a covering map, then there exists a pair $(Z, g_Z) \in \mathcal{C}(X)$ and a homeomorphism $h : W \to Z$ such that $g_Z \circ h = f$.

Proof

(1) By (m) above if there were a homeomorphism $h : Y \to Z$ such that $g_Z \circ h = g_Y$, then $c_Z = h \circ c_Y$. It follows from 8.4(b)(ii) and

the fact that (c_Y,Y) and (c_Z,Z) both are in $AB(PX)$, that $(c_Y,Y) = (c_Z,Z)$. Thus $Y = Z$ and $c_Y = c_Z$. Hence $c_Y \circ g_Y = c_X = c_Y \circ g_Z$ so by 8.4(i) $g_Y = g_Z$. Thus $(Y,g_Y) = (Z,g_Y)$ in contradiction to hypothesis.

(2) By 6.11(d) there is a covering map $k : PX \to W$ such that $f \circ k = c_X$. By 8.4(b)(iii) there is a pair $(c_Z,Z) \in AB(PX)$ and a homeomorphism $h : W \to Z$ such that $h \circ k = c_Z$. Arguing as we did in (m) above, we conclude that

$$g_Z \circ h \circ k = g_Z \circ c_Z = c_X = f \circ k,$$

and as k is surjective $g_Z \circ h = k$.

■

The above theorem says that $C(X)$ is formed by picking precisely one representative out of each "equivalence class" of covers of X, where "equivalence" is defined as in 8.4(j). Thus $C(X)$ has been defined in precisely the same way as $E(X)$ was in 4.1. As remarked after 8.4(b), an extension of a space X is really an ordered pair (i,eX) where $i : X \to eX$ is a dense embedding of X in a space eX; extensions (i_1,e_1X) and (i_2,e_2X) are equivalent if there is a homeomorphism $h : e_1X \to e_2X$ such that $h \circ i_1 = i_2$. This is completely analogous to our definition of equivalent covers in 8.4(j), and the construction of $E(X)$ is parallel to that of $C(X)$. Note that it is possible for (Y,g_Y) and (Z,g_Z) to be distinct members of $C(X)$ even though Y and Z are homeomorphic - the homeomorphism (call it h) will not satisfy $h \circ g_Y = g_Z$. Again, compare this to the analogous situation for extensions in 4AB.

We now define a partial order on $\mathcal{C}(X)$.

(o) <u>**Definition**</u>. Let X be a space, and let (Y,g_Y) and (Z,g_Z) belong to $\mathcal{C}(X)$. We say that (Y,g_Y) is **projectively smaller** than (Z,g_Z) — denoted $(Y,g_Y) \leqslant (Z,g_Z)$ — if there exists a surjection $f \in C(Z,Y)$ such that $g_Y \circ f = g_Z$.

Although we have used the symbol $\leqslant$ to denote both a partial order on $\mathcal{D}(X)$ and a relation on $\mathcal{C}(X)$, there will be no ambiguity about its meaning.

(p) <u>**Theorem**</u>. Let X be a space. Then

(1) $\leqslant$ is a partial order on $\mathcal{C}(X)$.

(2) The map $(Y,g_Y) \mapsto (c_Y,Y)$ is an order isomorphism from $\mathcal{C}(X)$ onto $\mathcal{D}(X)$. Thus $\mathcal{C}(X)$ is a complete lattice.

<u>Proof</u>

(1) Obviously $\leqslant$ is reflexive and transitive. To show $\leqslant$ is antisymmetric, suppose $(Y,g_Y) \leqslant (Z,g_Z)$ and $(Z,g_Z) \leqslant (Y,g_Y)$. Then there exist $f : Y \to Z$ such that $g_Y = g_Z \circ f$, and $k : Z \to Y$ such that $g_Z = g_Y \circ k$. Now argue as in 8.4(e) and apply 8.4(i) to show that f and k are homeomorphisms. By 8.4(n)(1) $(Z,g_Z) = (Y,g_Y)$.

(2) Define $\sigma : \mathcal{C}(X) \to \mathcal{D}(X)$ by setting $\sigma((Y,g_Y)) = (c_Y,Y)$. Obviously σ is onto, and if $(c_Y,Y) = (c_Z,Z)$, then $Y = Z$ and $c_Y = c_Z$. Hence $g_Y \circ c_Y = c_X = g_Z \circ c_Z = g_Z \circ c_Y$ and so $g_Y = g_Z$ as c_Y is surjective. Hence σ is one-to-one. It follows from 8.4(m) that $(Y,g_Y) \leqslant (Z,g_Z)$ iff $\sigma((Y,g_Y)) \leqslant \sigma((Z,g_Z))$ and so σ is an order isomorphism. As lattice properties are determined by order properties, it follows that

$(C(X),\leq)$ and $(D(X),\leq)$ are lattice isomorphic. Thus by 8.4(h) $(C(X),\leq)$ is a complete lattice.

■

It is evident that the largest element in $(C(X),\leq)$ can be taken to be (PX,Π_X) and the smallest element can be taken to be (X,id_X).

We have now defined an order structure on the set of all covers of a given space X and shown that it is a complete lattice. This is analogous to our achievement in Chapter 4, in which we showed that the set of extensions of a given space is a complete upper semilattice (see 4.1(g)). The next logical step is to attempt to mimic our achievements in Chapter 5 by developing a theory of "covering properties".

(q) **Definition**. Let P be a topological property.

(1) A space X is called **P-coverable** if there is a space Y with P and a covering map $f : Y \to X$.

(2) The set $\{(Y,g_Y) \in C(X) : Y \text{ has } P\}$ is denoted by $CP(X)$. (Note that by 8.4(n)(2) X is P-coverable iff $CP(X) \neq \emptyset$.)

(3) P is called a **covering property** if, whenever X is a P-coverable space, $CP(X)$ has a smallest member in $(C(X),\leq)$. This smallest member is denoted by $(c_P X,k_P)$ and is called the **minimum P-cover** of X. (It is characterized by the following property: if Y has P and $f : Y \to X$ is a covering map, there is a continuous surjection (and hence by 8.4(d) a covering map) $g : Y \to c_P X$ such that $k_P \circ g = f$.)

Evidently P-coverable spaces are analogous to the P-regular spaces defined in 5.2(a), and covering properties are analogous to the extension properties defined in 5.3(a). As a familiar example, if P is the property of being extremally disconnected, then P is a covering property and $(c_P X, k_P) = (PX, \Pi_X)$

In 5.3(c) we were able to characterize extension properties as being closed-hereditary, productive topological properties. Unfortunately there seems to be no analogous characterization of covering properties in terms of familiar, easily-described topological constructions. The best way to search for such an analogy is to rephrase 5.3(c) in category-theoretic terms and then try to interpret in topological terms the "dual" category-theoretic statement about the category of spaces and covering maps. This is done in Chapter 9; specifically, juxtapose 9.4(b), 9.5(d), 9.6(e), 9.7(b)(5), and 9M. When we do this, we are led to the following concept.

(r) **Definition**. Let $\{X_\alpha : \alpha \in A\}$ be a set of spaces. A space X together with a set of perfect irreducible continuous surjections $f_\alpha : X_\alpha \to X$ is called a **covering coproduct** of $\{X_\alpha : \alpha \in A\}$ if for every space Y and family of covering maps $g_\alpha : X_\alpha \to Y$ there exists a covering map $h : X \to Y$ such that $h \circ f_\alpha = g_\alpha$ for each $\alpha \in A$.

Note that since f_α is surjective, if such an h exists it must be unique.

(s) **Lemma**. Let X (with functions $\{f_\alpha : \alpha \in A\}$) and Y (with functions $\{g_\alpha : \alpha \in A\}$) both be covering coproducts of a set $\{X_\alpha : \alpha \in A\}$ of spaces. Then there is a homeomorphism $h : X \to Y$ such that $h \circ f_\alpha = g_\alpha$ for each $\alpha \in A$. In this sense covering coproducts are unique if they exist.

<u>Proof</u>. By definition there exist covering maps $h : X \to Y$ and $k : Y \to X$ such that $h \circ f_\alpha = g_\alpha$ and $k \circ g_\alpha = f_\alpha$ for each $\alpha \in A$. Thus $h \circ k \circ g_\alpha = id_Y \circ g_\alpha$ and $k \circ h \circ f_\alpha = id_X \circ f_\alpha$ for each $\alpha \in A$. As f_α and g_α are surjective it follows that $h \circ k - id_Y$ and $k \circ h = id_X$. Thus h and k are homeomorphisms.

■

The term "covering coproduct" was used because it is just the coproduct of a set of spaces in the category of spaces and covering maps (see 9M).

(t) **<u>Theorem</u>**. The following are equivalent for a topological property P.

(1) P is a covering property.

(2) If $\{X_\alpha : \alpha \in I\}$ is a set of spaces with P, and if X together with covering maps $f_\alpha : X_\alpha \to X$ is the covering coproduct of $\{X_\alpha : \alpha \in I\}$, then X has P. (In other words, "P is closed under the formation of covering coproducts".)

(3) If X is a P-coverable space, and if $\wedge CP(X) = (Y,g)$ (where the infimum is taken in $(C(X),\leq)$, then Y has P.

<u>Proof</u>

(1) $\Longleftrightarrow$ (3) Obviously if $CP(X)$ has a smallest member in

$\mathcal{C}(X)$ it will be $\wedge\mathcal{CP}(X)$; thus such a smallest member exists iff Y has P (where $(Y,g) = \wedge\mathcal{CP}(X)$).

(3) ⇒ (2) Under the hypothesis of (2), and by 8.4(n)(3), for each $\alpha \in I$ there is a homeomorphism $h_\alpha : Y_\alpha \to X_\alpha$ such that $g_\alpha \circ h_\alpha = f_\alpha$, and $(Y_\alpha, g_\alpha) \in \mathcal{C}(X)$. It is routine to show from the definition of the covering coproduct that $(X, id_X) = \wedge\{(Y_\alpha, g_\alpha) : \alpha \in I\}$. As each Y_α has P, evidently $(X, id_X) = \wedge\mathcal{CP}(X)$. Hence by (3) X has P.

(2) ⇒ (3) It is easy to check that $\wedge\mathcal{CP}(X)$ is the covering coproduct of $\mathcal{CP}(X)$, and the result quickly follows.

■

It is easy to see that the covering coproduct of a set of spaces exists iff the spaces are coabsolute. As noted above, it is not clear how to construct the infimum of a subset of $\mathcal{C}(X)$ in terms of "well-known" constructions like subspaces, products, quotients, etc. However, there do exist several techniques for producing covering properties. We will describe one of these now, and refer the reader to 8G for an example.

(u) __Definition__. A **cover-generating operator** (abbreviated **cg-operator**) is a class Q of ordered pairs (U,X) whose second component is a space X and whose first component is an open set U of X, with this property: if $f : Z \to X$ is a covering map and $(U,X) \in Q$ then $(f^{\leftarrow}[U], Z) \in Q$.

Informally, then, a cg-operator assigns to each space X a subset of $\tau(X)$ in such a way that this subset is "preserved inversely by covering maps". One way of defining a cg-operator Q is as follows: if P is a topological property that is inversely preserved by covering maps (i.e., if $f : X \to Y$ is a covering map and Y has P then X has P), then we can define a pair (U,X) to be in Q if U is an open subset of X and U has P. However, other "natural" examples of cg-operators exist; one is implicitly suggested in (aa) and (ab) below.

As one might expect, cg-operators can be used to generate covering properties as follows.

(v) __Theorem__. Let Q be a cg-operator and let P_Q be defined as follows: a space X has P_Q iff $c\ell_X U \in B(X)$ whenever $(U,X) \in Q$. Then P_Q is a covering property.

__Proof__. First we show that P_Q is indeed a topological property. If $h : X \to Y$ is a homeomorphism then it is evident that $(U,X) \in Q$ iff $(h[U],Y) \in Q$. Thus $c\ell_X U \in B(X)$ when $(U,X) \in Q$ iff $c\ell_Y h[U] \in B(Y)$ when $(h[U],Y) \in Q$. Thus P_Q is a topological property if P_Q contains a space with at least two points (see 4.1(j)).

Let X be a space. Since all open sets of PX have open PX-closures (as PX is extremally disconnected), $(PX,\Pi_X) \in CP_Q(X)$; so, P_Q contains a space with more than one point. By 8.4(t) above it suffices to show that the "first component" of the infimum of $CP_Q(X)$ has P_Q.

Let $\wedge CP_Q(X) = (Z,g_Z)$ and suppose that Z does not have P_Q. Then there exists an open subset U of Z such that $(U,Z) \in Q$ and $c\ell_Z U \notin B(Z)$. Let $T = c\ell_Z U \oplus c\ell_Z(Z\setminus c\ell_Z U) = (c\ell_Z U \times \{1\}) \cup (c\ell_Z(Z\setminus c\ell_Z U) \times \{2\})$ (see 1.2(h)). Define $k : T \to Z$ by $k(t,i) = t$ for each $(t,i) \in T$ ($i = 1$ or 2). One easily checks that k is a covering map but not a homeomorphism; so, without loss of generality $(T, g \circ k) \in \mathcal{C}(X)$ (see 8.4(n)(2)). Evidently $(T, g_Z \circ k) > (Z, g_Z)$ in $(\mathcal{C}(X), \geqslant)$. Now suppose $(Y, g_Y) \in \mathcal{C}P_Q(X)$. There exists a covering map $h : Y \to Z$ such that $g_Z \circ h = g_Y$. As $(U,Z) \in Q$ it follows that $(h^{\leftarrow}[U], Y) \in Q$ and so $c\ell_Y h^{\leftarrow}[U] \in B(Y)$. Now define $j : Y \to T$ as follows:

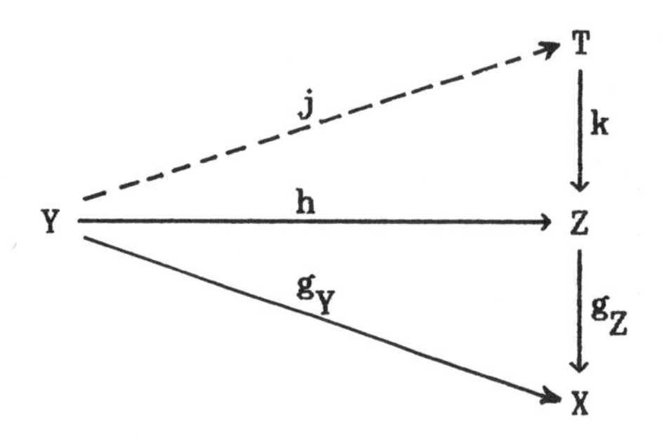

$j(y) = (h(y),1)$ if $y \in c\ell_Y h^{\leftarrow}[U]$, and $j(y) = (h(y),2)$ if $y \in c\ell_Y h^{\leftarrow}[Z \backslash c\ell_Z U]$. Since $c\ell_Y h^{\leftarrow}[U] \in B(Y)$, j is well-defined and is easily verified to be a covering map. Thus $(Y, g_Y) \geqslant (T, g_Z \circ k)$ in $\mathcal{C}(X)$. Hence $(T, g_Z \circ k) \leqslant \wedge \mathcal{C}P_Q(X) = (Z, g_Z)$ and we have obtained a contradiction. Thus P_Q is a covering property.

■

If Q is the class $\{(U,X) : U$ is open in X and X is a space$\}$, then P_Q is extremal disconnectedness. If Q is the class $\{(U,X) : U$ is a σ-compact open subset of the space X$\}$, then P_Q is the class of spaces whose σ-compact open sets have clopen closures. Other examples abound.

Just as we concentrated on Tychonoff and zero-dimensional extension properties in Chapter 5, we will now turn our attention to Tychonoff and zero-dimensional covering properties. We begin by considering the sublattice of Tychonoff (respectively, zero-dimensional) members of $C(X)$.

(w) __Definition__. Let X be a space. The set $\{(Y,g_Y) \in C(X)$: Y is Tychonoff (respectively, zero-dimensional)$\}$ is denoted by $C_T(X)$ (respectively, $C_0(X)$).

Note that if X is regular then PX is strongly zero-dimensional (see 6.4, 6.2(c), and 6.11(h)), and if X is not regular then $C_T(X) = \varnothing$ (see 1.8(h)). Thus $C_T(X) \neq \varnothing \neq C_0(X)$ iff X is regular. Surprisingly, if X is Tychonoff it does not follows that $C(X) = C_T(X)$ (see 8H). Also, if X is regular it does **not** follow that $C_T(X)$ is a sublattice of $C(X)$ - it is possible for (Y,g_Y) and (Z,g_Z) to belong to $C_T(X)$ while $(Y,g_Y) \wedge (Z,g_Z) \in C(X) \setminus C_T(X)$, where the infimum is taken in the complete lattice $(C(X),\geq)$ (see 8I). However, if X is Tychonoff and we restrict the partial order on $C(X)$ to $C_T(X)$, we find that $C_T(X)$ is a complete lattice, although it follows from the above remarks that the infimum in $C_T(X)$ of a subset of $C_T(X)$ need not always be the same as its infimum in $C(X)$ (see 8M). Similar remarks hold for $C_0(X)$.

(x) __Theorem__

(1) Let X be a space, P an extension property, and suppose $\varnothing \neq S \subseteq CP(X)$. Then the supremum in $(C(X),\geq)$ of S belongs to $CP(X)$.

(2) If $\emptyset \neq S \subseteq C_T(X)$ (respectively, $\emptyset \neq S \subseteq C_0(X)$) then $\bigvee S \in C_T(X)$ (respectively, $C_0(X)$).

<u>Proof</u>

(1) If $S = \{(Y_i,f_i) : i \in I\}$, construct $\bigvee S = (T,h)$ as in the proof of 8.4(f); T is a subspace of $\Pi\{Y_i : i \in I\}$ and hence is P-regular. For each $i \in I$ the function $\Pi_i' : T \to Y_i$ constructed in the proof of 8.4(f) is a covering map. Thus by 5.9(c) T has P, and so $(T,h) \in CP(X)$.

(2) If P is "Tychonoff" or "zero-dimensional" this follows from (1).

■

(y) **<u>Corollary</u>**

(1) Let P be an extension property and let $X \in P$. Then $CP(X)$, with the order inherited from $C(X)$, is a complete lattice. In particular, PX has P.

(2) Let X be a Tychonoff (respectively, zero-dimensional) space. Then $C_T(X)$ (respectively, $C_0(X)$), with the order inherited from $C(X)$, is a complete lattice.

<u>Proof</u>. (1) follows from (x) above and 2.1(e); (2) follows immediately from (1).

■

(z) **<u>Definition</u>**. A topological property P is a **Tychonoff covering property** if P is possessed only by Tychonoff spaces and if, for each Tychonoff space X, the infimum in $(C_T(X),\geq)$ of the set of

P-coverings of X is a P-covering of X. As before, we denote that infimum by $(c_P X, k_P)$ and call $c_P X$ the **minimum P-cover** of X. **Zero-dimensional covering properties** are defined analogously.

If P is a Tychonoff covering property, note that the set of P-coverings of X is a subset of $C_T(X)$ and so we can form its infimum in $C_T(X)$. It is evident that 8.4(v) above can be adapted to produce Tychonoff and zero-dimensional covering properties, since the direct sum of two Tychonoff (respectively, zero-dimensional) spaces is Tychonoff (respectively, zero-dimensional). We now examine in more detail a somewhat different example of a Tychonoff covering property.

(aa) **Definition**. A Tychonoff space is a **quasi-F-space** if each dense cozero-set of X is C^*-embedded in X.

(ab) **Theorem**. Being a quasi-F-space is a Tychonoff covering property.

Proof. We must produce a minimum quasi-F-cover of each Tychonoff space X (as described in 8.4(q)). Now PX is extremally disconnected and Tychonoff as X is regular (see 6.11(h)) so every dense open subset of PX is C^*-embedded in PX (see 6.2(c)). Thus (PX, Π_X) is a quasi-F cover of X. Let (Z,g) be the infimum (in $C_T(X)$) of the set of quasi-F covers of X (which we have just seen is nonempty). We will argue as in 8.4(v) to show that Z is a quasi-F-space.

Suppose C is a dense cozero-set of Z that is not C^*-embedded in Z. Let $\beta j : \beta C \to \beta Z$ extend the embedding map $j : C \to Z$, let T

$= (\beta j)^{\leftarrow}[Z]$, and let $k = \beta j|T$. Using 1.8(f) and the fact that C is dense in Z, we see that k is a covering map from T onto Z, and $(T,g\circ k) \geq (Z,g)$ in $C_T(X)$. Note that C is C^*-embedded in T but not in Z, so $g\circ k$ is not a homeomorphism; hence $(T,g\circ k) \neq (Z,g)$.

Now let (Y,h) be a quasi-F cover of X. There exists a covering map $f : Y \to Z$ such that $g\circ f = h$. By 1.4(h) and 6.5(b)(4) $f^{\leftarrow}[C]$ is a dense cozero-set of Y and hence is C^*-embedded in Y. Let $F = f|f^{\leftarrow}[C]$. Then F maps $f^{\leftarrow}[C]$ onto C and extends continuously to $\beta F : \beta Y \to \beta T$ (since $\beta Y = \beta(f^{\leftarrow}[C])$ and $\beta T = \beta C$). Now $\beta j\circ \beta F|f^{\leftarrow}[C] = \beta f|f^{\leftarrow}[C]$ so by 1.6(d) $\beta j\circ \beta F = \beta f$. Hence by 4.2(f)

$$Y = (\beta f)^{\leftarrow}[Z] = (\beta F)^{\leftarrow}[(\beta j)^{\leftarrow}[Z]] = (\beta F)^{\leftarrow}[T].$$

Thus $\beta F|Y$ is a covering map from Y onto T and $g\circ k\circ(\beta F|Y) = h$. Evidently T is Tychonoff and $(Y,h) \geq (T,k\circ g)$ for each quasi-F cover (Y,h) of X. Thus $(Z,g) \geq (T,k\circ g)$ in $C_T(X)$ which contradicts the result in the previous paragraph. Thus Z is quasi-F and (Z,g) is the minimum quasi-F cover of X.

■

In the very interesting problem 8J we outline how to construct the quasi-F-cover of a compact space X as a space of ultrafilters on a sublattice of $R(X)$. In general, however, we do not know of a method of representing P-covers as spaces of ultrafilters on a suitably chosen lattice.

In this section we have completed our program of showing that the set $C(X)$ of covers of a space X has a "parallel structure" to the structure of the set $E(X)$ of extensions of X. Both absolutes and extensions are intimately connected with maximal filters on certain lattices of subsets of the originial space, and the "new points" that are built - the points of the absolute or the points of the outgrowth of the extension - are just such maximal filters. Although the theory of covers is not (as this is written) nearly as well developed as the theory of extensions, the analogy between what was done is this section and what was done in Chapters 4 and 5 is striking.

8.5 Completions of C(X) vs. C(EX)

In 2.5(b) we saw that for each poset $(P,\leqslant)$ there exists a complete lattice D(P) (whose partial order we denote by $\subseteq$) and a one-to-one function $j : P \to D(P)$ possessing these properties:

(a) j is an order isomorphism from $(P,\leqslant)$ onto $(j[P],\subseteq)$;

(b) if $z \in D(P)$ then $z = \bigwedge\{j(x) : x \in P \text{ and } j(x) \supseteq z\} = \bigvee\{j(x) : x \in P \text{ and } j(x) \subseteq z\}$.

We called $(D(P),\subseteq)$ the **Dedekind-MacNeille completion** of $(P,\leqslant)$ and indicated that it is "unique up to isomorphism" (see 2.5(b) for a precise formulation of this).

We then constructed the **conditional completion** $(K(P),\subseteq)$ of $(P,\leqslant)$ as follows: delete the smallest (respectively, the largest) member of D(P) from D(P) iff P has no smallest (respectively, largest) member, and denote the resulting set as K(P). The poset $(K(P),\subseteq)$ (which is a lattice if $(P,\leqslant)$ is; see 2.5(c)(2)) is the conditional completion of P, and its properties are described in 2.5(c).

As we saw in 1.3 and 2.1(f)(5), if X is a Tychonoff space then $(C(X),\leqslant)$ is a poset (in fact a lattice-ordered ring). We will prove below that $(C(X),\leqslant)$ is conditionally complete iff X is extremally disconnected, and that $f \to f \circ k_X$ is an order isomorphic embedding of C(X) into C(EX). Hence for each Tychonoff space X, C(X) can be embedded in a natural fashion in the conditionally complete lattice C(EX). This raises the question of whether C(EX) is (up to isomorphism) the conditional completion of C(X) (via the mapping $f \to f \circ k_X$). The purpose of this section is to characterize those X for which this is the case.

(a) **Theorem**. The following are equivalent for a Tychonoff space X:

(1) C(X), with the usual ordering, is a conditionally complete lattice and

(2) X is extremally disconnected.

Proof

(1) ⇒ (2) Let V be a nonempty open subset of X. As X is Tychonoff, for each $x \in V$ there exists $f_x \in C(X)$ such that $\underline{0} \leqslant f_x \leqslant \underline{1}$, $f_x(x) = 1$, and $f_x[X \setminus V] \subseteq \{0\}$. As $\{f_x : x \in V\} = S$ is bounded above by the constant function $\underline{1}$, by hypothesis S has a supremum in C(X), which we denote by h. As $\underline{1} \geqslant h \geqslant f_x$ for each $x \in V$, it is evident that $h[V] = \{1\}$. Thus by 1.6(d) $h[c\ell_X V] = \{1\}$. Suppose that $x_0 \in X \setminus c\ell_X V$ and that $h(x_0) = \delta > 0$. As h is continuous and X is regular there exists an open set W of X such that $x_0 \in W \subseteq c\ell_X W \subseteq X \setminus c\ell_X V$ and $h[c\ell_X W] \subseteq [\delta/2,+\infty)$. As X is Tychonoff

there exists $g \in C(c\ell_X W)$ such that $g[bd_X W] \subseteq \{0\}$, $g(x_0) = -\delta/2$, and $-\underline{\delta/2} \leqslant g \leqslant \underline{0}$. Define $k \in F(X,\mathbb{R})$ as follows:

$$k(x) = \begin{cases} h(x) \text{ if } x \in c\ell_X(X \backslash c\ell_X W) \\ \\ h(x) + g(x) \text{ if } x \in c\ell_X W \end{cases}$$

Now k is well-defined as $g[bd_X W] = \{0\}$, and $k \in C(X)$ by 1.6(b). If $y \in c\ell_X(X \backslash c\ell_X W)$ then $k(y) = h(y) \geqslant f_x(y)$ for each $x \in V$, and if $y \in c\ell_X W$ then $k(y) = h(y) + g(y) \geqslant \delta/2 + (-\delta/2) = 0 = f_x(y)$ for each $x \in V$. Note that $k(x_0) = h(x_0) - \delta/2$; thus k is an upper bound of S that is strictly less than h, contradicting the definition of h. It follows that $h[X \backslash c\ell_X V] = \{0\}$. Thus $c\ell_X V = X \backslash h^{\leftarrow}(0)$, and so $c\ell_X V$ is open. Hence X is extremally disconnected.

(2) $\Rightarrow$ (1) Let X be extremally disconnected and let $\{f_i : i \in I] = S \subseteq C(X)$ be bounded above by $g \in C(X)$. For each rational number q, we let $U(q) = c\ell_X \cup \{f_i^{\leftarrow}[(q,+\infty)] : i \in I\}$. Evidently $U(q) \in B(X)$ since X is extremally disconnected. Furthermore, if $q_1 < q_2$ then $U(q_1) \supseteq U(q_2)$.

We claim that if $x \in U(q)$ then $g(x) \geqslant q$. For suppose $g(x) < q$; then there exists $\delta > 0$ such that $g(x) < q - \delta$. Thus $g^{\leftarrow}[(-\infty,q-\delta)] \cap [\cup\{f_i^{\leftarrow}[(q,+\infty)] : i \in I\}] \neq \emptyset$; so, there exists $j \in I$ and $y \in X$ such that $g(y) < q - \delta < q < f_j(y)$, which contradicts the assumption that g is an upper bound of S. Thus $\sup\{q \in \mathbb{Q} : x \in U(q)\}$ exists in $\mathbb{R}$ for each $x \in X$. We define $h \in F(X,\mathbb{R})$ by:

$$h(x) = \sup\{q \in \mathbb{Q} : x \in U(q)\}.$$

The above argument shows that $h \leqslant g$ for each upper bound g of S. Furthermore, let $i \in I$ and $x \in X$ and suppose that $f_i(x) > h(x)$. Find a rational $q \in (h(x),f_i(x))$ and note that it follows from the definition of h that $x \notin U(q)$, while the fact that $x \in f_i^{\leftarrow}[(q,+\infty)]$ implies that $x \in U(q)$. Hence $h \geqslant f_i$ for each $i \in I$, and so h is an upper bound (in $F(X,\mathbb{R})$) for S. If we show that $h \in C(X)$ it will follow that $\vee S$ exists in $C(X)$ (and is h).

Let $A(q) = X \backslash U(q)$. Then $A(q) \in \mathcal{B}(X)$, $q_1 < q_2$ implies $A(q_1) \subseteq A(q_2)$, and it is easily checked that $h(x) = \inf\{q \in \mathbb{Q} : x \in A(q)\}$. It thus follows from 1.10(e) that $h \in C(X)$. A symmetric argument shows that any nonempty subset of $C(X)$ that is bounded below has an infimum in $C(X)$, and so $C(X)$ is conditionally complete.

■

(b) **Proposition**. Let X be a Tychonoff space and define $\mu : C(X) \to C(EX)$ as follows: $\mu(f) = f \circ k_X$. Then μ is an order isomorphism (and a ring isomorphism) from $C(X)$ onto a subset of $C(EX)$.

Proof. It is easily checked that $\mu(f+g) = \mu(f) + \mu(g)$ and that $\mu(fg) = \mu(f)\mu(g)$. If $f \leqslant g$ then $f(k_X(\alpha)) \leqslant g(k_X(\alpha))$ for each $\alpha \in EX$; so, $\mu(f) \leqslant \mu(g)$. Conversely, if $\mu(f) \leqslant \mu(g)$ and $x \in X$, then there exists $\alpha \in EX$ such that $k_X(\alpha) = x$. Thus $f \circ k_X(\alpha) \leqslant g \circ k_X(\alpha)$ and $f(x) \leqslant g(x)$. Hence $f \leqslant g$, and so μ is both one-to-one and an order isomorphism.

■

Thus for each Tychonoff space X, the map $f \to f \circ k_X$ embeds $C(X)$ as a subset of the conditionally complete lattice $C(EX)$. We now characterize those spaces X for which C(EX) is the conditional completion of C(X). The following concept was introduced in 6U(2).

(c) <u>**Definition**</u>. A Tychonoff space X is a **weak cb space** if, whenever $\{A_n : n \in \mathbb{N}\}$ is a decreasing sequence of regular closed subsets of X with empty intersection, there exists a decreasing sequence $\{Z_n : n \in \mathbb{N}\}$ of zero-sets of X such that $A_n \subseteq Z_n$ for each $n \in \mathbb{N}$ and $\cap\{Z_n : n \in \mathbb{N}\} = \emptyset$.

The following theorem provides two classes of weak cb spaces. Recall that a space is **countably paracompact** if every countable open cover of it has a locally finite open refinement.

(d) <u>**Theorem**</u>

(1) Every normal countably paracompact space is weak cb.

(2) Every pseudocompact space is weak cb.

<u>Proof</u>

(1) It is well-known that a normal space X is countably paracompact iff for each decreasing sequence $\{A_n : n \in \mathbb{N}\}$ of closed sets of X with empty intersection there exists a decreasing sequence $\{G_n : n \in \mathbb{N}\}$ of open sets of X with empty intersection such that $A_n \subseteq G_n$ for each $n \in \mathbb{N}$ (see 8K). So, let $\{A_n : n \in \mathbb{N}\}$ be a decreasing sequence of regular closed subsets of X with empty intersection, and choose $\{G_n : n \in \mathbb{N}\}$ as above. As X is normal, there exists for each $n \in \mathbb{N}$ an $f_n \in C(X)$ such that $\underline{0} \leqslant f_n \leqslant \underline{1}$, $f_n[A_n] \subseteq \{0\}$, and $f_n[X \backslash G_n] \subseteq \{1\}$. Then $\{Z(\Sigma\{f_i^2 : 1 \leqslant i \leqslant n\}) : n \in \mathbb{N}\}$ is the required sequence of zero-sets witnessing the fact that X is

weak cb.

(2) There is no decreasing sequence of nonempty regular closed subsets of a pseudocompact space whose intersection is empty (see 1.11(e)).

■

(e) **Theorem**. The following are equivalent for a Tychonoff space X:

(1) X is weak cb,

(2) if $F \in C(EX)$ there exists $f \in C(X)$ such that $f \circ k_X \geqslant F$,

(3) if $G \in C(EX)$ and $G(\alpha) > 0$ for each $\alpha \in EX$, then there exists $f \in C(X)$ such that $0 < (f \circ k_X)(\alpha) < G(\alpha)$ for each $\alpha \in EX$, and

(4) $C(EX)$ is the conditional completion of $C(X)$ via the embedding $f \to f \circ k_X$.

Proof

(1) ⇒ (2) Let $F \in C(EX)$ and put $A_n = k_X[c\ell_{EX}F^{\leftarrow}[(n,+\infty)]]$ for each $n \in \mathbb{N}$. Since $\{c\ell_{EX}F^{\leftarrow}[(n,+\infty)] : n \in \mathbb{N}\}$ is a decreasing sequence of clopen subsets of EX with empty intersection, it follows from 6.6(e)(9) and that $\{A_n : n \in \mathbb{N}\}$ is a decreasing sequence of regular closed subsets of X with empty intersection. As X is weak cb there exists $\{g_n : n \in \mathbb{N}\} \subseteq C(X)$ such that $A_{n+1} \subseteq Z(g_{n+1}) \subseteq Z(g_n)$ for each $n \in \mathbb{N}$, $g_n \geqslant \underline{0}$, and $\bigcap\{Z(g_n) : n \in \mathbb{N}\} = \varnothing$. For each n define f_n as follows:

$$f_n = \underline{1} - ((\vee\{\underline{n}\ g_i : 1 \leqslant i \leqslant n\}) \wedge \underline{1}).$$

Evidently $f_n \in C(X)$ (see 1B) and $f_n \geqslant \underline{0}$.

We claim that if $x \in X$ then there exists an open set $V(x)$ in X such that $\{n \in \mathbb{N} : f_n[V(x)] \neq \{0\}\}$ is finite. To see this, note that as $\cap\{Z(g_n) : n \in \mathbb{N}\} = \emptyset$ there exists $k \in \mathbb{N}$ such that $g_k(x) \neq 0$. Thus there exists $j \in \mathbb{N}$ such that $g_k(x) > 1/j$ on an open neighborhood $V(x)$ of x. Thus if $n \geqslant j$ and $y \in V(x)$, it follows that $(\underline{n}\ g_k)(y) \geqslant 1$. Hence if $n \geqslant \max\{j,k\}$ then $(\vee\{\underline{n}\ g_i : 1 \leqslant i \leqslant n\})(y) \geqslant 1$ and so $f_n(y) = 0$ for each $y \in V(x)$. Our claim follows.

Now define $f \in F(X,\mathbb{R})$ as follows:

$$f(x) = 2 + \Sigma\{f_n(x) : n \in \mathbb{N}\}.$$

If $x \in X$ then by the above claim $f_n(x) = 0$ for all but finitely many $n \in \mathbb{N}$, so $f(x)$ is well-defined. Furthermore, it follows from that claim and from 1.6(a) that f is continuous.

Let $\alpha \in EX$. Then either $F(\alpha) \leqslant 2$, in which case $f \circ k_X(\alpha) \geqslant 2 \geqslant F(\alpha)$, or else there exists $n \in \mathbb{N}$ such that $\alpha \in c\ell_{EX}F^{\leftarrow}[(n,+\infty)]\backslash c\ell_{EX}F^{\leftarrow}[(n+1,+\infty)]$. In that case $n \leqslant F(\alpha) \leqslant n+1$ and $k_X(\alpha) \in A_n$. Thus if $j \leqslant n$ then $k_X(\alpha) \in Z(g_j)$ and it follows that

$$\begin{aligned} f_j(k_X(\alpha)) &= 1 - ((\vee\{\underline{j}\ g_i(k_X(\alpha)) : 1 \leqslant i \leqslant j\}) \wedge 1) \\ &= 1 - 0 \\ &= 1 \end{aligned}$$

Thus $f(k_X(\alpha)) \geqslant 2 + \Sigma\{1 : 1 \leqslant j \leqslant n\} = n + 2 > F(\alpha)$. It follows that $f \circ k_X \geqslant F$.

(2) $\Rightarrow$ (3) Suppose $G \in C(EX)$ and $G(\alpha) > 0$ for each $\alpha \in$

EX. Then $1/G \in C(EX)$ and $1/G > \underline{0}$. By (2) there exists $g \in C(X)$ such that $g \circ k_X \geqslant 1/G$. If $x \in X$ then there exists $\alpha \in EX$ such that $k_X(\alpha) = x$, so $g(x) = g \circ k_X(\alpha) \geqslant 1/G(\alpha) > 0$. Thus $1/g \in C(X)$ and $1/g > \underline{0}$. Thus $1/g \circ k_X \in C(EX)$, and if $\alpha \in EX$ then $0 < (1/g \circ k_X)(\alpha) \leqslant G(\alpha)$. Hence $1/g$ is the required f.

(3) $\Rightarrow$ (1) Let $\{A_n : n \in \mathbb{N}\}$ be a decreasing sequence of regular closed subsets of X with empty intersection. Without loss of generality assume $A_1 = X$ and $A_n \backslash A_{n+1} \neq \varnothing$ for $n \in \mathbb{N}$. It follows from 6.6(e)(3,8) that $\{\lambda(A_n) \cap EX : n \in \mathbb{N}\}$ is a decreasing sequence of clopen sets of EX with empty intersection. Define $G \in F(EX,\mathbb{R})$ by

$$G[\lambda(A_n)\backslash\lambda(A_{n+1})] = \{1/n\} \text{ for each } n \in \mathbb{N}.$$

Then G is a well-defined continuous function on EX, since $\bigcap\{\lambda(A_n) \cap EX : n \in \mathbb{N}\} = \varnothing$ implies $EX = \bigcup\{\lambda(A_n)\backslash\lambda(A_{n+1}) : n \in \mathbb{N}\} \cap EX$. By hypothesis there exists $f \in C(X)$ such that $f(x) > 0$ for each $x \in X$, and $f \circ k_X \leqslant G$. If $x \in A_n$ then $x = k_X(\alpha)$ for some $\alpha \in \lambda(A_n) \cap EX$ (see 6.6(e)(3)); so, $f(x) = f \circ k_X(\alpha) \leqslant G(\alpha) \leqslant 1/n$. Thus $A_n \subseteq f^{\leftarrow}[[0,1/n]]$. Let $Z_n = f^{\leftarrow}[[0,1/n]]$. Then $Z_n \in Z(X)$ (see 1.4(j)), $A_n \subseteq Z_n$ for each $n \in \mathbb{N}$, and $\bigcap\{Z_n : n \in \mathbb{N}\} = f^{\leftarrow}(0) = \varnothing$.

(2) $\Rightarrow$ (4) Let $F \in C(EX)$. To show that C(EX) is the conditional completion via the embedding $f \to f \circ k_X$, we must show that $\{f \circ k_X : f \in C(X) \text{ and } f \circ k_X \geqslant F\} \neq \varnothing \neq \{f \circ k_X : f \in C(X) \text{ and } f \circ k_X \leqslant F\}$ and that $F = \bigwedge\{f \circ k_X : f \in C(X) \text{ and } f \circ k_X \geqslant F\} = \bigvee\{f \circ k_X : f \in C(X) \text{ and } f \circ k_X \leqslant F\}$ (see 2.5(c)). Now (2) essentially says that the above two sets are nonempty. We will verify that F =

$\bigwedge\{f \circ k_X : f \in C(X) \text{ and } f \circ k_X \geq F\}$. The other verification is essentially the same.

To show that $F = \bigwedge\{f \circ k_X : f \in C(X) \text{ and } f \circ k_X \geq F\}$ we must show that if $H \in C(EX)$ and $F \not\geq H$, then there exists $h \in C(X)$ such that $h \circ k_X \geq F$ but $h \circ k_X \not\geq H$. If $H \in C(EX)$ and $F \not\geq H$, there exists $\alpha_0 \in EX$ such that $H(\alpha_0) - F(\alpha_0) > 0$. As $H - F \in C(EX)$, there exists $A \in R(X)\setminus\{\emptyset\}$ and $\delta > 0$ such that $H(\alpha) - F(\alpha) > \delta$ for each $\alpha \in \lambda(A) \cap EX$. By (2) there exists $f \in C(X)$ such that $f \circ k_X \geq H \vee F$. Let $r_0 = \inf\{f \circ k_X(\alpha) - H(\alpha) : \alpha \in \lambda(A) \cap EX\}$. As $H \geq F$ on $\lambda(A) \cap EX$ it follows that r_0 exists and is non-negative. By the definition of r_0 it is evident that $S = c\ell_{EX}[\lambda(A) \cap (f \circ k_X - H)^{\leftarrow}[(r_0 - \delta/4, r_0 + \delta/4)]]$ is a nonempty clopen set of EX. It follows from 6.6(e)(8) that $\text{int}_X k_X[S] \neq \emptyset$. If $x_0 \in \text{int}_X k_X[S]$, evidently $x_0 \in \text{int}_X A$ and there exists $\gamma_0 \in S$ with $x_0 = k_X(\gamma_0)$. Thus $f \circ k_X(\gamma_0) - H(\gamma_0) \leq r_0 + \delta/4$. We can choose $g \in C(A)$ as follows: $g(x_0) = -r_0 - \delta/3$, $g[\text{bd}_X A] = \{0\}$, and $-r_0 - \delta/3 \leq g(x) \leq 0$ for each $x \in A$. Consider the function $h \in F(X,\mathbb{R})$ defined as follows:

$$h(x) = \begin{cases} f(x) \text{ if } x \in c\ell_X(X\setminus A) \\ f(x) + g(x) \text{ if } x \in A \end{cases}$$

As $g[\text{bd}_X A] = \{0\}$, h is unambiguously defined, and by 1.6(b) $h \in C(X)$.

We now claim that $h \circ k_X \geq F$ but $h \circ k_X \not\geq H$. First note that $h \circ k_X(\gamma_0) = h(x_0) = f(x_0) + g(x_0) = f(x_0) - r_0 - \delta/3 < f \circ k_X(\gamma_0) - r_0 - \delta/4 \leq H(\gamma_0)$; so, $h \circ k_X \not\geq H$ as claimed. If $\alpha \in$

$EX \cap \lambda(c\ell_X(X\setminus A))$, then $k_X(\alpha) \in c\ell_X(X\setminus A)$ (see 6.6(e)(3)); so, $h \circ k_X(\alpha) = f(k_X(\alpha)) \geqslant F(\alpha)$ by definition of f. If $\alpha \in EX \cap \lambda(A)$ then $k_X(\alpha) \in A$; so,

$$h(k_X(\alpha)) = f(k_X(\alpha)) + g(k_X(\alpha))$$

$$\geqslant f(k_X(\alpha)) - r_0 - \delta/3$$

$$\geqslant f(k_X(\alpha)) - (f(k_X(\alpha)) - H(\alpha)) - \delta/3$$

(by definition of r_0)

$$= H(\alpha) - \delta/3$$

$$> F(\alpha) \text{ (by definition of A).}$$

Thus $h \circ k_X \geqslant F$ as required, and the proof is complete.

(4) ⇒ (2) If C(EX) is the conditional completion of C(X) via the embedding $g \to g \circ k_X$, it follows from 2.5(c) that if $F \in C(EX)$ then there exists $f \in C(X)$ such that $f \circ k_X \geqslant F$.

■

Further properties of weak cb-spaces, appear in 6U and 8L. We close by noting the fact that C(EX) is the conditional completion of C(X) for a space X does **not** imply that if $F, G \in C(EX)$ and $F < G$ then there exists $f \in C(X)$ such that $F \leqslant f \circ k_X \leqslant G$. In fact, this happens only if X is extremally disconnected.

(f) **Theorem**. Let X be a Tychonoff space. The following are equivalent:

(1) X is extremally disconnected, and

(2) if F, G $\in$ C(EX) and F $<$ G then there exists f $\in$ C(X) such that $F \leqslant f \circ k_X \leqslant G$.

Proof

(1) $\Rightarrow$ (2) Obvious.

(2) $\Rightarrow$ (1) If (1) fails, there is an A $\in$ R(X) such that $bd_X A \neq \emptyset$. Define F, G $\in$ C(EX) as follows:

$$F[\lambda(A) \cap EX] = \{0\} \qquad F[EX \setminus \lambda(A)] = \{1\}$$

$$G[\lambda(A) \cap EX] = \{0\} \qquad G[EX \setminus \lambda(A)] = \{2\}$$

If $F \leqslant f \circ k_X \leqslant G$ for some f $\in$ C(X), then $f[A] = \{0\}$ and $f[c\ell_X(X \setminus A)] \subseteq [1,+\infty)$ giving contradictory requirements on the value of f(x) for any $x \in bd_X A$. Thus (2) fails.

■

Chapter 8 - Problems

8A. <u>The commutativity of the Iliadis absolute with one-point H-closed extensions</u>. The reader should work through 7D before attempting this.

(1) Prove that the following are equivalent for any space X.

(a) $|\beta(EX)\setminus EX| = 1$.

(b) $|\kappa X \setminus X| = 1$.

(c) X is not H-closed, and if U and V are disjoint open subsets of X then at least one of $c\ell_X U$ and $c\ell_X V$ is H-closed. (See 7P.)

Henceforth let X be a locally H-closed, non-H-closed space.

(2) Prove that $E(X^{\#}) = (EX)^{\#}$ iff condition (c) of (1) is satisfied by X. (See 7D(4) and 8.2(c).)

(3) Prove that X is locally H-closed iff EX is locally compact.

(4) Prove that $E(X^{+}) = (EX)^{+}$ iff the set of non-isolated points of EX is compact. (See 7.6(j) and 7D(3).)

(5) Prove that $E(X^{+}) = (EX)^{+}$ iff the set of non-isolated points of X is Θ-closed in κX. (See 7Q(1) and 6.9(e).)

8B. <u>"$P(hX) = h(PX)$" can be ambiguous.</u> In this problem we show that the two interpretations of the equation $P(hX) = h(PX)$ discussed prior to 8.2(e) are not always equivalent.

(1) If A is a non-compact subset of $\beta\mathbb{N}\setminus\mathbb{N}$, show that $\sigma(\mathbb{N} \cup A)$ is not compact. (Hint: See 7B(5).)

Define the space Y to be the free union

$$(\{1\} \times \beta\mathbb{N}) \oplus (\{2\} \times \mathbb{N}) \oplus (\{3\} \times (\mathbb{N} \cup A))$$

If X is a space, we define an H-closed extension αX of X as follows:

(a) if X is not homeomorphic to Y, define αX to be σX and

(b) define αY to be $(\{1\} \times \beta\mathbb{N}) \oplus (\{2\} \times \sigma(\mathbb{N} \cup A)) \oplus (\{3\} \times \beta\mathbb{N})$.

(2) Prove that $\sigma Y = (\{1\} \times \beta\mathbb{N}) \cup (\{2\} \times \beta\mathbb{N}) \cup (\{3\} \times \sigma(\mathbb{N} \cup A)$.

(3) Prove that the function $h : \alpha Y \to \sigma Y$ defined by

$h(1,\alpha) = (1,\alpha)$ if $\alpha \in \beta\mathbb{N}$,

$h(2,\alpha) = (3,\alpha)$ if $\alpha \in \sigma(\mathbb{N} \cup A)$, and

$h(3,\alpha) = (2,\alpha)$ if $\alpha \in \beta\mathbb{N}$

is a homeomorphism, but that αY and σY are not equivalent extensions of Y.

(4) Prove that if X is any space, then there is a perfect

continuous irreducible surjection $f : \alpha(PX) \to \alpha X$ such that $(P(\alpha X), \Pi_{\alpha X}) \sim (\alpha(PX), f)$. (Hint: Use 8.2(d,h).)

(5) Let $\gamma\mathbb{N}$ denote the one-point compactification of $\mathbb{N}$, and let $X_0 = (\{1\} \times \gamma\mathbb{N}) \oplus (\{2\} \times \mathbb{N}) \oplus (\{3\} \times (\mathbb{N} \cup A))$. Prove that $P(\alpha X_0)$ and $\alpha(PX_0)$ are non-equivalent extensions of PX_0 (here, as usual, we identify PX_0 with $\Pi_{\alpha X_0}^{\leftarrow}[X_0]$).

Thus one interpretation of $P(\alpha X_0) = \alpha(PX_0)$ is true, while the other is false.

8C. <u>The commutativity of P and κ</u>. In this problem we present an alternative method of proving that $\kappa(PX) = P(\kappa X)$ (see 8.2(f)) for any space X. See 6.11(j) for a more direct proof. As in 8.2(f), we interpret PX to be $\Pi_{\kappa X}^{\leftarrow}[X]$.

(1) Show that $\Pi_{\kappa X}^{\leftarrow}[X]$ is open in $P(\kappa X)$.

(2) Let T be an extension of X in which X is hypercombinatorially embedded (see 7O). Prove that if $f : S \to T$ is a perfect irreducible continuous surjection, then $f^{\leftarrow}[X]$ is hypercombinatorially embedded in S. (Hint: Use 6.5(d).)

(3) Prove that $\Pi_{\kappa X} | P(\kappa X) \setminus PX$ is a bijection onto $\kappa X \setminus X$, and that $P(\kappa X) \setminus PX$ is discrete.

(4) Prove that $\kappa(PX) = P(\kappa X)$. (Hint: Use 7O(6).)

8D. The commutativity of P and σ. In this problem we present an alternative method of proving that $\sigma(PX) = P(\sigma X)$ (see 8.2(f)) for any space X. See 8.2(h) for a slightly more elegant proof. As in 8.2(f), we interpret PX to be $\Pi_{\sigma X}^{\leftarrow}[X]$. **Warning**: this exercise contains a number of messy computations.

(1) Let $\mathcal{S} \in \sigma(PX) \setminus PX$ (i.e., $\mathcal{S}$ is a non-convergent open ultrafilter on $\Pi_{\sigma X}^{\leftarrow}[X]$), and define $\mathcal{S}^{\#}$ as follows:

$$\mathcal{S}^{\#} = \{U : U \text{ is open in } \sigma X \text{ and } (0U) \cap PX \in \mathcal{S}\}.$$

(Here 0U is the set of (convergent) open ultrafilters on σX that contain U; see 6.8(c).) Show that $\mathcal{S}^{\#}$ is an open ultrafilter on σX that converges to a point of $\sigma X \setminus X$; i.e., show that $\mathcal{S}^{\#} \in P(\sigma X) \setminus PX$. (Hint: 6.8(d,f) may be useful.)

(2) Define a function $F : \sigma(PX) \to P(\sigma X)$ as follows:

$$F | PX = id_{PX} \text{ and}$$
$$F(\mathcal{S}) = \mathcal{S}^{\#} \text{ if } \mathcal{S} \in \sigma(PX) \setminus PX.$$

Prove that F is a one-to-one function.

(3) Prove that if U is open in σX then

$$F^{\leftarrow}[0U] = o_{\sigma(PX)}(0U \cap PX) \quad \text{(see 7.1(a,c))}.$$

(4) Let W be an open subset of X. Show that:

(a) $\Pi_{\sigma X}^{\leftarrow}[0_{A\sigma X}W] \cap PX$ is dense in $0(o_{\sigma X}W) \cap PX$ and

(b) $$o_{\sigma(PX)}(0(o_{\sigma X}W) \cap PX)\backslash PX = o_{\sigma(PX)}(\Pi_{\sigma X}^{\leftarrow}[o_{\sigma X}W] \cap PX)\backslash PX.$$

(5) Let W be an open subset of X. Prove that

$$F^{\leftarrow}[\Pi_{\sigma X}^{\leftarrow}[o_{\sigma X}W]] = o_{\sigma(PX)}(\Pi_{\sigma X}^{\leftarrow}[W]).$$

(6) Prove that $F : \sigma(PX) \to P(\sigma X)$ is a continuous surjection. (To show that F is onto, note that $F[\sigma(PX)]$ is H-closed (why?))

(7) Show that a closed nowhere dense subset of PX is closed in $P(\sigma X)$. (Hint: Use 6.5(d) and 7.2(b)(4).)

(8) Let Y be extremally disconnected and let Z be an extremally disconnected H-closed extension of Y such that closed nowhere dense subsets of Y are closed in Z. Let $G : \sigma Y \to Z$ be a continuous bijection such that $G|Y = id_Y$. Prove that G is an open function (and hence a homeomorphism).

(9) Prove that $P(\sigma X) \equiv_{PX} \sigma(PX)$ by showing that F is the required homeomorphism.

8E. <u>More about ultrarealcompact spaces</u>. Let P be a Tychonoff or zero-dimensional extension property. Prove that $\mathbb{N}$ has P iff every ultrarealcompact P-regular space has P. (Note that this adds to the list of equivalent conditions in 8.3(g). Ultrarealcompact spaces were introduced in 5AA. Use 5.9(c).)

8F. $\gamma_P(EX)$ vs. $E(\gamma_P X)$ for extension properties possessed by spaces failing to be almost realcompact. Suppose that P is a Tychonoff or zero-dimensional extension property with these properties: (i) there is a P-regular space whose cardinality is not Ulam-measurable and that does not have P and (ii) there is a space with P that is not almost realcompact.

It is shown in this problem that there exists a P-regular space that satisfies 8.3(c)(5) but not 8.3(f)(10). This shows that 8.3(h) cannot be strengthened much.

Let S be a P-regular space whose cardinality is not Ulam-measurable and that does not possess P, and let Z' be a space with P that is not almost realcompact.

(1) Let $Z = EZ'$. Show that Z has P and is not realcompact.

(2) Show that there exists $s \in \beta_P S \setminus S$ such that $\beta_P S \setminus \{s\}$ does not have P. (Hint: Use 5.9(e).)

(3) Let $L = (\{s\} \times \beta_P S) \cup (\beta_P S \times \{s\})$, viewed as a subspace of $\beta_P S \times \beta_P S$. Let $K = \beta_P S \times \{0,1\}$, $K_0 = \beta_P S \times \{0\}$, and $K_1 = \beta_P S \times \{1\}$. Define ϕ : $K \to L$ by:

$$\phi(x,0) = (s,x) \text{ and}$$
$$\phi(x,1) = (x,s).$$

Show that K and L are compact P-regular spaces and ϕ is a perfect irreducible continuous surjection.

(4) Let $X = (K \times \upsilon Z) \setminus [(\{(s,0)\} \cup K_1) \times (\upsilon Z \setminus Z)]$

and $Y = (L \times \upsilon Z)\setminus[\phi[K_1] \times (\upsilon Z\setminus Z)]$ (regarded as subspaces of $K \times \upsilon Z$ and $L \times \upsilon Z$ respectively). Prove that X and Y are both P-regular.

(5) Let $X_0 = (K_0 \times \upsilon Z)\setminus(\{(s,0)\} \times (\upsilon Z\setminus Z))$ and $X_1 = K_1 \times Z$. Prove that $\gamma_P X = \gamma_P X_0 \oplus X_1$.

(6) Suppose that $\mathbb{N}$ has P. Prove that $\gamma_P X_0 \subseteq K_0 \times \upsilon Z$. (Hint: Use 5AD to show that $\upsilon(K_0 \times Z) = K_0 \times \upsilon Z$. Now assume that $p \in \beta(K_0 \times \upsilon Z)\setminus(K_0 \times \upsilon Z)$ and let βj extend the inclusion map $j : K_0 \times \upsilon Z \to K_0 \times \beta Z$. Use 5.3(g) and 5G(3) to produce $f \in C(Z,\mathbb{N})$ such that $(\beta f \circ \Pi_{\beta Z} \circ \beta j)(p) = \infty$. (Here $\mathbb{N}^* = \mathbb{N} \cup \{\infty\}$ is the one-point compactification of $\mathbb{N}$). Now use 5.3(g) and 5.9(f).)

(7) Prove that $\gamma_P X_0 \subseteq K_0 \times \upsilon Z$. (Hint: If $\mathbb{N}$ does not have P, infer that every space with P is countably compact; see 5R(3,6,11). Now conclude that $K \times \upsilon Z = \beta(K \times Z)$ (see 5F(4) and 4AG(8)).)

(8) Prove that $\gamma_P X_0 = K_0 \times \upsilon Z$. (Hint: If there exists $z \in \upsilon Z\setminus Z$ such that $((s,0),z) \notin \gamma_P X_0$, consider $(K_0\setminus\{(s,0)\}) \times \{z\}$.)

(9) Prove that $(\phi[K_0] \times \upsilon Z) \cup Y \subseteq \gamma_P Y \subseteq L \times \upsilon Z$.

(10) Show that Y satisfies 8.3(c)(5). (Hint: Let $\{G_n : n \in \mathbb{N}\}$ be a countable locally finite family of open sets of Y. For each $n \in \mathbb{N}$ let $U_n = \cup\{G_i \cap (L \times Z) : i \geq n\}$. Verify that $\cap\{c\ell_Y U_n : n \in \mathbb{N}\} = \emptyset$. Let $H_n = c\ell_Z \Pi[U_n]$, where Π is the projection from $L \times Z$ onto Z. Use 1M and the fact that Z is extremally disconnected to show that $\{H_n : n \in \mathbb{N}\}$ is a decreasing sequence

of nonempty clopen subsets of Z with empty intersection. Use 5F(3) and then observe that $c\ell_{\Upsilon_P Y} U_n \subseteq L \times c\ell_{\upsilon Z} H_n$ (see (9)).)

(11) Show that if Y satisfies 8.3(f)(10), and if g is a perfect continuous surjection from the P-regular space T onto Y, then $Pg : \Upsilon_P T \to \Upsilon_P Y$ is a countably biquotient surjection. (Hint: Use 6.11(d).)

(12) Show that Y does not satisfy 8.3(f)(10). (Hint: Let $f = \phi \times id_{\upsilon Z} | X$. Then f is a perfect map from X onto Y. Choose $z_0 \in \upsilon Z \setminus Z$ and let $p_0 = (\phi((s,0)), z_0) \in \upsilon Y \setminus Y$. Use (5) and (8) to identify $\Upsilon_P X$ as $(K_0 \times \upsilon Z) \oplus (K_1 \times Z)$. Show that $\{K_0 \times \upsilon Z\}$ is a countable (in fact finite) open cover of $Pf^{\leftarrow}(\Pi_0)$ in $\Upsilon_P X$, but that $Pf[K_0 \times \upsilon Z]$ contains no neighborhoods of p_0 in $\Upsilon_P Y$.)

8G. Basically disconnected covers of compact spaces

(1) Let Q be the class $\{(U,X) : U$ is an open F_σ-subset of the regular space $X\}$. Show that Q is a cg-operator (see 8.4(u)).

(2) Prove that if X is normal and Q is as above, then X has the property P_Q (see 8.4(v)) iff X is basically disconnected (see 6K).

(3) Let X be compact. Prove that $(c_{P_Q} X, k_{P_Q})$ is the minimum basically disconnected cover of X. Then show that if $\mathcal{Q}$ is the σ-completion of the subalgebra of $R(X)$

generated by $\{c\ell_X C : C \in coz(X)\}$, then $(c_{P_Q} X, k_{P_Q})$ is equivalent (as a cover of X) to $(S(\mathcal{Q}), k_{\mathcal{Q}})$ (see 6H). (The **σ-completion** of $\mathcal{Q}$ is defined to be the intersection of all the σ-complete subalgebras of $R(X)$ that contain $\mathcal{Q}$ (see 6K).)

(4) Show that $c_{P_Q}(\beta\mathbb{N}\setminus\mathbb{N})$ is neither $\beta\mathbb{N}\setminus\mathbb{N}$ nor $E(\beta\mathbb{N}\setminus\mathbb{N})$. (Hint: Compare the size of $\mathcal{Q}$ for $X = \beta\mathbb{N}\setminus\mathbb{N}$ and the size of $R(\beta\mathbb{N}\setminus\mathbb{N})$; refer to (4) in §23 of [Si] in case of difficulty.)

8H. <u>A non-Tychonoff cover of a Tychonoff space</u>. In this problem we produce a Tychonoff space Y and a cover (X,f) of Y such that X is not Tychonoff. The reader may wish to review 2T before attempting this.

(1) Prove that the point (ω_1, ω_1) is a P-point (see 1W) of the product space $M = (\omega_1+1) \times (\omega_1+1)$, and conclude that if $f \in C(M)$ then f is constant on a neighborhood of (ω_1, ω_1).

(2) Let $P = M\setminus\{(\omega_1, \omega_1)\}$. Prove that $\beta P = M$. (Hint: Use 2.6(q)(6), 4AG(8) and 4.5(p)(3).) Infer that P is pseudocompact.

(3) Prove that if $h \in C(P)$ then there exists $\alpha_0 < \omega_1$ such that $h((\alpha, \omega_1)) = h((\omega_1, \alpha))$ for each $\alpha > \alpha_0$.

(4) Let $n \in \mathbb{N}$ and $F_n = M \times \{1,2,\ldots,n\}$ (thus F_n is the free union of n copies of M). Define a partition K_n of F_n as follows:

$$\begin{aligned} K_n &= \{\{(\omega_1,\alpha,i)\ ,\ (\alpha,\omega_1,i+1)\} : i = 1 \text{ to } n-1\} \\ &\cup \{\{(\omega_1,\omega_1,i) : i = 1 \text{ to } n\}\} \\ &\cup \{\{p\} : p \notin [(\{\omega_1\} \times \{\omega_1+1\}) \cup ((\omega_1+1) \times \{\omega_1\})] \\ &\times \{1,\ldots,n\}\}. \end{aligned}$$

(In effect we are "gluing n copies of M together" by gluing the right edge of the ith copy to the top edge of the $(i+1)^{st}$ copy for $i = 1$ to $n - 1$.) Given K_n the quotient topology induced by the map ϕ_n that maps each $x \in F_n$ to the member of K_n that contains it. Prove that K_n is a Hausdorff space.

(5) Let $L_n = K_n \setminus \{\phi_n(\omega_1,\omega_1,1)\}$ and $X = (\oplus\{L_n : n \in \mathbb{N}\}) \cup [0,1]$. Let

$$\begin{aligned} U(n,0) &= \emptyset, \\ U(n,1) &= \operatorname{int}_{K_n} \phi_n[P \times \{1\}], \\ U(n,k) &= \operatorname{int}_{K_n} \phi_n[P \times \{k,k+1\}] \text{ for } 2 \leqslant k \leqslant n-1, \text{ and} \\ U(n,n) &= \operatorname{int}_{K_n} \phi_n[P \times \{n\}]. \end{aligned}$$

For each $t \in [0,1]$ let $\{G_i(t) : i \in \mathbb{N}\}$ be a neighborhood base at t in [0,1]. Then let

$$H(m,i,t) = \cup\{\cup\{U(n,[ns]) \cup G_i(t) : s \in G_i(t)\} : n \geqslant m\}$$

where [ns] is the largest integer no larger than ns. Topologize X as follows: open subsets of $\oplus\{L_n : n \in \mathbb{N}\}$ are open in X, and if $t \in [0,1]$ then $\{H(m,i,t) : m \in$

$\mathbb{N}$, $i \in \mathbb{N}\}$ is a neighborhood base at t in X. Show that this defines a Hausdorff topology on X.

(6) Let $g \in C(X)$. Prove that g is constant on $[0,1]$, and hence X is not Tychonoff. (Hint: Use (3).)

(7) Let Y denote the quotient space of X obtained by collapsing $[0,1]$ to a point p, and let f denote the associated quotient map. Prove that Y is Tychonoff. (Hint: First show that if $y \neq p$ then y is completely separated from any closed set disjoint from it. Now show that if p belongs to the open subset H of Y, there exists $m \in \mathbb{N}$ such that

$$[0,1] \cup [\cup\{U(n,[ns]) : s \in [0,1] \text{ and } n \geqslant m\}] \subseteq f^{\leftarrow}[H].$$

If $n \geqslant m$ show that $U(n,k) \subseteq f^{\leftarrow}[H]$ for each $k \leqslant n$. Then verify that $L_n \subseteq f^{\leftarrow}[H]$ and note that L_n is clopen in Y.)

(8) Prove that X is regular and that $(X,f) \in \mathcal{C}(Y) \setminus \mathcal{C}_T(Y)$.

8I. A non-Tychonoff infimum of two Tychonoff spaces. In this problem we produce a regular non-Tychonoff space J(X) and covers $(Y_i,f_i) \in \mathcal{C}_T(J(X))$ $(i = 1,2)$ such that $(Y_1,f_1) \wedge (Y_2,f_2) = (J(X),id_{J(X)}) \in \mathcal{C}(J(X)) \setminus \mathcal{C}_T(J(X))$ (where the infimum is taken in the lattice $\mathcal{C}(J(X))$). Our construction is closely related to that discussed in 1Y and 2S. We will use the notation of 1Y.

(1) Let X be a Tychonoff, non-normal space (such as the "Tychonoff plank" defined in 2R). Let H and K be disjoint closed subsets of X that cannot be put inside disjoint open subsets of X. Assume that neither H nor K contains any isolated points of X. Show that the function q defined in 1Y(3) is irreducible.

(2) Define three equivalence relations E_1, E_2, and E on Z (defined as in 1Y(2)) as follows:

xE_1y if $x = y$ or if $x = (h,i)$, $y = (h,i+1)$, $h \in H$, and i is even;

xE_2y if $x = y$ or if $x = (k,i)$, $y = (k,i+1)$, $k \in K$ and i is odd; and

xEy if $x = y$ or if xE_1y or if xE_2y.

Let Y_i be the quotient space Z/E_i $(i = 1,2)$.

(3) Prove that Y_i is Tychonoff $(i = 1,2)$.

(4) Define $f_i : Y_i \to J(X)$ by letting $f_i(z)$ be the unique E-equivalence class containing z. Show that $(Y_i,f_i) \in \mathcal{C}_T(J(X))$ and conclude that in $\mathcal{C}(J(X))$, $(Y_1,f_1) \wedge (Y_2,f_2) = (J(X),id_{J(X)})$. Note by 1Y(5) that J(X) is not Tychonoff.

8J. <u>The minimum quasi-F-cover of a compact space</u>. In 8.4(ab) we showed that every Tychonoff space X has a unique minimum quasi-F-cover $(QF(X),\phi_X)$. In this problem we show that if X is compact, then QF(X) can be represented as the "space of ultrafilters" on the sublattice of $R(X)$ consisting of closures of interiors of zero-sets of X. Throughout this problem X denotes a compact space.

(1) Let L be a sublattice of the lattice of closed subsets of X. Let $L^{\#} = \{c\ell_X \text{int}_X L : L \in L\}$. Prove that $L^{\#}$ is a sublattice of the lattice $(R(X),\subseteq)$, and if L is a base for the closed sets of X then so is $L^{\#}$.

(2) Let S be a sublattice of $(R(X),\subseteq)$ and let $T(S)$ denote the set of ultrafilters on S. If $S \in S$ let $S^* = \{\alpha \in T(S) : S \in \alpha\}$. Show that $\{S^* : S \in S\}$ is a closed base for a compact topology τ on $T(S)$.

(3) Prove that $(T(S),\tau)$ is Hausdorff if the following condition (*) holds.

(*) If $A, B \in S$ and $A \wedge B = \emptyset$ there exist $C, D \in S$ such that $A \wedge C = B \wedge D = \emptyset$ and $C \vee D = X$.

(4) Let S be a sublattice of $(R(X),\subseteq)$ that is a base for the closed subsets of X and satisfying (*). If $\alpha \in T(S)$ let $\phi(\alpha) = \bigcap\{S \in S : S \in \alpha\}$. Prove that $|\phi(\alpha)| = 1$ and thus ϕ defines a covering map from $T(S)$ onto X.

(5) Prove that if $C \in \text{coz } X$ and $V \in \text{coz } C$ then $V \in \text{coz } X$. (Hint: If $C = X \setminus Z(f)$ and $V = C \setminus Z(g)$, where $f \in C^*(X)$ and $g \in C^*(C)$, define $h \in F(X,\mathbb{R})$ by

$$h(x) \begin{cases} = 0 \text{ if } x \in Z(f) \\ = f(x)g(x) \text{ if } x \in C. \end{cases}$$

(6) Prove that $T(Z(X)^{\#})$ is Hausdorff. (Hint: If $Z_1, Z_2 \in Z(X)$ and $c\ell_X int_X Z_1 \wedge c\ell_X int_X Z_2 = \emptyset$, conclude that $X \setminus (Z_1 \cap Z_2)$ is a dense cozero-set C of X. Find $V_1, V_2 \in coz\ C$ such that $Z_i \cap C \subseteq V_i$ $(i = 1,2)$ and $V_1 \cap V_2 = \emptyset$. Now consider $X \setminus V_1$ and $X \setminus V_2$ and use (5)).

(7) Show that $(T(Z(X)^{\#}), \phi)$ is a cover of X (where ϕ is as in (4)).

We now show that $(T(Z(X)^{\#}), \phi)$ is the minimum quasi-F-cover of X. Henceforth we denote $T(Z(X)^{\#})$ by K.

(8) If $C \in coz\ K$, show that there exists $V \in coz\ X$ such that $\phi^{\leftarrow}[V]$ is a dense subset of C. (Hint: As C is Lindelöf it is the union of countably many basic open subsets of K. Associate with these a $V \in coz\ X$.)

(9) Use (8) to show that if $S \in Z(K)$ then there exists $Z \in Z(X)$ such that $c\ell_K int_K S = c\ell_K \phi^{\leftarrow}[int_X Z]$.

(10) Prove that $A \rightarrow \phi[A]$ is a lattice isomorphism from $Z(K)^{\#}$ onto $Z(X)^{\#}$. (Use 6.5(d)(3).)

(11) Prove that if $Z \in Z(X)$ then $(c\ell_X int_X Z)^* = c\ell_K \phi^{\leftarrow}[int_X Z]$.

(12) Prove that a Tychonoff space Y is a quasi-F-space iff whenever $Z_1, Z_2 \in Z(Y)$ and $int_Y Z_1 \cap int_Y Z_2 = \emptyset$,

then $c\ell_Y\text{int}_Y Z_1 \cap c\ell_Y\text{int}_Y Z_2 = \varnothing$. (Hint: If $\text{int}_Y Z_1 \cap \text{int}_Y Z_2 = \varnothing$ and Y is quasi-F, let $C = Y \setminus (Z_1 \cap Z_2)$ and show that $C \cap \text{int}_Y Z_1$ and $C \cap \text{int}_Y Z_2$ are completely separated in C. For the converse use (5) and 4.6(h).)

(13) Prove that K is a quasi-F-space. (Hint: Use (9), (10), (11), and (12).)

(14) Prove that up to equivalence (K,ϕ) is the minimum quasi-F-cover of X. (Hint: If (Y,f) is another quasi-F-cover of X, let $S = \{(\alpha,Y) \in K \times Y : \phi(\alpha) = f(y)\}$ and argue as in 6.11(d). Use (9) and 6.5(b)(4) also.)

(15) Prove that $K = EX$ iff $Z(X)^{\#} = R(X)$. Infer that if X has countable cellularity (see 2N(6)), then $K = EX$.

(16) Prove that the quasi-F-cover of the ordinal space $\omega_1 + 1$ is the one-point compactification of $\times_0 D(\omega_1)$ (see the remarks preceding 5.10(d)).

8K. Countably paracompact spaces

(1) Let X be a space. Prove that the following are equivalent:

(a) X is countably paracompact,

(b) if $\{U_n : n \in \mathbb{N}\}$ is a countable open cover of X, there exists a countable open cover $\{V_n : n \in \mathbb{N}\}$ of X such that $c\ell_X V_n \subseteq U_n$ for each n, and

(c) if $\{F_n : n \in \mathbb{N}\}$ is a decreasing sequence of closed subsets of X with empty intersection, then there exists a decreasing sequence $\{G_n : n \in \mathbb{N}\}$ of open subsets of X with $F_n \subseteq G_n$ and $\bigcap\{c\ell_X G_n : n \in \mathbb{N}\} = \varnothing$.

(Hint: To show (a) $\Rightarrow$ (c), let $\mathcal{C}$ be a locally finite open refinement of $\{X \setminus F_n : n \in \mathbb{N}\}$. If $W \in \mathcal{C}$ let $g(W) = \min\{n \in \mathbb{N} : W \subseteq X \setminus F_n\}$, $V_n = \bigcup\{W \in \mathcal{C} : g(W) = n\}$, and $G_n = \bigcup\{V_i : i \geqslant n\}$. If $p \in X$ find an open neighborhood $U(p)$ of p and a finite subset $\mathcal{F}(p)$ of $\mathcal{C}$ such that $\mathcal{F}(p) = \{C \in \mathcal{C} : C \cap U(p) \neq \varnothing\}$. Let $k(p) = \max\{g(W) : W \in \mathcal{F}(p)\} + 1$, and verify that $p \notin c\ell_X G_{k(p)}$. To show (c) $\Rightarrow$ (a), let $\{U_i : i \in \mathbb{N}\}$ be an open cover of X and let $F_n = X \setminus \bigcup\{U_i : 1 \leqslant i \leqslant n\}$. Let $V_1 = U_1$ and $V_i = U_i \cap G_{i-1}$ if $i \geqslant 2$; now consider $\{V_i : i \in \mathbb{N}\}$.)

(2) Prove that a space X is countably compact iff it is feebly compact and countably paracompact.

(3) Let X be normal. The following are equivalent:

(a) X is countably paracompact and

(b) if $\{F_n : n \in \mathbb{N}\}$ is a decreasing sequence of closed subsets of X with empty intersection, there exists decreasing sequence $\{G_n : n \in \mathbb{N}\}$ of open subsets of X with $F_n \subseteq G_n$ and $\bigcap\{G_n : n \in \mathbb{N}\} = \varnothing$.

(Hint: For (b) ⇒ (a), note that F_n and $X \setminus G_n$ are completely separated.)

(5) Prove that an almost realcompact, normal, countably paracompact space is realcompact. (Hint: See 6U(2).)

8L. Weak cb spaces vs. weakly δ-normally separated spaces. Recall (see 1R) that a Tychonoff space X is weakly δ-normally separated if each regular closed subset of X is completely separated in X from every zero-set of X disjoint from it.

(1) Prove that each weak cb space is weakly δ-normally separated. (Hint: Let $A \in R(X)$, $Z \in Z(X)$, and assume $A \cap Z = \emptyset$. Suppose $Z = Z(h)$ where $h \in C(X)$. Let $G = \chi_{E(X) \setminus \lambda(A)} + h \circ k_X$ and produce an f as in 8.5(e). Now consider h/f.)

(2) Let $f : X \to Y$ be a perfect continuous open surjection. Prove that if Y is weak cb then so is X.

(3) Prove that if X is a weak cb space, then so is $X \times [0,1]$. (Hint: Use 1M.)

(4) Prove that if $X \times [0,1]$ is weakly δ-normally separated, then X is weak cb. (Hint: If $\{A_n : n \in \mathbb{N}\}$ is a decreasing sequence of regular closed subsets of X with empty intersection, let $H_k = (A_k \times [0,1]) \cup (X \times [1/k+1,1])$ and set $B = \cap\{H_k : k \in \mathbb{N}\}$. Prove that $B \in R(X)$. Find $F \in C(X \times [0,1])$ such that $F[B] = \{0\}$ and $F[X \times \{0\}] = \{1\}$. Let $Z_n = \{x \in X : F(x,1/n) = 0\}]$.

(5) Prove that a Tychonoff space X is weak cb iff $X \times$

[0,1] is weakly δ-normally separated.

8M. Infima and minimal elements of $C(X)$

(1) Let $f : X \to Y$ be a covering map from X to Y, let $p \in Y$, and let F be a finite set such that $\emptyset \neq F \subseteq f^{\leftarrow}(p)$. Let $Z = X/F$, i.e., Z is the quotient space of X with F identified to a point, say q. Let $g : X \to Z$ denote the quotient function and define $h : Z \to Y$ by $h(z) = f(z)$ if $z \neq q$ and $h(q) = p$.

(a) Prove that $f = h \circ g$ and g is a continuous surjection.

(b) Using 8.4(d), conclude that g and h are covering functions.

(c) If X is regular (respectively, Tychonoff or zero-dimensional), prove that Z is regular (respectively, Tychonoff or zero-dimensional).

(2) Let X be a regular space. Prove that $C_T(X)$ (respectively, $C_0(X)$) has a minimal element in $C_T(X)$ (respectively, $C_0(X)$) iff X is Tychonoff (respectively, zero-dimensional).

(3) Let Y be the Tychonoff space described in 8H and let $(X,f) \in C(Y)$ be its non-Tychonoff cover. Let $S = \{(Z,g) \in C_T(Y) : (Z,g) \geqslant (X,f)\}$. Show that (X,f) is the infimum of S in $C(Y)$. (Hint: Use (1).)

Note: In (3), we have constructed an example of a nonempty family $S \subseteq C_T(Y)$ such that the infimum of S in $C_T(Y)$ is not the same as the infimum of S in $C(Y)$.

CHAPTER 9

Categorical Interpretations of Absolutes and Extensions

9.1 Introduction

In virtually every branch of abstract mathematics the entities studied are sets endowed with some "structure" (e.g., a topology or a set of algebraic operations), together with "structure-preserving" functions between such sets. It is therefore not surprising that there are many similarities among the various constructions and techniques used in different branches of abstract mathematics, or within a single branch of mathematics. One theme of this book has been the development of a general theory emphasizing the similarities among the various instances of one class of topological constructions, namely extensions. (To a much lesser extent we have studied covers in the same way.) Thus, in Chapter 5, we showed that the Stone-Čech compactification, the maximum zero-dimensional compactification, and the Hewitt realcompactification are all specific examples of the same phenomenon.

The branch of mathematics which studies the abstract properties of "sets with structures" and "structure-preserving functions" is category theory. Category theory provides a tool by which many parallel techniques used in several branches of mathematics can be linked and treated in a unified manner. By providing a language in which a diversity of important mathematical concepts can be formulated and compared, category theory ties together individual techniques and places them in a light in which it is possible to pinpoint common ideas and results.

This text would not be complete without studying category theory, as the two main focal points of the book - extensions and absolutes - can be formulated in terms of the language of categories. The reader is **warned** that the category theory presented here is a one-sided view from a topological perch.

The first five sections of the chapter develop some of the basic concepts of category theory and characterize these concepts in some particular topological categories. Included in the fourth section is a categorical formulation of Stone's Duality Theorem between Boolean algebras and compact zero-dimension spaces. In the last three sections, the theories of extensions, covers, and absolutes are translated into categorical terms.

9.2 Categories, functors, natural transformations, and subcategories

As noted above, the main objects in mathematics are usually sets with an enriched structure, e.g., topology, order, group operation, ..., and functions with certain properties, e.g., continuous, order-preserving, homomorphism, ..., between the main objects. Roughly speaking, a category consists of sets with structure with structure-preserving functions, a functor is a function between categories, and a natural transformation is a link connecting functors. The formal definitions and examples of these concepts are given in this section.

(a) **Definition**. A **category** $\mathcal{C}$ consists of two classes - a class ob($\mathcal{C}$) of **objects** and a class of **morphisms** - obeying the following axioms:

(i) With each morphism f there are associated two objects (not necessarily distinct) called the **domain** of f and **codomain** of f. If A and B are objects, then Hom(A,B) is used to denote the class of morphisms whose domain is A and whose codomain is B. Hom(A,B) is a set. The notation "$f : A \to B$", in this context, means "$f \in \text{Hom}(A,B)$". When there is more than one category being used and there is a possibility of confusion, "Hom(A,B)" will be denoted by "$\text{Hom}_{\mathcal{C}}(A,B)$".

(ii) Let A, B, C, and D be objects of $\mathcal{C}$. If $f \in \text{Hom}(A,B)$ and $g \in \text{Hom}(B,C)$, there is a unique morphism $g \circ f \in \text{Hom}(A,C)$, called the **composition** of g with f. Composition is associative, i.e., if $h \in \text{Hom}(C,D)$, then $h \circ (g \circ f) = (h \circ g) \circ f$.

(iii) For each object A, there is an **identity morphism** $1_A \in \text{Hom}(A,A)$ with the following properties: if B is an object and $f \in \text{Hom}(A,B)$ (respectively, $g \in \text{Hom}(B,A)$), then $f \circ 1_A = f$ (respectively, $1_A \circ g = g$). (Up to now, the identity function on a topological space X has been denoted by id_X. In order to make our notation conform to standard category-theoretic usage, we will use "id_X" and "1_X" interchangeably in this chapter if X is a topological space.)

In category theory, it is useful to use a diagram as an aid in understanding a proof or definition. For example, let $\mathcal{C}$ be a category, A, B, C and D be objects of $\mathcal{C}$, and $f : A \to B$, $g : A \to C$, $h : B \to C$, $j : C \to D$, and $m : B \to D$ be morphisms in $\mathcal{C}$ such that $h \circ f = g$ and $m = j \circ h$.

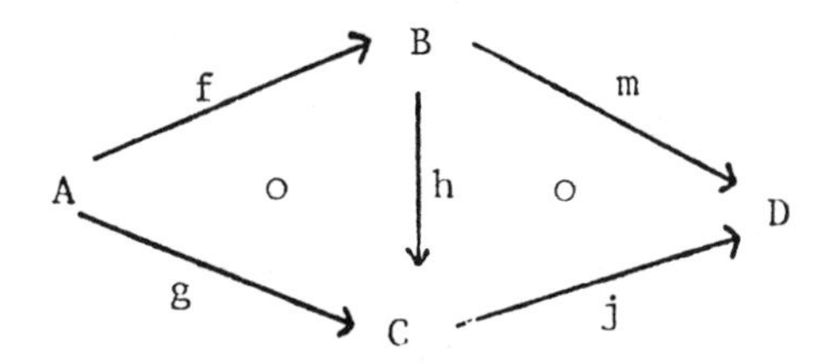

The above diagram is useful in analyzing the properties of the compositions $h \circ f = g$ and $j \circ h = m$. The small circle in the left-hand portion of the diagram (the portion containing the objects A, B, and C and the morphisms f, g, and h) denotes that $g = h \circ f$ which is expressed by saying that the left-hand part of the diagram **commutes**. Likewise, the right-hand part of the diagram commutes meaning that $m = j \circ h$. Since all parts of the diagram commute, we simply say the diagram commutes. The usefulness of the diagram is the motivation it provides in establishing that $m \circ f = j \circ g$. This is easily verified as follows: $j \circ g = j \circ (h \circ f) = (j \circ h) \circ f = m \circ f$. Since composition is associative in categories, we usually will write compositions without parentheses; so, the previous proof would be written as follows: $j \circ g = j \circ h \circ f = m \circ f$.

(b) **Examples**

(1) A category which is fundamental in mathematics is the category of sets and functions (we will use this brief description of categories instead of resorting to more accurate but long-winded descriptions such as "the category whose class of objects is the class of all sets and whose class of morphisms is the class of functions").

The "composition" of morphisms hypothesized in (a)(ii) is just ordinary composition of functions, and the "identity morphism" hypothesized in (a)(iii) is just the identity function on the set. The category of sets and functions is denoted as SET.

Most familiar categories have as their objects sets with additional structure, e.g., algebraic, topological, or both, and as their morphisms some or all of the "structure-preserving" functions from one object to another. In these categories, the "composition" of (a)(ii) and the "identity morphism" of (a)(iii) are the same as defined in SET. **In this book, all categories will be of this type.**

(2) The category of Hausdorff spaces and θ-continuous functions is denoted as θHAUS. To verify (a)(ii) and (a)(iii), we must show that the identity function is "structure-preserving" and that the composition of two "structure-preserving" functions is also "structure-preserving". In this case, the identity function is θ-continuous and the composition of θ-continuous functions is θ-continuous (see 4.8(h)(1)). In θHAUS, the additional "structure" on the sets is a topology.

(3) The category of Boolean algebras and Boolean algebra homomorphisms is denoted as BA. The objects of BA are sets with an algebraic structure.

(4) The category of zero-dimensional spaces and continuous functions is denoted as ZD, and the category of compact spaces and continuous functions is denoted as CPT. Both of these categories have topological structures on the sets which are objects.

Our next step is to define a functor.

(c) <u>**Definition**</u>. Let $\mathcal{B}$ and $\mathcal{C}$ be categories. A **covariant functor** (respectively, **contravariant functor**) from $\mathcal{B}$ to $\mathcal{C}$ is a correspondence F which assigns to each object B of $\mathcal{B}$, a unique object in $\mathcal{C}$, denoted as FB, and to each morphism f in $\mathcal{B}$, a unique morphism in $\mathcal{C}$, denoted as F(f), and which satisfies these three properties:

(1) If $f : A \to B$ is a morphism in $\mathcal{B}$, then $F(f) : FA \to FB$ (respectively, $F(f) : FB \to FA$) is a morphism in $\mathcal{C}$.

(2) For each object B of $\mathcal{B}$, $F(1_B) = 1_{FB}$.

(3) For morphisms $f : A \to B$ and $g : B \to C$ of $\mathcal{B}$, $F(g \circ f) = F(g) \circ F(f)$ (respectively, $F(g \circ f) = F(f) \circ F(g)$).

The word **functor** will mean either a covariant functor or a contravariant functor. The fact that F is a functor from category $\mathcal{B}$ to a category $\mathcal{C}$ is denoted by writing $F : \mathcal{B} \to \mathcal{C}$.

(d) <u>**Examples**</u>

(1) Consider the correspondence $F : \Theta\text{HAUS} \to \text{SET}$ defined by $FX = X$, i.e., if X is a space, then FX is the underlying set of X, and for a Θ-continuous function $f : X \to Y$ in ΘHAUS, $F(f) : FX \to FY$ is defined by $F(f)(x) = f(x)$. So, $F(f) = f$. Clearly F satisfies (c)(2) and (c)(3) and, hence, is a covariant functor. Sometimes F is called the **forgetful** functor as F "forgets" the topological structure of the space and the "structure-preserving" property of the Θ-continuous function.

(2) Another functor G from ΘHAUS to SET is defined as follows: for each space X, let $GX = \text{Hom}(X, \mathbf{2})$ (Hom as defined in ΘHAUS). If X and Y are spaces and $f : X \to Y$ is a Θ-continuous function, define $G(f) : GY \to GX$ by $G(f)(g) = g \circ f$. Now G satisfies

(c)(2) and (c)(3) and is a contravariant functor. Sometimes G is denoted as Hom(_,$\mathbf{2}$).

(3) Consider the functor B : ZD → BA defined as follows: $BX = B(X)$, i.e., the Boolean algebra of clopen sets of the zero-dimensional space X, and if f ∈ $\mathrm{Hom}_{ZD}(X,Y)$ define $B(f) \in \mathrm{Hom}_{BA}(B(Y),B(X))$ by $B(f)(U) = f^{\leftarrow}[U]$ for $U \in B(Y)$. Now, B satisfies (c)(2) and (c)(3) and is a contravariant functor.

(4) For a category $\mathfrak{a}$, the **identity functor** I : $\mathfrak{a} \to \mathfrak{a}$ is defined by IA = A for each object A ∈ $\mathfrak{a}$ and I(f) = f for each morphism f of $\mathfrak{a}$. The identity functor is covariant.

(5) A functor similar to the one defined in (1) is the "discrete functor" D : ΘHAUS → ΘHAUS defined as follows: for a space X, DX is defined to be the underlying set of X with the discrete topology and for a Θ-continuous function f : X → Y, where X and Y are spaces, D(f) : DX → DY is defined by D(f)(x) = f(x). Since DX has the discrete topology, D(f) is continuous. Clearly, D satisfies (c)(2) and (c)(3); D is a covariant functor.

(6) Another example of a covariant functor is β_0 : ZD → CPT where $\beta_0(X)$ is the maximal zero-dimensional compactification $\beta_0 X$ of the zero-dimensional space X ($\beta_0 X$ is constructed in 4.7) and if f ∈ $\mathrm{Hom}_{ZD}(X,Y)$, then $\beta_0(f)$ is defined to be the function $\beta_0 f$ defined in 4.7(d)(2) (recall that $\beta_0 f$ is the unique continuous function extending f). If X is a zero-dimensional space and $1_X : X \to X$, then $\beta_0(1_X) = 1_{\beta_0 X}$ by the uniqueness of the continuous extension of 1_X. Likewise, if f : X → Y and g : Y → Z are morphisms in ZD, then $\beta_0(g) \circ \beta_0(f)$ and $\beta_0(g \circ f)$ are continuous extensions of $g \circ f : X \to Z$. So, by 4.7(d)(2), $\beta_0(g \circ f) = \beta_0(g) \circ \beta_0(f)$. This shows that β_0 is a covariant functor.

We now define the link connecting functors together.

(e) **Definition**. Let $\mathcal{B}$ and $\mathcal{C}$ be two categories and F : $\mathcal{B} \to \mathcal{C}$ and G : $\mathcal{B} \to \mathcal{C}$ be covariant (respectively, contravariant) functors. A **natural transformation** η from F to G, denoted as η : F $\to$ G, is a correspondence which assigns to each object B of $\mathcal{B}$, an unique morphism $\eta(B)$: FB $\to$ GB in $\mathcal{C}$ such that for each morphism f : A $\to$ B in $\mathcal{B}$, $G(f) \circ \eta(A) = \eta(B) \circ F(f)$ (respectively, $\eta(A) \circ F(f) = G(f) \circ \eta(B)$).

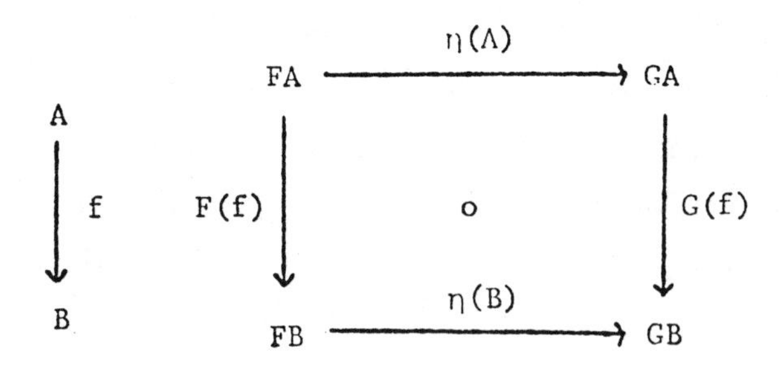

(the diagram illustrates the covariant functor case.)

(f) **Examples**

(1) Let I : ΘHAUS $\to$ ΘHAUS be the identity functor defined in (d)(4) and D : ΘHAUS $\to$ ΘHAUS the discrete functor defined in (d)(5). Define η : D $\to$ I as follows: for each space X, let $\eta(X)$: DX $\to$ IX be the identity function. Since DX has the discrete topology, $\eta(X)$ is a morphism in ΘHAUS. Also, if f : X $\to$ Y is a morphism in

ΘHAUS, then it easily follows that $I(f) \circ \eta(X) = \eta(Y) \circ D(f)$.

(2) Let $\beta : ZD \to CPT$ be the functor defined as follows: for each zero-dimensional space X, let $\beta(X) = \beta X$ (see 4.6) and for $f \in \mathrm{Hom}_{ZD}(X,Y)$, $\beta(f) = \beta f$ is the unique continuous extension of f (see 4.2(d)). As in (d)(6), β is a covariant functor. Now, define $\eta : \beta \to \beta_0$ in this manner: for each object X in ZD, there is a unique continuous function $\eta(X) : \beta X \to \beta_0 X$ which is the identity function on X (see 4.2(d)). If $f : X \to Y$ is a morphism in ZD, then $\beta_0(f) \circ \eta(X)$ and $\eta(Y) \circ \beta(f)$ agree on X; so, by the uniqueness of continuous extensions (see 4.1(b)), $\beta_0(f) \circ \eta(X) = \eta(Y) \circ \beta(f)$.

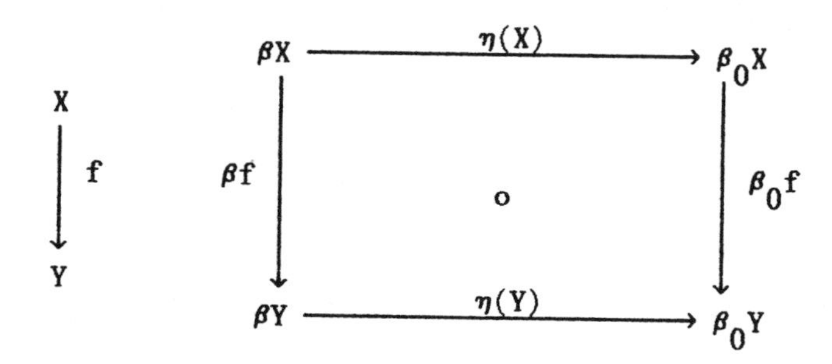

We conclude this section with the definition of subcategory.

(g) <u>Definition</u>

(1) A category $\mathcal{A}$ is a **subcategory** of a category $\mathcal{C}$ if each object of $\mathcal{A}$ is an object of $\mathcal{C}$, each morphism of $\mathcal{A}$ is a morphism of $\mathcal{C}$, the composition of two morphisms in $\mathcal{A}$ is the same as the composition in $\mathcal{C}$, and the identity morphism of an object in $\mathcal{A}$ is the same identity morphism in $\mathcal{C}$.

(2) A subcategory $\mathcal{A}$ of a category $\mathcal{C}$ is said to be **full** if for

objects A and B of $\mathcal{Q}$, $\mathrm{Hom}_{\mathcal{Q}}(A,B) = \mathrm{Hom}_{C}(A,B)$.

(h) **Example**. Both of the categories ZD and CPT are full subcategories of ΘHAUS as Θ-continuous functions into regular spaces are continuous by 4.8(h)(3). On the other hand CPT is not a subcategory of ZD as there are compact spaces which are not zero-dimensional.

9.3 Topological categories

In this section, topological categories are defined. Already, in the first section, three examples (ΘHAUS, ZD, and CPT) were presented; more examples will be described in this section.

(a) **Definitions**

(1) A **topological category** C is a subcategory of ΘHAUS with these two properties:

(i) if $A \in ob(C)$ and B is homeomorphic to A, then $B \in ob(C)$,

(ii) if $A, B \in ob(C)$ and $f : A \to B$ is a homeomorphism, then $f \in \mathrm{Hom}_C(A,B)$, and

(iii) $ob(C)$ contains a space with more than one point.

(2) The category of all Hausdorff spaces and all continuous functions is a topological category and denoted as HAUS.

(3) The full subcategory of HAUS consisting of all the Tychonoff spaces is denoted as TYCH, and the full subcategory of ZD consisting of compact spaces is denoted as ZDCPT. Both TYCH and

ZDCPT, as well as ZD and CPT, are topological categories.

(4) If $\mathcal{Q}$ is a topological subcategory of HAUS, then ob($\mathcal{Q}$) is a class *P* of Hausdorff spaces where the topological property *P* is defined as follows: a space X has property *P* (i.e., X is a *P*-space) iff X $\in$ ob($\mathcal{Q}$). Conversely, if *P* is a topological property, then the full subcategory of HAUS whose objects are the *P*-spaces is denoted by <*P*>. The full subcategory of <*P*> consisting of all the subspaces of products of *P*-spaces is denoted as <Reg(*P*)>, and <K(*P*)> denotes the full subcategory of <*P*> consisting of all the closed subspaces of products of *P*-spaces. Note that this notation conforms with the notation introduced and used in Chapter 5.

In defining a topological category, we must ensure that the proposed class of morphisms is closed under compositions. The composition of two continuous functions is continuous and the composition of two θ-continuous functions is θ-continuous but the composition of two compact (see 1.8(a)(1)) continuous functions need not be compact (see 9A). So, the class of spaces and compact, continuous functions is **not** a category. On the other hand, the class of spaces and perfect, continuous functions is a category; see 1.8(e).

We remind the reader that almost all of the categories treated in this book are topological; the main topological categories are θHAUS, HAUS, TYCH, ZD, CPT and ZDCPT.

9.4 Morphisms

At this point we are ready to interpret categorical concepts in our main topological categories. Since categories consist of objects and morphisms, for a topological property to have a formulation in terms of categorical language, we must be able to characterize the property in terms of functions. In this section, we start this process by translating various types of morphisms in topological categories.

(a) **Definition**. Let $\mathcal{C}$ be a category and A and B be objects of $\mathcal{C}$. A morphism m : B → A is a **monomorphism** if whenever C ∈ ob($\mathcal{C}$) and f, g ∈ Hom(C,B), the equality $m \circ f = m \circ g$ implies that f = g.

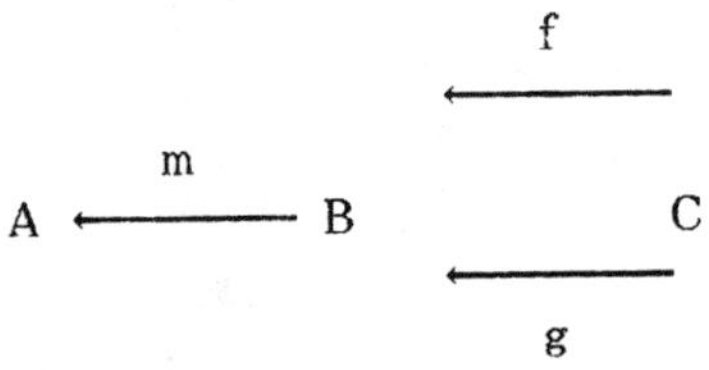

$$A \xleftarrow{\ m\ } B \underset{g}{\overset{f}{\leftleftarrows}} C$$

Each definition and theorem in category gives rise to a "dual concept" which is obtained by reversing the arrows in the diagram illustrating the definition or theorem. The dual to the concept of a monomorphism is the concept of an epimorphism. Explicitly:

(b) **Definition**. A morphism e : A → B in a category $\mathcal{C}$ is an **epimorphism** if whenever C ∈ ob($\mathcal{C}$) and f, g ∈ Hom(B,C), the equality $f \circ e = g \circ e$ implies f = g.

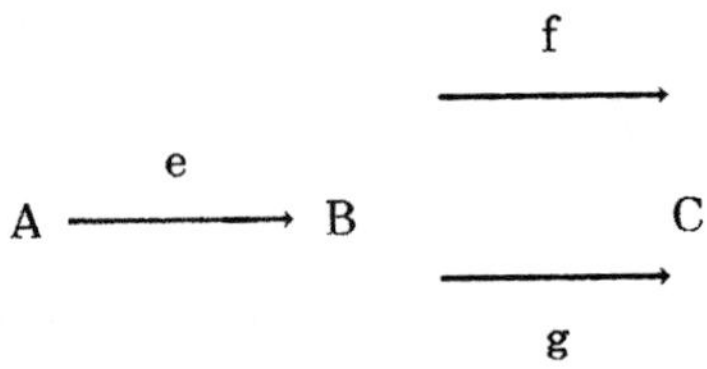

$$A \xrightarrow{\ e\ } B \underset{g}{\overset{f}{\rightrightarrows}} C$$

(c) **Example**. In SET, the monomorphisms are the one-to-one functions (injections) and the epimorphisms are the onto functions (surjections); this is easy to establish and is left as a problem (see 9B) to the reader.

(d) **Theorem**

(1) The monomorphisms in HAUS, in TYCH, and in ZD are the one-to-one continuous functions.

(2) The epimorphisms in HAUS, in TYCH, and in ZD are the continuous functions whose image is dense.

Proof

(1) For each of the categories in question, it is obvious that a one-to-one continuous function between objects of the categories is a monomorphism. Conversely, suppose X and Y are spaces and $m : Y \to X$ is continuous but not one-to-one; then, there are elements $y_1 \neq y_2$ in Y such that $m(y_1) = m(y_2)$. Let Z be the singleton space $\{z\}$. Define $f_i : Z \to Y$ for $i = 1, 2$ by $f_i(z) = y_i$. Then Z, f_1, and f_2 belong to each of the above categories, $f_1 \neq f_2$, and $m \circ f_1 = m \circ f_2$; so, m is not a monomorphism.

(2) Let X, Y, and Z be spaces in one of the three categories, let $e : X \to Y$ be a continuous function such that $e[X]$ is dense in Y, and $f_i : Y \to Z$ be a continuous function for $i = 1, 2$ such that $f_1 \circ e = f_2 \circ e$. Since $f_1 | e[X] = f_2 | e[X]$ and Z is Hausdorff, it follows that $f_1 = f_2$ by 1.6(d). Conversely, suppose $e : X \to Y$ is continuous but $c\ell_Y e[X] \neq Y$. Consider the product space $Y \times \mathbf{2}$ ($\mathbf{2} = \{0,1\}$ has the discrete topology) and the relation $R = \{((y,j), (y,i)) : y \in Y,\ i, j \in \mathbf{2}$, and $y \notin c\ell_Y e[X]$ implies $i = j\}$ on Y. Evidently, R is an

equivalence relation on Y. Let W denote the set of equivalence classes. For each equivalence class $[(y,j)] \in W$, $[(y,j)] = \{(y,j)\}$ if $y \notin c\ell_Y e[X]$ or $[(y,j)] = \{(y,0),(y,1)\}$ if $y \in c\ell_Y e[X]$. Let $q : Y \times \mathbf{2} \to W$ be the quotient function, i.e., $q(y,j) = [(y,j)]$ for each $(y,j) \in Y \times \mathbf{2}$, and let W have the quotient topology induced by q. (In effect, W is obtained by "gluing two copies of Y together along $c\ell_Y e[X]$.") It is left as a problem (see 9B) to verify that if Y is Hausdorff (respectively, Tychonoff, zero-dimensional), then so is W. For $i \in \mathbf{2}$, define $j_i : Y \to Y \times \mathbf{2}$ by $j_i(y) = (y,i)$ and let $f_i = q \circ j_i$. For $i \in \mathbf{2}$, clearly j_i is continuous and hence f_i is continuous. If $x \in X$, note that $(f_0 \circ e)(x) = [(e(x),0)] = [(e(x),1)] = (f_1 \circ e)(x)$ as $e(x) \in c\ell_Y e[X]$. So, $f_0 \circ e = f_1 \circ e$. However, $f_0 \neq f_1$ since there is some $y \in Y \setminus c\ell_Y e[X]$ for which $f_0(y) \neq f_1(y)$. This shows that e is not an epimorphism.

■

The algebraic concept of isomorphism, the topological concept of homeomorphism, and the set-theoretic concept of a bijection are analogous ideas. Two isomorphic groups have the "same" group-theoretic structure, and two homeomorphic topological spaces have the "same" topological structure. In category theory, the idea of "sameness", is captured by the concept of an "isomorphism"; the emphasis, however, is shifted from the similar structure of the objects to the properties of the morphisms connecting them.

(e) **Definition**. Let $\mathcal{C}$ be a category and A and B be objects of $\mathcal{C}$. Let $f : A \to B$ be a morphism in $\mathcal{C}$.

(1) A morphism $g : B \to A$ is a **right** (respectively, **left**) **inverse**

of f if $f \circ g = 1_B$ (respectively, $g \circ f = 1_A$).

(2) An **isomorphism** is a morphism that has both a left inverse and a right inverse.

It is easy to check that the isomorphisms in the category of groups and group homomorphisms are precisely the group isomorphisms. In BA they are precisely the Boolean algebra isomorphisms (see 9B), in ΘHAUS they are precisely the Θ-homeomorphisms (defined in 4.8(g)(4)), and in HAUS, TYCH, ZD, CPT, or ZDCPT they are precisely the homeomorphisms.

Now, we list some of the basic properties of isomorphisms.

(f) <u>**Theorem**</u>. Let A and B be objects of a category $\mathcal{C}$ and $f : A \to B$ a morphism. Then:

(1) If f has a right inverse g and a left inverse h, then $g = h$.

(2) If f is a monomorphism and has a right inverse, then f is an isomorphism.

(3) If f is an epimorphism and has a left inverse, then f is an isomorphism.

<u>Proof</u>

(1) Note that $g = 1_A \circ g = h \circ f \circ g = h \circ 1_B = h$.

(2) Let g be a right inverse of f. Since $f \circ 1_A = f = 1_B \circ f = f \circ g \circ f$ and f is a monomorphism, it follows that $1_A = g \circ f$ and g is a left inverse of f.

(3) (Note that (3) is dual to (2); so its proof is dual to (2). We write out the proof of (3) so that the reader may observe the correspondence between a proof and its dual. Henceforth, we shall

not include the dual proofs.) Let h be a left inverse of f. Since $1_B \circ f = f = f \circ 1_A = f \circ h \circ f$ and f is an epimorphism, it follows that $1_B = f \circ h$ and h is a right inverse of f.

■

(g) **Definition**. Let $\mathcal{C}$ be a category and $f : A \to B$ a morphism in $\mathcal{C}$. If g is both a left inverse and a right inverse of f, then g is called an **inverse** of f. (By (f), inverses are unique.) The inverse of f, if it exists, is denoted by "$f^{\leftarrow}$".

(h) **Definition**. A subcategory $\mathcal{A}$ of a category $\mathcal{C}$ is called a **replete subcategory** of $\mathcal{C}$ if whenever $A \in ob(\mathcal{A})$, $B \in ob(\mathcal{C})$, and $j : A \to B$ is an isomorphism in $\mathcal{C}$, then $B \in ob(\mathcal{A})$ and j is an isomorphism in $\mathcal{A}$.

As examples, note that HAUS is not a replete subcategory of ΘHAUS since there are Θ-homeomorphisms that are not homeomorphisms (see 4.8(t)); however, TYCH is a replete subcategory of HAUS.

(i) **Definition**

(1) Let $\mathcal{B}$ and $\mathcal{C}$ be categories, $F, G : \mathcal{B} \to \mathcal{C}$ be functors, and $\eta : F \to G$ a natural transformation. If for each object B of $\mathcal{B}$, $\eta(B)$ is an isomorphism, then η is called a **natural isomorphism**; we write $F \simeq G$ or $\eta : F \simeq G$ if we need to emphasize the particular natural isomorphism.

(2) If $\mathcal{A}$, $\mathcal{B}$, and $\mathcal{C}$ are categories and $F : \mathcal{A} \to \mathcal{B}$ and $G : \mathcal{B} \to \mathcal{C}$ are functors, then $G \circ F$ is used to denote the functor from $\mathcal{A}$ to $\mathcal{C}$ defined as follows: for each object A of $\mathcal{A}$, $(G \circ F)A = G(FA)$ and

for a morphism $f : A \to B$ in $\mathcal{Q}$, $(G \circ F)(f) = G(F(f))$. It is easy to verify that $G \circ F : \mathcal{Q} \to \mathcal{C}$ is a functor. Note that if both G and F are contravariant functors, then $G \circ F$ is a covariant functor.

(3) Two categories $\mathcal{B}$ and $\mathcal{C}$ are said to be **equivalent** if there are functors $F : \mathcal{B} \to \mathcal{C}$ and $G : \mathcal{C} \to \mathcal{B}$ such that $I_{\mathcal{B}} \simeq G \circ F$ and $I_{\mathcal{C}} \simeq F \circ G$ where $I_{\mathcal{B}}$ and $I_{\mathcal{C}}$ are the identity functors on $\mathcal{B}$ and $\mathcal{C}$, respectively (see 9.2(d)(4)).

(j) Proposition. The two categories ZDCPT and BA are equivalent.

Proof. Let $\mathcal{B}$: ZDCPT $\to$ BA be the functor defined in 9.2(d)(3). Define S : BA $\to$ ZDCPT by defining SA to be the Stone space of the Boolean algebra A (see 3.2(c)) and if $f \in \mathrm{Hom}_{BA}(A,B)$, let $S(f) = \lambda(f)$ as defined in 3.2(e). Now S is a contravariant functor. Let I and J denote the identity functor on ZDCPT and BA, respectively. For each space X of ZDCPT, let $\nu(X) = \nu_X$ which is a homeomorphism by 3.2(h). By 3.2(k)(1), $\nu : I \to S \circ \mathcal{B}$ is a natural isomorphism. For each Boolean algebra A of BA, let $\lambda(A) = \lambda_A$ which is a Boolean isomorphism by 3.2(d)(3). By 3.2(k)(2), $\lambda : J \to \mathcal{B} \circ S$ is a natural isomorphism. Hence, ZDCPT and BA are equivalent categories.

■

As proven in (d), the monomorphisms of HAUS, TYCH and ZD are the one-to-one continuous functions. For a topological property P, the objects of $\langle K(P) \rangle$ are precisely the homeomorphic copies of the closed subspaces of products of P-spaces. So, it will be useful to have a category-theoretic description, if possible, of closed embeddings. The categorical concept of an "extremal monomorphism" provides us with such a description.

(k) **Definitions**. Let $\mathcal{C}$ be a category and A and B be objects of $\mathcal{C}$.

(1) A monomorphism $m : A \to B$ is an **extremal monomorphism** in $\mathcal{C}$ if, for each object C of $\mathcal{C}$, and epimorphism $e : A \to C$, the existence of a morphism $h : C \to B$ such that $h \circ e = m$ implies that e is an isomorphism in $\mathcal{C}$.

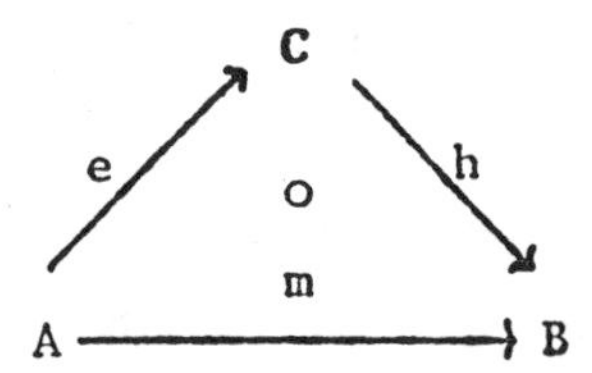

(2) The categorial dual to extremal monomorphism is "extremal epimorphism". An epimorphism $e : B \to A$ is an **extremal epimorphism** in $\mathcal{C}$ if, for each object C of $\mathcal{C}$ and monomorphism $m : C \to A$, the existence of a morphism $h : B \to C$ such that $m \circ h = e$ implies that m is an isomorphism in $\mathcal{C}$.

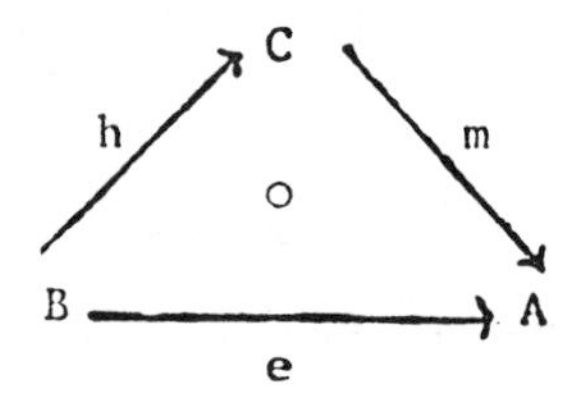

(l) **Theorem**. Let $\mathcal{C}$ be one of the categories HAUS, TYCH, or

ZD. The extremal monomorphisms (respectively, epimorphisms) in C are precisely the closed embeddings (respectively, quotient functions).

Proof. The proof for extremal epimorphisms is left as a problem (see 9B) for the reader. Here is the proof for extremal monomorphisms. Let $m : X \to Y$ be an extremal monomorphism in C, and let $Z = c\ell_Y m[X]$. Define $e : X \to Z$ by $e(x) = m(x)$. Let $h : Z \to Y$ be the inclusion function. Note that Z is an object in C and e and h are morphisms in C. Since $m = h \circ e$, e is an isomorphism. But, in C, isomorphisms are homeomorphisms. Thus, m is a closed embedding.

Conversely, let $m : X \to Y$ be a closed embedding where X and Y are objects in C. By (d), m is a monomorphism in C. Suppose Z is an object in C, $e : X \to Z$ is an epimorphism in C, i.e., $Z = c\ell_Z e[X]$ by (d), and $h : Z \to Y$ is a morphism in C such that $m = h \circ e$. Let W be the subspace $m[X]$ of Y. Then W is an object in C. Since $e[X]$ is dense in Z, it follows that $(h \circ e)[X] = m[X]$ is dense in $h[Z]$. Hence, $h[Z] \subseteq c\ell_Y m[X]$. But as m is closed, $m[X] = c\ell_Y m[X] \supseteq h[Z]$. So, we can define the function $s : Z \to m[X]$ by $s(z) = h(z)$ for $z \in Z$. Let $g : m[X] \to Y$ be the inclusion function and define $j : X \to m[X]$ by $j(x) = m(x)$. Evidently, j is a homeomorphism as m is an embedding. We have that $h = g \circ s$ and $m = g \circ j$, which implies that $g \circ s \circ e = h \circ e = m = g \circ j$. By (d), g is monomorphism; hence, $j = s \circ e$. As j is a homeomorphism, $\overleftarrow{j}$ is a morphism and $1_X = \overleftarrow{j} \circ j = (\overleftarrow{j} \circ s) \circ e$. Thus, e has a left inverse, so by (f)(3), e is an isomorphism. This shows that m is an extremal monomorphism.

■

In topological categories, continuous functions can be factored into the composition of two morphisms one of which is an extremal monomorphism or epimorphism.

(m) **Proposition**. Let X and Y be spaces and $f : X \to Y$ be a continuous function.

(1) Let Z be the subspace $c\ell_Y f[X]$ of Y, $e : X \to Z$ defined by $e(x) = f(x)$, and $m : Z \to Y$ the inclusion function. Then e is an epimorphism in HAUS, m is an extremal monomorphism in HAUS, and $f = m \circ e$.

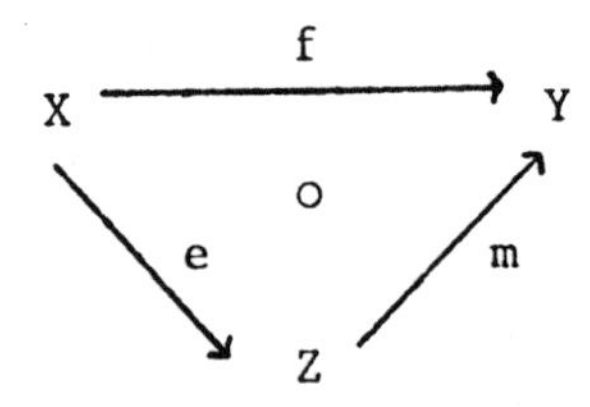

(2) Let $Z = f[X]$, and define $e : X \to Z$ by $e(x) = f(x)$ for $x \in X$. Let Z have the quotient topology induced by e, and let $m : Z \to Y$ be the inclusion function. Then Z is a Hausdorff space, e is an extremal epimorphism in HAUS, m is a monomorphism in HAUS, and $f = m \circ e$.

<u>Proof</u>. The proof is left as a problem (9B) for the reader. ∎

9.5 Products and Coproducts

Let P be a topological property. In 5.3(c), we proved that an object of <Reg(P)> has a maximum P-extension iff P is closed-hereditary and productive. Let C be one of HAUS, TYCH, and ZD. Now, by 9.4(l) a topological property P is closed-hereditary iff whenever X, Y ∈ ob(C), Y has P, and m : X → Y is an extremal monomorphism, then X has P. In this section, we develop the categorical interpretation of the topological concept of "product" and its categorical dual, the concept of a "coproduct". This will allow us to characterize the "extension properties" of Chapter 5 in categorical terms.

(a) **Definition**. Let C be a category and $\{A_a : a \in I\}$ a family of objects of C. An object B of C, together with a family $\{\Pi_a : a \in I\}$ (respectively, $\{j_a : a \in I\}$ of morphisms of C, is called a **product** (respectively, **coproduct**) of $\{A_a : a \in I\}$ in C if

(1) for each $a \in I$, $\Pi_a : B \to A_a$ (respectively, $j_a : A_a \to B$)

and

(2) if C is an object of C and $\{f_a : a \in I\}$ is a family of morphisms in C where $f_a : C \to A_a$ (respectively, $f_a : A_a \to C$) for each $a \in I$, then there is an unique morphism $g : C \to B$ (respectively, $g : B \to C$) such that $\Pi_a \circ g = f_a$ (respectively, $g \circ j_a = f_a$) for each $a \in I$.

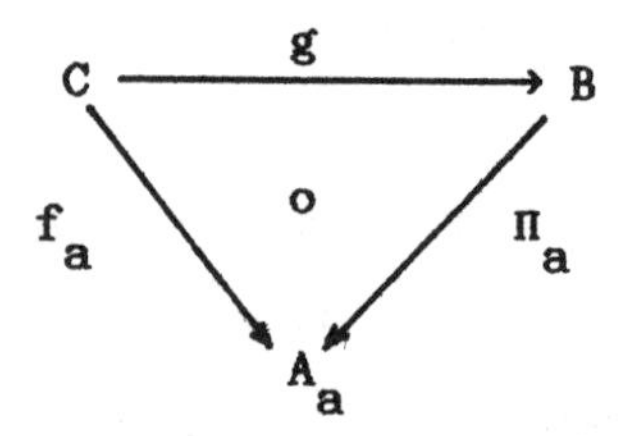

The product (respectively, coproduct) of the family $\{A_a : a \in I\}$, if it exists, is denoted as $\Pi\{A_a : a \in I\}$ (respectively, $\amalg\{A_a : a \in I\}$). Although this notation is the same as that used for Cartesian products of sets, no confusion will ensue. (Note that although a product consists of an object together with a set of morphisms, we will, as above, often abuse terminology and refer to the "object part of the product" as "the product".)

Categorical products and coproducts are unique in the following sense:

(b) **Proposition**. Let C be a category and $\{A_a : a \in I\}$ a family of objects in C. If $(B, \{\pi_a\}_{a\in I})$ and $(B', \{\pi_a'\}_{a\in I})$ (respectively, $(B, \{j_a\}_{a\in I})$ and $(B', \{j_a'\}_{a\in I})$) are products (respectively, coproducts) of $\{A_a : a \in I\}$ in C, then there is an isomorphism $h : B \to B'$ (respectively, $h : B' \to B$) such that $\pi_a = \pi_a' \circ h$ (respectively, $h \circ j_a' = j_a$) for each $a \in I$.

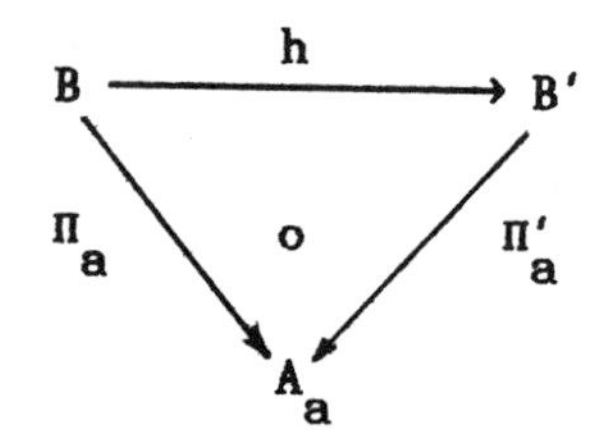

Proof. Only the proof for the product case is given. Since $(B, \{\Pi_a\}_{a\in I})$ is a product of $\{A_a : a \in I\}$ in C, there is a unique morphism $g : B' \to B$ such that $\Pi_a' = \Pi_a \circ g$ for each $a \in I$. Similarly, there is a unique morphism $g' : B \to B$ such that $\Pi_a = \Pi_a' \circ g$ for each $a \in I$. Now $g \circ g' : B \to B$ is a morphism such that $\Pi_a \circ (g \circ g') = \Pi_a$ for each $a \in I$. But $\Pi_a \circ 1_B = \Pi_a$ for each $a \in I$. By the uniqueness condition, $g \circ g' = 1_B$. Similarly, $g' \circ g = 1_{B'}$. So, g and g' are isomorphisms.

■

Let P be a topological property. When we say P is productive, we mean that the topological product of P-spaces is a P-space (see 1.7(h)). So, we can investigate the existence of categorical products in $<P>$ and, if they exist, compare them with the topological products. The next result shows that these two concepts of products coincide when P is productive and that the corresponding concepts of coproducts coincide when P is closed under the formation of topological sums.

(c) **Proposition**. Let P be a topological property which is productive (respectively, closed under the formation of topological sums). Then the product (coproduct) in $<P>$ of a set of objects in $<P>$ exists and is the topological product (respectively, sum) together with the set of coordinate projection functions (respectively, the corresponding set of inclusion functions).

Proof. The proof is left as problem 9C.

(d) **Corollary**. Products (respectively, coproducts) exist in HAUS, TYCH, and ZD and are just the topological products (respectively, sums) together with set of coordinate projection functions (respectively, inclusion functions).

Proof. This is a consequence of (c) and the observations that if P is the topological property of being Hausdorff (respectively, Tychonoff or zero-dimensional), then P is productive and closed under the formation of topological sums and $\langle P \rangle$ = HAUS (respectively, TYCH of ZD).

■

Likewise, products exist in CPT and ZDCPT; also, coproducts exist in CPT and ZDCPT even though the infinite topological sum of nonempty compact spaces is not compact (see 9C). Thus the converse to (c) is false. Also, for a topological property P, products may exist in $\langle P \rangle$ without P being productive.

(e) **Example**. Let P be the topological property of being discrete. Now, $\mathbf{2}$ (discrete topology on a doubleton) has property P, but $\mathbf{2}^{\mathbb{N}}$ is the Cantor space and is not discrete. So, P is not productive. However, if $\{A_a : a \in I\}$ is a family of objects of $\langle P \rangle$ and B is the underlying set of $\Pi\{A_a : a \in I\}$ with the discrete topology, then $(B, \{\Pi_a\}_{a \in I})$ is the categorical product of $\{A_a : a \in I\}$ in $\langle P \rangle$. (See 9C for a generalization of this example.)

(f) **Example**. Let pHAUS be the category of Hausdorff spaces with p-maps as the morphisms (see 7.6). (Note that it follows immediately from the definition of p-maps that the composition of two p-maps is a p-map.) Let κ denote the cardinal 2^{c} and for each $\alpha < \kappa$ let $X_\alpha = \mathbb{N}$. We will show that the category pHAUS is not closed under the formation of products by showing that the product of the set $\{X_\alpha : \alpha < \kappa\}$ of objects of pHAUS does not exist.

Suppose that the product in pHAUS of $\{X_\alpha : \alpha < \kappa\}$ did exist. Denote it by P, and let $p_\alpha : P \to X_\alpha$ (for each $\alpha < \kappa$) be the set of morphisms in pHAUS witnessing the fact that P is the product in pHAUS of $\{X_\alpha : \alpha < \kappa\}$. As p-maps are continuous, each p_α is continuous. Suppose that Y is a Hausdorff space and that $f_\alpha : Y \to X_\alpha$ is a continuous function for each $\alpha < \kappa$. Since X_α is discrete, each f_α is open and hence a p-map (see 7.6(d)). Hence by the definition of "product", there must exist a p-map f (of necessity continuous) from Y to P such that $p_\alpha \circ f = f_\alpha$ for each $\alpha < \kappa$. By the characterization and uniqueness of products in HAUS, it follows that P would have to be the product in HAUS of $\{X_\alpha : \alpha < \kappa\}$, i.e., the topological product $\Pi\{X_\alpha : \alpha < \kappa\}$ (which we will denote by X), and each p_α would have to be the projection function Π_α.

Let $Y = \mathbb{N}$, and let $f_\alpha : Y \to X_\alpha$ be the identity map for each $\alpha < \kappa$. If X were the category-theoretic product of $\{X_\alpha : \alpha < \kappa\}$ in pHAUS, there would have to be a p-map $f : \mathbb{N} \to X$ such that $\Pi_\alpha \circ f = f_\alpha$ for each $\alpha < \kappa$. Each open set of X contains 2^κ points, so by 4.1(c) $c\ell_X f[\mathbb{N}]$ is a nowhere dense subset of X. It follows from 7.6(e) that $c\ell_X f[\mathbb{N}]$ is H-closed (and hence compact). Thus $\Pi_\alpha[c\ell_X f[\mathbb{N}]]$ is a compact subset of X_α and hence is finite (for each $\alpha < \kappa$). Thus $(\Pi_\alpha \circ f)[\mathbb{N}]$ is finite. But $(\Pi_\alpha \circ f)[\mathbb{N}] = f_\alpha[\mathbb{N}]$

$= \mathbb{N}$ by our choice of f_α, which is a contradiction. Thus there is no such p-map f, and $\{X_\alpha : \alpha < \kappa\}$ has no product in pHAUS.

■

9.6 Reflective and epireflective subcategories

In this section, the theory of extension properties that was developed in Chapter 5 is shown to be a part of the theory of "epireflective subcategories" of categories.

(a) **Definition**. Let $\mathcal{C}$ be a category. A full, replete subcategory $\mathfrak{a}$ of $\mathcal{C}$ is called **reflective** if, for each object B of $\mathcal{C}$, there is an object rB in $\mathfrak{a}$ and a morphism $r_B : B \to rB$ in $\mathcal{C}$ with this property: if A is an object of $\mathfrak{a}$ and $f : B \to A$ is a morphism (in $\mathcal{C}$), there exists a unique morphism $\mathfrak{a}f : rB \to A$ (in $\mathfrak{a}$) such that $f = (\mathfrak{a}f) \circ r_B$.

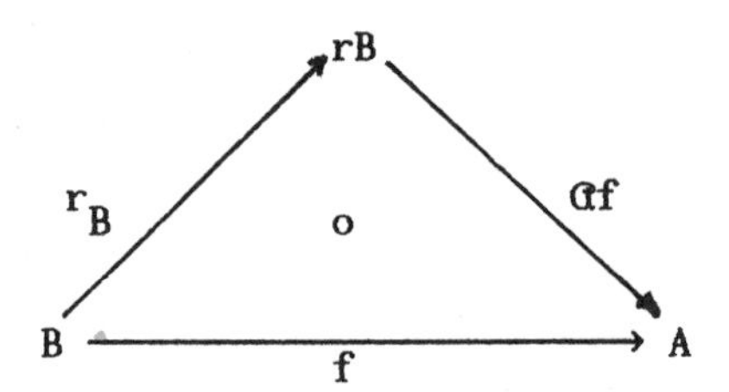

If r_B is an epimorphism (respectively, monomorphism) for each object B of $\mathcal{C}$, then $\mathfrak{a}$ is called an **epireflective** (respectively, monoreflective) subcategory of $\mathcal{C}$. The object rB is the **$\mathfrak{a}$-reflection** (or $\mathfrak{a}$-epireflection, $\mathfrak{a}$-monoreflection) of B, and r_B is called the **reflection**

morphism.

The dual concepts of reflections and epireflections are the terms "coreflections" and "mono-coreflections", respectively, which are introduced in the next section.

(b) <u>Examples</u>

(1) The category CPT is an epireflective subcategory of TYCH. The CPT-epireflection of a Tychonoff space X is its Stone-Čech compactification βX and the reflection morphism r_X is just the inclusion function from X into βX.

(2) The category ZDCPT is an epireflective subcategory of ZD. The ZDCPT-epireflection of a ZD space X is $\beta_0 X$ (discussed in 4.7), and the reflection morphism r_X is the inclusion function from X into $\beta_0 X$.

(3) Let HC be the category of H-closed spaces and continuous functions. Now, HC is an epireflective subcategory of pHAUS. The HC-epireflection of a space X is its Katetov H-closed extension κX, and the reflection morphism r_X is the inclusion function from X into κX (see 7.6(b)).

(4) Let P be a topological property which is productive and closed-hereditary. By 5.3(c), P is an extension property. It follows that $\langle P \rangle$ is an epireflective subcategory of $\langle \mathrm{Reg}(P) \rangle$. The $\langle P \rangle$-epireflection of a P-regular space X is $\gamma_P X$, and the reflection morphism r_X is the inclusion function from X into $\gamma_P X$. So, the theory of extension properties developed in Chapter 5 is a special case of the theory of epireflective subcategories.

We can describe epireflective subcategories using the language of functors as follows.

(c) **Theorem**. Let $\mathcal{A}$ be a reflective subcategory of a category $\mathcal{C}$. For each object C of $\mathcal{C}$, let rC be the reflected object of $\mathcal{A}$ and let $r_C : C \to rC$ be the $\mathcal{A}$-reflection. For each object C of $\mathcal{C}$, let $RC = rC$, and for each morphism $f : C \to D$ of $\mathcal{C}$, let $R(f) = \mathcal{A}(r_D \circ f)$, i.e., R(f) is the unique morphism from rC to rD such that $R(f) \circ r_C = r_D \circ \mathcal{A}f$.

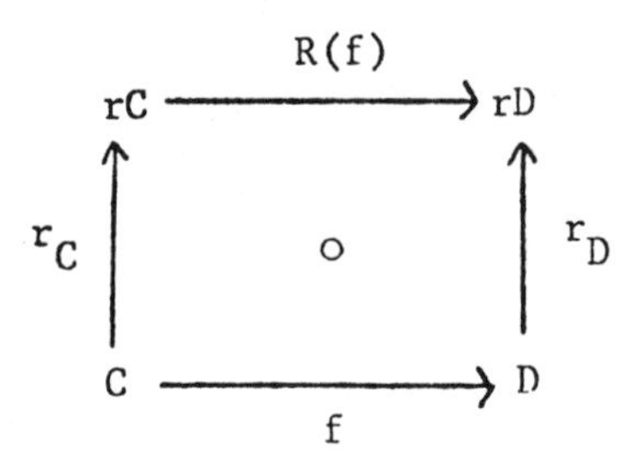

Then R is a functor.

Proof. First note that since $1_{RC} \circ r_C = r_C = r_C \circ 1_C$, it follows by the uniqueness of $R(1_C)$ that $R(1_C) = 1_{RC}$. Let $f : C \to D$ and $g : D \to E$ be morphisms of $\mathcal{C}$. Since $R(g) \circ R(f) \circ r_C = R(g) \circ r_D \circ f = r_E \circ g \circ f = R(g \circ f) \circ r_C$ it follows from the uniqueness of $R(g \circ f)$, that $R(g) \circ R(f) = R(g \circ f)$. So, $R : \mathcal{C} \to \mathcal{A}$ is a covariant functor.

■

If $\mathcal{A}$ is an epireflective subcategory of a category $\mathcal{C}$, then the $\mathcal{A}$-reflection is unique in the following sense.

(d) **Proposition**. Let $\mathcal{Q}$ be an epireflective subcategory of a category $\mathcal{C}$. For each object B of $\mathcal{C}$, let rB and sB be $\mathcal{Q}$-epireflections with the reflection morphisms of r_B and s_B, respectively. Then:

(1) There is an isomorphism $h : rB \to sB$ such that $s_B = h \circ r_B$.

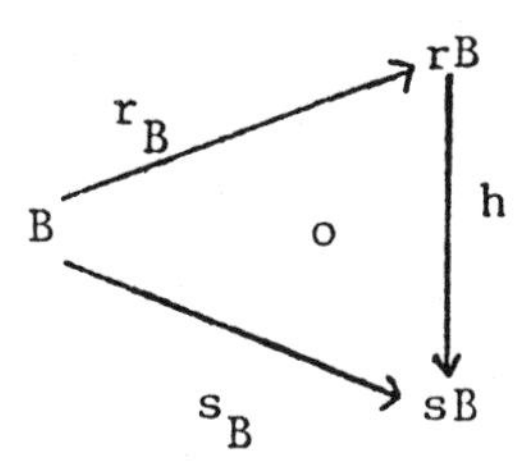

(2) If $A \in \mathcal{Q}$, then $r_A : A \to rA$ is an isomorphism.

Proof

(1) By hypothesis, there are morphisms $\mathcal{Q}s_B : rB \to sB$ and $\mathcal{Q}r_B : sB \to rB$ such that $s_B = (\mathcal{Q}s_B) \circ r_B$ and $r_B = (\mathcal{Q}r_B) \circ s_B$. Thus, $1_{rB} \circ r_B = r_B = (\mathcal{Q}r_B) \circ (\mathcal{Q}s_B) \circ r_B$. Since r_B is an epimorphism, it follows that $1_{rB} = (\mathcal{Q}r_B) \circ (\mathcal{Q}s_B)$. Similarly, $1_{sB} = (\mathcal{Q}s_B) \circ (\mathcal{Q}r_B)$. Thus, both $\mathcal{Q}s_B$ and $\mathcal{Q}r_B$ are isomorphisms.

(2) Clearly, A is the $\mathcal{Q}$-epireflection of A with the reflection morphism being 1_A. Since $r_A \circ 1_A = r_A$, it follows that $\mathcal{Q}r_A = r_A$. By the proof of (1), $\mathcal{Q}r_A$ $(= r_A)$ is an isomorphism.

■

The next result illustrates the similarity between the extension properties developed in Chapter 5 and the theory of epireflective subcategories.

(e) **Theorem**. Let $\mathcal{A}$ be a epireflective subcategory of the category $\mathcal{C}$. The following are true:

(1) If the product in $\mathcal{C}$ of a set of objects of $\mathcal{A}$ exists then the product belongs to $\mathcal{C}$.

(2) If $A \in ob(\mathcal{A})$, $C \in ob(\mathcal{C})$, and $m : C \to A$ is an extremal monomorphism in C, then $C \in ob(\mathcal{A})$.

Proof. For each object C of $\mathcal{C}$, let rC be the $\mathcal{A}$-epireflection and let $r_C : C \to rC$ be the epireflection morphism. To prove (1) let $\{A_a : a \in I\}$ be a set of objects of $\mathcal{A}$ and let $(B, \{\Pi_a\}_{a \in I})$ be its product in $\mathcal{C}$. For each $a \in I$, there is a unique morphism $\mathcal{A}\Pi_a : rB \to A_a$ such that $\Pi_a = (\mathcal{A}\Pi_a) \circ r_B$.

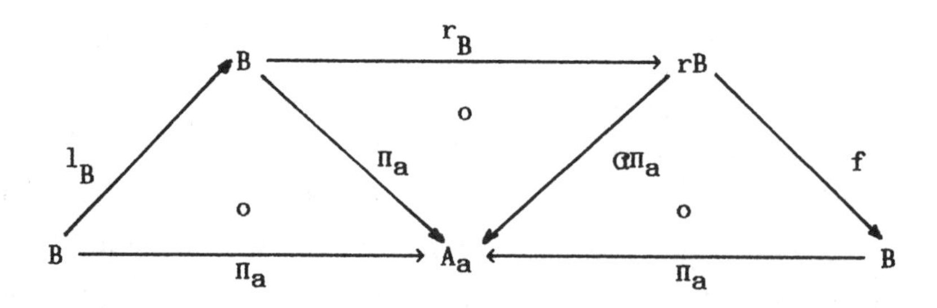

By the definition of products, there is a unique morphism $f : rB \to B$ such that $\Pi_a \circ f = \mathcal{A}\Pi_a$ for each $a \in I$. Combining these, we see that $\Pi_a = \Pi_a \circ (f \circ r_B)$ for each $a \in I$. But by the definition of product, there is a unique morphism $g : B \to B$ such that $\Pi_a \circ g = \Pi_a$ for each $a \in I$, and evidently the g that "works" is 1_B. It follows that $f \circ r_B = 1_B$. As r_B is an epimorphism with a left inverse, by 9.4(f), r_B is an isomorphism in $\mathcal{C}$. Since $\mathcal{A}$ is a replete subcategory of $\mathcal{C}$, B is an object in $\mathcal{A}$. To prove (2), first note there is a unique morphism $\mathcal{A}m : rC \to A$ such that $m = \mathcal{A}m \circ r_C$.

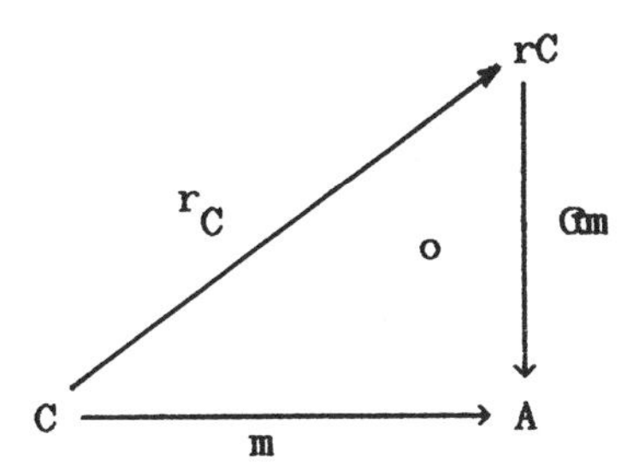

As m is an extremal monomorphism and r_C is an epimorphism, it follows that r_C is an isomorphism. Since $\mathcal{A}$ is a replete subcategory of $\mathcal{C}$, C is an object of $\mathcal{A}$.

■

The next result is a nice consequence of (e).

(f) **Theorem**. Let P be a topological property such that $\langle \mathrm{Reg}(P) \rangle$ is one of HAUS, TYCH, or ZD. Then $\langle P \rangle$ is an epireflective subcategory of $\langle \mathrm{Reg}(P) \rangle$ iff P is an extension property.

Proof. If P is an extension property, then, by (b)(4), $\langle P \rangle$ is an epireflective subcategory of $\langle \mathrm{Reg}(P) \rangle$. Conversely, suppose $\langle P \rangle$ is an epireflective subcategory of $\langle \mathrm{Reg}(P) \rangle$. By 9.5(d), $\langle \mathrm{Reg}(P) \rangle$ (= HAUS, TYCH, or ZD) is closed under the formation of products. By (e), $\langle P \rangle$ is closed under the formation of products. Since products in $\langle P \rangle$ are the same as the products in $\langle \mathrm{Reg}(P) \rangle$ which are the topological products, it follows that P is productive. Let X be a P-space and Y a closed subspace of X. The inclusion function $m : Y \to$

X is an extremal monomorphism by 9.3(l). Since Y is an object of <Reg(P)> and $m = \mathfrak{a} m \circ r_Y$, it follows that r_Y is an isomorphism. Hence, Y is a P-space. This shows that P is closed-hereditary. By 5.3(c), P is an extension property.

■

We now give an example of an epireflective subcategory <P> of HAUS where <Reg P> is not HAUS.

(g) **Example**. Let P be the property of being Tychonoff; then <P> = <Reg P> = TYCH. Let X be a space and $[x] = \bigcap\{f^{\leftarrow}(f(x)) : f \in C(X)\}$ for each $x \in X$. It is immediate that $\{[x] : x \in X\}$ is a partition, denoted as rX, of X. The function $r_X : X \to rX$, defined by $r_X(x) = [x]$, is the usual quotient function, but we do **not** place the quotient topology on rX. If $f \in C(X)$ and $y \in [x]$, then $f(x) = f(y)$; so for each $f \in C(X)$, there is a unique function $f' : rX \to \mathbb{R}$ such that $f = f' \circ r_X$. We place the weak topology on rX induced by $\{f' : f \in C(X)\}$ (see 1.7(a)). It easily follows that r_X is a continuous surjection. Suppose $x, y \in X$ and $[x] \neq [y]$. Then there is some $f \in C(X)$ such that $f(x) \neq f(y)$ which implies that $f'([x]) \neq f'([y])$. Since $f' : rX \to \mathbb{R}$ is continuous, it follows that rX is Hausdorff. If $g \in C(rX)$, then $f = g \circ r_X \in C(X)$ and, hence, $g = f'$. This shows that $C(rX) = \{f' : f \in C(X)\}$, and by 1.4(e,j), it follows that rX is Tychonoff. Let Y be a Tychonoff space and $h : X \to Y$ a continuous function. Let $x, y \in X$ such that $h(x) \neq h(y)$. Since Y is Tychonoff, there is some $g \in C(Y)$ such that $g(h(x)) \neq g(h(y))$. As $g \circ h \in C(X)$, $r_X(x) = [x] \neq [y] = r_X(y)$. So, the correspondence $k : rX \to Y$

defined by $k([x]) = h(x)$ is a function. Also, $h = k \circ r_X$. Next, we want to show that k is continuous. By 1.4(e,j), the topology on Y is generated by the zero-sets of Y. Let $g \in C(Y)$. Then $g \circ h \in C(X)$ and for some $f \in C(rX)$, $g \circ h = f \circ r_X$. Since $f \circ r_X = g \circ h = (g \circ k) \circ r_X$ and r_X is an epimorphism in HAUS, it follows that $f = g \circ k$. So, $f^{\leftarrow}(0) = k^{\leftarrow}(g^{\leftarrow}(0))$. This shows that k is continuous. It is easy to check that if $d : rX \to Y$ is a function (not necessarily continuous) and $h = d \circ r_X$, then $d = k$. This completes the proof that TYCH is an epireflective subcategory of HAUS. Note that, in general, the TYCH-reflection r_X is not an embedding function. So, the epireflective theory includes more than the extension theory developed in Chapter 5.

9.7 Coreflections

In this section, the concept of a "coreflection", the dual notion of reflection, is defined and developed. The "minimum P-covers" discussed in 8.4 (see 8.4(q)) are shown to be examples of "coreflections".

(a) **Definition**. Let C be a category. A full, replete subcategory $\mathcal{A}$ of C is called **coreflective** if, for each object B of C, there is an object sB in $\mathcal{A}$ and a morphism $s_B : sB \to B$ in C with this property: if A is an object of $\mathcal{A}$ and $f : A \to B$ is a morphism in C, there is a unique morphism $\mathcal{A}f : A \to sB$ (in $\mathcal{A}$) such that $s_B \circ \mathcal{A}f = f$.

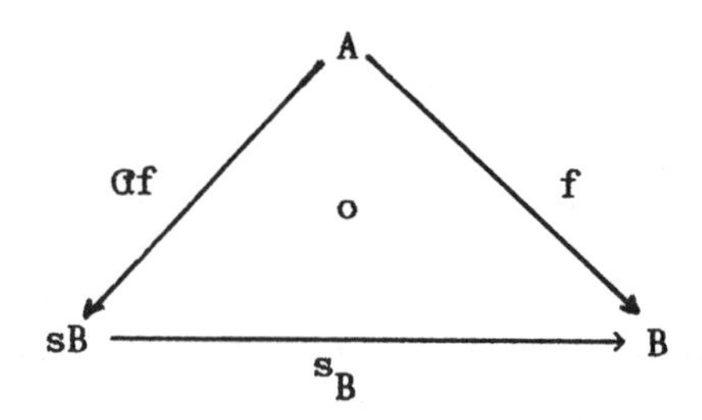

If s_B is a monomorphism (respectively, epimorphism) for each B of C, then $\mathcal{a}$ is called a **mono-coreflective** (respectively, **epi-coreflective) subcategory** of C. The object sB is called the **$\mathcal{a}$-coreflection** (or **$\mathcal{a}$-monocoreflection, $\mathcal{a}$-epi-coreflection)** of B, and s_B is called the **coreflection morphism**. Note that a coreflective subcategory of C is of necessity a full subcategory of C.

We now list a number of examples of coreflections.

(b) <u>**Examples**</u>

(1) Let D be the full subcategory of all discrete objects of HAUS. For each space X, let sX be X with the discrete topology and let $s_X : sX \to X$ be the identity function. If D is a discrete space and $f : D \to X$ a function, then since s_X is a bijection, there is a unique morphism $\mathcal{a}f : D \to sX$ such that $s_X \circ \mathcal{a}f = f$.

(2) Let kCPT denote the full subcategory of HAUS whose objects are k-spaces (see 1T). If X is a space, then kX is a k-space by 1T(4) and the identity function $id_X : kX \to X$ is the coreflection morphism. If Y is a k-space and $f : Y \to X$ is continuous, then the function $kf : kY \to kX$ defined by $(kf)(y) = f(y)$ is continuous by 1T(8). But $kY = Y$ by 1T(4); so, $kf : Y \to kX$. Clearly, $id_X \circ kf = f$. Since

id_X is a bijection, kf is the unique function $g : Y \to kX$ such that $id_X \circ g = f$.

(3) Let P denote the full subcategory of HAUS of P-spaces (defined in 1W). For each space X, let X_δ be the P-space defined in 1W(6); X_δ has the underlying set of X with the topology generated by the G_δ-sets of X. The identity function $id_X : X_\delta \to X$ is the coreflection morphism. If Y is a P-space and $f : Y \to X$ is a continuous function, then $f_\delta : Y \to X_\delta$ defined by $f_\delta(y) = f(y)$ is a continuous function by 1W(7) and $id_X \circ f_\delta = f$. As id_X is a bijection, f_δ is the only function $g : Y \to X_\delta$ satisfying $id_X \circ g = f$.

(4) Let ED be the full subcategory of HAUS consisting of extremally disconnected spaces. Each space X has a Hausdorff absolute PX (defined in 6.11) which is extremally disconnected and a perfect, irreducible continuous surjection $\Pi_X : PX \to X$. If $f : Y \to X$ is continuous, then there is a continuous function $g : PY \to PX$ such that $\Pi_X \circ g = f \circ \Pi_Y$ by 6G(2). If Y is also extremally disconnected, then Π_Y is a homeomorphism by 6.5(d)(4). If $h = g \circ \Pi_Y^{\leftarrow}$, then $h : Y \to PX$ is a continuous function and $\Pi_X \circ h = f$. We **almost** have a proof that ED is coreflective in HAUS. The problem is the issue of the uniqueness of a map h such that $\Pi_X \circ h = f$. The argument used in the previous three examples is not available as Π_X is not necessarily one-to-one. The key here is whether the continuous function $f : Y \to X$ is a c-map (defined in 6G(3)). In general, this is false. For example, let X be the unit interval, $Z = X \oplus X$, $m : Z \to X$ defined by $m[X \times \{0\}] = \{0\}$ and $m(x,1) = x$ for $x \in X$, and $f = m \circ \Pi_Z : PZ \to X$. The perfect continuous surjection f is not a c-map and by 6G(3), there is more

than one continuous function $h : PZ \to EX$ such that $\Pi_X \circ f = h$. If we let oHAUS denote the category of Hausdorff spaces and open continuous functions and oED the full subcategory of oHAUS of all extremally disconnected spaces, then by 6G(4), oED is a coreflective subcategory of oHAUS. Unfortunately, the composition of two c-maps need not be a c-map; see 6G(5). Hence, we cannot generalize this result to "the category of Hausdorff spaces and c-maps," as there is no such category.

(5) Let P be a covering property (see 8.4(q)(3)), and let $[i\hat{P}]$ denote the category of all P-coverable spaces and covering maps. It is clear that the composition of two irreducible continuous surjections is also irreducible, continuous and onto; by 1.8(e), the composition of two perfect functions is perfect. So, $[i\hat{P}]$ is a category. (The notation "[ ]" is extended in 9.8(i).) Let $[iP]$ denote the full subcategory of $[i\hat{P}]$ consisting of P-spaces.

Let X be a P-coverable space, Y a P-space, and $f \in C(Y,X)$ a covering map. As (Y,f) is a cover of X and Y is a P-space, there is a continuous function $h : Y \to c_P X$ such that $k_P \circ h = f$. By 8.4(d), h is a covering map. If $h' : Y \to c_P X$ is a continuous function such that $k_P \circ h' = f$, then by 8.4(d), h' is a covering map. Since $f = k_P \circ h = k_P \circ h'$, by 8.4(i), $h' = h$.

Thus, $[iP]$ is a coreflective subcategory of $[i\hat{P}]$. So, the theory of covers developed in 8.4 is a special case of the theory of coreflections. Since the theory of extensions is a special case of the theory of epireflections (see 9.6(b)(4)) and since coreflections and reflections are categorically dual concepts, we have shown that the theory of covers and the theory of extensions are in some sense categorically dual theories. Note that if P is extremal disconnectedness, then every space is P-coverable, $c_P X = PX$ for each space X, and $[i\hat{P}]$ is the category of Hausdorff spaces and covering maps. Other examples of coreflective subcategories are given in 9D and 9E.

Another way of showing that ED is not a coreflective subcategory of HAUS with the Hausdorff absolute as the coreflection is to use the next result.

(c) **Proposition**. Let $\mathcal{C}$ be HAUS, TYCH, or ZD and $\mathcal{a}$ a coreflective subcategory of $\mathcal{C}$. If $\mathcal{a}$ contains a nonempty space, then $\mathcal{a}$ is both mono-coreflective and epi-coreflective in $\mathcal{C}$.

Proof. For each space X in $\mathcal{C}$, let sX be the $\mathcal{a}$-coreflection and $s_X : sX \to X$ be the coreflective morphism. We will show that s_X is onto and one-to-one. Let Y be a nonempty object of $\mathcal{a}$. Let $x \in X$ and $f : Y \to X$ be defined by $f(y) = x$ for all $y \in Y$. Now, f is continuous and there is a continuous function $h : Y \to sX$ such that $s_X \circ h = f$. This shows that $x \in s_X[sX]$ and that s_X is onto. To show that s_X is one-to-one, let $f, g : Y \to sX$ be continuous functions such that $s_X \circ f = s_X \circ g$. By 9.4(d), we need to show that $f = g$. There is a unique continuous function $h : sY \to sX$ such that $s_X \circ h = s_X \circ f \circ s_Y$. Since $s_X \circ f \circ s_Y = s_X \circ g \circ s_Y$, it follows that $h = f \circ s_Y = g \circ s_Y$. As s_Y is an epimorphism, it follows that $f = g$.

■

In particular, if $\mathcal{a}$ is a coreflective subcategory of HAUS, X is a space, sX the $\mathcal{a}$-coreflection and s_X is the $\mathcal{a}$-coreflection morphism, then s_X is a bijection and, hence, sX is X with a finer topology.

(d) **Proposition**. Let $\mathcal{Q}$ be a coreflective subcategory of $\mathcal{C}$. For each object C in $\mathcal{C}$, let sC be the coreflection of C and $s_C : sC \to C$ the coreflection morphism. Let SC = sC for each object C of $\mathcal{C}$ and $S(f) = \mathcal{Q}(f \circ s_C)$ for each morphism $f : C \to D$ in $\mathcal{D}$. Then S is a functor from $\mathcal{C}$ to $\mathcal{Q}$.

Proof. Note that S(f) is the unique morphism $g : sC \to sD$ such that $s_D \circ g = f \circ s_C$. By the uniqueness property (see 9.6(c)) it follows that S is a functor.

■

(e) **Proposition**. Let $\mathcal{C}$ be a category closed under the formation of coproducts. Let $\mathcal{Q}$ be a coreflective subcategory of $\mathcal{C}$. Then the following are true:

(1) the coproduct in $\mathcal{C}$ of a set of objects of $\mathcal{Q}$ belongs to $\mathcal{Q}$ and

(2) if $A \in ob(\mathcal{Q})$, $C \in ob(\mathcal{C})$, and $e : A \to C$ is an extremal epimorphism in $\mathcal{C}$, then $C \in ob(\mathcal{Q})$.

Proof. The proof of this result is the dual of the proof of 9.6(e)

■

A nice application of (e) is this next result.

(f) **Proposition**. Let P be a topological property. Then $\langle P \rangle$

is a coreflective subcategory in HAUS iff P is closed under the formation of topological sums and the Hausdorff quotient of a P-space is a P-space.

Proof. The proof of one direction is immediate from (e), 9.5(c), and 9.4(l). Conversely, suppose P is closed under the formation of topological sums and the Hausdorff quotient of a P-space is a P-space. Let X be a space, and let $S = \{(A,f) : A$ is a P-space, the underlying set of A is a subset of X, and $f : A \to X$, defined by $f(a) = a$ for $a \in A$, is continuous$\}$. Clearly, S is a set. Let $Y = \oplus\{A : (A,f) \in S\}$. So, Y is a P-space. Define $g : Y \to X$ by $g(x,s) = x$ for $s = (A,f) \in S$ and $x \in A$. Let Z be the underlying set of $g[Y]$ with the quotient topology induced by g. Let $h : Y \to Z$ (defined by $h(y) = g(y)$) be the associated quotient map. Let $m : Z \to X$ be the inclusion function; now, by 9.4(m)(2), Z is the Hausdorff quotient of Y. So, Z is a P-space. Finally, we will show that (Z,m) is a $\langle P\rangle$-coreflection of X. Let T be a P-space and $k : T \to X$ a continuous function. By 9.4(m)(2), $k = j \circ e$ where $e : T \to R$ ($R = k[T]$ with the quotient topology) and $j : R \to X$ is the inclusion function. Now R is a P-space, and $s = (R,j) \in S$. So, the inclusion function $i : R \to Y$ defined by $i(r) = (r,s)$ is continuous. Now, $h \circ i \circ e : T \to Z$ is continuous, and for $t \in T$, $(m \circ (h \circ i \circ e))(t) = (m \circ h)(k(t),s) = g(k(t),s) = k(t)$. So, $k = m \circ (h \circ i \circ e)$. If $d : T \to Z$ is continuous and $k = m \circ d$, then $m \circ (h \circ i \circ e) = m \circ d$. By 9.4(d), m is a monomorphism; so, $h \circ i \circ e = d$. Thus, there is a **unique** continuous function $d : T \rightarrow Z$ such that $m \circ d = k$. This completes the proof that (Z,m) is the $<P>$-coreflection of X.

■

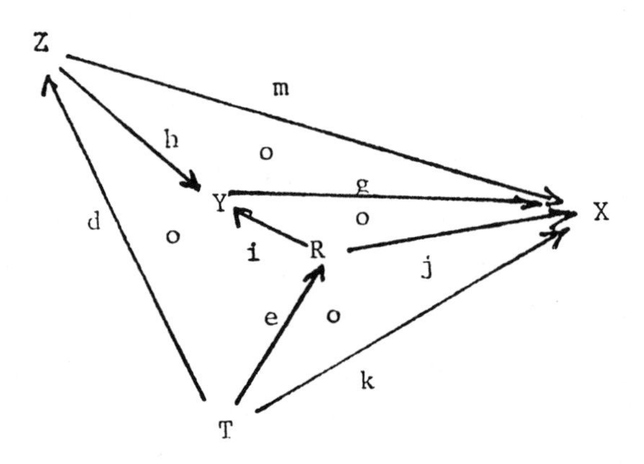

Since $[iP]$ is a coreflective subcategory of $[i\hat{P}]$ (see (b)(5) for notation), we know by (e) that the class of objects in $[iP]$ is closed under the formation of coproducts and extremal epimorphic images in $[i\hat{P}]$. Furthermore, in view of (f) above we might hope that these properties would in fact characterize the coreflective subcategories of $[i\hat{P}]$. In 9M we investigate these questions.

9.8 Projective covers

In this section, the theory of absolutes is shown to be a special instance of the theory of projective covers of categories. In certain topological categories, absolutes are "projective" objects and the absolute and absolute map of a fixed space is a "projective cover" of the space. We will determine the "projective" objects and "projective covers" in a number of topological categories.

(a) **Definition**. Let $\mathcal{C}$ be a catgeory and $\mathcal{M}$ a class of morphisms of $\mathcal{C}$.

(1) An object P of $\mathcal{C}$ is **$\mathcal{M}$-projective** if for B, C $\in$ ob($\mathcal{C}$), morphism $f : B \to C$ of $\mathcal{M}$, and morphism $g : P \to C$ in $\mathcal{C}$, there exists a morphism $h : P \to B$ in $\mathcal{C}$ such that $f \circ h = g$.

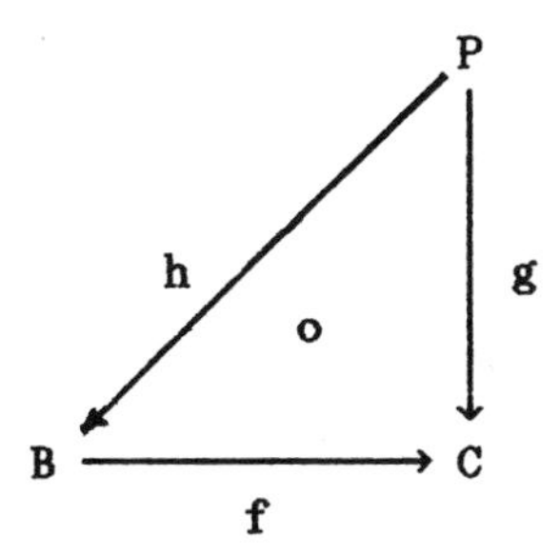

If $\mathcal{C}$ is a topological category and $\mathcal{M}$ is the class of all surjections in $\mathcal{C}$, an $\mathcal{M}$-projective object in $\mathcal{C}$ is simply called a **projective object** in $\mathcal{C}$.

(2) Let C be an object in a topological category $\mathcal{C}$. A pair (P,k) is called a **projective cover** of C in $\mathcal{C}$ if P is a projective object in $\mathcal{C}$ and $k : P \to C$ is an irreducible surjection in $\mathcal{C}$.

Our first goal in this section is to prove the "uniqueness" of projective covers.

(b) **Proposition**. Let $\mathcal{C}$ be a topological subcategory of HAUS and $C \in ob(\mathcal{C})$. If (P,k) and (P',k') are projective covers of C, then there is a homeomorphism $h : P \to P'$ such that $k' \circ h = k$.

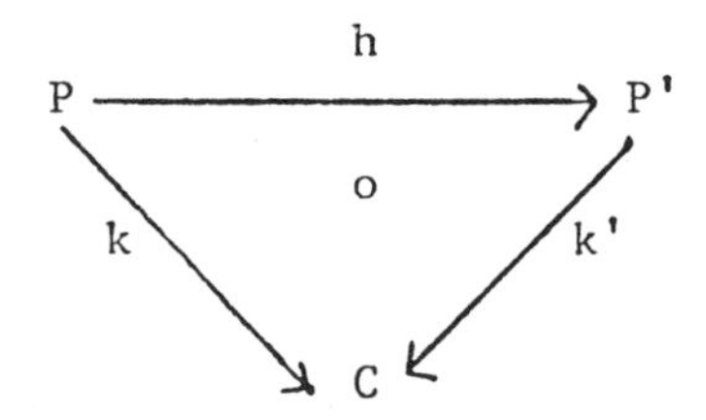

Proof. Since k and k′ are surjections, there are continuous functions $f : P \to P'$ and $g : P' \to P$ such that $k' \circ f = k$ and $k \circ g = k'$. Now, $k \circ 1_P = k = k \circ (g \circ f)$ and $k' \circ 1_{P'} = k' = k' \circ (f \circ g)$. By 8.4(i) $g \circ f = 1_P$ and $f \circ g = 1_{P'}$. So, f is a homeomorphism.

■

(c) **Definition**. Let C be an object of a topological category $\mathcal{C}$. Two projective covers (P,k) and (P',k') of $\mathcal{C}$ are said to be **equivalent** if there is a homeomorphism $h : P \to P'$ such that $k' \circ h = k$. We denote this by writing $(P,k) \approx (P',k')$. **In this book equivalent projective covers of a space are identified and considered to be the same.**

For most topological categories, the projective objects are a subclass of the class of extremally disconnected spaces.

(d) **Theorem**. Let $\mathcal{C}$ be a topological category satisfying these two properties:

(1) If X is an object of $\mathcal{C}$, then $X \oplus X$ is an object of $\mathcal{C}$ and the projection function : $X \oplus X \to X$ (defined by $(p,j) \to p$ for $j \in \{0,1\}$ and $p \in X$) is a morphism of $\mathcal{C}$.

(2) If X is an object of $\mathcal{C}$ and Y is a regular closed subset of X, then Y is an object of $\mathcal{C}$ and the inclusion function $Y \to X$ is a morphism of $\mathcal{C}$.

Then each projective object of $\mathcal{C}$ is extremally disconnected.

<u>Proof</u>. Let P be a projective object of $\mathcal{C}$. Let $f : P \oplus P \to P$ be the projection function onto P. Let U be an open subset of P and $V = P \backslash c\ell_P U$. Now, $X = ((c\ell_P U) \times \{0\}) \cup ((c\ell_P V) \times \{1\})$ is a regular closed subset of $P \oplus P$. So, $X \in \mathrm{ob}(\mathcal{C})$ and the inclusion function $i : X \to P \oplus P$ is a morphism of $\mathcal{C}$. Also, $f \circ i : X \to P$ is a surjection. As P is a projective object, there is a morphism

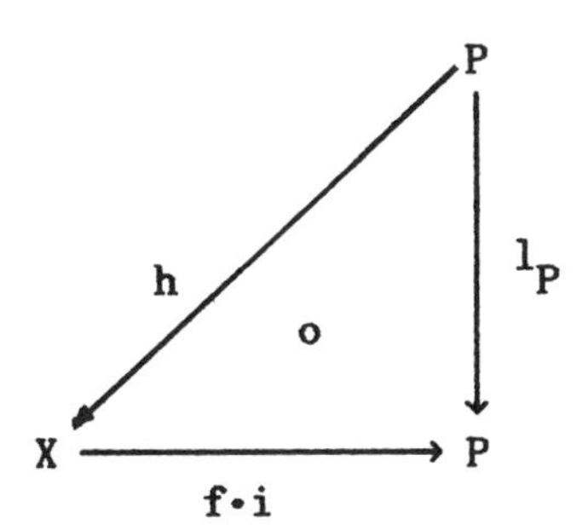

$h : P \to X$ such that $f \circ i \circ h = 1_P$. Since $U \cap c\ell_P V = \varnothing$, it follows that $h[U] = U \times \{0\}$, and by 4.8(s)(3), $h[c\ell_P U] \subseteq c\ell_X(U \times \{0\}) = (c\ell_P U) \times \{0\}$. Similarly, $h[c\ell_P V] \subseteq (c\ell_P V) \times \{1\}$. As $h[c\ell_P U \cap c\ell_P V] \subseteq ((c\ell_P U) \times \{0\}) \cap ((c\ell_P V) \times \{1\}) = \varnothing$, it follows that $c\ell_P U \cap$

$c\ell_P V = \emptyset$. This shows that $c\ell_P U$ is open and completes the proof that P is extremally disconnected.

■

Now, we shall show that discrete spaces in some topological categories are projective.

(e) **Proposition**. Let X and Y be spaces, $f : X \to Y$ a surjection, D a discrete space, and $g : D \to Y$ a function. There is a one-to-one continuous function $h : D \to X$ such that $f \circ h = g$.

Proof. For each $d \in D$, let $x_d \in f^{\leftarrow}(g(d))$; now, define $h : D \to X$ by $h(d) = x_d$. Since D is a discrete space, h is continuous. Clearly, $h \circ f = g$ and h is one-to-one.

■

We are now ready to identify the projective objects and projective covers in a number of topological categories.

(f) **Theorem**. The projective objects of CPT are precisely the extremally disconnected spaces and the projective cover of a compact space X is (EX, k_X).

Proof. By (d), we only need to show that a compact extremally disconnected space is projective. Since all continuous functions between compact spaces are perfect, all morphisms of CPT are perfect and continuous. The first part of the result now follows from 6.11(d). If X is an object of CPT, then EX is a compact, extremally disconnected space by 6.6(e)(1) and 6.9(b)(1), $k_X : EX \to X$ is an irreducible continuous surjection by 6.6(e)(5,6), and (EX,k_X) is a projective cover of X by 6.7(b).

■

The projective objects for HAUS are quite different than those of CPT.

(g) **Theorem**. The projective objects of HAUS, TYCH, or ZD are precisely the discrete spaces.

Proof. The proofs for HAUS, TYCH, and ZD are very similar and only the proof for HAUS is given. By (e), every discrete space is a projective object of HAUS. Conversely, suppose X is a projective object in HAUS. Let D be a discrete space such that $|D| = |X|$, and let $f : D \to X$ be a bijection. By the projective property of X, there is a continuous function $g : X \to D$ such that $g \circ f = 1_X$. As f is a bijection, it follows that g is a bijection. So, X is discrete.

■

Because of the similarities between compact spaces and H-closed spaces, we would suspect the projective objects of HC (defined in 9.6(b)(3)) to be similar to those of CPT as opposed to those of HAUS. However, this is not the case. Before presenting the next result, the reader needs to observe that a discrete space is H-closed iff it is finite.

(h) **Theorem**. The projective objects of HC are precisely the finite spaces.

Proof. As finite spaces are discrete, it follows from (e) that finite spaces are projective in HC. Conversely, let P be a projective object in HC. Let I be the unit interval with the usual topology and let $x \in P$. Now, $P \times I$ is H-closed. By 7S(2), the space Y, whose underlying set if the same as $P \times I$ and whose topology is generated by $\tau(P \times I) \cup \{P \times (0,1] \cup \{(x,0)\}\}$, is H-closed. Also, it is clear that $f : P \to P \times I$ defined by $f(y) = (y,0)$ and $g : Y \to P \times I$ defined by $g(z) = z$ for each $z \in Y$ are continuous. As g is onto and P is projective, there is a continuous function $h : P \to Y$ such that $g \circ h = f$. As f and g are one-to-one functions, it follows that $h(y) = (y,0)$ for each $y \in P$. Since $P \times (0,1] \cup \{(x,0)\}$ is open in Y, $h^{\leftarrow}[P \times (0,1] \cup \{(x,0)\}] = \{x\}$ is open in P. This shows that each point of P is isolated and that P is discrete.

■

(i) **Remark** The difference between the categories HAUS, HC, TYCH, and ZD where the projective objects are the discrete spaces and the category of CPT where the projective objects are the extremally disconnected spaces is not in the objects but in the morphisms of the categories; this is noted in the next result. First, we need to expand on some notation introduced in 9.7(b)(5). For a topological property P, let $[P]$ denote the category whose objects are P-spaces and whose morphisms are perfect, continuous functions. Note that the composition of perfect functions is perfect by 1.8(e), so $[P]$ is indeed a category. If P is the property of being Hausdorff, Tychonoff, or ZD, then $[P]$ is denoted as [HAUS], [TYCH], or [ZD], respectively. The category whose objects are P-spaces and whose morphisms are perfect, Θ-continuous functions is denoted as $[\Theta P]$; it follows from 1.8(e) and 4.8(h)(1) that the compositions of two perfect Θ-continuous functions is perfect and Θ-continuous. Hence, $[\Theta P]$ is indeed a category. On the other hand, it may be true that $[\Theta P]$ is not a replete subcategory of $[\Theta\text{HAUS}]$, as the isomorphisms in $[\Theta P]$ are precisely the Θ-homeomorphisms. For example, $[\Theta\text{TYCH}]$ is not a replete subcategory of $[\Theta\text{HAUS}]$; this follows from the fact that the Tychonoff space $\beta\mathbb{N}$ is Θ-homeomorphic to the non-Tychonoff space $\kappa\mathbb{N}$ (see 4.8(t)).

(j) **Theorem**. Let P be a topological property which is productive and closed hereditary. Then:

(1) the projective objects of $[P]$ are precisely the extremally disconnected spaces of P and

(2) each object X of $[P]$ has a projective cover which is (PX, Π_X).

Proof. If all of the P-spaces are singletons or the empty space, then (1) and (2) follow easily. Suppose there is a P-space with more than one point. Then $\mathbf{2}$ is a P-space. If X has property P, then $X \times \mathbf{2}$ is homeomorphic to $X \oplus X$; hence $X \oplus X$ is a P-space. Also, the projection function from $X \oplus X$ to X, defined by $(p,j) \to p$ for $j = 0,1$ and $p \in X$, is perfect. So, (d)(1) is satisfied. As P is closed hereditary, (d)(2) is satisfied. Hence, by (d), each projective object in $[P]$ is extremally disconnected. Conversely, let E be an extremally disconnected, P-space. Let Y and X be P-spaces and $f : E \to X$ and $g : Y \to X$ be continuous perfect functions such that g is onto.

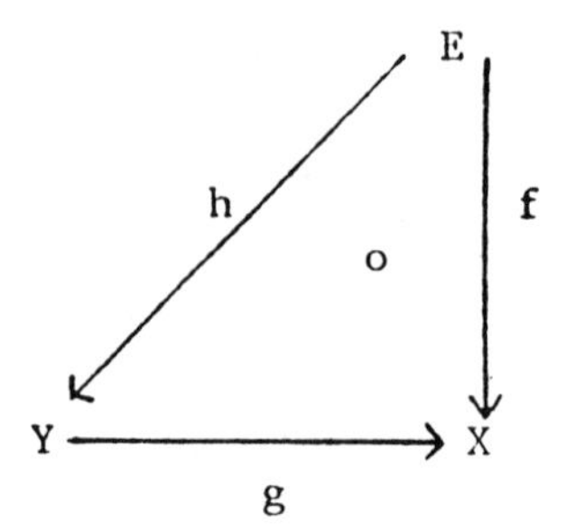

By 6.11(d), there is a perfect, continuous function $h : E \to Y$ such that $g \circ h = f$. This completes the proof of (1). To prove (2), let X be a P-space. By 8.4(y), PX is a P-space. Since $\Pi_X : PX \to X$ is a perfect, irreducible continuous surjection, it follows that (PX,Π_X) is the projective cover of X.

■

(k) **Corollary**. The projective objects in [HAUS], [TYCH], and [ZD] are precisely the extremally disconnected objects and the projective cover of a space X is (PX,Π_X).

Now, we will turn our attention to [ΘP] for various topological properties P. Since a Θ-continuous function into a regular space is continuous, it follows that [ΘTYCH] (respectively, [ΘZD]) is the same category as [TYCH] (respectively, [ZD]). So, for the remainder of this section, we will restrict our attention to two categories -- [ΘHAUS] and [ΘHC] -- which contain nonregular spaces.

(1) **Proposition**. Let P be a topological property which satisfies the following:

(1) if X is a P-space, then so is $X \oplus X$,

(2) P is regular closed hereditary, i.e., if A is a regular closed subspace of the P-space X, then A is a P-space,

(3) spaces which are Θ-homeomorphic to P-spaces are also P-spaces, in particular, $[\Theta P]$ is a replete subcategory of $[\Theta\text{HAUS}]$, and

(4) every compact space is a P-space.

Then a projective object of $[\Theta P]$ is compact and extremally disconnected.

Proof. Let P be a projective object in $[\Theta P]$. By (d), P is extremally disconnected. Let $j : P(s) \to P$ be the identity function. (Recall that P(s) denotes the semiregularization of P.) As j is a perfect, Θ-continuous surjection and P is projective, there is a perfect Θ-continuous function $h : P \to P(s)$ such that $j \circ h = 1_P$.

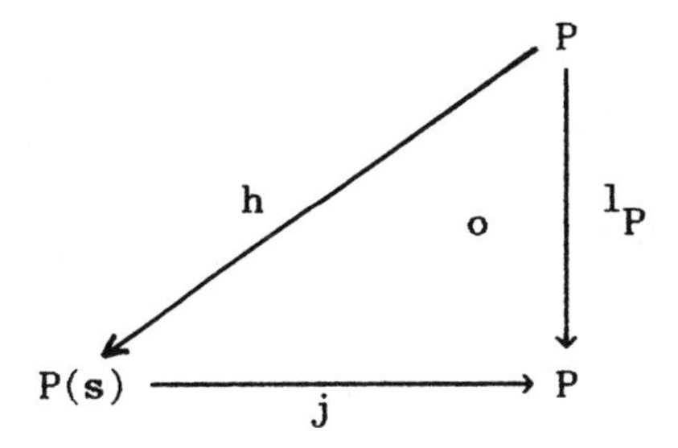

Since 1_P and j are the identity functions, it follows that h is the identity function. As P(s) is P with a coarser topology, h is continuous. However, h is a closed bijection; so, h is a homeomorphism and P is semiregular. It follows from 6.4 that P is an extremally disconnected, Tychonoff space and from 6F(5), there is an infinite discrete space D and an embedding function $f : P \to \beta D$ such that $f[P] \subseteq \beta D \setminus D$. Let $\beta^+ D$ be the corresponding simple extension defined in 7.1(g). The identity function $j : \beta^+ D \to \beta D$ is a continuous bijection and it is easy to show that j is a Θ-homeomorphism (see 7C(2)). Thus, $\beta^+ D$ is a P-space. Define $e : P \to \beta^+ D$ by $e = (j^{\leftarrow} | f[P]) \circ f$. By 4.8(h)(1,4) e is Θ-continuous. Now, e is one-to-one, $e[P] \subseteq \beta^+ D \setminus D$, and $\beta^+ D \setminus D$ is a closed, discrete subspace of $\beta^+ D$, so it follows that e is a perfect, Θ-continuous function from P into $\beta^+ D$. As P is a projective object, there is a perfect, Θ-continuous function $h : P \to \beta D$ such that $e = j^{\leftarrow} \circ h$. So, $j^{\leftarrow} \circ f = j^{\leftarrow} \circ h$. As $j^{\leftarrow}$ is one-to-one, $f = h$.

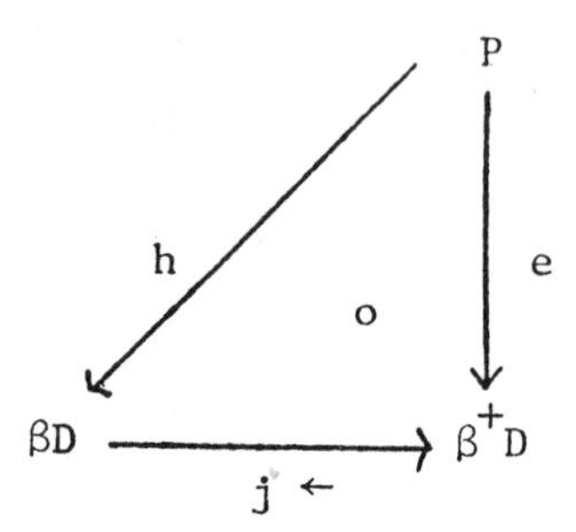

So, P is homeomorphic to $h[P]$ and $h[P]$ is closed in βD. This shows that P is compact.

■

The next result reduces our investigation from characterizing projective objects in many categories to only one category.

(m) **Proposition**. Let P be a topological property satisfying 9.8(l)(1,2,3) and

(4′) every H-closed space is a P-space.

Then the projective objects of $[\Theta P]$ are precisely the projective objects of $[\Theta \text{HAUS}]$.

Proof. If X is a projective object of $[\Theta\text{HAUS}]$, then X is a compact extremally disconnected space by 9.8(l). So, X is a P-space. Since $[\Theta P]$ is a full subcategory of $[\Theta\text{HAUS}]$, it follows that X is a projective object of $[\Theta P]$. Conversely, suppose X is a projective object of $[\Theta P]$. Then X is a compact extremally disconnected space by 9.8(l). Let Y be a space (not necessarily a P-space) and $f : X \to Y$ a perfect, Θ-continuous function. Let $i : Y \to \kappa Y$ be the inclusion function. So, $i \circ f : X \to \kappa Y$ is a Θ-continuous function, and by 4O(3), $i \circ f$ is perfect. Now κY is a P-space and so is $E(\kappa Y)$ by 6.9(b)(1). Since X is a projective object in $[\Theta P]$ and $k_{\kappa Y} : E(\kappa Y) \to \kappa Y$ is a perfect, Θ-continuous surjection, there is a perfect Θ-continuous function $h : X \to E(\kappa Y)$ such that $k_{\kappa Y} \circ h = i \circ f$. It follows that $h[X] \subseteq k_{\kappa Y}^{\leftarrow}[Y]$. By 6.9(a), $(EY, k_Y) \sim (k_{\kappa Y}^{\leftarrow}[Y], k_{\kappa Y} | k_{\kappa Y}^{\leftarrow}[Y])$. Thus, there is a function $g : X \to EY$ such that $k_Y \circ g = f$. Evidently g is perfect as h is, and by 6.5(b)(4) and 4.8(h)(5), g is Θ-continuous. Let Z be a space (not necessarily a P-space) and $m : Z \to Y$ a perfect, Θ-continuous surjection. By 6.7(b), there is a perfect, Θ-continuous function $n : EY \to Z$ such that $m \circ n = k_Y$.

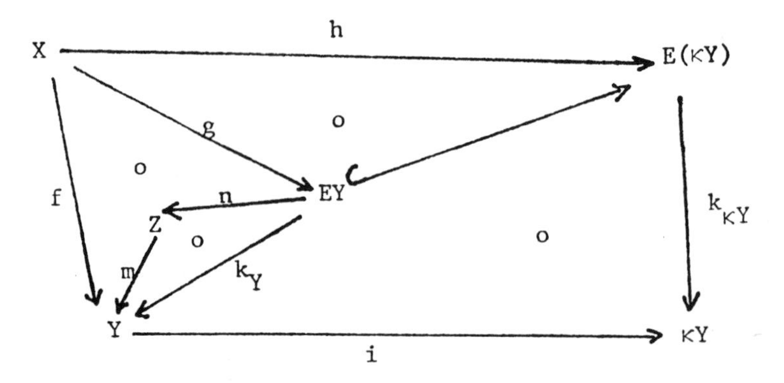

Thus, $n \circ g : X \to Z$ is a perfect Θ-continuous function and $m \circ (n \circ g) = k_Y \circ g = f$. This completes the proof that X is a projective object in [ΘHAUS].

■

(n) **Theorem**. The projective objects of [ΘHAUS] are precisely the finite spaces.

Proof. Since finite spaces are compact and discrete and since continuous functions with compact domain are perfect, it follows by (e) that finite spaces are projective objects in [ΘHAUS]. Assume X is an infinite projective object in [ΘHAUS]. By 9.8(l), X is a compact, extremally disconnected space. As X is infinite, X has a discrete infinite subspace N and by 6L(3,6), N is C^*-embedded in X. By mapping N onto the rationals of [0,1] and extending to X, there is a perfect continuous function $f : X \to \mathbb{R}$ such that $f[X] = [0,1]$. Consider $Y = \mathbb{R} \times \mathbb{Z}$ where $\mathbb{Z}$ is the set of all integers; $U \subseteq Y$ is defined to be open if $(r,0) \in U$ for $r \in \mathbb{R}$ implies there is some $\epsilon > 0$ and $n \in \mathbb{N}$ such that $B(r,n,\epsilon) \subseteq U$ where $B(r,n,\epsilon) = \bigcup\{[r,r+\epsilon) \times \{m\} : m \geqslant n\} \cup [\bigcup\{(r-\epsilon,r] \times \{m\} : m \leqslant -n\}]$. Now Y is a semiregular space. The function $f_0 : X \to Y$, defined by $f_0(x) = (f(x),0)$ for each $x \in X$, is Θ-continuous and by 4O(3) is perfect. Let $Z_0 = c\ell_Y(\mathbb{R} \times \mathbb{N})$, $Z_1 = c\ell_Y(Y \setminus Z_0)$, and $Z = Z_0(s) \oplus Z_1(s)$, and define $g : Z \to Y$ by $g(r,i) = r$ for $r \in Z_i$ and $i = 0,1$. Clearly, g is a perfect irreducible, Θ-continuous surjection. However, note that for $i = 0,1$, $Z_i(s)$ is a Tychonoff space and $(\mathbb{R} \times \{0\}) \times \{i\}$ is a copy of the Sorgenfrey line (see 1H) in $Z_i(s)$. As X is a projective object in [ΘHAUS], there is a perfect Θ-continuous function $h : X \to Z$ such that $g \circ h = f_0$. Since Z is regular, h is continuous and, hence, $h[X]$ is a compact subspace of Z. Since $f_0 = g \circ h$, $h[X] \subseteq (\mathbb{R} \times \{0\}) \times \{0\} \cup (\mathbb{R} \times \{0\}) \times \{1\}$. So, $h[X] \cap Z_i(s)$ is a compact subspace of the Sorgenfrey line $(\mathbb{R} \times \{0\}) \times \{i\}$ for $i = 0,1$. By 1H(7), $h[X] \cap Z_i(s)$ is countable for $i = 0,1$;

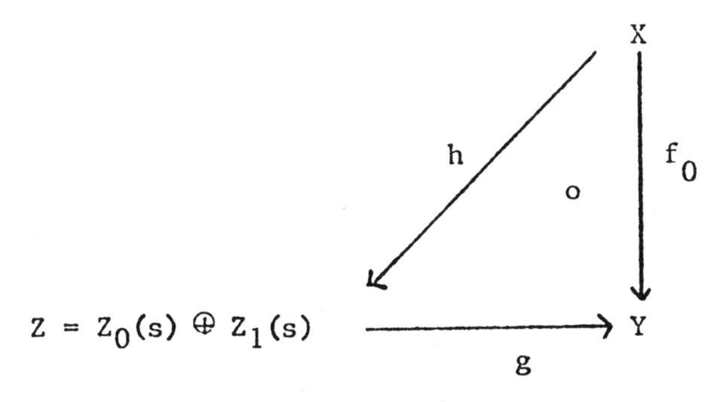

hence, $h[X]$ is countable. This implies $g \circ h[X]$ $(= f_0[X])$ is countable. This yields a contradiction as $|f_0[X]| = |[0,1]| > \aleph_0$. Thus, the space X is finite.

■

In view of (m), we have the next result.

(o) **Corollary**. If P is a topological property satisfying:

(1) if X is a P-space, so is $X \oplus X$,

(2) P is regular-closed hereditary,

(3) the Θ-homeomorphic image of a P-space is a P-space,

and

(4) every H-closed space has property P,

then the projective objects of $[\Theta P]$ are precisely the finite spaces.

In particular, the projective objects of $[\Theta HC]$ are precisely the finite spaces. If P is the property of being H-closed and Urysohn, then P satisfies 9.8(1)(1,2,3,4) but not 9.8(o)(4). As shown in 9H, there are infinite projective objects in $[\theta P]$.

CHAPTER 9 — Problems

9A. Compact Functions. Recall that a function $f \in F(X,Y)$, where X and Y are spaces, is compact if $f^{\leftarrow}(y)$ is compact for each $y \in Y$. Show that the composition of compact, continuous functions is not necessarily compact. (Hint: Consider a continuous bijection from a noncompact space onto a compact space and the continuous function from this compact space to a singleton space.)

9B. Monomorphisms, Epimorphisms, and Isomorphisms

(1) In SET, show that the monomorphisms (respectively, epimorphisms) are the one-to-one (respectively, onto) functions.

(2) Consider the functor $\mathcal{B} : \mathrm{ZD} \to \mathrm{BA}$. Find a monomorphism $f : X \to Y$ in ZD such that $\mathcal{B}(f)$ is not a monomorphism. (Hint: Use 3.2(j)(2).)

(3) If $F : \mathcal{B} \to \mathcal{C}$ is a functor between the categories $\mathcal{B}$ and $\mathcal{C}$ and f is an isomorphism in $\mathcal{B}$, show that F(f) is an isomorphism in $\mathcal{C}$.

(4) Let W be the space described in the proof of 9.4(d)(2). Prove that W, with the quotient topology induced by q, is Hausdorff and if Y is Tychonoff or zero-dimensional, then so is W.

(5) Show that categorical isomorphisms in BA are precisely

the Boolean algebra isomorphisms.

(6) Let C be an object in a category $\mathcal{C}$ and $e_C : C \to C$ a morphism with this property: if $f : A \to C$ and $g : C \to B$ are morphisms of $\mathcal{C}$, then $e_C \circ f = f$ and $g \circ e_C = g$. Prove that $1_C = e_C$. (Hint: Use 9.4(f)(1).)

(7) Prove that the extremal epimorphisms in HAUS, TYCH, and ZD are precisely the quotient functions.

(8) Prove 9.4(m).

9C. Products and Coproducts.

(1) Let P be a topological property. If P is productive (respectively, closed under the formation of topological sums), prove that products (respectively, coproducts) exist in $\langle P \rangle$ and coincide with the topological product (respectively, sum) together with the set of coordinate projection functions (respectively, the corresponding set of inclusion functions).

(2) Let $\mathcal{C}$ be a category in which products (respectively, coproducts) exist. Let $\mathfrak{a}$ be a coreflective (respectively, reflective) subcategory of $\mathcal{C}$. Prove that products (respectively, coproducts) exist in $\mathfrak{a}$.

(3) Prove that products and coproducts exist in CPT and ZDCPT. (Hint: For products, use (1); for coproducts, use (2) and 9.6(b)(1,2).)

9D. Sequential Spaces. Let X be a space and $\mathcal{C} = \{A \subseteq X :$ if $(x_n)_n$ is a sequence contained in A and $(x_n)_n$ converges to a point $p \in X$, then $p \in A\}$.

(1) Show that $\{X \setminus A : A \in \mathcal{C}\}$ is a Hausdorff topology on X.

The set X with the topology $\{X \setminus A : A \in \mathcal{C}\}$ is denoted by sX. A space X is called **sequential** if X = sX. Let SEQ denote the full subcategory of HAUS consisting of all sequential spaces.

(2) (a) Show that $\tau(sX) \supseteq \tau(X)$.

(b) For $U \subseteq X$, show that $U \in \tau(sX)$ iff for each sequence $(x_n)_n$ in X which converges to some point in U, there is some $N \in \mathbb{N}$ such that $\{x_n : n \geq N\} \subseteq U$.

Let $r_X : sX \to X$ denote the identity function; by (2)(a), r_X is continuous.

(3) Let $(x_n)_n$ be a sequence in X and $p \in X$.

(a) Prove that $(x_n)_n$ converges to p in X iff $(x_n)_n$ converges to p in sX.

(b) Use (a) to conclude that s(sX) = sX; in particular,

sX is a sequential space.

(4) (a) Show that first countable spaces are sequential.

(b) Show that sequential spaces are k-spaces (see 1T).

(5) (a) Show that the quotient of a sequential space is also sequential.

(b) Show that the topological sum of sequential spaces is sequential.

(c) Let Y be a sequential space and $S = \{\{x_n : n \in \mathbb{N}\} \cup \{p\} : (x_n)_n$ is a sequence in Y converging to $p\}$; in particular, each $A \in S$ is a compact metric subspace of Y. Let $Z = \oplus\{A : A \in S\}$ and define $f : Z \to Y$ by $f(x,A) = x$ where $x \in A \in S$. Prove that Z is a metric space and f is a quotient function.

(d) Prove that a space is sequential iff it is the quotient image of a metric space.

(6) Let $f : Y \to X$ be a continuous function where Y is a sequential space. Define $h : Y \to sX$ by $h(y) = f(y)$ (recall that sX and X have the same underlying set).

(a) Prove that h is continuous and $r_X \circ h = f$.

(b) If $g : Y \to sX$ is a function such that $r_X \circ g = f$, show that $g = h$.

(c) Using (a) and (b), conclude that SEQ is a coreflective subcategory of HAUS with sX being the coreflection of a space X via the coreflection function r_X.

(7) Let $Y = \{(0,0)\} \cup \{(n,m) : n,m \in \mathbb{N}\}$ and $C_n = \{(n,m) : m \in \mathbb{N}\}$ for $n \in \mathbb{N}$. Define $\tau(Y) = \{U \subseteq Y :$ if $(0,0) \in U$, then $C_n \setminus U$ is finite for each $n \in \mathbb{N}\}$. Show the following:

(a) Y is a zero-dimensional space with $Y \setminus \{(0,0)\}$ a discrete subspace,

(b) Y is a sequential space, and

(c) Y is not first countable at (0,0).

9E. Locally Connected Spaces. A space is **locally connected** if each point has a neighborhood base of connected neighborhoods.

(1) (a) Prove that a space X is locally connected iff the components of all the open subspaces of X are open. (For a space X and a point $p \in X$, recall that $C(p) = \bigcup\{A \subseteq X : A$ is connected and $p \in A\}$, the component of p, is the largest connected subset of X containing p and is closed.)

(b) If X is locally connected, $p \in X$, and U is an open set such that $p \in U$, prove, using (a), that there is an open, connected set V such that $p \in V \subseteq U$.

(c) Use (a) to prove that the components of a locally connected space are clopen and that a locally connected space is the topological sum of connected spaces.

(d) Consider the subspace $X = (\{1/n\} \times [0,1] : n \in \mathbb{N}\} \cup ([0,1]) \times \{0\}) \cup \{(0,1)\}$ of $\mathbb{R}^2$. Prove that X is connected but is not locally connected at the point $(0,1)$. (Hint: Use this well-known fact that about connectedness: if $A \subseteq B \subseteq c\ell_Y A \subseteq Y$ where A is a connected subspace of Y, then B is also connected.)

Note that even though a locally connected space is the topological sum of connected spaces by 1(c), the converse is not necessarily true as a connected space may not be locally connected (see 1(d)). However, by 1(a), it follows that locally connected spaces are precisely the spaces which are the topological sums of a locally connected, connected space.

(e) Show that the continuous image of a locally connected space is not necessarily locally connected. (Hint: Consider the space described in 1(d) with the discrete topology and consider the identity function from the discrete space to the space X.)

(2) (a) Prove that the quotient image of a locally connected space is also locally connected.

(b) Let $\{X_a : a \in A\}$ be a nonempty family of nonempty spaces. Prove that $\Pi\{X_a : a \in A\}$ is locally connected iff X_a is locally connected for each $a \in A$ and X_a is connected for each $a \in A \setminus F$ for some finite set $F \subseteq A$. (Hint: Use 2(a), 1.7(d)(1), the fact that open functions are quotient, and the fact that the product of nonempty spaces is connected iff each coordinate space is connected.)

(c) Prove that local connectedness is open hereditary but not closed hereditary. (Hint: To show local connectedness is not closed hereditary, consider a convergent sequence in $\mathbb{R}$.

(d) Prove that the topological sum of locally connected spaces is locally connected.

(e) Use 9.7(f) to conclude that the full subcategory of HAUS whose objects are all the locally connected spaces is a coreflective subcategory of HAUS.

(3) (a) Prove that a space X is connected iff for each open cover $\mathcal{C}$ of X and every pair of points x, y

$\in X$, there is a subfamily $\{U_1,\ldots,U_n\}$ of C such that $x \in U_1$, $y \in U_n$, and $U_i \cap U_{i+1} \neq \emptyset$ for $1 \leq i < n$. (Hint: Suppose the condition fails. Choose $x,y \in X$ and an open cover C to witness this failure and inductively define $U^n(x)$ as follows: $U^1(x) = \bigcup\{U \in C : x \in U\}$ and given $U^n(x)$, let $U^{n+1}(x) = \bigcup\{U \in C : U \cap U^n(x) \neq \emptyset\}$. Now, prove that $\bigcup_{\mathbb{N}} U^n(x)$ is clopen.)

(b) Prove that a space X is connected and locally connected iff for every open cover C of X and every pair of points $x, y \in X$, there is a finite family $\{V_1,\ldots,V_n\}$ of connected open subsets of X such that $x \in V_1$, $y \in V_n$, $V_i \cap V_{i+1} \neq \emptyset$ for $1 \leq i < n$, and each V_i is contained in some element of C.

(c) Prove that a space is compact and locally connected iff every open cover has a finite refinement of compact, connected sets.

(d) Let X be a connected, locally connected and locally compact space. Prove that every pair of points of X is contained in some compact, connected subspace of X.

9F. Reflective Subcategories

(1) If $\mathcal{A}$ is a reflective subcategory of $\mathcal{B}$ and $\mathcal{B}$ is a reflective subcategory of a category $\mathcal{C}$, prove that $\mathcal{A}$ is a reflective subcategory of $\mathcal{C}$.

(2) Prove that every mono-reflective subcategory $\mathcal{A}$ of a category $\mathcal{B}$ is epireflective.

9G. Compact Zero-dimensional Reflections of Tychonoff Spaces. Let X be a space. For $x \in X$, recall that $C(x) = \cup\{A \subseteq X : A$ is connected and $x \in A\}$ is called the component of x, that C(x) is connected and closed, and that $\{C(x) : x \in X\}$ forms a partition of X.

(1) If X is compact, prove that $C(x) = \cap\{A \subseteq X : x \in A,$ A is clopen$\}$ for each $x \in X$. (Hint: Let $Q(x) = \cap\{A \subseteq X : x \in A,$ A is clopen$\}$. First show that $C(x) \subseteq Q(x)$ (true for all spaces X) and that Q(x) is closed. To show $Q(x) \subseteq C(x)$, it is necessary and sufficient to show that Q(x) is connected. Assume Q(x) is not connected and use the normality of X to find disjoint open subsets U and V of X such that $Q(x) \subseteq U \cup V$, $x \in U \cap Q(x)$, and $V \cap Q(x) \neq \emptyset$. Using the compactness of X, find a clopen set W of X such that $Q(x) \subseteq W \subseteq U \cup V$. Obtain a contradiction by showing that $x \in W \cap Q(x) \subset Q(x)$.)

(2) If X is compact and $C(x) \subseteq U$ where U is an open set, show there is a clopen set W such that $C(x) \subseteq W \subseteq$

U. (Hint: Use (1) and the compactness of X.)

Let zdX denote $\{C(x) : x \in X\}$ and define $z_X : X \to zdX$ by $z_X(x) = C(x)$.

(3) If X is compact, prove that zdX with the quotient topology induced by z_X is a zero-dimensional Hausdorff space. (Hint: First show that $z_X^{\leftarrow}[z_X[U]] = U$ for each clopen subset U of X; use this fact to conclude that $z_X[U]$ is clopen in zdX. Now use (2) and normality of X to show that zdX is Hausdorff.)

(4) If X is compact, Y is a compact, zero-dimensional space, and $f \in C(X,Y)$, prove there is a unique continuous function $g : zdX \to Y$ such that $g \circ z_X = f$. (Hint: First prove that $f^{\leftarrow}(f(x)) \supseteq C(x)$, for each $x \in X$, since $f[C(x)]$ is connected. Now show the existence of a function $g : zdX \to Y$ such that $g \circ z_X = f$. Use that zd_X has the quotient topology to show that g is continuous.)

(5) Use (4) to show that the category ZDCPT is an epireflective subcategory of CPT where the epireflection of a compact space X is zdX and z_X is the reflection morphism.

(6) Use (5) and 9.6(b)(1) to show that the category ZDCPT is an epireflective subcategory of TYCH where the epireflection of a Tychonoff space X is $zd(\beta X)$ and $z_X \circ j_X$ is the reflection morphism (j_X denotes the inclusion function from X into βX).

9H. Projective Objects in H-closed, Urysohn Spaces. Let HCU (respectively, ΘHCU) denote the full subcategory of HAUS (respectively, ΘHAUS) consisting of H-closed, Urysohn spaces. By 4.8(h)(6), 4.8(s)(5), and 4K(7), the Θ-homeomorphic image of an H-closed, Urysohn space is also H-closed and Urysohn. So, ΘHCU is a replete subcategory of ΘHAUS.

In this problem, the goal is to determine the projective objects of HCU, ΘHCU, [HCU], and [ΘHCU] (see 9.8(i) for the definition of [ ]). The following facts are useful in this problem set: (i) a space X is H-closed and Urysohn iff X(s) is compact (follows from 4.8(h)(8), 4K(7), and 4.8(k)); (ii) let $f \in F(X,Y)$ where X and Y are spaces and define $f(s) \in F(X(s),Y(s))$ by $f(s)(x) = f(x)$ for $x \in X$. Then f is Θ-continuous iff f(s) is Θ-continuous (follows from 4.8(h)(7)); (iii) an extremally disconnected space is Urysohn (follows from 6.2(b)(5)).

(1) Use 6.11(d) and 9.8(d) to show that the projective objects of [HCU] are precisely the extremally disconnected, H-closed spaces.

(2) Show that the projective objects of HCU are precisely the finite spaces. (Hint: Show that the proof of 9.8(h) works for HCU, i.e., that $P \times I$ and Y are Urysohn spaces whenever P is Urysohn.)

(3) (a) Let X, Y and Z be H-closed, Urysohn spaces, $f : X \to Y$ a Θ-continuous surjection, $g : Z \to Y$ a

Θ-continuous function, and Z an extremally disconnected space. Use 9.8(f) to find a function $h \in F(Z,X)$ such that $h(s) : Z(s) \to X(s)$ is continuous and $f \circ h = g$.

(b) Use (a) to show that the projective objects of ΘHCU are precisely the H-closed, extremally disconnected spaces.

(4) Show that the projective objects of [ΘHCU] are precisely the compact, extremally disconnected spaces. (Hint: Let Z be a compact, extremally disconnected space, Y and X be H-closed, Urysohn spaces, and $f : Z \to Y$ and $g : X \to Y$ be perfect, Θ-continuous functions such that g is onto. Since $Z(s) = Z$ and Y(s) and X(s) are compact spaces, the Θ-continuous functions $f(s) : Z \to Y(s)$ and $g(s) : X(s) \to Y(s)$ are perfect and continuous. Use 9.8(f) to find a function $h : Z \to X$ such that $g \circ h = f$ and $h(s) : Z \to X(s)$ is perfect and continuous. Using that the identity function from X(s) to X is perfect and Θ-continuous, show that $h : Z \to X$ is perfect and Θ-continuous.)

9I. Projective Objects in Hausdorff and H-closed Spaces. In 9.8, we identified the projective objects for the categories HAUS, [HAUS], [ΘHAUS], HC, and [ΘHC] but not for the categories ΘHAUS, ΘHC, and [HC].

(1) Show that the projective objects of ΘHAUS are precisely the discrete spaces. (Hint: Use the proof of 9.8(g) and show that the Θ-homeomorphic image of a discrete space is discrete.)

(2) Use 6.11(d) and 9.8(d) to show that the projective objects of [HC] are precisely the extremally disconnected, H-closed spaces.

(3) Show that the projective objects in ΘHC are precisely the finite spaces. (Hint: Let X be a projective object in ΘHC. Use 9.8(d) to show that X is extremally disconnected. Next show that X(s) is also a projective object in ΘHC. Since X(s) is compact (see 9H(1) and (3)) and every Θ-continuous function with compact domain is perfect (see 4O(3)); show that X(s) is a projective object in [ΘHC] and apply 9.8(o).)

9J. Pullbacks. Let X, Y_1 and Y_2 be objects in a category C and $f_i : Y_i \to X$, i = 1,2, be morphisms in C. The pair $(P,\{p_1,p_2\})$ is a **pullback** of f_1 and f_2 if (i) P is an object in C; (ii) $p_i : P \to Y_i$, i = 1,2, are morphisms in C; (iii) $f_1 \circ p_1 = f_2 \circ p_2$; and (iv) if $(Q,\{q_1,q_2\})$ is a pair satisfying (i), (ii), and (iii), there is a unique morphism $h : Q \to P$ in C such that $p_i \circ h = q_i$ for i = 1,2.

The following diagram is called the **pullback square**.

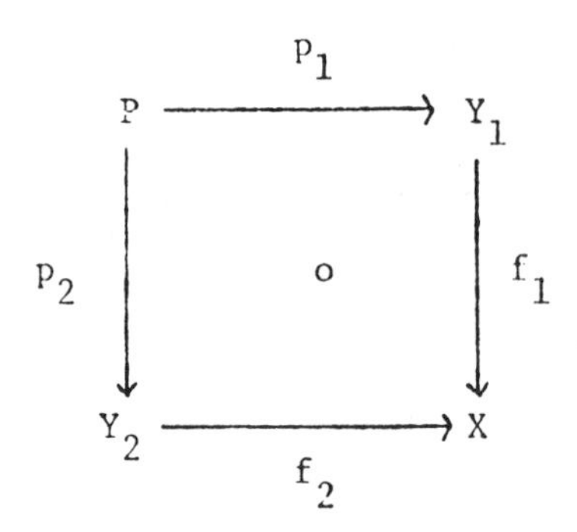

(1) Let $f_i : Y_i \to X$, $i = 1,2$ be morphisms and objects in SET, $P = \{(y_1,y_2) \in Y_1 \times Y_2 : f_1(y_1) = f_2(y_2)\}$, and $p_i = \Pi_i | P$, $i = 1,2$, where $\Pi_i : Y_1 \times Y_2 \to Y_i$ is the usual coordinate projection. Prove that $(P,\{p_1,p_2\})$ is the pullback in SET of f_1 and f_2.

(2) Let $f_i : Y_i \to X$, $i = 1,2$, be morphisms and objects in ΘHAUS. Let P, p_1, p_2 be defined as in (1). Prove that $(P,\{p_1,p_2\})$ is the pullback in ΘHAUS of f_1 and f_2.

(3) Let $f_i : Y_i \to X$, $i = 1,2$, be functions where X, Y_1, Y_2 are spaces and $(P,\{p_1,p_2\})$ be the pullback in SET (described in (1)). When we need to consider P as a space, we refer to the subspace topology inherited from the product space $Y_1 \times Y_2$. So we have this pullback square:

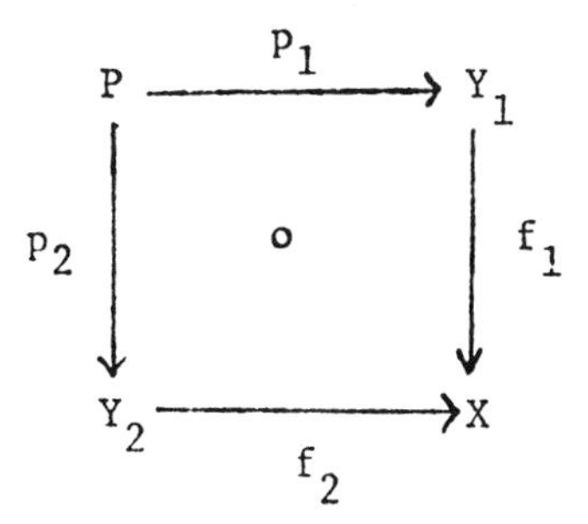

In particular, note that p_1 and p_2 are continuous even though f_1 and f_2 may not be continuous.

(a) If f_2 is onto (respectively, compact), prove that p_1 is onto (respectively, compact).

(b) If f_2 is perfect and f_1 is continuous, prove that p_1 is closed.

(4) Let X and Y be spaces, E an extremally disconnected space, $f : X \to Y$ a perfect surjection (not necessarily Θ-continuous) and $g : E \to Y$ a continuous function.

(a) Show there is a continuous function $h : E \to X$ such that $f \circ h = g$. (Hint: Apply (3), 6.5(c), and 6.5(d)(4).)

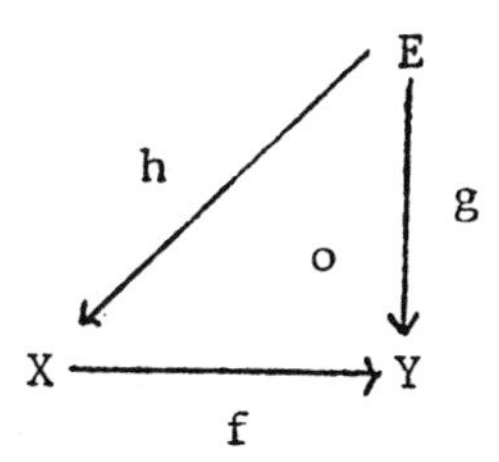

(b) If f is continuous and g is perfect, show that the function h defined in (a) is perfect.

Note that in 4(a) we obtained a continuous function without f necessarily being continuous. The reader might find it interesting to compare the pullback technique in 4(a) with the proof of 6.11(d)(1). A solution to 4(a) is a solution to 6G(1).

9K. Quasi-F-spaces as Projective Objects. In 8.4(ab) we showed that every Tychonoff space has a minimum quasi-F-cover. In 8J we denoted the minimum quasi-F-cover on X by $(QF(X), \phi_X)$ and constructed the space QF(X), in the case where X is compact, as a space of ultrafilters on a sublattice of $R(X)$. In this problem we show that $(QF(X), \phi_X)$ is the projective cover of the compact space X in a certain category. The reader is advised to work through 8J before attempting this problem.

(1) Let X be a Tychonoff space and (as in 8J) define $Z(X)^{\#}$ to be $\{c\ell_X \mathrm{int}_X Z : Z \in Z(X)\}$. Call a covering map $f : X \to Y$ **z-irreducible** if $A \to f[A]$ is a lattice isomorphism from $Z(X)^{\#}$ onto $Z(Y)^{\#}$. Prove that the composition of two z-irreducible maps is z-irreducible; thus, show that the class of compact spaces and z-irreducible maps forms a category denoted as ziCPT.

(2) Let X, Y, and Z be Tychonoff spaces and let $f : X \to Y$ and $g : Y \to Z$ be continuous functions. If $g \circ f$ and g are z-irreducible maps, prove that f is a z-irreducible map. (Hint: Start by using 8.4(d).)

(3) Prove that the projective objects in [ziCPT] are precisely the quasi-F-spaces. (Hint: For one direction, if X, Y, and Z are compact and $f : X \to Y$ and $g : Z \to Y$ are z-irreducible, consider $g \circ \phi_Z$ and use (2), 8J(10), and the minimality of QF(Z). For the other direction, suppose that U is a dense cozero-set of the compact space X that is not C^*-embedded in X. Let $j : U \to X$ be the inclusion map, and consider the maps $\beta j : \beta U \to X$ and $id_X : X \to X$.)

(4) Prove that if X is compact, then $(QF(X), \phi_X)$ is its projective cover in ziCPT.

Note. Some, but not all, of these results can be generalized from the category ziCPT to the category [ziTYCH] (whose definition is obvious); see the Notes.

9L. Projective Objects in kHAUS

(1) Verify that the composition of k-maps (defined in 1T(6)) is a k-map.

Let kHAUS denote the category of spaces and k-maps. The goal of this problem is to identify the projective objects of kHAUS. For a space X, the identity function from kX to X is denoted as j_X; recall that j_X is a k-map (see 1T(7)).

(2) Let Y be a projective object in kHAUS.

(a) Use 9.8(d) to show that Y is extremally disconnected. (Hint: Be careful! For a space X, it must be shown that the projection function $X \oplus X \to X$ is a k-map and if A is a regular closed subspace of X, the inclusion function $A \to X$ is a k-map.)

(b) Show that Y is a k-space. (Hint: Obtain a k-map $h : Y \to kY$ such that $j_Y \circ h = id_Y$ and show that h is a homeomorphism.)

(3) Show that extremally disconnected k-spaces are projective objects in kHAUS. (Hint: Let Z be an extremally disconnected k-space, $f : X \to Y$ be a k-map surjection, and $g : Z \to Y$ a k-map. Use 1T(9) to obtain perfect, continuous functions $kf : kX \to kY$ and $kg : kZ \to kY$ such that $j_Y \circ kf = f \circ j_X$ and $j_Y \circ kg = g \circ j_Z$. Note that j_Z is a homeomorphism and that kf is a surjection. Use 6.11(d) or 9J(4) to obtain a perfect, continuous function $h : kZ \to kX$ such that $kf \circ h = kg$. Show that $m = j_X \circ h \circ j_Z^{\leftarrow}$ is a k-map such that $f \circ m = g$.)

(4) (a) If $f \in C(X,Y)$ is a k-map surjection and Y is a k-space, prove that X is a k-space. (Hint: Use 1T(9) to show that $kf : kX \to kY$ is a perfect, continuous surjection. Since $f \circ j_X = kf$, apply 8.4(d) to conclude that the continuous bijection j_X is perfect and, hence, a homeomorphism.)

(b) For each space X, show that $(P(kX), j_X \circ \Pi_{kX})$ is the projective cover of X in kHAUS. (Hint: Use (4)(a) to show that P(kX) is a k-space, (1) to show that $j_X \circ \Pi_{kX}$ is a k-map, and the fact that j_X is a bijection, together with the irreducibility of Π_{kX} to show that $j_X \circ \Pi_{kX}$ is irreducible.)

9M. Coreflective Subcategories of $[iP]$. Let P be a topological property. As in 9.7(b)(5), we let $[i\hat{P}]$ denote the category of P-coverable spaces and covering maps, and $[iP]$ the subcategory of P-spaces and covering maps.

(1) Prove that every morphism in $[i\hat{P}]$ is both an epimorphism and a monomorphism. (Hint: The first assertion is immediate; for the second, see 8.4(i).)

(2) Prove that the extremal epimorphisms and the isomorphisms of $[i\hat{P}]$ are precisely the homeomorphisms. (Hint: In 9.4(k)(2), let $h = id_B$.)

(3) Let S be a set of objects of $[i\hat{P}]$. Prove that the following are equivalent:

(a) S has a coproduct in $[i\hat{P}]$ and

(b) there exists an extremally disconnected space E such that $S \subseteq AB(E)$ and S is bounded below in $(AB(E), \leqslant)$ (see 8.4(t)).

(4) Prove that if $\{X_\alpha : \alpha \in I\}$ is a set of objects of $[i\hat{P}]$ possessing a coproduct in $[i\hat{P}]$, then that coproduct is a space B, together with a set of covering maps $f_\alpha : X_\alpha \to B$, such that B is the infimum in $C(B)$ of $\{(X_\alpha, f_\alpha) : \alpha \in I\}$.

(5) Prove that the following are equivalent:

(a) $[iP]$ is a coreflective subcategory of $[i\hat{P}]$ and

(b) if the coproduct in $[i\hat{P}]$ of a set S of objects of $[iP]$ exists, then that coproduct is the coproduct of S in $[iP]$. (Hint: See 8.4(t).) This shows that 9.7(f) holds for subcategories of $[i\hat{P}]$.)

NOTES

Chapter 1. Much of the material on C(X), C^*-embedding and C-embedding, and normal spaces (which appears in 1.3, 1.4, 1.9, and 1.10) was first treated in a systematic manner in the fundamental text by Gillman and Jerison [GJ]. The more recent text by Walker [Wa] discusses many of the same ideas.

Our treatment of complete separation via separating chains is patterned after that given in Mandelker and Johnson [MJ]. Theorem 1.8(i) is a slight generalization of a corresponding result for Tychonoff spaces (which appears as 4.2(g)) due to Henriksen and Isbell [HI].

The spaces introduced in 1I and 1N are discussed in [GJ]; the space Ψ of 1N is attributed to Isbell by Gillman and Jerison. The spaces appearing in 1H, 1V, and 1X are discussed in most advanced topology texts (e.g., [En]). The result in 1R(3) is due to Zenor [Z_1]. The "Jones machine" of 1Y is due to F. Burton Jones and appears in [Jo]. The space discussed in 1U is due to Stephenson; it and similar examples appear in [Ste]. Most pseudocompact spaces X appearing in the literature possess a dense subspace each infinite subset of which has a limit point in X, and are pseudocompact by virtue of this property. An example of a pseudocompact space lacking this feature can be found in 5.3 and 5.4 of [GS].

Chapter 2. Basic references for lattice theory are the books by Birkhoff [Bir] and Gratzer [Gr]. Most of the material about regular open sets, semiregular spaces, and the semiregularization of a space is contained in Halmos' book [Hal] and Katětov's papers [Ka_1, Ka_2, Ka_3, Ka_4]. The concept of a filter is an important tool and is used in this book to construct extensions and absolutes of spaces. Most of the results about filters presented here are quite standard (see [Wi]). The Birkhoff-Stone representation theorem and the Dedekind-MacNeille completions are common topics in lattice theory texts, e.g., see [Gr].

Before trying to prove a statement, a topologist usually checks its validity in some nice class of spaces, e.g., compact Hausdorff spaces, metric spaces, or linearly ordered spaces. Ordered spaces appeal to topologists because of their similarity to subspaces of the reals. Additional information about ordered spaces can be found in the monographs by Maurice [Ma] and Nachbin [Na]. The ordinal spaces, special cases of ordered spaces, provide a good source of examples; see 2R, 2S, and 2T.

Cardinal topological invariants are useful in classifying and studying topological spaces. A few basic cardinal invariants are introduced in 2N and used in 2O to establish the well-known theorem by Arhangel'skii. Extensive development of cardinal invariants and the results relating them are contained in [Ju_2, Ju_3]. More information about set theory, including the proofs of many of the results in 2.6, can be located in [CN, De, Du, HrJ, Ku, Mo].

Chapter 3. Basic references on the theory of Boolean algebras are Halmos [Hal] and Sikorski [Si]. The "Stone duality" theory developed in 3.2 is due to Marshall Stone [Sto]. A detailed and lucid discussion of Martin's Axiom and some of its consequences appears in Kunen [Ku], and also in Fremlin [Fre]; we refer the reader to these for further references to set-theoretic topics. Topological consequences of Martin's axiom are discussed in Rudin [Ru_1]. Theorem 3.5(l) is due to Juhasz [Ju_1], and 3V is taken from Bell-Ginsburg [BG]. More detailed discussions of P(κ), SL(κ), and BF(κ) (which appear in 3.5(o), 3.5(q), and 3U respectively) appear in [Ku] and [Fre]. Most of the material on the product of *ccc* spaces in 3T appears in [Gal].

Chapter 4. Discussions of arbitrary extensions appear in [Ba_7, Fo_1, Fo_2, Sa_1, Sa_2, Th]. Taimanov's result [Ta] (4.1(m)) plays a central role in this chapter. In 1930, Tychonoff [Ty] showed that subspaces of compact Hausdorff spaces are precisely the completely regular, Hausdorff (i.e., Tychonoff) spaces. One-point compactifications were developed by Alexandroff [Al_1] in 1924; Lubben [Lu] established that for a Tychonoff space X, $K(X)$ is a complete lattice iff X is locally compact. The problem of finding necessary and sufficient conditions on a Tychonoff space X for which $K(X)$ is a lattice remains unsolved; partial results about when $K(X)$ is a lattice are contained in [Chn, Shi, Tz, Un, VF].

In 1939, Wallman [Wal] constructed a T_1 compactification of a space using the lattice of closed sets. This construction was extended by Banaschewski [Ba_4, Ba_5], Fan and Gottesman [FGo], Frink [Fri], and Šanin [Sa_1, Sa_2]. Additional information about Wallman compactifications can be found in [AS_1, AS_2, Bil_1, Brk, Chn, Hag_1, Nj, SS_1, Stn]. The problem of whether every compactification of a Tychonoff space is Wallman was resolved in 1977 when Ul'janov [Ul_2]

showed that for each cardinal α such that $2^{\alpha} \geq \aleph_2$, there is a compactification of a discrete space of cardinality α which is not Wallman. This result, together with a result by Bandt [Ban], shows that every compactification of every separable Tychonoff space is Wallman iff the continuum hypothesis holds. The relationship between Wallman and Gelfand techniques is developed in 4G. Additional material about Gelfand compactifications can be found in [Bil_2, GJ, GK]. Other methods of generating compactifications are presented in problems 4F, 4W, and 4AA (see [Ty, FG, NW, Sm]).

Various methods of constructing the Stone-Čech compactification (other than those developed in the body of Chapter 4) appear in problems 4Y and 4AC (see [Al_2, Fo_2, Sa_3, GJ]). The original articles by Stone [Sto] and Cech [Ce] as well as the books by Chandler [Chn], Gillman and Jerison [GJ], and Walker [Wa] are excellent sources for additional information about the Stone-Čech compactification.

The largest zero-dimensional compactification of a zero-dimensional space was constructed by Banaschewski in [Ba_1]; other information can be found in [McC, MG, Wo_7]. The relationship between Boolean algebras and zero-dimensional compact spaces which is developed in Chapter 3 is used to generate all of the zero-dimensional compactifications in problem 4I. In 4V, a zero-dimensional space X is constructed for which $\beta X > \beta_0 X$.

An interesting topic not discussed in this book is the relationship between $\beta X \setminus X$, where X is a locally compact space, and $K(X)$. Magill [Mag] proved that if X and Y are locally compact spaces, $\beta X \setminus X$ is homeomorphic to $\beta Y \setminus Y$ iff $K(X)$ is lattice isomorphic to $K(Y)$. This elegant result has been extended in many directions, see [Chn, KT, MG, Por_2, Ra, Thr_1, Thr_2, Thr_3, Wo_1].

The concept of H-closed spaces was introduced in 1924 by Alexandroff and Urysohn [AU]; the theory of H-closed spaces was advanced significantly by the 1940 paper of Katětov [Ka_1]. In 1929, Tychonoff [Ty] proved that every space can be embedded in an H-closed space (not necessarily densely) and Fomin [Fo_1, Fo_2], Katětov [Ka_1], Šanin [Sa_1, Sa_2], and Stone [Sto] showed that every space has an H-closed extension. Additional information about H-closed spaces and the Katětov H-closed extension can be found in [Ba_2, Bo_1, Bo_2, BPS, CF, D'A, DiP, DoP, Fl_2, He_2, HS_1, Il_1, Ka_2, Ka_4, Li, MR, Ob_2, Ob_4, Por_1, PT, PW_4, SV, SW, Ve_3, Ve_7, Vel_1, Vel_2, Vel_3].

The sources for problems 4AD and 4AE are [Hec, PW_3, Wei]. Additional information about the products of pseudocompact spaces and Stone-Čech compactifications (problems 4AF and 4AG) can be found in [Gli, Ste, Tam, Wa]. The existence of remote points without set-theoretic axioms beyond the ZFC axioms was established independently by Chae and Smith [CS] and van Douwen [vD_1]; the results presented in 4AH are taken, with some variation, from these two papers. Other good sources of information about remote points are the papers [vDvM, Do_1, Do_2, Gat, Pe_1, Pe_2].

<u>Chapter 5</u>. Most of the fundamental results on extension properties, such as 5.3(c) and 5.9(e), are due to Herrlich and van der Slot [HvdS]. Many of the more recent results, especially those involving the notion of P-pseudocompactness found in 5S, can be found in [Wo_8] and in [Brn_1]. S. Mrowka is responsible for most of the early work on E-compactness; much of it appears in [Mr]. Theorem 5.8(c) is due to Herrlich and Strecker [HS_1]. The construction in

5.10(d) appears in [Wo_2].

The Hewitt realcompactification was introduced by Hewitt [Hew]. An excellent, although algebraically oriented, introduction to υX appears in Chapter 8 of [GJ]; also see the book by Weir [We], which contains an excellent bibliography of work done on the subject up to 1975. Theorems 5.11(h) appears in [FG_1], and 5.11(i) appears in [FG_2]. Our proof of 5.11(l) (together with 5.11(k)) is due to Mack [Mac_2], and does not to our knowledge appear in the literature. The result in 5.11(l) is due to Zenor [Z_2], who proved it by a quite different method from that appearing herein.

Problem 5B is taken from [Wo_2]. In [He_1] Herrlich proves that there is no space E such that the class of E-completely regular spaces contains all regular spaces; this improves 5C. Problem 5E(6) appears in [Che]. The results in 5H are due to Broverman, and appear in [Brn_1] and [Brn_2]; 5J is also due to Broverman (see [Brn_1] and [Brn_3]). Problems 5K, 5S, 5T, 5U, 5W, and 5X are all taken from [Wo_8]. Problems 5N and 5O are from [Hu_1], and 5P is an adaptation of [My]. The results in 5R appear both in [Wo_8] and [GS].

Many authors have investigated the problem of when $\upsilon(X \times Y) \equiv_{X\times Y} \upsilon X \times \upsilon Y$. The results in 5AD are early theorems drawn from [Co] (which is also the source for 5V). In [Hu_2], Hušek shows that unlike the corresponding situation for β (see 4AG(7)), there is no topological property P such that $\upsilon(X \times Y) \equiv_{X\times Y} \upsilon X \times \upsilon Y$ iff $X \times Y$ has P; he does this by exhibiting spaces X, Y, and Z such that

$$\upsilon(X \times Y \times Z) \equiv_{X\times Y\times Z} \upsilon(X \times Y) \times \upsilon Z,$$

while

$$\upsilon(X \times Y \times Z) \neq_{X \times Y \times Z} \upsilon X \times \upsilon(Y \times Z).$$

The reader is referred to [Oh_1] for recent results, and an extensive bibliography, concerning this problem.

Problem 5Y is taken from [Brn_1]. Ultrarealcompact spaces (see 5AA) were introduced by Ohta [Oh_2].

Chapter 6. The study of absolutes was initiated by Gleason [Gl], who constructed the absolute of a compact space X as the Stone space of $R(X)$. During the next six years this construction was extended, for a variety of spaces and by a variety of methods, by a number of authors. In [Po_1] and [Po_2] Ponomarev constructed the absolute of a space as an inverse limit, first for paracompact spaces and then for arbitrary (Hausdorff) spaces; cf. 6Z. Flaschmeyer [Fl_1] generalized Gleason's original construction and produced the construction of EX discussed in 6.6. Iliadis [Il_2] constructed EX in the manner described in 6.8. Ponomarev (in [Po_2]) seems to have originated the word "absolute". The term "Iliadis absolute", which we have adopted, had come into general use by Eastern European and Soviet authors by the 1970's.

Banaschewski [Ba_7], and (slightly later) Mioduszewski and Rudolf [MR], independently constructed the space PX discussed in 6.11; following Vermeer [Ve_3], we have dubbed the pair (PX,Π_X) the "Banaschewski absolute" for this reason. The "commutativity of κ and P", which is established in 6.11(j), is first treated in [MR].

The notion of an absolute has been generalized to non-Hausdorff spaces, notably by Blaszczyk ([Bl_2], [Bl_3]), Šapiro [Sap], and Ul'janov

$[Ul_1]$. As "all spaces in this book are Hausdorff", we have not discussed this construction. A good expository account of the absolute of a non-Hausdorff space appears in [PS].

Several comprehensive survey articles concerning absolutes have appeared over the years; we mention those by Iliadis and Fomin [IF], Ponomarev and Šapiro [PS], and Woods $[Wo_9]$. The lengthy article by Efimov [Ef], although not a survey article, is also worth consulting.

Much of the basic material on extremally disconnected spaces appearing in 6.2 can be found in [GJ].

Many of the results in 6F appear in [Ef]. Fleissner [Fle] has shown that it is relatively consistent with ZFC that every normal space of character no greater than $2^{\aleph_0}$ is collectionwise Hausdorff, so the supposition in 6F(11) that $2^{\aleph_0} = 2^{\aleph_1}$ cannot be dropped. The results in 6G appear in §6 of [MR], and generalize similar results appearing in [HJ]. Basic subalgebras (see 6H) were introduced in $[Wo_3]$. Vermeer's thesis is the source of 6J (see 3.2.5 of $[Ve_3]$). Most of the results in 6N were discovered independently by several authors; Scott [Sc] seems to have been the first. The proof outlined in 6N is unpublished and due to Ortwin Förster. Isbell [Is] proved 6O(2); the proof here is modelled after that appearing in 12H of [GJ]. Van Douwen $[vD_2]$ proved the results in 6R.

Almost realcompact spaces were introduced by Frolik [Fro]; the equivalence in 6U(1) is due to Nancy Dykes [Dyk], while the construction of aX outlined towards the end of 6U is due to Woods $[Wo_6]$. The derivation of the properties of the example presented in 6V appears in Kato [Ka].

Example 6W was communicated to us by Vermeer $[Ve_2]$.

Problem 6X is taken from [PW_1]. Problem 6Y comes from [Wo_4, Wo_5]. Bellamy [Be], and independently Woods [Wo_1], proved 6AA(9); the proof outlined here comes from [Wo_1].

The question of the existence of P-points of $\beta X \setminus X$ has a long history. W. Rudin [Ru_2] proves 6AC(4) when $X = \mathbb{N}$, and the proofs indicated in 6AD were communicated to us by Alan Dow. In the late 1970's, S. Shelah produced a model of set theory in which $\beta\mathbb{N} \setminus \mathbb{N}$ has no P-points; see the paper by Wimmers [Wim] for a detailed description of Shelah's work.

<u>Chapter 7</u>. The largest H-closed extension of an arbitrary space, the Katětov extension, was introduced in 4.8. In this Chapter, other types of H-closed extensions are introduced and developed. The complete upper semilattice of H-closed extensions of a fixed space has an extra degree of freedom when compared to the complete upper semilattice of compactifications of a Tychonoff space. In particular, when X is a Tychonoff space, each compactification in $K(X)$ induces a partition (actually, an upper semi-continuous partition - see 4Q) of compact subsets of $\beta X \setminus X$ and distinct compactifications of X induce distinct partitions of $\beta X \setminus X$. In a similar fashion, each H-closed extension of a space X induces a partition of compact subsets of $\sigma X \setminus X$ (the details are in section 7.4), but distinct H-closed extensions of X may induce the <u>same</u> partition of $\sigma X \setminus X$. This additional flexibility of H-closed extensions requires an investigation of simple and strict extensions (done in 7.1) and an investigation of when two H-closed extensions induce the same partition of the remainder of the Fomin extension (done in 7.7). A good source of information concerning simple and strict extensions is a paper [Ba_6] by Banaschewski. Most

of the material for 7.7 is contained in the papers [Por_2, PVo_1, PVo_2].

When introduced to H-closed extensions, the reader might guess that the cardinality of the set of H-closed extensions giving rise to the same partition would be no greater than the cardinality of the set of partitions of compact subsets of the remainder of the Fomin extension. Surprisingly, this is false as verified in 7F(5) for an infinite discrete space (see [Fe_2, Fe_3]).

For a fixed Tychonoff space X, there are a number of ways of generating each member of $K(X)$, e.g., the Gelfand method presented in 4.5 and the proximity method presented in 4Z and 4AA. In 1956, Tychonoff (see [Al_3]) asked if there was a method, corresponding to the proximity method, of generating each member of $H(X)$, when X is a fixed space. For a space X, a proximity on X is a subset $\rho \subseteq \mathbb{P}(X) \times \mathbb{P}(X)$ which satisfies certain properties (see 4Z and 4AA). The idea was to modify the properties so that it would be possible to generate all of $H(X)$. Now the maximum number of modified proximities on an infinite space X would be no more than $|\mathbb{P}(\mathbb{P}(X) \times \mathbb{P}(X))| = |\mathbb{P}(\mathbb{P}(X))|$. But when X is infinite and discrete, $|H^{\Theta}(X)| = |\mathbb{P}(\mathbb{P}(\mathbb{P}(X)))|$ as proven in 7.7(e). So, it is not possible to generate even those H-closed extensions which are in a one-to-one correspondence with partitions of compact subsets of $\sigma X \backslash X$. The cardinality results about $H(X)$ and $H^{\Theta}(X)$ are contained in [Fe_2, Fe_3, PVo_1]. The problems of generating all of the Θ-isomorphic H-closed extensions, all strict H-closed extensions, and all H-closed extensions were solved respectively by Fedorčuk [Fe_1], Carlson and Votaw [CV, Vo], and Ovespjan [Ov_1, Ov_2].

The Katětov and Fomin extensions play important roles in studying H-closed extensions. The projective maximum property (see

4.8(n)) of the Katětov extensions is most useful as a common connecting point for all the other H-closed extensions. The facts that each H-closed extension can be linked to a partition of compact subsets of the remainder of the Fomin extension and that the remainder of the Fomin extension is a zero-dimensional space are useful in generating and classifying H-closed extensions. Basic facts about the Fomin extension can be found in [Bl_1, Fl_3, Fo_1, Fo_2, JF, Ov_3, Ov_4, PVo_1, PVo_2].

Additional information about one-point H-closed extensions and locally H-closed spaces is located in [Bl_1, Ob_4, Por_2]. Minimal Hausdorff spaces were introduced in 1940 by Katětov [Ka_1]. The results of 7.5(a)-(d) are contained in [Bo_1, Ka_1]. Banaschewski [Ba_3], Fomin [Fo_1, Fo_2], and Šanin [Sa_1, Sa_2] established 7.5(e)(1), and 7.5(e)(2) is established in [Li, SW]. For a semiregular space X, the bijection between $H^{\Theta}(X)$ and $M(X)$ (7.7(e)(5)) is unexpected and interesting.

The concept of a p-map was introduced by Harris, and the results 7.6(a)-(c) are contained in [Har]. The result 7.6(d) is contained in [HS_1, Li, PT]. The rest of the results in 7.6 are from a paper by Woods [Wo_{10}]. The result 7.6(g)(3) gives a characterization of H-closed subspaces of an H-closed space using p-maps; the use of "p-maps" makes this an external characterization. A problem that is still unsolved is to find an internal characterization of H-closed subspaces of an H-closed space. The result 7.5(g)(4) reduces the problem to finding an internal characterization of nowhere dense, H-closed subspaces of an H-closed space. Additional results about p-maps are in [Ov_5].

The results in 7B are contained in [Ka_4, PT, PVo_2]. Most of the results of 7H are from [Ve_3, VW]. Problem 7I is taken from [PW_2]. An important source of information about H-closed extensions with relatively zero-dimensional remainders is the paper [Fl_3] by Flachsmeyer, and the use of hypercombinatorial embeddings to characterize the Katětov and Fomin extensions is developed in [Ka_2]. The extension of Taimanov's result (7Q(2)) is by Velichko [Vel_1], and the characterization of Martin's Axiom in terms of Θ-closed subsets (7Q(3)) is by Dickman and Porter [DiP]. Problem 7U is taken from [PW_4]. Many of the results in 7V are believed to be new.

Other sources of information concerning H-closed extensions and minimal Hausdorff spaces are [Ba_2, Bel, BL, BM, DoP, Fl_2, He_2, He_3, Ho, MP, Ob_3, Pa, PV, PVW, PW_5, Rud, Sa_3, Ti_1, Ti_2, Ti_3, Ve_1, Ve_4, Vel_3].

Chapter 8. The commutativity results that appear in 8.2(d)(3) and 8.2(h) were proved in [PVW]; 8.2(d) first appeared in [PVo_2, DiP]. The example given in 8B is due to Vermeer [Ve_8]. The proof outlined in 8C that $P(\kappa X) = \kappa(PX)$ appears in [PVW], although (as remarked in the Notes for Chapter 6) this result was originally proved using different methods by Mioduszewski and Rudolf [MR]. The unpublished "direct proof" given in 8D that $P(\sigma X) = \sigma(PX)$ is due to Porter, Vermeer, and Woods.

Theorem 8.3(i), together with the earlier results in 8.3 preparatory to it, are due to Ohta [Oh_2]. So are the results in 8F. The characterization in 8.3(l) of the circumstances under which $E(\upsilon X) = \upsilon(EX)$ is due to Hardy and Woods [HW].

Section 8.4 is, for the most part, a modification of the "theory of expansions" developed by Vermeer in his doctoral dissertation [Ve_3]. Our treatment of these ideas is somewhat different from his, and the introduction of the posets AB(E) and $D(X)$ is new. The notion of a "covering coproduct" (and its category-theoretic treatment in Chapter 9) is also new. The construction of suprema in AB(E) in 8.4(f) appears in [Hag_2]. The construction in 8.4(u) is due to Vermeer [Ve_5].

The construction of the minimum basically disconnected cover described in 8G is due independently to Vermeer [Ve_6] and Zaharov and Koldunov [ZK_2]. Vermeer also constructs the minimum basically disconnected cover of an arbitrary Tychonoff space by using inverse limits. Recently, a method has been developed to construct covers in a manner similar to the construction of Wallman compactifications; see [HVW_2].

The example in 8H is due to Chaber [Chb], while 8I is a modification (using the Jones machine) of an example in [Ve_3].

The minimum quasi-F cover of a compact space X was first constructed by Dashiell, Hager, and Henriksen [DHH], using both inverse limits and maximal ideal spaces of certain "completions" of the ring C(X). Huijsmans and de Pagter [HdP] constructed the quasi-F cover of a compact space X as a space of certain ideals of C(X). Vermeer [Ve_5] proved (using inverse limits) than in arbitrary Tychonoff space has a minimum quasi-F cover; however, the method used in 8.4(ab) to prove this is new. The construction described in 8J, together with a wealth of other information about quasi-F covers, appears in [HVW_1]. A vaguely similar construction is due to Zaharov and Koldunov [ZK_1].

Most of the material in 8.5 is adapted from [MaJ]. Our treatment is somewhat more self-contained than that appearing in

[MaJ], and does not make use of the ring of normal upper semi-continuous functions utilized in [MaJ]. Weak cb spaces are introduced in [MaJ]. The characterization of weak cb spaces given in 8L is taken from [Mac_1]; it is analogous to Dowker's well-known theorem that $X \times [0,1]$ is normal iff X is normal and countably paracompact.

Chapter 9. In this chapter, category theory is used to place the two main concepts of the book - extensions and absolutes -- in a more general mathematical framework. Sections 9.1 - 9.5 contain the basic terminology of category theory applied to the topological concepts needed for the remainder of the chapter. The reader interested in learning more categorical topology should consult the texts by Herrlich and Strecker [HS_3] and Walker [Wa, Ch. 10] and the surveys by Herrlich [He_4, He_5].

The purpose of 9.6 and 9.7 is to show that the theory of extensions is a part of the epireflective theory of categories and the theory of covers is a part of the coreflective theory of categories. In these two sections, the categorical language links together a number of concepts developed in the first eight chapters. Problems 9D and 9E give additional examples of topological coreflections and 9G presents another example of a topological reflection. Problem 9M shows that for a topological property P, the category $[iP]$ is a coreflective subcategory of $[i\hat{P}]$ iff $[iP]$ is closed under formations of coproducts and extremal images in $[i\hat{P}]$; this result, we believe, is new.

The definition of projective objects in 9.8(a)(1) is not a categorical definition as M is the class of all "surjectives". This definition is consistent with the spirit of this book as the primary focus is topological and not categorical; however, there are a number of categories, e.g., CPT and HC, in which "surjection" can be replaced by the categorical term "epimorphism". Gleason [Gl] characterized the projective objects of CPT (see 9.8(f)), and Liu [Li] characterized the projective objects of HC (see 9.8(h)). The result 9.8(j) is from [Ba_7]. The projective objects of HCU and [HCU] (see 9H) were characterized in [Bl_3, MR, Ve_1]. The results 9.8(l) and (m), which we believe are new, extend results in [VW]; 9.8(n) is from [VW]. The projective objects of [ΘHCU] (respectively, kHAUS) are characterized in 9H (respectively, 9L); these results appear in [VW] (respectively, [Wat]). The reference for 9K, and the results in 9K which can be generalized, [HVW_1].

Additional references for this chapter are [Dyc, Har, HS_2, Por_3, Por_4, Pu, Wo_9, Wo_{10}].

BIBLIOGRAPHY

Note: The bibliographic items are ordered alphabetically by the first author's last name rather than by the symbol denoting the item. For example, AU precedes AS_1 because Alexandroff precedes Alo.

Al_1 Alexandroff, P.S., Uber die Metrisation der im kleinen topologischen Räume, Math. Ann. **92**(1924), 294-301.

Al_2 Alexandroff, P.S., Bikompakte Erweiterungen topologischer Räume, Mat. Sb. N. S. **5**(47)(1939), 403-423.

Al_3 Alexandroff, P.S., Some results in the theory of topological spaces, obtained within the last twenty-five years, Uspehi Mat. Nauk **15**(1960), 25-95 = Russian Math. Surveys **15**(1960), 23-83.

AU Alexandroff, P.S. and Urysohn, P., Zur Theorie der topologischen Raüme, Math. Ann. **92**(1924), 258-266.

AS_1 Alo, R.A. and Shapiro, H.L., Normal bases and compactifications, Math. Ann. **175**(1968), 337-340.

AS_2 Alo, R.A. and Shapiro, H.L., A note on compactifications and semi-normal bases, J. Austral. Math. Soc. **8**(1968), 102-108.

Ba_1 Banaschewski, B., Uber null dimensionale Räume, Math. Nachr. **13**(1955), 129-140.

Ba_2 Banaschewski, B., On the Katětov and Stone-Čech extensions, Canad. Math. Bull., **2**(1959), 1-4.

Ba_3 Banaschewski, B., Uber Hausdorffsch-minimale Erweiterungen von Räumen, Arch. Math. **12**(1961), 355-365.

Ba_4 Banaschewski, B., Normal systems of sets, Math. Nachr. **24**(1962), 53-57.

Ba_5 Banaschewski, B., On Wallman's method of compactification, Math. Nachr. **27**(1963), 105-114.

Ba_6 Banaschewski, B., Extensions of topological spaces, Canad. Math. Bull. **7**(1964), 1-22.

Ba_7 Banaschewski, B., Projective covers in categories of topological spaces and topological algebras, Proc. Kanpur Topology Conf. 1968, Academic Press, New York, 1971, 63-91.

Ban Bandt, C., On Wallman-Shanin compactifications, Math. Nachr. **77**(1977), 333-351.

BG Bell, M.G. and Ginsburg, J., First countable Lindelöf extensions of uncountable discrete spaces, Canad. Math. Bull. **23**(1980), 397-399.

Be Bellamy, D.P., A non-metric indecomposable continuum, Duke Math. J. **38**(1971), 15-20.

Bel Bel'nov, V.K., Classification of Hausdorff extensions, Vestnik Moskov. Univ. Ser. I. Mat. Meh. **24**(1969), 23-29 = Moscow Univ. Math. Bull. **24**(1969), 17-20.

BPS Berri, M.P., Porter, J.R., and Stephenson, R.M., A survey of minimal topological spaces, Proc. Kanpur Topology Conf. 1968, Academic Press, New York, 1971, 93-114.

Bil_1 Biles, C.M., Wallman-type compactifications, Proc. Amer. Math. Soc., **25**(1970), 363-366.

Bil_2 Biles, C.M., Gelfand and Wallman-type compactifications, Pac. J. Math., **35**(1970), 267-278.

Bir Birkhoff, G., Lattice theory (3rd edition), American Mathematical Society, Providence, Rhode Island, 1967.

Bl_1 Blaszczyk, A., On locally H-closed spaces and the Fomin H-closed extension, Coll. Math. **25**(1972), 241-253.

Bl_2 Blaszczyk, A., Extremally disconnected resolutions of T_0 spaces, Colloq. Math. **32**(1974), 57-68.

Bl_3 Blaszczyk, A., On projectiveness in H-closed spaces, Colloq. Math. **32**(1975), 185-192.

BL Blaszczyk, A. and Lorek, U., A classification of H-closed extensions, Coll. Math. **39**(1978), 29-33.

BM Blaszczyk, A. and Mioduszewski, J., On factorization of maps through τX, Coll. Math. **23**(1971), 45-52.

Bo_1 Bourbaki, N., Espaces minimaux et espaces complétement séparés, C. R. Acad. Sci. Paris **212**(1941), 215-218.

Bo_2 Bourbaki, N., General Topology, Addison-Wesley, Reading, Mass., 1966.

Brk Brooks, R.M., On Wallman compactifications, Fund. Math. **60**(1967), 159-173.

Brn_1 Broverman, S., Topological extension properties, Doctoral dissertation, The University of Manitoba, Winnipeg, Canada, 1976.

Brn_2 Broverman, S., $\mathbb{N}$-compactness and weak homogeneity, Proc. Amer. Math. Soc. **62**(1977), 173-176.

Brn_3 Broverman, S., Some classes of Θ-compactness, Canad. Math. Bull., **29**(1986), 54-59.

CV Carlson, S.C. and Votaw, C., Para-uniformities, para-proximities, and H-closed extensions, Rocky Mountain J. Math., **16**(1986), 805-836.

Ce Čech, E., On bicompact spaces, Ann. of Math. **38**(1937), 823-844.

Chb Chaber, J., Remarks on open-closed mappings, Fund. Math. **74**(1972), 192-208.

CS Chae, S.B. and Smith, J.H., Remote points and G-spaces, Topology Appl. **11**(1980), 243-246.

Chn Chandler, R.E., Hausdorff Compactifications, Marcel Dekker, New York, 1976.

CF Chevally, C. and Frink, O., Bicompactness of Cartesian products, Bull. Amer. Math. Soc. **47**(1941), 612-614.

Che Chew, K.P., A characterization of ℕ-compact spaces, Proc. Amer. Math. Soc. **26**(1970), 679-682.

Co Comfort, W.W., On the Hewitt realcompactification of a product space, Trans. Amer. Math. Soc. **131**(1968), 107-118.

CN Comfort, W.W. and Negrepontis, S., The Theory of Ultrafilters, Springer-Verlag, New York, 1974.

D'A D'Aristotle, A.J., A note on H-closed extensions of a product, Canad. Math. Bull. **19**(2)(1976), 149-153.

DHH Dashiell, F., Hager, A., and Henriksen, M., Order-Cauchy completions of ring and vector lattices of continuous functions, Canad. J. Math. **32**(1980), 657-685.

De Devlin, K.J., Fundamentals of Contemporary Set Theory, Springer-Verlag, New York, 1974.

DiP Dickman, R.F. and Porter, J.R., Θ-closed subsets of Hausdorff spaces, Pac. J. Math. **59**(1975), 407-415.

vD_1 van Douwen, E.K., Remote points, Diss. Math. **188**(1981), 1-45.

vD_2 van Douwen, E.K., Transfer of information about $\beta\mathbb{N}\setminus\mathbb{N}$ via open remainder maps, unpublished manuscript.

vDvM van Douwen, E.K. and van Mill, J., Spaces without remote points, Pac. J. Math. **105**(1983), 67-75.

Do_1 Dow, A., Products without remote points, Topology Appl. **15**(1983), 239-246.

Do_2 Dow, A., Remote points in large products, ibid., **16**(1983), 11-17.

DoP Dow, A. and Porter, J.R., Cardinalities of H-closed spaces, Topology Proc. **7**(1982), 27-50.

Du Dugundji, J., Topology, Allyn and Bacon, Boston, Mass., 1967.

Dyc Dyckhoff, R., Factorization theorems and projective spaces in topology, Math. Z. **127**(1972), 256-264.

Dyk Dykes, N., Mappings and realcompact spaces, Pac. J. Math. **31**(1969), 347-358.

Ef Efimov, B.A., Extremally disconnected spaces and absolutes, Trudy Moscovsk. Mat. Obshch **23**(1970), 235-276 = Trans. Moscow Math. Soc. **23**(1970), 243-285.

En Engelking, R., General Topology, Polish Scientific Publishers, Warsaw, 1977.

FGo Fan, K. and Gottesman, N., On compactifications of Freudenthal and Wallman, Nederl. Akad. Wetensch, Proc. Ser. A **55**(1962), 504-510.

Fe_1 Fedorčuk, V.V., On H-closed extensions of Θ-proximities, Mat. Sbornik **89**(131)(1972), 400-418 = Math. USSR Sbornik **18**(1972). 407-424.

Fe_2 Fedorčuk, V.V., On a problem of A.N. Tychonoff, Dokl. Akad. Nauk SSSR **210**(1973), 1297-1299 = Soviet Math. Dokl. **14**(1973), 912-915.

Fe_3 Fedorčuk, V.V., A problem of A.N. Tychonoff on the classification of H-closed extensions, Fund. Math. **86**(1974),

69-90.

FG_1 Fine, N. and Gillman, L., Extensions of continuous functions in $\beta\mathbb{N}$, Bull. Amer. Math. Soc. **66**(1960), 376-381.

FG_2 Fine, N. and Gillman, L., Remote points in $\beta\mathbb{R}$, Proc. Amer. Math. Soc. **13**(1962), 29-36.

FGL Fine, N., Gillman, L. and Lambek, J., Rings of quotients of rings of functions, McGill University Press, 1966.

Fl_1 Flachsmeyer, J., Topologische Projektivraüme, Math. Nachr. **26**(1963), 57-66.

Fl_2 Flachsmeyer, J., H-abgeschlossene Räume als schwach-stetige Bilder kompacter Räume, Math. Z. **91**(1966), 336-343.

Fl_3 Flachsmeyer, J., Zur Theorie der H-abgeschlossenen Erweiterungen, ibid., **94**(1966), 349-381.

Fle Fleissner, W.G., Normal Moore spaces in the constructible universe, Proc. Amer. Math. Soc. **46**(1974), 294-298.

Fo_1 Fomin, S.V., Extensions of topological spaces, C.R. (Doklady) Acad. Sci. URSS (N.S.) **32**(1941), 114-116.

Fo_2 Fomin, S.V., Extensions of topological spaces, Ann. of Math. **44**(1943), 471-480.

Fre Fremlin, D.H., Consequences of Martin's Axiom, Cambridge Tracts in Mathematics Vol. 84, Cambridge University Press, London, 1984.

Fri Frink, O., Compactifications and semi-normal spaces, Amer. J. Math., **86**(1964), 602-607.

Fro Frolik, Z., A generalization of realcompact spaces, Czech. Math. J. **13**(88)(1963), 127-138.

Gal Galvin, F., Chain conditions and products, Fund. Math., **108**(1980), 33-48.

Gat Gates, C.L., Some structural properties of the set of remote points of a metric space, Canad. J. Math. **32**(1980), 195-209.

GK Gelfand, I. and Kolmogoroff, A., On rings of continuous functions on topological spaces, Dokl. Akad. Nauk SSSR **22**(1939), 11-15.

GJ Gillman, L. and Jerison, M., Rings of continuous functions, Princeton, Van Nostrand, 1960.

GS Ginsburg, J. and Saks, V., Some applications of ultrafilters in topology, Pac. J. Math. **57**(1975), 403-417,

Gl Gleason, A.M., Projective topological spaces, Ill. J. Math. **2**(1958), 482-489.

Gli Glicksburg, I., Stone-Čech compactifications of products, Trans. Amer. Math. Soc. **90**(1959), 369-382.

Gr Gratzer, G., Lattice Theory, Freeman and Co., San Francisco, 1971.

dG de Groot, J., Groups represented by homeomorphism groups I, Math. Ann. **138**(1959), 80-102.

Hag_1 Hager, A., On inverse-closed subalgebras of C(X), Proc. Lond. Math. Soc. (3)**19**(1969), 233-257.

Hag_2 Hager, A., Isomorphisms of some completions of C(X), Topol. Proc. **4**(1979), 402-435.

Hal Halmos, P.R., Lectures on Boolean Algebras, Van Nostrand Mathematical Studies #1, Van Nostrand, New York, 1963.

HW Hardy, K. and Woods, R.G., On c-realcompact spaces and locally bounded normal functions, Pacific J. Math. **43**(1972), 647-656.

Har Harris, D., Katětov extension as a functor, Math. Ann. **193**(1971), 171-175.

Hec Hechler, S.H., On some weakly compact spaces and their products, Topology Appl. **5**(1975), 83-93.

HI Henriksen, M. and Isbell, J.R., Some properties of compactifications, Duke Math. J. **25**(1958), 83-106.

HJ Henriksen, M. and Jerison, M., Minimal projective extensions of compact spaces, Duke Math. J. **32**(1965), 291-295.

HVW_1 Henriksen, M., Vermeer, J. and Woods, R.G., Quasi-F-covers of Tychonoff spaces, Trans. Amer. Math. Soc., to appear.

HVW_2 Henriksen, M., Vermeer, J. and Woods, R.G., Wallman covers of compact spaces, to appear.

He_1 Herrlich, H., Wann sind alle stetigen Abbildungen in Y konstant?, Math. Zeit. **90**(1965), 152-154.

He_2 Herrlich, H., T_ν-Abgeschlossenheit und T_ν-Minimalität, Math. Z. **88**(1965), 285-294.

He_3 Herrlich, H., Nicht alle T_2-minimale Räume sind von 2 Kategorie, ibid., **91**(1966), 185.

He_4 Herrlich, H., Categorical topology, Gen. Topol. Appl. **1**(1971), 1-15.

He_5 Herrlich, H., Categorical topology. 1971-1981, Mathematik Arbeitspapiere Nr. 24, Universität Bremen, 1981.

HvdS Herrlich, H. and van der Slot, J., Properties which are closely related to compactness, Indag. Math. **29**(1967), 524-529.

HS_1 Herrlich, H. and Strecker, G., H-closed spaces and reflective subcategories, Math. Ann. **177**(1968), 302-309.

HS_2 Herrlich, H. and Strecker, G., Coreflective subcategories in general topology, Fund. Math. **73**(1972), 199-218.

HS_3 Herrlich, H. and Strecker, G., Category Theory, Allyn and

Bacon, Boston, 1973.

Hew Hewitt, E., Rings of real-valued continuous functions I, Trans. Amer. Math. Soc. **64**(1948), 45-99.

Ho Holsztyński, W., Minimal Hausdorff spaces and T_1-compacta, Dokl. Akad. Nauk SSSR **178**(1968), 24-26 = Soviet Math. Dokl. **9**(1968), 18-20.

HrJ Hrbacek, K. and Jech, T., Introduction to Set Theory, Marcel Dekker, New York, 1978.

HdP Huijsmans, C. and de Pagter, B., Maximal d-ideals in a Riesz space, Canad. J. Math. **35**(1983), 1010-1029.

Hu_1 Hušek. M., The class of k-compact spaces is simple, Math. Zeit. **110**(1969), 123-126.

Hu_2 Hušek, M., Hewitt realcompactifications of products, Colloq. Math. Janos Boylai **8**(1974), 427-435.

Il_1 Iliadis, S., Characterization of spaces by H-closed extensions, Dokl. Akad. Nauk SSSR **149**(1963), 1015-1018 = Soviet Math. Dokl. **4**(1963), 501-504.

Il_2 Iliadis, S., Absolutes of Hausdorff spaces, Dokl. Akad. Nauk SSSR **149**(1963), 22-25 = Soviet Math. Dokl. **4**(1963), 295-298.

IF Iliadis, S. and Fomin, S., The method of centered systems in the theory of topological spaces, Uspekhi Math. Nauk. 21(1966), 47-76 = Russian Math. Surveys **21**(1966), 37-62.

Is Isbell, J.R., Zero-dimensional spaces, Tohoku Math. J. **7**(1955), 1-8.

Jo Jones, F.B., Hereditarily separable, non-completely regular spaces, Topology Conf., Virginia Polytechnic Inst. and State U. 1973, Springer Lecture Notes in Mathematics Vol. **375**(1974), 149-152.

Ju_1 Juhasz, I., Martin's axiom solves Ponomorev's problem, Bull. Acad. Polon. Sci. Sér. Sci. Math. Astronom. Phys. **18**(1970), 71-74.

Ju_2 Juhasz, I., Cardinal Functions in Topology, Mathematische Centrum Tracts No. 34, Amsterdam, 1971.

Ju_3 Juhasz, I., Cardinal Functions in Topology - Ten Years Later, Mathematische Centrum, Amsterdam, 1980.

KT Kannan, V. and Thrivikraman, T., Lattices of Hausdorff compactifications of a locally compact space, Pac. J. Math. **57**(1975), 441-444.

Ka_1 Katětov, M., Uber H-abgeschlossene und bikompakte Raüme, Cas. Mat. Fys. **69**(1940), 36-49.

Ka_2 Katětov, M., On H-closed extensions of topological spaces, ibid, **72**(1947), 17-32.

Ka_3 Katětov, M., A note on semiregular and nearly regular spaces, ibid, **72**(1947), 97-99.

Ka_4 Katětov, M., On the equivalence of certain types of extension of topological spaces, ibid, **72**(1947), 101-106.

Kat Kato, A., Union of realcompact spaces and Lindelöf spaces, Canad. J. Math. **31**(1979), 1247-1268.

Ku Kunen, K., Set theory; an introduction to independence proofs, Studies in Logic and the Foundations of Mathematics Vol. 102, North-Holland, Amsterdam, 1980.

La Lambek, J., Lectures on rings and modules, Blaisdell, Toronto, 1966.

Li Liu, C-T., Absolutely closed spaces, Trans. Amer. Math. Soc. **130**(1968), 86-104.

Lu Lubben, R.G., Concerning the decomposition and amalgamation of points, upper semi-continuous collections, and topological extensions, Trans. Amer. Math. Soc. **49**(1941), 410-466.

Mac_1 Mack, J., Countable paracompactness and weak normality properties, Trans. Amer. Math. Soc. **148**(1970), 265-272.

Mac_2 Mack, J., private communication.

MaJ Mack, J. and Johnson, D.G., The Dedekind completion of C(X), Pac. J. Math. **20**(1967), 231-243.

Mag Magill, K.D., Jr., The lattice of compactifications of a locally compact space, Proc. Lond. Math. Soc. (3)**18**(1968), 231-244.

MG Magill, K.D., Jr. and Glasenapp, J.A., 0-dimensional compactifications and Boolean rings, J. Aust. Math. Soc. **8**(1968), 755-765.

MJ Mandelker, M. and Johnson, D.G., Separating chains in topological spaces, J. Lond. Math. Soc. (2), **4**(1972), 510-512.

Ma Maurice, M.A., Compact ordered spaces, Mathematische Centrum Tracts No. 6, Amsterdam, 1964.

McC McCartney, J.R., Maximal zero-dimensional compactifications, Proc. Cambridge Philos. Soc. **68**(1970), 653-661.

MP McCoy, R.A. and Porter, J.R., Baire extensions, Topol. Appl. **7**(1977), 39-58.

MR Mioduszewski, J. and Rudolf, L., H-closed and extremally disconnected Hausdorff spaces, Dissert. Math. **66**(1969), 1-55.

Mo Monk, J.D., Introduction to Set Theory, McGraw-Hill, New York, 1969.

Mr Mrowka, S., Further results on E-compact spaces I, Acta Math. **120**(1968), 161-185.

My Mysior, A., The category of all zero-dimensional realcompact spaces is not simple, Topol. Appl. **8**(1978), 259-264.

Na Nachbin, L., Topology and Order, Van Nostrand, New York, 1965.

NW Naimpally, S.A. and Warrack, B.D., Proximity Spaces, Cambridge Press, London, 1970.

Nj Njåstad, O., On Wallman-type compactifications, Math. Z. **91**(1966), 267-276.

Ob_1 Obreanu, F., Filtre Deschise, Acad. Repub. Pop. Romîne Bul. Sti. Ser. Mat. Fiz. Chim. **2**(1950), 1-5.

Ob_2 Obreanu, F., Spatti Absolut Inchise, ibid. **2**(1950), 21-25.

Ob_3 Obreanu, F., Spatti Separate Minimale, An. Acad. Repub. Pop. Romîne, Sect. Sti. Mat. Fiz, Chem. Ser. A **3**(1950), 325-349.

Ob_4 Obreanu, F., Spatti Local Absolut Inchise, ibid. **3**(1950), 375-394.

Oh_1 Ohta, H., Local compactness and Hewitt realcompactifications of products II, Topol. Appl. **13**(1982), 155-165.

Oh_2 Ohta, H., Topological extension properties and projective covers, Canad. J. Math. **34**(1982), 1255-1275.

Ov_1 Ovsepjan, S.G., On a new method for the construction of extensions of topological spaces, Dokl. Akad. Nauk SSSR **206**(1972), 819-822 = Soviet Math. Dokl. **13**(1972), 1330-1334.

Ov_2 Ovsepjan, S.G., Construction of all Hausdorff and all H-closed extensions of topological spaces, Dokl. Akad. Nauk SSSR **224**(1975), 764-767 = Soviet Math. Dokl. **16**(1975), 1301-1305.

Ov_3 Ovsepjan, S.G., On the sequential closure of a topological space in its Hausdorff extensions, Dokl. Akad. Nauk SSSR **231**(1976), 806-809 = Soviet Math. Dokl. **17**(1976), 1650-1654.

Ov_4 Ovsepjan, S.G., A new method in the theory of extensions of topological spaces, Uspekhi Mat. Nauk **34**(1979), 174-178 = Russian Math. Surveys **34**(1979), 199-203.

Ov_5 Ovsepjan, S.G., Extension of continuous mappings and epireflectivity of the category of H-closed spaces, Dokl. Akad. Nauk SSSR **225**(1980), 286-289 = Soviet Math. Dokl. **22**(1980), 692-696.

Pa Parovicenko, I.I., On suprema of families of H-closed extensions, Dokl. Akad. Nauk SSSR **193**(1970), 1241-1244 = Soviet Math. Dokl. **11**(1970), 1114-1118.

Pe_1 Peters, T.J., G-spaces: products, absolutes and remote points, Topology Proc. **7**(1982), 119-146.

Pe_2 Peters, T.J., Dense homeomorphic subspaces of X* and of (EX)*, ibid., **8**(1983), 285-301.

Po_1 Ponomarev, V.I., On paracompact spaces and their continuous mappings, Dokl. Akad. Nauk SSSR **143**(1962), 46-49 = Soviet Math. Dokl. **3**(1962), 347-350.

Po_2 Ponomarev, V.I., The absolute of a topological space, Dokl. Akad. Nauk SSSR **149**(1963), 26-29 = Soviet Math. Dokl. **4**(1963), 299-302.

PS Ponomarev, V.I. and Sapiro, L.B., Absolutes of topological spaces and their continuous maps, Uspekhi Math. **31**(1976), 121-136 = Russian Math. Surveys **21**(1976), 138-154.

Por_1 Porter, J.R., On locally H-closed spaces, Proc. Lond. Math. Soc. (3), **20**(1970), 193-204.

Por_2 Porter, J.R., Lattices of H-closed extensions, Bull. Acad. Polon. Sci. Sér. Sci. Math. Astronom. Phys. **22**(1974), 831-837.

Por_3 Porter, J.R., Extension function and subcategories of HAUS, Canad. Math. Bull. **18**(4)(1975), 587-590.

Por_4 Porter, J.R., Categorical problems in minimal spaces, Proc. Mannheim Categorical Top. Conf. 1975, Lecture Notes in Math. **540**(1976), 482-500.

PT Porter, J.R. and Thomas, J., On H-closed and minimal Hausdorff spaces, Trans. Amer. Math. Soc. **138**(1969), 159-170.

PV Porter, J.R. and Vermeer, J., Spaces with coarser minimal Hausdorff topologices, Trans. Amer. Math. Soc. **289**(1985), 59-71.

PVW Porter, J.R., Vermeer, J., and Woods, R.G., H-closed extensions of absolutes, Houston J. Math. **11**(1985), 109-120.

PVo_1 Porter, J.R. and Votaw, C., H-closed extensions I, Topol. Appl. **3**(1973), 211-224.

PVo_2 Porter, J.R. and Votaw, C., H-closed extensions II, Trans. Amer. Math. Soc. **202**(1975), 193-209.

PW_1 Porter, J.R. and Woods, R.G., Minimal extremally disconnected Hausdorff spaces, Topol. Appl. **8**(1978), 9-26.

PW_2 Porter, J.R. and Woods, R.G., Ultra-Hausdorff H-closed extensions, Pac. J. Math. **82**(1979), 399-411.

PW_3 Porter, J.R. and Woods, R.G., Feebly compact spaces, Martin's axiom, and "Diamond", Topology Proc. **9**(1984), 105-121.

PW_4 Porter, J.R. and Woods, R.G., Accumulation points of nowhere dense sets in H-closed spaces, Proc. Amer. Math. Soc. **93**(1985), 539-542.

PW_5 Porter, J.R. and Woods, R.G., When all semiregular H-closed extensions are compact, Pac. J. Math., **120**(1985), 179-188.

H7:Pu Purisch. S., Projectives in the category of ordered

spaces, Proc. Charlotte Top. Conf. 1974, Studies in Topology, Acad. Press (1975), 467-478.

Ra Rayburn, M.C., On Hausdorff compactifications, Pac. J. Math. **44**(1973), 707-714.

Ru_1 Rudin, M.E., Lectures in set-theoretic topology, C.B.M.S. Regional Conference Series in Mathematics, Vol. 23(1974).

Ru_2 Rudin, W., Homogeniety problems in the theory of Čech compactifications, Duke Math. J. **23**(1956), 409-419.

Rud Rudolf, L., Θ-continuous extensions of maps on τX, Fund. Math. **74**(1972), 111-131.

Sa_1 Šanin, N.A., On special extensions of topological spaces, C.R. (Doklady) Acad. Sci. URSS (N.S.) **38**(1943), 6-9.

Sa_2 Šanin, N.A., On separation in topological spaces, ibid., **38**(1943), 110-113.

Sa_3 Šanin, N.A., On the theory of bicompact extensions of topological spaces, ibid. **38**(1943), 154-156.

Sap Šapiro, L.B., On absolutes of topological spaces and continuous mappings, Dokl. Akad. Nauk SSSR **226**(1976), 523-526 = Soviet Math. Dokl. **17**(1976), 147-151.

Sc Scott, B.R., Pseudocompact metacompact spaces are compact, Topology Proc. **4**(1979), 577-588.

Shi Shirota, T., On systems of structures of a completely regular space, Osaka Math. J. **41**(1950), 131-143.

Si Sikorski, R., Boolean algebras (2nd ed.), Springer-Verlag, New York, 1964.

Sm Smirnov, J.M., On proximity spaces, Mat. Sb. (N.S.) **31**(73)(1952), 543-574 = Amer. Math. Soc. Transl. (2) **38**(1964), 5-35.

So Solomon, R.C., A Hausdorff compactification that is not regular Wallman, Topol. Appl. **7**(1977), 59-63.

SS Steiner, A.K. and Steiner, E.F., Wallman and z-compactifications, Duke Math. J. **35**(1968), 269-276.

Ste Steiner, E.F., Wallman spaces and compactifications, Fund. Math. **61**(1968), 295-304.

Ste Stephenson, R.M., Jr., Pseudocompact spaces, Trans. Amer. Math. Soc. **134**(1968), 437-448.

Sto Stone, M.H., Applications of the theory of Boolean rings to general topology, Trans. Amer. Math. Soc. **41**(1937), 375-481.

SV Strecker, G.E. and Viglino, G., Co-topology and minimal Hausdorff spaces, Proc. Amer. Math. Soc. **21**(1969), 569-574.

SW Strecker, G.E. and Wattel, E., On semi-regular and minimal Hausdorff embeddings, Indag. Math. **29**(1967), 234-237.

Ta Taimanov, A.D., On extension of continuous mappings of topological spaces, Mat. Sb. **31**(1952), 459-463.

Tam Tamano, H., A note on the pseudocompactness of the product of two spaces, Mem. Coll. Sci. Univ. of Kyoto, Ser. A, Math. **33**(1960), 225-230.

Thr_1 Thrivikraman, T., On the lattices of compactifications, J. Lond. Math. Soc. **4**(1972), 711-717.

Thr_2 Thrivikraman, T., On compactifications of Tychonoff spaces, Yokohama Math. J. **20**(1972), 99-106.

Thr_3 Thrivikraman, T., On Hausdorff quotients of spaces and Magill's theorem, Monat. für Math. **76**(1972), 345-355.

Th Thron, W.J., Topological Structures, Holt, Rinehart and Winston, New York, 1966.

Ti_1 Tikoo, M.L., Remainders of H-closed extensions, Doctoral dissertation, Univ. of Kansas, 1984.

Ti_2 Tikoo, M.L., The Banaschewski-Fomin-Šanin extension μX, Topol. Proc. **10**(1985), 187-206.

Ti_3 Tikoo, M.L., Remainders of H-closed extensions, Topol. Appl. **23**(1986), 117-128.

Ty Tychonoff, A., Uber die topologische Erweiterung von Räumen, Math. Ann. **102**(1930), 544-561.

Tz Tzung, F.-C., Sufficient conditions for the set of Hausdorff compactifications to be a lattice, Pac. J. Math. **77**(1978), 565-573.

Ul_1 Ul'janov, V.M., On compactifications satisfying the first axiom of countabliliy and absolutes, Mat. Sb. **98**(140)(1975), 223-254 = Math USSR Sbornik **27**(1975), 199-226.

Ul_2 Ul'janov, V.M., Solution of a basic problem on compactifications of Wallman type, Dokl. Akad. Nauk SSSR **233**(1977), 1056-1059 = Soviet Math. Dokl. **18**(1977), 567-571.

Un Unlü, Y., Lattices of compactifications of Tychonoff spaces, Topol. Appl. **9**(1978), 41-57.

Vel_1 Velichki, N.V., On extension of mappings of topological spaces, Sibirsk. Mat. Ž. **6**(1965), 64-69 = Amer. Math. Soc. Transl. (2) **92**(1970), 41-47.

Vel_2 Velichko, N.V., H-closed topological spaces, Mat. Sb. (N.S.) **70**(112), (1966), 98-112 = Amer. Math. Soc. Transl. 78(2)(1968), 103-118.

Vel_3 Velichko, N.V., On the theory of H-closed spaces, Sibirsk. Mat. Ž. **8**(1967), 754-763 = Siberian Math. J. **8**(1967), 569-575.

Ve_1 Vermeer, J., Minimal Hausdorff and compact-like spaces, Top.

Struc. II (1979), 271-282, Math. Centr. Amsterdam.

Ve_2 Vermeer, J., Two generalizations of normality, Topology and measure III, Proc. Conf., Vitte/Hiddensee 1980, Part 2 (1982), 329-338.

Ve_3 Vermeer, J., Expansions of H-closed spaces, Doctoral dissertation, Vrije Universiteit, Amsterdam, 1983.

Ve_4 Vermeer, J., Embeddings in minimal Hausdorff spaces, Proc. Amer. Math. Soc. **87**(1983), 533-536.

Ve_5 Vermeer, J., On perfect irreducible pre-images, Topology Proc. **9**(1984), 173-189.

Ve_6 Vermeer, J., The smallest basically disconnected preimage of a space, Topol. Appl. **17**(1984), 217-232.

Ve_7 Vermeer, J., Closed subspaces of H-closed spaces, Pac. J. Math. **118**(1985), 229-247.

Ve_8 Vermeer, J., An example concerning H-closed extensions of absolutes, manuscript.

VW Vermeer, J. and Wattel, E., Projective elements in categories with perfect Θ-continuous maps, Canad. J. Math. **33**(1981), 872-884.

VF Visliseni, J. and Flaksmaier, J., Power and construction of the structure of all compact extensions of a completely regular space, Dokl. Akad. Nauk SSSR **165**(1965), 258-260 = Soviet Math. Dokl. **6**(1965), 1423-1425.

Vo Votaw, C., H-closed extensions as para-uniform completions, Doctoral dissertation, University of Kansas, 1971.

Wa Walker, R.C., The Stone-Čech compactification, Springer, New York, 1974.

Wal Wallman, H., Lattices and topological spaces, Ann. of Math.

(2) **39**(1938), 112-126.

Wat Wattel, E., Projective objects and k-mappings, Report No. 72, Vrije Universiteit, Amsterdam, 1977.

We Weir, M.D., Hewitt-Nachbin spaces, North-Holland Math. Studies, American Elsevier, New York, 1975.

Wei Weiss, W., Countably compact spaces and Martin's axiom, Canad. J. Math. **30**(1978), 243-249.

Wh Wheeler, R.F., Topological measure theory for completely regular spaces and their projective covers, Pac. J. Math. **82**(1979), 565-584.

Wi Willard, S., General Topology, Addison-Wesley, Reading, Mass., 1970.

Wim Wimmers, E.L., The Shelah P-point independence theorem, Israel J. Math. **43**(1982), 28-48.

Wo_1 Woods, R.G., Certain properties of $\beta X \setminus X$ for σ-compact X, Doctoral dissertation, McGill University, Montreal, 1969.

Wo_2 Woods, R.G., Some $\aleph_0$-bounded subsets of Stone-Čech compactifications, Israel J. Math. **9**(1971), 545-560.

Wo_3 Woods, R.G., A Boolean algebra of regular closed subsets of $\beta X \setminus X$, Trans. Amer. Math. Soc. **154**(1971), 23-36.

Wo_4 Woods, R.G., Co-absolutes of remainders of Stone-Čech compactifications, Pac. J. Math. **37**(1971), 545-560.

Wo_5 Woods, R.G., Ideals of pseudocompact regular closed sets and absolutes of Hewitt realcompactifications, Topol. Appl. **2**(1972), 315-331.

Wo_6 Woods, R.G., A Tychonoff almost realcompactification, Proc. Amer. Math. Soc. **43**(1974), 200-208.

Wo_7 Woods, R.G., Zero-dimensional compactifications of locally compact spaces, Canad. J. Math. **26**(1974), 920-930.

Wo_8 Woods, R.G., Topological extension properties, Trans. Amer. Math. Soc. **210**(1975), 365-385.

Wo_9 Woods, R.G., A survey of absolutes of topological spaces, Topological Structures II, Mathematische Centrum Tract 116 (1979), 323-362 (Amsterdam).

Wo_{10} Woods, R.G., Epireflective subcategories of Hausdorff categories, Topol. Appl. **12**(1981), 203-220.

ZK_1 Zaharov, V.K. and Koldunov, A.V., The sequential absolute and its characterizations, Dokl. Acad. Nauk SSSR **253**(1980), 280-283 = Soviet Math. Dokl. **22**(1980), 70-74.

ZK_2 Zaharov, V.K. and Koldunov, A.V., Characterizations of the σ-cover of a compact, Math. Nachr. **107**(1982), 7-16.

Z_1 Zenor, P., A note on Z-mapping and WZ-mappings, Proc. Amer. Math. Soc. **23**(1969), 273-275.

Z_2 Zenor, P., Certain subsets of products of Θ-refinable spaces are realcompact, Proc. Amer. Math. Soc. **40**(1973), 612-614.

LIST OF SYMBOLS

Note: Each symbol is listed alphabetically. Thus $\cup$ ("union") is listed under "U". After each symbol there appears a brief description of its meaning, followed by a location in the book. The location refers to the section or problem of first occurrence or definition of the symbol.

A

$a(\mathcal{F})$	adherence of a filter base $\mathcal{F}$, 2.3(c)(3)
$\aleph_\alpha$	cardinal number, 2.6(l)
$a_\Theta(\mathcal{F})$	$\cap\{c\ell_\Theta F : F \in \mathcal{F}\}$, 4.8(r)
$\mathfrak{a}P$	the class of almost P-spaces, 6T
$\mathfrak{a}PT$	the class of Tychonoff almost P-spaces, 6T
$a_T(\mathcal{U})$	the adherence in T of a filter $\mathcal{U}$ on a subspace X of T, preceding 6.9(c)
AB(E)	set of (equivalence classes of) pairs (c_Y,Y) consisting of covering map c_Y from a fixed space E to Y, 8.4(b)

B

$\mathcal{B}(X)$	Boolean algebra of clopen subsets of X, 1.5(a)
$BF(\kappa)$	combinatorial consequence of MA, 3U
βX	Stone-Čech compactification of X, 4.2(b)
βf	Stone extension of $f : X \to Y$ to $\beta f : \beta X \to$

	βY, 4.2(d)
f^{β}	if $f \in C^*(X)$, f^{β} is the extension to $C^*(\beta X)$, 4.5(r)
$\beta_0 X$	maximum zero-dimensional compactification of X, following 4.7(b)
$\beta_0 f$	extension of $f : X \to Y$ (zero-dimensional spaces) to $\beta_0 f : \beta_0 X \to \beta_0 Y$, 4.7(d)
BA	category of Boolean algebras and Boolean algebra homomorphisms, 9.2(b)(3)

C

$\|X\|$	cardinality of a set X, 1.1(c)
χ_A, $\chi_{A,X}$	characteristic function; $\chi_{A,X}(x) = 1$ if $x \in A$, $\chi_{A,X}(x) = 0$ if $x \notin A$, 1.1(d)
C(X,Y)	set of continuous functions from X to Y, 1.2(c)(1)
C(X)	$C(X,\mathbb{R})$, 1.2(c)(2)
coz(f)	$X \setminus Z(f)$; the cozero set of f, 1.4(c)
$c\ell_X$	closure with respect to space X, 1.2(d)(5)
$C^*(X)$	bounded real-valued continuous functions on X, 1.9(a)
$c(\mathcal{F})$	set of convergent points of a filter base $\mathcal{F}$, 2.3(c)(3)
$\chi(p,X)$	character at p, 2N(4)
$\chi(X)$	character, 2N(4)
c(X)	cellularity of X, 2N(6)
ccc	countable chain condition, 2N(6), 3.5(b)

a', A'	complement of a or A as elements in a Boolean algebra, 3.1(d)(1)
$a \backslash b$	$a \wedge b'$ in a Boolean algebra, 3.1(f)
$\langle\langle S \rangle\rangle$	smallest complete subalgebra of a Boolean algebra containing S, following 3.4(b)
CH	continuum hypothesis, 3.5(a)
$\hat{x}$	$\{y \in P : y \leqslant x\}$, P a poset, 3.5(f)(2); also linear functional on $\mathfrak{A} \subseteq C^*(X)$, 4.5(l)
$c_B X$	Wallman compactification via Boolean algebra B, 4.7(a)
$c\ell_\Theta A$	Θ-closure of A, 4.8(r)
c_Z	covering map from space E onto Z, 8.4(b)
$C(X)$	set of (equivalence classes of) covers of X, 8.4(l)
$(c_P X, k_P)$	minimum P-cover of the space X, 8.4(q)
$CP(X)$	set of P-covers of X, 8.4(q)
$C_T(X)$	set of Tychonoff covers of X, 8.4(w)
$C_0(X)$	set of zero-dimensional covers of X, 8.4(w)
CPT	category of compact spaces and continuous functions, 9.2(b)(4)
$\amalg\{A_a : a \in I\}$	category-theoretic coproduct of the set $\{A_a : a \in I\}$ of objects, 9.5(a)

D

$D(\alpha)$	discrete space of cardinality α, 1.2(a)
Δ_Y	diagonal of Y (i.e., $\{(y,y) \in Y \times Y : y \in Y\}$), 1C
$d(X)$	density character, 1G

$a \Delta b$	symmetric difference in a Boolean algebra, 3L
$\mathcal{D}(X)$	$\{(c_Y,Y) \in AB(PX) : (c_Y,Y) \geqslant (c_X,X)\}$, 8.4(g)
DJ(X)	quotient of Jones space of X, 1Y(6)

E

$\exp(\lvert X \rvert)$	cardinality of power set of X, 1.1(c)
e_F	evaluation map, 1.7(i)
$\mathcal{E}(X)$	set of all extensions of X (up to equivalence), 4.1(d)(2)
EX	Iliadis absolute of X, 6.6(b)
E′X	absolute of X as a set of open ultrafilters, 6.8(a)
$E_{\mathfrak{a}}X$	convergent ultrafilters on a basic subalgebra $\mathfrak{a}$ of $\mathcal{R}(X)$, 6H
ED	category of extremally disconnected spaces and continuous functions, 9.7(b)(4)

F

F(X,Y)	set of all functions from X to Y, 1.1(a)
$f \mid A$	restriction of function f to A, 1.1(b)
$f^{\leftarrow}$	inverse of function f, 1.1(g)
F_σ-set	union of countably many closed sets, 1.2(g)
$f^*[\mathcal{C}]$	$\{Y \setminus f[X \setminus C] : C \in \mathcal{C}\}$, 1.8(c)
$\lvert f \rvert$	$\lvert f \rvert(x) = \lvert f(x) \rvert$ if $f \in C(X)$, following 1.3(b)
$\hat{\mathcal{F}}$	filter generated by filter base $\mathcal{F}$, 2.3(c)(1)
f_s	function f : X → Y viewed as having domain

	X(s), 2.2(g)

G

G_δ-set	intersection of countably many open sets, 1.2(g)
Γ	Nemitskii plane, 1I
$\langle E \rangle$	subalgebra of Boolean algebra generated by E, 3.1(j)
GCH	generalized continuum hypothesis, 3.5(a)
$\mathcal{GB}(X)$	set of regular subrings of C*(X), following 4.5(p)
$\gamma_P X$	maximum P-extension of P-regular space X, 5.1(c)
$\gamma_E X$	maximum E-compact extension, following 5.4(a)
$\gamma_{\mathbb{N}} X$	maximum $\mathbb{N}$-compact extension, 5#, 5F
g_Y	covering map in the pair (Y,g_Y) onto X, 8.4(k,l)

H

$\mathcal{H}$	class of H-closed spaces, 4.8(m)
$\mathcal{H}(X)$	set of H-closed extensions of X (up to equivalence), 4.8(m)
$\mathcal{H}^\theta(X)$	set of largest members of θ-equivalence classes of H-closed extensions of X, 7.7(a)
$\mathcal{H}^\#(X)$	set of strict H-closed extensions of X (i.e., $\mathcal{H}^\#(X) = \{Y^\# : Y \in \mathcal{H}(X)\}$), 7C
$\mathcal{H}^+(X)$	set of simple H-closed extensions of X (i.e., $\mathcal{H}^+(X) = \{Y^+ : Y \in \mathcal{H}(X)\}$), 7C
$\mathcal{H}_0(X)$	set of fully disconnected H-closed extensions of

	X, 7I
Hom(A,B)	set of morphisms from A to B, 9.2(a)
$Hom_{\mathfrak{a}}$(A,B)	set of morphisms from A to B in category $\mathfrak{a}$, 9.2(a)
HAUS	category of Hausdorff spaces and continuous functions, 9.3(a)(2)
HCU	category of H-closed, Urysohn spaces and continuous functions, 9H
HC	category of H-closed spaces and continuous functions, 9.6(b)(3)

I

$\cap\mathcal{C}$	intersection of all sets in a set $\mathcal{C}$ of sets, 1.1(h)
id_A	identity function in F(A,A), 1.1(j)
$\subset$	proper set inclusion, 1.1(k)
$\subseteq$	set inclusion, 1.1(k)
$int_X A$	interior of A in X, 1.2(d)(5)
f^{-1}	algebraic inverse of $f \in C(X)$ (i.e., $f^{-1}(x) = \frac{1}{f(x)}$), 1.3(a)
$f \wedge g$	infimum of functions in C(X), 1.3(b)
$\wedge B$	infimum of a subset B of a poset, 2.1(c)(5)
$\wedge$	lattice infimum, 2.1(c)(7)
(a,b),[a,b), (−∞,a], ...	interval notation in posets, 2.1(c)(2,4)
I(X)	set of isolated points of X, preceding 6.9(e), 7.6(k)

1_A	identity morphism, 9.2(a)
$[i\hat{P}]$	category of all P-coverable spaces and covering maps (where P is a covering property), 9.7(b)(5)
$[iP]$	category of all spaces with P and covering maps (where P is a covering property), 9.7(b)(5)

J

J(X)	Jones machine applied to X, 1Y

K

kX	k-space coreflection of X, 1T
kf	k-space coreflection of $f : X \to Y$, 1T(8)
K	class of compact spaces, 4.2 (second paragraph)
$K(X)$	set of compactifications of X (up to equivalence), 4.2(a)
K_0	class of compact, zero-dimensional spaces, 4.7 (first paragraph)
$K_0(X)$	set of zero-dimensional compactifications of X, 4.7(a)
κX	Katětov H-closed extension, 4.8(o)
$K(P)$	class of P-compact spaces, 5.2(a)
k_X	map from EX onto X, 6.6(d)
k_X'	map from E'X onto X, 6.8(e)
κf	extension to $\kappa X \to \kappa Y$ of the p-map $f : X \to Y$, 7.6(c)
$\kappa_F D$	modification of topology of κD, where D is

	discrete, 7F(1)
$\kappa_0 X$	maximum fully disconnected, H-closed extension of X, 7I
$\kappa_{\mathcal{B}} X$	simple H-closed extension of X generated with b-basis $\mathcal{B}$, 7N
kCPT	category of k-spaces and continuous functions, 9.7(b)(2)

L

L(X)	Lindelöf degree of X, 2N(7)
λ	$\lambda(a) = \{\alpha \in S(B) : a \in \alpha\}$ where B is Boolean algebra and $a \in B$, 3.2(a)

M

MA(κ)	κ-Martin's axiom, 3.5(h)
MA	Martin's axiom, 3.5(h)
$m_{\mathfrak{a}} X$	maximal ideal space of $\mathfrak{a} \subseteq C^*(X)$, 4.5(f)(4), 4.5(i)
mX	maximum m-bounded extension of X, preceding 5.10(d)
μX	$(\kappa X)(s)$, the Banaschewski-Fomin-Šanin semiregular extension of the semiregular space X, 7.5(f)
$\mathcal{M}(X)$	set of minimal Hausdorff extensions of a semiregular space X, 7.5(f)

N

$\mathbb{N}$	set of positive integers (used as index set and as a countable discrete space), 1.2(b)
$\mathcal{N}(p)$	set of neighborhoods of p, 1.2(d)(6)
$\mathcal{N}(p,X)$	set of neighborhoods of p in X, 1.2(d)(6)
$\neg S$	negation of the statement S, 3.5 (second paragraph)
$\|\cdot\|$	norm, 4.5 (second paragraph)

O

$f \leqslant g$	order structure on C(X), 1.3(b)
ord(A)	set of ordinals whose cardinality is $\lvert A \rvert$, 2.6(h)(4)
ω, ω_0	smallest infinite ordinal, 2.6(e)(3)
ω_γ	smallest ordinal with cardinality $\aleph_\gamma$, 2.6(j)
O^p, $O_Y{}^p$	filter trace of an open neighborhood system of a point p in a space Y on a dense subspace X, 4.1(c), 7.1(a)(1)
OU, O(U)	set of all open ultrafilters on X containing the open set U, 6.8(c)
oU, $o_Y U$	$\{p \in Y : U \in O^p\}$ where X is dense in Y and U is open subset of X, 7.1(a)(2)
$\lhd$	order on $M(X)$, 7G
ob($\mathcal{C}$)	class of objects of a category $\mathcal{C}$, 9.2(a)

P

$\mathbb{P}(X)$	power set of set X, 1.1(c)
$\Pi\{X_i : i \in I\}$	product of sets $\{X_i : i \in I\}$, 1.1(f) (also see $\Pi\{A_i\ i \in I\}$ below)
$\langle x_i \rangle_{i\in I}$	point $f \in \Pi\{X_i : i \in I\}$ with $f(i) = x_i$, 1.1(f)
$\Pi\{A_i : i \in I\}$	category-theoretic product of objects $\{A_i : i \in I\}$, 9.5(a) (also see $\Pi\{X_i : i \in I\}$ above)
Π_j	projection function from $\Pi\{X_i : i \in I\}$ onto X_j, 1.1(f)
Π_S	projection function from $\Pi\{X_i : i \in I\}$ onto $\Pi\{X_i : i \in S\}$, 1.1(f) (also see Π_X listed below)
Π_X	map from PX onto X, 6.11(b) (also see Π_S listed above), following 6.11(e)
$f \cdot g$	product of functions, 1.3(a)
αf	product of scalar and function, 1.3(a)
$\Pi\{f_i : i \in I\}$	product of $\{f_i : X_i \rightarrow Y_i : i \in I\}$ ("product function"), 1.7(f)
P	topological property, 4.1(j)
ψ	extension of $\mathbb{N}$ by a maximal almost disjoint family, 1N
$\alpha + 1$	successor ordinal of α, 2.6(e)
m^+	successor cardinal of m, preceding 2.6(m)
$\pi w(X)$	π-weight of X, 2N(5)
$x \perp y$	x and y are not compatible elements of a poset, 3.5(b)(1)
P(X)	set of P-points of X, 1W (also see P(Y) below)
$P(\kappa)$	combinatorial consequence of MA, 3.5(q)

P(Y)	partition of $\sigma X \setminus X$ induced by $f : X \to Y (Y \in \mathcal{H}(X))$, 7.4 (first paragraph) (also see P(X) above)
$\mathcal{P}(X)$	set of all extensions of X with property $\mathcal{P}$, 4.1(j)(2)
$\mathcal{P}f$	extension of $f : X \to Y$ to $\Upsilon_{\mathcal{P}} X$ $(Y \in \mathcal{P})$, 5.1(a)
$\mathcal{P}_E$	the property of being homeomorphic to E, 5.4(a)
PX	Banaschewski absolute, 6.11(b)
Y^+	extension Y of X with the simple extension topology, 7.1(g)
X^+	one-point H-closed extension of the locally H-closed space X, 7D
$\mathcal{P}_Q$	$X \in \mathcal{P}_Q$ if $cl_X U \in \mathcal{B}(X)$ whenever $(U,X) \in Q$, 8.4(v)
$\langle\mathcal{P}\rangle$	category of Hausdorff spaces with property $\mathcal{P}$ and continuous functions, 9.3(a)(4)
pHAUS	category of Hausdorff spaces and p-maps, 9.5(f)
$[\mathcal{P}]$	category of Hausdorff spaces with property $\mathcal{P}$ and perfect, continuous functions, 9.8(i)

Q

$\mathbb{Q}$	set of rational numbers, 1.2(a)
$(QF(X), \phi_X)$	minimum quasi-F cover of X, 8J

R

$\mathbb{R}$	set of real numbers, 1.2(a)

$\underline{r}$	constant function on X whose only value is $r \in \mathbb{R}$, 1.2(c)(3)
$\mathcal{R}(X)$	set of regular closed sets of X, 2.2(b)
$\mathcal{RO}(X)$	set of regular open sets of X, 2.2(b)
$\mathrm{Reg}(\mathcal{P})$	class of $\mathcal{P}$-regular spaces, 5.2(a)
rC	reflection of $C \in ob(\mathcal{C})$ in reflective subcategory of $\mathcal{C}$, 9.6(a)
r_C	reflection morphism from C to rC, 9.6(a)

S

$A \subseteq B$	A is a subset of B, 1.1(k)
$A \subset B$	A is a proper subset of B, 1.1(k)
$\oplus\{X_i : i \in I\}$	sum or free union of spaces $\{X_i : i \in I\}$; $\oplus\{X_i : i \in I\} = \cup\{X_i \times \{i\} : i \in I\}$, 1.2(h)(1)
$\oplus\{f_i : i \in I\}$	sum of $\{f_i : X_i \to Y_i : i \in I\}$ ("sum function"), 1.2(h)(2)
$f + g$	sum of functions, 1.3(a)
$f \vee g$	supremum of functions in C(X), 1.3(b)
spt(f)	support of f, 1.4(d)
$\Sigma(c)$	Σ-product subspace based at $c \in \Pi\{X_a : a \in I\}$, 1X
$\vee B$	supremum of subset B of a poset, 2.1(c)(5)
$\vee$	lattice supremum and infimum, 2.1(c)(7)
$\alpha + 1$	successor ordinal of α, 2.6(e)(1)
S(B)	Stone space of Boolean algebra B, 3.2(a,c)
$SL(\kappa)$	combinatorial consequence of MA, 3.5(o)
$[S]^2$	set of doubletons of a set S,2N(1), 3T

$S(A)$	set of ultrafilters on a Wallman base L that contain the set A, 4.4(a)(2)
$Y^{\#}$	extension Y of X with the strict extension topology, 7.1(d)
σX	Fomin H-closed extension of X, 7.2 (first paragraph)
$X^{\#}$	one-point H-closed extension of the locally H-closed space X, 7D
SET	category of sets and function, 9.2(b)(1)
sB	coreflection of B $\in$ ob(C) in a coreflective subcategory of C, 9.7(a)
s_B	coreflection morphism from sB to B, 9.7(a)

T

$2^{\|X\|}$	cardinality of power set of X, 1.1(c)
2^{α}	cardinality of power set of X if $\|X\| = \alpha$, 1.1(c)
$\mathfrak{2}$	two point discrete space {0,1}, 1.2(a)(1)
$\tau(X)$	set of open subsets of X, 1.2(d)(2)
τ	topology on X in (X,τ), 1.2(d)(4)
τ_{δ}	P-space coreflection topology obtained from τ, 1W(6)
$\tau(s)$	semiregularization topology, 2.2(e)(2)
$\tau(\leqslant)$	order topology on a linearly ordered set, 2.5(e)(1)
$\tau^{\#}$	strict extension topology on an extension of X, 7.1(d)
τ^{+}	simple extension topology on an extension of X,

	7.1(g)
$\Theta C(X,Y)$	set of Θ-continuous functions from X to Y, 4.8(g)(3)
θX	Gleason space of X (i.e., $S(R(X))$), 6.6(a)
$\theta' X$	set of all open ultrafilters on X, 6.8(c)
ΘHAUS	category of Hausdorff spaces and Θ-continuous functions, 9.2(b)(2)
ΘHCU	category of H-closed Urysohn spaces and Θ-continuous functions, 9H
$[\Theta P]$	category of Hausdorff spaces with P and perfect Θ-continuous functions, 9.8(i)
TYCH	category of Tychonoff spaces and continuous functions, 9.3(a)(3)

U

$\cup$	Union, 1.1(h)
$\cup C$	union of all sets in a set C of sets, 1.1(h)
[0,1]	closed unit interval, 1.2(a)
υX	Hewitt realcompactification, 5.5(c)
υf	extension of $f \in C(X)$ to υX, 5.10(b)

W

$w(X)$	weight of X, 2N(3)
$w_L X$	Wallman compactification of X generated by Wallman base L, 4.4(a), and preceding 4.4(d)

X

X^I	product of $\|I\|$ copies of X, 1.7(c)
X_δ	P-space coreflection of X, 1W
X(s)	semiregularization of X, 2.2(e)(2)

Y

Y(P)	largest H-closed extension of X corresponding to a partition of $\sigma X \setminus X$ into compact sets, 7.4(a)

Z

Z(f)	zero-set of $f \in C(X)$, 1.4(b)
$Z(X)$	set of zero-sets of X, 1.4(b)
ZFC	Zermelo-Fraenkel set theory with Axiom of Choice, 3.5 (first paragraph)
ZD	category of zero-dimensional spaces and continuous functions, 9.2(b)(4)
ZDCPT	category of zero-dimensional, compact spaces and continuous functions, 9.3(a)(3)
ziCPT	category of compact spaces and z-irreducible covering maps, 9K

A

AB(X), 8.4(b)
    order structure on, 8.4(c), (f)
absolute, §6.1, 6.6(a)
    (see Iliadis absolute and Banaschewski absolute)
    and measure theory, §6.1
accumulation point (of nowhere dense set), 7U
a(F), 2.3(c)(3)
    vs. c(F), 2.3(f)
    for ultrafilters, 2.3(h)
    vs. continuous maps, 2.3(i)
    vs. z-filters, clopen filters, 2.3(m)
$\aleph_\alpha$, 2.6(l)
almost compact, 5U
    implies pseudocompact, 5U
    and Tychonoff extension properties, 5U
    EX is iff, 8A
almost H-closed space, 7P, 8A
almost P-space, 6T
almost realcompact space, 6U
    vs. realcompact, 6U, 6V
    examples, 6V
    properties, 6U, 6V, 8.3(i)
Archimedian field, 4.5(f)
    characterizations, 4.5(g), 4D
Arhangel'skii's theorem, 2O
atom, 3.2(l)
    vs. isolated point, 3.2(m)
atomic Boolean algebra, 3.2(l)
    Stone space of, 3.2(m)(2)
aX, 6U

B

Banaschewski absolute, §6.1, §6.11, 6.11(b), 6.11(e)
relation to EX, 6.11(c), (h)
equivalence to, 6.11(f), (g), (h), (j), 6E
of H-closed spaces, 6.11(i)
of sums and products, 6A
as an inverse limit, 6Z
and H-closed extensions, 8.2(d), (f), (h), (i), 8B
vs. κX, 6.11(j), 8C
vs. μX, 8.2(i)
vs. σX, 7B, 8.2(h), 8D
Banaschewski-Fomin-Šanin extension, 7.5(f)
example, 7.7(g)
as an element of $\mathcal{M}(X)$, 7.5(g), 7.7(f)
vs. βX, 7B
vs. κX, 7B
vs. (PX,Π_X), 8.2(i)
and remote points, 7V(9)
vs. σX, 7.7(g), 7B
Baire space, 7T
base, 1.2(d)
for Tychonoff space, 1.4(e)
b-base, 7N
basic subalgebra, 6H
applications, 6I, 6Y
basic sublattice, 6S
basically disconnected spaces, 6K
properties, 6K
are F-spaces, 6L(3)
vs. P-space, 6L(3)
as cover, 8G

βX, 4.2(b) (also see Stone-Čech compactification)
X is C*-embedded in, 4.2(e), 4.6(j)
characterizations of, 4.5(p)(3), 4.6(g)
extension of continuous functions to, 4.2(c), (e)
extension of perfect functions to, 4.2(f), (g)
closed base for, 4.6(h)
zero-sets of, 4.6(d)
and z-ultrafilters on X, 4.4(h), et seq.
and CR-filters on X, 4Y
and maximal ideals of C*(X), 4.5(p), 4.6(a), (e)
and maximal ideals of C(X), 4AC
as a one-point compactification, 4A
as a Wallman compactification, 4.4(h)
as a Gelfand compactification, 4.5(p)
when X is discrete, 4U, 6F
remote points of 4AH
extremally disconnected when, 6.2(c)
as a reflection in CPT, 9.6(b)
and EX, 6.9(b)
vs. $\beta_0 X$, 4.7(g)
vs. κX, σX, μX, 7B
$\beta X \setminus X$
co-absolutes of, 6Y
as a continuum, 6AA
properties, 5.11(h), (i), 6L(2)
$\beta X \setminus \upsilon X$, 5Z, 6AB
$\beta \mathbb{N}$, 4.8(t)
vs. $\kappa \mathbb{N}$, 4L
C*-embedded subspaces of, 6F(9)
separable subspaces of, 6Q
βf, 4.2(d)
vs. f^{β}, end of 4.5
explicit description, 4.6(k), (l)

$\beta_0 X$, after 4.7(b)
X is z-embedded in, 4.7(f)
vs. βX, 4.7(g), (j)
and clopen ultrafilters on X, 4.7(c)
extension of continuous functions to, 4J
as a reflection in ZDCPT, 9.6(b)
$\beta_0 f$, 4.7(d)
B-filter, 3.1(q), (r)
BF(κ), 3U
topological consequences of, 3V
bijection, 1.1(e)
binding family, 4W
binomial expansion, 4C
Birkhoff-Stone theorem, 2.4(g)
proof, 2I
Boolean algebra, 3.1(d)
examples, 3.1(e), 3A, 3B, 3M, 3N
complete, 3.1(o)
completion of, 3.4(d), 3J
σ-complete, 6K
σ-completion, 8G
identities in, 3.1(g)
atomic, 3.2(l), 3.2(m)
countable atomless, 3.3(g)
free, 3K
subalgebra of, 3.1(h), (k), §3.4, 3C, 3J, 6H, 6I
generated subalgebra, 3.1(j), (k)
factor or quotient algebra, 3L, 3M, 3N, 6Y
Stone space of, §3.2, 3.2(c), 3N(4), (also see Stone space)
Boolean homomorphism, isomorphism, 3.1(m)
Boolean monomorphism, preceding 3.4(d)
bonding map (in inverse limit), 2U
bounded

poset, 2.1(c)(6), 2B
subset of ordinal, 2.6(p)
B-ultrafilter, 3.1(q), (r)
characterization, 3.1(r)
in Stone space, 3.2(a), (b), (c)
$B(X)$, 1.5(a)
and zero-dimensional spaces, 1.5(b)
vs. $B(\beta_0 X)$, 4J
for extremally disconnected X, 6.2(a)

C

Cantor space, prior to 3.3(f)
topological characterization, 3.3(f)
irreducible image of, 6I
Cantor κ-space, 3K
cardinal functions, 2N
cardinal numbers, 2.6(l)
sums, products, 2.6(m)
exponentiation, 2.6(n)
arithmetic of, 2.6(o)
regular, singular, 2.6(p)
category, 9.2(a)
examples, 9.2(b)
topological, 9.3(a)
$c_B X$, after 4.7(a)
ccc (countable chain condition)
in Boolean algebras, 3.5(b)(4), 3K(5), 3M3, 3N(3), 3R(7)
in posets, 3.5(b)(3), 3.5(c)
in product spaces, 3R, 3S, 3T
in topological spaces, 2N(6), 3.5(b)(5), 3K(5), 3P, 3R, 3S, 3T
cellularity, 2N(6) (also see *ccc*)

vs. density character, 2N(6)
vs. weight in metric spaces, 2N(6)
C-embedding, 1.9(b)
vs. C*-embedding, 1.9(c), 1.9(j)
and compact spaces, 1.9(k)
in normal spaces, 1.10(g)
in υX, 5.10(b)
in Σ-products, 4R(9)
C*-embedding, 1.9(b) (also see C*-$\mathbb{N}$-embedding)
sufficient conditions for, 1.9(h), (i), 4.6(h), (j)
in normal spaces, 1.10(g), 1O
in βX, 4.2(e), 4.6(j)
in H-closed extensions, 4M
in extremally disconnected spaces, 6.2(c)
in F-spaces, 6L
in products, 4AG
in P-spaces, 4AG(6)
c(F), 2.3(c)(3)
vs. a(F), 2.3(f)
vs. z-filters, clopen filters, 2.3(m)
cg-operator, 8.4(u)
and covering properties, 8.4(v)
character (of a space), 2N(4)
vs. density character, 2N(4)
vs. cardinality of a space, 2N(4), 2O
characteristic function, 1.1(d)
C.I.P., see countable intersection property
clopen set, 1.5(a)
and characteristic functions, 1.5(c)
closed function, 1.2(e)
closed at a point, 7V(9)
closed-hereditary, 1.2(f)
extension properties are, 5.3(c)

closed ultrafilter (also see ultrafilter), 2.3(c)(4)
in normal spaces, 2Q
vs. z-ultrafilters, 2Q, 2R(5)
in δ-normally separated spaces, 2Q(4)
closure, 1.2(d)
c-map, 6G(3)
C*-ℕ-embedding, 1L
co-absolute spaces, 6Y
examples, 6Y(8), 6.10(f)
cofinal set, confinality, 2.6(p)
collectionwise Hausdorff, 6F(11)
commuting diagram, preceding 9.2(a)
compact function, 1.8(a)
and irreducible with H-closed domain, 7H
not closed under composition, 9A
compactifications, beginning of 4.2, 4F (also see separate listings of following topics)
Fan-Gottesman, 4W
Freudenthal, 4X
Gelfand, 4.5; 4.5(m), 4G
maximum zero-dimensional, prior to 4.7(c)
one-point, 4.3, 4A
partial order on family of, 4.2(a)
via proximities, 4AA
Stone-Čech, 4.2(b), 4.6, 4Y
Wallman (of Wallman type), 4.4, 4A(2), 4B, 4G
compactness
of closed nowhere dense subsets, 7.6(j), (k)
vs. H-closed, 4.8(k)
in linearly ordered spaces, 2.5(g)
vs. normal, preceding 1.10(b)
preserved by perfect maps, 1.8(d).
vs. Tychonoff, 1.10(i)

and upper semicontinuous decompositions, 4Q
compatible (members of a poset), 3.5(b)(1)
complement (in a Boolean algebra), 3.1(b)
uniqueness of, 3.1(c), (d)
complete
lattice, semilattice, 2.1(d)
conditionally, 2.5(a)
Boolean algebra, 3.1(o)
subalgebra of Boolean algebra, 3.4(d), (b)
Boolean homomorphism, 3.4(a)(2)
complete ring of quotients, 6.1
complete separation, 1.9(d)
conditions equivalent to, 1.9(f), (g)
and C*-embedding, 1.9(h)
and compact sets, 1.9(i)
and normal spaces, 1.10(g)
and separating chains, 1.10(f)
completion
of a Boolean algebra, 3.4(d), (e), (f)
conditional, 2.5(c), 8.5(e)
of C(X), 8.5
Dedekind-MacNeille, 2.5(b) et seq.
of a lattice, 2.5(b), (c)
minimal, 3.4(h), (j), 3J, §6.1
composition (of morphisms), 9.2(a)
conditional completion, 2.5(c)
of C(X), §8.5; 8.5(e)
continuum hypothesis (CH), 3.5(a)
generalized, 3.5(a)
and product spaces, 3T
and P-points of $\beta X \setminus X$, 6AC
contravariant functor, 9.2(c)
examples, 9.2(d)(2), (3)

coproduct, 9.5(a)
vs. coreflective subcategories, 9.7(e)
vs. topological sum, 9C
copseudocompact, 5W
coreflective subcategory, 9.7(a)
examples, 9.7(b)
vs. covering properties, 9.7(b)(5), 9M
properties, 9.7(c) to (f)
coregular spaces, 5.9(g)
almost compact, 5U
extensions, 5T, 5U
countable chain condition, see *ccc*
countable intersection property (C.I.P.), 2P(1), 5.10(a)
countable π-weight, preceding 6.10(d); also see 2N(5)
and absolutes, 6D
and P-points, 6AD
and remote points, 4AH(6)
countably compact, 1P
vs. compact, 2.6(q)
vs. feebly compact, 1N, 1P(3)
and perfect, 3.5(l)(2), 4AE
products, 1P(4)
vs. pseudocompact, 1.11(d), 1N, 1P(3), 5(S)
and separable, 4AD
vs. U-compact, 5R
countably paracompact, 8K
covariant functor, 9.2(c)
examples, 9.2(d), 9.6(c)
covering coproduct, 8.4(r)
uniqueness, 8.4(s)
vs. coreflective subcategories, 9.7(b)(5)
covering map, 8.4(a)
properties, 8.4(d), (i), (k)

covering property (also see Tychonoff, zero-dimensional covering property) 8.4(q), 9M
characterizations, 8.4(t)
examples, 8.4(v) et seq., 8G
covers, 8.4(a)
equivalence of, 8.4(j)
examples, 8.4(v), 8G, 8H, 8J
family of, 8.4(l)
cozero-sets, 1.4(c)
as a base, 1.4(e)
closed under countable unions, finite intersections, 1.4(i)
C*-embedding of, 6L
$\mathcal{C}_0(X)$, 8.4(w)
as a complete lattice, 8.4(y)
order structure on, 8.4(x)
c_pX, 8.4(q)
$\mathcal{CP}(X)$, 8.4(q)
as a complete lattice, 8.4(y)
order structure, 8.4(x)
CR-filter, 4Y
$\mathcal{C}_T(X)$, 8.4(w)
as a complete lattice, 8.4(y)
order structure on, 8.4(x), 8I
$\mathcal{C}(X)$, 8.4(l)
properties, 8.4(n)
order structure, 8.4(o), (p)
relation to $\mathcal{D}(X)$, 8.4(p)
as a lattice, 8M
C(X), 1.2(c)
conditional completion of, 8.5
as a lattice, 2.1(f)(5), 1.3(b)
maximal ideal space of, 4AC
as a ring, 1.3(a)

C*(X), 1.9(a)
as a ring, 1.9(a) et seq.
as a lattice, 2.1(f)(5)
maximal ideal space of, 4.5(p)
C(X,Y), 1.2(c)

D

Dedekind-MacNeille completion, 2.5(b) et seq.
of a Boolean algebra, 3.4(j), 6.1
of C(X), 8.5
δ-normal separation, 1R
and closed ultrafilters, z-ultrafilters, 2Q(4)
vs. countable compactness, 1R
vs. normal, 2Q(5)
vs. pseudocompactness, 1R
vs. realcompactness, 5Q
Δ-system, Δ-system lemma, 3R
De Morgan's laws, 3.1(g), 3F
dense subset
of a Boolean algebra, 3.4(d)(4), (i), 3J, 6.10(b)
of C(X), 3Q
of a poset, 3.5(f)(1), (g), 3Q
density character, 1G, 2N(2)
vs. cardinality of space, 2N(2)
vs. character, 2N(4)
and C*(X), 1G
of products, 3O
directed set, 2U
distributive lattice, 2.4(e), (f), (g)
duality
for Boolean algebras, and Stone spaces, §3.2

for posets, 2A
$D(X)$, 8.4(g)
properties, 8.4(k), (m)
as a lattice, 8.4(h)
relation to $C(X)$, 8.4(p)

E

$E_Q X$, 6H
E-compactness, 5.4(a)
examples, §5.5
vs. m-boundedness, 5B
vs. m-compactness, 5O
vs. one-point extensions, 5AC
vs. realcompact zero-dimensional, 5P
E-completely regular, 5.4(a)
E-open characterization, 5.4(c)
examples, 5.5(a), (b), (e), 5C, 5AE
E-embedded subspace, 5.4(e)
embedding, 1.2(e)
vs. extremal monomorphism, 9.4(l)
hypercombinatorial, 7O
into compact spaces, 4.2(b)
into H-closed spaces, 4.8(n), §7.1, §7.2
into products of extensions, 4.1(g)
Embedding theorem, 1.7(j)
E-open, 5.4(b)
epicoreflective subcategory, 9.7(a)
epimorphism, 9.4(b)
characterizations, 9.4(d)
examples, 9.4(c), 9B
extremal, 9.4(k), (l), (m), 9.7(e), 9B

properties, 9.4(f)
epireflective subcategory, 9.6(a)
examples, 9.6(b), (g)
vs. extension properties, 9.6(f)
properties, 9.6(d), (e), 9F
equivalence in AB(E), preceding 8.4(b)
equivalence to absolute, 6.7(c), 6.9(b), 6.11(f), 8.1(a), (b)
equivalent categories, 9.4(i)(3)
examples, 9.4(j)
equivalent covers, 8.4(j)
equivalent extensions, 4.1(d)
homeomorphic but not equivalent, 4AB
vs. absolutes, 6.9(b), 6.11(j)
evaluation map, 1.7(i)
and compactifications, 4.2(c), 4F
as an embedding, 1.7(j)
and extensions, 4.1(g)
$E(X)$, 4.1(d)(2)
complete lattice iff, 7.3(g)
as complete upper semilattice, 4.1(g)
and inverse limits, 4P
projective maximum in, 4.1(h)
projective minimum in, 7.3(e)
EX, §6.1, 6.6(b); see Iliadis absolute
E′X, 6.8(a) et seq, 6E(2)
extension of a function, 4.1(a)(1)
to absolutes, 6G
to βX, 4.2(c), 4.6(g), (l)
to $\beta_0 X$, 4.7(d)
criteria for, 4.1(l), (m), (n)
and E-compactness, 5.4(d), (e)
to $\gamma_p X$, 5.1(a)
to κX, 4.8(n), 7.6(b)

for perfect functions, 1.8(i), 1D, 4.2(f), (g)
uniqueness, 1.6(d)
to υX, 5.5(c)(2), 5.10(b)
extension of a space, beginning of Chapter 4
compact, §4.2
equivalent, 4.1(d)
H-closed, 4.8(m) et seq., Chapter 7
non-equivalent but homeomorphic, 4AB
simple, 7.1(g)
strict, 7.1(d)
T_1, 4T
extension property (also see Tychonoff extension property, zero-dimensional extension property), 5.3(a)
characterizations, 5.3(c)
closed-hereditary, 5.3(b)
and compact spaces, 5L
and discrete spaces, 5AB
vs. epireflective subcategories, 9.6(f)
maximum, 5X
and ordinal spaces, 5J
productive, 5.3(b)
properties of, 5.9(a), (b), (c), (d), (e), 7J
extremally disconnected space, §6.1, 6.2(a)
are basically disconnected, 6K
and C*-embedding, 6.2(c)
characterizations, 6.2(b)
form coreflective subcategory, 9.7(b)(4)
and C(X), 8.5(a)
examples, §6.3, 6.10(d), 6C
minimal, 6X
and open maps, 6R
P-points of, 6O
products of, 6P

and projective objects, 9.8(d)
and regularity, §6.4
and Stone-Čech compactification, 6.2(c)
and Stone space, 6.2(d)
and zero-dimensionality, §6.4

F

F_σ-set, 1.2(d)
factor algebra (of Boolean algebra), 3L, 3M, 3N, 4U, 6Y
Fan-Gottesman compactification, 4W
f^β, 4.5(r)
vs. βf, end of 4.5
feebly compact (also see pseudocompact), 1.11(a)
characterizations, 1.11(b)
and countable π-weight, 4AD
vs. countably compact, 1N, 1P
examples, 1N, 1U (also see pseudocompact)
and metacompact, 6N
and products, 4AF
vs. pseudocompact, 1.11(d), 1U
filter, 2.3(a), 2H
adherence of, 2.3(c)(3)
on a Boolean algebra, 3.1(q), (r)
convergent, 2.3(c)(3)
CR-filter, 4Y
fixed, free, 2.3(c)(3)
open, clopen, closed, regular open, z-, 2.3(c)(4)
p-filter (see also proximity), 4AA
on a poset, 3.5(d)
filter base, 2.3(a)
fixed filter, 2.3(c)(3)

Fomin extension, §7.2 (also see σX)
vs. absolutes, 8.2(d), (h), 8D
characterization, 7O
compact subsets of, 7.2(c), 7.4(c), 7E
of locally H-closed spaces, 7.3(b), (c)
properties, 7.2(b), (c), 7.4(a), (c)
of quasiregular spaces, 7T
free Boolean algebra, 3K
free filter, 2.3(c)(3)
free generators, 3K
Freudenthal compactification, 4X
F-space, 6L
basically disconnected spaces are, 6L(3)
vs. extremally disconnected, 6L(8)
vs. products, 6P
fully disconnected spaces, 7I
functor (see covariant, contravariant functor), 9.2(c)
forgetful, 9.2(d)(1)
identity, 9.2(d)(4)

G

G_δ-set, 1.2(g)
in βX, 5.11(b)
of $\beta X \setminus \upsilon X$, 5Z
in non-normal spaces, 1I(4), 1N
in normal spaces, 1K
vs. zero-set, 1K
Gelfand compactification, §4.5, 4.5(n)
all compactifications as, 4.5(q)
βX as, 4.5(p)
vs. Wallman compactifications, 4G

of extensions, 4P
inverse system, 2U
irreducible function, §6.1; 6.5(a) (also see below)
properties, 6.5(b)
irreducible Θ-continuous function
vs. c-maps, 6G(4)
onto extremally disconnected spaces, 6.5(d)
induces Boolean algebra isomorphism, 6.5(d)
vs. perfect function, 6B(2), (3)
and products, 6A
properties, 6.5(d), 8.4(i)
and sums, 6A
isomorphism
of Boolean algebras, 3.1(n), 3.2(f)
in categories, 9.4(e), 9B
vs. homeomorphisms in linearly ordered spaces, 2.5(f), 2J
of lattices, 2.4(b), (c)
natural, 9.4(i)

J

join (in a lattice), 2.1(c)(7)
Jones machine, 1Y
applied to Tychonoff plank, 2S
J(X), 1Y

K

N, 4.8(q), (t)
vs. βN, 4L
X, preceding 4.8(n)
vs. βX, 7B(2), (8)

Gelfand theorem, 4.5(m)
generated Boolean algebra, 3.1(j), (k), 3I, 3K
generated filter, 2.3(c)
g.l.b., see greatest lower bound
Gleason space, 6.6(a)
Glicksberg's theorem, 4AG(7)
greatest lower bound, 2.1(c)(5)

H

Hausdorff extension property, 5.8(a)
characterization, 5.8(c)
H-closed space, 4.8(a)
characterizations, 4.8(b), 7.1(i), (j)
not closed-hereditary, 4.8(f)
closed subspaces of, 4N
vs. compact, 4.8(c), (d)
examples, 4.8(d), (f), (q), 7.1(d), §7.2, 7.5(f), 7H, 7S, 7T, 7U(9)
and Martin's axiom, 7Q, 7T
maximal, submaximal, 7M
minimal, §7.5
products, 4.8(l)
regular closed subspaces of, 4.8(e)
semiregularization of, 4.8(h)(8), 7.5(b)(3), (d)
vs. Urysohn, 4.8(k)
H-closed subspace, 4S, 7.6(g)(4), 7D(7), 7H(4)
hereditary properties, 1.2(f)
Hewitt realcompactification (see υX), 5.5(c)
Hilbert space, 1V, 5P(14), (15)
homeomorphism vs. equivalence, 4AB
homomorphism,
vs. B-filters, 3G

of Boolean algebras, 3.1(m), (n), 3G, 6Y
into complete Boolean algebras, 3.1(p)
onto factor algebras, 3L(3)
induced on Boolean algebra by continuous function, 3.2(i), (j), (k)
join, meet, 2.4(b), (c)
of lattices, 2.4(b), (c), 3D
of posets, 2.4(b), (c)

H-set, 4N
and absolutes, 6J

$H_\theta(X)$, 7I

$H^\theta(X)$, 7.7(a)
as complete upper semilattice, 7.7(e)(3)
vs. $\mathcal{M}(X)$, 7.7(e)(4)
vs. simple extensions, 7C

$H(X)$, 4.8(m)
cardinality of, 7.3(f), 7.4(b), 7F
is complete lattice iff, 7.3(g)
is a complete upper semilattice, 7.3(g)
vs. partitions of σX, 7.4(b)
projective maxima in, 4.8(n)
projective minima in, 7.3(e)
and semiregularizations, 7.5(b)(4), 7.7(f)

hypercombinatorial embedding, 7O
in κX, σX, 7O
and perfect irreducible preimages, 8.2(g)

I

ideal (in a Boolean algebra), 3L
examples, 3M, 3N, 4U, 6Y

identity function, 1.1(j)

identity functor, 9.2(d)(4)

identity morphism, 9.2(a)

Iliadis absolute, §6.1, 6.6(b)
almost compact when, 8A
of $\beta X\backslash X$, 6Y
cardinality of, 6M
compact subsets of, 6J
embeddings in βD, 6F
equivalence to, 6.7(c), 6.9(a), (b), (d), (f), 6.10(a)
examples, 6.10(a), (e), (f), 6W
vs. extension properties, §8.3
generalizations, 6H, 6S; also see covers
and H-closed extensions, 8.2(a), (b), (c), (d)
and Hewitt realcompactification, 8.3(l)
isolated points of, 6.9(e)
vs. Katětov, Fomin, Banaschewski-Fomin-Šanin extensions,
non-normality of, 6W
vs. one-point H-closed extensions, 8A
preservation of topological properties by, 6B
of products, 6A, 6P
properties of, 6.6(d), 6.7(b), 6.9(a), (b)
of a space with countable π-weight, 6.10(d)
as a space of open ultrafilters, §6.8, 6E
and Stone-Čech compactification, 6.9(b)
of sums, 6A
uniqueness of, 6.7(a)

indecomposable continuum, 6AA

interior, 1.2(d)

interval (in a linearly ordered set), 2.5(e)

inverse (category-theoretic), 9.4(e), (g)
uniqueness, 9.4(f)

inverse limit, 2U
Banaschewski absolute as, 6Z
and compactness, 2U

characterization, 7O
extension of continuous functions to, 4.8(n), 7.6(b)
extremally disconnected when, 6.2(b)
vs. EX, 8.2(d)
vs. μX, 7B, 7.5(f)
as projective maximum of H(X), 4.8(n)
vs. PX, 6.11(j), 8C
properties of, 4.8(p), 5.8(b)
vs. σX, 7B, 7P(8)
is Urysohn when, 7B
Katětov H-closed extension, 4.8(o); also see κX
kHAUS, 9L
k-map, 1T(6)
K_0(X), beginning of 4.7
vs. Boolean bases for X, 4I
as complete upper semilattice, 4.7(a)
projective maximum in, following 4.7(b)
k_p, 8.4(q)
k-spaces, 1T
form coreflective subcategory, 9.7(b)(2), 9L
vs. sequential spaces, 9D(4)
kX, 1T
k_X, §6.1, 6.6(b) et seq.
properties, 6.6(e)
k'_X, 6.8(e)
K(X), 4.2(a)
as complete upper semilattice, 4.2(a)
complete lattice iff, 4.3(e)
vs. H(X), preceding 7.4(b)
projective maximum of, 4.2(b)
and upper semicontinuous decompositions, 4Q

L

lattice, 2.1(c)(7); also see semilattice
basic sublattice, 6S
vs. Boolean algebra, 3.1(e)(6)
complete, 2.1(d), 2.5(a), (g)
completion of, 2.5(b), §8.5
C(X) as, 2.1(f)(5)
distributive, 2.4(e), (f)
examples, 2.1(f), 2.2(c), 4.3(e), 7.3(g), 8.4(h), (p), (y)
homomorphism on a, 2.4(b)
identities in, 2C
isomorphism, 2.4(b)
partial order on, 2D
set representation of, 2.4(d)
subset of, 2K
sublattice, 2.4(a)
topological representation of, 3.1(a)
lattice-ordered ring, 1.3
C(X) as; 1B(3)
least upper bound, 2.1(c)(5)
left inverse (see inverse)
L-filter, etc. (see filter)
Lindelöf degree, 2N(7)
vs. character and cardinality, 2O(5)
Lindelöf space
with uncountable discrete subspace, 3V
zero-dimensional, 4.7(i)
linearly ordered set, 2.5(a)
completion of, 2.5(b)(3), (4)
rationals characterized as, 2M
topology on, 2.5(e)
linearly ordered space, 2.5(e), 2J

compactness in, 2.5(g)
compactifications of, 2.5(h)
are hereditarily normal, 2.5(n)
homeomorphisms of, 2.5(f)
open sets in, 2.5(g)
subspaces of, 2L
are Tychonoff, 2.5(i)
locally compact spaces, beginning of 4.3
characterizations, 4.3(e)
one-point compactifications of, 4.3(b), (c), (f), 4A
locally connected spaces, 9E
locally finite family, 1.2(k)(l)
and continuous functions, 1.6(c)
locally H-closed space, 7.3(a)
Fomin extension of, 7.3(b)
and *E*(X), *H*(X), 7.3(e), (g)
and one-point H-closed extensions, 7.3(e), 7D
and products, 7D(10)
loset, 2.5(a) (see linearly ordered set)
lower bound (in a poset), 2.1(c)(3)
lower semilattice, 2.1(d), 2D
l.u.b., see least upper bound
Lusin set, 3V

M

MA, MA(κ), see Martin's axiom
Martin's axiom, κ-Martin's axiom, 3.5(h)
and Boolean algebras, 3.5(i)
combinatorial consequences of, 3.5(n), (p), (r), 3U
equivalences to, 3.5(i), (m), 7Q, 7T(6)
topological equivalent, 3.5(m)

topological consequences of, 3.5(l), 3S, 3V, 4AD, 4AE, 6F(9), 6AB, 6AD, 7Q, 7T(6)
$m_{\Omega}X$, 4.5(f)(4)
as a compactification of X, 4.5(m), (o), (p)
constructed using evaluation map, 4E
as a topological space, 4.5(i), (j), (n), (o), (p)
maximum (in a poset), 2.1(c)(6)
maximal ideal space
of C*(X), 4.5(p)
of C(X), 4AC
of regular subalgebras of C*(X), 4.5(m), (o)
maximum *P*-extension, 4.1(c), Chapter 5
m-bounded, 5.6(c)
extensions, 5.10(d)
as *P*-pseudocompactness, 5S
properties, 5B
m-compact, 5N
and zero-dimensional, 5N
vs. E-compact, 5O
measurable cardinals (also see Ulam-measurable cardinals), 2P
and P-points, 6O
and products, 6P
and realcompactness, 5.11(l), (m), 5I
meet (in a lattice), 2.1(c)(7)
metacompact, 1.2(l)
and δ-normally separated, 5Q
and feebly compact, 6N
vs. realcompact, 5.11(m), 5Q
and pseudocompact, 6N
metric spaces
zero-sets of, 1.4(g)
minimal extremally disconnected spaces, 6X
minimal Hausdorff space, beginning of §7.5

characterization, 7.5(a)
closed subspaces of, 7.5(e)(2)
dense subspaces of 7.5(e)(1)
examples, 7.5(c), 7S(5), 7T
and products, 7.5(b)(2)
and semiregularizations, 7.5(b)(3)
minimum (in a poset), 2.1(c)(6)
minimum P-cover, 8.4(q)
monomorphism
for Boolean algebras, preceding 3.4(d)
category-theoretic, 9.4(a)
characterizations, 9.4(d)
examples, 9.4(c), (d), 9B
properties, 9.4(f)
morphism, 9.2(a)
μX (see Banaschewski-Fomin-Šanin extension), 7.5(f)
properties, 7.5(h)
M(X), 7.5(f)
alternate order on, 7G
vs. H(X), 7.7(e)
largest element in, preceding 7.5(g), 7.7(f)
no projective maximum, 7.5(g)
vs. strict extensions, 7C

N

natural transformation, 9.2(e)
examples, 9.2(f)
ℕ-compactness, 5.5(a) (also see E-compact)
associated maximum extension, 5E
characterization, 5E(6)
and discrete spaces, 5I

P-pseudocompactness for, 5S
vs. realcompact zero-dimensional, 5G
and weakly homogeneous spaces, 5H
$\mathbb{N}$-embedding, 1L
Nemitskii plane, 1I
normal space, 1.10(a)
absolute not, 6W
and C*-embedding, C-embedding, 1.10(g), 1O
characterizations, 1.10(g)
closed G_δ-sets in, 1K
generalizations, 1R
not productive, 1H(5), 2T
realcompact when, 5.11(l), 8K(5)
vs. Tychonoff, 1.10(h), 1O, 2R, 2T
z-ultrafilters, closed ultrafilters on, 2Q
normed ring, 4.5(b)
nowhere dense, 7U

O

object (in a category), 9.2(a)
ω, 2.6(e)
ω_γ, after 2.6(j)
topologized as linearly ordered space, 2.6(q)
is C-embedded in $\omega_\gamma + 1$, 2.6(q)
compact subsets of, 2.6(q)
σ-compact, countably compact subspaces of, 2.6(q)
Stone-Čech compactification of, 4A
one-point compactification, §4.3
and $K(X)$, 4.3(e)
as Stone-Čech compactification, 4A
uniqueness of, 4.3(f)
as Wallman compactification, 4A

one-point H-closed extension, §7.3
and EX, 8A
family of, 7D
and locally H-closed space, 7.3(b)
non-uniqueness of, 7.3(d)
open function, 1.2(e)
and c-maps, 6G(4)
onto extremally disconnected spaces, 6R
and p-maps, 7.6(d)
open-hereditary, 1.2(f)
open ultrafilter (also see ultrafilter), 2.3(c)(4)
0^p, 7.1(a)
order
homomorphism, 2.4(b)
linear, 2.5(a)
topology, 2.5(e) (also see linearly ordered space)
ordered ring, 4.5(f)
ordered topological space (also see linearly ordered space), 2.5(e), 2J
is hereditarily normal, 2.5(n)
ordinal number, 2.6(c) to (k)
finite, infinite, 2.6(i), (j), (k)
initial, 2.6(i), (j), (k)
limit, 2.6(e)
successor, 2.6(e)
vs. well-ordered set, 2.6(h)
0(U), 6.8(c)
properties of, 6.8(d)
oU, 7.1(a)(2)
properties of, 7.1(c), (e), 7.2(b)

P

[*P*], 9.8(i)
  projective objects in, 9.8(j)
paracompact space, 1.2(l)
partially ordered ring, 4.5(f)
partially ordered set (also see poset), 1.1(i), 2.1(a)
partition, 1.2(j)
  into clopen sets, 3.3(d)
P-closed, 7J
P-compact, 5.2(a)
  examples, 5.2(a)
p-cover, 7.6(a)(1)
P-cover, 8.4(q)
  examples, 8G, 8J
P-coverable, 8.4(q)
P-embedded subspace, 5.3(e)
  P-compact extension of, 5.3(f)
perfect function, 1.8(a)
  and absolutes, 6.1, 6.6(e), 6.11(c)
  and almost realcompactness, 6U
  characterization using open covers, 1.8(c)
  composition of, 1.8(e), 8.4(d)
  extensions of, 1.8(i), 4.2(f), (g)
  from extremally disconnected spaces, 6.11(d)
  images of remainders under, 1.8(i), 4.2(f), (g)
  inverse images of compact subsets under, 1.8(d)
  vs. irreducible perfect functions, 6B
  non-preservation of realcompactness by, 6U, 6V
  non-preservation of semiregularity by, 2.2(k)
  preservation of covering properties by, 1J
  preservation of regularity by, 1.8(h), 1J
  product of, 1.8(g)
  projection maps are iff, 1M
  restrictions of, 1.8(f)

perfect space, 3.5(k)(1)
    and compact, 3.5(l)(3)
perfectly normal space, 3.5(k)(2), 6AB
    and countably compact, 4AE
P-extension, 4.1(j)
    maximum, see maximum P-extension, 4.1(j), Chapter 5
P-filter, p-ultrafilter, 3.5(d)
pHAUS, 9.5(f)
π-base, 2N(5), 3Q, 7V
π-base, 4AH, 7V
π-weight, 2N(5) (also see countable π-weight)
    of absolutes, 6B(4)
    vs. density character, 2N(5)
    of strict extensions, 7A
    vs. weight, 2N(5)
Π_X, 6.11(b)
    properties, 6.11(c)
$P(\kappa)$, 3.5(q)
    combinatorial consequences, 3.5(r), (t), 3U
    topological consequences, 3.5(s), 3V, 4AD, 4AE
p-maps, 7.6(a)(2)
    characterization, 7.6(b)
    with discrete domain, 7.6(k)
    vs. embeddings, 7.6(d), (g), (h)
    examples, 7.6(d)
    and κX, 7.6(b), (h)
    with Lindelöf or Urysohn range, 7R
    and regular closed sets, 7.6(f)
po-cover, 7I(11)
point finite, 1.2(k)
po-map, 7I(11)
poset, 1.1(i), 2.1(a)
    by inclusion, 2.1(b)(1)

subsets of, 2K
P-point, 1W
in $\beta X \setminus X$, 6AC, 6AD
in extremally disconnected space, 6O
P-pseudocompact, 5S
P-regular space, 5.2(a)
characterization, 5.2(b)
and compact spaces, 5M(2)
examples, 5.2(a)
vs. P-compact, 5.3(c)
P-compact extension of, 5.3(d)
P-embedded subspaces of, 5L(1)
and P-pseudocompactness, 5S
subspaces of, 5M
product, (also see product space)
category-theoretic, 9.5(a), (b), (c), (f), 9.6(e), 9C
of compact subsets, 1.7(g)
of extremally disconnected spaces, 6P
of functions, 1.7(e), 1.8(g)
of H-closed spaces, 4.8(l)
of normal spaces, 1H(5), 2T
of perfect functions, 1.8(g)
of semiregular spaces, 2.2(j)
of sets, 1.1(f)
of spaces, 1.7(c), 1F
of Θ-continuous functions, 4O
of Urysohn spaces, 4K
product spaces (also see product), 1.7(c), 1E
properties, 1F
density character of, 3O
and extensions, 4.1(k)
feeble compactness and pseudocompactness in, 4AF
as Stone-Čech compactification, 4AG

productive properties, 1.7(h)
projection function, 1.1(f)
projective cover, 9.8(a)(2)
properties, 9.8(d)
uniqueness, 9.8(b)
projective maximum, 4.1(h)
vs. maximum P-extension, 5A
projective minimal, 7D(8)
projective minimum, 7.3(e), 7D(9)
projective object, 9.8(a)(1)
in CPT, 9.8(f)
in HAUS, 9.8(g)
in HC, 9.8(h)
in [HC], 9I
in kHAUS, 9L
in [P], 9.8(j)
in ΘHAUS, 9I
in [ΘHAUS], 9.8(n)
in ΘHC, 9I
in ΘHCU, 9H
in [ΘP], 9.8(l), (o)
in TYCH, 9.8(g)
in ZD, 9.8(g)
in ziCPT, 9K
projectively larger (extensions), 4.1(f)
projectively smaller (covers), 8.4(c), (o)
proximity, 4Z
and compactifications, 4AA
proximity topology, 4Z(4)
pseudocompact, 1.11(c); (also see feebly compact)
absolutes are iff, 6B
vs. almost compact, 5U
$\beta X \setminus X$ is when, 5.11(i)

characterizations, 1.11(e), 1Q(7), 4AF(5)
continuous surjections preserve, 1Q(1)
vs. countably compact, 1N, 1Q(8), 1R(2), 2R
examples, 1N, 2.6(q), 2R, 2S, 2T, 4R(9)
vs. feebly compact, 1.11(d), 1U
generalizations, 5S
hereditary properties, 1Q(2), (9)
and metacompact, 6N
and metrizable, 1Q(5)
and ℕ-compact, 5G
and normal, 1R, 2T
product spaces, 4AF, 4AG(7), (8)
P-space, 1W(4), 4AG(6)(e)
and realcompact, 5F
regular closed subsets, 1Q(2), 5V
and Σ-products, 4R(q)
Ψ, 1N
P-space, 1W
countable subsets are C*-embedded, 4AG(6)(d)
feebly compact, 1W(4)
pseudocompact, 4AG(6)(e)
X_δ is, 1W(7)
and z-closed maps, 4AG(5)
is zero-dimensional iff, 1W
zero-sets in, 1W(2), 4AG(5)
pullback, 9J
PX (see Banaschewski absolute), 6.11(b)
$P(X)$, 4.1(j)(2)
complete upper semilattice iff, 4.1(k)

Q

$\mathbb{Q}$, 1.2(a)
characterizations, 2M
$\mathbb{Q}$-indexing, 1.10(c)
and continuous functions, 1.10(e)
quasi-F-cover, 8J
quasi-F-space, 8.4(aa)
and covering properties, 8.4(ab)
quasiregular space, 7T
quotient map, 9.4(l), 9B

R

realcompactness, 5.5(c)
and absolutes, 6T, 6U
vs. almost realcompact, 6V
and $\beta X \setminus X$, 5.11(i)
characterizations, 5.10(c), 5.11(c)
for discrete space, 5I
of extensions, 5.5(c)(2), 5F(3), (8)
vs. metacompact, 5.11(m), 5O
vs. normal Θ-refinable, 5.11(l)
and pseudocompact, 5F(5)
vs. ultrarealcompact, 5AA
and zero-dimensionality, 5G, 5P, 5W
refinement, 1.2(j)
of compact zero-dimensional spaces, 3.3(b)
reflection morphism, 9.6(a)
reflective subcategory, 9.6(a)
examples, 9.6(b), 9G
properties, 9.6(c), (d), (e), 9F

regular cardinal, 2.6(p)(2)
regular closed set, 2.2(b) (also see $R(X)$)
regular open set, 2.2(b) (also see $RO(X)$)
regular point, 7U
regular space
    preservation by perfect functions, 1.8(h), 1J
    not Tychonoff, 1Y
regular subring of C*(X), 4.5(k), 4E
    vs. compactification of X, 4.5(m), (q), 4E(4)
    and embeddings in products, 4E
relatively zero-dimensional remainder, 4X, 7N
remainder, 4.1(a), 5T
    of βX, 5.11(h), (i), 5H, 5T, 5U, 5W, 5AA
    of κX, 4.8(p)(2), 7P
    of σX, §7.4, 7E
    of υX, 5.11(b), (c), 5T, 5Z
    relatively zero-dimensional, 4X, 7N
remote points, 4AH
    vs. accumulation points, 7U(2)
    generalizations, 7V
replete (class of spaces), 4.1(j)
representation (of a lattice), 2.4(d)
right inverse (see inverse)
rimcompact, 4X
ring of subsets, 2.4(a)
    as Wallman base, 4.4(a)
root (of a Δ-system), 3R
$RO(X)$, 2.2(b)
    as a Boolean algebra, 3.1(e)(3)
    as a complete lattice, 2.2(c)
    as a completion, 3.4(e)
    for extremally disconnected X, 6.2(b)
    is isomorphic to $R(X)$, 3B

R(X), 2.2(b)
as a Boolean algebra, 3.1(e)(4)
as a complete lattice, 2.2(c)
for extremally disconnected X, 6.2(b)
is isomorphic to RO(X), 3B
relation to Z(X), 5.11(h)

S

semilattice, 2.1(d)
AB(E) as, 8.4(f)
E(X) as, 4.1(g)
K(X) as, 4.2(a)
P(X) as, 4.1(j)
semiregular space, 2.2(e)
embedding T_2 spaces in, 2G
examples of, 2.2(h), (k)
extremally disconnected iff, §6.4
generalizations, 7L
and H-closed, 7.5(a)
hereditary properties, 2.2(i)
and perfect maps, 2.2(k)
products, 2.2(j)
vs. regular space, 2.2(g), (h)(1)
semiregularization, 2.2(e)
of absolutes, 6.11(c)
is minimal Hausdorff when, 7.5(b)
and Θ-equivalence, 7.7(d)
separate points, 4.5(d)
separates points and closed sets, 1.7(i)(2)
separating chain, 1.10(b)
and complete separation, 1.10(f)

and C(X), 1.10(e), 1S
and $\mathbb{Q}$-indexing, 1.10(d)
sequential spaces, 9D
S-equivalence, 7C
vs. $\dot{\theta}$-equivalence, 7C(2)
Σ-products, 1X, 4R
σX, preceding 7.2(a) (also see Fomin extension)
vs. βX, 7B(1), (5), 7.7(g)
vs. κX, 7B, 7P(8)
vs. μX, 7.7(g), 7B(1)
σ-completion, 6K
simple extension, 7.1(g)
number of, 7C(4)
properties, 7.1(h), (j)
vs. strict extension, 7.1(h), 7C
simply generated (extension property), 5.4(a)
singular cardinal, 2.6(p)(2)
SL(κ), 3.5(o) (also see P(κ))
combinatorial equivalence, 3.5(r)
consequences, 3.5(s), (t), 3U, 3V
Sorgenfrey line, 1H
stable ultrafilter, 8.2(j)
and points of υX, 8.2(k)
Stone space (of Boolean algebra), 3.2(c)
βD as, 4U
Cantor set as, 3.3(g)
clopen sets of, 3.2(d)
of complete Boolean algebra, 6.2(d), (e)
continuous functions on, 3.2(f), (j)
of factor algebras, 3L, 3M, 3N, 4U
Gleason space as, 6.6(a)
of σ-complete Boolean algebra, 6K
topological properties, 3.2(d)

Stone-Čech compactification, see βX
Stone-Weierstrass theorem, 4.5(e)
strict extension, 7.1(d)
    characterizations, 7A, 7L(9)
    number of, 7C(4)
    properties, 7.1(h), (j)
    semiregular extension is, 7.1(e)(4)
    vs. simple extension, 7.1(h), 7C
strongly zero-dimensional, 4.7(h)
    characterization, 4.7(j)
    vs. zero-dimensional, 4.7(g), (h), 4V
subalgebra (of a Boolean algebra), 3.1(h), (k), §3.4, 3C, 3J, 6H, 6I
    intersection of, 3C
    regular, 3.4(d)(3)
    union of, 3E
subcategory, 9.2(g)
    coreflective, 9.7(a)
    examples, 9.2(h)
    full, 9.2(g)
    reflective, epireflective, 9.6(a) et seq.
    replete, 9.4(h)
submaximal space, 7M(7)
sum, 1.2(h)
    of absolutes, 6A
surjection, 1.1(e)

T

Taimanov theorem, 4.1(m)
    consequences, 4.2(h)
Tamano's theorem, 4AF(7)
Θ-closure, 4.8(r)

properties, 4.8(s), 7Q
Θ-continuous function, 4.8(g) (also see irreducible Θ-continuous function)
and compact sets, 4O
into dense subspace, 4.8(h)(5)
examples, 4.8(i), 6.6(e)(5)
and H-closed spaces, 4.8(h)(6)
products, 4O
properties, 4.8(h), 4O
into regular spaces, 4.8(h)(3)
restrictions of, 4.8(h)(4)
Θ-equivalence (of H-closed extensions), 7.7(a)
characterization, 7.7(d)
Θ-homeomorphism, 4.8(g)(4)
example, 4.8(t)
Θ-refinable, 5.11(j)
vs. realcompact, 5.11(l), 5Q
[ΘP] (for various P), 9.8(i)
ΘX (see Gleason space), 6.6(a)
Θ′X, 6.8(c)
topological category, 9.3(a)
topological property, 4.1(j)
transfinite construction, 2.6(g)
transfinite induction. 2.6(f)
$\mathfrak{z}$-embedding, 1L, 4.7(e)
and $\beta_0 X$, 4.7(f)
characterizations, 1L
vs. ℕ-embedding, 1L(5)
Tychonoff covering property, 8.4(z)
examples, 8.4(ab)
Tychonoff extension property, 5.6(a)
vs. absolute, §8.3, 8F
and almost compact spaces, 5U

characterizations, 5.6(b), 5.9(e)
and co-pseudocompactness, 5X
and countably compact subspaces, 5R
and discrete spaces, 5AB
examples, 5.6(d), §5.10, 5B, 5D, 5N, 5R, 5W, 5AA, 6T, 6U
and maximum extension, 5.9(f)
and one-point extensions, 5AC
properties, 5.9(a), (f), (h)
pseudocompactness classes for, 5W
Tychonoff plank, 2R
Tychonoff space, 1.4
vs. compact, 1.10(i)
vs. normal, 1I, 1N
vs. regular, 1Y
Tychonoff spirals, 2S

U

U-compactness, 5R
vs. countable compactness, 5R
vs. (p,S)-compactness, 5AC(2)
Ulam-measurable cardinal, 2P
and ℕ-compactness, 5I
and realcompactness, 5I, 5AD
ultrafilter, 2.3(c), 2H (also see open, clopen, z-ultrafilter, etc.)
characterizations, 2.3(d), (e)
on a poset, 3.5(d), (e)
ultrarealcompact space, 5AA
properties, 8E
upper bound (in poset), 2.1(c)(3)
upper semicontinuous decomposition, 4Q, and preceding 7.4(b)
upper semilattice, 2.1(d), 2D (also see semilattice)

υX, 5.5(c)
and absolutes, 8.2(l)
characterization, 5.10(b)
compact iff, 5F(4)
and products, 5AD
properties, 5F, 5V, 5Z, 6AB
regular closed subsets of, 5V
Urysohn-closed space, 7J
Urysohn's extension theorem, 1.9(h)
Urysohn space, 4.8(j)
compact iff, 7.5(b)
properties, 4K, 7J(1), 7R(2)

W

Wallman base, 4.4(a)
examples, 4.4(b), (h), 4.7(b), 4A(2)
for normal space, 4B
properties, 4B
and Wallman compactification 4.4(d)
Wallman compactification, 4.4(d) (also see $w_L X$)
weak c.b. space, 6U(2), 8.5(c)
and C(X), 8.5(e)
examples, 8.5(d)
vs. weak δ-normal separation, 8L
weak δ-normal separation, 1R
and pseudocompactness, 1R
vs. weak c.b., 8L
weak topology, 1.7(a)
and product spaces, 1.7(c)
weakly Lindelöf spaces, 3P
Weierstrass M-test, 1B(6)

weight of a space, 2N(3)
well-ordered set, 2.6(a)
functions on, 2.6(h)
$w_L X$ (Wallman compactification of X re L), 4.4(a)(2), (d)
(also see Wallman base)
βX as, 4.4(h)
$\beta_0 X$ as, 4.7(b), (c)
as a compactification of X, 4.4(d), (f)
vs. Gelfand compactifications, 4G
one-point compactification as, 4A
zero-dimensional compactifications as, 4.7(b)
woset, 2.6(a) (see well-ordered set)

X

X_δ, 1W(7)
vs. Hewitt realcompactification, 5F(7)

Z

z-closed mapping, 4AF
z-embedding, 1.9(l)
vs. C*-embedding, 1.9(m)
zero-dimensional, 1.5(a)
compactifications, 4.7
and m-compact, 5O
and realcompact, 5P, 5S
vs. strongly zero-dimensional, 4.7(g), (h), 4V
vs. totally disconnected, 1V
implies Tychonoff, 1.5(c)
zero-dimensional covering property, 8.4(z)

zero-dimensional extension property, 5.7(a)
vs. absolute, §8.3
associated maximum extension, 5.9(b)
characterizations, 5.7(b), 5.9(e)
and compact subspaces, 5L
copseudocompact, 5W, 5X
and discrete spaces, 5AB
examples, 5D, 5G, 5J
properties, 5.9(a), (f), (h)
vs. Tychonoff extension properties, 5D, 5M
zero-sets, 1.4(b) (also see Z(X))
base of, 1.4(e)
non C*-embedded, 1N(4)
vs. clopen set, 1.5(c), 1W(2)
countable intersections of, 1.4(i)
finite unions, intersections, 1.4(i)
vs. G_δ-sets, 1K
inverse image of, 1.4(j)
in metric spaces, 1.4(g)
in normal spaces, 1K
in P-spaces, 1W(2), 4AG(6)
ZFC, preface, introduction to 3.5
independent of, introduction to 3.5
z-irreducible map, 9K
z-ultrafilter, 2.3(c)(4) (also see ultrafilter)
vs. closed ultrafilters, 2Q, 2R(5)
vs. maximal ideals of C*(X), 4.6(f)
as points of βX, 4.4(h)
P-stable, 5Y
$Z(X)$, 1.4(b) (also see zero-sets)
vs. βX, 5.11(h)
not countably complete, 2E
as a lattice, 1.4(i), 2.1(f)(4)
relation to $Z(\upsilon X)$, 5.11(g)